Y0-BIR-342

SPACE SYSTEMS AND THEIR INTERACTIONS WITH EARTH'S SPACE ENVIRONMENT

Edited by
Henry B. Garrett
and
Charles P. Pike
Air Force Geophysics Laboratory
Hanscom Air Force Base
Bedford, Massachusetts

Volume 71
PROGRESS IN
ASTRONAUTICS AND AERONAUTICS

Martin Summerfield, Series Editor-in-Chief
New York University, New York, New York

Published by the American Institute of Aeronautics and Astronautics

American Institute of Aeronautics and Astronautics
New York, New York

Library of Congress Cataloging in Publication Data
Main entry under title:

Space systems and their interactions with Earth's space environment.

(Progress in astronautics and aeronautics; v. 71)
Includes bibliographies and index.
1. Space vehicles—Design and construction. 2. Space environment. 3. Astronautics—Environmental aspects. I. Garrett, Henry B. II. Pike, Charles P. III. American Institute of Aeronautics and Astronautics. IV. Series.
TL507.P575 vol. 71 [TL875] 629.ls [629.47'l] 80-15518
ISBN 0-915928-41-8

Table of Contents

Authors

Environmental Effects of Space Systems: A Review

D. M. Rote
Argonne National Laboratory, Argonne, Ill.

Effects of Microwave Beams on the Ionosphere

Lewis M. Duncan
Los Alamos Scientific Laboratory, Los Alamos, N. Mex.

Upper Atmosphere Modifications Due to Chronic Discharges of Water Vapor from Space Launch Vehicle Exhausts

Jeffrey M. Forbes
Boston University, Boston, Mass.

Modification of the Ionosphere by Large Space Vehicles

Michael Mendillo
Boston University, Boston, Mass.

Argon-Ion Contamination of the Plasmasphere

Y. T. Chiu, J. M. Cornwall, J. G. Luhmann, and Michael Schulz
The Aerospace Corporation, El Segundo, Calif.

Magnetospheric Modification by Gas Releases from Large Space Structures

Richard R. Vondrak
SRI International, Menlo Park, Calif.

Spacecraft Charging: A Review

H. B. Garrett
Air Force Geophysics Laboratory, Hanscom Air Force Base, Bedford, Mass.

Spacecraft Charging During Eclipse Passage

H. B. Garrett and D. M. Gauntt
Air Force Geophysics Laboratory, Hanscom Air Force Base, Bedford, Mass.

Occurrence of Arcing and Its Effects on Space Systems

Joseph E. Nanevicz and Richard C. Adamo
SRI International, Menlo Park, Calif.

Surface Discharge Effects

K. G. Balmain
University of Toronto, Toronto, Canada

Preface

Sputnik alerted America over twenty years ago to the dawn of the space age, and Neil Armstrong's moon walk captured the nation's attention more than ten years ago. With great anticipation we await the beginning of the Space Shuttle era and, with it, the Orbiter's capability for regularly launching massive payloads. In this volume we focus on space environment issues that will be associated with our increased utilization of space in the Space Shuttle era. Papers presented here cover aspects of spacecraft-environment interaction phenomena currently believed to be critical to the design and development of Shuttle era space systems. The intent of this volume is to serve as a reference source for spacecraft engineers and for space scientists working on environmental questions related to these systems and on the Space Transportation System itself.

The volume is organized into five topical areas including: the effects of space operations on the Earth's environment; the effects of spacecraft charging on space systems; the effects of space radiation on spacecraft systems; the interactions between the space environment and large dimension, high-power spacecraft; and the effects of the environment on the structural integrity of space vehicles. Each topical area is introduced by a review paper which is followed by specific technical papers. It should be noted that, as recently as five years ago, many of the effects considered were not known or were considered to be of no practical interest. With the advent of planning for physically large (>50m), high-power (>1-kW) systems, many of these environmental interactions have become relevant issues. Since the perceived direction for utilization of the Space Shuttle Orbiter includes deploying such large, high-power spacecraft, this volume's arrival is very timely.

Figure 1 contains data on the predicted increase in the size of space structures through the 1980's,[1] whereas, in Fig. 2, the projected increase in complexity vs susceptibility to radiation damage of electronic microcircuit components is plotted.[2] Similar growth curves have been projected for the number of satellites in near-Earth orbit and for the increase in rocket exhaust contaminants. Based on these projections, we feel that the space community may be approaching critical decision points regarding the problems of the space environment's effects on space systems and, conversely, the system's effects on the space environment. The topical areas covered

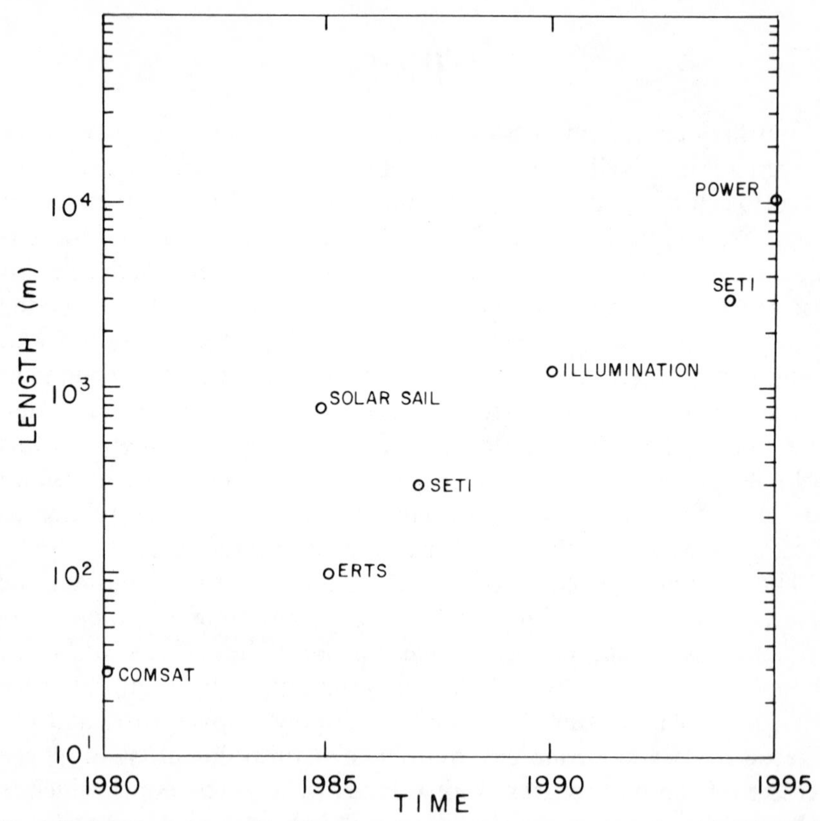

Fig. 1 Projected maximum dimension of various space systems as a function of time (adapted from Ref. 1).

in this volume address these issues. The conclusions to be drawn in each area are summarized below.

In Chapter I, the effects of space operation on the Earth's space environment, a critical emerging issue is the growing public awareness and concern over the environmental effects of high technology systems on the environment. As exemplified by public concern over the supersonic transport plane's effects on ozone, the degradation of air quality by automobile exhaust emissions, and the leakage of radiation and waste from nuclear plants, societal considerations can now play a key role in legislative actions concerning these systems. The environmental impact of systems can become a crucial decision point for their use or nonuse. Therefore, will the increased utilization of space offered by the Space Transportation System alter

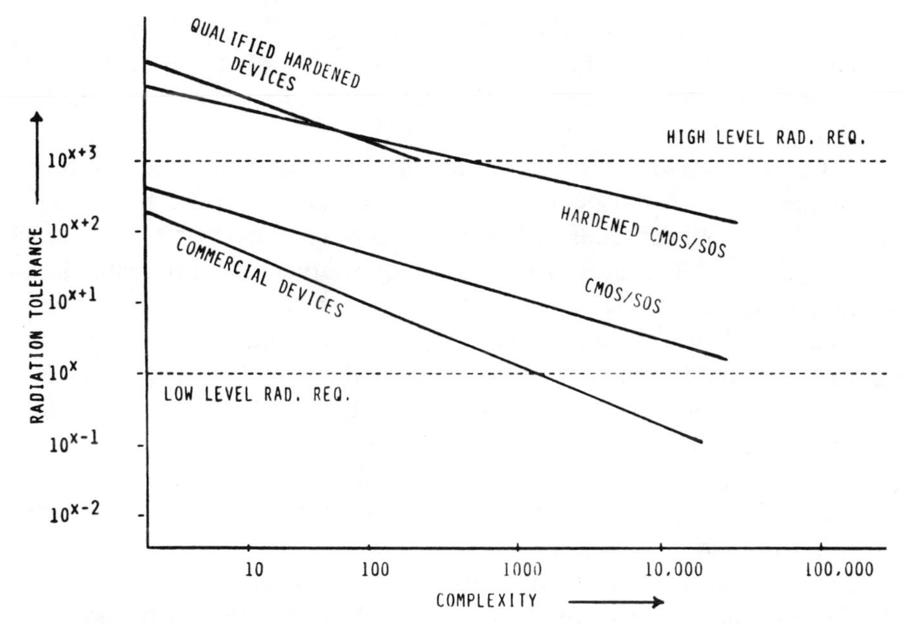

Fig. 2 Projected decrease in radiation tolerance of monolithic semiconductor devices as a function of complexity.[2]

the nature of the space environment? In our examination of the effects on the environment, little evidence is presented to indicate that Shuttle operations could have long-term effects on the space environment or measurable effects on the Earth's weather. It may be possible, however, if extremes of space system operations occur, that the space environment could be significantly perturbed on a short time scale. This statement is made, however, in the context of our present knowledge. We feel, based on the papers in Chapter I, that there exists a serious need for further study to improve our knowledge of the space environmental impact of space systems.

Chapter II, spacecraft charging, is a new technical development area. Specifically, in the last decade the complexity of spacecraft systems and the desire to reduce their power requirements has led to the development of electronic circuits which are vulnerable to low voltage impulses—impulses which are injected into the satellite system following electrical discharges between adjacent satellite surfaces. The electrical discharges result from the buildup on spacecraft surfaces of static charge from the space environment. In this chapter, calculations of charge buildup and techniques to mitigate charge buildup are covered. The papers demonstrate that, by using spacecraft charging computer codes, the spacecraft charging phenomena (not, however, the actual arcing process which is still not well characterized) can be estimated and corrective spacecraft design measures taken.

One of the first major environmental effects to be observed on a satellite system was high-energy particle damage to the Explorer 1 satellite. Engineers rapidly found means of circumventing the effects of the Van Allen natural radiation environment by using shielding and radiation hardening of electronic components. For example, solar panels were oversized to compensate for solar cell degradation due to radiation damage. It was not until the 1960's, when massive releases of radiation from high-altitude nuclear tests occurred, that problems associated with radiation damage to spacecraft became acute. Engineering methods were again found to compensate for these effects. With the latest advances in microcircuitry and the trend toward very large, lightweight space structures with complex signal microprocessors to control them, the effects of radiation damage may once again become a basic factor in satellite system design goals. Radiation damage effects and means of circumventing these effects on future systems are presented in Chapter III.

As implied in Fig. 1, the physical size and power requirements of satellite systems will grow in the next decade. These systems will have problems unique to their size. For the first time, satellites may become so large that environmental effects on one portion of the satellite may be independent of effects on another portion. High operating voltages are associated with such systems as the Solar Power Satellite which could, potentially, suffer parasitic power loss through the plasma. The scale size of space structures can significantly change the physical assumptions made in calculating environmental effects. Even the attitude thrusters, used to control large structures, could damage the structure. In Chapter IV, the proposition clearly emerges that careful consideration must be given to environmental factors early in the design of large, high-power systems if they are to be successful.

Chapter V describes possible systems limiting effects due to the environment. These range from structural vibrations to actual collisions between satellites. Important questions peculiar to large, high-powered systems are brought out. For example, is there a limit to the physical size that a structure can be built? Are there limits on the size of a structure such that gravitational, thermal, and electrostatic forces become so great as to prevent us from compensating for their effects? Likewise, is it possible that the number of spacecraft launches and the associated density of space debris will become so great as to seriously limit the lifetime of future systems through spacecraft collisions? These issues are addressed in a quantitative fashion in this chapter.

We believe that the topical areas in this volume present a forward-looking view of some of the environmental problems expected in the Shuttle era. Furthermore, the direction of the space program, through developments both in basic science and engineering, could be significantly affected by these issues. We hope that this volume will contribute toward a better understanding of the totality of space systems environmental interactions.

References

[1] Hagler, T., Patterson, H. G., and Nathan, C. A., "Learning to Build Large Structures in Space," *Astronautics & Aeronautics,* Vol. 12, 1977, pp. 51-57.

[2] Buchanan, B. L., "Keynote Address: SOS Technology and Military Electronics," Institute of Electrical and Electronics Engineers Silicon on Sapphire Workshop, 1978.

Henry B. Garrett
Charles P. Pike
April 1980

Progress in Astronautics and Aeronautics

VOLUMES	EDITORS
*1. **Solid Propellant Rocket Research.** 1960	Martin Summerfield *New York University*
2. **Liquid Rockets and Propellants.** 1960	Loren E. Bollinger *The Ohio State University* Martin Goldsmith *The Rand Corporation* Alexis W. Lemmon Jr. *Battelle Memorial Institute*
3. **Energy Conversion for Space Power.** 1961	Nathan W. Snyder *Institute for Defense Analyses*
*4. **Space Power Systems.** 1961	Nathan W. Snyder *Institute for Defense Analyses*
5. **Electrostatic Propulsion.** 1961	David B. Langmuir *Space Technology Laboratories, Inc.* Ernst Stuhlinger *NASA George C. Marshall Space Flight Center* J. M. Sellen Jr. *Space Technology Laboratories, Inc.*

*Now out of print

6. Detonation and Two-Phase Flow. 1962	S. S. Penner *California Institute of Technology* F. A. Williams *Harvard University*
7. Hypersonic Flow Research. 1962	Frederick R. Riddell *AVCO Corporation*
8. Guidance and Control. 1962	Robert E. Roberson *Consultant* James S. Farrior *Lockheed Missiles and Space Company*
***9. Electric Propulsion Development.** 1963	Ernst Stuhlinger *NASA George C. Marshall Space Flight Center*
***10. Technology of Lunar Exploration.** 1963	Clifford I. Cummings and Harold R. Lawrence *Jet Propulsion Laboratory*
11. Power Systems for Space Flight. 1963	Morris A. Zipkin and Russell N. Edwards *General Electric Company*
12. Ionization in High-Temperature Gases. 1963	Kurt E. Shuler, Editor *National Bureau of Standards* John B. Fenn, Associate Editor *Princeton University*
13. Guidance and Control—II. 1964	Robert C. Langford *General Precision Inc.* Charles J. Mundo *Institute of Naval Studies*
14. Celestial Mechanics and Astrodynamics. 1964	Victor G. Szebehely *Yale University Observatory*
***15. Heterogeneous Combustion.** 1964	Hans G. Wolfhard *Institute for Defense Analyses* Irvin Glassman *Princeton University* Leon Green Jr. *Air Force Systems Command*

16. **Space Power Systems Engineering.** 1966	George C. Szego *Institute for Defense Analyses* J. Edward Taylor *TRW Inc.*
17. **Methods in Astrodynamics and Celestial Mechanics.** 1966	Raynor L. Duncombe *U. S. Naval Observatory* Victor G. Szebehely *Yale University Observatory*
18. **Thermophysics and Temperature Control of Spacecraft and Entry Vehicles.** 1966	Gerhard B. Heller *NASA George C. Marshall Space Flight Center*
*19. **Communication Satellite Systems Technology.** 1966	Richard B. Marsten *Radio Corporation of America*
20. **Thermophysics of Spacecraft and Planetary Bodies: Radiation Properties of Solids and the Electromagnetic Radiation Environment in Space.** 1967	Gerhard B. Heller *NASA George C. Marshall Space Flight Center*
21. **Thermal Design Principles of Spacecraft and Entry Bodies.** 1969	Jerry T. Bevans *TRW Systems*
22. **Stratospheric Circulation.** 1969	Willis L. Webb *Atmospheric Sciences Laboratory, White Sands, and University of Texas at El Paso*
23. **Thermophysics: Applications to Thermal Design of Spacecraft.** 1970	Jerry T. Bevans *TRW Systems*
24. **Heat Transfer and Spacecraft Thermal Control.** 1971	John W. Lucas *Jet Propulsion Laboratory*

25. Communications Satellites for the 70's: Technology. 1971	Nathaniel E. Feldman *The Rand Corporation* Charles M. Kelly *The Aerospace Corporation*
26. Communications Satellites for the 70's: Systems. 1971	Nathaniel E. Feldman *The Rand Corporation* Charles M. Kelly *The Aerospace Corporation*
27. Thermospheric Circulation. 1972	Willis L. Webb *Atmospheric Sciences Laboratory, White Sands, and University of Texas at El Paso*
28. Thermal Characteristics of the Moon. 1972	John W. Lucas *Jet Propulsion Laboratory*
29. Fundamentals of Spacecraft Thermal Design. 1972	John W. Lucas *Jet Propulsion Laboratory*
30. Solar Activity Observations and Predictions. 1972	Patrick S. McIntosh and Murray Dryer *Environmental Research Laboratories, National Oceanic and Atmospheric Administration*
31. Thermal Control and Radiation. 1973	Chang-Lin Tien *University of California, Berkeley*
32. Communications Satellite Systems. 1974	P. L. Bargellini *COMSAT Laboratories*
33. Communications Satellite Technology. 1974	P. L. Bargellini *COMSAT Laboratories*
34. Instrumentation for Airbreathing Propulsion. 1974	Allen E. Fuhs *Naval Postgraduate School* Marshall Kingery *Arnold Engineering Development Center*
35. Thermophysics and Spacecraft Thermal Control. 1974	Robert G. Hering *University of Iowa*

36. Thermal Pollution Analysis. 1975	Joseph A. Schetz *Virginia Polytechnic Institute*
37. Aeroacoustics: Jet and Combustion Noise; Duct Acoustics. 1975	Henry T. Nagamatsu, Editor *General Electric Research and Development Center* Jack V. O'Keefe, Associate Editor *The Boeing Company* Ira R. Schwartz, Associate Editor *NASA Ames Research Center*
38. Aeroacoustics: Fan, STOL, and Boundary Layer Noise; Sonic Boom; Aeroacoustics Instrumentation. 1975	Henry T. Nagamatsu, Editor *General Electric Research and Development Center* Jack V. O'Keefe, Associate Editor *The Boeing Company* Ira R. Schwartz, Associate Editor *NASA Ames Research Center*
39. Heat Transfer with Thermal Control Applications. 1975	M. Michael Yovanovich *University of Waterloo*
40. Aerodynamics of Base Combustion. 1976	S. N. B. Murthy, Editor *Purdue University* J. R. Osborn, Associate Editor *Purdue University* A. W. Barrows and J. R. Ward, Associate Editors *Ballistics Research Laboratories*
41. Communication Satellite Developments: Systems. 1976	Gilbert E. LaVean *Defense Communications Engineering Center* William G. Schmidt *CML Satellite Corporation*
42. Communication Satellite Developments: Technology. 1976	William G. Schmidt *CML Satellite Corporation* Gilbert E. LaVean *Defense Communications Engineering Center*

43. Aeroacoustics: Jet Noise, Combustion and Engine Noise. 1976	Ira R. Schwartz, Editor *NASA Ames Research Center* Henry T. Nagamatsu, Associate Editor *General Electric Research and Development Center* Warren C. Strahle, Associate Editor *Georgia Institute of Technology*
44. Aeroacoustics: Fan Noise and Control; Duct Acoustics; Rotor Noise. 1976	Ira R. Schwartz, Editor *NASA Ames Research Center* Henry T. Nagamatsu, Associate Editor *General Electric Research and Development Center* Warren C. Strahle, Associate Editor *Georgia Institute of Technology*
45. Aeroacoustics: STOL Noise; Airframe and Airfoil Noise. 1976	Ira R. Schwartz, Editor *NASA Ames Research Center* Henry T. Nagamatsu, Associate Editor *General Electric Research and Development Center* Warren C. Strahle, Associate Editor *Georgia Institute of Technology*
46. Aeroacoustics: Acoustic Wave Propagation; Aircraft Noise Prediction; Aeroacoustic Instrumentation. 1976	Ira R. Schwartz, Editor *NASA Ames Research Center* Henry T. Nagamatsu, Associate Editor *General Electric Research and Development Center* Warren C. Strahle, Associate Editor *Georgia Institute of Technology*
47. Spacecraft Charging by Magnetospheric Plasmas. 1976	Alan Rosen *TRW Inc.*

48. Scientific Investigations on the Skylab Satellite. 1976	Marion I. Kent and Ernst Stuhlinger *NASA George C. Marshall Space Flight Center* Shi-Tsan Wu *The University of Alabama*
49. Radiative Transfer and Thermal Control. 1976	Allie M. Smith *ARO Inc.*
50. Exploration of the Outer Solar System. 1977	Eugene W. Greenstadt *TRW Inc.* Murray Dryer *National Oceanic and Atmospheric Administration* Devrie S. Intriligator *University of Southern California*
51. Rarefied Gas Dynamics, Parts I and II (two volumes). 1977	J. Leith Potter *ARO Inc.*
52. Materials Sciences in Space with Application to Space Processing. 1977	Leo Steg *General Electric Company*
53. Experimental Diagnostics in Gas Phase Combustion Systems. 1977	Ben T. Zinn, Editor *Georgia Institute of Technology* Craig T. Bowman, Associate Editor *Stanford University* Daniel L. Hartley, Associate Editor *Sandia Laboratories* Edward W. Price, Associate Editor *Georgia Institute of Technology* James G. Skifstad, Associate Editor *Purdue University*
54. Satellite Communications: Future Systems. 1977	David Jarett *TRW Inc.*

55. Satellite Communications: Advanced Technologies. 1977	David Jarett *TRW Inc.*
56. Thermophysics of Spacecraft and Outer Planet Entry Probes. 1977	Allie M. Smith *ARO Inc.*
57. Space-Based Manufacturing from Nonterrestrial Materials. 1977	Gerard K. O'Neill, Editor *Princeton University* Brian O'Leary, Assistant Editor *Princeton University*
58. Turbulent Combustion. 1978	Lawrence A. Kennedy *State University of New York at Buffalo*
59. Aerodynamic Heating and Thermal Protection Systems. 1978	Leroy S. Fletcher *University of Virginia*
60. Heat Transfer and Thermal Control Systems. 1978	Leroy S. Fletcher *University of Virginia*
61. Radiation Energy Conversion in Space. 1978	Kenneth W. Billman *NASA Ames Research Center*
62. Alternative Hydrocarbon Fuels: Combustion and Chemical Kinetics. 1978	Craig T. Bowman *Stanford University* Jørgen Birkeland *Department of Energy*
63. Experimental Diagnostics in Combustion of Solids. 1978	Thomas L. Boggs *Naval Weapons Center* Ben T. Zinn *Georgia Institute of Technology*
64. Outer Planet Entry Heating and Thermal Protection. 1979	Raymond Viskanta *Purdue University*
65. Thermophysics and Thermal Control. 1979	Raymond Viskanta *Purdue University*

66. Interior Ballistics of Guns. 1979	Herman Krier *University of Illinois at Urbana-Champaign* Martin Summerfield *New York University*
67. Remote Sensing of Earth from Space: Role of "Smart Sensors." 1979	Roger A. Breckenridge *NASA Langley Research Center*
68. Injection and Mixing in Turbulent Flow. 1980	Joseph A. Schetz *Virginia Polytechnic Institute and State University*
69. Entry Heating and Thermal Protection. 1980	Walter B. Olstad *NASA Headquarters*
70. Heat Transfer, Thermal Control, and Heat Pipes. 1980	Walter B. Olstad *NASA Headquarters*
71. Space Systems and Their Interactions with Earth's Space Environment. 1980	Henry B. Garrett and Charles P. Pike *Hanscom Air Force Base*

(Other volumes are planned.)

Chapter I—Effects of Space System Operations on the Space Environment

ENVIRONMENTAL EFFECTS OF SPACE SYSTEMS: A REVIEW

D. M. Rote*
Argonne National Laboratory, Argonne, Ill.

Abstract

This review and the papers in this section focus on the effects of large space systems, primarily the Satellite Power System (SPS), on the upper atmosphere. From 56-500 km, the major contaminant sources are SPS microwave transmissions and rocket effluents. Although no significant effects have yet been found for microwave transmissions, deposition of rocket effluents causes compositional changes, most of which appear to be associated with the release of large amounts of water. From 500-36,000 km, rocket effluents and ion engine contaminants (primarily Ar^+) could alter magnetospheric and plasmaspheric structure and dynamics. One of the major impacts of these alterations could be perturbation of Van Allen radiation belt stability, leading to changed radiation hazards to materials and personnel and to modification of high energy particle precipitation events. The ambient density falls rapidly and the potential for significant environmental alteration increases as one goes outwards from the Earth's surface. And, the further from the Earth's surface, the less certain our knowledge of environmental change processes.

I. Introduction

The unique properties of the Earth's space environment make it an attractive place for a broad spectrum of scientific, commercial, industrial, and military activities (see, for example, Ref. 1). Current development of the Space Shuttle and other advanced space transportation systems is evidence that the uses of space will continue to proliferate. Such proliferation has resulted in increased concern about preserving the very unique properties of space that stimulated its exploitation in the first place. For example, the gradual accumulation of objects in space poses safety hazards that will require careful management. Thus, the identification and investigation

---

*Head, Air Resources Section, and Director of Atmospheric Effects Assessment, Satellite Power Systems Program.

of the effects of proposed new uses of space are assuming greater importance. In particular, the potential impacts of the Space Shuttle program and the proposed Satellite Power System (SPS) are receiving considerable attention, because these projects represent unprecedented increases in the use of the space environment.

It is important, therefore, to gain some perspective on the proposed scope of these two programs. (Figure 1 is a schematic diagram of the various regions of the atmosphere to help orient the reader.) The Space Shuttle will go into operation during the 1980's and will provide a general-purpose transportation system for equipment and personnel between the Earth's surface and low Earth orbits. Up to 60 launches per year are anticipated. One of the early uses of this system could be for SPS technology verification.[2] A recently completed environmental impact statement provides a detailed discussion of the potential environmental impacts, including both lower and upper atmospheric effects, of the Space Shuttle program.[3]

Glaser has proposed that the SPS produce baseload power for the utility electric power grid.[4] A reference design has

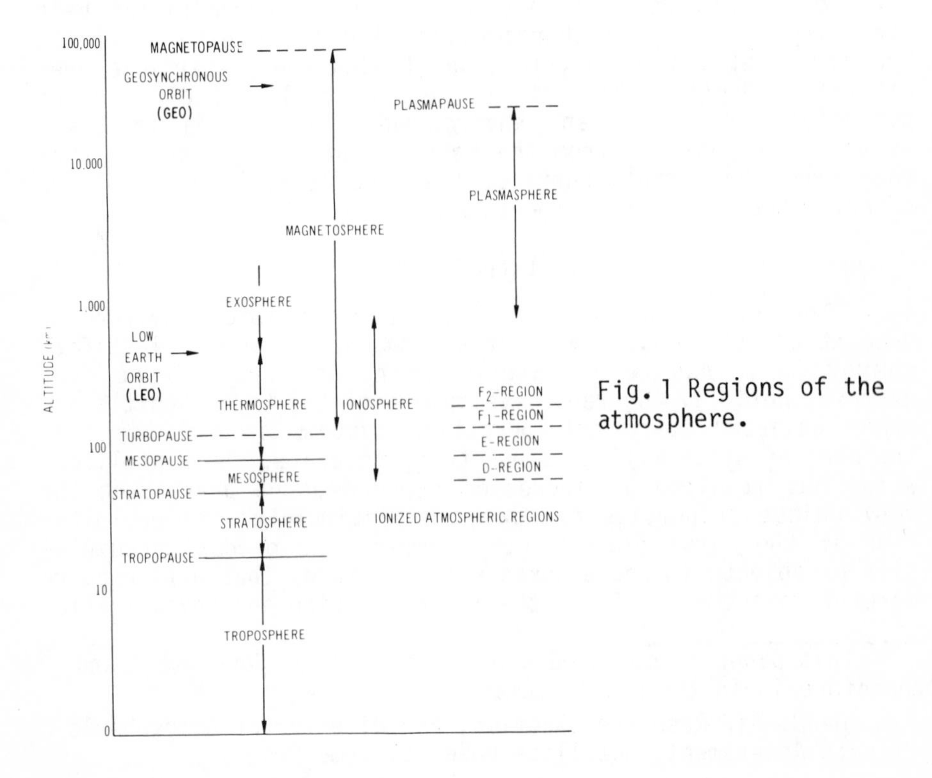

Fig. 1 Regions of the atmosphere.

been developed by the National Aeronautics and Space Administration (NASA) and is being used in a Department of Energy (DOE)-sponsored program to evaluate the environmental and socioeconomic impacts of the SPS, as well as its technical feasibility.[5] Briefly, the NASA reference design utilizes 60 satellites in geosynchronous orbit to collect solar energy, convert it to microwave energy, and transmit it to the Earth's surface, where receiving antennas (rectennas) rectify the microwave radiation to DC power. Each satellite delivers 5 GW of power to the grid. The total system is projected for operation around the year 2020 and is expected to supply approximately 20% of that year's total U.S. electrical energy demand.

An international SPS could conceivably be much larger and contribute significantly to the projections by Dupas and Claver[6] as to the fraction of global demand for electric power that could be satisfied by all forms of solar energy. In other words, during the current assessment of the NASA reference design -- each satellite of this system has a projected 30-year lifetime -- one should not lose sight of the possibility of a larger, international system operating into the indefinite future.

The types of rockets required for space transportation for the NASA reference design are listed in Table 1. The initial construction phase involves building base stations at low Earth orbit (LEO) and geosynchronous earth orbit (GEO).

Table 1 Space vehicle types

Name	Abbreviation	Propellants	Operating, region, km
Heavy lift launch vehicle	HLLV	LCH_4/LO_2 1st stage	0-56
		LH_2/LO_2 2nd stage	56-120
		LH_2/LO_2 circularization and deorbit	450-500
Personnel launch vehicle	PLV	Details not available; probably same as above	0-500
Cargo orbit transfer vehicle	COTV	Argon ions and electrons LH_2/LO_2	500-36,000
Personnel orbit transfer vehicle	POTV	LH_2/LO_2	500-36,000

The space components of the system (solar satellites and microwave generators and transmitters) would be constructed in space from materials delivered to the LEO base station and then transferred to the GEO construction site. The transportation requiremenmts for this initial construction phase are given in Fig. 2.

Figure 3 shows the scenario for construction of two 5-GW satellites per year. Note that two photovoltaic options

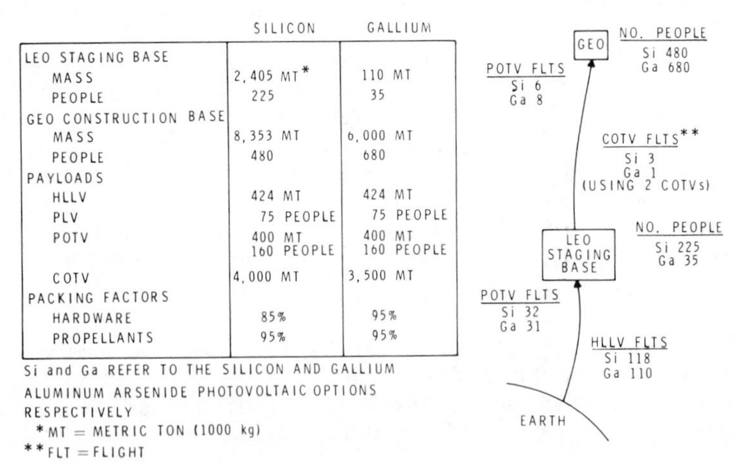

	SILICON	GALLIUM
LEO STAGING BASE		
MASS	2,405 MT*	110 MT
PEOPLE	225	35
GEO CONSTRUCTION BASE		
MASS	8,353 MT	6,000 MT
PEOPLE	480	680
PAYLOADS		
HLLV	424 MT	424 MT
PLV	75 PEOPLE	75 PEOPLE
POTV	400 MT 160 PEOPLE	400 MT 160 PEOPLE
COTV	4,000 MT	3,500 MT
PACKING FACTORS		
HARDWARE	85%	95%
PROPELLANTS	95%	95%

Si and Ga REFER TO THE SILICON AND GALLIUM ALUMINUM ARSENIDE PHOTOVOLTAIC OPTIONS RESPECTIVELY

* MT = METRIC TON (1000 kg)

** FLT = FLIGHT

Fig. 2 Scenario for buildup of construction bases needed for construction of power satellites (Source: Ref. 5).

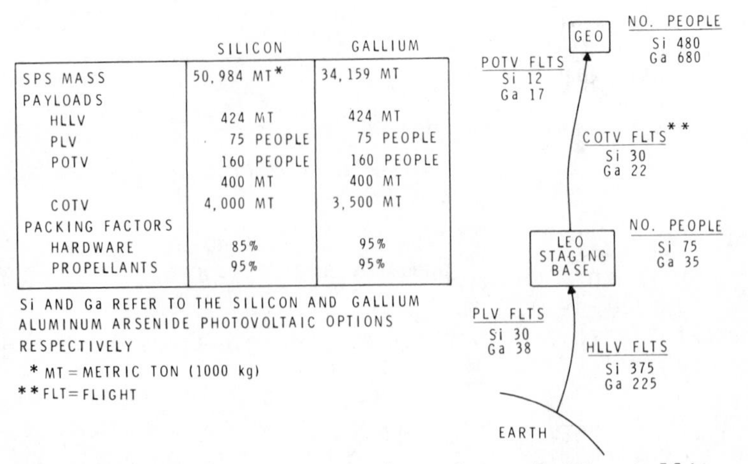

	SILICON	GALLIUM
SPS MASS	50,984 MT*	34,159 MT
PAYLOADS		
HLLV	424 MT	424 MT
PLV	75 PEOPLE	75 PEOPLE
POTV	160 PEOPLE 400 MT	160 PEOPLE 400 MT
COTV	4,000 MT	3,500 MT
PACKING FACTORS		
HARDWARE	85%	95%
PROPELLANTS	95%	95%

Si AND Ga REFER TO THE SILICON AND GALLIUM ALUMINUM ARSENIDE PHOTOVOLTAIC OPTIONS RESPECTIVELY

* MT = METRIC TON (1000 kg)

** FLT = FLIGHT

Fig. 3 Scenario for construction of two 5-GW satellites per year (Source: Ref. 5).

are indicated: (1) the use of silicon cells (Si) and (2) the use of gallium aluminum arsenide cells (Ga). As many as 400 heavy lift launch vehicle (HLLV) flights per year are required to construct 60 satellites over a 30-year period.

The major space impacts of the Space Shuttle and the SPS derive from: 1) transmission of microwave radiation through the atmosphere, 2) deposition of rocket effluents and ablated materials in the atmosphere, and 3) structures in orbit. Each of the above produces an array of effects.

This introductory review and the other papers in this section discuss a broad range of impact issues already identified in connection with the Space Shuttle and the SPS. However, the major emphasis is on the SPS because of its potentially greater, overall impact.

It should be noted that, where possible, some indication of the probability of occurrence of the various effects and their importance has been given. However, in view of the limited state of our present knowledge regarding the natural state of the upper atmosphere and its variations and, more importantly, the paucity of relevant data regarding perturbations caused by large space systems, the probability of occurrence of the effects and their significance should be regarded as uncertain or unknown unless specifically stated to the contrary.

In summary, the following review discusses our state of knowledge of those cause and effect relationships in the upper atmosphere that are regarded as most important. Section II discusses the effects of transmitting power through space. The impacts of rocket effluents and ablated materials being injected into various regions of the upper atmosphere are addressed in Sec. III. Section IV describes some of the anticipated effects of large space structures in the magnetosphere.

II. Microwave Effects

The greatest benefit of a system designed to collect solar energy in space for subsequent transmission to the Earth's surface is the essentially constant exposure (except for occultations) to direct solar radiation (1350 W/m^2) as opposed to the diurnally varying, attenuated solar radiation received on the Earth's surface (150-570 W/m^2). Several proposals have been made for the transmission of energy from the space environment to the Earth's surface. These include solar radiation reflected from large mirrors, infrared laser beams, and microwave beams. Each of these systems has its own technical feasibility and environmental problems, as well as its particular advantages and disadvantages.

The advantage of microwaves over the other alternatives is the relative transparency, even during storm conditions, of the atmosphere to the anticipated frequency band around 2.45 GHz. Hence, by far the greatest attention to technical feasibility and atmospheric effects has been devoted to the microwave system as described in the previously mentioned NASA reference design. Therefore, this introductory review and the detailed discussion in the paper by Duncan emphasize the microwave alternative. The reader interested in the other alternatives can consult some of the recent literature for information and additional references. The various laser options have been reviewed by Walbridge,[7] and the March 1979 issue of Astronautics and Aeronautics has several articles related to this topic.

The major concern regarding the transmission of microwaves through the atmosphere is interaction with the ionosphere. Modifications to the ionosphere could affect the propagation of both the SPS microwave and reference pilot beams, as well as the propagation of other frequency bands used by terrestrial and satellite communication systems. Duncan's contribution reviews this subject and focuses on recent results of ionospheric heating experiments conducted at Arecibo in Puerto Rico. Duncan also discusses a number of microscopic and collective plasma phenomena connected with the propagation of high energy electromagnetic radiation through the ionosphere in an attempt to identify the phenomena most likely to be influenced by the transmission of microwave power beams from GEO to the Earth's surface. However, before going any further, it will be worthwhile to review briefly the relevant aspects of the SPS reference design.

The reference design contemplates 60 satellites, each capable of delivering 5 GW of power at the bus bar. The microwave components of the system include the transmitting antenna, microwave beam, rectenna, and pilot reference beam. The pilot reference beam is used to maintain phase coherence and is transmitted from the rectenna center to the satellite transmitter center. Its frequency is 2450 MHz, which is in the middle of the governmental and nongovernmental, industrial, medical, and scientific (IMS) user band (2450 ± 50 MHz). This frequency happens to be at the upper limit of the Eastern European IMS band (2375 ± 50 MHz). As pointed out by Duncan, it is also a compromise between increasing absorption by hydrometeors in the atmosphere at higher frequencies and increasing interactions with the ionosphere at lower frequencies, although these variations are fairly slow in both cases.

The power beam has a Gaussian shape with a center power flux density of 22,000 W/m^2 at the transmitting antenna (1-km diam) and 230 W/m^2 at the rectenna. The rectenna is roughly 10 km in diameter and has a power flux density at its edge of 10 W/m^2. The rectenna's shape is adjusted according to the angle of incidence of the power beam. The maximum power flux density at the rectenna, which is approximately equal to that in the ionosphere, represents a compromise between exceeding the threshold for enhanced electron heating in the ionosphere and further increasing land use requirements.

The two main phenomena identified in Duncan's paper are ohmic heating, which converts a small fraction of the beam energy into thermal energy of free electrons in the ionosphere, and a plasma instability called thermal self-focusing, which results in plasma density and beam density irregularities.

Ohmic heating is a lower ionospheric phenomenon that results in increased temperatures of local free electrons. These temperatures were predicted to be on the order of several hundred degrees Kelvin. A small increase in electron density was also predicted because of a reduction in the charge-recombination rate of electrons and ions. However, no significant telecommunications or climatic impacts from this small density change were expected.

An experiment to verify these predictions was conducted at Arecibo. However, because of frequency limitations, a scaling law saying that ohmic heating scales inversely as the square of the driving frequency was employed to simulate SPS effects. The heating experiment resulted in a 100 K temperature increase. According to Duncan, most of the discrepancy between theory and experiment has been resolved. One reason for the low experimental value was thought to be a heating pulse that was too short in duration (nine milliseconds) to reach the maximum value. In any case, before reliable estimates of SPS heating effects can be made, it will be necessary to verify the scaling procedures used to extrapolate to the SPS reference design parameters.

The thermal self-focusing instability mentioned above is an F-region plasma effect that can give rise to large-scale variations in both the ambient plasma and in the beam profile. The threshold for exciting this instability is about 50 W/m^2 for the SPS frequency of 2450 MHz and is therefore, according to Duncan, well below the proposed peak power density of 230 W/m^2. Hence, one can expect self-focusing to occur. Depending on the degree of induced density fluctuations, this instability could cause significant modifications to high frequency (HF) and pilot reference beam propagation. It has not yet been possible to verify these effects with existing experimental facilities. However, as reported by Duncan, planned upgrading of the Arecibo and Platteville facilities will permit simulations of SPS power beam effects. Therefore, Duncan concludes that "no significant telecommunications or climatic effects have yet been experimentally demonstrated."

III. Rocket Effluents and Ablated Materials Effects

Direct deposition of rocket effluents can, of course, occur throughout the entire space environment. However, the principal exhaust releases of the SPS are expected to occur in three altitude ranges: 56-124 km, 450-500 km, and 500-36,000 km. Hence, for convenience, the upper atmosphere has been divided into the following three domains:

Domain A 56-124 km,

Domain B 124-500 km (LEO $\simeq$ 450-500 km), and

Domain C 500-36,000 km (GEO $\simeq$ 36,000 km).

A. Sources

Domain A Sources. As shown in Table 1, the major source of rocket exhaust in Domain A is the second stage of the HLLV, with the primary combustion products being H_2O and H_2 in the ratio of approximately 2:1. According to the NASA reference design, the HLLV engines burn between about 56 and 124 km as shown in Fig. 4. Because of the shape of the HLLV trajectory, a substantial fraction of the total effluent is injected between 110 and 124 km, i.e., just above the turbopause as shown in Figs. 1 and 5. The personnel launch vehicle (PLV) also burns its engines in this region, but details regarding its trajectory are not available. In any case, the HLLV exhaust will dwarf that of the PLV.

The other main source of pollutants in Domain A is vehicle reentry. Reentry results in ablation of material from the vehicles and heat shields and in production of nitric oxide (NO). Estimates by Park suggest that the mass of NO produced should equal about 22% of the mass of the vehicle.[8] With peak production at around 70 km, the NO should be distributed between 55 and 100 km.[9]

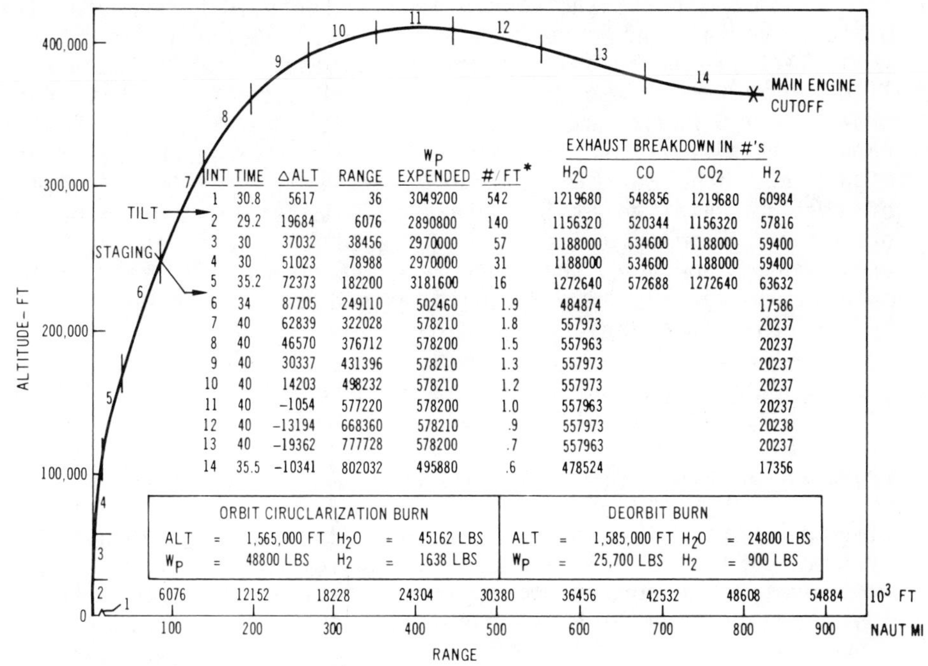

INT	TIME	ΔALT	RANGE	W_P EXPENDED	#/FT*	H_2O	CO	CO_2	H_2
1	30.8	5617	36	3049200	542	1219680	548856	1219680	60984
2	29.2	19684	6076	2890800	140	1156320	520344	1156320	57816
3	30	37032	38456	2970000	57	1188000	534600	1188000	59400
4	30	51023	78988	2970000	31	1188000	534600	1188000	59400
5	35.2	72373	182200	3181600	16	1272640	572688	1272640	63632
6	34	87705	249110	502460	1.9	484874			17586
7	40	62839	322028	578210	1.8	557973			20237
8	40	46570	376712	578200	1.5	557963			20237
9	40	30337	431396	578210	1.3	557973			20237
10	40	14203	498232	578210	1.2	557973			20237
11	40	−1054	577220	578200	1.0	557963			20237
12	40	−13194	668360	578210	.9	557973			20238
13	40	−19362	777728	578200	.7	557963			20237
14	35.5	−10341	802032	495880	.6	478524			17356

ALTERNATE TRAJECTORIES CAN BE CONSIDERED WITH LOWER INSERTION ALTITUDES IF ENVIRONMENTAL CONSIDERATIONS DEEM NECESSARY

* #/FT=WEIGHT OF PROPELLANT EXPENDED PER FOOT OF ALTITUDE CHANGE

Fig. 4 SPS heavy lift launch vehicle trajectory and exhaust products data (Source: Ref. 5).

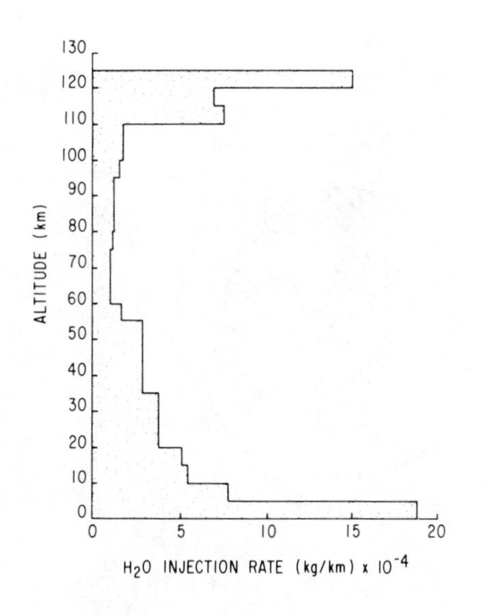

Fig. 5 HLLV water vapor injection rate versus altitude (Source: Ref. 18).

A less significant source of pollutants in this region is waste material and construction debris. Assuming that the equivalent of one percent of the total mass of two 5-GW satellite systems would be lost each year, Whitten estimates that this source would amount to about 10^6 kg/yr of material ranging in size from fine dust to relatively large objects that might reach the Earth's surface intact.[10] Approximately half of this material might be metallic, with the remainder being oxides of aluminum or silicon. Based on estimates by Park and Menees, however, the annual injection rate of meteoritic material (4×10^7 kg/yr) would exceed this man-made source by more than an order of magnitude.[11]

Domain B Sources. The major sources of effluents in Domain B are the circularization and deorbit burns of the HLLVs and PLVs, as well as similar burns of the personnel orbit transfer vehicles (POTVS). In each case, the propellant is LH_2/LO_2. One can also expect minor injections of combustion products from the space-shuttle-type orbit maneuvering system (OMS) and reaction control system (RCS) engines that will presumably burn monomethylhydrazine and nitrogen tetroxide. In addition to these chemical rockets, the electric ion

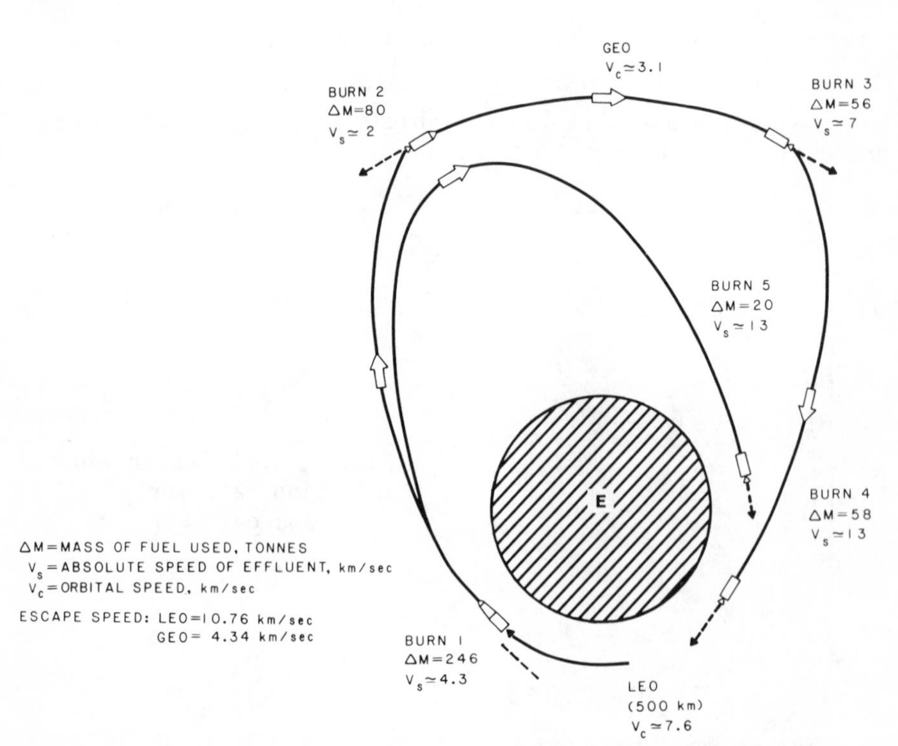

Fig. 6 POTV effluent deposition (Source: Ref. 13).

engines of the cargo orbit transfer vehicle (COTV) deposit some argon ions and electrons in the region near Domain B, although most of these effluents are injected at higher altitudes.

Domain C Sources. There are five engine burns in all for each POTV round trip between LEO and GEO. Figure 6 illustrates a typical scenario with nominal fuel consumption at each burn. The scenario includes a deorbit burn of the first stage rocket at LEO, stage separation and circularization burn of the second stage rocket at GEO, deorbit burn of the second stage at GEO, circularization of the second stage at LEO, and circularization of the first stage at LEO. In addition to the two burns at GEO, other effluent sources in Domain C include the electric ion propulsion system and the chemical rockets used on the COTV for control during eclipses of the sun by the Earth.

Perturbation Factors. In order to obtain a rough idea of the relative importance of these injections, one can compare the accumulation of these man-made injectants with the natural abundance of the material in each region. In such comparisons, cognizance must be taken of the natural processes that "turn over" ambient and injected material in each region, as well as those mechanisms that operate exclusively to remove the injectants or reduce them to ambient species and unperturbed levels. Taking account of such processes is never easy. Hence, such comparisons should be regarded as tentative. An attempt to make such comparisons with respect to SPS-related injections in the stratosphere and mesosphere was made by Brubaker.[12] More recently, Bauer made similar comparisons and extended them to Domains A, B, and C.[13] Brubaker defined a "perturbation factor" PF as:

$$PF = \frac{\text{injection rate (mass/yr) x residence time (yr)}}{\text{ambient loading (mass)}}$$

As defined, PF is a nondimensional quantity that can be evaluated for a particular material for each domain of space. Obviously, this evaluation requires estimates of both residence time for the material in the domain and ambient loading. Bauer's results for all three domains for the major SPS transportation system injectants are given in Table 2. From this table one can conclude that the injection of hydrogen (actually H_2O and H_2) in Domain A may be important and that injection of hydrogen atoms, argon ions (Ar^+), and energy is likely to be significant in Domain C. One failing of this type of comparison is that it does not take into account interdomain transport of injectants, e.g., transfer of hydrogen atoms to Domain B from Domain A. A second inadequacy is that the

Table 2 Perturbation factor (adapted from Ref. 13)

Altitude Range of Injection	Injectant	Injection rate, molecules/yr	Residence time	Ambient loading, molecules	PF[a]
70-120 km	H_2O/H_2	7.4×10^{34} (H atoms)	10 days	3.3×10^{34} (H atoms)	0.07
	NO	2×10^{33}	4 days	3.5×10^{33}	0.005
450-500 km	H_2O/H_2	1×10^{33} (H atoms)	ice: 5 minutes water: 2×10^{-4} yr	1×10^{30}	0.006
LEO to GEO	H_2O/H_2	1.3×10^{33} (H atoms)	1-10 days	4.4×10^{32}	0.01-0.1
	Ar^+	3.7×10^{32}		4.4×10^{25}	>> 1 ($\simeq 10^5$)
	Energy	14.5 MT-HE[b]/yr		7-70 MT-HE	0.0006-0.6

[a]Perturbation factor, PF = (injection rate)x(residence time)/(ambient loading). This is a dimensionless quantity that gives a measure of the significance of an injectant in a given altitude range. The numbers are only suggestive, because the concept of "residence time" is not always well defined.

[b]MT-HE = megaton of high explosive.

comparison may not be between the most relevant materials. In Domain B, for example, the comparison refers to hydrogen atoms. However, if the hydrogen atoms enter initially as H_2O or H_2, they are much more important to the phenomenon of ionospheric depletion than if they enter as hydrogen atoms.

For convenience this review will address the potential impacts of these rocket effluents according to the most relevant domain, which may not necessarily be the altitude range of initial injection.

B. Effects

Domain A Effects. As can be seen from Fig. 1, Domain A begins at the lower boundary of the mesosphere (50 km), includes the mesopause (80-90 km), which is the coldest portion of the atmosphere, and ends at the nominal second stage HLLV engine cutoff around 124 km. This domain also coincides with the D- and lower part of the E-regions of the ionosphere. Thus, it can be thought of as a transition region between the dense sea of neutrals below, which are governed by molecular collisions and the longer wavelength portion of the electromagnetic spectrum (UV and longer), and the rarified mixture of ionized and neutral species above, which is governed by long-range interactions and largely (although not entirely) by shorter wavelength radiation.

Although this domain may not be regarded by some readers as part of the "space environment," it is nevertheless a region of considerable importance to both the terrestrial and the outer space environments. Its neutral and ionized constituents partake in a rich variety of complex chemical and physical phenomena, many of which are very poorly understood. It is well-known, for example, that water, which is present only as a trace substance, has not been adequately measured[14] and that its role in the overall chemistry of the region is not understood well. With respect to the lower ionosphere, Sechrist has remarked that:[15]

> ...without doubt, the D-region is the least understood portion of the ionosphere, and it offers the scientific community a stimulating and challenging assortment of strikingly interdisciplinary unsolved problems.

However, since this domain seems to be the dumping ground for second stage effluents, some attempt must be made to evaluate the potential impacts in spite of the present lack of knowledge.

During the past year or two, efforts by several investigators to address Domain A effects have gotten underway. The researchers include Mendillo and Forbes of Boston University (see their contributions in this volume), Zinn and Sutherland from Los Alamos Scientific Laboratory, Turco from R and D Associates, and Bernhardt et al. of Stanford University. In addition, several workshops designed to identify atmospheric effects in this domain (and others) have been conducted, including: Workshop on Ionospheric and Magnetospheric Effects,[16] Workshop on Stratospheric and Mesospheric Effects,[17] and, most recently, Workshop on Modification of the Upper Atmosphere by Satellite Power System (SPS) Propulsion Effluents.[13]

The major effects identified thus far are:

1) Compositional perturbations in the vertical profiles of neutral chemical constituents, particularly trace substances such as ozone, water vapor, nitric oxide, and ablation products (materials ablated from rocket components during reentry).

2) Lower ionospheric (D- and lower E-region) compositional, density, and electrical conductivity changes.

3) Response of the atmosphere to local thermal energy inputs from rocket engine burns and microwave energy absorption.

4) Formation of mid-latitude noctilucent clouds and their influence on upper atmospheric dynamics and albedo.

5) Climatic effects associated with changes in composition, cloud cover, and energy injection.

6) Changes in electromagnetic and radiative properties, including enhanced radiative cooling, airglow, and alteration of radio wave propagation (sporadic-E, VLF propagation, etc.).

Clearly, these effects are not all distinct. An attempt to show plausible interrelationships is shown in Fig. 7. The initiating causes appear in the small boxes to the left, and the final impacts are shown in large ellipses. The indicated relationships should be regarded as suggestive because of our limited understanding. Two major problem areas are immediately apparent from this figure. First, will the intermediate and ultimate impacts indeed result from the rocket effluents? And, second, will these effects be significant, either by themselves or in comparison with naturally occurring phenomena?

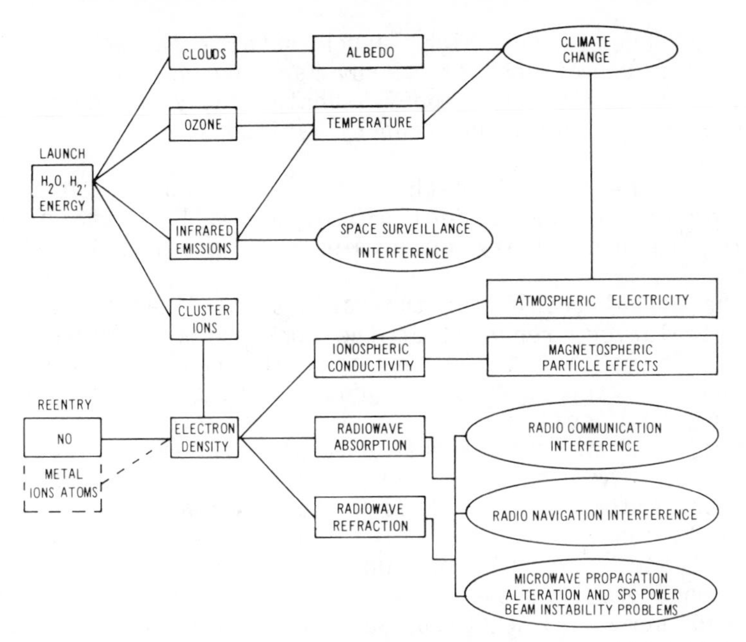

Fig. 7 Rocket effluent effects in Domain A (56-124 km).

The three workshops cited above[13,16,17] and Mendillo's and Forbes' contributions to this volume have addressed all of these issues, at least in a preliminary way. In most cases, the general finding of these efforts is that our ability to estimate the nature and extent of the above listed effects is severely limited by our lack of knowledge of the unperturbed atmospheric chemical constituents and dynamics. Bearing this in mind, an attempt has been made to summarize our present understanding of each of the above effects.

a) Compositional Perturbations. Participants in the Workshop on the Stratospheric and Mesospheric Impacts of Satellite Power Systems in September, 1978, concluded that:

1) The most significant perturbations arise in the mesosphere from exhaust injections of water vapor and from nitrogen oxides produced by reentry. These effects may not be negligible.

2) Based on preliminary one-dimensional model calculations and model-independent considerations, a global increase in the water vapor concentration of 1-2% at 50 km and 4-5% at 80 km may be expected for a launch rate of about 700 HLLV flights per year.

3) Due to the high launch rates expected, a significant "corridor effect" (a narrow band around the globe) may result. This effect could modify substantially the composition perturbations at launch and reentry latitudes.

4) The effect on the total, vertically integrated ozone concentration is negligible; a small fraction of 1% according to preliminary one-dimensional model calculations.

The actual results of the various one-dimensional global model calculations reported at the workshop are summarized in Fig. 8. The shaded areas represent the range of values reported by the various modeling groups. It should be stressed that, due to uncertainties in emission rates, rocket trajectories, and various ranges of applicability, the modeling results should be regarded as very preliminary. In particular, only annual emissions from HLLVs up to 80 km were considered in the calculations. Since the bulk of the emissions occur above this altitude (Fig. 5), one would expect these results to change considerably above 80 km. Brubaker has very recently completed some calculations using a simplified modeling approach that takes into account emissions up to the second stage engine cutoff altitude.[18] His results (shown in Fig. 9) appear to agree quite well with the earlier one-dimensional model calculations up to about 70 km; above that altitude, however, they show much larger changes in H_2O concentration than would be expected based on the emission distribution shown in Fig. 5.

Forbes has recently completed some calculations for the entire Domain A, and his work is reported in his contribution

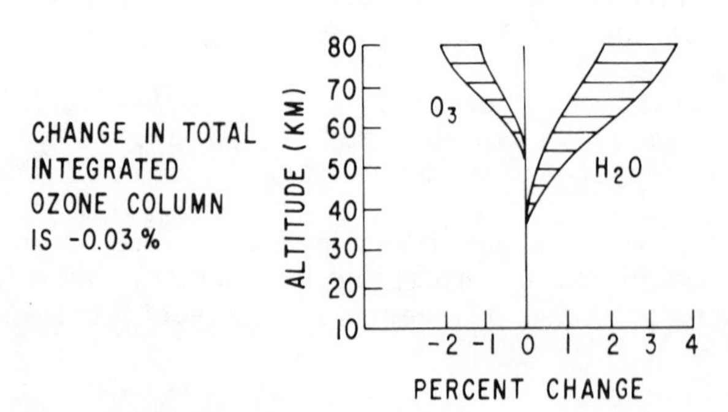

Fig. 8 One-dimensional model estimates of effect of rocket exhaust emissions on stratospheric/mesospheric ozone and water distributions (Source: Ref. 18).

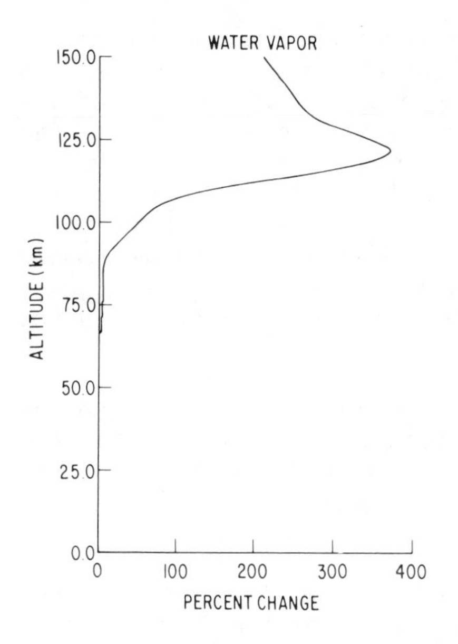

Fig. 9 Percent change in ambient water vapor concentration (Source: Ref. 18).

to this volume. The work is especially relevant to the issue of "corridor effects" referred to in the workshop conclusions, since he estimated the extent of the perturbation in water vapor concentration required to produce "measurable" effects. Forbes found that the net water vapor concentration would have to be >100 ppm (one to two orders of magnitude greater than ambient values) for measurable effects to occur. Such high water vapor concentrations could lead, according to Forbes, to at least a doubling of hydroxyl radical (OH^-) concentrations below 100 km and to a net radiative cooling because H_2O radiates in the infrared without absorbing much sunlight. The high H_2O concentration requirement for significant radiative cooling arises, according to Forbes, from relatively low collision frequencies, which makes the transfer of energy from translational to excited states of H_2O inefficient. Since, as Forbes' calculations show, the size of the area in which the concentration of H_2O exceeds 100 ppm is limited to about 20,000 km^2, large-scale effects are not expected to be significant.

It should be noted that some experimental evidence for a cooling effect has been reported by Chenurnoy and Charina.[19] They studied variations in hydroxyl band IR emission before, during, and after a noctilucent cloud display and concluded that higher H_2O concentrations before formation lead to enhanced OH^- emission, causing local cooling. After formation, the water vapor is condensed into cloud particles which, according to these authors, reduces the IR emission.

The enhanced production of OH^- radicals by photolysis of H_2O is important, because these radicals are involved in an atomic oxygen loss mechanism and because they are one of the most intense emitters of IR in the airglow spectrum. The atomic hydrogen released by the photolysis of H_2O also is important, as it is involved in the dominant ozone loss mechanism between about 75 and 95 km. However, a significant fraction of the hydrogen atoms is likely to diffuse upward into the thermosphere. The consequences of this added upward hydrogen atom flux into the thermosphere will be discussed later.

The production of NO during vehicle reentry is expected to have little effect on mesospheric ozone.[20] In any case, this production mechanism is expected to be too small relative to natural sources to affect stratospheric NO_x levels significantly.

b) Lower Ionospheric Effects. The Workshop on Ionospheric and Magnetospheric Effects of Satellite Power Systems held in August, 1978,[16] reached the following conclusions regarding D- and lower E-region effects. Water injection tends to convert the region from one dominated by light molecular ions to one dominated by heavier water cluster ions that recombine more rapidly with free electrons. This leads to a reduction in the daytime electron density. On the other hand, the injection of NO by reentering vehicles might lead to an enhancement of daytime ionization, since NO ionization by solar UV radiation ordinarily dominates free electron production anyway. At night, scattering of UV radiation in the geocorona by hydrogen atoms, which ordinarily maintains the lower E-region nighttime ionization, is somewhat enhanced. This results from the photodissociation of H_2O and the diffusion of hydrogen atoms into the geocorona. Forbes estimates that with water vapor concentrations $\geq$100 ppm, a 75% reduction in daytime ionization levels between 70 and 100 km could occur.

c) Energy Injection Effects. Preliminary estimates of energy injection rates from both the rockets and the SPS microwave beam are quite small when compared with solar UV radiation absorbed by ozone.[21] Hence, temperature changes resulting from these energy inputs are expected to be negligible. Forbes estimates that one might expect to find some "noticeable" disturbances within a radius of 6 km of the rocket trajectory because of thermal energy and momentum transfer from the rocket engines.[22] In particular, he suggests that both shocks and gravity waves may be detectable as has been the case for past launches of large rockets. The significance of such disturbances is unclear at this time, but Forbes has suggested that archived data should be studied to attempt to correlate launches of large rockets with the occurrence of sporadic-E or other electromagnetic effects.

Another source of energy is chemical energy released by reaction of rocket effluents with ambient constituents. The potential magnitude and impact of this source of energy probably is small, but it has not yet been evaluated. Turco points out that such an evaluation is difficult, because the dynamics of the unperturbed mesosphere are only poorly understood at present.[20]

d) Noctilucent Clouds and Other Condensation Phenomena. Noctilucent clouds are thin clouds formed at an altitude of approximately 85 km, which is in the coldest region of the atmosphere (mesopause). These clouds are generally visible in the summer northern latitudes after sunset when the observer is in the Earth's shadow and the clouds can still reflect the direct rays of the sun. (For an excellent discussion of this type of cloud accompanied by some beautiful pictures, the interested reader is encouraged to see the early paper by G. Witt.[23] For additional sightings and morphology, see the paper by Fogle and Haurwitz.[24]

The possibility of increased noctilucent cloud formation near the mesopause, especially at lower latitudes, is one of the more curious topics to arise in connection with water vapor releases from rockets. Ellsaesser points out that these clouds have been the subject of scientific investigation for nearly a century. However, it is only within the last few years that their nature has become at least partially understood.[14] Current theories, according to Ellsaesser, favor a crystalline ice composition covering a meteoritic dust nucleus or, alternatively, a hydrated-metallic-ion nucleus.

The conditions favorable to the formation of these clouds include an extremely cold mesopause, as well as sufficient water vapor, and an abundance of numerous condensation nuclei. Ellsaesser suggests that these conditions are met in the summer in the high-latitude mesopause region, and it is there that such clouds have been observed.[14] Noctilucent clouds are not seen at latitudes below about 45 deg, except for artificial ones correlated with SCOUT missile launches from Pt. Mugu at latitude 34 deg[25] and with French sounding rockets at latitude 45 deg.[26]

The occurrence of "artificial" noctilucent clouds at low latitudes is curious in view of Forbes' theoretical discussion.[22] Forbes notes that in order for ice clouds to form, the ambient temperature T_a must be less than the frost-point temperature T_{fp}, which is lower during the summer than in the winter at all mesopause latitudes. Below 60 deg latitude, $T_a \gtrsim T_{fp}$ only for water vapor concentrations >100 ppm, even in the summer. (Note that ambient water vapor concentrations are highly uncertain but are in the range of 1-10 ppm.) Below 40 deg latitude, the water vapor concentration would have to exceed 1000 ppm. In the case of the HLLV, these concentrations are estimated by Forbes to occur over areas 20,000 km^2 and 200 km^2, respectively.

Another very important point raised by Forbes is that sufficiently large updrafts must exist to support the ice crystals during their growth. The required updraft is about 1 m/sec for 0.1 μm spherical ice particles. Forbes suggests that since such updrafts occur only at high latitudes during the summer, this is consistent with the observation of noctilucent clouds only under such conditions. However, and this appears to be critical, Reid has shown that the growth of needle- or disk-shaped crystals does not require such large updrafts.[27] Therefore, it may still be theoretically possible to form persistent middle-latitude noctilucent clouds. This may explain, at least in part, the controversial mid-latitude sightings of noctilucent clouds that were associated with small rocket engine burns.

In comparing theoretical studies (such as those of Forbes and others cited by him) with observations of clouds formed by high-altitude, water vapor releases, it is important to keep in mind that there may be more than one process involved depending on the mode of release and, in particular, on the initial temperature and density of the released fluid. Reported water vapor release experiments are discussed by Mendillo, both in this volume and in a workshop proceedings publication.[28] These experiments are of several types: 1) those involving the rupture of water tanks carried aloft, 2) those involving the production of water vapor as a by-product of an explosion, and 3) those involving rocket exhaust.

The question of the early fate of releases by these various modes was addressed at the LaJolla Workshop in June, 1979, by Bernhardt,[29] Park,[30] and Mendillo[28]. In particular, Park points out that, in the case of rocket exhaust, rapid adiabatic expansion of the plume outside the nozzle into the near vacuum of space leads to rapid condensation of the water vapor within a fraction of a second. This implies that an initial condensation of at least a part of the water vapor occurs essentially independent of altitude in Domains A, B, and

C, above 70 km altitude. Park suggests that as much as 90% of the water vapor in rocket exhaust should condense to the solid state. Bernhardt, on the other hand, points out that several factors can influence the degree of condensation of released vapor. These include the initial specific enthalpy (C_pT/ρ) of the vapor, the expansion geometry, and the exhaust constituents (which, in turn, depend upon the rocket fuel and especially on whether it is solid or liquid). Bernhardt's calculations indicate that while the low specific enthalpy of the Lagopedo releases resulted in approximately half of the mass of the water vapor condensing, the relatively high specific enthalpy of typical rocket exhaust is expected to yield much lower condensate fractions (5-15%). In the observations reported by Meinel et al.[25] and Benech and Dessens,[26] the rockets were solid-fueled, and the exhaust contained a substantial mass fraction of aluminum oxide particles on the order of 5μ in diameter as well as water vapor. Hence, independent of the condensation of the exhaust water vapor, the aluminum oxide particles could act as nuclei to stimulate growth of ice crystals from the ambient moisture. The injected water vapor would, of course, provide additional moisture to enhance the growth process.

Mendillo concludes in this volume and in Ref. 28 that there is ample evidence to indicate that at least 50% of the water vapor injected by all three methods listed above goes into the rapid formation of exhaust clouds that are probably composed of ice crystals. He also suggests that such clouds become visible within less than 1 sec of release. In addition, the ratio of water vapor to ice crystal mass formed in each case probably depends on the injection mechanism as well as the altitude. Finally, he points out that whereas evidence from the high-water experiments of 1963 and the Lagopedo experiments of 1977 suggest that 80-100% of the water injected via explosives or storage tank ruptures quickly condenses, rocket exhaust injections may result in significantly lower fractions of condensate ranging from 50-100%. The uncertainties are apparently quite large.

Consequently, one would expect that at least some fraction of the total water injected by the rockets will enter Domain A (or Domains B and C) initially as ice crystals. In view of this, two questions come to mind. First, what happens to all of these ice crystals, and what role do they play in the formation of persistent clouds, such as noctilucent clouds? And, second, what happens to the remaining water vapor?

To answer the second question first, those molecules that escape the rapid condensation process are available for subsequent cloud formation, for reaction with ambient neutral and ion species, for photodissociation and ionization, and for transport out of the injection region by ambient winds or diffusion. Some of those water molecules and the hydrogen atoms resulting from photodissociation diffuse upward into the upper ionosphere and thermosphere. The fate of these species is discussed in a subsequent section.

The first question is more relevant to the present section. The ice crystals formed rapidly in the rocket exhaust plume follow ballistic trajectories until they are slowed by collisions to ambient terminal falling speed, unless they are completely reevaporated by energy absorbed through collisions or from solar radiation. The terminal velocity is a rapidly changing function of the neutral density and quickly drops to about 1 m/sec at 85 km for spherical ice crystals and less for other shapes. If updrafts approximately equal to the terminal velocity persist and the crystals are in thermodynamic equilibrium with the ambient air, the crystals could remain to form persistent clouds.

Thus, provided that conditions are favorable, the initially formed ice crystals and the water vapor that escaped the initial condensation are available to form persistent clouds. If the conditions are not favorable, only the initial condensation will be observed, and that should dissipate in a relatively short time. In most cases, however, available data do not allow one to distinguish exactly what is going on when clouds have been observed. The two French rocket launches that triggered the formation of clouds are excellent examples; both the initial and then the persistent condensation were observed. The following quote from Benech and Dessens is self-explanatory:[26]

> While this stage (third stage burned from 159 km down to 77 km) was burning, the rocket left a visible exhaust trail behind it; the trail disappeared after some seconds except on a short length of the trajectory; this small segment (later identified as lying between 79 and 92 km) of the tail became more and more visible, and at its upper and lower extremities two trails expanded in opposite directions.

The cloud remained visible until it was no longer illuminated by the setting sun. The authors stated that the rocket produced 260 g water vapor, 520 g hydrogen chloride, and 1600 g aluminum oxide on a 10-km-long trajectory. Had this cloud been more uniformly visible along the entire trajectory, it might have been explained as either the sunlight being reflected from the aluminim oxide particles or the rapid condensation of exhaust vapor within the plume.

Returning now to the possibility of SPS rockets forming noctilucent clouds, Forbes concludes that no possibility for formation exists in the winter hemisphere, whereas the possibility for local cloud formation does exist in the summer hemisphere.[22] Cloud formation or natural cloud induced growth presumably would be enhanced by the formation of heavy water cluster ions that can serve as ice nuclei and by the enhanced local radiative cooling by the added water vapor. Forbes, however, does not think that such cloud formation is likely to be a global problem in the middle latitudes.

On the other hand, the participants in the Workshop on Stratospheric and Mesospheric Impacts of Satellite Power Systems were not quite as certain in their evaluation of the problem and stated:

> It is not possible at present to assess the likelihood of an increase in noctilucent cloud formation due to water injection in the mesosphere. Better estimates of the expected increase in the water concentration as a function of latitude are required, as is better information about existing mesospheric temperatures and water concentrations.

Finally, based on the observations described by Mendillo in this volume and our present understanding, the initial condensation of at least 50% of the water vapor is quite likely to occur. What happens to the water vapor and ice crystals thereafter seems to be a fairly open question.

e) Climatic Effects. No clear-cut indications exist that rocket exhaust emissions in Domain A will cause weather or climatic effects. However, in view of the unprecedented increase in space vehicle activity anticipated and the obvious consequences if such effects did arise, this subject must be treated with considerable care. To date, a number of possible atmospheric effects have been defined that could potentially lead to inadvertant weather and/or climate modification. These include but are not limited to: (1) formation of persistent, large-scale clouds that can alter the radiation balance, (2) compositional changes that can modify the thermal structure and/or dynamics of the atmosphere, and (3) alterations of atmospheric electricity that can influence thunderstorm processes.

The participants in the Workshop on the Stratospheric and Mesospheric Impacts of Satellite Power Systems reached the consensus that:[17]

1) It is uncertain whether an increase in noctilucent cloud formation would have a detectable effect on the tropospheric climate; the coupling between mesospheric phenomena and climate is poorly understood.

2) Other direct climatic effects of the expected composition perturbations are probably negligible; changes in surface temperatures are estimated to be much less than 0.1 K.

Based on his recent work as discussed in this volume, Forbes concludes that while small-scale atmospheric effects are possible over Domain A regions of $\lesssim$ 20,000 km^2, it is unlikely that such effects will have a significant global influence. On the other hand, Forbes regards his work as preliminary. The challenge to the scientific community of answering this question in a more definitive manner remains.

Somewhat more speculative mechanisms for coupling the upper atmosphere to tropospheric weather and climate have been suggested by various investigators. Hines, for example, has suggested that changing the mode of planetary wave propagation in the mesosphere may modulate tropospheric weather.[31] Such changes could conceivably be brought about by compositional changes that alter the reflectivity of such waves at the mesopause.

Another example of a speculative mechanism is related to the recently proposed mechanisms being developed to explain the so-called solar-weather effect.[32] The atmospheric electricity mechanism of Markson may be relevant to the present problem.[33] His view is that, if the global electric circuit is modified by, for example, modulation of the ionizing radiation that controls the conductivity of the atmosphere above thunderstorms, then the thunderstorm processes themselves could be altered. Unfortunately, the role of atmospheric electricity in thunderstorm processes is not well understood and how the use of space might modify the conductivity to influence this mechanism needs further study.

f) Changes in Electromagnetic and Radiative Properties. The radiative characteristics of Domain A most subject to change will be: 1) the screening of the lower portion of Domain A from direct solar Lyman alpha and beta radiation by the deposition of large amounts of water vapor between 110 and 125 km, 2) the scattering of this same radiation in the geocorona with subsequent enhanced flux reaching Domain A at nighttime, 3) the airglow in the IR part of the spectrum, and 4) the UV absorption profile corresponding to the ozone profile. While all of these phenomena are likely to be changed, at least to some extent, by implementation of the SPS, the magnitude of the changes and their significance have not been established beyond preliminary results.

The main electromagnetic properties subject to modification are the plasma density profiles and the electrical conductivity distributions and their spatial and temporal fluctuations. Changes in the plasma density (electron and ion concentrations) already have been discussed. Since the conductivity is directly proportional to the density of positively and negatively charged particles, a measurable reduction in the daytime conductivity due to ion-cluster formation (which enhances recombination) and screening of solar Lyman alpha radiation is expected, at least over regions $\simeq$ 20,000 km^2. Similarly, one would expect the reentry of space vehicles to lead to an increase in conductivity and that the diffusion of hydrogen atoms into the geocorona would lead to an increase in the nighttime conductivity in Domain A. In addition, the ablated material from space vehicle reentry and space debris could lead to striations or thin layers of greatly enhanced conductivity (1- or 2-km thick), called sporadic-E, in the region of 100-110 km. Another source of conductivity change would be modifications in the magnetospheric structure brought about by injection of rocket effluents in Domain C or even Domain B. If such structural modifications were to influence the frequency of occurrence or the intensity of high-energy-particle-precipitation events, then the ionospheric conductivity could be significantly altered.

The potential consequences of changing the conductivity distribution have not been investigated in detail. However, based on discussions conducted at workshops,[16,34] it seems plausible that such changes could lead to alteration of radio wave propagation and possibly to large-scale changes in electric fields and current systems. For example, the high-latitude conductivity distribution affects the high-latitude electric field which, in turn, influences the equatorial electric field, which can affect the occurrence of spread-F, a form of plasma density irregularity affecting radio wave propagation. The high-latitude conductivity distribution also is important because this region of the ionosphere completes the electrical circuit that couples the atmosphere auroral zone to the outer magnetosphere. The currents that flow through this circuit (referred to as Birkeland currents and electrojets) undergo large fluctuations during magnetic substorms and have been known to cause damaging current surges in power transmission lines and long telephone lines.[35,36] (For a review of a number of occurences of tripping circuit breakers, refer to a paper by Slothower and Albertson.[37]) Thus, alteration of the auroral zone conductivity could modify the morphology of this current system and influence the occurrence or intensity of terrestrial current surges, perhaps moving them to more populated areas.

The conductivity of Domain A is also important for communication systems, especially those using VLF. Finally, Markson has speculated that, since the lower ionosphere is part of the global atmospheric electric circuit through which currents are driven by thunderstorms, large-scale perturbations in the conductivity, especially if they reach down to the middle stratosphere, may influence thunderstorm processes and therefore weather and climate.[33]

Domain B Effects. Domain B (124-500 km) is subjected to: (1) rocket effluents and their by-products, both of which diffuse upward from Domain A, (2) effluents injected directly by orbit maneuvers in Domain B, and (3) effluents and their by-products that fall or diffuse downward from Domain C.

Included within this region (Fig. 1) are the upper portion of the E-region and most of F-region of the ionosphere. The E-region extends from about 90-150 km and is formed principally by soft, solar X rays. The main positive ions are O_2^+, NO^+, and O^+. Even though the free-electron density drops by one to two orders of magnitude from daytime to nighttime, the density remains high enough during the night to cause reflection of radio waves. Historically, this was the region of the ionosphere discovered to be responsible for the first long-range transmissions of radio signals. The E-region also constitutes a transition zone between regions dominated by molecular ions and those dominated by atomic ions. Approaching the top of the E-region, the $O+$ ion concentration becomes increasingly larger and begins to dominate in the F-region above 150 km. This is a key feature of the E-region, because the electron-ion recombination rate is much greater in the presence of molecular ions. This point will be returned to later.

The F-region is divided into the F_1-region, extending from about 150-200 km, and the F_2-region, extending out into the plasmasphere beyond 1000 km. The F_1 electron density profile reaches a peak or shoulder during the daytime and undergoes a quite large reduction during the nighttime, as shown in Fig. 10.[38] The F_2-region contains the highest electron density profile peak ($\simeq 10^6$ electrons/cm^3), which undergoes a reduction by about a factor of 10 during the nighttime period. The F_2-region positive ion population is dominated by O^+ ions which recombine rather slowly with electrons. Extreme ultraviolet solar radiation is the principal source of ionizing radiation, and maximum production occurs around 160 km. However, the electron density continues to increase beyond 160 km because above this altitude the reduction in the electron loss rate is greater than the reduction in the ionization rate.

As shown in Fig. 1, the F-region overlaps both the neutral atmosphere (thermosphere) and the lower part of the magnetosphere. As the altitude increases from 150 to 500 km, the mean free path increases from 40 to 80,000 m. Therefore, the Earth's magnetic field plays an increasingly important role in determining the mean charged-particle motion. Collisions with the neutral particles cause the ionosphere, on the average, to corotate with the rest of the atmosphere around the Earth. This net corotation is superimposed on the spiral motions of individual charged-particles along magnetic field lines brought about by the VxB force. In addition to these motions, short-term imbalances in the basic driving forces can give rise to large changes in particle motion relative to the corotating frame of reference. Neutral wind patterns, for example, are highly variable and difficult to predict but are very important in determining the morphology of the ionospheric "holes," which are strongly influenced by net charged-particle drifts relative to the mean corotational motion brought about by changes in ionospheric electric field patterns. "Holes" are discussed in more detail later.

a) Fate of the Rocket Effluents. The fate of the rocket effluents in Domain B is illustrated schematically in Fig. 11. The main impacts are: 1) enhanced airglow, which may influence satellite-based surveillance systems; 2) reduction in plasma (electron and ion) densities, which could interfere with radio communications and navigation; and 3) thermal expansion, which could lead to increased satellite drag.

Ultimately, rocket exhaust molecules will contribute hydrogen atoms that will escape from the Earth's atmosphere, either at present or possibly at enhanced escape rates. If the rate of escape is not enhanced, there would be a buildup in ambient hydrogen atoms to a new equilibrium concentration. This greater hydrogen atom concentration could lead to important neutral atmosphere effects, possibly including changes in

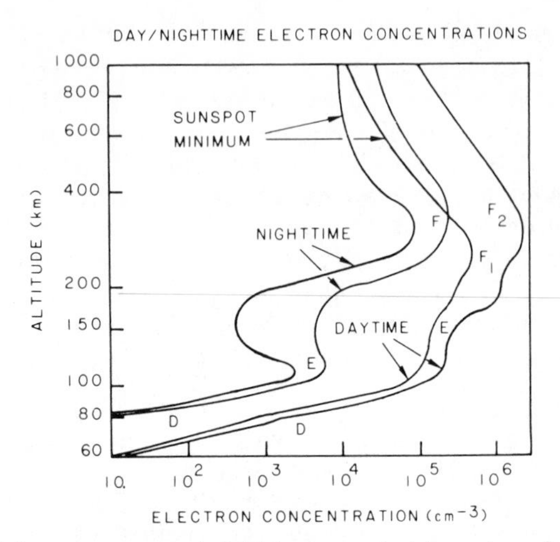

Fig. 10 Daytime versus nighttime variation in electron concentrations (Source: Ref. 38).

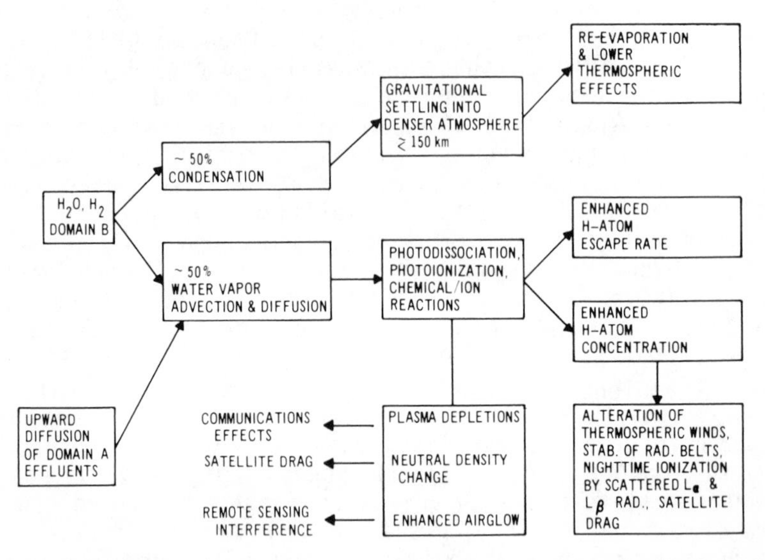

Fig. 11 The fate of rocket effluents in Domain B (124-500 km).

the thermospheric wind patterns, changes in stability of the radiation belts from the expanded geocorona[39] (see Vondrak's contribution to this volume), and increases in satellite drag. The significance of these effects remains to be evaluated.

b) Ionospheric Depletions. The exhaust molecules themselves (H_2O and H_2) will cause the dominant F_2-region ions (O^+) to be replaced by molecular ions (e.g., H_2O^+)

that recombine very rapidly with electrons (see Mendillo's contribution to this volume). This depletion or "ionospheric hole" can extend far beyond the local source of injected molecules and thus may affect radio communication and navigation systems over a wide geographical area. The perturbation may be extended to the conjugate ionosphere (at the opposite end of the geomagnetic field line passing through a hole) via draining effects on a plasmaspheric flux tube and/or by allowing enhanced photoelectron escape from the depleted region. This enhanced escape would occur because the electron collision frequency would be reduced.

It has now been established that rocket exhaust products deposited in the F_2-region will cause such ionospheric depletions. Although the mechanisms involved are reasonably well understood in general terms, specific details require further investigation. In particular, the influence of such holes on radio communications remains uncertain.

Chacko and Mendillo at Boston University (see Mendillo's contribution to this volume) have undertaken a detailed investigation of archived data in an attempt to uncover as much information as possible on the effects of rocket launches on the upper atmosphere. To date, their reappraisal of all papers on rocket effects published prior to the Skylab launch (1973) indicates the following:

1) Rapid injection of gases into the F-region produces localized, short-lived regions of depleted free-electron density by physically pushing the ambient plasma aside. The spatial and temporal extent of such a "snowplow" effect was determined by using the time required for the injectants to expand to ambient densities. Thus, these effects are altitude-dependent, with time scales ranging from a few seconds to minutes and spatial scales up to several kilometers from the rocket nozzle. No significant consequences have been associated with these effects.

2) Effects from chemical interactions between the injectants and the atmosphere were never considered in conjunction with pre-Skylab observations of rocket effects. However, these effects did turn out to be the cause of longer-term and spatially-extended effects, as observed in the Skylab launch.

3) Even extremely large injections of H_2O into the 100- to 150-km region distinguished themselves only by the size of the ice clouds produced.

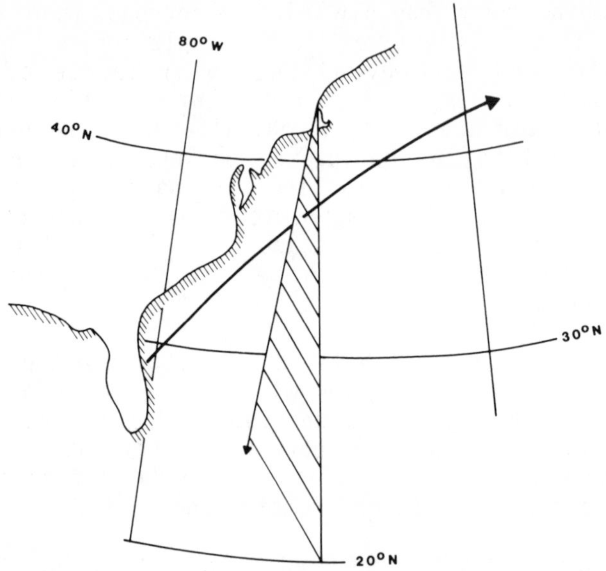

Fig. 12 Intersection of the ray path from the ATS-3 satellite to the Sagamore Hill Observatory with the rocket trajectory from Kennedy Space Center into the ionosphere (Source: Ref. 40).

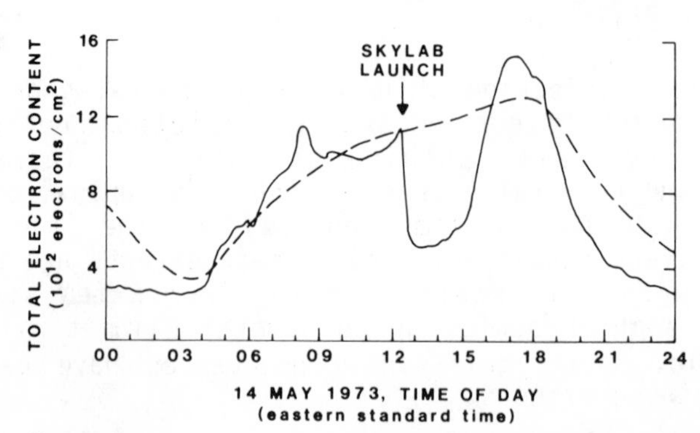

Fig. 13 "Hole" seen in the total electron content measurement of the ionosphere by the Sagamore Hill Observatory (Source: Ref. 40).

The May 14, 1973, launch of Skylab I with a Saturn V rocket marked the first time, according to Chacko and Mendillo, that a substantial amount of rocket exhaust was injected into the F_2-region of the ionosphere. As has been mentioned, this region is distinguished by the relative absence of molecular ions that are responsible for the relatively lower free-electron equilibrium content in the F_1 region. As described in detail by Mendillo et al., it was a pure accident that the trajectory of that rocket happened to intercept the radio-signal ray-path connecting the ATS-3 communication satellite with the ground-based observatory at Sagamore Hill (Fig. 12).[40] The deep depression in the total electron content (TEC) observed at Sagamore Hill after the rocket passed the intersection point is shown in Fig. 13.

Based on observations made at other ground-based observatories, whose lines-of-sight with ATS-3 were intercepted by the resulting depleted region, Mendillo et al. reported that the "ionospheric hole" had a radius of about 1000 km and lasted for about four hours. More recently, in an attempt to simulate or, more appropriately, "match" the observed data of that ionospheric depletion, Zinn and Sutherland reported that the hole observed by ground-based stations could have drifted out of the lines-of-sight of the observatories along magnetic field lines.[41] Further, if typical thermospheric winds existed at the time (no observations were in progress at the time), the residual rocket exhaust products not consumed in the initial hole-making process could have been transported to adjacent ionospheric regions to rapidly recombine additional electron-ion pairs. This effect of winds or movement of neutral atmospheric constituents, which are far more abundant than the ions and electrons, would also result in the effective displacement of the hole out of the lines-of-sight of the observatories.

The first successful attempt to produce an ionospheric depletion under controlled experimental conditions occurred in September, 1977.[42] Project Lagopedo involved two rocket-borne experiments, Lagopedo uno and duo. Each rocket carried 88 kg of high explosives and an instrument package that was separated from the parent rocket prior to detonation of the explosive. The detonation products included about 30 kg of H_2O, 16 kg of CO_2, and 20 kg of N_2.[43] These experiments generally confirmed theoretical predictions of an initial period, on the order of several seconds in duration, during which the ambient plasma would be swept away over a distance of less than 1 km by the rapidly expanding detonation cloud, followed by a much larger-scale plasma depletion process lasting for at least 30 minutes and extending over a radius of 30 km or more. In addition to the rocket-borne instrument package, which sampled the disturbed area for about two to three minutes, ground-based instruments monitored the "hole" for about 30 minutes.

A major finding was the importance of the suppression of participation of water molecules in the electron-ion recombination process because of initial formation of ice crystals followed by subsequent gravitational settling. The ice crystals then reevaporate and/or sublime at lower altitudes where the presence of the water is less significant because of the natural abundance of molecular ions. While precise details regarding this suppression mechanism are not yet available, it seems clear from the analysis of ground-based optical measurements that ice crystal clouds did form.[44] (Due to the time of detonation, the cloud from only one experiment was visible.) Particle radii were deduced to be 0.1 m.[41] The fraction of the water that condensed remains uncertain. Such condensed-phase exhaust clouds have been observed in high-altitude rocket burns as discussed earlier and by Molander and Wolfhard in regard to the translunar injection burn of Apollo 8 from its parking orbit at 185 km.[45]

Until now, a number of hole formation effects have been identified as plausible, but few have been experimentally verified. Some of these are summarized briefly below.

Large Scale Reductions in Electron Density. Based on work already referenced, there exists at least the possibility for large-scale, perhaps even global-scale, permanently depleted electron-density zones. (It should be noted, however, that the size of the ionospheric F-region depletion does not scale with the number of the reactive exhaust molecules (H_2, H_2O, CO_2).[46]) This possibility was considered at both the workshops cited earlier.[13,16] For example, one plausible scenario discussed at the Bauer workshop was that the daily HLLV launches would lead to an electron-depleted zonal region centered on Kennedy Space Center extending over the altitude region of 150-500 km, with a N-S extent of 1000-3000 km and uniformly distributed around the globe. In view of the uncertainties involved, however, the state of knowledge was felt to be inadequate to evaluate the likelihood of forming such a zone of depletion.

Nevertheless, the consequences of such large-scale, permanent changes in ionospheric morphology could be quite serious, especially for those using the ionosphere as an inexpensive means of conducting long-range communications in the 3-30 MHz radio band.[47] Hence, it is anticipated that this subject will receive considerable attention in the near future.

Draining of the Plasmasphere. The plasmasphere (Fig. 1) might be drained to some extent because of the depletion of F-region plasma at the foot of the magnetic flux tube. Downward diffusion of plasma parallel to Earth's magnetic field lines will attempt to fill in the F-region hole. If large (excess) amounts of H_2O remain at the foot of the flux tube, it is possible that the entire flux tube will be drained of plasma. If the plasmasphere is not precisely corotating with the Earth, the flux tube depletion process could deplete plasma from a large longitudinal sector of the magnetic shell that intersects the F-region hole. Slower cross-magnetic-shell diffusion could lead to removal of plasma from shells far from the shell that intersects the F-region hole. Therefore, a large part of the plasmasphere could be affected. The cold, light-mass plasma normally present in the plasmasphere leads to pitch angle diffusion of Van Allen belt energetic particles. Absence of some of this cold plasma might increase the lifetime of Van Allen particles, which would increase the energetic particle radiation dosage of satellites.

Generation of Plasma Instabilities. Both the ionospheric depletions just described and disturbances caused by the passage of acoustic gravity waves generated in the wake of a rocket or by its rapidly expanding exhaust cloud can be sources of plasma instabilities or irregularities in the equatorial ionosphere. Such irregularities may give rise to a form of radio-wave interference called spread-F. The main characteristic of spread-F is that it yields a widened echo trace on the ionosonde record (radio frequency reflections versus altitude). The widened echo trace results from the alignment of the irregularity structures along the magnetic field lines, which are horizontal in the equatorial region. Spread-F occurs mainly at night when there is no ionizing radiation from the sun. In equatorial regions, since the Earth's magnetic field is horizontal, ions cannot diffuse downward to lower altitudes to replenish the electrons that recombine more rapidly with the denser ions. Hence, during the nighttime, a sharp ion gradient can form in the vertical. A region of higher ion density sitting on top of a region of lower ion density and held in that position by the horizontal magnetic field is unstable and favors the upwelling of ion bubbles that, in turn, cause the so-called spread-F irregularities that can scatter HF, VHF, and UHF waves. The occurrence of spread-F as a result of rocket-produced ionospheric depletion was predicted by Ossakow et al.[48] and by Anderson and Bernhardt.[49] During the Spacelab-2 mission (currently scheduled for 1981), plans call for monitoring the ionosphere during a specially designed burn of the Space Shuttle orbital engines near the geomagnetic equator in order to look for the predicted spread-F.

Spread-F cannot occur in mid-latitude regions by the previously described mechanism because of the vertical component of the Earth's magnetic field. However, other irregularities are possible at the edge of an F-region depletion. Also, the reduction of the conductivity in the E- or F-region may promote irregularities, since a high conductivity will "short-out" the E-fields responsible for the irregularities. Ionospheric irregularities promote scintillations in satellite VHF waves passing through the disturbed regions, which could affect radio navigation and communication systems.

According to Bernhardt and daRosa, one constructive effect of ionospheric holes is the possibility of using the hole as a focusing lens for HF radio waves for radio astronomical observations.[50] Bernhardt has suggested that the ionospheric holes also would defocus VLF waves. So far no observations of this latter phenomenon have been reported.[51]

Airglow Enhancement. Release of water in the thermosphere leads to a dissociative-recombination process that results in the production of OH^{-1} radicals in excited states. The degree of vibrational and electronic excitation of these radicals is presently uncertain.[20] However, the emission produced during de-excitation could enhance the natural background daytime and nighttime airglow in the near infrared part of the spectrum. Pongratz has reported that very little OH^{-1} vibrational emission in the 9.4 and 5.1 micron bands was observed in the Lagopedo Experiments.[44] Hence, the significance of this airglow is questionable.

Additional sources of airglow include excited oxygen atoms and molecules. Both excited species presumably can be produced by reaction of rocket exhaust with ambient constituents. For example, enhanced emission from singlet-state oxygen atoms was observed in the Lagopedo experiments.[42] However, according to Pongratz, it appeared that these emissions could be accounted for by CO_2 chemistry rather than water chemistry, for both were injected in these experiments.[44]

In any event, increases in visible and near-IR daytime or nighttime airglow along with reflectivity due to ice crystals could lead to the reduced effectiveness of satellite-borne surveillance and remote sensing systems.

Changes in the Hydrogen Atom Cycle of the Upper Atmosphere. The injection of large amounts of H_2O or H_2 may ultimately lead to the addition of two hydrogen atoms per molecule to the thermosphere. Recent calculations by Zinn and Sutherland have indicated that the number of hydrogen atoms that can reach the upper thermosphere as a result of injection of H_2O and H_2 during the second-stage burn of an HLLV from 60 to 120 km is larger than previously thought.[41] Specifically, a computation was run representing a portion of an HLLV second-stage burn trajectory covering an altitude range of 118-123 km and a ground distance of 300 km. Total emissions were 2.5 x 10^{31} H_2O and 8.5 x 10^{30} H_2 molecules, or a total of 6.7 x 10^{31} hydrogen atoms. In the course of 24 hr, they found that some H_2O and H_2 diffused up to the F_2-region where it reacted with O^+ ions. During the daytime, since the upward molecular diffusion rate is relatively slow, fresh O^+ ions were produced by photoionization about as fast as old ones were destroyed. However, the reaction sequence does not destroy the hydrogen atoms that are produced; consequently, they tend to accumulate. During the first 30 hr after launch, about 1.3 x 10^{31} hydrogen atoms were produced at the expense of about 20% of the exhaust molecules. This means that over a five-day period, the number of hydrogen atoms produced would equal about 15% of the total thermospheric hydrogen atom inventory above the mesopause. Of course, HLLVs will be launched at the rate of one or two per day, so that the total hydrogen atom production after five days would about equal the total hydrogen atoms naturally present. As Zinn points out, however, the real question is how rapidly the H_2O molecules are injected relative to the rate at which they naturally migrate upward from the lower atmosphere across the mesopause (85 km).[52]

The natural flux of H_2O molecules across the mesopause leads to a rather large natural flux of hydrogen atoms upward through the thermosphere to the exosphere, where they replace those hydrogen atoms that naturally escape from the Earth's gravitational field into outer space. Therefore, the problem is understanding this natural process better and then examining the effects of the perturbations caused by the exhaust products.

One by-product of Zinn's second-stage HLLV burn simulation is that, for a nighttime launch as opposed to the noontime launch discussed above, the O^+ ions and also the free electrons are not replenished by photoionization. Hence, upward diffusion from below 120 km of H_2O and H_2 to the F_2-region will produce some ionospheric depletion. Zinn's calculation indicates that the F_2-peak electron density would be reduced by about 30% by this mechanism. This should be compared to the factor-of-three depletion predicted by the direct injection during the HLLV circularization burn.

Heat Balance Effects of Rocket Effluents. The thermodynamic effects of the rocket effluents are complicated by several heating and cooling mechanisms that have not been fully evaluated. First, the rocket exhaust will enter the ambient atmosphere with a large amount of streaming and thermal energy, both of which will be rapidly dissipated during the initial expansion of rocket exhaust plume into the ambient environment. The effluents themselves (mostly H_2O and H_2) will then undergo interactions with solar radiation and ambient constituents and will release additional energy to the local environment. This additional energy will be partitioned between the kinetic and excitation energy of the reaction products. The former will lead to thermospheric temperature increases, while the latter could lead to enhanced IR and visible emissions and, therefore, to some degree of cooling. Some of these interactions are not yet fully understood. For example, one of the important sets of reactions involved in this partitioning of the available energy is the dissociative-recombination of H_2O^+ with an electron (e^-):

$$H_2O^+ + e^- \rightarrow OH + H$$
$$\rightarrow H_2 + O$$
$$\rightarrow H + H + O$$

Unfortunately, we do not know the branching ratios, states of excitation of the products, or the reaction rates.[13] On the other hand, since the IR emissivity of H_2O is orders of magnitude larger than that of the normal constituents (O, H, N_2), the injection of substantial numbers of H_2O molecules could lead to enhanced radiative cooling.

In addition to neutral atmospheric thermal effects, plasma temperatures also may be modified. Model calculations show that ionospheric depletions produce 2000 K and 50 K electron temperature enhancements during daytime and nighttime, respectively.[51] The mechanism for this temperature rise is reduced cooling of electrons onto ions. In other words, since the ion concentration is depleted, the remaining electrons cannot be cooled as readily. Thus, photoelectrons are not effectively cooled in the depleted region, which produces electron temperature enhancements and nonlocal heating of the F-layer at both hemispheres. This, in turn, leads to a thermal expansion of the plasma. However, neutral atmospheric temperatures are not expected to be significantly affected by these electron temperature changes, because the neutral density is a thousand times greater than the electron density in the F-region.[51]

Domain C Effects. Domain C includes the plasmasphere (region of ionized, rarified gas) and the magnetosphere. Figure 14 shows a very simplified cross-sectional view of these

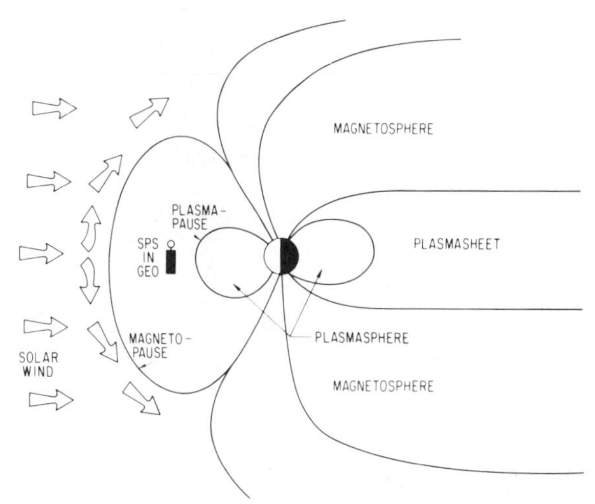

Fig. 14 Cross section of magnetosphere and plasmasphere.

regions in a plane perpendicular to the Earth's orbital plane. The orbital plane of the SPS satellite orbit is perpendicular to the page. The rather dramatic departure from a simple magnetic dipole field results from the interaction of the Earth's field with the stream of charged particles that constitutes the solar wind. In addition to distorting the field lines, this interaction plays an important, but not completely understood role in various auroral and perhaps other processes that influence upper atmospheric dynamics and terrestrial radio communications. Recently, considerable speculation has been given to the role that such interactions may play in tropospheric weather. In fact, the so-called solar-weather effect was the subject of a scientific symposium/workshop entitled Solar-Terrestrial Influences on Weather and Climate, held at Ohio State University, July 24-28, 1978.[32]

The two SPS-related sources of disturbance of the natural plasmasphere and magnetosphere are: (1) the rocket effluents used to propel the COTVs and the POTVs from LEO to GEO and (2) the presence of the satellite structures themselves (including debris). Figure 15 illustrates these sources of disturbance and some potential impacts. The effects associated with rocket exhaust in the plasmasphere are described in detail in the Chiu et al. contribution to this volume. Vondrak addresses outer magnetospheric effects including those related to large space structures.[34] The Chiu et al. paper and some related work are summarized briefly in the following section, and Vondrak's work is summarized in a later section.

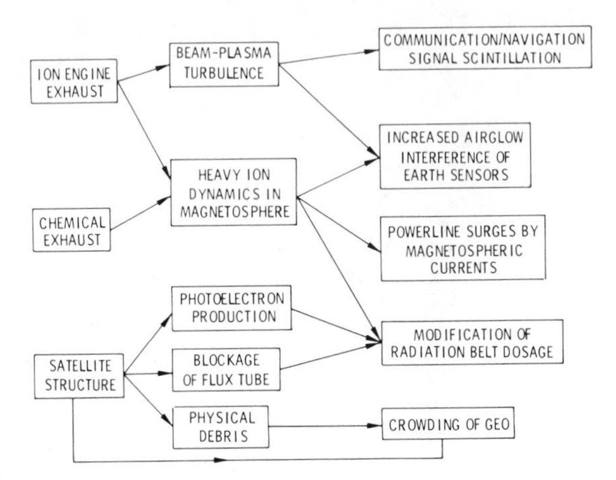

Fig. 15 Identified causes and effects of magnetospheric impact (Source: 54).

Fate of the Rocket Effluents. The present construction scenario designed by NASA calls for the construction of two 5-GW satellites per year over the next 30 years.[5] This requires 22-30 COTV round trips per year between LEO (500 km) and GEO (6.6 Earth radii). Following a slowly unraveling spiral trajectory, each COTV trip out will take about 130 days and the return trip another 30 or more days. Therefore, the total amount of propellant expelled will decrease with increasing radial distance. About 80% will be injected within the plasmasphere, which extends out to about four Earth radii (4Re). As already indicated in Table 2, the number of Ar^+ ions injected per year (and their total mass) is expected to exceed the total number of particles (and their mass) in the entire plasmasphere. In addition, the kinetic energy of these ions is orders of magnitude greater than the ambient particle thermal energy and can be compared to the energy dumped into the near-Earth magnetosphere during a moderately quiet solar year ($\simeq 6 \times 10^{23}$ ergs/yr).[39] It seems plausible that a mass and energy addition of this magnitude is likely to modify significantly both the structure and dynamics of the magnetospheric plasma and even affect the coupling to the ionosphere and lower atmosphere. Unfortunately, unlike the F-region where both accidental and planned modification experiments have yielded relevant information, the phenomenology of Domain C effects is largely unknown. Some experiments, such as the barium releases mentioned by Chiu et al. in this volume do reveal important effects. However, precise applicability of these experiments is not always obvious. The areas to be resolved are: 1) what happens to the Ar^+ ion beam (accompanied by neutralizing elections) after injection? and 2) what influence do the Ar^+ ions have on the plasmasphere?

The first question has been addressed by dividing the fate of the Ar^+ ions into two phases. The initial phase begins with the injection, perpendicular to the geomagnetic field lines and accompanied by some lateral spreading, and continues for a period on the order of 10-100 sec while the beam maintains some degree of integrity. This is a very critical phase of the Ar^+ ion residence in the geomagnetic field. At the risk of drastically over-simplifying the situation, two plausible scenarios are suggested for the initial phase:

1) Most of the beam's streaming energy is inelastically transferred to the magnetospheric plasma or neighboring regions, and the mean beam motion ceases within the ambient plasma. The structure of the ion engine plasma cloud would resemble a sheet stretched out in the poleward directions. This plasma sheet would contain relatively low energy or "cold" ions and electrons that are still more energetic than the ambient plasma particles.

2) The bulk of the beam remains intact, penetrates through the geomagnetic field, and escapes into outer space. The ions making up the outer portions of the beam are separated from the beam and become trapped in the geomagnetic field as relatively "hot" or energetic ions and accompanying electrons.

Chiu et al. in this volume give a detailed description of the first scenario where the beam is stopped within about 1000 km of the injection point. They consider a relatively narrow beam (initial angular width = 10 deg) and two energy options, 3.5 keV and 0.5 keV.

Curtis and Grebowsky, on the other hand, present the case for the second option and consider a higher-energy (5-keV) beam.[53] They find that the bulk of the beam traverses the plasmasphere in about 100-200 sec. However, they also find that a nonnegligible fraction of the beam is deposited with relatively high energy in the plasmasphere.

A third scenario has been proposed by Liemohn et al. in their contribution to this volume. This scenario which is, in some respects, between the two scenarios described previously, contemplates the ion beam as spreading geometrically with an angular width 15 deg from a point source. In this case, the beam pressure would drop to ambient pressure in less than 10 km. Once this happens, acording to Chiu, the density is not sufficient to maintain a polarization electric field to enable the beam to propagate across field lines.[54] Hence, the Ar^+ ions would become trapped as individual 3-5-keV (i.e., hot) particles. The critical difference between this alternative and the first scenario described above is that the ion exhaust emerges from a point source and spreads out rapidly, losing its identity as a plasma beam. Hence the Ar^+ ions enter the plasmasphere with relatively high energy. By contrast, Chiu's model assumes an array of ion engines such that the ion exhaust emerges as a beam with little transverse growth and interacts with the ambient plasma and magnetic field as an entity. Since much of the beam's streaming energy is lost to the ambient plasma, the Ar^+ ions have significantly lower energies than in Liemohn's case when the beam finally disperses.

Following the initial beam-magnetosphere interaction phase, the particles enter the final phase. The lifetime in this second and final phase is determined by three major types of processes: 1) coulomb scattering with electrons, 2) wave-particle interactions, and 3) charge-exchange interactions. The first two processes transfer energy to the ambient plasma and can lead to modifications of both the plasma temperature and structure. The wave-particle interactions also can give rise to interference with radio frequency communications systems. The third process turns the Ar^+ ions into Ar neutrals, whereupon they may escape from the geomagnetic field and either enter the lower atmosphere or leave it altogether. These processes are described in greater detail by Chiu et al. and Curtis and Grebowsky.[53]

The effectiveness of these types of processes in limiting the lifetime of the Ar^+ ions in the plasmasphere on the one hand, and the consequences with respect to the dynamic structure of the plasmasphere on the other, depend to a certain extent on what happens in the first phase of beam motion. After the initial phase, the density and velocity distribution of the Ar^+ ions must be elucidated.

Curtis and Grebowsky have argued that, since the beam density cannot fall below a certain critical number density (n_c) in the plasmasphere, it will not be stopped.[53] This critical density is defined by:

$$n_c = B^2/4\pi M_A c^2$$

where:

M_A = Ar mass
c = velocity of light
B = geomagnetic field strength

These authors also note that the beam velocity is "transalfvénic" in the plasmasphere, i.e., greater than the propagation speed of Alfvén waves (transverse plasma waves that propagate along B-field lines). This being the case, only Ar^+ ions that make up the outer portions of the beam are able to be pealed off as the beam streams through the geomagnetic field. As emphasized by Curtis and Grebowsky, this fraction of the Ar^+ ion beam, however, is still a nonnegligible contribution of the plasmasphere.[53]

Chiu et al., on the other hand, have noted that the critical density argument used by Curtis and Grebowsky doesn't apply to the nonvacuum case of the plasmasphere. In fact, they computed the Alfvén velocity for field lines or L-values as a function of radial distance from the center of the Earth. (The L-value is a dimensionless parameter used to label a magnetic field line or the family of field lines that make up a magnetic shell. It is also an adiabatic invariant of the particle motion along the field line. A magnetic shell intersects the equatorial plane at a radial distance $R/Re \simeq L$ from the center of the Earth. The higher the L-value, the larger the radial distance at the equator and the higher the latitude at which the field line enters the atmosphere. See Ref. 55 for a detailed discussion of the meaning of L.) Chiu et al. found that, for all plasmaspheric field lines except very close to LEO, the Alfvén speed exceeds the beam speed (150 km/sec for a 3.5 keV beam) by at least an order of magnitude! Hence, in contrast to Curtis and Grebowsky, it would seem that the beam is subalfvénic and stoppable. It could turn out that both authors are partially correct and that the stopping phenomenology is critically dependent on the specifics of the ion beam design, i.e., both its streaming velocity and the distribution of transverse velocity components, as well as on ambient parameters such as the L-value and plasma density.

b) Effects of Deposition of Argon Ions and Energy. Whether or not the beam is in fact "stopped," both groups of authors agree that a nonnegligible fraction of the Ar^+ ions will be deposited in the geomagnetic plasma. Chiu et al. assume all of it will be deposited with 80% in the plasmasphere, while Curtis and Grebowsky find the maximum fraction of Ar^+ ions deposited to be $\simeq 10\%$ for the smaller L-values (1-2).[53] Even in the case of a smaller fraction of deposited Ar, the energy density will still exceed that of the ambient plasma by two orders of magnitude according to Curtis and Grebowsky. Thus, regardless of the fraction stopped, a large source of free energy is still deposited in the plasmasphere. This free energy can be transferred to the ambient constituents by coulomb scattering of electrons, which will cause the ambient temperature to increase, or via wave-particle interactions. According to Chiu et al., the latter interactions can contribute to ambient plasma heating, modification of the ring-current and radiation belt particles, and radio wave scintillation.

Particle-wave interactions include a number of different types of processes in which turbulent waves are first generated by the Ar^+ ions. This turbulence can either suppress or enhance the growth of natural plasma instabilities. Some of these instabilities are associated with the precipitation of high energy protons or electrons from the ring-current and radiation belts. Following their beam stopping model, Chiu et al. have argued that the net energy effect of adding the slow Ar^+ ions is a suppression of the natural plasma instability that is credited with periodically dumping relativistic electrons from the inner Van Allen Belt. Curtis and Grebowsky, in keeping with their scenario, argue that the fast Ar^+ ions will generate turbulence components that will be in resonance with a plasma instability, which will lead to dumping of high energy protons.[53] In either case, the consequences could be significant.

If Chiu et al. are correct, suppression of relativistic electron precipitation events will lead to a buildup of electron radiation levels by as much as a factor of two or three. If Curtis and Grebowsky are correct, increased dumping of high energy protons will create RF disturbances and will affect ionizing radiation levels as far down as the stratosphere. However, as pointed out by Aiken, the magnitude of these proton precipitation events is limited by the rate of repopulation of inner belt protons.[56] Since the major source of repopulation is beta decay of cosmic ray neutrons, repopulation is a process that would require as long as 100 years.

Aiken also suggested that since precipitation of electrons from the outer radiation belt (4-6Re) will cause atmospheric ionization at mid- and higher-latitudes, possible resonant interactions that could lead to enhanced precipitation of these particles (whose repopulation presumably is fairly rapid) should be looked into.[56]

Chiu has also pointed out that some of the Ar^+ ions may be recycled and accelerated to relatively high energies (hundreds of MeV).[39] This could substantially increase the high atomic number radiation levels that are hazardous to space workers. (The level of relativistic electrons also is significant for degradation of spacecraft instrumentation.)

Vondrak has suggested in this volume that an argon-contaminated plasmasphere may alter the process of mid-latitude ionospheric trough formation and coupling processes between the ionosphere and magnetosphere. There is a possibility, although rather speculative, that multiple plasmapauses may be formed at the equatorial plane. This could alter the magnetospheric convection pattern and, consequently, the location of the auroral oval. This, in turn, could induce power line surges at population centers in the event of magnetospheric substorms. The probability of occurrence of these phenomena is unknown at this time, but their consequences could be important. Power line surges resulting from natural magnetic substorms have been reported in the literature as discussed earlier. Therefore, the connection between natural disturbances of the magnetosphere and terrestrial impacts exists. What is lacking is any concrete evidence whether man-made disturbances will be of sufficient magnitude to cause similar impacts or whether the location of the auroral oval can be shifted by such disturbances.

Also worth mentioning is the potential increase in airglow due to precipitated 500-eV Ar^+ and to 5-eV Ar^+ thermalized electrons that may impact space-based optical sensors for military, scientific, weather, and earth resources application. The probability of occurrence as well as the severity of the impacts, if any, are unknown.

IV. Effects of Large Space Structures in the Magnetosphere

In the previous section, Ar^+ ion injections were reviewed, with particular emphasis on the fate of these ions in the plasmasphere. This region is approximately toroidal in shape and extends from the top of the ionosphere out to $\simeq$ 4Re in the equatorial plane. In terms of the magnetic shell parameter defined earlier, the region is roughly bounded by the magnetic shells corresponding to $1.2 \leq L \leq 4$.

Beyond the plasmapause (the outer boundary of the plasmasphere) is the region referred to as the outer magnetosphere. Its dynamic structure is strongly influenced by the solar wind and the associated inner-planetary magnetic field. Vondrak's paper in this volume contains a detailed description of this region, its structure, and the sources of contamination associated with large space structures. These sources and their effects are summarized briefly below:

1) The satellite structures in GEO will serve as sinks for particles striking them. Each of the 60 SPS satellites,

for example, will cover an area of $\approx$ 55 km^2 and be distributed along an equatorial line over the U.S. on the magnetic shell defined by an L value of about 6.6. Hence, one would expect that flux tubes passing through these satellites and perhaps even the whole region of the magnetic shell could suffer some plasma reduction. This, according to Vondrak, would be analogous to the "sweeping" of the Jovian radiation belts by the Galilean satellites.

2) The satellites will become charged and, consequently, will accelerate ambient charged particles. This could lead to alteration of the ambient plasma and to enhanced radiation damage over and above that expected by unaccelerated particle bombardment. (The satellites will reside in the outer portion of the outer Van Allen Belt.)

Project SCATHA (Spacecraft Charging at High Altitudes), which is the subject of another section in this volume, also is relevant to this issue. For example, SCATHA experiments have demonstrated that electric potentials established on spacecraft can accelerate and decelerate plasma components to tens of keV. These experiments have also indicated that klystron operations can be adversely influenced by the plasma environment, which may be modified by gaseous and particulate emissions in GEO.

3) The SPS satellite structure will dissipate most of the solar input energy as either reflected light or infrared radiation, since the proposed solar cells are expected to be, at most, 17-19% efficient. In addition, the DC-to-RF converters will dissipate approximately 1.2 GW of waste heat, probably at relatively high temperatures (> 500°C). The effects of this radiation, as well as the effects of RF radiation on the astronomical community, could be important.[57]

4) Satellite structural members will be exposed continuously to direct solar and cosmic radiation and are expected to be sources of photoelectrons and other charged and neutral particles. The ejected neutral particles could form relatively long-lived, orbiting, dust clouds or rings near GEO.

5) Finally, the structures will be sources of various neutral contaminant gases (as well as Ar^+ ions and electrons from attitude control and station-keeping thrusters). Sources of neutral gases include: (a) chemically-fueled attitude control and station-keeping thrusters used during eclipses of the sun by the Earth, (b) leakage from pressurized chambers including living quarters, containment vessels, etc., and (c) satellite surface erosion and outgassing of volatiles.

In addition to these satellite structural sources, there will be both chemical and ion propellants injected near GEO from the POTV circularization and deorbit burns, from COTV operations, and from thrusters required in space construction operations and maneuvers. Based on exhaust and escape velocity, it is anticipated that the POTV circularization burn neutral effluents will be gravitationally trapped, while those from the deorbit burn will escape into outer space. The other sources of neutral gases probably will have relatively low exhaust speed and be more randomly directed. Thus, they probably will be trapped and may form gas clouds in the GEO environment. The minimum gas release rates needed to produce significant effects have been evaluated by Vondrak in this volume. Garrett and Forbes also have performed some calculations relevant to this topic.[58] They suggest that it may be possible to measure an actual contaminant cloud (Xe^+ ions) during the 1979-1980 time frame by using the GEOS II satellite.

Finally, although it is not part of the SPS reference design, space operations between GEO and lunar orbit (a region referred to as "cislunar space") also may have complicating environmental consequences. Vondrak has suggested that released gases in this region could form a ring around the Earth at approximately lunar-orbital distance similar in structure to the Jovian ring associated with gaseous emissions from the Jovian satellite IO.[34] A total leakage rate $\geq$ 600 kg/sec could lead to significant effects, such as the alteration of the solar wind flow pattern and, consequently, the structure of the outer magnetosphere. However, this value is probably well above that expected from space operations on the scale of the reference design.

V. Conclusion

This introductory review summarized briefly some of the main points of the papers in this section. The review was organized by altitude regime as follows:

Domain A 56-124 km,

Domain B 124-500 km (LEO $\simeq$ 450-500 km)

Domain C 500-36,000 km (GEO $\approx$ 36,000 km), and

GEO and beyond.

Specific contaminant sources were determined for each of these regimes, and the environmental effects of the sources were assessed.

In Domains A and B, the major sources are SPS microwave transmission and rocket effluents. Although no significant effects have yet been found for microwave transmissions, the deposition of rocket effluents is known to cause compositional changes. Current work has focused on evaluating the duration and spatial extent of these perturbations. Most appear to be associated with the release of large amounts of H_2O in regions where it is not usually found.

In Domain C, along with chemical effluents from rockets, ion engine contaminants (anticipated to be primarily Ar^+) are expected to be a significant factor in altering magnetospheric and plasmaspheric structure and dynamics. Although still quite controversial, one of the major impacts of these alterations may be to perturb the stability of the Van Allen radiation belts. This could lead to changes in radiation hazards to space equipment and space workers, as well as alteration of high energy particle precipitation events.

In the outer magnetosphere, current knowledge of environmental changes is limited. The space environment is very low in density and energy, and the environmental loading can be quite severe. It is not clear, however, how major alterations in this environment, which are known to vary by several orders of magnitude naturally, could affect the global climatic system. Although a connection between solar activity and climate is believed to exist through subtle changes in geomagnetic activity, this phenomenon is not at all certain or well understood. Thus, further research will be required in this region.

In summary, the ambient density falls rapidly and the potential for significant environmental alterations increases as one goes outwards from the Earth's surface. Correspondingly, as the altitude increases our current knowledge of environmental change processes and their importance decreases. Hence, the reader is cautioned that, unless specifically indicated to the contrary, either here or in the papers that follow, the probability of occurrence of the potential atmospheric effects of large space systems and their significance for the terrestrial environment should be regarded as uncertain or unknown.

Acknowledgment

The diligent efforts of the editors, R.F. Bryans, AIAA, and M.W. Tisue, Argonne National Laboratory, to make this chapter more attractive and readable and the patient typing of the manuscript by L. Kaatz are gratefully acknowledged. In addition, the author wishes to express his appreciation to the other contributors to this part of the book, who carefully reviewed this chapter. In particular, the helpful assistance and comments of H.B. Garrett and C.P. Pike are sincerely appreciated.

This work was supported by the Satellite Power System Project Office of the U.S. Dept. of Energy.

References

[1]Stine, G.H., "Industry Goes to Space," Omni, April 1979, pp. 37-41, 100-102.

[2]Grumman Aerospace Corp., "Utilization of Shuttle Sortie Missions for Satellite Power Systems Technology Verification," prepared for National Aeronautics and Space Administration, March 1978.

[3]NASA, Space Shuttle Program Environmental Impact Statement, April 1978.

[4]Glaser, P.E., "Power from the Sun: Its Future," Science, Vol. 162, Nov. 1968, pp. 857-861.

[5]DOE, Satellite Power System (SPS) Concept Development and Evaluation Program, Reference System Report, DOE/ER-0023, Oct. 1978.

[6]Dupas, A., and Claverie, M., Centre National de la Recherche Scientifique, personal communication, 1979.

[7]Walbridge, E.W., "Laser Satellite Power Systems," U.S. Dept. of Energy (in press), 1980.

[8]Park, C., "Equivalent-Cone Calculation of Nitric Oxide Production Rate During Space Shuttle Entry," Atmospheric Environment (in press), 1979.

[9]Rakich, J.V., Bailey, H.E., and Park, C., "Computation of Nonequilibrium Three-Dimensional Inviscid Flow Over Blunt-Nosed Bodies at Supersonic Speeds," AIAA Paper 75-835, June 1975.

[10]Whitten, R., see comments in Sec. 2.2.3 in Bauer, 1979.

[11]Park, C., and Menees, G.P., "Odd Nitrogen Production by Meteoroids," Journal of Geophysical Research, Vol. 83, Aug. 1978, pp. 4029-4035.

[12]Brubaker, K.L., Preliminary Assessment for the Satellite Power System (SPS): Vol. 2, Detailed Assessment, U.S. Dept. of Energy, DOE/ER-0021/2, Oct. 1978, pp. 86-90.

[13]Bauer, E., "Workshop on Upper Atmospheric Environmental Effects Arising from Propulsion Effluents Associated with Placing Large Payloads on Orbit: Satellite Power System Environmental Assessment," U.S. Dept. of Energy (in press), 1979.

[14]Ellsaesser, H., see comments in Secs. 2.3 and 2.4 in Bauer, 1979.

[15]Sechrist, C.F., Jr., "The Ionospheric D Region," The Upper Atmosphere and Magnetosphere, National Academy of Sciences, Washington, D.C., 1977, pp. 102-116.

[16]Rote, D.M., "Report on the Workshop on Ionospheric and Magnetospheric Effects of Satellite Power Systems, U.S. Dept. of Energy (in press), 1979.

[17]Brubaker, K.L., "Proceedings of Workshop on Stratospheric and Mesospheric Impacts of Satellite Power Systems," U.S. Dept. of Energy (in press), 1979.

[18]Brubaker, K.L., Argonne National Laboratory, personal communication, 1979.

[19]Chenurnoy, V.N., and Charina, G.A., "Semi-Annual Temperature Variations in the Thermosphere and Their Relation with Atmospheric Circulation and Mesospheric Clouds," Geomagnetic Aeronomy, Vol. 17, 1977, pp. 58-60.

[20]Turco, R., see comments in Sec 2.8 in Bauer, 1979.

[21]Brubaker, K.L., "Energy Injection," Preliminary Environmental Assessment for the Satellite Power System (SPS), U.S. Dept. of Energy (in press), 1979.

[22]Forbes, J., see comments in Sec. 2.7 in Bauer, 1979.

[23]Witt, G., "Height, Structure, and Displacements of Noctilucent Clouds," Tellus, Vol. 14, 1962, pp. 1-18.

[24]Fogle, B., and Haurwitz, B., "Noctilucent Clouds," Space Science Review, Vol. 6, 1966, pp. 279-340.

[25]Meinel, A.B., Middlehurst, B., and Whitaker, E., "Low-Latitude Noctilucent Cloud of 15 June 1963," Science, Vol. 141, Sept. 1963, pp. 1176-1178.

[26]Benech, B., and Dessens, J., "Mid-Latitude Artificial Noctilucent Clouds Initiated by High-Altitude Rockets," Journal of Geophysical Research, Vol. 79, March 1974, pp. 1299-1301.

[27]Reid, G.C., "Ice Clouds at the Summer Polar Mesopause," Journal of Atmospheric Science, Vol. 32, March 1975, pp. 523-535.

[28]Mendillo, M., see comments in Sec. 2.5 in Bauer, 1979.

[29]Bernhardt, P.A., see comments in Sec. 2.5.4 in Bauer, 1979.

[30]Park C., see comments in Sec. 3.1.3 in Bauer, 1979.

[31]Hines, C.O., "A Possible Mechanism for the Production of Sun-Weather Correlations," Journal of Atmospheric Science, Vol. 31, 1974, pp. 589-591.

[32]McCormac, B.M., and Seliga, T.A., Solar Terrestrial Influences on Weather and Climate, Reidel Publishing Co., Boston, Mass., 1979.

[33]Markson, R., see comments in Sec. 3.1 in Rote, 1979.

[34]Vondrak, R.R., "Environmental Impact of Space Manufacturing," AIAA Paper 77-539, May 1977.

[35]Anderson, C.W., III, Lanzerotti, L.J., and MacLennan, C.G., "Outage of the L4 System and the Geomagnetic Disturbances of 4 August 1972," The Bell System Technical Journal, Vol. 53, Nov. 1974, pp. 1817-1837.

[36]Albertson, V.D., and Van Baelen, J.A., "Electric and Magnetic Fields at the Earth's Surface Due to Auroral Currents," IEEE Transactions on Power Apparatus and Systems, Vol. 89, April 1970, pp. 578-584.

[37]Slothower, J.C., and Albertson, V.D., "The Effects of Solar Magnetic Activity on Electric Power Systems," Journal of the Minnesota Academy of Science, Vol. 34, No. 2, 1967, pp. 94-100.

[38]Carrigan, A.L., and Skrivanek, R.A., "Aerospace Environment," USAF Cambridge Research Laboratories, L.G. Hanscom Field, Bedford, Mass, 1974.

[39]Chiu, Y., see comments in Sec. 4 in Bauer, 1979.

[40]Mendillo, M., Hawkins, G.S., and Klobuchar, A., "Sudden Vanishing of the Ionospheric F Region Due to the Launch of Skylab," Journal of Geophysical Research, Vol. 80, June 1975, pp. 2217-2228.

[41]Zinn, J., and Sutherland, C.D., "Effects of Rocket Exhaust Products in the Thermosphere and Ionosphere, Los Alamos Scientific Laboratory Report No. LA-UR-79-2750, Oct. 1979, submitted for publication to Space Solar Power Review, Pergamon Press, Inc., Elmsford, N.Y., 1979.

[42]Pongratz, M.B., and Smith, G.M., "The Lagopedo Experiments - An Overview," paper #SA15, American Geophysical Union 1978 Spring Meeting, Miami, Fla., March 1978.

[43]Sjolander, G.W., and Szuszczewicz, E.P., "Chemically Depleted F_2 Ion Composition: Measurements and Theory," Journal of Geophysical Research, Vol. 84, No. A8, 1979, pp. 4393-4399.

[44]Pongratz, M.B., Los Alamos Scientific Laboratory, personal communication, 1979.

[45]Molander, R.C., and Wolfhard, H.G., "Explanation of Large Apollo-8 (J-2) Rocket Plume Observed during Trans-Lunar Injection," Institute for Defense Analysis Note N-610, IDA Log #HG 69-9734, Feb. 1969.

[46]Mendillo, M., Herniter, B., and Rote, D., "Modification of the Aerospace Environment by Large Space Vehicles," Journal of Spacecraft and Rockets (in press), May/June 1980.

[47]Sailors, D., U.S. Navy Communications group, Naval Ocean Systems, personal communication with E. Bauer, 1979.

[48]Ossakow, S.L., Zalesak, S.T., and McDonald, B.D., "Ionospheric Modification: An Initial Report on Artificially Created Equatorial Spread-F," Geophysical Research Letters, Vol. 5, No. 8, 1978, pp. 691-694.

[49]Anderson, D.N., and Bernhardt, P.A., "Modeling the Effects of an H_2 Gas Release on the Equatorial Ionosphere," Journal of Geophysical Research, Vol. 83, 1978, pp. 4777- 4790.

[50]Bernhardt, P.A., and daRosa, A.V., "A Refracting Radiotelescope," Radio Science, Vol. 12, 1976, pp. 327-336.

[51]Bernhardt, P.A., "Environmental Effects of Plasma Depletion Experiments," draft report, International Council of Scientific Unions Committee on Space Research, Paris, France, Oct. 1978.

[52]Zinn, L., Los Alamos Scientific Laboratory, personal communication, 1979.

[53]Curtis, S.A., and Grebowsky, J.M., "Changes in the Terrestrial Atmosphere-Ionosphere-Magnetosphere System Due to Ion Propulsion for Solar Power Satellite Placement," submitted to Space Solar Power Review, Pergamon Press, Inc., Elmsford, N.Y, Dec. 1979.

[54]Chiu, Y., The Aerospace Corp., personal communication, 1979.

[55]Stone, E.C., "The Physical Significance and Application of L, B_0, and R_0 to Geomagnetically Trapped Particles," Journal of Geophysical Research, Vol. 68, No. 14, July 1963, pp. 4157-4166.

[56]Aiken, A., see comments in Sec. 4.4.2 in Bauer, 1979.

[57]Stokes, G.M., and Ekstrom, P.A., "Workshop on Satellite Power Systems Effects on Optical and Radio Astronomy," U.S. Dept. of Energy (in press), 1979.

[58]Garrett, H.B., and Forbes, J.M., "Time Evolution of Ion Contaminant Clouds at Geosynchronous Orbit," Geophysical Research Letters (in press), 1979.

EFFECTS OF MICROWAVE BEAMS ON THE IONOSPHERE

Lewis M. Duncan*

Los Alamos Scientific Laboratory, Los Alamos, N. M.

Abstract

This is a review of the effects associated with the propagation of intense microwave beams through the ionosphere. Collisional damping of the microwave beam in the lower ionosphere will significantly enhance the local free electron temperatures. Experimental observations of this enhanced electron heating are in general agreement with the theoretical models. In addition, thermal self-focusing of electromagnetic waves in the ionosphere can produce variations in the beam power flux density and create large-scale electron density irregularities. These large-scale irregularities also may trigger the formation of small-scale plasma striations. Again, experimental results support theoretical models of this phenomenon. These investigations of the dominant physical processes involved in microwave propagation through the ionosphere are applicable to the environmental impacts assessment of the proposed solar-power satellite microwave power-transmission system. Ionospheric modifications can lead to the potentially enhanced telecommunications and climate impacts.

Introduction

The ionosphere is commonly defined as that portion of Earth's upper atmosphere where sufficient numbers of free electrons exist so as to affect radio wave propagation. Electromagnetic radiation propagating through this region is collisionally damped by the free electrons. For microwave frequencies, the fraction of wave energy absorbed by the plasma is expected to be relatively small. However, the resulting ohmic heating of the plasma can affect the local ionospheric thermal budget significantly. Strong electromagnetic radiation can initiate rapid enhancements in electron temperature, and also affect ionospheric densities and structure. The direct results of collisional damping of the electromagnetic wave by the free electrons are called resistive heating effects.

---

*Staff Member.

In addition, differential ohmic heating of the plasma gives rise to electron temperature gradients, convective plasma motions, and macroscopic thermal forces capable of exciting secondary plasma instabilities. The large-scale ionospheric responses to these dynamic thermal forces can be described generally as collective plasma phenomena.

The study of electromagnetic wave interactions with plasmas is motivated by its relevance to ionospheric modification research,[1] laser-fusion plasma heating,[2] and investigations of diverse astrophysical phenomena.[3] In particular, this study finds application in the environmental impact assessment of ionosphere-microwave beam interactions associated with the microwave power transmission system of the solar-power satellite concept.[4] Numerous telecommunications systems rely on ionospheric reflections or transionospheric propagation as part of their communications signal path. Any system that can modify the ionosphere significantly has the potential to produce wide-ranging telecommunications interference. In addition, the role of the ionosphere in solar-terrestrial coupling and climate change is not well understood. As a result, modification of the ionosphere by electromagnetic radiation is of general concern. This paper reviews the dominant physical processes involved in the propagation of intense microwave beams through the ionosphere.

Resistive Heating Effects

Solar photoionization in the ionosphere produces free electrons with an effective temperature usually exceeding that of the background neutral gas. As electrons gain energy through solar photoionization, they also lose energy by collisions with the much heavier atoms and molecules of this background gas. The electron temperature is therefore an energy balance between these heating and cooling processes.

The collisional heating and cooling interactions of the ionospheric plasma are all dependent on the electron temperature. Under certain conditions, the rate of heating may temporarily dominate the normal cooling losses, initiating a rapid increase in the electron temperature that continues until compensating cooling processes develop, limiting the temperature rise. This enhanced electron heating can affect the electron-ion recombination rates, changing ionospheric densities, or drive secondary nonlinear ionospheric interactions, further disturbing the ambient plasma. These disturbances can produce potentially serious telecommunications impacts. This section investigates the threshold conditions required to excite enhanced electron heating in the ionosphere through the collisional damping of electromagnetic radiation.

Electron Heating Theory

The electron energy balance equation in the ionosphere can be expressed as

$$\frac{dU}{dt} = Q^+ - Q^- \tag{1}$$

where dU/dt is the change in electron energy with time, Q+ is the heat source function, and Q^- describes the volume heat losses. For a Maxwellian electron energy distribution,

$$\frac{dU}{dt} = \frac{d}{dt} \int \frac{1}{2} mv^2 F \, d^3v \tag{2}$$

$$= \frac{3}{2} n_e k \frac{dTe}{dt} \quad (\text{erg/cm}^3\text{-s}) \tag{3}$$

where m is the electron mass, v is the electron velocity, f is the electron distribution function, n_e is the electron number density, k is Boltzmann's constant, and dTe/dt is the change of electron temperature with time. Clearly whenever the energy input exceeds cooling losses, the electron temperature must increase. Conversely, in the absence of additional heat sources, the electron temperature will relax to its ambient level.

As an electromagnetic wave propagates through a plasma, free electrons respond to the wave's oscillating electric field. If, while under the action of this wave, the electron suffers a collision, it will scatter out of the electric field, taking with it a part of the wave's energy. This collisional damping of the electromagnetic wave results in ohmic heating of the plasma. The rate of energy input to the atmospheric free electrons resulting from the absorption of microwave or underdense radio-frequency radiation can be attributed entirely to this ohmic heating process. The corresponding heat source function is given by

$$Q^+ = \frac{E^2}{8\pi} \frac{f_p^2}{f^2} (\nu_{ei} + \nu_{en}) \tag{4}$$

where E is the wave electric field amplitude, f_p is the local plasma frequency, f is the electromagnetic wave frequency, and ν_{ei} and ν_{en} are the electron-ion and electron-neutral collision frequencies. In the lower ionosphere, the electron-neutral collision frequency dominates and can be approximated by[5]

$$\nu_{en} = \sum_{x}^{N_2, O_2, O} \nu_{ex} \quad (5)$$

with

$$\nu_{e,N_2} = (2.33 \times 10^{-11}) n\langle N_2\rangle [1 - (1.21 \times 10^{-4}) Te] Te \quad (s^{-1}) \quad (6a)$$

$$\nu_{e,O_2} = (1.82 \times 10^{-10}) n\langle O_2\rangle [1 + (3.6 \times 10^{-2}) Te^{1/2}] Te^{1/2} \quad (s^{-1}) \quad (6b)$$

and

$$\nu_{e,O} = (8.2 \times 10^{-10}) n\langle O\rangle Te^{1/2} \quad (s^{-1}) \quad (6c)$$

where $n\langle m\rangle$ is the neutral number density (cm^{-3}) and Te is the electron temperature (K).

To maintain their energy balance, electrons lose energy through collisions with the atoms and molecules of the background gas. In the collision-dominated lower ionosphere, thermal conduction is not an important cooling mechanism. The most effective kinds of energy transfer collisions are inelastic interactions with O_2 and N_2, producing rotational and vibrational excitation, and collisions with atomic oxygen, producing excitation of hyperfine levels of the ^3P ground state. For neutral temperature Tn, the electron energy loss rates corresponding to these individual processes are approximately[5]

$$Q^-|_{N_2 \text{ rotational}} = 3.2 \times 10^{-26} n_e n\langle N_2\rangle (Te - Tn)/Te^{1/2} \quad (erg/cm^3\text{-}s) \quad (7)$$

$$Q^-|_{O_2 \text{ rotational}} = 1.1 \times 10^{-25} n_e n\langle O_2\rangle (Te - Tn)/Te^{1/2} \quad (erg/cm^3\text{-}s) \quad (8)$$

and

$$Q^-|_{N_2 \text{ vibrational}} = 2.08 \times 10^{-16} n_e n\langle N_2\rangle \{1 - \exp[3200(\frac{1}{Te} - \frac{1}{Tn})]\} \quad (erg/cm^3\text{-}s) \quad (9)$$

where

$$5.715 \times 10^{-8} \exp(-3353/Te), \quad Te \leq 1000 \text{ K}$$

$$A = \begin{cases} 2.0 \times 10^{-7} \exp(-4605/Te), & 1000 < Te < 2000 \text{ K} \\ 2.53 \times 10^{-6}\, Te^{1/2} \exp(-17620/Te), & Te \geq 2000 \text{ K} \end{cases} \tag{10}$$

and

$$Q^{-}\Big|_{\substack{O_2 \\ \text{vibrational}}} = 1.6 \times 10^{-23} n_e n\langle O_2 \rangle B \ (\text{erg/cm}^3\text{-s}) \tag{11}$$

where

$$B = 0, \quad Te \leq 350 \text{ K} \tag{12a}$$

$$B = 1.3 \times 10^{-4} \exp[2.532 \times 10^{-2}(Te - 350)] \quad 350 < Te \leq 500 \text{ K} \tag{12b}$$

$$B = 1.933 \times 10^{-3}\{1 + 9.11 \times 10^{-2}(Te - 500) + 2 \exp[7.68 \times 10^{-3}(Te - 500)]\} \quad 500 \leq Te < 1000 \text{ K} \tag{12c}$$

$$B = 0.135\{1 + 5.185 \times 10^{-3}(Te - 1000) + \exp[2.8 \times 10^{-3}(Te - 1000)]\} \quad 1000 < Te \leq 1400 \text{ K} \tag{12d}$$

$$B = 0.83 + 1.95 \times 10^{-3}(Te - 1400) \quad 1400 < Te \leq 2000 \text{ K} \tag{12e}$$

$$B = 2 + 2.75 \times 10^{-3}(Te - 2000) \quad 2000 < Te \leq 3000 \text{ K} \tag{12f}$$

$$B = 4.75 + 2.925 \times 10^{-3}(Te - 300) \quad 3000 < Te \leq 5000 \text{ K} \tag{12g}$$

$$B = 10.6 \exp[1.59 \times 10^{-4}(Te - 5000)] \quad Te > 5000 \text{ K} \tag{12h}$$

and

$$Q^{-}\Big|_{O(^3P)} = 4.69 \times 10^{-24} n_e n\langle O \rangle (Te - Tn)/Tn, \ \text{erg/cm}^3\text{-s} \tag{13}$$

The electron temperature is a sensitive balance between heating processes and cooling interactions. Equations (4-6) show that, as the electron temperature increases, the electron-neutral collision frequency rises, thereby further increasing the ohmic heating. This self-amplifying process must be balanced by electron cooling losses, which also become more efficient as the electron temperature increases above its ambient value.

For close-to-ambient electron temperatures in the lower ionosphere, the principal cooling mechanism is rotational excitation of N_2 and O_2. Above about 100 km, excitation of hyperfine levels in atomic oxygen becomes important, depending on the concentration profile. As the electron temperature increases to several times ambient, vibrational excitation of N_2 and O_2 provides substantial cooling losses and ultimately saturates the heating.

The time it takes a plasma to self-consistently reach an equilibrium between these competing processes is called the heating time scale. After the ionosheric wind has swept the plasma beyond the heating beam, the electron temperature relaxes to its ambient level on a cooling time scale not very different from its normal heating counterpart. In the lower ionosphere, both heating and cooling equilibria are reached in a few milliseconds or less.

For normal ionospheric heating, the balance between the heat input and cooling loss is continuously converging on some stable equilibrium. For this "stable" heating, the rate of change of electron temperature is decreasing at all times. However, for a sufficiently large heat source function, this criterion is not always satisfied. When the rate of electron temperature change is increasing as a function of time, a condition of "thermal runaway" temporarily exists. When this occurs, the electron temperature and the heating time constant become nonlinearly dependent on the incident electromagnetic wave intensity. A description of this enhanced electron-heating process and predictions for ionospheric heating by the solar-powered satellite microwave power beam are presented in the next section.

Enhanced Electron Heating

Holway and Meltz,[6] investigating the effects of strong radiowave heating of free electrons in the lower ionosphere, first introduced the concept of an electron temperature runaway. Subsequent studies,[7] including comprehensive models for the dominant collisional cooling processes, have been performed to predict the enhanced electron-heating thresholds, time dependences, and saturation limits, as functions of the initial ionospheric conditions and the heating wave frequency and power density. These theoretical results can be compared to experimental radiowave heating observations of the ionosphere and used to predict the ionospheric effects of proposed systems such as the solar power satellite (SPS) microwave power transmission beam.

The SPS concept proposes to collect solar energy in space and beam this energy via microwaves to ground-based receiving antennas, where it can be converted to electrical power. A

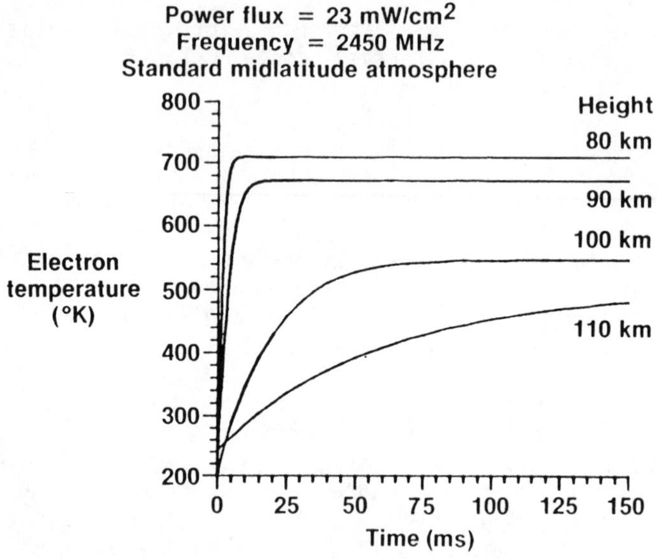

Fig. 1 Theoretical predictions of microwave heating in the lower ionosphere (from Meltz.[7]).

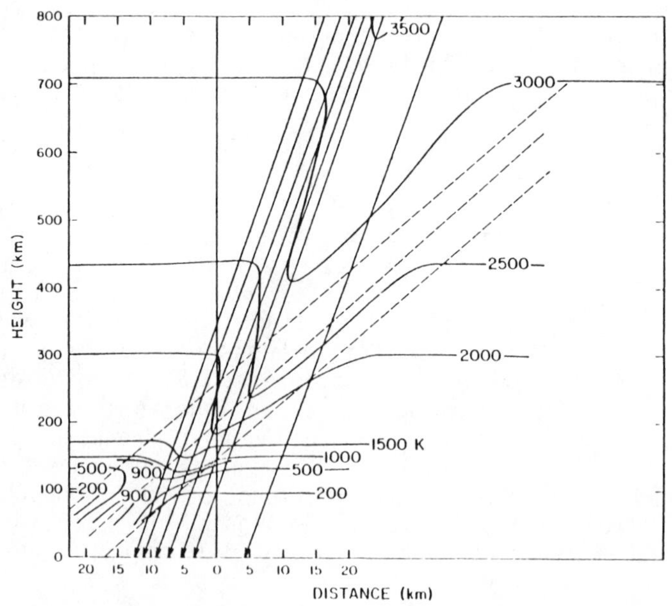

Fig. 2 Contours of theoretically calculated electron temperature over Boulder, Colo. for microwave heating by the solar-power satellite power beam. The dashed lines give the beam direction, and the light solid lines indicate the geomagnetic-field direction (from Perkins and Roble.[7]).

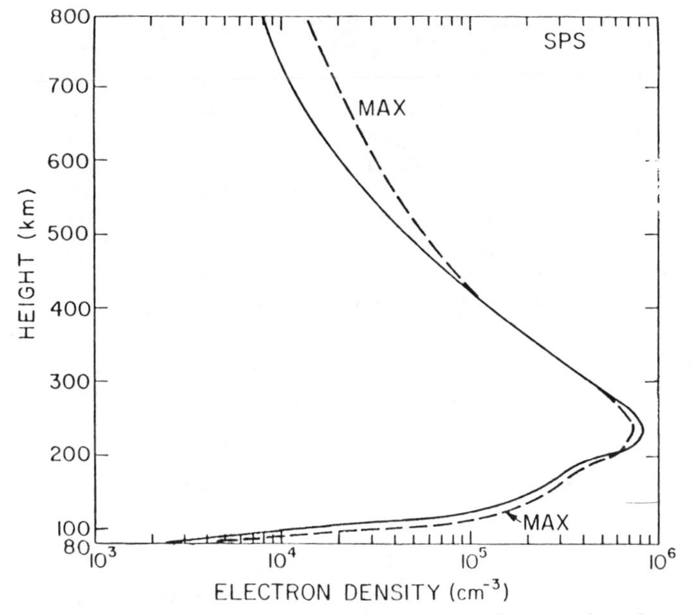

Fig. 3 Theoretically calculated profiles of electron density. The solid curve represents the undisturbed electron-density profile; the dashed curve is the electron-density profile determined after 3 hr of the SPS microwave heating (from Perkins and Roble.[7]).

network of these space power stations, each generating 5 to 10 GW of power, could make a substantial contribution toward satisfying future energy needs.

The microwave power beam responsible for transmitting energy from space to the ground-based rectenna is designed to operate at 2.45 GHz. This frequency represents a compromise between the ionosphere-microwave interactions, which are more easily excited at lower frequencies, and the increased scattering losses from atmospheric hydrometeors such as rain and hail, which occur at higher frequencies. Although no major changes in this operating frequency are anticipated, small changes may be made to reduce interference effects on other electromagnetic systems. An alternative option of laser power transmission also is being considered.

To avoid nonlinear ionospheric interactions, the maximum power flux density in the SPS microwave beam was originally limited to 23 mW/cm^2. Therefore, to deliver 5 GW of power, the downcoming beam would have to be at least 5 km in diameter. In addition, this downcoming microwave power beam is directed onto the ground-based rectenna by an upgoing pilot beam. This retrodirective system is designed to operate at a frequency close to that of the power beam, and is centered in the rectenna. Thus, the upgoing pilot beam propagates through the same ionospheric plasma as the downcoming power beam. Any disturbances generated in the ionosphere could lead to pilot-beam scintillation or scattering, resulting in wandering or defocusing of the microwave power beam.

Most of the ionospheric changes induced by the SPS microwave beam are believed to be restricted to an area near the beam. Just as increased electron temperatures cause enhanced collisional damping of the microwave power beam, absorption of other radio waves propagating through the local heated region also is increased. In addition, thermal forces deriving from this ohmic heating of the ionospheric plasma are capable of driving secondary instabilities. The resultant collective plasma phenomena will be discussed in greater detail.

Enhanced electron heating is predominantly a lower ionosphere effect. For SPS microwave beam parameters, the predicted electron temperatures are shown in Fig. 1 as a function of altitude and time. The electron temperature apparently increases within the beam by a factor of 3 to 4. Predicted changes in electron temperature throughout the ionosphere are shown in Fig. 2. In addition to direct temperature effects, the ohmic heating causes a decrease in the rate of electron recombination, resulting in a general slight daytime enhancement in the electron number density. Predictions of the ionospheric density modification due to microwave heating are presented in Fig. 3. No significant telecommunications impacts are expected to accompany these relatively small changes in ionospheric density.

The threshold power flux density for exciting enhanced electron heating is altitude dependent but can be approximated at 20-30 mW/cm^2. However, even for smaller power fluxes, significant increases in electron temperature can occur, as seen in Fig. 2. Predicted electron temperatures and the electron heating rate at 90 km are shown in Fig. 4 for two different power fluxes of the SPS beam. Thermal runaway temporarily exists for the heating at 30 mW/cm^2. However, even for the SPS power level of 23 mW/cm^2, significant heating develops. In addition, these results show that the heating saturation limit is reached only after tens of milliseconds, much longer than normal heating time scales. Recent experimental tests of this enhanced electron heating theory are described in the next section.

Experimental Observations

Initial tests of the enhanced electron heating theory were made in two series of experimental studies using the 430-MHz radar system at the Arecibo Observatory (National Astronomy and Ionosphere Center).[8] The rapid electron heating and cooling initially predicted for the lower ionosphere suggested that pulsed heating experiments using high peak powers would be sufficient to initiate enhanced electron heating. The 430-MHz radar system at Arecibo operates with 2.5-MW peak pulse power and a maximum 10-ms pulse length with a 6% duty cycle. Coupled with the gain of the 305-m Arecibo reflector, this system delivers ~ 15 W/m^2 (in the center of the radar beam) to 100-km altitude. This is roughly twice the estimated threshold for exciting enhanced electron heating.

The 430-MHz radar system also serves as the principal ionospheric diagnostic in this experiment. Performing as an incoherent backscatter radar, a sensitive receiver system detects the radar signal power backscattered by free electrons in the ionosphere, recorded as a function of time. In this manner, ionospheric electron density altitude profiles were measured before and after a long (0.4- to 9-ms) radar heating pulse. On these short time scales, electron number density variations are

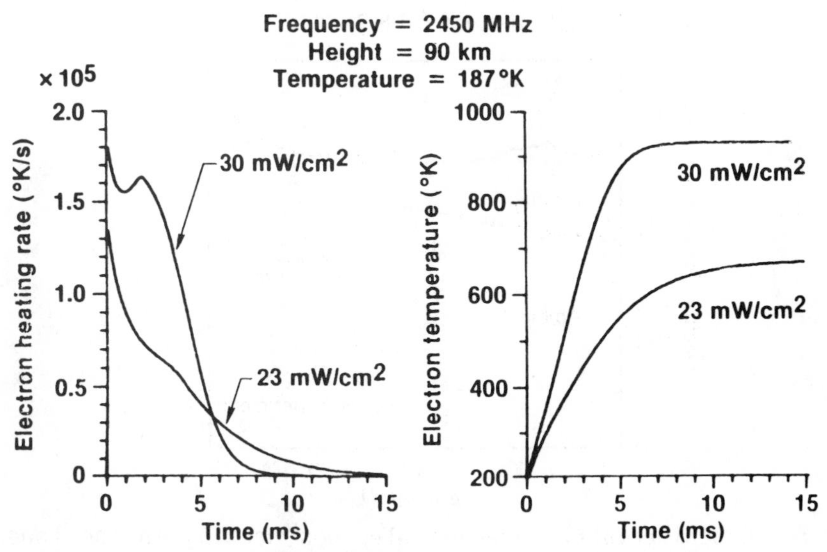

Fig. 4 Theoretical predictions of enhanced electron heating. a) Predicted electron temperature at 90 km as a function of time for power flux densities of 23 and 30 mW/cm^2. b) Electron heating rate for the same conditions.

very small. As a result, differences in these "hot" and "cold" profiles can be interpreted as due to changes in the electron scattering cross section. Because of the known temperature dependence of this cross section, the effective electron heating averaged across the beam could be unambiguously determined. The heating and short diagnostic radar pulses were separated slightly in frequency to avoid signal contamination, with an altitude resolution of 3 km. Figure 5 shows the results of this experiment.

Although more than 100 K increases in electron temperature were typically observed, this was much less than the predictions of enhanced electron heating theory at the time. This discrepancy was resolved by explicitly computing the heating time dependence as a function of power and accurately treating the radar power distribution across the heating and diagnostic beam. Using these improvements, agreement between theory and observations is good, as shown in Fig. 5. The remaining differences at the lowest altitudes result from complications in interpreting the radar backscatter data when the local plasma Debye length and the radar wavelength become comparable.

By delaying the radar diagnostic pulse with respect to the heating pulse, it was possible to measure the ionospheric

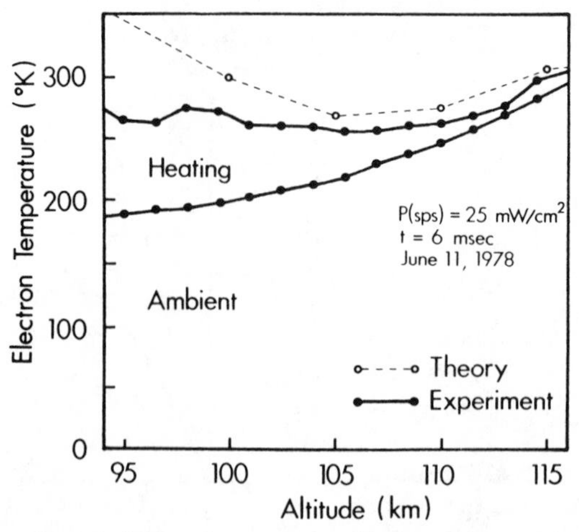

Fig. 5 Experimentally observed electron heating in the lower ionosphere. The solid curves represent the electron temperature for ambient and heated conditions. The dashed curve is the corresponding theoretical prediction. The power is expressed frequency-scaled to the SPS microwave beam.

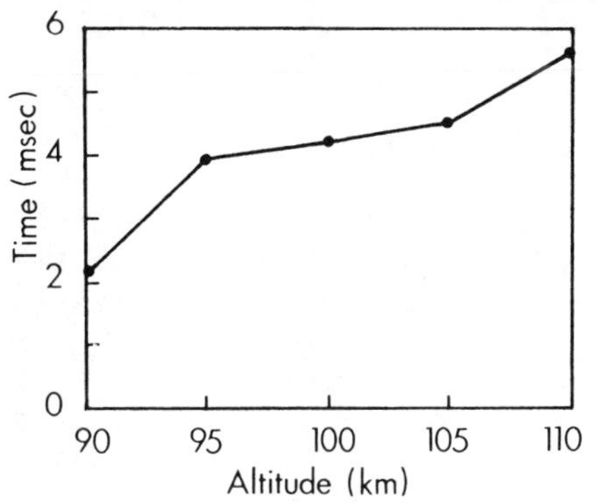

Fig. 6 The experimentally measured ionospheric electron cooling time constants as a function of altitude.

cooling time scales as a function of altitude (Fig. 6). The measurement accuracy is about a factor of two and confirms the predictions of the electron cooling models.

Although these experiments usefully provided the first test of enhanced electron heating theory, the results are limited by the restriction to 10-ms pulsed heating. As shown in the current theoretical heating predictions, this heating would not have achieved equilibrium throughout the beam.

Future experiments are designed to produce continuous, SPS-equivalent heating within the same spot size through the ionosphere. As shown in Eq. (4), ionospheric heating scales vary inversely as the square of the radio wave frequency. As a result, ionospheric heating equivalent to that of the SPS microwave beam may be achieved at lower radiated powers by heating at lower wave frequencies. This scaling is valid as long as the radio wave does not excite resonant plasma interactions, or roughly $f > 2f_p$. Other high-frequency ionospheric studies have produced both resonant and ohmic plasma heating effects; application of these results to the SPS environmental assessment must distinguish the responsible mechanism producing the observed effects. Future ionospheric heating studies to simulate SPS ionosphere-microwave interactions are planned, using HF facilities at Arecibo, Puerto Rico, and Platteville, Colo.

Past experimental programs at these facilities could produce power flux densities in the F-region of approximately 20 $\mu W/m^2$ at Arecibo and 50 $\mu W/m^2$ at Platteville for heating frequencies of 3 to 12 MHz. Recent improvements to the Arecibo facility have increased its maximum tranmitted power flux by about an order of magnitude over the same frequency range. Nevertheless, even with the benefit of frequency scaling, these facilities cannot produce SPS-equivalent heating throughout the ionosphere. Definitive experimental tests await additional upgrading at both sites.

Collective Plasma Phenomena

Differential ohmic heating of the ionosphere produces electron-temperature gradients that can cause convective plasma motions and excite plasma instabilities. This large-scale plasma behavior is driven by macroscopic thermal forces, as opposed to the microscopic kinetics of resistive heating effects. The resulting changes in ionospheric structure and electron density can induce radio wave scintillation and scattering, affecting a wide range of telecommunications systems. This section investigates a plasma phenomenon leading to ionospheric irregularities and striations and estimates its threshold and magnitude for conditions relevant to the solar-power satellite microwave beam.

Thermal Self-Focusing Theory

The collective plasma process that has received the most attention with respect to potential ionosphere-microwave interactions is the phenomenon of thermal self-focusing. Small natural density fluctuations in the ionosphere cause a variation in the plasma index of refraction. As a result, an electromagnetic wave propagating through the plasma is slightly focused and defocused, with the local electric field intensity increased as the incident wave refracts into regions of comparatively underdense plasma. Differential ohmic heating of the plasma gives rise to a temperature gradient, driving plasma from the focused region and amplifying the initial density perturbation. This self-focusing instability continues until hydrodynamic equilibrium is reached, creating large-scale ionospheric irregularities similar to natural spread-F conditions.

The theory of thermal self-focusing is developed separately for underdense and overdense plasma heating. Underdense ionospheric heating corresponds to electromagnetic radiation with frequency greater than the ionosphere's maximum plasma frequency so that the radio wave propagates through the ionosphere. The ionospheric critical frequency is typically 3 to 10 MHz; therefore, the ionosphere is underdense for the SPS microwave beam. Overdense heating occurs when the radio wave frequency is smaller than the ionosphere's critical frequency so that the wave is reflected in the ionosphere. Past HF experiments were almost entirely of this kind.

Before further discussing thermal self-focusing, it is appropriate to mention some equivalent terminologies that can be found in the open literature. Thermal self-focusing also has been referred to as thermal diffusive scattering, thermally driven stimulated Brillouin scattering or as collisionally coupled, purely growing, parametric instability in ionospheric modification theories. In nonlinear optics, it is called stimulated Rayleigh scattering; as applied to laser technology, it is described as beam filamentation. The effect of the self-focusing instability, which we call ionospheric irregularities or plasma striations, frequently is referred to as "hot spots" in laser-plasma coupling, "solitons," in astrophysical and controlled fusion research, and "cavitons" in some laboratory plasma studies. Nevertheless, despite subtle differences, the physics of self-focusing is essentially the same for each of these applications.

Thermal self-focusing does not have an absolute threshold in the usual sense. The focusing process organizes the plasma into large-scale density irregularities or striations. The threshold for creating these striations depends on the striation width λ. As the electric field intensity of the incident radiation increases, the resulting self-focused striations become narrower. The threshold power flux density for underdense thermal self-focusing in the ionosphere is given by[9]

$$E_o^2 c/8\pi = D^3 (n_m k T_e\ c)(\lambda_o \lambda_{mfp}^2/L^3)(n_c/n_m)^2 \qquad (14)$$

where E_o is the electric field amplitude of the incident wave $D = k_\perp^2 L/2k_o$; $k_\perp$ is the wave vector perpendicular to the radio wave propagation direction; L is the ionosphere scale height $k_o = 2\pi f/c$; n_m is the maximum plasma density $\lambda_o = c/f$; λ_{mfp} is the electron mean free path; and n_c is the critical electron density at $f_p = f$. In terms of experimental parameters, this threshold can be expressed as

$$P_{thresh}(W/cm^2) = (3.8 \times 10^{11})(D/L)^3 (T_e^5 f^3/f_p^6) \qquad (15)$$

For a reasonable spatial growth rate (D = 5) and typical ionospheric conditions, the SPS self-focusing threshold is approximately 50 W/m^2, well below the proposed microwave power flux density of 230 W/m^2. Thus, beam self-focusing is expected to occur for the solar-power satellite. The resulting striation width is approximately 500 m, with induced density fluctuations on the order of a few percent or less. Another study, assuming an initial condition that low-level spatially coherent density fluctuations are already naturally present in the ionosphere, predicts density variations of up to 80%.[10]

In both studies, the SPS microwave beam is predicted to generate large-scale ionospheric irregularities. The environmental and system impacts of this process will depend on the degree of beam self-focusing, the size of the resulting large-scale density striations, and the magnitude of the density fluctuations within the irregularities. Secondary instabilities may develop along the striation density gradients. Ionospheric striations may cause scintillation of radio waves using transionospheric propagation through the perturbed region, including the SPS pilot beam. In addition, hf radio waves will undergo multiple reflections in the striated region, resembling a natural spread-F environment. Recent experimental observations of self-focusing electromagnetic waves in the ionosphere are described in the next section.

Experimental Observations

Thermal self-focusing of high-frequency electromagnetic radiation has been observed in overdense ionospheric modification experiments. The overdense self-focusing instability develops with theoretically predicted threshold fields, scale lengths, and growth rates.[11] Associated density irregularities form on time scales of seconds to tens of seconds and decay on time scales of minutes. Above the hf wave-reflection height, these striations are extended along the geomagnetic field for at least 100 km.

For overdense self-focusing, the theoretical threshold can be expressed as[9]

$$E_0^2 c/8\pi = (7.5 \times 10^{10}) D^4 Te^5 / f n_m L^3 \tag{16}$$

For typical ionospheric parameters and $f_{HF} = 6$ MHz,

$$P_{thresh}(W/m^2) \simeq D^4 (.4\ \mu W/m^2) \tag{17}$$

exceeded in the F-region by both the Platteville ($\sim 50\ \mu W/m^2$) and Arecibo ($\sim 120\ \mu W/m^2$) hf facilities. The threshold striation scale size in the magnetic meridian can be found from[12]

$$P_{HF}(W/m^2) \cong (1.9 \times 10^{12}) Te^5 \sin^2\theta / L f_{HF}^3 \lambda_\perp^2 \tag{18}$$

where θ is the angle between the wave-propagation direction and the magnetic field. For typical experimental conditions,

$$P_{HF}(W/m^2) \cong 38 \lambda_\perp^{-2} \tag{19}$$

The hf power flux at 250 km is estimated at 30 $\mu W/m^2$ for this experiment, yielding a north-south striation width of $\lambda_\perp$

= 1.1 km. In the east-west plane, the striation widths are predicted to be approximately 500 m.[13] However, these theories derive only the instability threshold conditions; a saturation theory is being developed.

Figure 7 is a schematic of a self-focusing experiment conducted at the Arecibo Observatory. Intense hf electromagnetic radiation incident as an overdense ionospheric plasma excites parametric instabilities, enhancing electron-plasma oscillations observable by incoherent backscatter radar. These instabilities continue to be the subject of intense experimental study. Of importance here is the fact that, above instability threshold, the strength of enhanced plasma waves directly depends on the local power of the pump electric field. In addition, because of exact frequency and wave-number matching conditions for both the parametric wave-plasma interaction and the radar incoherent backscatter process, these enhanced waves are detected at only one altitude. As a result, systematic scanning of the narrow radar beam across the ionospheric interaction region of the enhanced plasma waves yields a two-dimensional cross section characteristic of the electric-field intensity. These maps of electric-field strength clearly show

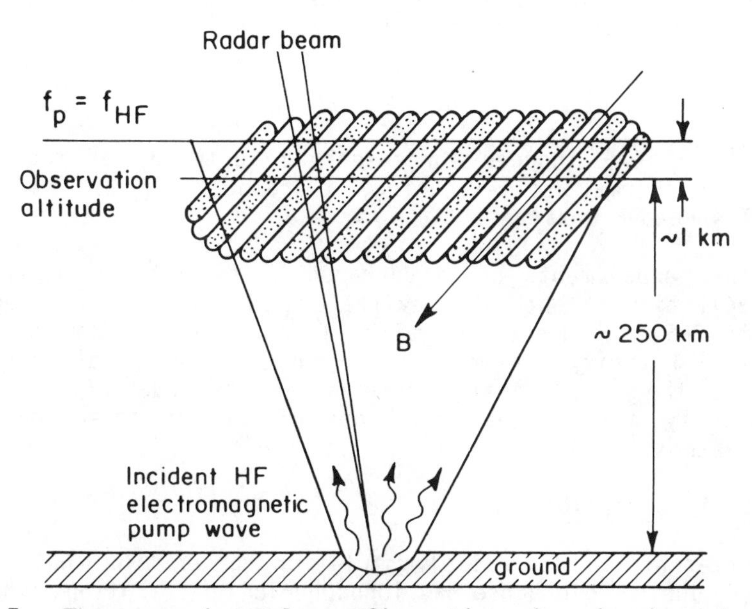

Fig. 7 The experimental configuration for incoherent backscatter radar mapping of ionospheric plasma striations, attributed to thermal self-focusing of incident HF electromagnetic radiation (from Duncan and Behnke.[12]).

self-focusing striations and large-scale structuring of the illuminated plasma.[12]

The results of a self-focusing experiment conducted at Arecibo from June 2-17, 1977 are shown in Fig. 8. When the radar beam is fixed (Fig. 8a), signal modulations are induced by the natural ExB drift of striations through the beam. Immediately afterward, rapid scanning of the radar beam (shown in Fig. 8b) detected a series of striations in the interaction region. Typical striation dimensions deduced from observations over many such scans are 0.8 to1.2 km in the north-south plane and 0.5 to 1.0 km in the east-west plane, in good agreement with thermal self-focusing theory. Striation velocities are on the order of 25 m/s, with components of 20 m/s to the east, and smaller than 15 m/s in the north-south direction. North-south velocity measurements are complicated by the strong spatial dependence of the striation width in the magnetic meridian plane.

The enhanced plasma waves detected in these striations are observed approximately 5-km higher in altitude than photoelectron enhanced plasma oscillations seen outside the focused regions.[14] If this difference is interpreted as a shift of the plasma-density profile, the magnitude of the density fluctuations in the striation can be estimated by

$$\Delta n_e \simeq \Delta x \; n_e/L \tag{20}$$

where $\Delta n_e \gtrsim 5\%$. Induced large-scale structuring of the ionosphere by thermal self-focusing, therefore, is much like a natural spread-F environment.

Future experiments using upgraded HF ionospheric modification facilities at both Platteville and Arecibo are designed to simulate SPS underdense thermal self-focusing. Improved diagnostics will verify the instability threshold and scaling laws, determine the irregularity geometry and secondary phenomena, and directly measure the striations effects on telecommunications systems.

Short-Scale Striations

Short-scale (meter-size) striation generation has been observed during overdense HF ionospheric modification experiments.[1,15] These irregularities can affect not only hf communications links, but also many vhf and uhf systems. Several theories have been proposed explaining the development of short-scale striations. All of these theories require resonant or parametric interactions between the exciting electromagnetic wave and the ionospheric plasma. These parametric interactions represent direct coupling of the electromagnetic wave energy into one or more of the natural plasma modes of oscillation. As a result, short-scale plasma striations are not expected to be

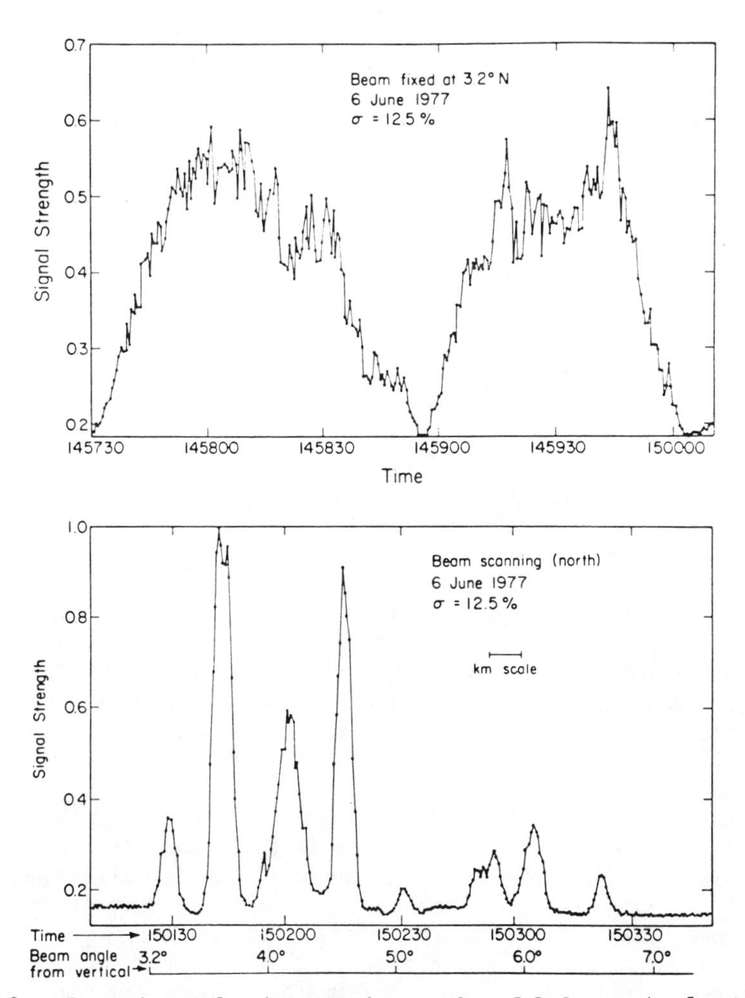

Fig. 8 Experimental observations of self-focused electromagnetic waves in the F-region ionosphere. a) Signal modulation induced by the natural dirft of striations through the fixed radar beam. b) A series of striations observed from rapid scanning of the radar beam across the interaction region, immediately following the drift measurements.

produced by the underdense SPS microwave beam. The proposed underdense hf ionospheric-heating experiments will demonstrate this effect if it occurs. The next section describes other potential plasma phenomena which have been investigated but which are considered unlikely for SPS ionosphere-microwave interactions.

Other Potential Effects

Although the phenomena believed to be of most importance to the SPS microwave beam interactions study have been discussed in the preceding sections, several other potential processes have been considered. Either because of frequency constraints or high instability thresholds, these effects are not expected to be excited by the SPS microwave beam. Several of these effects are described briefly in the next sections.

Parametric and Resonant Plasma Interactions

Parametric excitation of ionospheric plasma waves can develop for multiple-frequency electromagnetic radiation. In this case, parametric interactions occur not only at the wave frequency, but also at resonances of the difference frequencies. This process has been treated theoretically[16] and observed in recent hf experiments at Arecibo.[17] These interactions may be important because the SPS microwave power beam does not operate at a single frequency. Parametric interactions might be excited by beat waves generated within the finite bandwidth of the downcoming beam. In addition, the upgoing pilot beam operates at a frequency slightly separated from the downcoming beam. Again, parametric instabilities could be produced if the difference frequency of these waves coincides with a multiple of a resonant ionospheric plasma frequency.

Stimulated scattering of electromagnetic waves in the ionosphere is another process that can disturb the ambient plasma significantly. In this case, natural plasma oscillations are enhanced by scattering of the electromagnetic radiation from the plasma waves. Stimulated diffusion scattering probably will not occur for the SPS microwave beam but may be generated in hf experiments designed to simulate ionosphere-microwave interactions. Plasma heating and density modifications and a decrease in hf power flux reaching the higher altitudes would be associated with this phenomenon. Careful experimental studies of multiple-frequency parametric interactions and stimulated scattering should resolve any questions about potential SPS microwave beam effects and application of the hf experimental results.

Thermal Self-Focusing in the Lower Ionosphere

In addition to thermal self-focusing of electromagnetic waves in the F-region of the ionosphere, an analogous phenomenon may occur in the collision-dominated lower ionosphere. In this process, the nonlinear radio wave focusing is driven by changes in the electron collision frequency.[18] In a sense, this is thermal self-focusing driven by enhanced electron heating. This process is more rapid than its F-region counterpart, developing on a time scale of tens of milliseconds. This effect may have been detected in a previous hf experiment; a detailed study is planned as part of the future hf experimental research program.

Electric Breakdown of the Neutral Gas

A radio wave propagating through a partially ionized plasma accelerates free electrons in its electric field. Under normal conditions, collisions of these electrons with neutrals of the background gas leads to ohmic heating of the plasma. However, for a sufficiently intense electric field, the ionization frequency of the accelerated electrons may increase rapidly, causing electrical breakdown of the neutral gas.[19] This may lead to runaway ionization of the 20- to 100-km region, affecting communications systems up to uhf frequencies. The threshold field required to initiate electric breakdown has been estimated to be above 4×10^6 mV/m; the maximum SPS electric field is approximately 4×10^5 mV/m. Intense focusing of the beam may increase the microwave electric field significantly; an increase in power density by a factor of 100 would be required to approach the theoretical threshold. Although focusing of the microwave beam to this extent is extremely improbable, a process leading to impact ionization of neutrals would be very important if it did occur. The frequency-scaled hf experiments will look for this effect. In addition, the possibility of determining the relevant threshold through plasma laboratory studies has been discussed.

Atypical Ionospheric Conditions

Studies of ionosphere-microwave interactions must investigate the physics of potential phenomena for both ideal ionospheric conditions and the disturbed ionosphere. These variations include natural diurnal, seasonal, and solar-cycle changes; atypical ionospheric occurrences such as sporadic-E and spread-F; and man-made perturbations, i.e., ionospheric depletions following large rocket launches. The last concern may be the greatest, since construction of the solar-power satellite would involve frequent launches of heavy-lift launch vehicles. Large ionospheric density depletions are predicted to accompany the SPS construction,[20] and the first solar-power satellites will be expected to operate in this modified environment. As a result, the study of ionosphere-microwave interactions must allow for large variations in the unheated ionospheric composition and structure.

Conclusions

We have reviewed numerous phenomena associated with propagation of intense electromagnetic radiation through the ionosphere. Within the solar-power satellite microwave power beam, electron temperature enhancements of several hundred degrees Kelvin are predicted in the lower ionosphere. Small changes in electron density will accompany this heating. In addition, thermal self-focusing of the microwave beam in the F-region ionosphere is expected to produce large-scale density irregularities. Secondary nonlinear interactions may develop in this perturbed region. The principal impact of these ionospheric changes will be on telecommunication systems. An experimental program to determine these effects quantitatively is underway.

In the study of ionospheric physics, theory has historically been relatively poor at predicting experimental results, although quick to explain observations. It is therefore premature to form conclusions based on the ionospheric studies to date. Theoretical investigations have identified several potentially serious SPS ionospheric impacts; limited experimental studies have supported these results. However, no significant telecommunications or climate effects have yet been demonstrated experimentally. Furthermore, to extrapolate the results of current ionospheric research to the SPS studies, we also must verify the scaling assumptions on which those extrapolations are based. All these considerations are incorporated into the ongoing research program to determine ionospheric effects of microwave beams.

Acknowledgments

This paper has benefited from informative discussions with W. E. Gordon, F. W. Perkins, G. Meltz, and C. M. Rush. The Arecibo Observatory is operated by Cornell University under contract to the National Science Foundation. This work was supported by the Department of Energy. Editorial services were provided by J. Stelzer.

References

[1]See, for example, Radio Science Vol. 9, 1974.

[2]Brueckner, K. A., "Laser Driven Fusion," IEEE Transactions in Plasma Science Vo. PS-1, No. 1, March 1973, pp. 13-26; Nuckolls, J., Emmett, J., and Wood, L., "Laser-Induced Thermonuclear Fusion," Physics Today Vol. 26, No. 8, Aug. 1973, pp. 46; Johnson, L. C., and Chu, T. K., "Measurements of Electron Density Evolution and Beam Self-Focusing in a Laser-Produced Plasma," Physical Review Letters Vol. 32, No. 10, August 1974, pp. 517-520.

[3]Nicholson, D. R., Goldman, M. V., Hoyng, P., and Weatherall, J. C., "Nonlinear Langmuir Waves During Type III Solar Radio Bursts," Astrophysical Journal Vol. 223, July 1978, pp. 605-619; Goldman, M. V., and Nicholson, O. R., "Vital Theory of Direct Langmuir Collapse," Physical Review Letters Vol. 41, No. 6, Aug. 7, 1978, p. 406.

[4]Brown, W. C., "Satellite Power Stations: A New Source of Energy?" IEEE Spectrum Vol. 10, March 1973, pp. 38-47; also Glaser, P. E., "Solar Power from Satellites," Physics Today Vol. 30, No. 2, Feb. 1977, p. 30.

[5]Banks, P. M., and Kockartz, G., Aeronomy, Acad. Press., New York, 1973; Rees M. H., and Roble, R. G., "Observations and Theory of the Formation of Stable Auroral Red Arcs," Review of Geophysical and Space Physics Vol. 13, No. 1, Feb. 1975, pp. 201-242; Schunk, R. W., and Nagy, A. F., "Electron Temperatures in the F-Region of the Ionosphere: Theory and Observations," Review of Geophysical and Space Physics Vol. 16, No. 3, August 1978, pp. 355-399.

[6]Holway, L. H., and Meltz, G., "Heating of the Laser Ionosphere by Powerful Radio Waves," Journal of Geophysical Research Vol. 78, No. 34, Dec. 1, 1973, pp. 8402-8408.

[7]Perkins, F. W., and Roble, R. G., "Ionospheric Heating by Radio Waves: Predictions for Arecibo and the Satellite Power Station," Journal of Geophysical Research Vol. 83, No. A4, April 1, 1978, pp. 1611-1624; Duncan, L. M., and Zinn, J., "Ionosphere-Microwave Interactions for Solar Power Satellite," Los Alamos Scientific Laboratory report LA-UR-78-758, March 1978; G. Meltz, SPS Environmental Assessment, Washington DC, June 1979.

[8]Duncan, L. M., "SPS Environmental Assessment," Washington, DC, June 1979; Coco, D. S., "An Experimental Study of Electron Thermal Runaway in the Lower Ionosphere," MS Thesis, Rice University, Houston, TX, May 1979.

[9]Perkins, F. W., and Valeo, E. J., "Thermal Self-Focusing of Electromagnetic Waves in Plasmas," Physical Review Letters Vol. 32, No. 22, June 1974, pp. 1234-1237.

[10]Drummond, J. E., "Thermal Stability of Earth's Ionosphere under Power Transmitting Satellites," IEEE Transactions of Plasma Science Vol. PS-4, No. 4, Dec. 1976, p. 228.

[11]Thome, G. D., and Perkins, F. W., "Production of Ionospheric Striations by Self-Focusing of Intense Radio Waves," Physical Review Letters Vol. 32, No. 22, June 1974, pp. 1238-1240.

[12]Duncan, L. M., and Behnke, R. A., "Observations of Self-Focusing Electromagnetic Waves in the Ionosphere," Physical Review Letters Vol. 41, No. 14, Oct. 2, 1978, p. 998.

[13]Cragin, B. L., and Fejer, J. A., "Generation of Large-Scale Field-Aligned Irregularities in Ionospheric Modification Experiments," Radio Science Vol. 9, No. 11, Nov. 1974, pp. 1071-1075; Cragin, B. L., Fejer, J. A., and Leer, E., "Generation of Artificial Spread-F by a Collisionally Coupled Purely Growing Parametric Instability," Radio Science Vol. 12, No. 2, March 1977, pp. 273-284.

[14]Muldrew, D. B., and Showen, R. L., "Height of the HF-Enhanced Plasma Line at Arecibo," Journal of Geophysical Research Vol. 82, No. 29, Oct. 1, 1977, pp. 4793-4804.

[15]Duncan, L. M., and Gordon, W. E., "Final Report, Ionosphere/Microwave Beam Interaction Study," Rice University, Houston, TX, Final Report NAS9-15212, September 1977.

[16]Cragin, B. L., Fejer, J. A., and Showen, R. L., "Theory of Coherent Parametric Instabilities Excited by Two or More Pump Waves," Geophysical Research Letters Vol. 5, No. 3, March 1978, pp. 183-186.

[17]Showen, R. L., Duncan, L. M., and Cragin, B. L., "Observations of Plasma Instabilities in a Multiple Pump Ionospheric Heating," Geophysical Research Letters Vol. 5, No. 3, March 1978, pp. 187-190.

[18]Gurevich, A. V., Milikh, G. M., and Shlyuger, I. S., "Nonlinear Thermal Focusing of Radio Waves in the Lower Ionosphere," Geomagnetics and Aeronautics Vol. 16, No. 4, 1977, pp. 366-369.

[19]A. V. Gurevich, "Nonlinear Phenomena in the Ionosphere," in Physics and Chemistry in Space Vol. 10, 1978, Springer, Berlin-New York.

[20]Zinn, J., Sutherland, C. D., Smith, G. M., and Pongratz, M. B., "Predicted Ionospheric Effects from Launches of Heavy Lift Rocket Vehicles for the Construction of Solar Power Satellites," Los Alamos Scientific Laboratory report LA-UR-78-1590, 1978.

UPPER ATMOSPHERE MODIFICATIONS DUE TO CHRONIC DISCHARGES OF WATER VAPOR FROM SPACE LAUNCH VEHICLE EXHAUSTS

Jeffrey M. Forbes*
Boston University, Boston, Mass.

Abstract

The influences of transport, photodissociation, and frequency of injection on the global redistribution of water deposited in the Earth's upper atmosphere by repeated launches of large rockets are investigated. Measurable environmental effects of the injected water are found to occur when the mesospheric water vapor mixing ratio χ exceeds 100 ppmv, which occurs over areas of order 20,000 km^2 in connection with possible future Satellite Power System activities. These effects include a) a 50% reduction in D-region ionization due to screening of L_α radiation by water; b) a 50% reduction of D-region ionization as a result of converting NO^+ to water cluster ions which possess more rapid recombination rates; c) a doubling of OH concentrations below 100 km; and d) a global doubling of nighttime E-region ionization due to L_α and L_β radiations geocoronally scattered by atomic hydrogen released by photolysis of H_2O. Mixing ratios of 10^3 ppmv necessary for the maintenance of clouds at the mesopause are reached only over areas of order 200 km^2.

I. Introduction

We are approaching an era in which the existence in space of large structures (e.g., solar collectors) and even colonies are real possibilities. One scenario for constructing such structures involves launching many large rockets to transport the necessary materials and personnel into space. It is possible that sufficiently frequent and profuse depositions of thruster effluents (mostly CO_2, H_2O, and H_2) might seriously modify the atmosphere, especially since CO_2 and H_2O are minor

---

*Research Associate, Dept. of Astronomy, also Research Scientist, Space Data Analysis Laboratory, Boston College, Chestnut Hill, Mass.

atmospheric constituents which nevertheless play an important role in the radiative budget of the planet. The possibility of severely modifying the upper atmosphere recently was dramatized by the "ionospheric hole" created during the launch of Skylab by chemical reactions of injected H_2O and H_2 with O^+ to produce molecular ions which rapidly recombined with electrons (Mendillo et al., 1975). However, possible modifications of the atmosphere below about 150 km because of chronic discharges of this type have received relatively little attention.

In this paper a simple time-dependent analytic formalism is developed for examining the competing effects of transport, photodissociation, and frequency of injection on the steady-state global distribution of water discharged from the exhausts of large space launch vehicles in the mesosphere and lower thermosphere. This formalism is used to predict (and in some cases only to suggest) some environmental effects that might be expected from such activities. The model (described in Section II) and sample simulations (see Section III) are specifically geared to predict the fate of effluents from the second stages of Heavy Lift Launch Vehicles (HLLV's) associated with construction of a Satellite Power System (SPS). These vehicles are expected to deposit significant amounts of water in the 70 - 120 km region of the atmosphere. The methodology developed herein is easily extrapolated to handle a) simultaneous launches at several longitudes (only one launch site is considered here), b) less serious discharges, for instance from Space Shuttle operations, and c) other exhaust effluents, such as CO_2, which are expected to contribute a much smaller perturbation on ambient concentrations.

In Section IV possible environmental effects associated with the perturbation water vapor mixing ratios computed in Section III are reviewed briefly. Physical processes addressed include mesospheric cloud formation, ozone depletions, subsequent thermospheric effects of atomic hydrogen produced by H_2O photolysis, depletions of D and E region ionization, enhanced airglow emissions, and modifications of the heat budget of the mesosphere and lower thermosphere.

II. Mathematical Model

Given the rough nature of various input data, the following model is considered adequate for a first approximation. Keep in mind that this approach is basically a sophisticated scale analysis, and quantitative inferences must be made within this context. We will consider the 70-120 km height range as a slab, or closed system; in other words, diffusion of H_2O to above 120 km or below 70 km is prohibited. This simplifying assumption can be crudely justified as follows: The eddy mixing rates become slow in the upper stratosphere, and parcels of air are exchanged with the mesosphere infrequently. For instance, if we consider that vertical diffusion velocities w_D are of order $\frac{D}{H}$, then w_D varies from about .1 km day^{-1} between 40-50 km to 1 km day^{-1} between 60-70 km, based on the vertical profiles of eddy diffusion coefficient given in Oliver et al. (1977). Photolysis of H_2O is extremely slow ($J \simeq 10^{-7}$sec^{-1}) below 70 km (Bowman et al., 1970). On the other hand, in the lower thermosphere, say 120 km, the vertical diffusion velocity is on the order of 10 km day^{-1}; water molecules, then, travel only about 20 km in one photochemical half-life (~2 day; Bowman et al., 1970). Any perturbation in the H_2O mixing ratio between 70 and 120 km can be considered an overestimate in the following model, since diffusion out of the 70-120 km region is neglected. As will become evident, the analytic simplicity of the model will permit future analyses to take these effects into account if warranted by the results of the present study. A suggested approach is outlined in Section V.

In a slab model the equation governing the height-integrated number density, $n = \int_{Z_1}^{Z_2} n' \, dZ$, in a cartesian co-ordinate system is

$$\frac{\partial n}{\partial t} + \bar{u}\frac{\partial n}{\partial x} + \bar{v}\frac{\partial n}{\partial y} - K_h\frac{\partial^2 n}{\partial x^2} - K_h\frac{\partial^2 n}{\partial y^2} = C_o\,\delta(x)\,\delta(y)\,f(t) - Jn \quad (1)$$

where n' = molecules m^{-3}
$\bar{u}$ = an average E-W wind speed
$\bar{v}$ = an average N-S wind speed
x = E-W coordinate
y = N-S coordinate
K_h = horizontal eddy diffusivity
J = rate of photolysis of H_2O
$C_o = N_o/(z_2 - z_1)$
N_o = total number of H_2O molecules injected between 70 and 120 km per launch
z_2 = 120 km
z_1 = 70 km

Values of $\bar{u}$ = 30 deg long day^{-1} and $\bar{v}$ = .5 deg lat day^{-1} are adopted, based on the models of Schoeberl and Strobel (1977) and Roble et al. (1977) at midlatitudes.

The horizontal eddy diffusivity K_h is a function of cloud scale, ranging from an isotropic value of ~10^6 cm^2 sec^{-1} for scales <100 m to ~10^{10} cm^2 sec^{-1} for scales >100 km in the mesosphere (Bauer, 1974). Actually, experimental data for the mesosphere exist only for scales less than about 10 km (corresponding to $K_h \simeq 10^9$ cm^2 sec^{-1}), so the latter value quoted for large scale sizes represents a crude extrapolation by the present author of Bauer's (1974) curves to larger scale sizes. In the present paper, a constant value of $K_h = 10^{10}$ cm^2 sec^{-1} is adopted. This represents an overestimate of the true horizontal eddy diffusivity (and hence also the rate of cloud growth) at early times of expansion ($t \lesssim 1$ hr). Since the results to be presented in Section III consider only water vapor distributions for times of 1 hr and 6 hr after injection, the assumption of $K_h = 10^{10} cm^2$ sec^{-1} is adequate for current purposes.

A value of J = .5 day^{-1} (Bowman et al., 1970) represents a reasonable day-night average value over the height range of interest, and an isothermal atmosphere with scale height H = 6.66 km is a good approximation in this height range. The total number of H_2O molecules injected per launch is taken to be 7.0×10^{31} molecules, evenly distributed between 70 and 120 km. This number is based on data provided in a Preliminary Environmental Assessment for the Satellite Power System (Doc. No. SPS-EA-001, 1978) performed under the auspices of Argonne National Laboratory for the Department of Energy. The source molecule distributions are assumed to be described by delta functions in the x and y directions. The time history of the water discharges is represented by equally-spaced delta functions with period T: $f(t) = \sum_{n=o}^{\infty} \delta(t-nT)$. For later reference, the Laplace transform of f(t) is

$$F(s) = \frac{1}{1 - e^{-sT}}$$

Equation (1) is solved by taking a Laplace transform in time:

$$N = \int_0^{\infty} ne^{-st}\, dt$$

and a double Fourier transform in x and y:

$$\eta = \int_{-\infty}^{\infty}\int_{-\infty}^{\infty} N \exp(-i\alpha x - i\beta y)\, dx\, dy$$

The resulting algebraic equation is solved for η:

$$\eta = \frac{C_o}{s+a}\left[1 - \frac{1}{e^{-ST}-1}\right]$$

where $a = i\alpha\bar{u} + i\beta\bar{v} + D\alpha^2 + D\beta^2 + J$. Inversion of η and N yields

$$n(x,y,t) = \sum_{m=o}^{M} \frac{C_o U(t_m)}{4\pi D t_m} \exp\left\{-\frac{(x-\bar{u}t_m)^2+(y-\bar{v}t_m)^2}{4Dt_m} - Jt_m\right\} \quad (2)$$

where $t_m = t-mT$, $M=t/T$, and $U(t-t_o)$ is the unit step function. This solution for $n(x,y,t)$ simply represents the superposition of M Gaussian distributions each spreading in time, advected $\bar{u}t$ in longitude and $\bar{v}t$ in latitude, and exponentially modulated by the H_2O photolysis rate.

The quantity n represents the height integral of n´ over the 70-120 km slab. A more useful parameter to evaluate possible modification effects would be the water vapor mixing ratio. Further, some idea of the height dependence of n´ would be desirable. There are two extreme cases one might consider. First, if n´ is assumed to be constant with height, then $n' = \frac{n}{z_2-z_1}$ and

$$\chi = n'/n_t = n \exp \frac{z-z_1}{H} / n_o(z_2-z_1) \quad (3)$$

where n_t is the total number density at z, and n_o is the value of n_t at $z=z_1$. On the other hand, if n´ is well mixed, then we can define

$$n' = n \exp\left(\frac{z_1-z}{H}\right)/H\left[1 - \exp\left(\frac{z_1-z_2}{H}\right)\right] \quad (4)$$

which is a hydrostatic distribution for n´ with scale height H where the height-integrated number density is the same as in Eq. (3), or (z_2-z_1) n´. This gives

$$\chi = n'/n_t = n/n_o \; H\left[1 - \exp\left(\frac{z_1-z_2}{H}\right)\right] \quad (5)$$

Comparison between Eqs. (3) and (5) indicates that mixing ratios given by Eq. (3) are roughly 10, 1, and .1 times those given by Eq. (5) at altitudes of 100, 85, and 70 km, respectively. Thus, both of these equations give approximately the same result between 80 and 90 km. If we consider that the time constant for diffusion, H^2/D is typically on the order of 1 day, then on time scales less than about a day the deposited water vapor has not yet had time to reach a mixed distribution, and application of Eq. (3) might be more appropriate. For the values of $\bar{u}$ and $\bar{v}$ given above, the injected material is advected 30 deg long and .5 deg lat within 1 day.

Note that a constant H_2O photolysis rate of $J = 0.5$ day^{-1} is assumed. A constant J with height is only valid if the injected H_2O diffuses to optically thin concentrations on a time scale short with respect to photodissociation (~2 days). Water vapor below 120 km is primarily dissociated by L_α radiation and wavelengths greater than 1724Å. The dissociation by wavelengths greater than 1724Å is roughly half that due to L_α above 70 km (Banks and Kockarts, 1973). Due to the absorption cross sections involved, optical thinness is a valid assumption for $\lambda > 1724$Å when $\chi \lesssim 1000$, whereas the 70-120 km region is optically thin to L_α only for $\chi \lesssim 100$. According to the model simulations in Section III, χ exceeds 100 over areas on the order of 20,000 km^2 in the vicinity of injection, so that the assumption that $J = .5$ day^{-1} is independent of height, time, latitude, and longitude can be violated over areas of this magnitude. Given the crudeness of the model, it is not expected that the results presented here would be affected significantly by a more detailed specification of the H_2O photolysis rate.

It is implicitly assumed that all of the water is in vapor form, at least within a short time after injection. Adiabatic expansion of the exhaust cloud into the tenuous upper atmosphere will cause sufficient cooling to freeze part of the water (see Bernhardt, 1976). Freezing-out of water released at high altitudes was, in fact, observed during the 1962 Saturn High Water Experiment when a tank containing 86,000 kg of water in a Saturn rocket was explosively ruptured at 105 km (Debus et al., 1964; Edwards et al., 1962). The ice particles fall at a rate determined by their size and weight and the ambient atmospheric density. Based on numbers given in Reid (1975) these ice particles will sublime in a time scale on the order of minutes, which is short compared with the photochemical lifetime of H_2O (~2 days). It can therefore be assumed that all of the H_2O is in vapor form for purposes of calculating the fate of the released water. At the other extreme, it is possible (but unlikely, see Section IV) that favorable conditions would exist to maintain clouds at the mesopause level. Such a situation would significantly change the physics of the problem from that envisioned here.

III. Model Simulations

The steady-state solution for χ obtained from Eq. (2) can be visualized as a series of 3-dimensional Gaussian-shaped pulses moving away from the point of injection at $\bar{u} = 30$ deg long day^{-1} and $\bar{v} = .5$ deg lat day^{-1}, the width of the pulses increasing as $\sqrt{t}$ due to diffusive expansion, and the peak amplitude decreasing as e^{-Jt}/t because of both diffusive expansion and photolysis of H_2O. Steady-state values of χ for T = .125, .25, 1.0, and 4.0 days at 1 hr and 6 hr after injec-

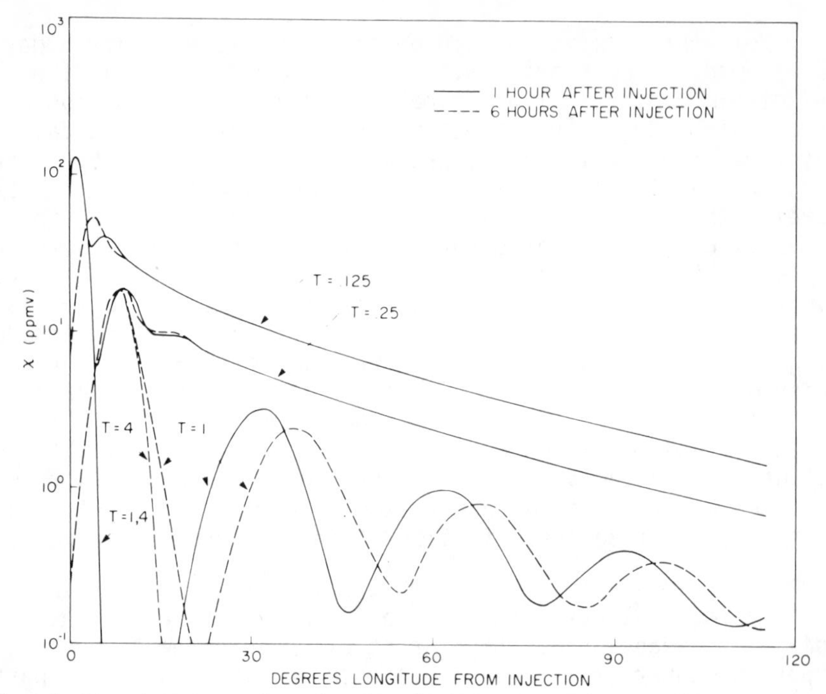

Fig. 1 Steady-state longitudinal distribution of χ at $\Delta t = 1$ hr and $\Delta t = 6$ hr after injection for T = .125, .25, 1.0, and 4.0 days. (The latitude varies along each curve and is given by $\bar{v}\ (\lambda/\bar{u})$ in deg lat, where $\bar{v} = .5$ deg lat day^{-1} and λ is deg long from injection.)

tion are plotted vs longitude in Fig. 1 and vs latitude in Fig. 2. As illustrated in Fig. 1, only for $T \gtrsim 1$ day do the pulses retain their identity without diffusing into one another. This is because the 10% width of the pulses is $\sqrt{4Dt \log_e 10}$ or ~ 10 $\sqrt{t}$ deg (where t is in days), whereas the peak-to-peak spacing is $\bar{u}T$ in longitude and $\bar{v}T$ in latitude. Thus, for T = 1 day, the peak-to-peak spacing is 30 deg long and .5 deg lat and the pulse width is ~ 10 deg at 1 day and ~ 30 deg at 3 days after injection, whereas for T = .2 day the pulse spacing is only 6 deg long and .1 deg lat. Note that χ is diminished to negligible values long before being advected one circuit (360 deg) around the Earth.

The combined effects of advection by winds, the high mixing rates characteristic of the mesosphere and lower thermosphere, and the relatively short lifetime of H_2O with respect to photodissociation (~2 days), act to prevent significant global or even regional steady-state buildups of water vapor concentrations between 70 and 120 km. In the following section, it is noted that a baseline value for measurable envi-

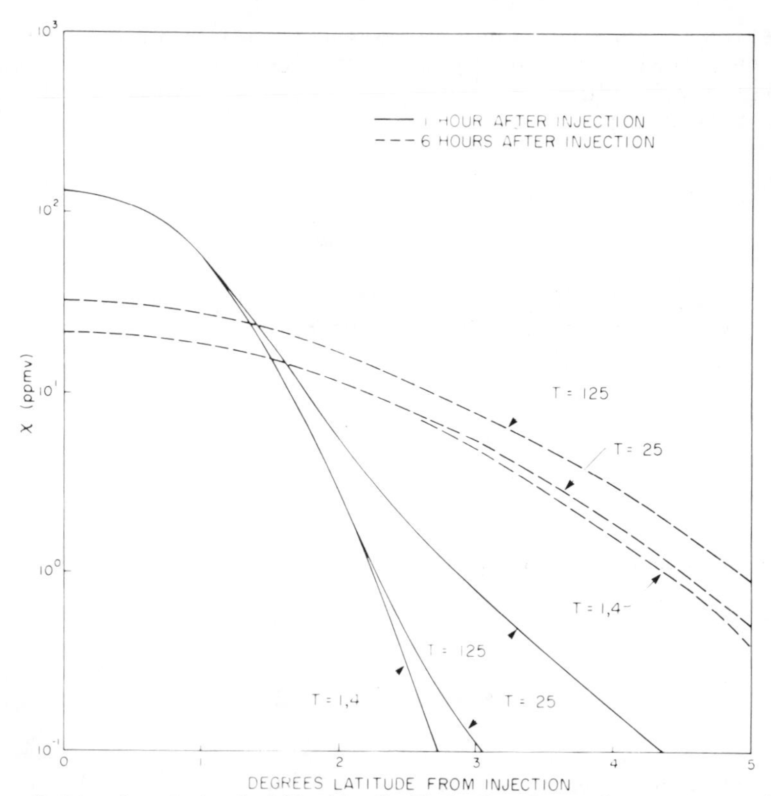

Fig. 2 Steady-state latitudinal distribution of χ at Δt = 1 hr and Δt = 6 hr after injection for T = .25, 1.0, and 4.0 days. (The longitudes corresponding to each curve are $\bar{u}\ \Delta t/24$ where $\bar{u}$ = 30 deg long day^{-1} and Δt is in hours.)

ronmental effects is χ=100 ppm between 80 and 90 km, which is only exceeded (see Figs. 1 and 2) within an area on the order of 20,000 km^2 (.5 deg lat × 2 deg long) around the point of release within 1 hr of injection for the nominal value of 7.0 × 10^{31} molecules of H_2O initially distributed evenly between 70 and 120 km. Around 100-120 km, χ could reach 10^3 ppmv over an area of this size, and 10^2 ppmv in an area on the order of 500,000 km^2 (5 deg lat × 10 deg long) within 6 hr of injection. [See Section II regarding the height dependence of χ for early (t<<1 day) and late (t>>1 day) times after injection.]

IV. Modifications of the Mesosphere and Lower Thermosphere

A. Ionospheric Effects

The possibility of severely modifying the ionosphere was recently dramatized by the injection of H_2 and H_2O into the

F-region during the launch of Skylab (Mendillo et al., 1975). In the normal F-region, the rate of electron loss L is limited by relatively slow ($\gamma \sim 10^{12} cm^{-3} sec^{-1}$) charge transfer and ion-atom interchange reactions:

$$O^+ + O_2 \rightarrow O_2^+ + O$$

$$O^+ + N_2 \rightarrow NO^+ + O$$

The injection of CO_2, H_2, or H_2O can greatly speed-up the conversion to molecular ions ($\gamma \sim 10^{-9} cm^{-3} sec^{-1}$):

$$O^+ + CO_2 \rightarrow O_2^+ + CO$$

$$O^+ + H_2O \rightarrow H_2O^+ + O$$

$$O^+ + H_2 \rightarrow OH^+ + H$$

which subsequently rapidly ($\alpha \sim 10^{-7} cm^{-3} sec^{-1}$) recombine:

$$O_2^+ + e^- \rightarrow O + O$$

$$H_2O^+ + e^- \rightarrow H + OH$$

$$OH^+ + e^- \rightarrow O + H$$

and hence deplete the F-region ionization. Since H_2O molecules around 120 km only diffuse about 20 km within their nominal 2-day photochemical lifetime (see Section II), relatively few H_2O molecules are expected to reach F-region altitudes from a HLLV injection between 70 and 120 km. (However, more quantitative verification of this statement is needed.) Therefore, we direct our attention in this section to possible D- and E-region modifications caused by the injection of water vapor.

In the upper D- and E-regions, O_2^+ and NO^+ are the dominant species and the rate of electron loss is already normally determined by dissociative recombination and hence follows a square loss law: $L = \alpha[e]^2$. However, there are at least two possible ways that the injection of water vapor below 120 km could affect ambient concentrations of ionization:

a) by screening the D-region of photoionizing radiation; and

b) by enhancing the formation of heavy water cluster ions which possess faster recombination rates than O_2^+ and NO^+.

These possibilities are examined here.

Table 1 Percent attenuation of L_α radiation reaching D region, because of screening by H_2O injected between 70 and 120 km

	Mixed distribution[a]		Height-independent distribution[a]	
Altitude, km	χ = 100	χ = 1000	χ = 100	χ = 1000
120	0.	0.	0.4	0.
110	1.	13.	24.	94.0
100	4.	34.	43.	98.6
90	10.	67.	57.	99.98
80	25.	94.	67.	100.
70	75.	100.	75.	100.

[a]Defined in Section II.

The dominant source of molecular ions in the 70 to 90 km region is photoionization of NO by L_α radiation:

$$NO + h\nu(\lambda=1216\mathring{A}) \rightarrow NO^+ + e^-$$

There is also a significant contribution due to photoionization of $O_2(^1\Delta)$:

$$O_2(^1\Delta) + h\nu(\lambda=1027\text{-}1118\mathring{A}) \rightarrow O_2^+ + e^-$$

The absorption cross section of H_2O at L_α is about 1.4×10^{17} cm^2 (Watanabe, 1958). Therefore, a H_2O column content of 10^{17} molecules cm^{-2} yields about 75% attentuation of L_α. For values of water vapor mixing ratio χ at 85 km (as defined in Section II), % attenuations of L_α radiation vs height are tabulated in Table 1 assuming either a mixed distribution or height-independent distribution of H_2O, each with the same column content of H_2O. It is evident that χ must exceed about 100 ppm for significant (~50%) attenuation of L_α radiation reaching the D-region. For $\chi \lesssim 10$, attenuation is less than 1% at all altitudes between 70 and 120 km. Note from Section III that χ exceeds 100 ppmv over areas on the order of 20,000 km^2. Also, since the absorption cross section of H_2O at L_β (1026Å) is almost 2×10^{17} cm^2, and between 1027 and 1118Å averages about 5×10^{-18} cm^2 (Watanabe, 1958), considerations discussed here also apply to photoionization of $O_2(^1\Delta)$.

As will be discussed in the following section, one of the net effects of H_2O photolysis is to create atomic hydrogen. The nature of the formalism adopted here precludes any prediction of the diffusion and redistribution of H atoms in the thermosphere. However, drawing on some computations by Reidy (1962), Kellogg (1964) estimates that at least 100 metric tons

(10^6g) of H must be added (to the natural abundance of ~50 metric tons) above 105 km to globally increase the attenuation of L_α radiation reaching the D-region from the normal 1.6% (see Purcell and Tousey, 1960) to 10%, and that the additional H must be replenished at least once a week to be maintained. This may be compared to the nominal 200 tons of H atoms introduced by every second stage HLLV exhaust. Since injection frequency will probably exceed 1 week^{-1} and H atoms will not have time to redistribute themselves uniformly over the globe in their 1-week thermospheric lifetime, a nominal 10% attenuation of L_α may be assigned to this global effect as a lower limit until more quantitative estimates are available.

In the D-region there exists a sharp transition somewhere between 75 and 85 km where NO^+ and O_2^+ are the dominant positive ions above, and water clusters of the type $H^+(H_2O)_n$ (n= 1~7) are dominant below (Narcisi et al., 1971; Reid, 1977; Chakrabarty et al., 1978). NO^+ and O_2^+ are precursor ions for reaction chains which lead to formation of $H^+(H_2O)_n$. Transition altitude for the $O_2^+ \rightarrow H^+(H_2O)_n$ chain occurs where O exceeds H_2O (Sechrist, 1977), since loss of O_4^+ via

$$O_4^+ + O \rightarrow O_2^+ + O_3$$

effectively short circuits the reaction chain. The transition from water cluster ions to O_2^+ and NO^+ moves upward by about 5 km for every order of magnitude increase in water vapor mixing ratio between 80 and 100 km. Transition altitude for the $NO^+ \rightarrow H^+(H_2O)_n$ chain also is dependent on the H_2O content, but it is sensitive to temperature since the low binding energy of intermediary cluster ions of the type $NO^+(H_2O)_m(N_2)$ indicates that thermal breakup must be important at D-region temperatures. According to Reid's (1977) calculations, the transition altitude increases from about 73 km for a "warm and dry" (1.2 ppm) atmosphere to over 90 km for a "cold and wet" (9 ppm) atmosphere. It does not appear unrealistic to assume that values of χ = 100 ppmv at 85 km would be accompanied by a near complete conversion of O_2^+ and NO^+ (with recombination coefficients $\alpha_1 \sim 7\ 10^{-7}$ cm^3 sec^{-1}) to hydrated ions ($\alpha_2 \sim 3\ 10^{-6}$; Reid, 1977) between 70 and 100 km. Assuming a square loss law ($L=\alpha[e]^2$) and steady state conditions, the corresponding reduction in electron density would be no more than $(\alpha_1/\alpha_2)^{1/2} \approx 0.5$. Combined with the screening effects expected at χ = 100 ppmv (see previous discussion), a nominal reduction of order 75% in ionization density between 70 and 100 km can be expected over areas on the order of 20,000 km^2.

In passing it is noted that ionization of NO below 100 km and O_2 above 100 km by geocoronally-scattered L_α and L_β radiations, respectively, are important mechanisms for maintaining the nighttime D- and E-regions (Swider, 1972). Addition of H to the thermosphere will tend to increase nighttime ionization levels at these altitudes. This effect should be examined more carefully since the difference between nighttime and daytime E-region conductivities is important for E- and F-region electrodynamic coupling mechanisms (Rishbeth, 1971a, 1971b) currently thought to explain features of the electric field morphology at low latitudes (Behnke and Hagfors, 1974).

B. Minor Neutral Chemistry and Airglow Emissions

Photolysis of water in the mesosphere can lead to a complex chain of reactions involving hydrogen-oxygen compounds. It is out of the scope of the present study to deal with the complexities of the minor neutral chemistry of the mesosphere. Below, some potentially important modifications are suggested, merely to provide some impetus to other researchers interested in evaluating the effects of the type of chronic discharges addressed in this paper.

Photodissociation of water vapor can proceed as

$$H_2O + h\nu \rightarrow H + OH^*$$

or

$$H_2O + h\nu \rightarrow H_2 + O(^3P)$$

The latter is relatively slow. Production of OH is potentially important since it provides an important loss mechanism for O:

$$O(^3P) + OH \rightarrow O_2 + H \qquad k = 4.2\ E\text{-}11\ cm^{-3}$$

In the near infrared, it is also one of the most intense emitters in the airglow spectrum. In fact, according to Bowman et al. (1970), the marked increase in $[O(^3P)]$ above 75-80 km is directly related to the decrease in [OH] above this height. A lower limit on the attainable values of [OH] due to this mechanism can be obtained by assuming that [O] remains unchanged and that steady-state conditions apply:

$$[OH] \sim \frac{J[H_2O]}{k[O]}$$

using $k = 4.2E\text{-}11\ cm^3\ sec^{-1}$, $J = .50E\text{-}05\ sec^{-1}$, and [O] profiles from Bowman et al. (1970), it is found that an H_2O volume mixing ratio greater than 100 ppmv will at least double normal OH concentrations below 100 km.

The 5577Å airglow originates in the $^1S \rightarrow {}^1D$ atomic oxygen transition between 90 and 100 km. Production of $O(^1s)$ is thought to occur via

$$O + O + \rightarrow O(^1s) + O_2$$

Since production of $O(^1s)$ is sensitive to $O(^3p)$ concentrations near 100 km, any reduçtions in $O(^3p)$ should be reflected in a reduction in the 5577Å airglow intensity. Near 100 km, the $O(^3p)$ distribution can be crudely approximated by balancing production by photodissociation of O_2:

$$O_2 + h\nu \rightarrow O + O \qquad (J' \sim 10^{-7}\ sec^{-1})$$

with loss by

$$O + O + M \rightarrow O_2 + M \qquad (k_1 \approx 10^{-32}\ cm^6\ sec^{-1})$$

and

$$O + OH \rightarrow O_2 + H \qquad (k_2 \approx 4 \times 10^{-11}\ cm^3\ sec^{-1})$$

The half-life of O with respect to either of these loss mechanisms is on the order of 200 days for normal concentrations at 100 km. An equation describing the photochemical concentration of O is:

$$[O]^2 + \frac{k_2[OH]}{k_1[M]}[O] - \frac{2J'[O_2]}{k_1[M]} = 0 \qquad (6)$$

Solving for [O], it is found that even a factor of 10 increase in OH concentration (to $10^3\ cm^{-3}$) would lead only to roughly a 10% reduction in [O]. Although this might produce a measurable reduction in the 5577Å green line intensity, this effect in itself is not particularly significant.

Between 70 and 120 km the major production mechanism for O_3 is

$$O(^3p) + O_2 + M \rightarrow O_3 + M$$

and loss of O_3 occurs by

$$H + O_3 \rightarrow OH^* + O_2$$

$$O_3 + h\nu \rightarrow O(^1D) + O_2$$

Reaction with H exceeds photolysis as the major loss mechanism for O_3 between about 75 and 95 km (Bowman et al., 1970). Thus, production of H by H_2O photolysis is likely to reduce O_3 concentrations between 70 and 100 km, although a significant fraction of the hydrogen would be expected to diffuse into the thermosphere. Production of OH by this mechanism would provide an additional contribution of the same order as that produced directly by H_2O photolysis.

C. Cooling and Cloud Formation at the Mesopause

The existence of noctilucent clouds at high latitudes during summer suggests the possibility that addition of large amounts of water at the cold (T=150-200K, depending on latitude) mesopause level (~85 km) could result in the maintenance of clouds in the mesosphere. The possible cooling effects of injecting H_2O in the mesosphere, and the conditions necessary for the maintenance of clouds at the mesopause level, are addressed in this section.

The 15μ band of CO_2 plays the dominant role in the radiation balance of the mesosphere; the contributions of H_2O (6.3μ, 80μ) and Ozone (9.6μ) are relatively unimportant assuming normal concentrations of these constituents (Kuhn and London, 1969). H_2O radiates in the infrared without absorbing much sunlight, so one would expect that injection of H_2O between 70 and 120 km would tend to have a net cooling effect. Due to the low collision frequencies, thermal emission processes are inefficient in controlling the heat balance in the lower thermosphere, so the maximum effect would be felt at the mesopause level (~ 85 km) or below. Since the mesospheric mass mixing ratio of CO_2 is about 300 ppm, one would expect that H_2O would have to reach comparable concentrations in order to cause a measurable effect on the local heat balance. In fact Kuhn and London (1969) show that increasing the mesospheric water content from 1 ppm to 100 ppm increases the cooling rate near the mesopause from less than .1 deg day^{-1} to ~1-2 deg day^{-1}, which can be compared to a CO_2 heating rate of ~4 deg day^{-1} at 85 km. Thus, any significant cooling effects at mesopause levels would occur only when $\chi \gtrsim 100$ ppmv, and hence only over areas on the order of 20,000 km^2 (see Section III). It does not appear that even a small perturbation of the global circulation could result from this effect. However, infrared emissions over small areas could be of interest to optical surveillance systems.

Studies of mesospheric cloud formation indicate that three conditions must be met in order to form and maintain clouds near the mesopause:

1) As discussed by Witt (1969), homogeneous nucleation is slow, and condensation nuclei are required for sufficiently fast growth of ice particles.

2) The ambient temperature must be less than the frost point temperature at which saturation occurs.

3) Sufficiently large vertical velocities might be required to ensure growth of ice to required dimensions ($\sim .1\mu$) for a given water vapor content.

Witt (1969) points out that an effective source of condensation nuclei might be large water cluster ions, since the coulomb forces act to lower the free-energy barrier against the formation of small droplets, allowing particle growth to proceed relatively easily. In a previous section it has been shown that injection of water vapor into the mesosphere in concentrations greater than 100 ppmv could convert all of the O_2^+ and NO^+ ions at 85 km to ions of the type $H^+(H_2O)_n$ (n=1 to 7), and thus provide an abundance of condensation nuclei.

As indicated by Kellog (1964), the terminal fall velocity at 85 km of particles with density 1g cm^{-3} and radius $.1\mu$ is about 1 $msec^{-1}$. Updrafts of 1 $msec^{-1}$ do not exist at midlatitudes, but only at high latitudes during summer (see Schoeberl and Stroble, 1978). This is consistent with the observed seasonal and latitudinal occurrence of noctilucent clouds. However, it is shown by Reid (1975) that growth of needle-shaped or disc-shaped particles is much faster than growth of spherical particles, which could relax the requirement of vertical velocities of order 1 $msec^{-1}$.

The ambient temperature must be less than the frost-point temperature at which saturation occurs in order for ice particles to grow and persist. The frost-point temperature can be computed from (Reid, 1975):

$$T_f = T_o/\left[1 - \left(\frac{kT_o}{L_v m_w}\right) \ln\left(\frac{p_i}{p_{io}}\right)\right] \qquad (7)$$

where L_v = latent heat of sublimation
m_w = mass of water molecule
k = Boltzmann's constant
p_i = saturation vapor pressure
p_{io} = known saturation vapor pressure at T_o = 273K.

Using Reid's (1975) value of $T_f \simeq 142K$ at 1.6 ppm (1 ppmv) at 85 km, the above equation can be used to extrapolate to other water vapor mixing ratios. In Fig. 3, the latitude variation of mesopause temperature during summer and winter (Newell, 1968) are plotted with frost point temperatures at 85 km indicated for 1, 10, 10^2, 10^3, and 10^4 ppmv. During winter, there does not seem to be any possibility of maintaining clouds,

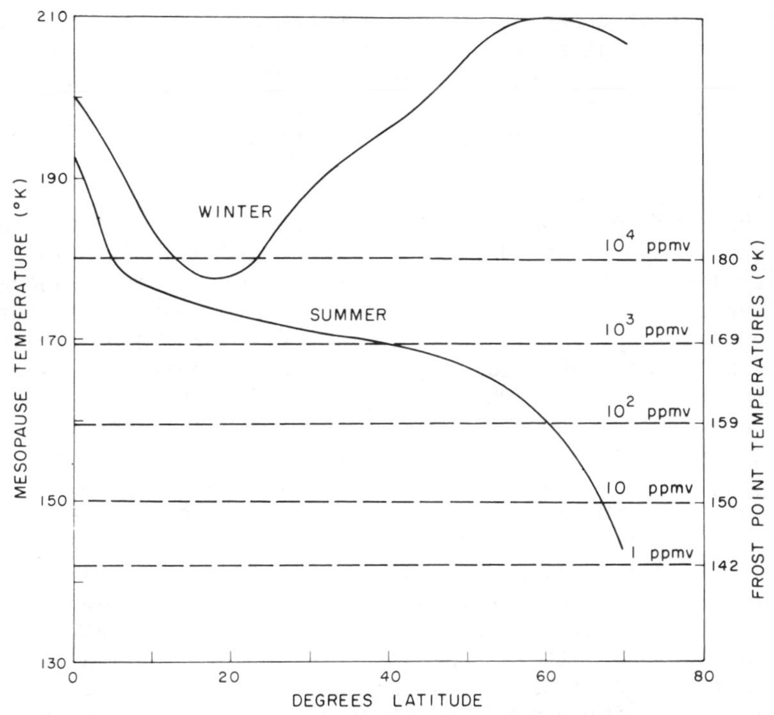

Fig. 3 Mesopause temperature vs latitude (from Newell, 1968) and frostpoint temperatures for various H_2O volume mixing ratios at 85 km.

even for mixing ratios as high as 10^4 ppmv. During summer, frost point temperatures exceed ambient temperatures above 67 deg N for mixing ratios on the order of 10 ppmv. At midlatitudes (30-45 deg), where SPS launch activities would likely occur, mixing ratios of order 10^3 ppmv would be required for the maintenance of clouds at the mesopause. Such conditions would only exist with very small areas (~ 200 km^2) in the vicinity of injection, and thus mesospheric clouds can be ruled out as an environmental hazard of SPS launch activities, unless the scenarios examined here represent gross underestimates with regard to frequency and magnitude of injected H_2O.

V. Discussion

The present calculations are intended more or less as a sophisticated scale analysis to sort out the competing effects of photodissociation, transport, and frequency of injection with regard to water injected by space launch vehicles. In particular, the present work is an attempt to arrive at some crude estimates of upper atmosphere modifications which might be expected from future SPS activities. It is shown, for instance, that transport by zonal winds and photodissociation play an important role in preventing significant build-ups of H_2O ($\chi \gtrsim 100$ ppmv) over wide areas ($\gtrsim 20{,}000$ km^2). Further, it is estimated that water vapor mixing ratios χ on the order of 100 ppmv or more are required to produce any measurable perturbations on the normal atmosphere.

Possible environmental effects quantitatively evaluated for $\chi > 100$ ppmv include: a) a 50% reduction in D-region ionization due to screening of L_α radiation by water (and a smaller contribution due to screening by thermospheric hydrogen produced by photolysis of the injected H_2O below 120 km, b) an additional 50% reduction of D-region ionization as a result of converting ambient NO^+ and O_2^+ ions to heavy water cluster ions which possess more rapid recombination rates, and c) at least a doubling of OH concentrations below 100 km. In addition, it is found that any radiative cooling produced by the injected H_2O would have a negligible effect on the general circulation of the mesosphere and lower thermosphere. At mid-latitudes mixing ratios on the order of 10^3 ppmv would be required to reach the frost-point temperatures necessary for the maintenance of clouds at the mesopause. Qualitatively, it is expected that atomic hydrogen released by photolysis of H_2O can increase the loss rate of ozone between 75 and 95 km, can significantly increase OH concentrations and accompanying airglow emissions, and also can act to increase nighttime E-region ionization by geocoronally-scattering L_α and L_β radiations after diffusing into the upper thermosphere. These affects of hydrogen released by H_2O photolysis may indeed comprise the most important upper atmosphere environmental impacts of discharging water in the 70 to 120 km height regime by SPS-type activities and should be investigated further.

A shortcoming of the present approach is the omission of an explicit calculation of vertical diffusion. In particular, the F-region ionization depletions that might result from H_2O injected below 120 km (which subsequently diffuses into the upper thermosphere) is unknown. Extension of the analytic formalism to handle vertical diffusion, if warranted by future SPS environmental assessments, can be achieved as outlined here.

The equation governing $n'(x,y,z,t)$, analagous to Eq. (1) for the height-integrated number density, $n(x,y,t)$, is

$$\frac{\partial n'}{\partial t} + \bar{u}\frac{\partial n'}{\partial x} + \bar{v}\frac{\partial n'}{\partial y} - K_h\frac{\partial^2 n'}{\partial x^2} - K_h\frac{\partial^2 n'}{\partial y^2} - \frac{\partial}{\partial z}K_v\left(\frac{\partial n'}{\partial z} + \frac{n'}{H}\right) = \delta(x)\,\delta(y)\,f(z)\,f(t) - Jn' \tag{8}$$

Proceeding in the same manner as in the solution of Eq. (1), we can apply a Laplace transform in t and Fourier transform in x and y, obtaining

$$\frac{\partial}{\partial z} K_v \left(\frac{\partial \eta}{\partial z} + \frac{\eta}{H}\right) - \left(i\alpha\bar{u} + i\beta\bar{v} + K_h\,\alpha^2 + K_h\,\beta^2 + s + J\right)\eta = f(z)\left[\frac{1}{e^{-ST}-1} - 1\right] \tag{9}$$

An adequate physical model would be to assume a 3-slab atmosphere ($z < 70$ km; 70 km $\leq z \leq$ 120 km; $z > 120$ km) where "effective" values are chosen for K_h, K_v, H, $\bar{u}$, $\bar{v}$, and J. These values remain constant within each slab but vary between slabs. Also, $f(z)$ would be constant between 70 and 120 km and would be zero for slabs 1 and 3. Such a model would simulate the very rapid diffusion and winds in the thermosphere, the very slow diffusion from the mesosphere into the stratosphere, and the long photochemical lifetime of H_2O below 70 km. A homogeneous solution corresponding to Eq. (9) would be of the form

$$\eta_h = A\,e^{r_1 z} + b\,e^{r_2 z} \tag{10}$$

and a particular solution is

$$\eta_p = \frac{f(z)\left[1 - \frac{1}{e^{-ST}-1}\right]}{i\alpha\bar{u} + i\beta\bar{v} + K_h\alpha^2 + K_h\beta^2 + S + J} \tag{11}$$

In a 3-slab model there would be 6 coefficients to be determined by applying boundary conditions at $z=o$, $z=\infty$, and two each at $z=70$ km and $z=120$ km. Then inverse Fourier and Laplace transforms would have to be applied to return to (x,y,z,t) space. Again, the numbers obtained from such a model would be crude, but probably adequate for the type of environmental assessments needed for SPS and Space Shuttle programs.

Acknowledgments

This work was supported in part by NASA contract NSG 7254 to Boston University.

References

Banks, P. M., and Kockarts, G., Aeronomy, Part A, Academic Press, New York, 1973, pp. 430.

Bauer, E., "Dispersion of Tracers in the Atmosphere and Ocean: Survey and Comparison of Experimental Data," Journal of Geophysical Research, Vol. 79, February, 1974, pp. 789-795.

Benhke, R. A. and Hagfors, T., "Evidence for the Existence of Nighttime F-Region Polarization Fields at Arecibo," Radio Science, Vol. 9, January, 1974, pp. 211-216.

Bernhardt, P. A., " The Response of the Ionosphere to the Injection of Chemically Reactive Vapors," Stanford Electronics Laboratories Tech. Rept. No. 17 (SEL-76-009), May, 1976.

Bowman, M. R., Thomas, L., and Geisler, J. E., "The Effect of Diffusion Processes on the Hydrogen and Oxygen Constituents in the Mesosphere and Lower Thermosphere," Journal of Atmospheric and Terrestrial Physics, Vol. 32, October, 1970, pp. 1661-1674.

Chakrabarty, D. D., Chakrabarty, P., and Witt, G., "An Attempt to Identify the Obscured Paths of Water Cluster Ions Build-up in the D-Region," Journal of Atmospheric and Terrestrial Physics, Vol. 40, April, 1978, pp. 437-442.

Debus, K. H., Johnson, W. G., Hembree, R. V., and Lundquist, C. A., "A Preliminary Review of the Upper Atmosphere Observations Made During the Saturn High Water Experiment," Proceedings of the XIIIth International Astronautical Congress, Springer Verlag, New York, 1964, pp. 182-196.

Edwards, H. D., Young, L. C., and Justus, C. G., "Analysis of Photographic Coverage of the Saturn SA-2 Water Experiment on April 25, 1962," Georgia Institute of Technology, Atlanta, Georgia, Tech. Rept. No. 1, Proj. A-637, 1962.

Kellogg, W. W., "Pollution of the Upper Atmosphere by Rockets-Space Science Review, Vol. 3, August, 1964, pp. 275-316.

Kuhn, W. R. and London, J., "Infrared Radiative Cooling in the Middle Atmosphere (30-110 km)," Journal of the Atmospheric Sciences, Vol. 26, March, 1969, pp. 189-204.

Mendillo, M., Hawkins, G. S., and Klobuchar, J. A., "A Sudden Vanishing of the Ionospheric F-Region Due to the Launch of Skylab," Journal of Geophysical Research, Vol. 80, June, 1975, pp. 2217-2228.

Murgatroyd, R. J. and Goody, R. M., "Sources and Sinks of Radiative Energy from 30 to 90 km," Quarterly Journal of the Royal Meteorological Society, Vol. 84, January, 1958, pp. 225-334.

Narcisi, R. S., Bailey, A. D., Della Lucca, L., Sherman, C., and Thomas, D. M., "Mass Spectrometric Measurements of Negative Ions in the D and Lower E Regions," Journal of Atmospheric and Terrestrial Physics, Vol. 33, August, 1971, pp. 1147-1159.

Newell, R. E., "The General Circulation of the Atmosphere Above 60 km," Meteorological Monograph, Vol. 9, 1968, pp. 98- 113.

Oliver, R. C., Bauer, E., Hidalgo, H., Gardner, K. A., and Wasylkiwskyj, W., "Aircraft Emissions: Potential Effects on Ozone and Climate," U.S. Dept. of Transportation, Wash., D.C., FAA-EQ-77-3, 1977.

Purcell, J. D. and Tousey, R., "The Profile of Solar Hydrogen Lyman-α," Journal of Geophysical Research, Vol. 65, January, 1960, pp. 370-372.

Reid, G. C., "Ice Clouds at the Summer Polar Mesopause," Journal of the Atmospheric Sciences, Vol. 32, March, 1975, pp. 523-535.

——————, "The Production of Water-Cluster Positive Ions in the Quiet Daytime D Region," Planetary and Space Science, Vol. 25, February, 1977, pp. 275-290.

Reidy, W., "Summary of Ultraviolet Missile Exhaust Gas Interactions," Geophysics Corporation of America, Bedford, Mass., Scientific Report No. 5, 1962.

Rishbeth, H., "The F-Layer Dynamo," Planetary and Space Science, Vol. 19, February, 1971a, pp. 263-267.

——————, "Polarization Fields Produced by Winds in the Equatorial F-Region," Planetary and Space Science, Vol. 19, March, 1971b, pp. 357-369.

Roble, R. G., Dickinson, R. E., and Ridley, E. C., "Seasonal and Solar Cycle Variations of the Zonal Mean Circulation in the Thermosphere," Journal of Geophysical Research, Vol. 82, December, 1977, pp. 4593-5504.

Schoeberl, M. R., and Strobel, D. F., "The Zonally-Averaged Circulation of the Middle Atmosphere," Journal of the Atmospheric Sciences, Vol. 35, April, 1978, pp. 577-591.

Sechrist, C. F., "The Ionospheric D-Region," The Upper Atmosphere and Magnetosphere, Upper Atmosphere Panel, Francis S. Johnson, Chairman, National Academy of Sciences, Wash., D.C., 1977.

Swider, W., "E-Region Model Parameters," Journal of Atmospheric and Terrestrial Physics, Vol. 34, October, 1972, pp. 1615-1626.

Watanabe, K., "Ultraviolet Absorption Processes in the Upper Atmosphere," Advances in Geophysics, Vol. 5, 1958, pp. 153-221.

Witt, G., "The Nature of Noctilucent Clouds," Space Research IX, Vol. 1969, Pergamon Press, Oxford, pp. 157-169.

Zimmerman, S. P., and Murphy, E. A., "Stratospheric and Mesospheric Turbulence," Dynamical and Chemical Coupling, edited by B. Grandal and J. A. Holtet, D. Reidel Co., Dordrecht, Holland, 1977, pp. 35-47.

MODIFICATION OF THE IONOSPHERE BY LARGE SPACE VEHICLES

Michael Mendillo*
Boston University, Boston, Mass.

Abstract

A brief history of rocket-induced perturbations upon the upper atmosphere is presented. The theory of "ionospheric hole" formation is described, stressing the role of a rapidly diffusing cloud of highly reactive rocket exhaust molecules interacting with the ionospheric plasma. Computer simulation results of this F-region modification problem show that carefully planned modification experiments can lead to significant advances in our understanding of the near-Earth plasma environment. These modification studies are of particular value in attempts to understand large-scale plasma dynamics, the thermal energy balance of a plasma, and the various modes by which plasma instabilities may be generated on a geophysical scale. The results also demonstrate that the F-region ionosphere will experience significant modification effects with virtually every in-orbit engine burn of the Space Shuttle and the proposed Heavy Lift Vehicles needed to construct Solar Power Satellites. Finally, a method of determining how to maximize (or minimize) ionospheric hole formation is detailed.

Introduction

A. Background

A little over two decades ago, when plans for the International Geophysical Year (IGY) of 1957-58 were underway, it was proposed to launch a small Earth satellite that would carry instruments capable of investigating conditions "just outside" the Earth's atmosphere (Abell, 1964). While the notion of a "top" to the terrestrial environment has now been replaced by the concept of "earthspace"(generally accepted to mean that the Earth's domain as a planet is defined by its "magnetosphere"), the spirit to explore that environment has not diminished. An important epoch in the history of human technology occurred on Oct.4, 1957, when Sputnik I marked the first

*Associate Dean, Graduate School.

successful launching of an artificial Earth satellite. The field of space science came of age with that mission; the atmosphere itself became a laboratory-in-space and in situ diagnostics probed its structure. As with all experiments, concerns were raised about probes affecting the medium to be measured and/or unintentionally modifying the surrounding regions. By January 1959, 11 space probes had been placed in various terrestrial, lunar, and solar orbits (four by the USSR and seven by the U.S). The 12th such mission, the U.S. Navy's Vanguard II, was launched in Feb. 1959, and led to the first openly documented upper atmosphere (ionospheric) disturbance associated with space flight activities (Booker, 1961). By the end of 1963, US agencies alone had launched their 100th rocket into the upper atmosphere. In the following year, W.W. Kellogg published a milestone review paper entitled Pollution of the Upper Atmosphere by Rockets. Its first sentence ("There have been so many deplorable examples of man's pollution of his environment that a conscious effort is being made in many quarters to forestall further cases.") is today, 15 years after publication, more in concert with scientific and public concerns than ever before.

The aim of Kellogg's paper was to alert the space science community to the many possible ways that rocket "pollutants" could have environmental impacts. The final impression left by the study was that the terrestrial atmosphere was sufficiently dense to absorb any conceivable shock that an aerospace technology might reasonably be expected to build, even to the point of the giant Saturn V rockets envisioned for the Apollo Program.

In the 15 years that have passed since Kellogg's assessment, very little evidence has been found to suggest that rocket effluents were, in fact, perilous to our environment. A reappraisal of all published accounts of rocket-induced modifications of the atmosphere since Sputnik-1 (Mendillo, 1979) reveals that hardly a dozen accounts exist describing specific aeronomic perturbations associated with the many hundreds of rocket launches that have occurred since 1957. There are good reasons for this:

1) The vast majority of rockets launched in the last two decades were relatively small ones, and thus the exhaust emissions amounted to no more than very minor additions of essentially trace species.

2) Of the large rockets (Saturns, Atlas/Centaurs, SL-4/SOYUZ) the overwhelming majority of launches carried payloads into low-Earth-orbit (h < 200 km) where the typical exhaust products (H_2O, H_2, CO_2, N_2, O_2) were again relatively inconspicuous additions to ambient conditions.

3) Large rocket launches made by U.S. agencies generally occur at the Kennedy Space Flight Center (KSFC), thereby insuring that rocket ascent trajectories occur over water. This condition makes atmospheric monitoring via ground-based facilities extremely difficult.

The main conclusion to be drawn from these points is that virtually all past rocket launches offered little reason, whether scientific or technological, to search systematically for the atmospheric perturbations caused by rocket effluents. The sole exception rests in the large-scale ionospheric perturbation observed during the launch of NASA's Skylab Workshop on May 14, 1973 (described in the following section). The purpose of this paper is to point out that new demands on current and future space transportation systems require engine burns in regions of the upper atmosphere where rocket effluents cause large-scale perturbation effects. In addition, recent advances in ionospheric modification studies strongly suggest that a wide variety of new scientific objectives may be examined by transforming what are essentially "events of opportunity" into important geophysical experiments.

Atmospheric Regions Susceptible to Rocket Exhaust Effects

Figure 1 presents a summary of terminology in which the Earth's atmosphere is divided into regions based upon properties of the neutral atmosphere as well as its ionized component (the ionosphere). While a rocket launch obviously can affect any region it traverses, the areas of prime concern to this study are those regions where the sun's radiation causes partial ionization to occur, that is, in the so-called D, E, F1,and F2 regions of the ionosphere. These regions contain a small fraction of the total mass of the atmosphere, and yet they are important because their electrically-charged components govern many aspects of radio communications. The positive ions found in the lower three regions are molecular ions (e.g., NO^+, O_2^+), while in the F2 region the dominant ion is atomic (O^+). Overall charge neutrality implies that in any given volume there are equal numbers of ions and electrons(e^-). As will be shown in later sections, the dominant cause of atmospheric perturbations due to rocket exhaust rests in the variety of <u>chemical reactions</u> that can occur between the exhaust material (usually molecular species, e.g., H_2O, H_2, CO_2) and the neutral and ionized components of the atmosphere.

In the F2-region of the ionosphere (h > 200 km), the dominant neutral species is atomic oxygen (O) and the major plasma components are O^+ and e^-. In this tenuous region the neutral concentrations are ~ 10^9 cm^{-3} or less, and the ionized concentrations are ~ 10^6 cm^{-3} or less. The introduction of large quantities of rocket exhaust molecules thus can represent a significant departure from ambient conditions, and the resultant atomic-molecular chemistry can lead to dramatic consequences. The bulk of this report deals with the specifics of this concern.

Below 200 km, the neutral atmosphere becomes increasingly dense and is composed almost entirely of molecular species. The plasma densities are orders of magnitude smaller than those found at F2-region heights, and the ions are predominantly molecular. In these lower portions of the ionosphere, molecular chemistry already is the dominant ionospheric process, and thus truly enormous amounts of rocket exhaust are required to overpower ambient processes, even for small spatial regions.

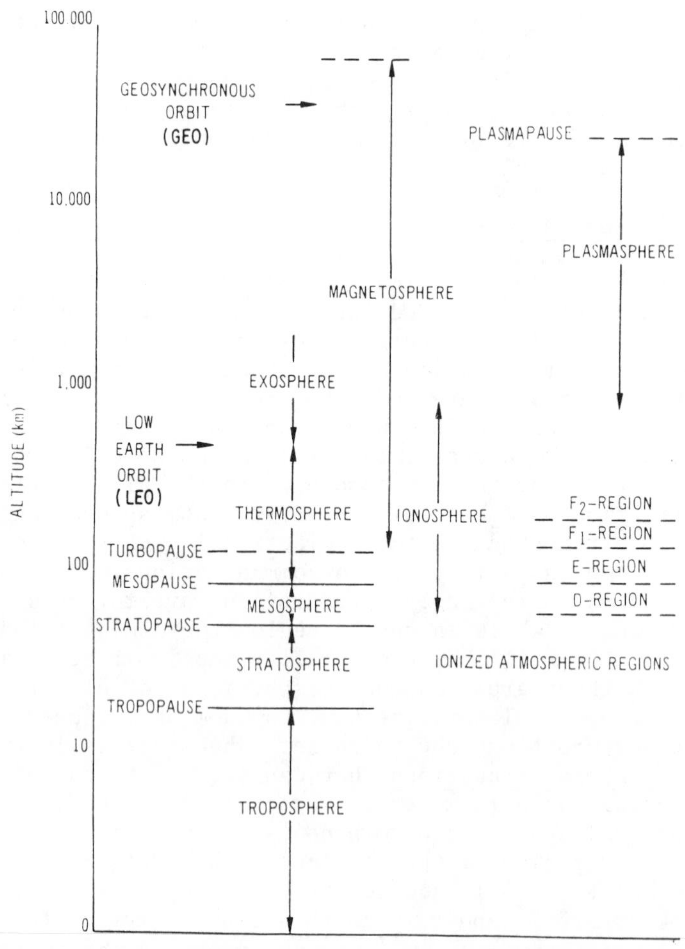

Fig. 1 Regions of the Earth's atmosphere.

Table 1 Observations of ionospheric disturbances caused by rocket launchings

Case	Rocket	Date	Altitude of engine shut-off, km	Effect	Observation Technique	Observer
1	Vanguard 2[a]	1959	F-region	F-region depletion	Vertical soundings	Booker(1961)
2	Scout[b]	1961	300	F-region depletion, E-region enhancement	Vertical soundings, Faraday rotation from rocket beacon	Jackson, et al.(1962)
3	Atlas(5 cases)[a]	1961	350	F-region depletion, E-region enhancement	Faraday rotation from rocket beacon	Stone, et al, (1964)
4	Saturn SA-9/ Pegasus A[a]	1965	500	F-region depletion, E-region enhancement	Vertical soundings	Felker & Roberts(1966)
5	Saturn 5 (Apollo 14)[a] (Apollo 15)[a]	1971 1971	≈190 ≈190	Traveling Ionospheric Disturbance	Vertical soundings	Arendt (1971) Arendt (1972)
6	Black Brant[c]	1971	35	F-region depletion	Vertical soundings	Reinisch(1973)
7	Saturn 5[a]	1973	442	Large-scale F-region depletion ('ionospheric hole')	Total electron measurements using geostationary satellite beacons	Mendillo, et al. (1975)
8	Saturn-5 (Apollo)[a] (Soyuz-19)[d]	1975	200	F-region traveling ionospheric disturbance, E-region enhancement	Vertical soundings	Bakai,et al. (1977)

[a]Kennedy Space Flight Center [b]Wallops Island [c]Eglin, Fla. [d]Baykonur Cosmodrone

As a result, very little data exists showing how, if at all, artificially-induced perturbations can be treated in these regions. The companion paper by Forbes (1980) considers, in detail, the D and E region perturbations possibilities capable of being induced by large rockets.

Past Studies of Ionospheric Disturbances Due to Rockets

While the discussion of rocket exhaust effects upon the Earth's upper atmosphere is not a new topic, past documentations of plume-associated disturbances have not been very great in number. Table 1 presents a chronological summary of published reports that suggest a causative link between observed ionospheric variations and rocket launches. Most of the rocket-induced aeronomic effects described in the early literature of the field were termed "localized" due to factors already mentioned: a) small amounts of injected material, b) injections into the very dense lower atmosphere, and c) poor diagnostics for detecting large-scale effects. In addition, the theoretical understanding of the processes involved was limited to discussions of rocket exhaust "snow-plow effects" upon the atmosphere, effects that by their very nature are limited to the region immediately surrounding a moving rocket.

A new class of ionospheric disturbance, the large-scale F-region hole, became apparent during the launch of Skylab in

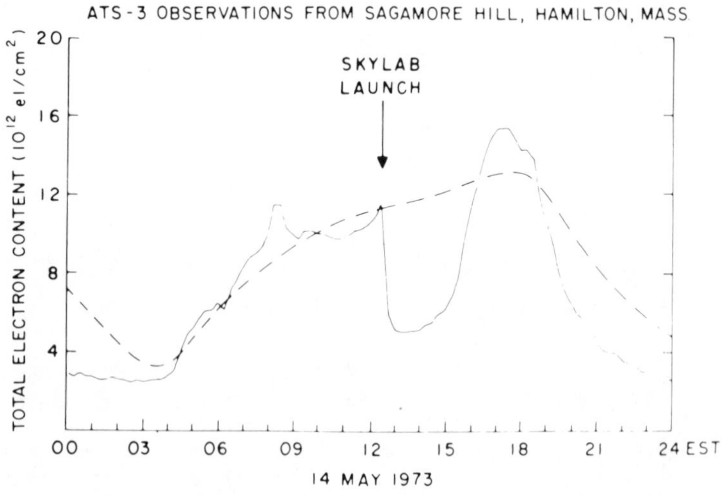

Fig. 2 Total Electron Content (TEC) data used to detect the "Skylab effect" on 14 May 1973. The dashed curve gives the anticipated diurnal TEC behavior based upon a monthly median prediction updated for geomagnetic storm effects (see Mendillo et al., 1975a,b).

1973. (See Fig.2 and Mendillo et al., 1975a,b.) The Saturn V rocket that launched NASA's Skylab Workshop was the last Saturn V used in the U.S. space program and the only one ever to have its main engines burning above 200 km. The resultant deposition of approximately 1000 kg/sec of exhaust into the 200-440 km altitude region initiated a rapid and large-scale depletion of the ionosphere to an extent never seen before. The artificially-created "ionospheric hole" amounted to nearly a 50% decrease in the total electron content (TEC) of the ionosphere over an area of approximately 1,000,000 km^2. Mendillo et al. (1975a,b) attributed the effect to the rapid expansion of an exhaust cloud of highly reactive molecules (water and hydrogen) that initiate a rapid recombination of the ionospheric plasma. In the following section the specifics of the ionospheric hole-making process are described.

Physical Processes Responsible for Ionospheric Holes

The main exhaust products of standard upper-stage rocket engines include molecular species of important ionospheric consequence. For example, a liquid oxygen-hydrogen engine (LH_2/LO_2), such as that used in the Saturn V vehicles, has an exhaust plume of water vapor H_2O and unused fuel H_2 in an abundance ratio of approximately 75%/25% by molecular count. An alternative propulsion scheme, such as that used by the Space Shuttle while in orbit (NASA, 1976), burns monomethylhydrazine (MMH) and nitrogen tetroxide N_2O_4 to produce, among other products, significant amounts of water H_2O, hydrogen H_2, and carbon dioxide CO_2. The introduction of H_2O, H_2, or CO_2 into the upper ionosphere ($h > 200$ km, where the dominant ion is O^+) causes a chemical transformation to molecular ions at rates 100 to 1000 times faster than occur with the naturally occurring molecules of nitrogen N_2 and oxygen O_2. These important reactions are:

$$O^+ + H_2O \xrightarrow{k_1} H_2O^+ + O, \quad k_1 = 2.4 \times 10^{-9}\ cm^3/sec \qquad (1)$$

$$O^+ + H_2 \xrightarrow{k_2} OH^+ + H, \quad k_2 = 2.0 \times 10^{-9}\ cm^3/sec \qquad (2)$$

$$O^+ + CO_2 \xrightarrow{k_3} O_2^+ + CO, \quad k_3 = 1.2 \times 10^{-9}\ cm^3/sec \qquad (3)$$

Once a molecular ion is formed, its dissociative recombination with an ambient electron occurs rapidly. (See Banks and Kockarts (1973) for a full treatment of normal ionospheric chemistry.) For the above cases, the reactions are:

$$H_2O^+ + e^- \xrightarrow{\alpha_1} OH + H \quad \alpha 1 = 3.0 \times 10^{-7} cm^3/sec \qquad (4)$$

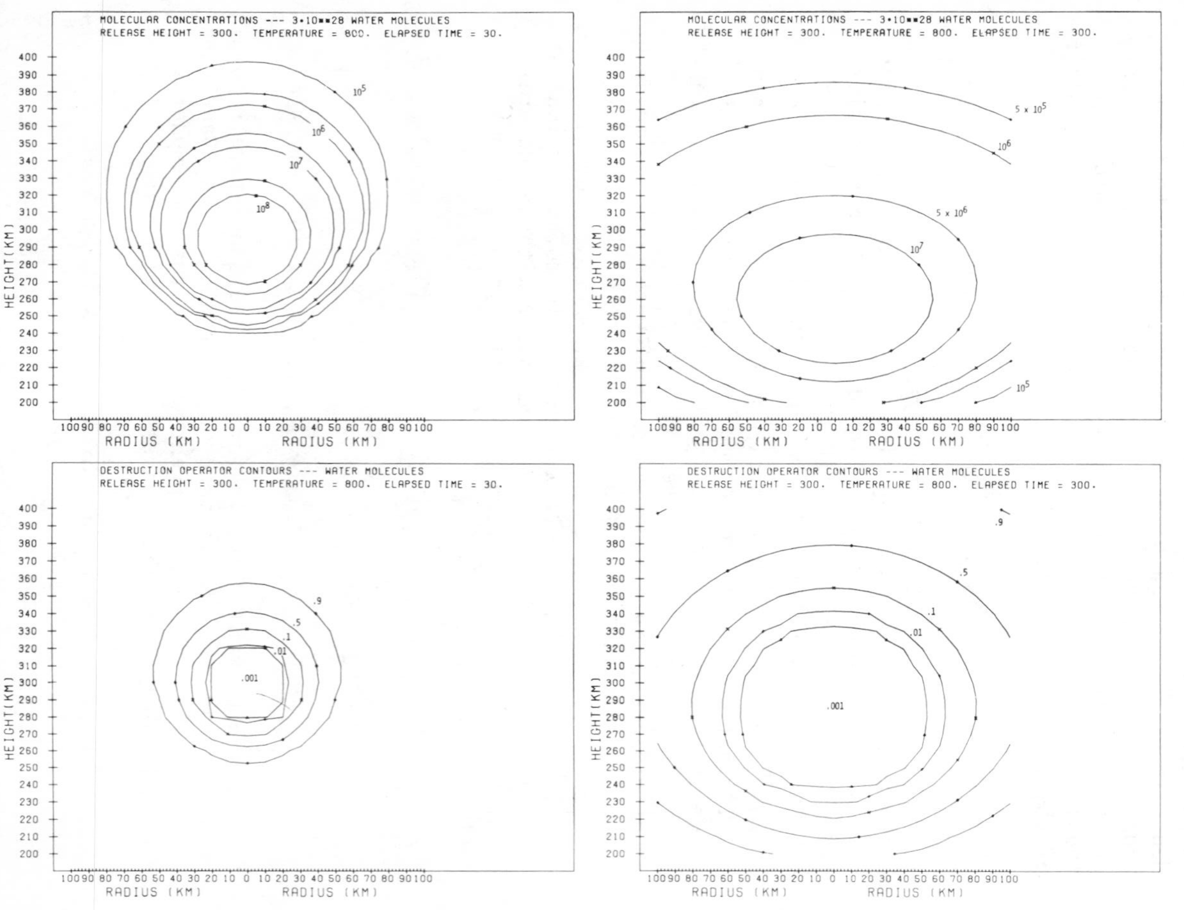

Fig. 3 Model simulation results for a point source release of 3 x 10^{28} water molecules into a 800K neutral atmosphere. The top two panels show molecular concentrations (in cm^{-3}) at 30 and 300 sec after the release. The bottom two panels give O^+ "destruction operators" defined as $D(t) = \{O^+ (t)\} / \{O^+ (t=0)\}$ at t = 30 and 300 sec. Note that the O^+ concentrations are not affected by H_2O concentrations less than $10^5 cm^{-3}$ (after Mendillo and Forbes, 1978).

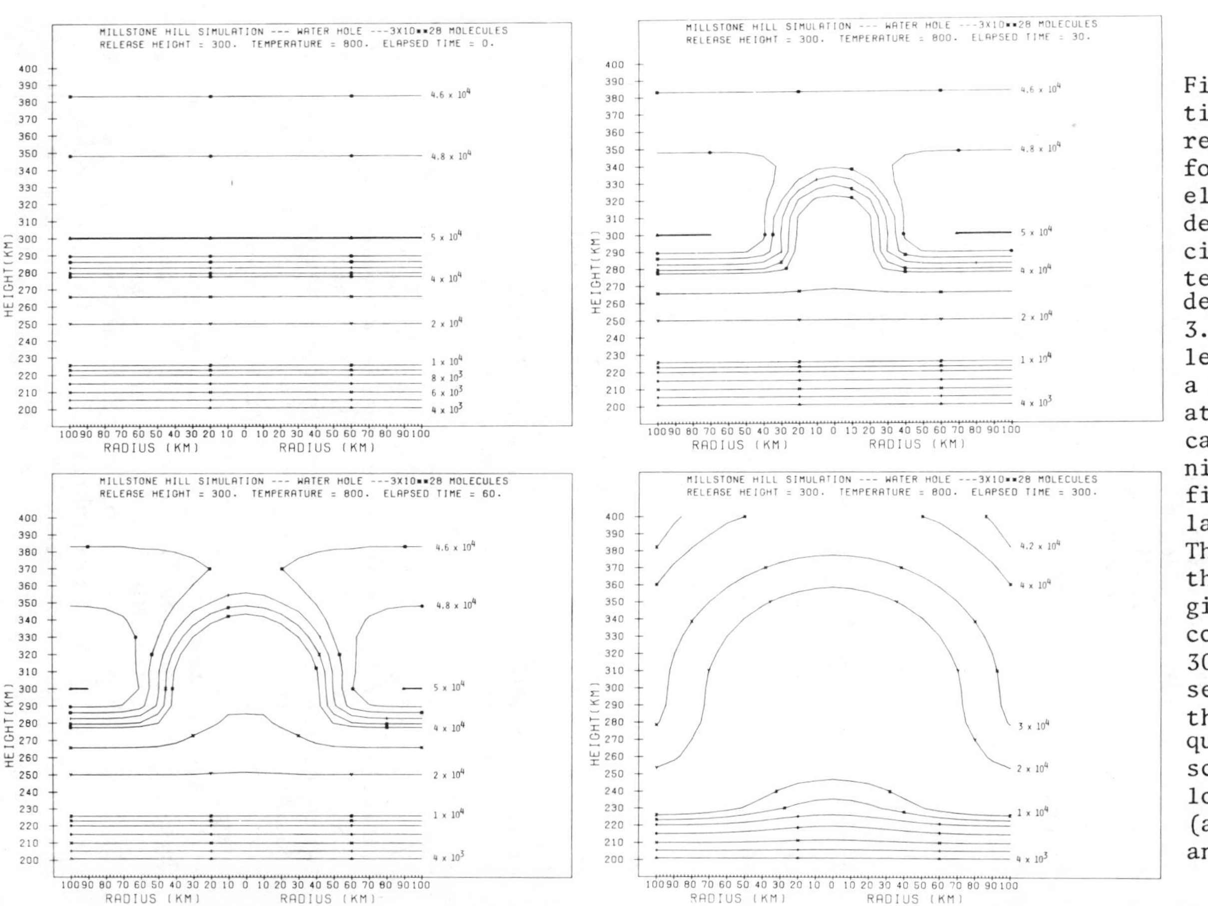

Fig. 4 A spatial/temporal representation for the F-region electron density depletions associated with a water release as depicted in Fig. 3. The upper left panel shows a Ne (h,r) array at t=0 for the case of a winter nighttime profile at a mid-latitude site. The remaining three panels give Ne(h,r) contours at t= 30,60 and 300 sec. Note that the hole forms quickly and descends into the lower F-region (after Mendillo and Forbes, 1978).

$$OH^+ + e^- \xrightarrow{\alpha_2} H + O, \alpha_2 = 1.0 \times 10^{-7} cm^3/sec \quad (5)$$

$$O_2^+ + e^- \xrightarrow{\alpha_3} O + O, \alpha_3 = 2.0 \times 10^{-7} cm^3/sec \quad (6)$$

The introduction of the highly reactive molecules H_2, H_2O, or CO_2 thus leads to the loss of ion-electron pairs. Mendillo and Forbes (1978) described a relatively simple formulation whereby the hole-making potential of a release scenario may be computed by forming a set of O^+ "destruction operators" for each molecular species involved. Figures 3 and 4 give an illustration of the technique for the case of 3×10^{28} water molecules (approximately 1 ton of water and similar to 1 sec of Saturn V exhaust). Figure 3 shows the molecular diffusion results and their resultant effect upon the O^+ distribution. Figure 4 shows the consequence such a release would have upon the electron densities of a typical mid-latitude winter nighttime ionospheric profile.

The creation of an ionospheric hole via the reactions described above will be accompanied by a significant amount of atmospheric airglow emissions. Anderson and Bernhardt (1978) have described the anticipated emission rates from the dissociative recombinations of OH^+ and H_2O^+ (Eqs. 4 and 5). These include radiations at 6300 Å and 5577 Å from excited states of atomic oxygen, and emissions at 3060 Å from the OH radical. Injections of CO_2 lead to dissociative recombinations of O_2 (Eqs. 3 and 6) which again yield emissions from excited states of atomic oxygen.

Once an ionospheric hole is created, its longevity depends on two main ionospheric replenishment processes: a) normal production of O^+ and e^- by solar ionizing radiations at extreme ultraviolet (EUV) and soft X-ray wavelengths, and b) plasma fluxes into the depleted region driven by normal diffusion, neutral winds, or electrodynamic (EXB) drifts. In summary, then, the ionospheric hole phenomenon may be considered to be the result of three coupled processes: 1) diffusion of an exhaust cloud of highly-reactive neutral molecules through a tenuous, multi-constituent atmosphere, 2) chemical reactions between the expanding cloud and the varies species (ionized and neutral) in the atmosphere, and 3) solar and/or dynamical replenishment processes. Figure 5 presents a summary of the overall nature of the ionospheric hole-making process together with several estimates of practical consequences resulting from the possible modifications induced (Rote, 1979).

All of the physical and chemical processes depicted in Fig.5 lend themselves to computer simulation techniques. Since the spatial/temporal availability of the injected molecules essentially determines all subsequent effects, a great deal of effort has been expended in describing the diffusion of molecules in the upper atmosphere (Bernhardt et al., 1978; Forbes and Mendillo, 1976; Schunk, 1978; Bernhardt, 1979a,b). The comprehensive treatment of Bernhardt (1979a,b) describes a variety of effects due to the coupling of rocket exhaust speed, vehicle orientation, and orbital location in the atmosphere. The studies by Bernhardt et al. (1975), Balko and Mendillo (1977), Mendillo and Forbes (1978), Anderson and Bernhardt (1979), and Zinn et al. (1978a,b, 1979) deal with the overall development of a hole. The models of Anderson and Bernhardt (1979) and Zinn et al. (1978) are particularly well-suited for studying the recovery and longevity aspects of a hole.

Scientific and Technologic Interest in Artificially-Created Ionospheric Holes

The Skylab event of May 1973 rekindled an interest in molecular release experiments that had languished since the mid-1960's. The effects caused by highly reactive molecules introduced into the F-region offer the opportunity to study a variety of atmospheric processes (e.g., molecular diffusion, neutral/plasma chemistry, airglow emissions, thermal energy balance, plasma transport, and electron density irregularity generation). In addition, the large-scale disruption of the ionosphere's vertical density structure, $N_e(h)$, has implications for radio propagation. The discussion of these issues by Mendillo et al. (1975a,b), Bernhardt et al. (1975), Papagiannis and Mendillo (1975), Bernhardt (1976), and Bernhardt and da Rosa (1977) led to the very successful set of Lagopedo modification experiments of Sept. 1977 in which rocket-borne payloads of H_2O and CO_2 were released into the F-region over Hawaii (Pongratz and Smith, 1978). A more extensive series of F-region modification experiments are to be performed via "dedicated burns" of the Space Shuttle engines as the orbiter passes near five ionospheric and radio astronomical observatories. The experiments are part of the Spacelab-2 mission currently scheduled for Jan. 1982 (Mendillo et al. (1978), Rosendhal (1978), Bernhardt et al.,(1978). A more immediate opportunity to observe large-scale ionospheric hole formation occurred during NASA's launch of satellite HEAO-C by an Atlas/Centaur rocket on 20 Sept. 1979 (Mendillo et al., 1979b).

Quite separate from the scientific objectives of the Lagopedo and Spacelab-2 plasma depletion experiments is a concern that the new generation of space transportation systems being developed and/or considered for the future require routine engine firings in the upper ionosphere and therefore the routine creation of large-scale ionospheric holes. The soon-to-be-operational Space Shuttle program will include orbital configurations usually in the 250-450 km altitude range, at precisely the heights where the ionospheric plasma densities (O^+ and e^-) reach their maximum values. The Space Shuttle's engine exhaust rates (NASA, 1976) are considerably smaller than those of a Saturn V rocket, and thus the very large spatial extent associated with the SKYLAB effect will not be found for the Space-Shuttle-induced holes.

The heavy-lift launch vehicles (HLLV's) and personnel orbital transfer vehicles (POTV's) required for the proposed Solar Power Satellite (SPS) system represent a very substantial increase over "conventional" Space Shuttle cargo and support-staff transportation modes. The regular transfer of material and personnel to low-Earth-orbit (LEO) and from LEO to geostationary Earth-orbit (GEO) suggests that a routine modification of the ionosphere will be a consequence of the fully implemented SPS concept. The ultimate spatial and temporal extent of the SPS-induced ionospheric holes will depend on the specifics of the launch vehicles to be designed and the orbital flight plans that evolve. Ample opportunity exists to influence these decisions by model/simulation studies, as suggested by some of the preliminary environmental assessment results presented by Zinn et al. (1978 a,b; 1979) and Mendillo et al. (1979) to be described in the following section.

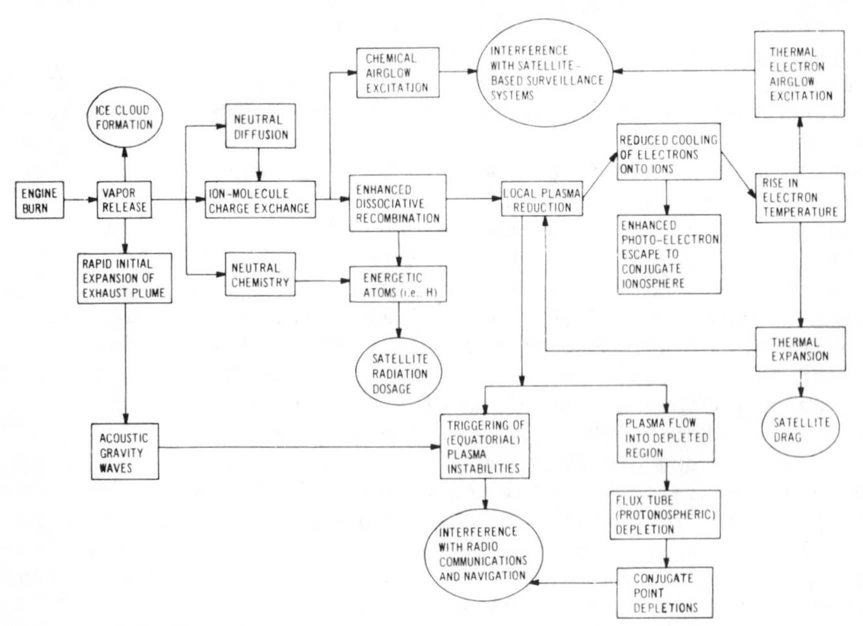

Fig. 5 Rocket effluent effects in upper ionosphere/F-region (above 200 km).

Table 2 Comparison of rocket exhaust rates for highly-reactive molecules capable of creating ionospheric holes (in units of 10^{25} molecules/sec)

Vehicle	H_2O	H_2	CO_2	Total
Saturn V Second Stage	3230	1320	---	4440
Atlas/ Centaur Stage	98	59	---	157
Space Shuttle OMS Engines	14	12	6	32

F-Region Hole Assessment Studies

Table 2 presents a summary of nominal molecular exhaust rates for the U.S. rocket vehicles (past and future) capable of creating large-scale F-region holes. The Saturn V type of effect, as seen during the Skylab launch, defines the maximum disturbance possible from "operational" vehicles. The Space Shuttle's orbital maneuvering subsystem (OMS) offers the smallest source of F-region perturbations for the near future. Zinn et al (1978a) alerted the aerospace community to the possible large-scale effects the proposed HLLV might induce as part of an SPS program. In partial response to those concerns, current plans for the nominal HLLV launch profile call for main engine shut-down below 125 km, with a subsequent orbit circularization burn near 450 km (NASA, 1978).

Mendillo et al. (1979) described a first-order way of assessing how each of these vehicles (Space Shuttle and HLLV) might affect the ionosphere in the vicinity of engine burns. A "source hole-making efficiency" was defined, given by

$$\varepsilon_s = \frac{\Delta N_{TOT}}{S_o} \times 100$$

where ΔN_{TOT} is the total number of electrons lost within the geographical area surrounding the source of injected molecules, S_o. The ε_s parameter is expressed in percent as a convenient way of describing the actual units of electrons lost/molecules released.

Figures 6, 7, and 8 give a sample electron density profile and ε_s results for OMS and HLLV engine burns at various heights in such an ionosphere. For the Space Shuttle case (a "small release"), the ε_s curve closely resembles the $N_e(h)$ curve in

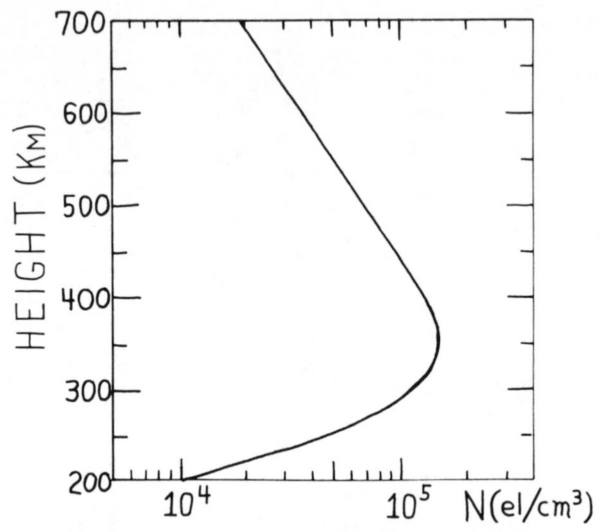

Fig. 6 A winter nighttime electron density profile, Ne(h), for mid-latitudes during a solar maximum year.

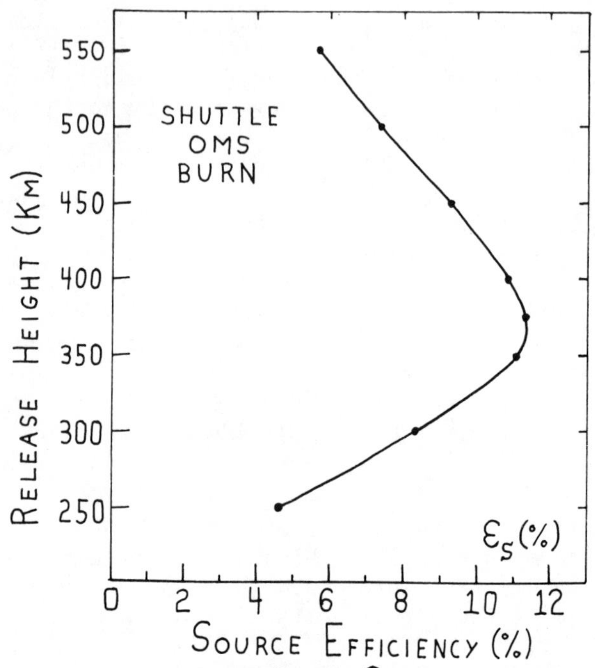

Fig. 7 Source efficiency results, ε_S = (total # e^- lost/total # molecule released) x 100, for the case of an 11-sec burn of the Space Shuttle OMS engines at various ionospheric heights- Total source = 3.2 x 10^{27} molecules.

Fig. 6. The most efficient release height, h (ε_{max}), is slightly above the height where the peak electron density occurs, h(N_{max}), an effect due to the gravitational settling of the molecules, as shown in Figs. 3 and 4. The fact that only 10% of the molecules released actually led to the loss of ion-electron pairs should not be confused with the fact that severe N_e depletions have occurred near the release point. Thus, while an "over-kill" situation exists near the release point, two factors limit the hole-making effectiveness of the molecules at more remote distances:

1) Vertical diffusion through an exponentially decreasing atmosphere results in the rapid escape of molecules before they can react with the O^+ in the topside ionosphere.

2) Horizontal diffusion through the Earth's tenuous atmosphere quickly results in molecular concentrations too low ($<10^5$ molecules/cm^3)to be of chemical importance.

Figure 8 presents the source efficiency factor, ε_s, results for equivalent point-source circularization burns by a single HLLV over the 250 - 550 km altitude range. Certain features distinguished these "large release" results from those in Fig. 7.

1) The efficiencies are more than a factor of three smaller than those found in the Space Shuttle simulation.

2) The efficiencies increase with altitude throughout the ionosphere, and thus no h(ε_{max}) is defined over the height range sampled.

It should be mentioned again that while efficiencies are low, the spatial extent of the resultant hole is obviously much larger than it is in the Space Shuttle case. The low ε_s values for the SPS case are caused by the same "over-kill" effects discussed with respect to the Shuttle. The curve in Fig. 8 does not go through a maximum simply because of the enormity of the SPS source function. Thus, even for an equivalent engine burn high in the topside ionosphere, diffusion and gravitational settling effects result in a significant number of molecules reacting with the ionospheric O^+ ions to form a hole.

Discussion and Conclusions

A brief history of rocket-induced perturbations upon the upper atmosphere has been presented. The theory of "ionospheric hole" formation was described stressing the role of a rapidly diffusing cloud of highly reactive rocket exhaust molecules interacting with the ionospheric plasma. This concept is in contrast to older theories in which the hole was thought to be formed by a rocket's exhaust physically displacing the ambient F-region components. Such "snow-plow" effects do, in fact, occur, but only close to the exhaust exit plane and only for time scales on the order of seconds. The degree to which the exhaust molecules may condense, and therefore be removed from the hole-making process, was not considered. Condensation is of most interest for the H_2O component of an exhaust plume since H_2 will not crystallize under upper atmospheric conditions. The end result of H_2O condensations is to scale down the source of H_2O molecules available for immediate chemical effects (Bernhardt, 1976), though subsequent evaporation may lead to longer-lived effects (Zinn et al., 1979).

Several groups have applied computer simulation techniques to the F-region modification problem with a general consistency of results obtained. All of the model results show that carefully planned modification experiments can lead to significant advances in our understanding of the near-Earth plasma environment. Modification studies can be of particular value in attempts to understand large-scale plasma dynamics, the

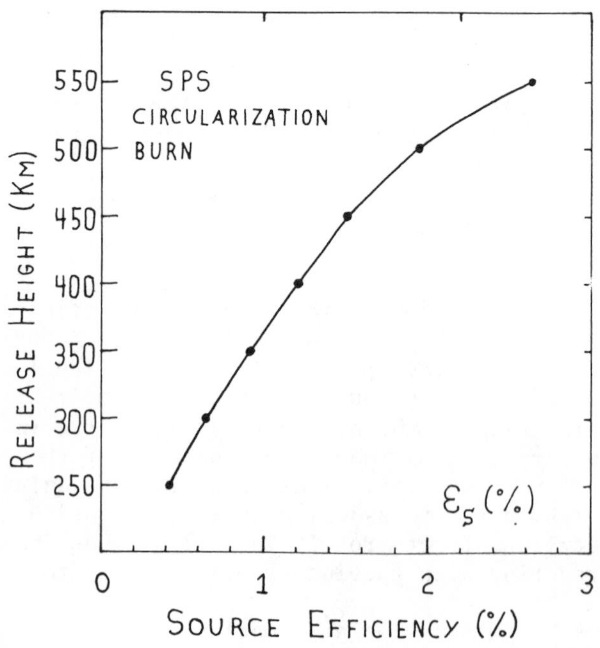

Fig. 8 Source efficiency results, ε_S(%), for the case of an equivalent point source circularization burn of the heavy-lift launch vehicles (HLLV) proposed for the Solar Power Satellite program. Total source = 10^{30} molecules.

thermal energy balance of a plasma, and the various modes by which plasma instabilities may be generated on a geophysical scale.

In terms of technological impact, it is important to realize that the F-region of the ionosphere will experience significant modification effects with virtually every in-orbit engine burn of the Space Shuttle and the proposed Heavy Lift Launch Vehicles needed to construct Solar Power Satellites. Recent simulation studies indicate that simple scaling arguments based on the number of highly reactive molecules contained in a rocket exhaust plume cannot be used to estimate the number of electrons that will be lost due to an engine burn. For a source efficiency factor defined as ε_s = number of electrons lost/number of molecules released, the rend clearly exists that ε_s gets smaller as the size of the release increases. (The spatial extent of the affected region is, of course, larger for larger injections.) These results may be generalized into two main conclusions for predicting rocket plume effects:

1) The maximum possible ionospheric perturbation would be associated with a horizontal engine burn situated slightly above the peak of the ionospheric $N_e(h)$ profile (e.g., in the 350-450 km range). Such a configuration points to the most efficient dispersal of a molecular exhaust plume for the purpose of creating ionospheric depletions.

2) The opposite case of minimizing the extent of an F-region ionospheric hole would be achieved by requiring as concentrated a horizontal engine burn as is possible and by locating that burn well below the peak of the $N_e(h)$ profile.

As the technology of the 21st century approaches, it is appropriate to review the concerns first raised by Kellogg (1964) in order that the great promise of laboratory-in-space plasma experiments and the possibility of large, useful structures in space do not cause inadvertent modification of our terrestrial domain.

References

[1]Abell, G., Exploration of the Universe, Holt, Rinehart and Winston, New York, 1964.

[2]Anderson, D.N. amd Bernhardt, P.A., "Modeling the Effects of an H_2 Gas Release on the Equatorial Ionosphere," Journal of Geophysical Research, Vol. 83, pp.4777-4790, Oct., 1978.

[3]Balko, B. amd Mendillo, M, "Finite Element Simulation Applied to the Transport of Neutral and Ionized Particles in the Earth's Upper Atmosphere," Astronomical Contributions of Boston University, Ser. III, No. 1, May 1977.

[4]Bank, P.M. and Kockarts, G., Aeronomy, Academic Press, New York, 1973.

[5]Bernhardt, P.A. Park, C.G. and Banks, P.M., "Depletion of the F_2 Region Ionosphere and Protonosphere by the Release of Molecular Hydrogen," Geophysical Research Letters, Vol. 2, Aug., 1975, pp. 341-344.

[6]Bernhardt, P.A., "The Response of the Ionosphere to the Injection of Chemically Reactive Vapors, Stanford University, SU-SEL-76-009, Tech. Rept. 17, 1976.

[7]Bernhardt, P.A. and deRosa, A.V., A Refracting Radiotelescope", Radio Science, Vol. 12, March, April 1977, pp. 327-336.

[8]Bernhardt, P.A., daRosa, A.V. amd Mendillo, M, "High Altitude Vapor Releases from the Space Shuttle," Transactions of the American Geophysical Union, Vol. 59, April 1978, 1162.

[9]Bernhardt, P.A., Three-Dimensional, Time-Dependent Modeling of Neutral Diffusion in a Nonuniform, Chemically Reactive Atmosphere," Journal of Geophysical Research, Vol. 84, March 1979 pp. 793-802.

[10]Booker, H.G., "A Local Reduction of F-Region Ionization Due to Missile Transit," Journal of Geophysical Research, Vol. 66, April, 1961, pp.1073-1079.

[11]Forbes, J.M. and Mendillo, M., "Diffusion Aspects of Ionoshperic Modification by the Release of Highly Reactive Molecules into the F-Region," Journal of Atmospheric Terrestrial Physics, Vol. 38, 1299, 1976.

[12]Kellogg, W.W., "Pollution of the Upper Atmosphere by Rockets", Space Science Reviews, June, 1964, pp.275-316.

[13]Mendillo, M., Hawkins, G.S. and Klobuchar, J.A., "A Large-Scale Hole in the Ionosphere Caused by the Launch of Skylab," Science, Vol. 187, January, 1975a, pp.343-346.

[14]Mendillo, M. and Forbes, J.M. "Artificially Created Holes in the Ionosphere," Journal of Geophysical Research, Vol. 83, Jan. 1978, pp. 151-162.

[15]Mendillo, M., Herniter, B. and Rote, D., "Modification of the Aerospace Environment by Large Space Vehicles," J. Spacecraft and Rockets, (in press, May-June, 1980).

[16]Mendillo, M. "Ionospheric Modifications Caused by Rocket Exhaust and Chemical Releases --- A Review of Observational Evidence and Theoretical Models,"DOE/Argonne National Lab Tech. Report, Submitted for publication, December 1979.

[17]NASA, MSFC/JSC, "Solar Power Satellite Baseline Review," NASA Headquarters, Washington, D.C., July, 1978.

[18]NASA, JSC, Space Shuttle, U.S. Government Printing Office, Washington, D.C., Stock No. 033-000-00651-9, 1976.

[19]Papagiannis, M.D. and Mendillo, M., "Low Frequency Radio Astronomy through an Artificially Created Ionospheric Window," Nature, Vol. 255, May, 1975, pp.42-44.

[20]Pongratz, M.B. and Smith, G.M. "The Lagopedo Experiments - An Overview," Transactions of the American Geophysical Union, Vol. 59, April 1978, 334.

[21]Rosendhal, J.D., "The Spacelab-2 Mission," Sky and Telescope, Vol. 55, June, 1978.

[22]Rote, D., "SPS Atmospheric Effects Review," U.S. Department of Energy, Washington, D.C.1979.

[23]Schunk, R.W., "On the Dispersal of Artificially-Injected Gases in the Night-time Atmosphere," Planetary Space Science, Vol. 26, June, 1978, pp.605-610.

[24]Zinn, J., Sutherland, C.D., Pongratz, M.B. and Smith, G.M., "Predicted Ionospheric Effects from Launches of Heavy Lift Rocket Vehicles for Construction of Solar Power Satellites," Los Alamos Scientific Lab., Doc. LA-UR-78-1590, June, 1978a, Dec., 1978b.

[25]Zinn, J., Sutherland, C.D., "Effects of Rocket Exhaust Products in the Thermosphere and Ionosphere," Los Alamos Scientific Lab., Doc. LA-UR-79-2750, October, 1979.

ARGON-ION CONTAMINATION OF THE PLASMASPHERE

Y. T. Chiu,[*] J. M. Cornwall,[†] J. G. Luhmann,[#] and Michael Schulz[§]
The Aerospace Corporation, El Segundo, Calif.

Abstract

Large-scale operation of argon-ion engines in space may give rise to global-scale modification of the magnetosphere. In this paper, we consider ion injectant effects of solar-powered orbit transfer operations of large payloads ($\sim 10^7$ kg) similar to that of the projected Satellite Power System. It is likely that the ion beam would interact and deposit its energy and mass in the magnetosphere. Magnetospheric heating may change the compositional distribution of thermal ions, thus causing enhancement of relativistic Van Allen radiation belt electrons. Effects upon the ring-current (auroral processes) also are discussed.

I. Introduction

As is pointed out in the review of environmental effects of space systems, transportation of large payloads, such as that of the proposed satellite power system (SPS), from low Earth orbit to higher space orbit (e.g., geosynchronous orbit) by ion engines implies substantial modifications of the magnetospheric environment. This is because the mass of exhaust plasma approaches or exceeds that of the local natural magnetospheric plasma. In this paper, we shall examine the basic physical scenarios of magnetospheric modification by this artificially injected plasma. The emphasis will be

---

*Research Scientist, Space Sciences Laboratory.

†Professor of Physics, Department of Physics, University of California, Los Angeles, Calif.

#Member of Technical Staff, Space Sciences Laboratory.

¶Research Scientist, Space Sciences Laboratory.

placed on identifying the physical mechanisms of the modifications rather than on a specific enumeration of space-based and Earth-based systems which may be affected by these modifications. A discussion of the effects on systems operations can be found in Chiu et al. (1979a).

Despite the title of this paper, the effects that we shall deal with are sufficiently general in that they are neither limited to argon ions nor to plasma exhaust products of ion engines alone. The latter is true because of the critical velocity phenomenon (Alfvén and Arrhenius, 1976; Angerth et al, 1962; Danielsson and Brenning, 1975; Möbius et al., 1979) which suggests that a neutral particle cloud moving with velocity relative to the ambient plasma greater than $v_c \equiv (|e| V_i/m)^{1/2}$ is apt to be ionized, where $|e| V_i$ is the ionization potential energy of the neutral particles of mass m. By noting the masses (~ 10-50 atomic numbers) and the ionization potentials (~1-10 eV) of the usual chemical rocket propellant elements, the reader can easily verify that their critical velocities v_c are ~ 10 km/sec. These velocities are frequently exceeded for cases of rocket propellants in the magnetosphere, where during tail injection events the background magnetospheric plasma may have speeds well above 100 km/sec. Therefore, if the critical velocity phenomenon observed in the laboratory is applicable in space, the operation of chemical rockets in the magnetosphere will generate ion clouds which behave in much the same manner as the argon-ion clouds to be dealt with in this paper. In this regard, the magnetospheric modifications considered here are applicable to large-scale injections of high-speed (> v_c) rocket exhaust in the magnetosphere, irrespective of whether the booster engines are ion engines or not. Although the Alfvén critical velocity phenomenon needs to be verified, some preliminary evidence for it may have been observed in the interactions between a lunar dust cloud and the solar wind (Lindeman et al.,1974).

The subject of magnetospheric modification by a massive plasma cloud is obviously very complicated and, to a major extent, unsolved. An attempt to trace the dynamical history of the artificial plasma is too difficult a task for the present paper although it eventually will have to be faced. Our efforts here will be limited to the initial phases of the problem; we ignore the question of the long-term fate of the artificially injected material.

Section II of this paper will present a simplified emission scenario of argon-ion engines for the proposed satellite power system which will be used as the standard for comparison. Sections III, IV, and V will address details of modifications due to interactions of the plasma beam with the magnetosphere, modifications of the plasmaspheric composition and density, and the modification of radiation belt dosage, respectively.

II. Argon-Ion Emission Models

The current technology of ion engines is still evolving (Kauffman, 1974; Byers and Rawlin, 1976); therefore, the parameters of argon-ion engine operations in space must be regarded as uncertain at present, although it is by now fairly firm that the most economical and environmentally safe propellant is argon. According to the reference system report of the satellite power system concept development

Table 1 Ion engine characteristics

	Option 1	Option 2
Specific impulse, sec	13000	5000
Ar^+ kinetic energy, keV/Ar^+	3.5	0.5
Ar^+ streaming speed, km/sec	130	50
Current density, amp/cm^2	2.5×10^{-2}	2.5×10^{-2}
Temperature, K	~1000[a]	~1000[a]
Beam density, cm^{-3}	$\sim 1.5 \times 10^{10}$	$\sim 4 \times 10^{10}$
Beam diameter at exit, cm	100	100
Beam spread at exit, deg	~10°	~10°
Number of engines required for SPS/COTV	~300	~800

[a]Kauffman (1974)

and evaluation program (DOE, 1979), argon-ion engines of specific impulse 13000 sec are projected to perform major propulsion and station-keeping duties for cargo orbit transfer vehicles as well as for the spacecraft at geosynchronous orbit. Other considerations (Byers and Rawlin, 1976), with perhaps less stringent requirements on projected advances on the technology of ion engines, assumed 5000 sec specified impulse as standard for comparison. Some projected characteristics of these two options of ion engine operation are listed in Table 1. Since the reader may easily scale the results of this paper by means of Table 1, we shall, henceforth, consider Option 2 only.

From Table 1, it is seen that the ion beam exhaust is a very dense but fairly cool plasma whose streaming kinetic energy (0.5 keV) far exceeds the thermal energy. Further, in order to propel the cargo orbit-transfer vehicles (COTV) to geosynchronous orbit, the argon plasma beam will be directed perpendicular to the geomagnetic field at the equatorial plane in the azimuthal direction (Fig. 1). The propagation of plasma beams perpendicular to the geomagnetic field entails very interesting dyn-

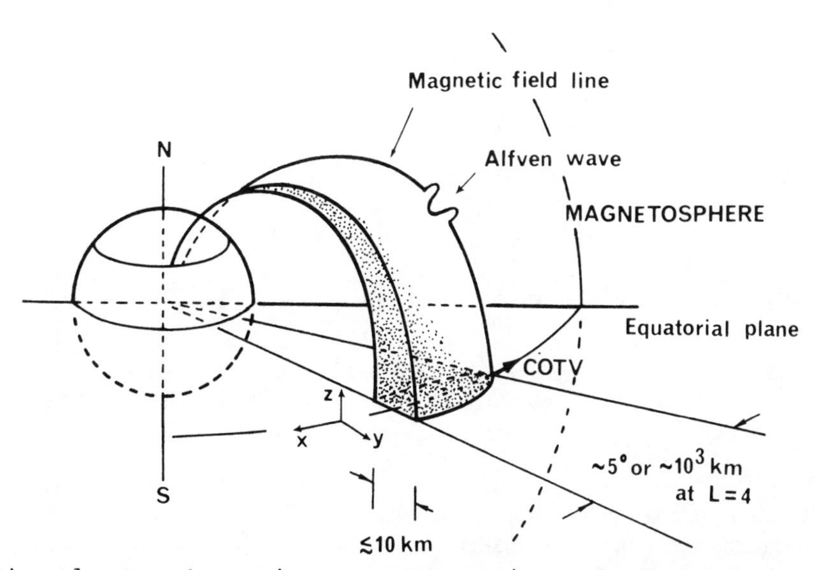

Fig. 1 A schematic representation of the cargo orbit transfer vehicle (COTV) argon-ion emission scenario at L = 4. (Ion beam trailing the COTV is roughly 1000 km long and ~10 km wide.)

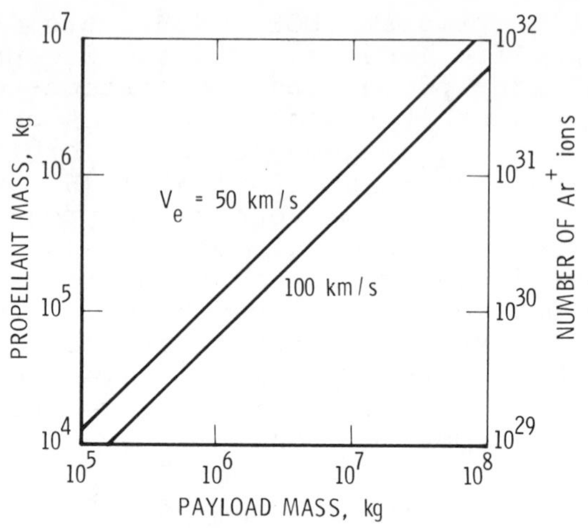

Fig. 2 Argon propellant mass necessary for the transport of different payload masses between 350 km altitude and synchronous altitude. (Orbital inclination is assumed to change from 0 to 28.5 deg during transport.)

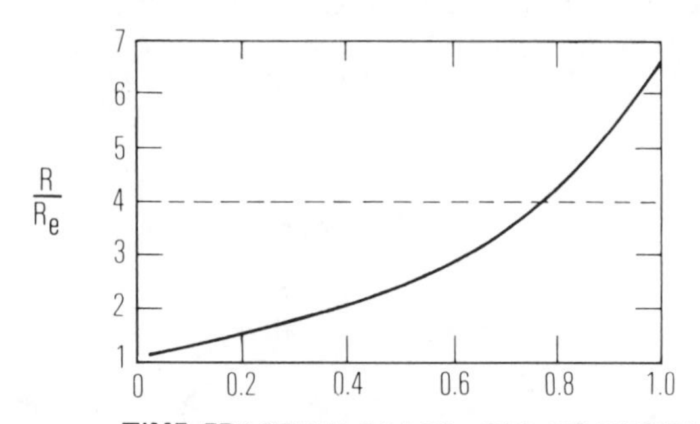

Fig. 3 Fraction of mission lifetime μ spent at various geocentric distances.

amical interactions with the magnetosphere-ionosphere system. This topic will be treated in the next section.

For consideration of emission parameters, Fig. 2 (taken from Chiu et al., 1979a) shows the relationship between payload mass and argon propellant mass needed to transport the payload from low Earth orbit (350 km altitude) to synchronous altitude, with an accompanying orbital plane change of 28.5 deg. Obviously, the amount of propellant required for a given payload depends on the ion-beam streaming speed. For satellite power system payload of ~ 10^7 kg, it will be necessary to expend ~ 10^6 kg of argon propellants for Option 2 in Table 1. This is ~ 1.5 x 10^{31} Ar^+ ions, roughly comparable to the total content of the natural plasmasphere and ionosphere above 500 km. The exhaust deposition rate in terms of the fraction μ of mission lifetime, which is nominally ~ 130 days, is shown as a function of geocentric radius R in Fig. 3 (Chiu et al., 1979a). Thus, 80% of the total propellant content is released in the plasmasphere, R ≤ 4 R_E. The number of Ar^+ released at a given geocentric distance for a payload mass of 10^7 kg is shown in Fig. 4; for comparison, the number of ambient electrons lying within a flux shell of thickness equal to twice the argon gyroradius at a given

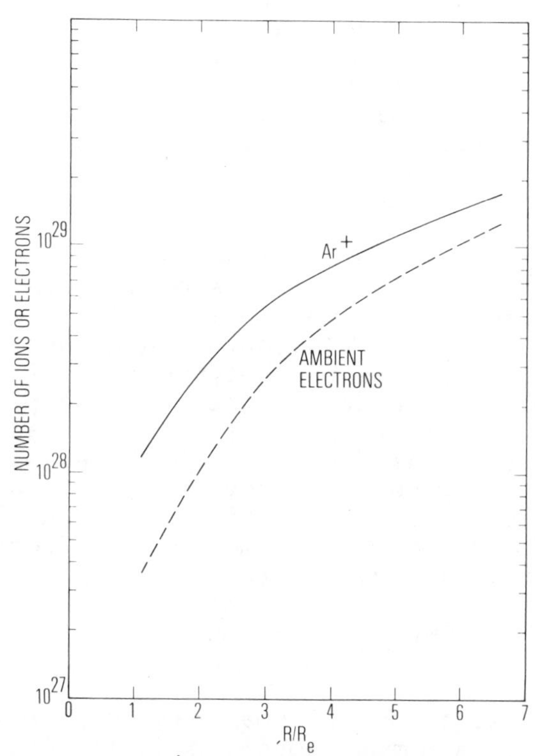

Fig. 4 Number of Ar^+ ions released at various geocentric distances for a payload mass of 10^7 kg, compared to the number of ambient electrons in a dipole field shell two argon gyroradii thick.

distance R is also shown. The energy content released into a given shell dominates the ambient energy content, however, since the streaming Ar^+ kinetic energy is considerably greater than the ambient thermal energy.

The fate of the injected Ar^+ depends largely on the plasma dynamics of the Ar^+ cloud interacting with the ambient magnetosphere. Some features of this dynamical interaction will be addressed in the next section. For the majority of the injected Ar^+ ions that are trapped in the plasmasphere, their energy (~ 500 eV per Ar^+) degrades via at least two routes. First, the propagation of the Ar^+ plasma across the geomagnetic field drives ionospheric currents which dissipate a substantial fraction of the Ar^+ streaming energy (see next section). Second, long range coulomb collisions with ambient electrons degrade the kinetic energy into ambient thermal energy without causing substantial changes in pitch angle because these coulomb collisions are primarily forward scatterings. For the energetic Ar^+ ions to be physically lost from the magnetosphere, they will have to suffer charge exchange collisions which would allow the neutral argon atom to escape the plasmasphere. Figure 5 shows the comparison of charge exchange lifetimes of Ar^+ in the plasmasphere ($L \geqslant 2$) with the thermalization lifetime due to coulomb collisions. Note that the charge exchange lifetimes (~ 100 hr) are comparable to the average duration between magnetospheric storms which typically define the lifetimes of the natural magnetospheric plasma. Thus, we expect that the injected Ar^+ plasma will substantially modify the magnetosphere, with a time constant roughly equal to several days, until a new magnetospheric storm sweeps out the accumulation, provided the storm mechanism remains effective under the modified circumstances.

III. Beam-Magnetosphere and Beam-Ionosphere Interactions

The physics of a plasma beam propagating transverse to a homogeneous <u>vacuum</u> magnetic field is very simple: if the beam is sufficiently dense that polarization currents can maintain the charge separation electric field necessary to satisfy $\vec{E} + (1/c)\vec{v}\times\vec{B}=0$ ($\vec{v}$ the beam velocity), the beam will

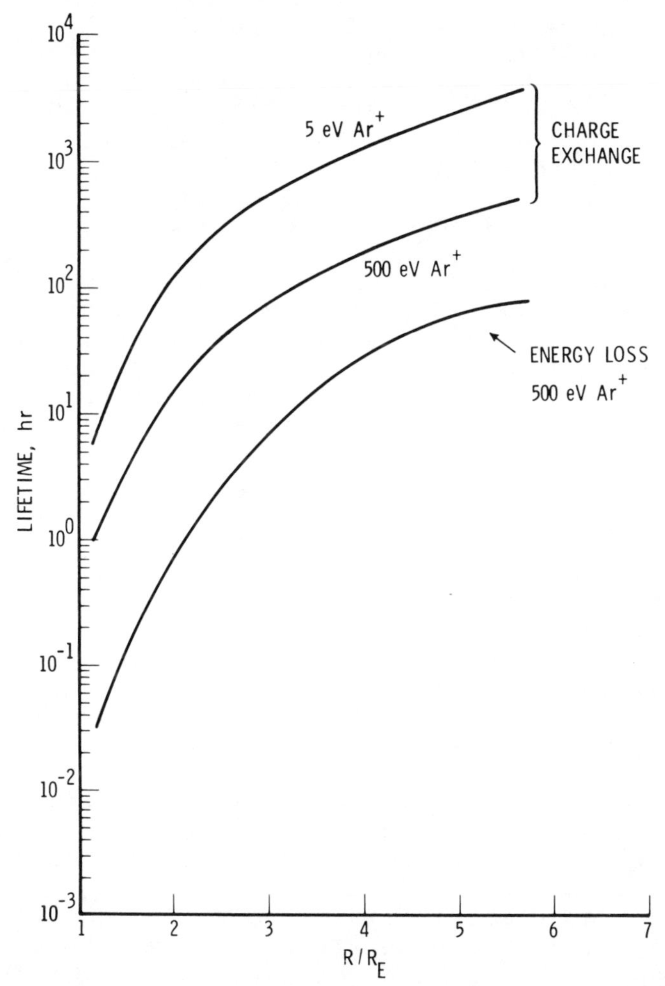

Fig. 5 Comparison of charge exchange and coulomb lifetimes of Ar^+ ions at various geocentric distances in the equatorial plane. (Charge exchange cross section used in calculation of lifetime was measured by Gilbody and Hasted, 1956. Plasmaspause is assumed to occur at 4 Re.)

propagate across the magnetic field. An alternative view of the effects of the polarization electric field $\vec{E}$ seen by a co-moving observer above is that, in the coordinate system of the stationary magnetic field outside of the cloud, the plasma cloud, under the force of $\vec{E}$, appears to be drifting with a velocity $\vec{v} = c(\vec{E}\times\vec{B})/B^2$. But $\vec{v}$ above is also the drift velocity of magnetic field lines in the cloud induced by the electric field $\vec{E}$; hence, the field lines in the cloud are said to be "frozen" into the plasma, drifting with velocity $\vec{v}$ relative to the field lines outside of the cloud. For this condition to apply, the beam density n_A must satisfy

$$4\pi\, n_A\, m_A\, c^2/B^2 \gg 1 \tag{1}$$

where m_A is the argon-ion mass (Curtis and Grebowsky, 1979). Numerically, Eq. (1) yields $(500\text{-}30000)\ n_A\ (\text{cm}^{-3}) > 1$ for $2 \leq L \leq 4$, which would seem to be well-satisfied for the beam parameters of Table 1; if so, the beam simply moves out of the magnetosphere to be dissipated in space.

Unfortunately, this conclusion is false. The magnetosphere, and particularly the ionosphere, cannot be considered in terms of a vacuum magnetic field. Plasmas in the magnetosphere, and especially in the ionosphere, act to short out the charge-separation electric field $\vec{E}$ and stop the beam in a distance on the order of 1000 km. Later in this section, we give some details of this picture, based on consideration of explosive barium release in the far magnetosphere under high $\beta = 8\pi\, p_\perp/B^2$, [i.e., $\beta > 1$] where $p_\perp$ is plasma cloud pressure perpendicular to the magnetic field (Pilipp, 1971; Scholer, 1970), although β for the argon ion beam is still much greater. A second "confirmation" of the action of an Alfvén shock to short out the polarization electric field can be found in the stopping of Starfish debris motion perpendicular to the magnetic field (Zinn et al., 1966); for this case $\beta > 1$ but again not as large as the case of the argon-ion beam.

It is sometimes claimed (Curtis and Grebowsky, 1979) that the argon-ion beam regime is trans-Alfvénic (i.e., $v_A \sim v$) and that the physics of beam stopping is different. However, actual calculation of plasmapheric Alfvén speeds, as in Fig. 6, based on a plasmasphere model which compares favorably with observations (Chiu et al., 1979), shows conclusively that the 3.5 keV argon beam is sub-Alfvénic by at least one order of magnitude. While the basic detailed physics of beam stopping requires further research, for our assessment task here we feel that the physical ideas of the above

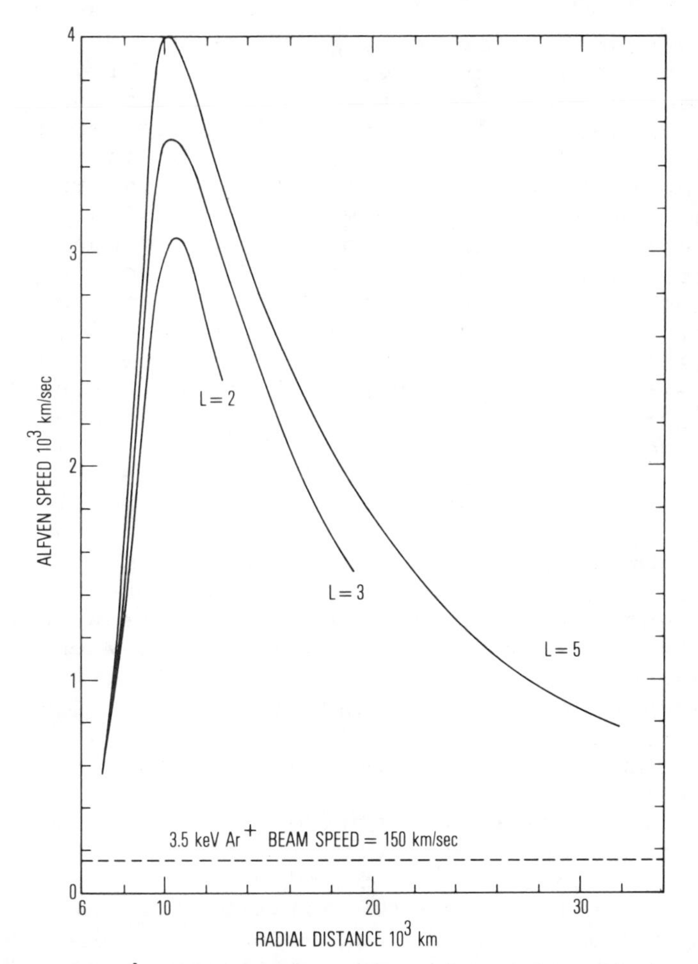

Fig. 6 Alfvén speed as a function of radial distance from the Earth's surface for geomagnetic field lines of various L-values.

reference are sufficiently sound to allow us to estimate beam stopping characteristics. As a consequence, the beam dissipates its energy in a rather localized region and may well deposit (by Joule heating) a substantial amount ($\sim 10^{14}$ ergs sec^{-1}) of energy in a relatively small area of the ionosphere, connected by magnetic field line to the argon beam (see Fig. 6), long before the majority of Ar^+ ions are physically present in the ionosphere.

Granted that the argon beam is stopped in the magnetosphere, it begins to act like a man-made ring current of ≤keV ions, with a residual pitch angle anisotropy left over from the initial injection nearly perpendicular to $\vec{B}$. (Just what energy and anisotropy are left after the beam-plasma interactions described in this section is not known at present; they need to be evaluated in order to give a more precise picture of their influences on systems such as increase in airglow and changes in radiation belt dosage levels.) This ring current acts much like a natural one, subject to charge exchange, coulomb scattering (see Section II), and wave-particle interactions. If charge exchange is the dominant loss process (coulomb scattering or other forms of energy loss does not remove argon), this ring current may have $\sim 10^{30}$ ions in it, comparable to the natural ring current. A major difference is that the argon ring current is mostly inside the plasmasphere while the natural ring current penetrates perhaps one Earth radius inside the plasmasphere with the rest outside. Nonetheless, there could be substantial overlap between the ring currents, especially during storm times when the plasmasphere is eroded and the natural ring current driven in by convection electric fields. The argon ring current can make a substantial contribution to the plasmaspheric pressure and hence to the currents which flow there (both across and along $\vec{B}$); this will significantly stiffen the plasmasphere to deformations associated with enhanced storm-time convection electric fields and diminish the strength of electrostatic radial diffusion. The result may be (other things unchanged) a buildup of ring-current particles and inhibition of inward transport of higher-energy particles.

Aside from the microscopic processes alluded to previously, there are many possible beam-plasma interactions. These involve not only the dense (1-100 cm^{-3}), relatively cold (≤ 1 eV) thermal plasma, but also ring-current and radiation-belt particles whose energies range from kiloelectron volts to megaelectron volts. A partial list of important processes (in order of increasing time scale would include: a) crossfield current-driven instabilities (time scale R_e/v ~milliseconds, where R_e is the electron Larmor radius, b) ion electrostatic modes (time scale $= \omega_A^{-1}$; ω_A is the Ar^+ plasma frequency $\approx 10^{-2}$/sec), and c) proton electromagnetic cyclotron (EMC) modes (time scale $\approx \Omega_p^{-1} \approx 0.1$ sec). These are important because they can cause local transfer of energy from the beam to the plasmasphere (first two modes), or because they affect natural magnetospheric processes. Thus the proton EMC mode is inhibited by heavy ions (Cornwall and Schulz, 1971). This may mean a buildup both of ring-current protons and of relativistic electrons which otherwise would be precipitated into the ionosphere by the EMC instability (see Section V). There are also magnetohydrodynamic instabilities of considerably larger time scale (e.g., Perkins, et al., 1973), which cause deformations of the beam (striations, etc.), but which probably do not transfer much energy from beam to plasmasphere. The free energy for these instabilities is not only the beam kinetic energy, but the energy associated with pitch-angle anisotropies and spatial gradients.

In the remainder of this section we give a brief description of the physics of stopping the beam and of dissipating some of its energy in the ionosphere. The basic physics is well-known, and was worked out by Scholer (1970) and Pilipp (1971) in connection with the HEOS-I release of an ionized barium cloud at L = 12 (Haerendel and Lust, 1970). This high-altitude release had, as will the argon engines, a high initial β (= $8\pi p_\perp/B^2$, where $p_\perp$ is the pressure perpendicular to the field lines of the injected plasma). The beam expands rapidly, in a direction perpendicular to $\vec{v}$ and to $\vec{B}$, to the point where $\beta \leqslant 1$. Of course, the beam also spreads without constraint (except for mirroring forces) along $\vec{B}$.

One could calculate the final beam spread y in the y direction (or $\vec{v} \times \vec{B}$ direction; see Fig. 1) using zero-Larmor-radius magnetohydrodynamics, that is, by equating P (including thermal pressure $n_A kT$ plus dynamic pressure (1/2) $n_A m_A v^2 \tan^2\theta$) to the asymptotic plasmaspheric pressure, which is essentially $B^2/8\pi$. Our flux conservation requires $n_A = n_{0A} A/\Delta z \Delta y$, where n_{0A} and A are the initial beam density and area (see Table 1) and Δz is the beam spread along the field. Assuming $\Delta z > \Delta y$, one finds that Δy is less than ~ 1 km for L > 4, much smaller than the argon

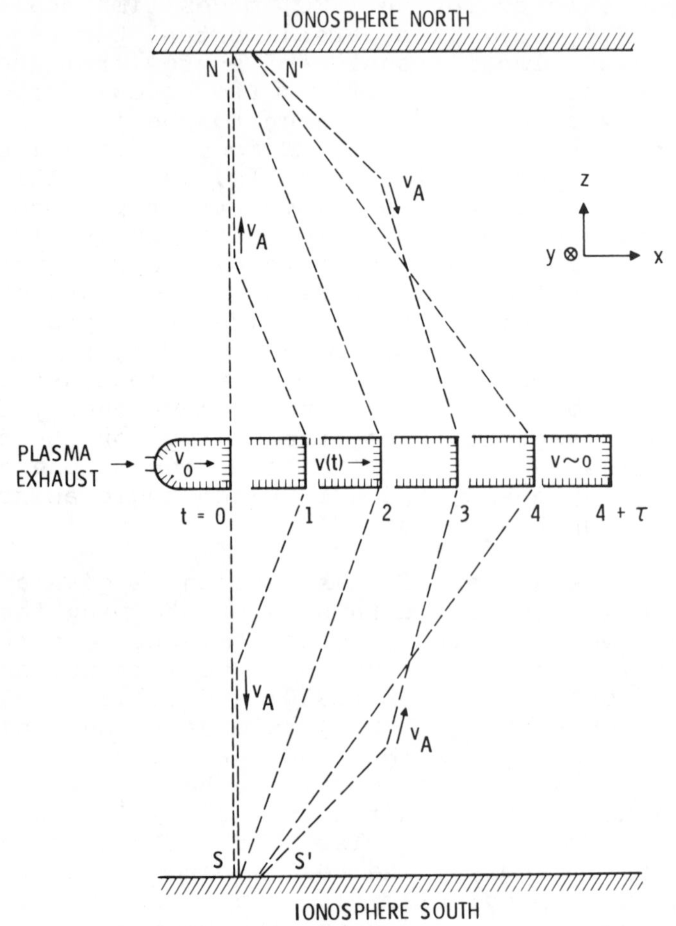

Fig. 7 Schematic of the beam-magnetosphere and beam-ionosphere interactions involving propagation and reflection of an Alfvén wave.

Larmor radius R_A. In effect, this calculation of the confinement of gyration centers tells us that Δy is of the order of the argon Larmor radius (40-80 km at L=4) because the gyration centers are confined to a Δy much smaller than the argon Larmor radius, at least for the first ten or so Larmor radii downstream from the nozzle; past this, the 10 deg angular divergence of the beam could produce substantially larger Δy [as long as Eq. (1) continues to hold].

In the first approximation, then, we have a beam of $\Delta y \approx R_A$ propagating across the Earth's field as shown in Fig. 7. In this figure, the dotted lines show schematically Earth's magnetic field lines at various times. The conditions $\vec{E} + (1/c)\, \vec{v}\times\vec{B} = 0$ means that these lines are frozen into the plasma beam at the equator; their distortion is an Alfvén wave (t = 1, in Fig. 6). At t = 2, the wave reaches the ionosphere where the foot of the field line slips because of the ionosphere's finite conductivity; the wave then reflects back to the beam (t = 3,4).

The field lines act somewhat like rubber bands, tending to retard the cloud. The physical mechanism is that the polarization charges responsible for $\vec{E}$ move along the field lines at the Alfvén speed v_A, accelerating magnetospheric plasma and transferring momentum out of the beam. Ultimately, the Alfvén wave reaches the ionosphere and drives dissipative Pedersen currents. (In the absence of dissipation, the argon beam would oscillate like a mass on a rubber band field line.)

Let M_A be the mass density of the argon beam, integrated along field lines passing through the beam:

$$M_A = \int dz\; n_A m_A \qquad (2)$$

When this mass density is equal to the mass per unit area incorporated by the Alfvén wave, namely $2v_A \tau n_0 m_p$, the beam is essentially stopped. Here τ is the time it takes the Alfvén wave to travel a distance $v_A\tau$, and $n_0 m_p$ is the magnetospheric mass density per unit volume. For the argon beam, $M_A \sim 5 \times 10^{-13}$ gm cm^{-2} at L = 4; if $v_A \sim 3 \times 10^7$ cm/sec^{-1} and $n_0 \sim 10^3$ cm^{-3}, this gives $\tau \approx$ few seconds. The beam's velocity behaves like $v \simeq v_0\, e^{-t/\tau}$, so the beam can only travel a distance on the order of $v_0\tau \lesssim 10^3$ km. In this example, the beam momentum is soaked up by magnetospheric plasma extending at most a few thousand kilometers down the field line on either side of the beam.

The ionosphere "feels" the beam after a time $(\ell/v_A) \sim$ tens of seconds (ℓ is the length of the field line). The electric field E_I imposed on the ionosphere differs from the $(1/c)\, \vec{v}\times\vec{B}$ field, mapped to the ionosphere, for two reasons: first, the electric field diminishes along the field line by a factor of order $e^{-\ell/v_A\tau}$, and second, the electric field is partly reflected at the ionosphere (to make the upgoing Alfvén wave at t = 3, 4 in Fig. 7). Scholer (1970) gives the relation

$$E_I = [2/(1+\chi)]\ E_{out} \tag{3}$$

$$\chi = 4\pi\Sigma_p\ v_A/c^2 \tag{4}$$

where E_{out} is the field just outside the ionosphere, E_I the field in the ionosphere, and Σ_p the ionospheric Pedersen conductivity. It is somewhat hazardous to make numerical estimates with exponential factors floating around, but it would be surprising if the ionospheric time scale for dissipation, τ_I, defined by

$$\tau_I = M_A\ v_0^2/2\Sigma_p E_I^2 \tag{5}$$

were very much greater than τ. Taking the two time scales equal, the rate at which beam energy is dissipated in the ionosphere is roughly $M_A v_0^2/2\tau \sim$ few ergs/cm^2 sec, and this might occur over an area of 10^3 -10^4 km^2 in the ionosphere. If these numbers are roughly right, a substantial fraction of the beam's energy is dissipated in the ionosphere; otherwise, the beam energy goes into formation of magnetospheric waves and particle energization. In any case, we argue that the whole beam power of ~ 5 x 10^{15} ergs/sec will be dissipated in the magnetosphere and ionosphere. Indeed, observations of barium releases in the far magnetosphere seem to qualitatively confirm the above considerations (Scholer, personal communications; Zinn et al.,1966).

By comparison, a natural geomagnetic storm dissipates energy at a rate ~ 10^{17} ergs/sec but only sporadically and over a much larger area. The dissipation rate per unit area for a storm is also a few ergs/cm^2 sec, comparable to the argon beam dissipation. The difference is that the argon beam will be present, day and night, for several months, while storms occur only every few days. The amount of naturally-occurring energy precipitated into the ionosphere over a span of time like four months (the ion-engine operational period), is not likely to exceed a few times 10^{22} ergs per month in moderately quiet times; this is on the order of magnitude of the energy released into the plasmasphere by the SPS ion engines.

IV. Modification of the Plasmasphere

In the previous sections it was found that the argon ions that are emitted by the ion engines will merge with the magnetospheric plasma rather than escape through the magnetopause. Here the effect of the addition of these heavy ions and the accompanying heat input to the inner magnetosphere is considered. As a first approximation it is assumed that the ionized argon can be treated as a constituent of the steady-state plasmasphere, with the extreme limits of the ionospheric concentration equal to the densities of hydrogen ions and oxygen ions at 500 km altitude. An earlier analytical model of the equilibrium plasmasphere derived by Chiu et al. (1976b) is used to predict the changes that will occur due to the presence of argon ions and to the heating of the ambient plasma by the thermalization of the argon ions.

Chiu et al. (1976b) found that the observed plasmaspheric density between an altitude of 500 km and the dipole field line L = 5 can be described adequately by the expression

$$n_j(s) = n_j(\ell)[B(s)\,T_j\,(\ell)/B(\ell)T_j(s)] \cdot \exp\int_{\ell}^{s} ds' \cdot$$

$$\cdot [q_j\,|e|\,E - m_j\,(GM_E/r^2 + \Omega_E^{\,2}\,r)\,\hat{r}'\cdot\hat{s}']/kT_j\,(s')$$

where

n_j	=	density of ion species j
s	=	distance along the dipole field line measured from the equator (s=0) to the ionosphere (s=ℓ)
B	=	magnetic field strength
q |e|	=	charge
E	=	electric field strength
m	=	ion mass
G	=	gravitational constant
M_E	=	mass of the Earth
Ω_E	=	angular rotation frequency of the Earth
k	=	Boltzmann's constant
r	=	geocentric radial coordinate

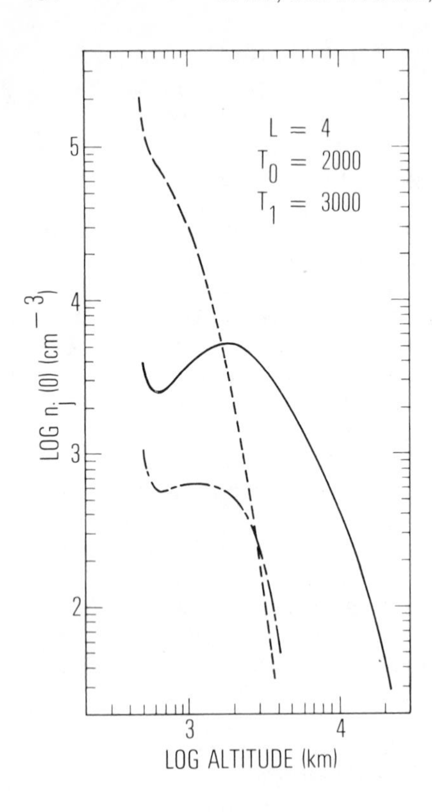

Fig. 8 Density distribution along the L = 4 dipole field line for the natural plasmasphere model of Chiu et al. (1979b).

The empirically determined temperature profile along a field line is given by

$$T(s) = T_0 + T_1 \left[\frac{L-1}{L_0}\right]^{\alpha} \left[\frac{\ell - s}{\ell}\right]^{\beta} \tag{7}$$

where T_0, T_1, L_0, α and β are constants with the following nominal values: $T_0 = 2000^{\circ}K$, $L_0 = 3$, $\alpha = 1/2$, $\beta = 1/2$. These authors considered a charge-neutral four constituent model composed of electrons, H^+, O^+, and He^+. Their result for L = 4 and the temperature parameters is shown in Fig. 8.

For the purpose of the present investigation, the minor constituent He^+ was omitted and argon ions were substituted as the fourth plasmasphere constituent in the model. As mentioned previously, the ionospheric (500 km) concentration of argon ions was set equal to the local concentration of either hydrogen ions or oxygen ions as nominal lower and upper limits. It must be recognized that these limits are at present arbitrary; eventually these limits are set by a detailed study of the fate of Ar^+ in the magnetosphere. Several different fractional temperature increases were assumed. The results of the calculations for L = 4 are shown in Fig. 9 for comparison with Fig. 8. It is seen from the left-hand panel of Fig. 9 that the mere addition of the argon ions will not upset the normal distribution of the plasmaspheric hydrogen and oxygen; however, as shown by the right-hand panel, the addition of heat by the argon-ion interaction with the ambient medium or by Joule heating can drastically alter the plasmasphere. At high altitudes, hydrogen can be replaced by the heavier constituents (oxygen or a combination of oxygen and argon). Some consequences of the possible changes in plasmasphere composition will be considered in the next sections. Before continuing the discussion, however, it should be pointed out that the calculation described above is a crude approximation of the actual situation, which is very far from steady state. The argon ions will be introduced onto different L shells at different times; plasma convection processes and radial diffusion can modify the spatial distribution of the argon which, because of the nature of its source, is already quite different from that of a normal plasmasphere constituent. In addition, interhemispheric flows, which may affect the plasmasphere as a whole were not included in the model. Work on including these effects is currently underway.

V. Ring-Current and Radiation-Belt Effects

The widespread use of ion engines in space carries certain implications for the high-energy ($E \gtrsim 10$ keV) particles that populate the Earth's magnetosphere. The effects of ion engines on such high-energy populations are necessarily indirect if the injected argon ions have energies of at most 500 eV. The high-energy particles, however, are subject to the presence of plasma turbulence in the magnetospheric environment, and the level of such turbulence may possibly be enhanced or reduced by the addition of argon ions from a space engine.

It is customary, for both kinematical and dynamical reasons, to distinguish between particles having $E \sim 10$-100 keV and those having $E \gtrsim 100$

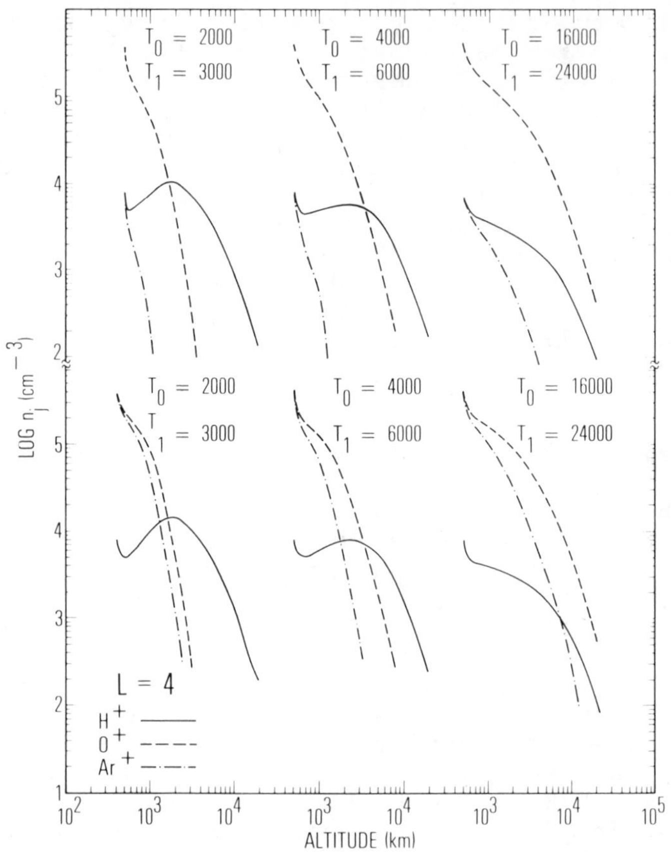

Fig. 9 Calculated L = 4 ion density distributions for various Ar^+ concentrations at the 500 km altitude boundary and several thermal models.

keV. The former are known as ring-current particles, since they account for the major share of diamagnetic disturbances observable with low-latitude magnetometers at the Earth's surface. The latter are known as radiation-belt particles, since they account for most of the radiation damage to hardened spacecraft. The ring current and the radiation belts consist of electrons, protons, and heavier ions trapped in the Earth's magnetic field. Such particles obey the laws of adiabatic charged-particle motion in the first approximation, and their iso-intensity profiles are toroidal in shape (e.g., White,1966).

The ring current and radiation belts are populated with energetic particles by virtue of the dynamical interaction between the Earth's magnetosphere and the solar wind. They are depleted of such particles mainly by interaction with plasma turbulence, which tends to violate the adiabatic invariants of charged-particle motion. Such violation allows particles to escape confinement by the Earth's magnetic mirror and to precipitate (i.e., deposit their energy) in the Earth's atmosphere and ionosphere. However, the interaction of geomagnetically trapped particles with plasma turbulence is very selective in that it is contingent upon a "resonance" between the gyration of a particle and the frequency of a wave in the turbulent spectrum (after one takes account of the Doppler shift associated with the motion of the particle through the plasma). Thus, different classes of turbulence are found to interact with different classes of particles.

Turbulence in a plasma can be created by various mechanisms. For example, the injection of a cold (i.e., monoenergetic) ion beam into a background plasma can easily lead to instability in various electrostatic wave modes (e.g., Hasegawa, 1975). The consequence of such instability is the rapid amplification of thermal noise to appreciable amplitude, such that the beam itself becomes diffused. In the case of ion engine beams, the question is whether the resulting plasma turbulence can interact resonantly with ring-current or radiation-belt particles. The answer to this question is not yet known since beam-plasma instabilities have not yet been investigated thoroughly in this context.

Another source of instability in the magnetospheric plasma is the velocity-space anisotropy intrinsic to ring-current particle distributions. Since the magnetic mirror points of ring-current protons and electrons are concentrated near the magnetic equator rather than distributed uniformly along a field line, there must be relatively more energy per degree-of-freedom associated with particle gyration than with translation of guiding centers along field lines. This condition can easily lead to instability in certain electromagnetic wave modes. The anisotropy of ring-current electrons can lead to instability of a field-guided wave with right-handed polarization, i.e., the so-called "whistler" wave mode. The anisotropy of ring-current ions can lead to instability of the analogous field-guided wave with left-handed polarization. Both instabilities require the wave frequency to be somewhat smaller than the corresponding particle gyrofrequency. Both instabilities cause velocity-space diffusion so as to reduce the anisotropy of the corresponding charged-particle species and reduce the lifetime of that species against precipitation into the Earth's atmosphere (Kennel and Petschek, 1966; Cornwall, 1966; Cornwall et al., 1970). Moreover, the unstable ion-cyclotron waves generated by the anisotropy of ring-current protons are resonant with relativistic radiation-belt electrons ($E \gtrsim 2$ MeV) and thus account for the observed precipitation of such electrons during the recovery phase of a magnetic storm (Thorne and Kennel, 1971; Vampola, 1971). It happens that the electromagnetic instabilities noted here are not effective at ring-current energies for protons outside the plasmasphere since the larger phase velocities attained there require a correspondingly larger proton energy for cyclotron resonance. Thus, the precipitation of relativistic electrons is contingent on the spatial coexistence of ring current and plasmasphere, which occurs only during the plasmaspheric expansion characteristic of the recovery phase of a magnetic storm.

The electromagnetic proton-cyclotron instability, however, is likely to be suppressed by the presence of substantial numbers of heavy ions, such as Ar^+ or O^+, in the magnetospheric plasma (Cornwall and Schulz, 1971). This means that the major mechanism for the depletion of relativistic electrons from the outer radiation belt is likely to be made inoperative by the widespread use of ion engines in space. The present population of such hazardous electrons is kept in balance by the occurrence of several large magnetic storms per year. Suppression of the proton-cyclotron instability might, therefore, result in a major enhancement of the radiation level from relativistic electrons within a year after argon-ion saturation of the plasmasphere.

The enhancement of relativistic-electron radiation can be analyzed within the framework of a simple model. Consider the model equation (e.g., Schulz,1974)

$$dI/dt = S - \lambda I \tag{8}$$

in which I represents the radiation intensity, λ the natural radiation-belt decay rate, and S the strength of a weak source of relativistic electrons. The decay rate λ is perhaps inversely proportional to the particle energy E, such that $1/\lambda \approx$ (E/1 MeV) x 10 days. The solution of Eq. (8) approaches the limit $I_{\infty} = S/\lambda$ as $t \to \infty$. Suppose, however, that I is reduced to zero by relativistic electron precipitation (REP) events which occur randomly in time. Thus, let exp $(-\Delta t/\tau)$ be the probability that a time interval of length Δt is free of REP events. The radiation intensity grows from zero in accordance with Eq. (8) after each REP event and so never quite reaches I_{∞}. A careful implementation of this model shows that the instantaneous probability for I to be in excess of some arbitrary threshold $I_0 < I_{\infty}$ at any given time is

$$P\ (I > I_0) = [1 - (I_0/I_{\infty})]^{1/\lambda\tau} \qquad (9)$$

and that $P(I > I_{\infty}) = 0$. The mean value of I, averaged over a time $\Delta t \gg \tau$, is given by

$$\bar{I} = [\lambda\tau/(1 + \lambda\tau)]\ I_{\infty} \qquad (10)$$

This is the natural state of affairs. The intensity of relativistic electrons in the vicinity of the plasmapause fluctuates in time in such a way that Eqs. (9) and (10) are satisfied. Suppose, however, that REP events are suppressed, e.g., by the addition of heavy ions to the plasmasphere. This environmental modification corresponds to the limit whereupon $\bar{I} = I_{\infty}$. The mean radiation intensity has been enhanced by a factor $(1 + \lambda\tau)/\lambda\tau$. Thus, if τ = 20 days (corresponding to the annual occurrence of about 18 REP events), the intensity of 2-MeV electrons will be enhanced by a factor 2 (relative to its natural mean value) by the addition of major quantities of Ar^+ or O^+ to the plasmasphere. The mean intensity of 4-MeV electrons will be enhanced by a factor ~ 3. With the addition of heavy ions to the plasmasphere, the radiation intensity ceases to fluctuate between zero and I_{∞}; it approaches I_{∞} as a permanent condition of the modified environment.

There is no danger that the population of ring-current protons would be similarly enhanced since these are subject to rapid charge exchange with ambient neutral hydrogen, as are ring-current helium and oxygen ions on a substantially longer time scale (Tinsley, 1976; Lyons and Evans, 1976). Moreover, the addition of argon plasma (or any other type of plasma) to the natural plasmasphere would tend to enhance the loss rate of ring-current electrons (Brice, 1970; 1971). One should not worry about the development of a charge imbalance in the magnetosphere on account of modified precipitation rates. There is ample cold plasma in the plasmasphere to balance charges through minor modifications of the already weak ambipolar electric field which has been included in the model calculations in the previous section.

The estimates of I and $P(I > I_0)$ are based on a highly idealized model of radiation-belt dynamics. Other sources of fluctuation are quite likely to broaden the probability distribution. For example, one might expect both λ and (especially) S in Eq. (8) to depend on geomagnetic activity and therefore on time. Moreover, the magnetosphere is subject to reversible (adiabatic) compressions that change the scales of energy and distance. Such compressions (caused by slow variations of solar-wind speed) are not of dynamical interest and so are omitted from Eqs. (8)-(10), but such adiabatic processes would have to be considered in formulating a model of the detailed time history of radiation dosage impacting a hardened spacecraft.

Suppressions of the electromagnetic ion-cyclotron wave model is illustrated in Fig. 10. The normalized growth rate γ/Ω_p is plotted as a function of normalized wave frequency ω/Ω_p for selected argon/hydrogen plasma-concentration ratios. For the parameters chosen it is clear that a ratio $N_{Ar}/N_H \geqslant 1$ is sufficient to suppress the proton-cyclotron branch altogether. The lower-frequency argon-cyclotron branch achieves a growth rate of at most 3% of the maximum proton-cyclotron growth rate attained in the absence of argon. Thus, the electromagnetic ion-cyclotron mode is very much stabilized overall by the addition of cold argon plasma to the medium. This conclusion is consistent with the results obtained by similar methods for the addition of even lighter ions such as Li^+ and He^+ (Märk, 1974).

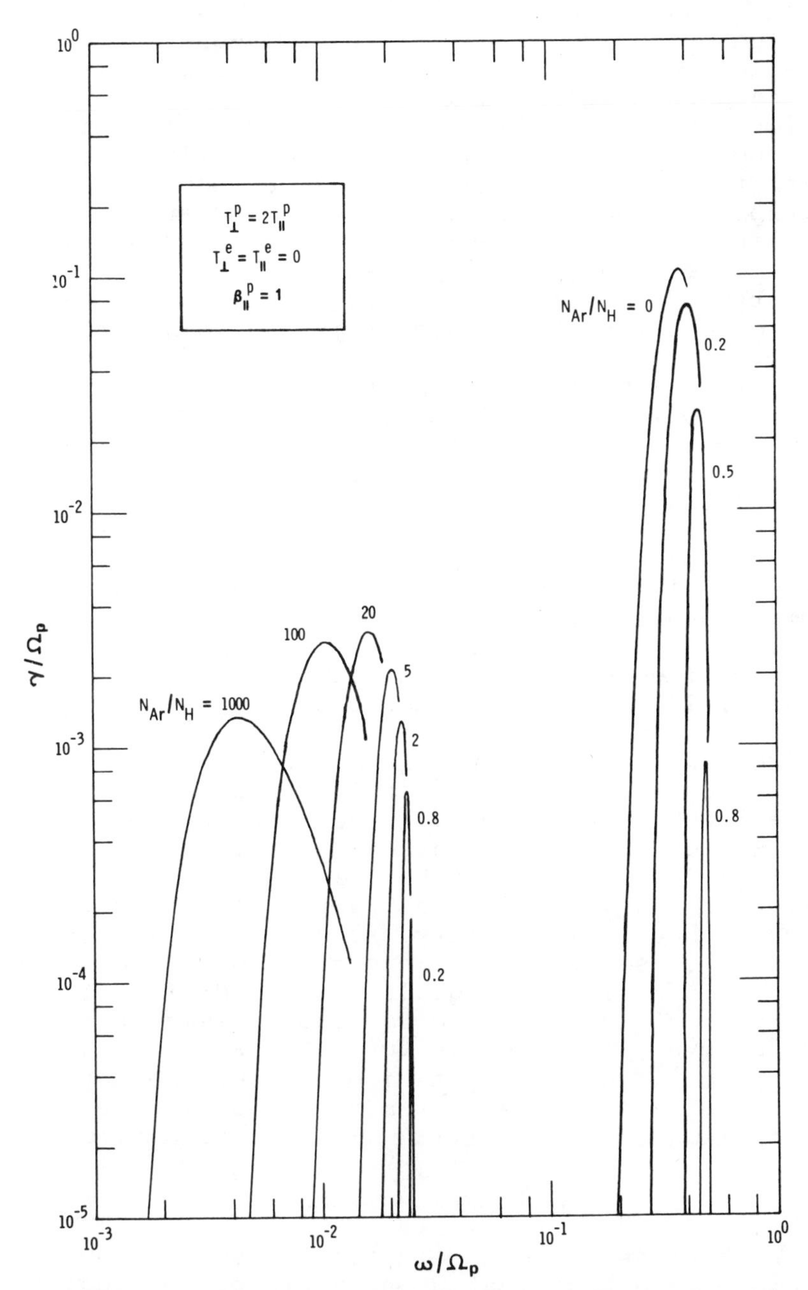

Fig. 10 Lowest-order estimates for the normalized growth rate as a function of normalized wave frequency, for various relative concentrations (N_{Ar}/N_H) of argon and hydrogen plasma. (Method of calculation is that of Märk, 1974.)

The foregoing discussion of ring-current and radiation-belt effects might have to be modified if the injected argon ions were to have energies ~ 3.5 keV (Option 1 Table 1). A beam of such ions might be stopped by the mechanism of Scholer (1970) and Pilipp (1971) before the ions became geomagnetically trapped. The result would be a cold argon plasma having the same effects as are outlined above. On the other hand, the injected 3.5-keV ions might be scattered out of the beam by some mechanism and come to constitute a plasma having a 3.5-keV temperature. In this latter case the injected argon plasma would contribute directly to the ring-current population, and the O^+ drawn up from the F-region would constitute the cold-plasma additive. The cold O^+ would tend to suppress the electromagnetic proton-cyclotron and helium-cyclotron instabilities of the natural ring current at frequencies below the oxygen gyrofrequency (e.g., Cornwall and Schulz, 1971). However, the hot argon plasma would become anisotropic with time because of charge exchange, and it would thus constitute an additional source of free energy for instability (see Cornwall, 1977). The argon-cyclotron instability would occur at frequencies somewhat below the argon gyrofrequency. Its growth rate would be enhanced by the presence of cold O^+ in the plasmasphere (see Cornwall and Schulz, 1971). The argon-cyclotron waves would tend to resonate with relativistic electrons (see Thorne and Kennel, 1971) and thus to facilitate the loss of such trapped electrons into the atmosphere. In summary, some of the plasma-dynamical consequences of injecting hot argon plasma ($E \gtrsim 3.5$ keV) may be qualitatively different from those of injcecting cold argon plasma ($E \lesssim 500$ eV). However, it is clear that in the case of cold-plasma injection by an ion engine, the intensity of relativistic-electron radiation in the Earth's magnetosphere will be substantially increased.

VI. Conclusions

We have assessed the initial phases of the magnetospheric modification scenario of ion-engine exhaust from the projected cargo orbit transfer vehicles of the satellite power system. Aside from summarizing the emission scenarios of the various options, we have dealt with several important aspects of the initial fate of the ion-engine exhaust plasma, using presently available information.

We point out at the outset that if the critical-velocity phenomenon observed in the laboratory is applicable in the magnetosphere, the problem of chemical rocket exhaust (neutral molecules) release in the magnetosphere is almost equivalent to the case of ion-engine plasma release that we have analyzed, because the chemical-rocket exhaust speed in the magnetosphere may exceed the critical velocity for ionization of the constituents.

We have examined the question of the propagation of the ion-engine exhaust plasma beam across the geomagnetic field. We show that it is fallacious to consider the problem as a plasma beam moving across a vacuum magnetic field. Even though the argon-ion beam has much higher $\beta = 8\pi p_{\perp}/B^2$ than artificial explosive release of ion clouds, the argon-ion beam, including both options, is definitely subAlfvénic in the plasmasphere. While details of the beam stopping effects needs further research, we are of the opinion that for the present assessment purposes, the basic physics of the referenced authors are essentially applicable. The plasma beam moves across the geomagnetic field and causes a disturbance of the magnetospheric plasma in the form of an Alfvén wave. The inertia of the natural magnetospheric plasma, driven by the Alfvén wave disturbance, represents a braking force on the motion of the beam. This reaction force is able to stop the ion beam in a distance of ~ 1000 km, according to scaling of studies conducted for HEOS-I release of barium in the far magnetosphere. The generation of an Alfvén wave disturbance by the plasma beam further implies that an electric field of magnitude comparable to the auroral electric field but associated with the Alfvén wave is applied to the ionosphere at the footprint of the magnetic flux tube of the plasma beam (Fig. 1). The Joule dissipation (heating) in the ionosphere by this electric field may be comparable to that in the auroral region during a magnetic storm.

We have roughly modelled the density and composition changes of the plasmasphere based on the above findings that the entire ion-engine exhaust will be stopped inside the magnetosphere. It is found that the mere addition of Ar^+ to the plasmasphere does not drastically change the composition of the plasmasphere, but the input of heat associated with the Ar^+ exhaust will have a tendency to change the plasmaspheric composition by allowing ionospheric O^+ to be forced up into the far plasmasphere where the radiation belts are located.

We have shown that an increase in the proportion of thermal ions of high mass, such as Ar^+ or O^+, can suppress the ion-cyclotron wave generation mechanism, i.e., the plasma instability associated with pitch-angle anisotropy of ring-current ions. Since in the natural state such ion-cyclotron waves cause the loss of radiation-belt electrons by pitch-angle scattering and thus limit the natural dosage level, the suppression of cyclotron-wave generation implies that the radiation belt dosage would build up to a higher level. Estimates, based on production and loss considerations applicable to the natural state of the radiation belts, indicate that ion cyclotron wave suppression may increase the radiation belt average relativistic electron dosage level by as much as a factor of two or three.

Our considerations here are limited to the application of present knowledge and by considerations of the initial fate of the plasma beam exhaust. Further investigations are needed before we can clearly define the eventual fate of the Ar^+ exhaust plasma.

Acknowledgments

This work was supported by the Department of Energy/Argonne National Laboratory contract No. 31-109-38-5075.

References

Alfvén, H. and Arrhenius, G., Evolution of the Solar System, NASA SP-345, 1976, Ch. 21, pp 383-391.

Angerth, B., Block, L., Fahlson, U., and Sopp, K., "Experiments with Partly Ionized Rotating Plasmas," Nuclear Fusion Supplement, 1962, 39-44, Part I.

Brice, N., "Artificial Enhancement of Energetic Particle Precipitation through Cold Plasma Injection: A Technique for Seeding Substorms?," Journal of Geophysical Research, Vol. 75, Sept. 1970, pp. 4890-4892.

Brice, N., "Harnessing the Energy in the Radiation Belts," Journal of Geophysical Research, Vol. 76, July 1971, pp. 4698-4701.

Byers, D. C. and Rawlin, V. K., "Electron Bombardment Propulsion System Characteristics for Large Space Systems," AIAA Paper 76-1039, Nov. 1976.

Chiu, Y. T., Luhmann, J. G., Ching, B. K., Schulz, M., and Boucher, D. J., "Magnetospheric and Ionospheric Impact of Large-Scale Space Transportation with Ion Engines," Aeronautics and Astronautics, accepted for publications, 1979a; also Aerospace Technical Report SAMSO-TR-79-3, 1979a.

Chiu, Y. T., Luhmann, J. G., Ching, B. K., and Boucher, D. J., Jr., "An Equilibrium Model of Plasmaspheric Composition and Density," Journal of Geophysical Research, Vol.84, March 1979b, pp. 909-916.

Cornwall, J. M., "Micropulsations and the Outer Radiation Zone," Journal of Geophysical Research, Vol. 71, May 1966, pp. 2185-2199.

Cornwall, J. M., Coroniti, F. V., and Thorne, R. M., "Turbulent Loss of Ring Current Protons," Journal of Geophysical Research, Vol. 75, Sept. 1970, pp. 4699-4709.

Cornwall, J. M. and Schulz, M., "Electromagnetic Ion-Cyclotron Instabilities in Multicomponent Magnetospheric Plasmas," Journal of Geophysical Research, Vol. 76, Nov. 1971, pp. 7791-7796; correction, Journal Geophysical Research, Vol. 78, Oct. 1973, pp. 6830.

Cornwall, J. M., "On the Role of Charge Exchange in Generating Unstable Waves in the Ring Current," Journal of Geophysical Research, Vol. 82, March 1977, pp. 1188-1196.

Curtis, S. A. and Grebowsky, J. M., "Changes in the Terrestrial Atmosphere-Ionosphere-Magnetosphere System Due to Ion Propulsion for Solar Power Satellite Placement," NASA TM 79719, Goddard Space Flight Center, unpublished, 1979.

Danielsson, L. and Brenning, N., "Experiment on the Interaction Between a Plasma and a Neutral Gas, II," Physics of Fluids, Vol. 18, June 1975, pp. 661-665.

DOE, "Satellite Power System (SPS) Concept Development and Evaluation Program," DOE/ER-0023, Jan. 1979.

Gilbody, H. B. and Hasted, J. B., "Anomalies in Adiabatic Interpretation of Charge Transfer Collisions," Proceedings of the Royal Society (London), Vol. A238, 1956, pp. 972-976.

Haerendel, G. and Lüst, R., "Electric Fields in the Ionosphere and Magnetosphere," Particles and Fields in the Magnetosphere edited by B.M. McCormac, Reidel, Dordrecht, Holland, 1970, pp. 213-228.

Hasegawa, A., Plasma Instabilities and Nonlinear Effects, Springer, Heidelberg, Germany, 1975, pp. 28-43.

Kauffman, H. R., "Technology of Electron-Bombardment Ion Thrusters," Advances in Electronics and Electron Physics, Vol. 36, edited by L. Marton, Academic Press, New York, 1974.

Lindeman, R. A., Vondrak, R. R., and Freemen, J. W., "The Interaction between an Impact-Produced Neutral Gas Cloud and the Solar Wind at the Luna Surface," Journal of Geophysical Research, Vol. 79, June 1974, pp. 2287-2296.

Lyons, L. R. and Evans, D. S., "The Inconsistency between Proton Charge Exchange and the Observed Ring Current Decay," Journal of Geophysical Research, Vol. 81, Dec. 1976, pp. 6197-6200.

Märk, E., "Growth Rates of Ion Cyclotron Instability in the Magnetosphere," Journal of Geophysical Research, Vol. 79, Aug. 1974, pp. 3218-3220.

Möbius, E., Boswell, R. W., Piel, A., and Henry, D., "Investigation of the Critical Velocity Phenomenon from SPACELAB," Transactions of the American Geophysical Union/EOS, Vol. 59, 1979, pp. 1162.

Perkins, F. W., Zabusky, N. J., and Doles, J. H., III., "Deformation and Striation of Plasma Clouds in the Ionosphere, 1," Journal of Geophysical Research, Vol. 78, Feb. 1973, pp. 697-709.

Pilipp, W. G., "Expansion of an Ion Cloud in the Earth's Magnetic Field," Planetary and Space Science, Vol. 19, July 1971, pp. 1095-1119.

Scholer, M., "On the Motion of Artificial Ion Clouds in the Magnetosphere," Planetary and Space Science, Vol. 18, June 1970, pp. 977-1004.

Schulz, M., "Particle Saturation of the Outer Zone: A Nonlinear Model," Astrophysics and Space Science, Vol. 29, Jan. 1974, pp. 233-242.

Thorne, R. M. and Kennel, C. F., "Relativistic Electron Precipitation during Magnetic Storm Main Phase," Journal of Geophysical Research, Vol. 76, July 1971, pp. 4446-4453.

Tinsley, B.A., "Evidence That the Recovery Phase Ring Current Consists of Helium Ions," Journal of Geophysical Research, Vol. 81, Dec. 1976, pp. 6193-6196.

Vampola, A. L., "Electron Pitch Angle Scattering in the Outer Zone During Magnetically Disturbed Times," Journal of Geophysical Research, Vol. 76, July 1971, pp. 4685-4688.

White, R. S., "The Earth's Radiation Belts," Physics Today, Vol. 19, Oct. 1966, pp. 25-36.

Zinn, J., Hoerlin, H., and Petschek, A. G., "The Motion of Bomb Debris Following the Starfish Test," Radiation Trapped in the Earth's Magnetic Field, edited by B. M. McCormac, Reidel, Dordrecht, Holland, 1966.

MAGNETOSPHERIC MODIFICATION BY GAS RELEASES FROM LARGE SPACE STRUCTURES

Richard R. Vondrak*
SRI International, Menlo Park, Calif.

Abstract

The deployment and operation of large structures in space will be accompanied by the release of gases into the Earth's space environment. For example, the launch of a spacecraft into low Earth orbit is accompanied by the deposition of large amounts of rocket exhaust into the atmosphere and ionosphere. Transfer to a higher orbit requires the release into the magnetosphere of rocket combustion products (or of energetic heavy ions if an electric propulsion engine is used). Even when the spacecraft is in final orbit, both the spacecraft itself and its attitude control system are potential sources of released gases. In the inner magnetosphere, gas releases from large space systems may alter the composition and thermal structure of the plasmasphere and the stability of the Van Allen radiation belts. Neutral gases released at even higher altitudes in the outer magnetosphere initially form a toroidal cloud around the Earth. After ionization, these gases may modify the plasma sheet, the magnetospheric current systems, and the magnetopause location.

Introduction

The deployment and operation of large space structures will be accompanied by the release of large amounts of gases into the Earth's space environment. These gas releases are of interest because they may interfere with the satellite system operation, and they may alter the natural environment. Environmental alteration is of concern for future space systems because those systems are expected to be much larger than systems in use at present. Several large systems have been proposed for deployment in the next two decades that require structures of kilometer dimensions operating in very high orbits where the Earth's plasma environment is quite tenuous.

---

*Senior Physicist, Radio Physics Laboratory.

Examples include systems for power generation (Glaser, 1968), communications (Bekey, 1979), scientific studies (Basler et al., 1978), industrialization (O'Neill, 1974, 1976), and national defense.

Gas releases generally occur during both the deployment and operation of any space system. The lift into low Earth orbit deposits large quantities of rocket exhaust products into the terrestrial atmosphere and ionosphere. If the satellite is to be deployed at higher altitudes, interorbital transfer requires the use of chemical rockets or electric propulsion systems that emit energetic ions. The operation of the satellite will release exhaust gases as reaction mass from the attitude control systems, effluents from the spacecraft pressurized volume, surface material eroded by micrometeorite abrasion, and photoelectrons emitted from the spacecraft surface.

Other papers in this volume describe the effects that lift to low earth orbit will have on the Earth's upper atmosphere (Rote, 1980; Forbes, 1980) and ionosphere (Mendillo, 1980). This paper concentrates on environmental effects due to deployment of a space system at high altitudes in the Earth's magnetosphere (see Fig. 1).

The inner magnetosphere is the region of the Earth's space environment where the Earth's magnetic field lines are approximately dipolar. This region is a complex one, consisting of many plasmas with different characteristics (see review by Garrett, 1979). The base of this region is the ionosphere; at higher altitudes the inner magnetosphere includes the plasmasphere, the geomagnetically-trapped Van Allen particles, and the ring current. The major sources of environmental modification in this region will be the transportation systems used for lifting payloads into low Earth orbit and for interorbital transfer. Ionospheric modification by the rocket exhaust from launch vehicles has been considered by Mendillo et al. (1980). They show that large depletions can occur in the F-region ionosphere and that alteration of plasmaspheric content and modification of the conjugate ionosphere may result. The alteration of the plasmasphere by argon ions from electric propulsion systems already has been evaluated (Chiu et al., 1979; Curtis and Grebowsky, 1979). They find that the transport of Solar Power Satellite (SPS) systems to geosynchronous orbit results in deposition of enough argon to substantially alter the thermal structure and composition of the plasmasphere. An important potential modification of the inner magnetosphere involves the geomagnetically-trapped Van Allen radiation (Chiu et al., 1979). The

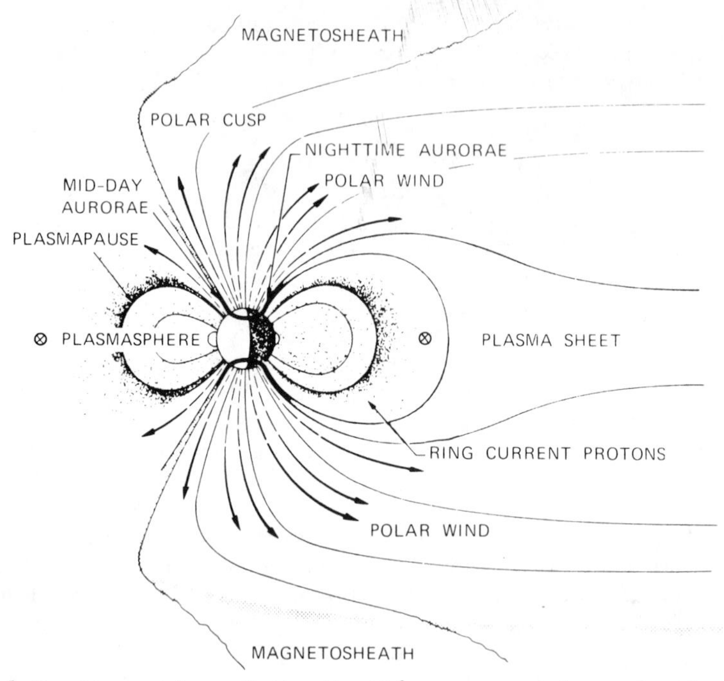

Fig. 1 Configuration of the Earth's magnetosphere in the noon-midnight meridian plane. (Geosynchronous orbit is indicated by the symbol ⊠.)

addition of heavy ions, such as argon, may stabilize the radiation belts and perhaps double the trapped radiation energy density. In contrast, the addition of light ions may precipitate the trapped Van Allen radiation. Such precipitation could have a significant effect on the terrestrial atmosphere and ionosphere, at least on a localized basis.

In the outer magnetosphere the plasma is not trapped on dipolar field lines, and plasma motion is determined by the ambient electric fields that change in response to variations in the solar wind parameters. The effects of gas releases in this region have been examined by Vondrak (1977) and Garrett and Forbes (1979).

This paper is organized such that the next section describes the expected sources of released gases from various space systems. Alteration of the magnetosphere by such gas releases is considered in the following sections. The last two sections identify some practical implications of modification of the magnetospheric environment and describe possible research studies and experiments that could improve our understanding of magnetospheric modification.

Gas Releases from Large Space Systems

In the present context, the term gas release is applied to any nonparticulate matter that is released during the deployment or operation of a space system. This includes exhaust products from the space transportation or attitude control system as well as gases and ions emitted from the spacecraft structure by degassing, erosion, or photoelectron emission. Further, the spacecraft structure can effectively act as a negative source (sink) of gases by absorbing the ambient particles. The sources and expected magnitudes of gas releases are summarized in Table 1 for various space systems. Some of these sources are discussed in more detail in this section.

The largest quantities of gas are released during the lift of space systems into Earth orbit. Fortunately, most of the rocket exhaust products are deposited at low altitudes where the atmosphere is dense. However, large space transportation systems do deposit significant quantities of exhaust, principally H_2O, H_2, and CO_2, into the ionosphere (Mendillo, 1980). Water is not normally present at these altitudes, and its chemical reactivity with the ionized components of the atmosphere results in a significant depletion of the ionosphere. For example, during the Skylab launch in 1973, exhausts from the Saturn V resulted in a 50% decrease in the total number of electrons within a 1000-km radius of the rocket launch path (Mendillo et al., 1975). Each flight of the Heavy-Lift Launch Vehicle (HLLV), designed for use in the postshuttle era, injects substantially more water into the ionosphere than the Skylab launch, but at a lower altitude. Zinn and coworkers (1979) estimate that F-layer depletion from each HLLV launch will be significant (a factor of four reduction) but not as pronounced as in the Skylab case.

Interorbital transfer deposits either rocket exhaust or energetic ions into the inner magnetosphere. The quantities listed in Table 1 are for the Cargo Orbital Transfer Vehicle (COTV) and Personnel Orbital Transfer Vehicle (POTV) designed for support of the SPS program. For this system, Chiu et al. (1979) have shown that the argon ions are substantial additions to the natural plasmasphere which has a total mass of only 10^6kg and a total energy content of 10^{31} eV. In contrast, the geomagnetically-trapped Van Allen radiation has a large energy content but is insignificant in terms of total mass.

Another type of interorbital transfer system that has been proposed is the mass-driver, an electromagnetic propulsion device that converts electrical energy into the kinetic

Table 1 Gas releases from large space structures

Source	Example	Emission Size[a]	Region
Lift to Earth orbit	HLLV	2×10^5 kg (H_2O)	Ionosphere (E-region)
	HLLV	3×10^4 kg (H_2O)	Ionosphere (F-region)
Interorbital transfer	COTV	1×10^5 kg (Ar^+)	Plasmasphere
	POTV	4×10^5 kg (H_2O)	Plasmasphere
Spacecraft systems	ACS(SPS)	2 gm/sec (Ar^+)	Geosynchronous orbit
Spacecraft structure	Degassing	10^{-2} gm/cm^2 -sec	Geosynchronous orbit
	Photoelectrons	3×10^{13} el/m^2 -sec	Geosynchronous orbit
	Micrometeorite erosion	10^{-14} gm/cm^2 -sec	Geosynchronous orbit
	Ion sputtering	10^{-17} gm/cm^2 -sec	Solar wind

[a]For propulsion systems, emission size is the total amount released in the indicated region. For spacecraft structure, the emission size is the rate per spacecraft surface area.

energy of high-velocity reaction mass (O'Neill, 1977). The reaction mass would be in the form of solid pellets, powder, or even liquid oxygen. Transfer of a 100,000 ton payload (equivalent to one SPS) from low Earth orbit to geosynchronous orbit would require the ejection of about 223,000 tons of reaction mass during a round trip time of 83 days. Although possible environmental effects have been noted in a preliminary manner (Vondark, 1977), no detailed assessment has yet been made.

The satellite structure itself is a source of gases in four ways: 1) exhaust gases or ions emitted by the attitude control system (ACS), 2) gas leakage, 3) surface erosion by meteoritic ablation and ion sputtering, and 4) cold photoelectron emission.

Gas leakage rates from large space structures are difficult to estimate. The Gemini and Apollo space capsules had passive leakage rates of 10 and 30 mg/sec, respectively (Kovar et al., 1969). These leakage rates were due to continuous diffusion through the cabin walls and outgassing through small leaks and do not include gas release from thruster firings and cabin ventings. These and Skylab leakage measurements indicate a minimum passive leakage rate of about 0.2 mg/m^2-sec for current-technology spacecraft. Improved sealing technology and system design during the next 30 years are expected to reduce the rate by not more than a factor of ten (Nishioka et al., 1973).

Attitude control systems for geosynchronous spacecraft currently use chemical rockets, and ion engines are planned for future systems. The baseline SPS system emits 55,000 kg of Ar^+ per year for each spacecraft. This is equivalent to a mean rate of 2 gm/sec. During eclipse periods of about 1 hr duration for 60 days each year, chemical rockets will burn at an average rate of about 10 gm/sec. Thus, Ar^+ is the most important contaminant to be expected from planned large space systems.

Meteoritic impact and energetic ion sputtering of spacecraft surfaces are another source of gases and particulate material. Solar wind sputtering of surface material occurs at the rate of about 0.04 Å/year (McDonnell et al., 1974), and micrometeorite ablation weathers the surface at the rate of about 10 Å/year (Thomas, 1974). The micrometeorite abrasion rate corresponds to approximately 500 gm/year for each km^2 of spacecraft surface area. This probably is not significant relative to other potential gas sources.

Satellites at geosynchronous orbit will emit photoelectrons of low energy (~10eV), in comparison to the ambient electron energies of typically 1 - 10 keV. Garrett and De Forest (1979) estimate the flux of photoelectrons that actually escape the spacecraft surface to be approximately 4×10^{-6} A/m^2. In addition to emitting these cold electrons, the spacecraft also is a negative source, or sink, of the ambient plasma in the magnetosphere. This occurs in two ways:

1) The structure can absorb particles that strike it (flux tube blockage). This is analogous to the sweeping of the Jovian radiation belts by the Galilean satellites (Thomsen, 1979). In the outer magnetosphere this is probably unimportant because the moon has a cross-sectional area of about 10^{17} cm^2 and yet no significant geophysical effects are known to result from its presence in the magnetosphere.

2) The satellite structure will charge, probably to a negative potential, and the ambient ion population will be accelerated to it.

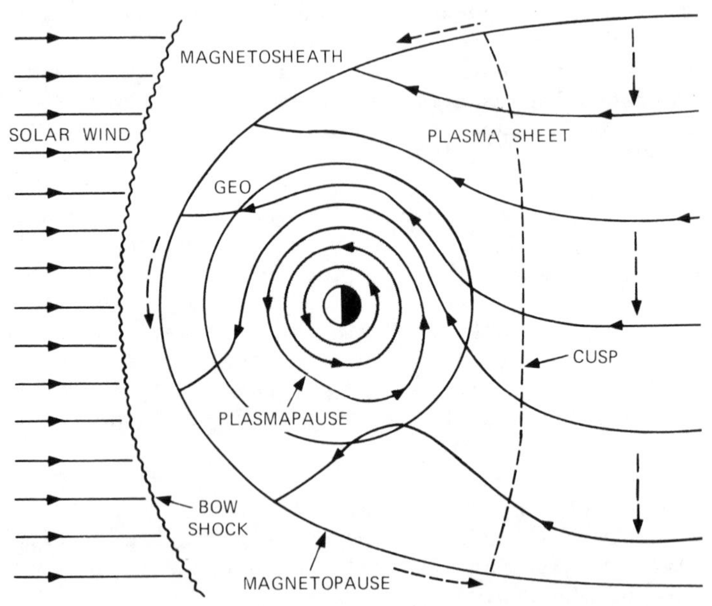

Fig. 2 Flow patterns of magnetospheric plasma in the equatorial plane. (Dashed arrows indicate electric-current directions. The location of geosynchronous orbit is labeled GEO. The plasmasphere (shaded) is a region of dense plasma corotating with the Earth. CUSP is the approximate limit of dipolar field lines.)

Fate of Released Gases

When gases are released at high altitudes in the magnetosphere, their motion is determined by whether they are neutral gases or ionized gases. Ionized gases flow in a convection pattern that is determined by the ambient electric and magnetic fields (see Fig. 2). Neutral gases released in Earth orbit will follow a trajectory determined by their initial velocity. If the thermal velocity of such gases is less than or comparable to the orbital velocity at the release point, then a ring of gases will be formed surrounding the Earth. Numerically, the criterion for gravitational trapping is

$$T \lesssim 2540\ m/r_o$$

where T is the gas temperature in K, m is the atomic or molecular weight, and r_o is the orbital distance in earth radii. At 300 K all gases except hydrogen and helium have a thermal velocity much smaller than the geosynchronous orbital velocity.

The time evolution of neutral gas clouds released by orbiting satellites is critically dependent upon the ionization lifetime of the gases. Prior to ionization, the motion is determined by gravitational forces; after ionization, plasma convection determines the subsequent motion. The ionization lifetime depends upon the location of release, because the total ionization rate is the sum of photoionization, electron impact, proton impact, and proton charge exchange. For example, the ionization lifetime of molecular nitrogen is approximately 21 days in the solar wind, but is much less within the magnetosphere where electron impact ionization is more important. A preliminary assessment of the motion of gas clouds released at geosynchronous orbit has been made by Garrett and Forbes (1979).

Modification of the Outer Magnetosphere

Once the gas release rate has been evaluated, the effects on the magnetosphere can be assessed by considering the contaminant abundance relative to the ambient plasma density. For example, the inner magnetosphere is characterized by dipolar magnetic field lines that co-rotate with the Earth and by a trapped plasma population. Within this region (known as the plasmasphere) the plasma temperature is low, but the equatorial density is high (10^3 to 10^5 cm^{-3}). In contrast, the plasma sheet in the outer magnetosphere has a small (≤ 2 cm^{-3}) plasma density, and plasma motion is controlled by both magnetic and electric fields. It is a dynamic region with rapid temporal variations (auroral substorms and geomagnetic storms) that are a magnetospheric response to changing solar wind conditions.

The ionized plasma resulting from gas clouds released in the vicinity of geosynchronous orbit may substantially modify the normal plasma sheet flow, at least on a local scale. In the plasma sheet the plasma flows towards the Earth under the influence of the cross-tail magnetospheric electric field. This flow is tenuous, consisting of about 1 proton/cm^3 moving earthward at a speed of 100 km/sec. This amounts to a flux of approximately 10^7 protons/cm^2-sec or 10^{-17} gm/cm^2-sec. The total mass flux integrated over the entire geomagnetic tail is about 1 kg/sec. The SPS attitude control system emits about 2 gm/sec; it is a substantial addition to plasma sheet flow over a scale length of about 4,000 km. Because the plasma sheet is the direct source of the aurora (and indirectly the source of the midlatitude ionospheric trough), this modification of the plasma sheet may have a significant effect on the terrestrial ionosphere. For example, disruption of the current in the geomagnetic tail is thought to be a trigger of auroral substorms. Artificial modification of this tail current by the addition of contaminant plasma may initiate or even inhibit auroral activity. Furthermore, any currents that may flow between the released ion cloud and the terrestrial ionosphere, as shown in Fig. 3, will result in heating of the high-latitude thermosphere.

Gases released at altitudes higher than geosynchronous orbit will spread great distances as neutrals, because electron impact ionization is smaller there than it is in the inner magnetosphere. Molecular oxygen or nitrogen released at lunar distance at a temperature of 300 K will form a ring (or torus) extending from 8 R_e (Earth radii) out to 77 R_e, with a thickness of 60 R_e perpendicular to the lunar orbital plane. The existence of an analogous ring around Saturn has been postulated by McDonough and Brice (1973). A similar structure around Jupiter, due to the escape of gases from the satellite Io, was observed by Pioneer 10 (Carlson and Judge, 1975). The lifetime of the gases will be nearly equal to their photo-ionization lifetime (10^6-10^7 sec), because once they are ionized the particles are lost to the solar wind. The moon will sweep up some of these gases, but since their ionization lifetime is comparable to their orbital period, such loss probably will not be significant. The mean equilibrium density is approximately equal to the product of the gas source rate times the ionization lifetime divided by the torus volume. Greater densities will, of course, occur near the release point.

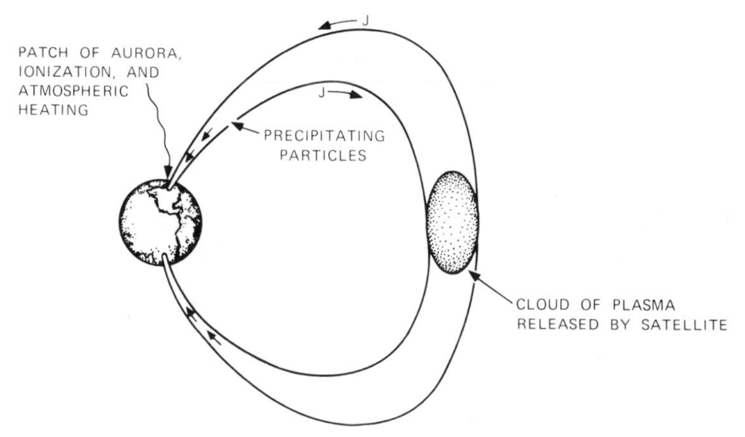

Fig. 3 Schematic view of the coupling between a gas release at satellite altitude and the high-latitude ionosphere.

Gas clouds released at geosynchronous and higher altitudes will pass, at least occasionally, into the solar wind and magnetosheath. The magnetosphere is a cavity formed in the solar wind because the solar wind dynamic pressure is less than the pressure of the Earth's magnetic field. However, a diversion of the solar wind also can be accomplished by a mass loading of the solar wind flow, as in the case of Venus which has a substantial ionosphere and lacks a strong magnetic field. The solar wind mass flow is 10^{-15} gm/cm^2-sec and will be diverted if the ion mass pick-up rate exceeds this amount. The entire frontal cross-sectional area of the magnetosphere is about 5×10^{19} cm^2, so that the total solar wind mass flux incident on the magnetosphere is only about 50 kg/sec. When ionized gases are deposited at a larger rate in the solar wind ahead of the magnetosphere, a substantial alteration of magnetospheric structure is expected (Fig. 4). If the ionized gases originate from photoionization of a torus of neutral gases, then diversion occurs when

$$(\rho v)_{sw} \lesssim n\ m\ L/\tau$$

where ρ and v are the solar wind density and velocity, n and m are the gas torus density and molecular mass, L is the torus thickness and τ is the photoionization lifetime. For a gas cloud at lunar distance, this occurs when the torus has an average density of about 10^{-20} gm/cm^3. This density would be the equilibrium value resulting from a neutral gas release rate of about 600 kg/s. If we assume that the gas source is passive leakage from pressurized space structures, then the total surface area of all structures should not exceed 3000 km^2. This is a very large surface area, although habitats of that size

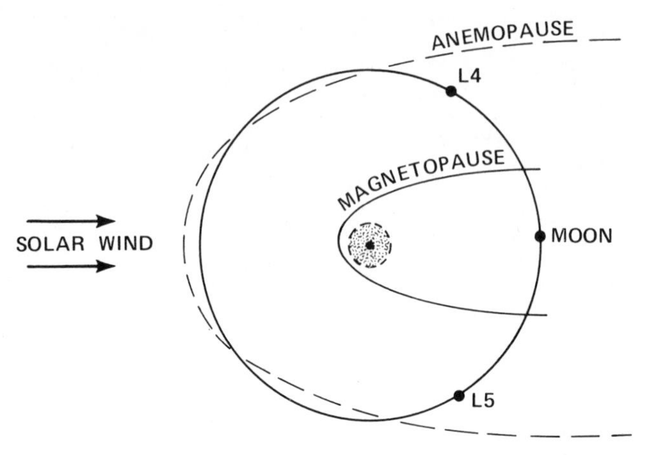

Fig. 4 Artificial anemopause formed by the release of gases at lunar distance. (At present the magnetopause is the location where solar wind flow is diverted by the geomagnetic field.)

have been proposed for deployment at lunar distance (O'Neill 1974, 1976). The upper limit for spacecraft deployment in cislunar space is probably less because gas sources other than passive leakage are expected.

Implications of Magnetospheric Modification

The processes described in the previous sections indicate that the deployment and operation of large space structures in high Earth orbit can modify the natural aerospace environment, at least on a local and possibly on a global scale. The practical implications of such modifications have not yet been evaluated fully. Some potential consequences that have been identified are listed in Table 2 and include the following:

1) Alteration of the Van Allen radiation belts will either increase or decrease the radiation exposure experienced by satellites in this region, depending upon the atomic mass of the released gases. If the addition of heavy ions increases the energy density of the radiation belts, satellites will require additional shielding. On the other hand, if the radiation belts are precipitated, less satellite damage will occur.

2) Scintillations produced by irregularities associated with directed plasma beams at high altitudes, such as ion electric propulsion systems, may adversely affect communication systems that use transionospheric propagation by satellite relay.

Table 2 Implications of environmental modification

Effect	Cause	User Relevance
Satellite radiation dosage	Alteration of Van Allen belts	Satellite shielding requirements
Ionospheric modification	Particle precipitation	Communication systems (reflected propagation)
Scintillation production	Magnetospheric plasma beams	Communication systems (transionospheric)
Thermospheric heating	Particle precipitation and joule heating	Satellite drag

3) The ionosphere, particularly at high latitudes, will be modified by particle precipitation. Increased auroral particle precipitation will affect communication systems that use ionospheric reflection techniques. Energetic particle precipitation currently affects communication systems by increasing high frequency absorption or altering very low frequency propagation characteristics.

4) Increased particle precipitation and intensification of high-latitude current systems will lead to thermospheric heating. Such heating at high latitudes causes the thermosphere to expand upwards and equatorwards. At a given altitude thermospheric density is increased, resulting in increased drag for low-altitude satellites and reduced accuracy of satellite ephemeris predictions.

Not included in this list, because of its speculative nature, is possible alteration of weather and climate by modification of the way in which the Earth's magnetosphere couples to the solar wind plasma. There is growing evidence that solar variations affect weather and climate (see review by Herman and Goldberg, 1978). Unfortunately, our understanding of these putative sun-weather links is so incomplete that it is impossible, at present, to estimate whether expected magnetospheric modification would have any significant effect on Earth's weather and climate.

Discussion

A definitive evaluation of the magnetospheric effects of large space systems is not easy because most regions are not well surveyed, and many of the physical processes are not understood. Measurements and improvements in magnetospheric models are required before any detailed analyses can be made. Of particular importance are experimental surveys of the natural plasma populations, electric and magnetic fields, and current systems that are present in the inner and outer magnetosphere. The dynamics of these regions, such as natural transport processes and wave/particle interactions, need to be measured accurately. Finally, there is a need for measurements and modeling of the atmospheric and ionospheric response to changes in the magnetosphere during geomagnetic storms and auroral substorms. Of particular importance are variations in the ionospheric electric fields, conductivity, and currents. Increased understanding of magnetic storms and auroral substorms will be important because these events result in the creation of plasma enhancements. Such naturally-occurring phenomena may be useful analogs from which one can assess the effects of artificial plasma injections.

Although measurements and modeling of the natural environment will be very useful, as long as knowledge of fundamental processes is still incomplete, progress in some areas will require active experiments. One of the few active releases at high altitudes was the U.S.-German cooperative experiment that released a 16 kg barium cloud at an altitude of 5 R_e in the equatorial plane (Brence et al., 1973). The ion cloud was observed to striate and to drift in the direction expected for the normal plasma flow at the location. Although no high-altitude magnetic or ionospheric disturbances were detected, at the time of release the ionosphere was recovering from a geomagnetic storm that may have obscured any effects produced by the barium release (Davis et al., 1973). Newer experiments planned for the future include FIREWHEEL (Haerendel et al., 1978) that will inject in the geomagnetic tail 160 kg of barium at 9.5 R_e and 40 kg of lithium at 7 R_e. These releases will serve as tracers of natural ion motions and may result in long-range perturbation of the magnetosphere and, possibly, the high-latitude ionosphere. Such experiments will help to identify magnetospheric processes and to evaluate the magnetospheric and ionospheric response to artificial plasma injection.

In summary, it appears certain that large space structures will alter the natural environment. However, it should be kept in mind that modification is not necessarily a deterioration of environmental quality. For example, the operation and reliability of space systems would benefit from the elimination of the Van Allen belts. In addition, magnetospheric modifications that increase the predictability of ionospheric variability could be very beneficial for users of long-distance communication systems. If there are harmful effects of any environmental modifications, early identification of their impact will allow an opportunity to design mitigation techniques. Finally, the economic benefits of space systems need to be compared to the cost of environmental degradation, and environmental effects of space systems need to be compared to those resulting from use of alternative Earth-based systems.

References

Basler, R., Johnson, G., and Vondrak, R., "Antenna Concepts for Interstellar Search Systems," Radio Science, Vol. 12, Sept. 1977, pp. 845-858.

Bekey, I., "Big Comsats for Big Jobs at Low User Cost," Astronautics & Aeronautics, Vol. 17, Feb. 1979, pp. 42-56.

Brence, W.A., Carr, R., Gerlach, J., and Neuss, H., "NASA/MPE Barium Cloud Project," Journal of Geophysical Research, Vol. 78, Sept. 1973, pp. 5726-5731.

Carlson, R.W., and Judge, D.L., "Pioneer 10 Ultraviolet Photometer Observations of the Jovian Hydrogen Torus: the Angular Distribution," Icarus, Vol. 24, 1975, pp. 395.

Chiu, Y.T., Cornwall, J.M., Luhmann, T.G., and Schulz, M., "Argon Ion Contamination of the Plasmasphere," Published elsewhere in this volume.

Chiu, Y., Luhmann, J., Ching, B., Schulz, M., and Boucher, D., "Magnetospheric and Ionospheric Impact of Large-Scale Space Transportation with Ion Engines," Space Sciences Lab., Rept. No. SSL 78 (3960-04)-3, to be published in Astronautics & Aeronautics, AIAA.

Curtis, S., and Grebowsky, J., "Changes in the Terrestrial Atmosphere-Ionosphere-Magnetosphere System Due to Ion Propulsion for Solar Power Satellite Placement," Journal of Geophysical Research (submitted for publication), 1979.

Davis, T., Stanley, G., and Boyd, J., "Geophysical Disturbance Environment During the NASA/MPE Barium Release at 5 Re on September 21, 1979," Journal of Geophysical Research, Vol. 78, Sept. 1973, pp. 5732-5735.

Forbes, J.M., "Upper Atmosphere Modifications Due to Chronic Discharges of Water Vapor from Space Launch Vehicle, published elsewhere in this volume.

Garrett, H.B., "Review of Quantitative Models of the 0-to 100-keV Near-Earth Plasma," Reviews of Geophysics and Space Physics, Vol. 17, May 1979, pp. 397-417.

Garrett, H.B., and DeForest, S., "Effects of a Time-Varying Photoelectron Flux on Spacecraft Potential," Journal of Geophysical Research (submitted for publication), 1979.

Garrett, H.B., and Forbes, J.M., "Time Evolution of Ion Contaminant Clouds at Geosynchronous Orbit," Geophysical Research Letters (submitted for publication), 1979.

Glaser, P.E., "Power from the Sun: Its Future, Science, Vol. 162, November 1968, pp. 857-886.

Haerendel, G., Hausler, B., Syocker, J., Hippman, H., and Paschmann, G., "Feurrad-Firewheel," Max-Planck-Institute Report, 1978.

Herman, R.R., and Goldberg, R.A., "Sun, Weather, and Climate," NASA SP-426, Washington, 1978.

Kovar, N., Kovar, R., and Bonner, G., "Light Scattering by Manned Spacecraft Atmospheres," Planetary and Space Science, Vol. 17, Feb. 1969, pp. 143.

Mendillo, M., "Modification of the Ionosphere by Large Space Vehicles," published elsehwere in this volume.

Mendillo, M., Hawkins, G.S., and Klobuchar, J.A., "A Large-Scale Hole in the Ionosphere Caused by the Launch of Skylab," Science, Vol. 187, 1975, pp. 343-345.

McDonnell, J.A.M., and Flavill, R.P., "Sputter Erosion on the Lunar Surface: Measurements and Features Under Simulated Solar He^+ Bombardment (Abstract)," Lunar Science-V, The Lunar Science Inst., Houston, 1974, pp. 478-479.

McDonough, T., and Brice, N., "A Saturnian Gas Ring and the Recycling of Titan's Atmosphere," Nature, Vol. 242, 1973, pp. 513-515.

Nishioka, K., Arno, R., Alexander, A., and Slye, R., "Feasibility of Mining Lunar Resources for Earth Use: Circa 2000 A.D.," NASA TMX-62267, Aug. 1973.

O'Neill, G.K., "The Colonization of Space," Physics Today, Vol. 27, Sept. 1974, pp. 32-38.

O'Neill, G.K., "The High Frontier: Human Colonies in Space," William Morrow and Co., New York, 1976.

O'Neill, G.K., "Mass Driver Reaction Engine as a Shuttle Upper Stage," Space Manufacturing Facilities, Vol. 2, edited by J. Grey, AIAA, New York, 1977.

Rote, D.M., "Environmental Effects of Space Systems," published elsewhere in this volume.

Thomas, G.E., "Mercury: Does Its Atmosphere Contain Water?" Science, Vol. 183, 1974, pp. 1197-1198.

Thomsen, M.F., "Jovian Magnetosphere-Satellite Interactions: Aspects of Energetic Charged Particle Loss," Reviews of Geophysics and Space Physics, Vol. 17, May 1979, pp. 369-379.

Vondrak, R.R., "Environmental Impact of Space Manufacturing," Space Manufacturing Facilities, Vol. 2, edited by J. Grey, AIAA, New York, 1977.

Zinn, J., Sutherland, C., and Pongratz, M., "Effects of Rocket Exhaust Products in the Thermosphere and Ionosphere," Los Alamos Scientific Laboratory Rept. LA-7926-MS, 1979.

Chapter II—Spacecraft Charging Interactions

SPACECRAFT CHARGING: A REVIEW

H. B. Garrett*
Air Force Geophysics Laboratory, Hanscom Air Force Base, Mass.

Abstract

The process of charge buildup on satellite surfaces is reviewed. In particular, the types of charging processes, the different charging models, and the effects of charging are described in a simplified manner in order to prepare the reader for the more detailed studies presented in other sections of this volume. Special emphasis is placed on fundamental concepts and on the space environment.

I. Introduction

Langmuir[1] published in 1924 what has remained the standard work on the problem of current flow to an object immersed in a plasma. The decades since his early work on plasma probes have seen the birth of the space age and the applications of his theories to complex satellite systems. Just as the need for more complex satellites has progressed, so the need for more detailed models of the spacecraft charging process has grown. The simple current-balance, single-probe models of Langmuir have given way to complex, circuit element descriptions and, ultimately, sophisticated programs capable of self-consistently calculating the trajectories of individual particles in the sheath surrounding a satellite.

The purpose of this review will be to present the basic principles of the charging process and to order the different types of processes and charging models into a coherent picture of the overall field of spacecraft charging. As many of the subjects are covered in detail in other sections of this volume, the review will provide only a brief description of the various aspects of charging. Where possible, simple examples of the basic charging phenomena will be advanced in order to illustrate specific points. In the first section the sources of currents to and from a spacecraft surface will be presented. In the next section, models of the space environment respon-

---

* Planetary Space Physicist.

sible for these currents (i.e., the 0-100 keV plasma) will be detailed. In the third section, the various types of spacecraft charging models will be introduced within the framework of the current sources and corresponding space environment, and specific examples will be described. The review will conclude with a brief discussion of the effects of charge buildup and means of mitigating the worst aspects of the problem.

II. Current Sources

The problems associated with charge buildup on spacecraft surfaces are, in general, the results of nature's attempt to bring about current and charge balance. The potential that develops between a spacecraft and the ambient medium is the result of current balance, whereas the arcing phenomenon is the result of charge balance. In this section, seven major charge (or current) sources will be described. The two most important sources to be described are ambient plasma fluxes and photoelectron emission due to sunlight. Closely related are the secondary and backscattered electron fluxes associated with impacting electrons and ions (secondary ion emission is insignificant). Currents generated by the movement of a structure across the ambient magnetic field or relative to the ambient ion population and by high energy (~ 10 keV and greater) electrons which deposit charge inside insulating surfaces are additional, more subtle sources. Finally, man-made sources such as plasma beams from ion thrusters or induced current flows to exposed surfaces with high potentials, which artificially enhance local charging rates, are also considered.

The major natural source of potentials of 10 kV or higher on satellite surfaces is the ambient space plasma. Although the actual space plasma is seldom representable in terms of a single temperature and density,[2] the Maxwell-Boltzmann distribution function is a useful starting point for describing the ambient plasma conditions that generate these large potentials. The distribution function (or probability density that a particle will have a certain velocity) f is, for an isotropic Maxwell-Boltzmann plasma

$$f\,(V_i) = n_i \left(\frac{m_i}{2\pi kT_i}\right)^{3/2} e^{-m_i v_i^2/2kT_i} \tag{1}$$

where

n_i = number density of species i
m_i = mass of species i
T_i = temperature of species i
v_i = velocity of species i
k = Boltzmann constant

The current flux to a surface is given by[3]

$$J_{i0} = q_i \int_0^{\infty} \vec{v}_i \cdot \vec{n}\, f d^3v = \frac{q_i n_i}{2} \left(\frac{2kT_i}{\pi m_i} \right)^{\frac{1}{2}} \quad (2)$$

where

J_{i0} = current density per unit area for 0 potential
$\vec{n}$ = unit normal to surface
q_i = charge on species i
d^3v = volume element in velocity space

In many cases where the ambient plasma is not well represented by a single Maxwell-Boltzmann distribution, a sum of two or more such distributions has proven useful in calculating the actual current to a spacecraft surface. Also, as will be shown in a subsequent section, when the effects of spacecraft potential and plasma anisotropies are considered, the simple distribution function in Eq. (1) is no longer valid, and the integration of Eq. (2) becomes difficult. In general, however, Eq. (2) (using a sum of two components or a slight modification of it) is accurate for most purposes. In the next section, appropriate values of T and n are tabulated.

The photoelectron current from a surface is a complex function of satellite material, solar flux, solar incidence angle, and satellite potential. In Figs. 1a and 1b, adapted from Grard,[4] are plots of the four functions necessary to describe the photoelectron current. In Fig. 1a is plotted the solar flux S(E) as a function of energy E (or wavelength). The exact details of the spectrum vary slightly with solar activity and can vary greatly if the sunlight reaching the spacecraft is severely attenuated by the atmosphere. Also shown in Fig. 1a are the electron yield per photon, W(E), and the total photoelectron yield H(E) [= W(E)S(E)], as a function of energy for aluminum oxide.[4] The total current for zero or negative satellite potential is given by

$$J_{PHO} = \int_0^{\infty} W(E)\, S(E)\, dE = \int_0^{\infty} H(E)\, dE \quad (3)$$

J_{PHO} is tabulated in Table 1 for a variety of materials.[4]

If the satellite is positively charged, the ambient photoelectron current is attracted to the surface. Since this return current is a function of potential and geometry, in order to accurately calculate it, the energy spectrum of the electrons for a given incident monochromatic photon must be known. Grard[4] has carried out these calculations for a variety of materials and different probe geometries. (See also

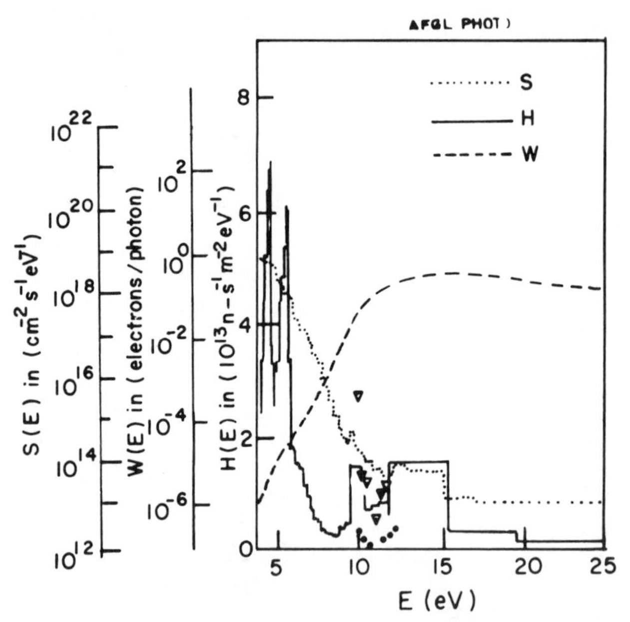

Fig. 1a Composite plot of W(E)(electron yield per photon), S(E) (solar flux), and their product H(E) (total photoelectron yield), as functions of energy E, for aluminum oxide; from Ref. 4.

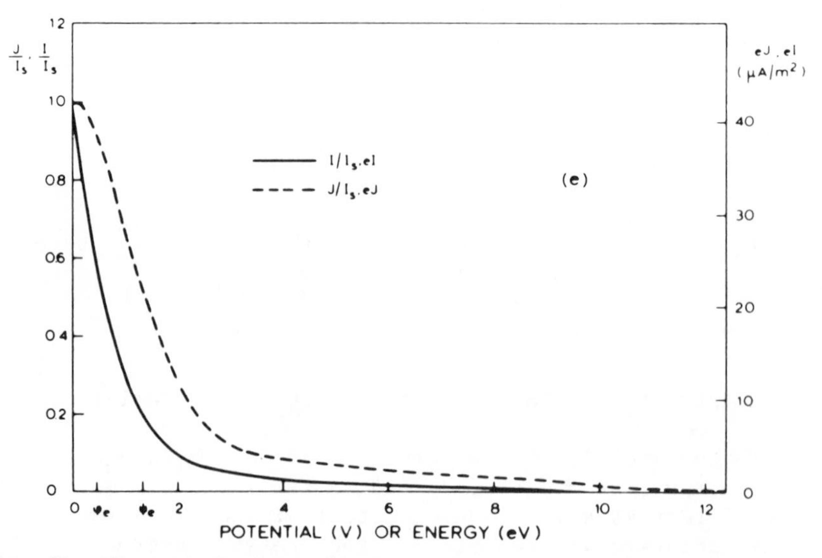

Fig. 1b The total photoelectron current as a function of positive satellite potential for planar (I) and point (J) probes; from Ref. 4.

Table 1 Photoelectron emission characteristics[a]

Material	Work function, eV	Saturation Flux, 10^{12} $n_e/sec\text{-}m^2$	Saturation Current Density, $\mu A/m^2$
Aluminum oxide	3.9	260	42
Indium oxide	4.8	190	30
Gold	4.8	180	29
Stainless Steel	4.4	120	20
Aquadag	4.6	110	18
LiF on Au	4.4	90	15
Vitreous carbon	4.8	80	13
Graphite	4.7	25	4

[a]From Ref. 4

Ref. 5) His results for aluminum oxide are plotted in Fig. 1b for a planar and a point probe as functions of positive probe potential.

As a satellite passes into the Earth's shadow, the sunlight is severely attenuated, greatly complicating the calculation of the photoelectron flux. An approximate attenuation curve has been derived, however, for estimating this flux[6]:

$$J_{PH}(X) = J_{PHO}\, e^{-e^{-(X-Z_o)/\delta Z}} \tag{4}$$

where

$J_{PH}(X)$ = photoelectron current density due to ray of light passing through atmosphere at height X

X = height above Earth's surface of ray path from satellite to a point on the solar disk

Z_o, δZ = experimentally determined parameters ($Z_o \sim 90$ km, $\delta Z \sim 40$ km)[6]

Equation (4) must be integrated over the entire solar disk, as seen from the satellite, to obtain the total photoelectron current density, J_{PH}.[6] The effects of this variation in photoelectron flux during eclipse passage will be discussed later.

Much as photons cause the emission of photoelectrons, the impact of ambient electrons and ions on a spacecraft surface generates backscattered and secondary electrons (backscattered and secondary ion fluxes are insignificant). Backscattered electrons are those ambient electrons reflected back from the surface (usually with some energy loss) while secondary electrons are electrons emitted as a result of energy deposition by incident electrons or ions. Each type has a characteristic emission spectrum.

The equation for the current density due to secondary emission, assuming an isotropic flux (and ignoring other angular variations) is[5]

$$J_{Si} = \frac{2\pi q_i}{m_i^2} \int_0^\infty dE' \int_0^\infty g_i(E',E) \quad \delta_i(E)\, f_i(E)\, dE \qquad (5)$$

where J_{Si} = secondary electron flux due to incident species i (usually assumed to be electrons e^- or protons H^+)

g_i = emission spectrum of secondary electrons due to incident species i of energy E

δ_i = secondary electron yield due to incident species i of energy E

E' = secondary electron energy

f_i = distribution function of incident particles at surface

Typical curves for δ_i for e^- and H^+ impacting on aluminum[5] are plotted in Figs. 2a and 2b. (More recent curves are available in Ref. 7). The function g_i is nearly independent of incident energy and incident particle species. The normalized curve[5] for aluminum is plotted in Fig. 2c and can be assumed for all incident particles.

In general, the integration of Eq. (5), even for a Maxwell-Boltzmann distribution, must be carried out numerically. As a gross approximation, for a negatively charged surface the secondary flux due to electrons is 40% of the incident flux; for H^+ on Al the value is more like 80% to 300%.[6] Thus, secondary currents cannot be ignored for a negatively charged surface. (For a positively charged surface, the secondary fluxes are approximately 0 since the average energy of a secondary electron is about 2 eV and is therefore reattracted).

For backscattered electrons, the current is given by

$$J_{BSE} = \frac{2\pi q_e}{m_e^2} \int_0^\infty dE' \int_{E'}^\infty B(E',E)\, E \quad f_e(E)\, dE \qquad (6)$$

where $B(E,'E) = G\left(\frac{E'}{E}\right)\frac{1}{E}$

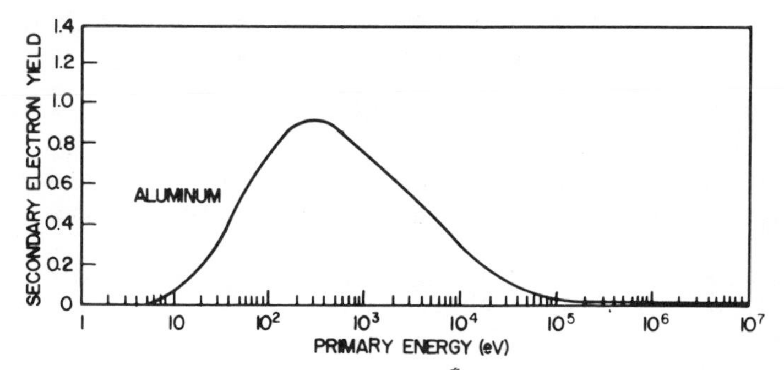

Fig. 2a Secondary electron yield δ_E, due to incident electrons of energy E impacting on aluminum; from Ref. 5.

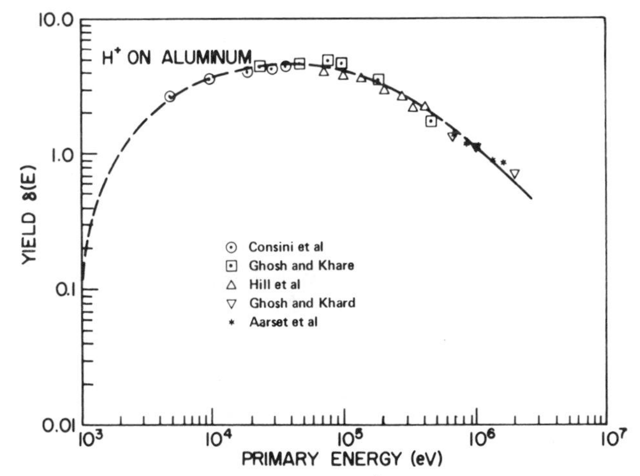

Fig. 2b Secondary electron yield δ_I, due to incident ions of energy E impacting on aluminum; from Ref. 5.

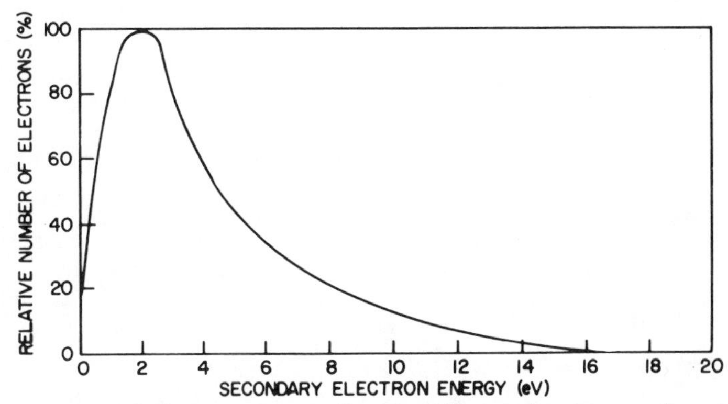

Fig. 2c Emission spectrum of secondary electrons due to incident electrons or ions impacting on aluminum; from Ref. 5.

G = percent of electrons scattered at fraction E'/E of the incident energy E

Sternglass[8] has published experimental measurements of backscatter parameters for various materials. A crude estimate of G is plotted in Fig. 3 for Al.[9] For negative potentials, the backscatter flux is roughly 20% of the incident flux.

For both secondary and backscattered currents, the actual values are dependent on angle of incidence which has been ignored in the preceding. Further, the secondary and backscatter properties of most materials are not well known. Currently, the lack of knowledge in this area is one of the major deficiencies in spacecraft-charging theory.

If a magnetic field is present, the electric field measured by an observer must be defined relative to a fixed coordinate system. That is, we can define an electric field $\bar{E}$ and magnetic field $\bar{B}$ in a given rest frame. In another reference frame moving at velocity $\bar{v}$ relative to the rest frame, the new electric and magnetic fields E' and $\bar{B}'$ are in a vacuum (i.e., components parallel to $\bar{v}$ remain unchanged):

$$\bar{E}' = \frac{\bar{E} + (\bar{v}/C) X \bar{B}}{\sqrt{1 - v^2/C^2}} \tag{7}$$

$$\bar{B}' = \frac{\bar{B} - (\bar{v}/C) X \bar{E}}{\sqrt{1 - v^2/C^2}}$$

where C = the speed of light

For all velocities we deal with, $v/C \ll 1$; then, including all components relative to $\bar{v}$:

$$\bar{E}' = \bar{E} + \left(\frac{\bar{v}}{C}\right) X \bar{B} \tag{8}$$

$$\bar{B}' = \bar{B}$$

This means that a satellite moving relative to a plasma (assumed to have $\bar{E} = 0$ in its rest frame) with velocity $\bar{v}_s$ will see an electric field in its rest frame given by

$$\bar{E}' = \frac{\bar{v}_s}{C} X \bar{B} \tag{9}$$

An Earth-orbiting satellite will see a maximum induced $\frac{\bar{v}}{C} X \bar{B}$

at low altitudes on the order of ~0.4 V/m. This electric field in the satellite rest frame can cause a variety of effects. Principally, however, it makes accurate electric field measurements difficult and, through ohmic loss due to the currents it drives, contributes to satellite drag and torque.[10,11] Of major concern here, however, is the induced electric field which leads to anisotropies in the satellite

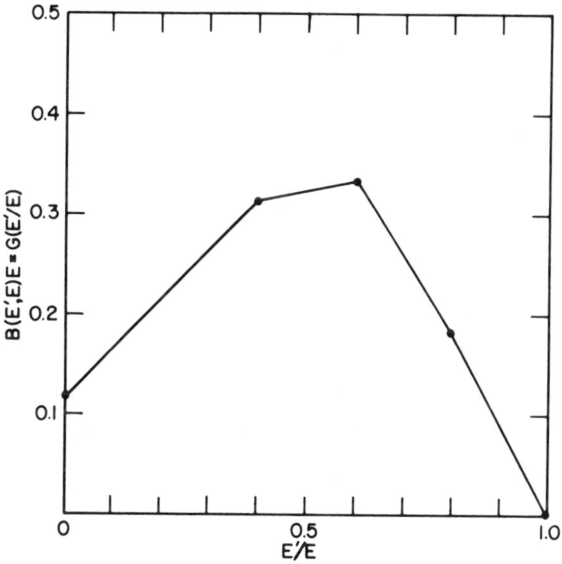

Fig. 3 G (E',E) as a function of (E'/E) where G is approximately the percentage of electrons scattered at a given energy E' as a result of an incident particle of energy E; from Ref.9.

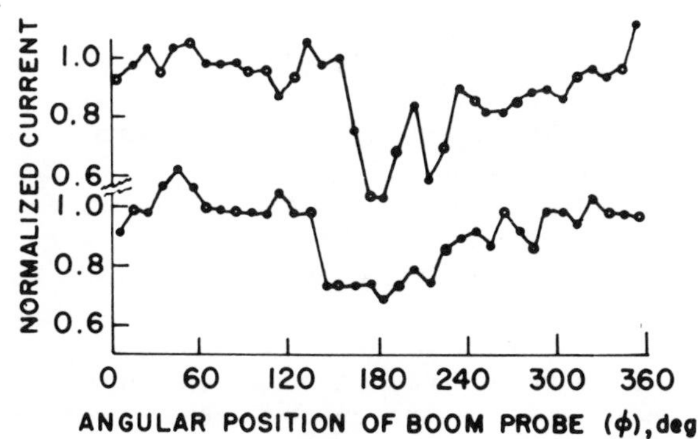

Fig. 4 Normalized electron current vs angular position of the plasma probe on Ariel 2; from Ref. 3. (The probe was approximately 150 cm from the center of the satellite; θ = 180 deg would correspond to the center of the satellite wake.)

sheath at low altitudes or for a large structure by perturbing the ambient currents flowing to its surface.

Much as a boat passing through water, a satellite can create a plasma wake. At an altitude below 800 km, a satellite has a velocity of ~ 7.5 km/sec decreasing to ~ 3 km/sec at geosynchronous orbit. Below ~ 4 R_E (Earth radii), in the equatorial plane, the plasma co-rotates with the Earth at a velocity of ~.5 km/sec increasing to 1-10 km/sec at geosynchronous orbit. The satellite and ambient plasma will therefore have a velocity relative to each other which can generate a wake. Fortunately, the electrons and ions also have a thermal velocity associated with them. At low altitudes, the average thermal velocity for electrons is about 600 km/sec, whereas it is only 10 km/sec for H^+. These increase to 60,000 km/sec and 1000 km/sec, respectively, at geosynchronous orbit where the ambient temperatures can be in excess of 10 keV. Thus, below 800 km wake effects may be significant for ions. These effects result in an asymmetry between the flux of ions to the leading edge and rear surface of the satellite. (The current difference is roughly proportional to the satellite-ion orbital velocity relative to each other.) Parker[12] has calculated the differential in surface voltages due to this asymmetry for a spacecraft with nonconducting surfaces. An example of wake structure at a distance of 150 cm behind the Ariel 2 satellite[13] is presented in Fig. 4. The effects of a wake are quite complex, and the reader is referred to Kasha,[10] Parker,[14] and Wildman[11] for details.

In a later section of this volume, Frederickson[15] considers the effects of charge deposition by energetic particles in dielectrics. Depending on the material's density and thickness, electrons of energy greater than 100 eV can penetrate it and deposit significant amounts of negative charge internally. If the electrons have MeV energies, then ~ MV/cm fields are necessary to significantly retard the incident flux. As Frederickson demonstrates, such fields are more than adequate to cause electrical breakdown. In Fig. 5 (from Frederickson's article) is plotted the potential versus depth in a PCV sample. Interestingly, grounding the surface can actually increase the potential gradient between the surface and interior of the sample.[15]

Finally, man-made sources of charging current are numerous and are being exploited to control spacecraft potentials. Examples are electron and ion beam systems which are capable of providing ~ keV particles at currents in excess of mA. Such systems have been flown on several satellites and rockets (ATS-5, ATS-6, GEOS I, GEOS II, P78-2, and ISEE-1 to name a

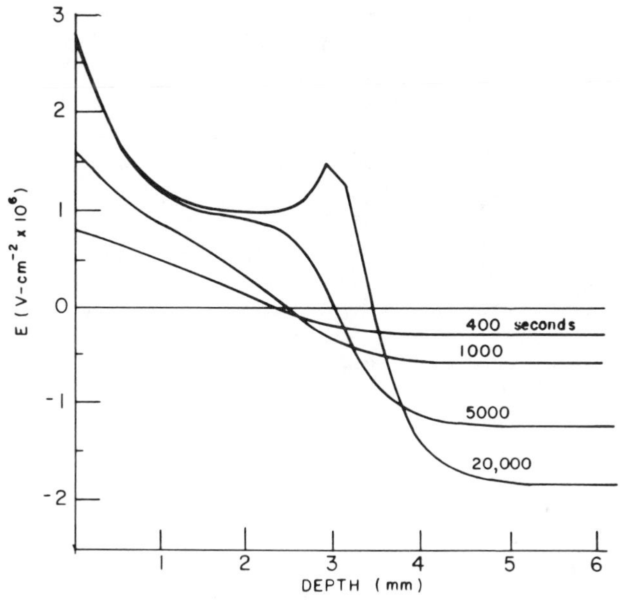

Fig. 5 The electric field measured at various depths in a dielectric as a function of exposure time to MeV electrons; from Ref. 15.

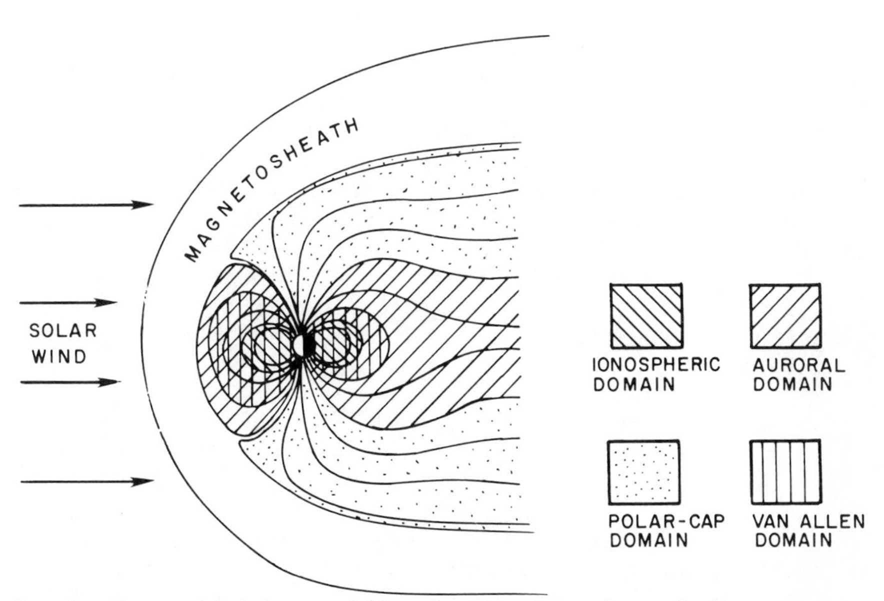

Fig. 6 Noon-midnight meridional cross section of the magnetosphere showing the approximate locations of the magnetospheric domains, from Ref. 24.

few). Several papers in this volume[16,17] treat the effects of these systems on the environment and their effects on satellite potential. Beam systems and any exposed surfaces having high potentials (such as the junctions between solar cells) alter the fields around a spacecraft and influence the current flow.[18-21] In some instances such field changes are capable of focusing ambient or beam particles on satellite surfaces as in the case of an electrostatic lens. Further, the ions from beam systems and other man-made contaminants, such as thrusters or outgassed materials from the satellite, are capable of supplying an artificial current source. More importantly, however, they may coat exposed surfaces, seriously altering material properties such as conductivity and thermal emissivity.[22]

III. The Ambient Environment

Of all the current sources considered in the following section, the most important for spacecraft charging calculations are the photoelectron and ambient fluxes. In this section, the fluxes to be expected in the Earth's magnetosphere (and to a lesser extent, those of Jupiter and Venus, and in the solar wind) will be discussed in terms of observations and models of the ambient environment. Garrett[23] reviewed the current status of our knowledge of the 0-100 keV near-Earth magnetosphere (the most important from a charging standpoint). In this section the main results of that study will be presented from the standpoint of spacecraft charging: the reader is referred to that review for a more detailed description. As in Garrett,[23] comments will be confined to the ionospheric and auroral domains as defined by Vasyliunas.[24] That is, only the low energy (0-100 keV) charged particle population in the plasmasphere and the near-Earth plasma sheet regions will be considered (see Fig. 6). Also, variations in ionic composition and pitch angle anisotropies will be ignored even though such variations are potentially important for spacecraft charge modeling. (See Young[25] for a treatment of ionic composition variations and DeForest[26] for a discussion of pitch angle anisotropies.)

Four types of quantitative models of the magnetospheric plasmas can be defined, depending upon the ratio of theoretical to empirical input to the model. The models of most use to the spacecraft charging community are those which consist of statistical compendiums of various parameters as functions of space, time, and geomagnetic activity. These statistical models require little theoretical input, relying primarily on actual measurements. Consideration of basic physical principles makes possible the derivation of analytic expressions

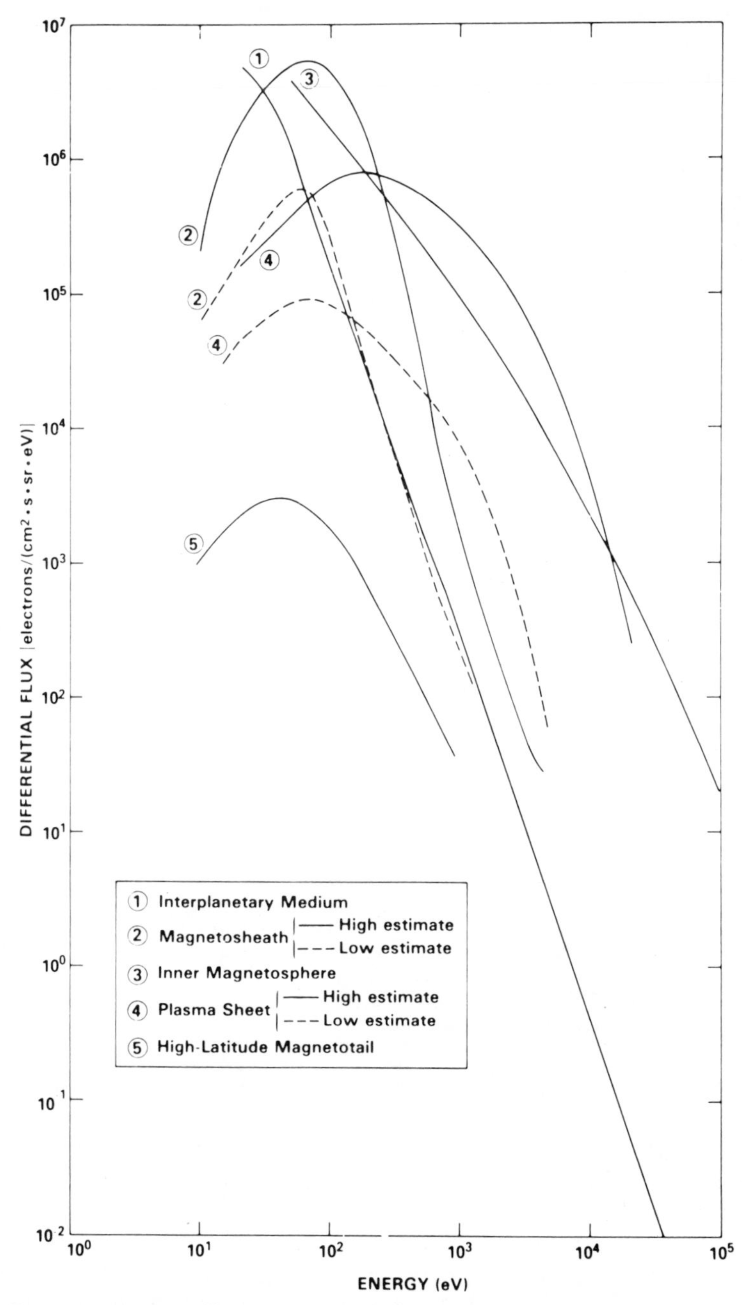

Fig. 7a Estimated "typical" spectra for the indicated magnetospheric and solar wind regions for 10 eV to 100 keV electrons; from Ref. 27).

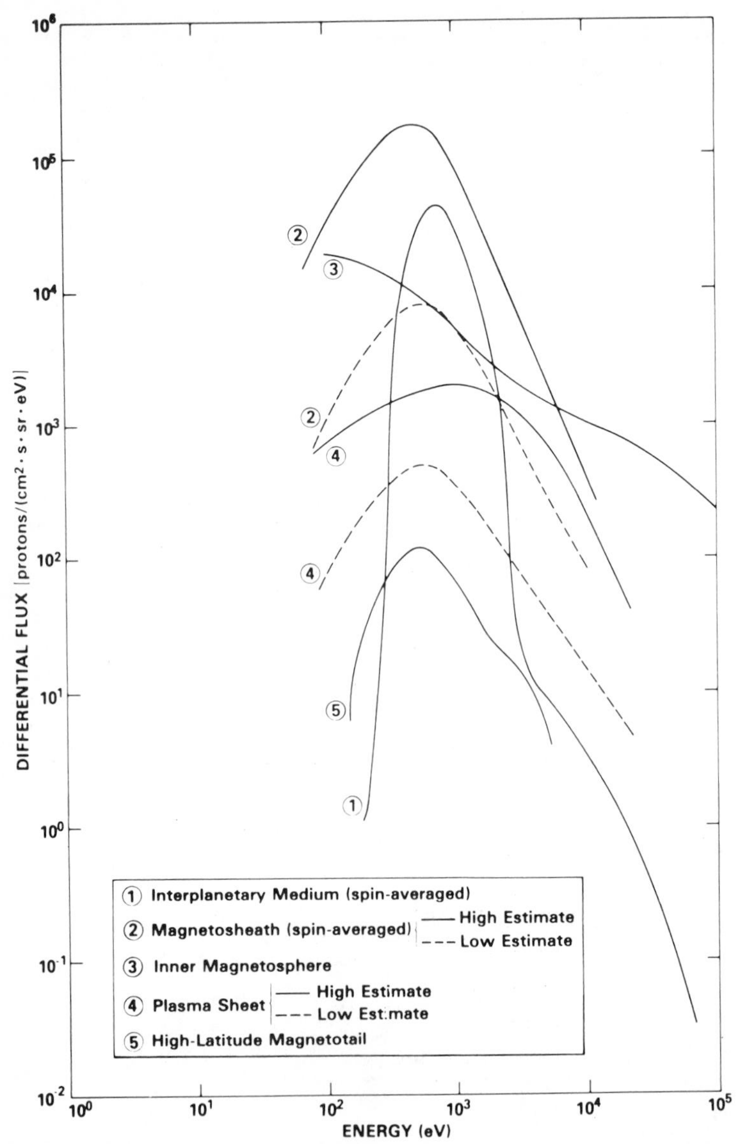

Fig. 7b Estimated "typical" spectra for the indicated magneto-spheric and solar wind regions for 10 eV to 100 keV protons; from Ref. 27.

capable of simulating changes in the environment; this is the second type of model. Of less importance per se to spacecraft charge modeling, but more important in allowing the scientist to order the magnetospheric particle data, are models which employ theory to predict trajectories of particles in static electric and magnetic field models. Finally, the most complete models from a theoretical standpoint are the full, three-dimensional, time-dependent models capable of taking into account time-varying particle injection events. Although this later category bears a close theoretical similarity to some of the models of the spacecraft sheath, it will be covered only briefly, the review being concentrated on the first two types of models.

Statistical models, as defined here, are compendiums or histograms of various plasma parameters based on actual data. The basic examples of this type of model are the composite or average distribution functions generated by Chan et al.[27] for various magnetospheric and solar wind regions (Fig. 7). Although such descriptions are particularly useful in predicting the typical potentials to be expected in various regions, a prohibitive number of distribution functions are needed as functions of time, spatial coordinates, and geomagnetic activity to adequately describe the magnetosphere. Instead, a description in terms of the first four plasma moments (density, number flux or current density, pressure or energy density, and energy flux), from which a number of the parameters of importance to the spacecraft charging problem (such as current and temperature) can be derived, has evolved as a compact means of describing the environment. Vasyliunas,[28] DeForest and McIlwain,[29] Su and Konradi,[30] and Garrett et al.,[2] all have carried out statistical studies of these parameters. In Fig. 8a, results from the analysis of ATS-5 and ATS-6 temperatures and currents are plotted as an example. In this figure, T(AVG) is the temperature obtained from dividing the energy density by the number density, and T(RMS) is the temperature derived from dividing the energy flux by the number flux. (In general, these are not equal and it is unclear which should be used in spacecraft charging analysis, although some evidence tends to indicate that T(RMS) is better.) In Fig. 8b the results of the statistical model are combined with a simple model of spacecraft charging in eclipse (see Eq. 31b) to predict the occurrence frequency of the potential in eclipse as a function of geomagnetic activity as represented by K_p.[31] This figure demonstrates that in a statistical sense, even a simple model can effectively predict spacecraft charging.

Methods have been developed for extending statistical models to other spatial positions by plotting the results from eccentric, inclined satellites in the same manner as the intensity plots of high-energy particles. A particularly good example is given in Fig. 9[32,33] for the low energy ions (200 eV $\lesssim E \lesssim$ 50 keV) in the generalized R-λ_m coordinate system.

Similar statistical studies (Refs. 34-38) have been conducted for the plasmasphere (see Fig. 10).

The major difficulty with statistical models from the spacecraft charging standpoint is their inability to include complex time variations while maintaining the proper phases between different parameters such as temperature and current. As the potential a satellite charges to may depend on the time history of the ambient plasma changes, this may be a serious omission. One solution to this problem is simply to provide several detailed but representative sequences[39] of actual plasma spectra. Unfortunately, from a practical standpoint such information lacks compactness and does not easily provide many of the parameters required by the spacecraft charging com-

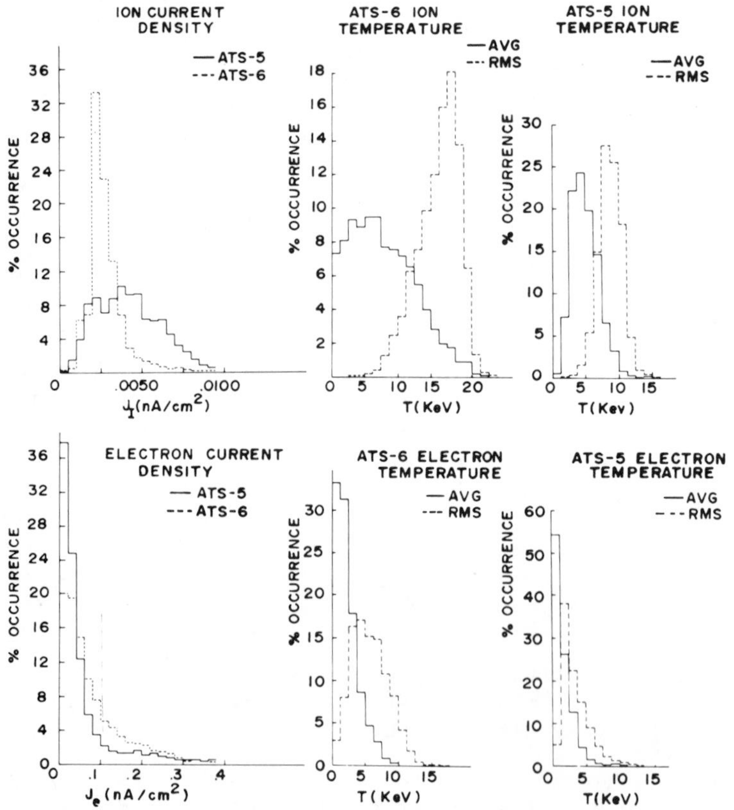

Fig. 8a Histograms of the occurrence frequencies of the electron and ion temperatures and current at geosynchronous orbit as measured by ATS-5 and ATS-6; from Ref. 2. T(AVG) is two-thirds the ratio of energy density to number density; T(RMS) is one half the ratio of particle energy flux to number flux.

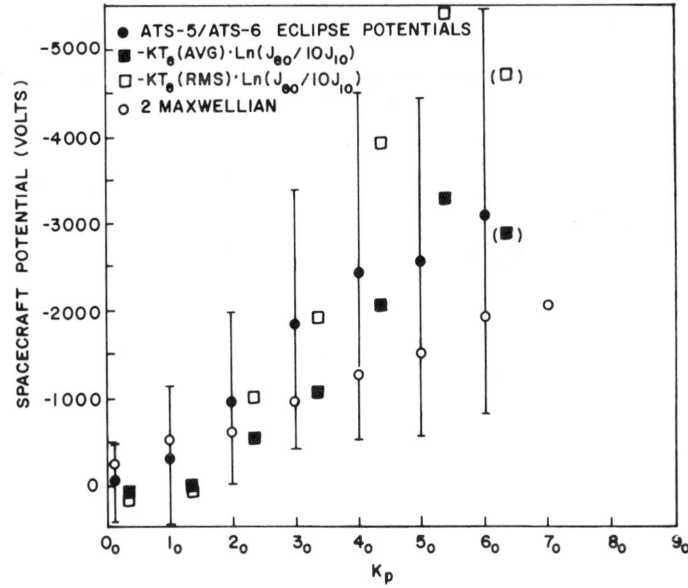

Fig. 8b Statistical occurrence frequency of observed variations of ATS-5 and ATS-6 satellite potentials in eclipse as a function of Kp (colored dots) and predicted potentials using the observed occurrence frequencies of the ambient current densities (J's) and electron temperature (T's); from Ref. 31.

munity. A solution to this problem of tradeoff between accuracy and massive amounts of data has been the introduction of analytic equations capable of modeling specific plasma variations. An example, useful for determining when a satellite is inside the midnight-dawn plasmapause (and hence likely to charge to only low positive potentials) rather than the hot plasmasheet (and likely to experience significant negative potentials), is the well known formula of Carpenter and Park[37] (see also Refs. 40-42) for the plasmapause location:

$$L_{pp} = 5.7 - 0.47\,\bar{K}_p \tag{9}$$

where L_{pp} is the plasmapause boundary (in Earth radii) and K_p is the maximum value of K_p in the preceding 12 hr.

Fairly detailed analytic formulas have been developed for many geosynchronous plasma parameters as functions of time, spatial position, and geomagnetic activity. Su and Konradi,[30] Garrett,[3] and Garrett and DeForest[43] all have empirically derived analytic expressions capable of defining the low-energy geosynchronous environment. Garrett and DeForest[43] have

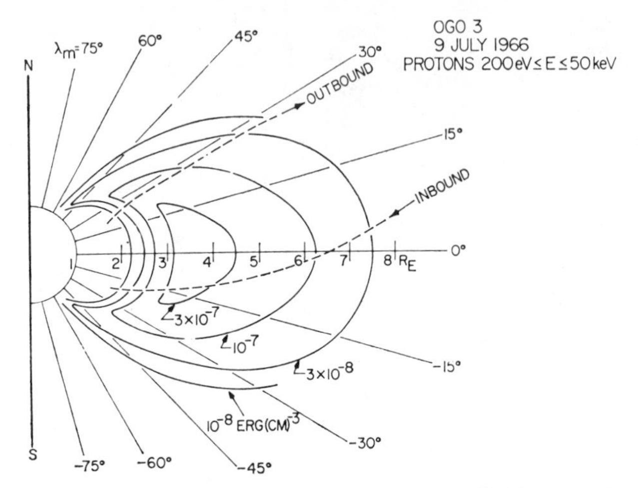

Fig. 9a Contour plot of constant proton (200 eV ≤ E ≤ 50 keV) energy densities in the R-λ_m coordinate system on July 9, 1966; from Ref. 32.

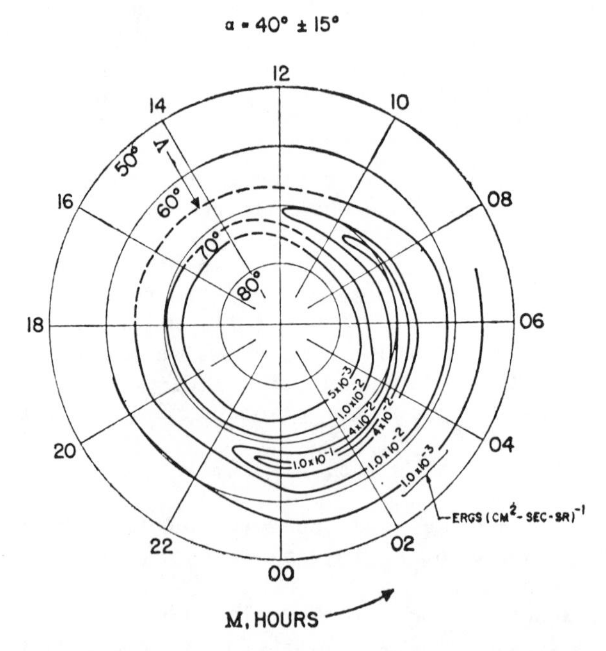

Fig. 9b Average contours of ≥ 5-keV electron energy fluxes over the northern auroral regions in terms of invariant latitude and local time; from Ref. 33.

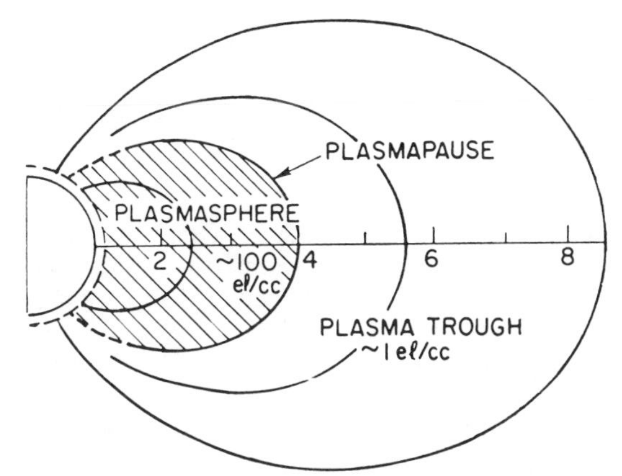

Fig. 10a Idealized meridional cross section of the magnetosphere near 1400 local time showing the location of the plasmasphere and plasma trough; from Ref. 34.

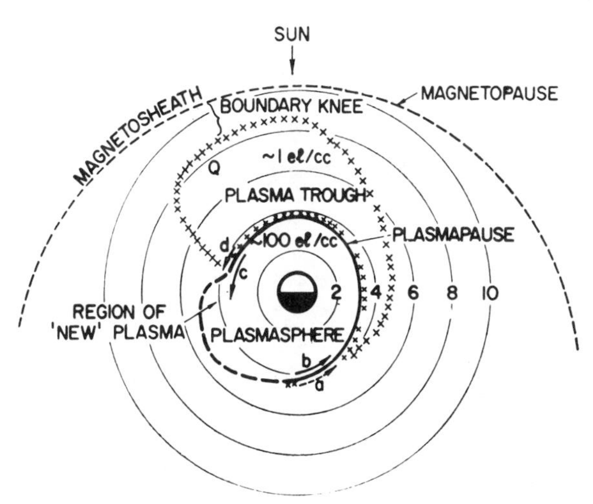

Fig. 10b Equatorial cross section of the magnetosphere illustrating the approximate locations of the plasmasphere and plasma trough in local time and with Earth radii; from Ref. 34.

developed a compact representation in which the first four moments of the distribution function at geosynchronous orbit are expressed in terms of equations linear in the geomagnetic index A_p (daily average of a_p) and varying diurnally and semidiurnally in local time, LT:

$$M_i(A_p,LT) = (a_0 + a_1 A_p)\left(b_0 + b_1 \cos\left(\frac{2\pi}{24}(LT + t_1)\right) + b_2 \cos\left(\frac{4\pi}{24}(LT + t_2)\right)\right) \tag{10}$$

where

M_i = moment i

$a_0, a_1, b_0, b_1, b_2, t_1, t_2$ = fitted parameters

Typical results are given in Fig. 11. An advantage of this model is that a two-Maxwellian (i.e., the sum of two distinct plasma components) distribution function can be derived directly from the four moments. Of particular importance to the spacecraft charging problem is that, although the model does not predict maximum charging conditions, it does allow accurate simulation of the average variations in the charging environment at geosynchronous orbit.[43] The model currently is being extended to lower altitudes.

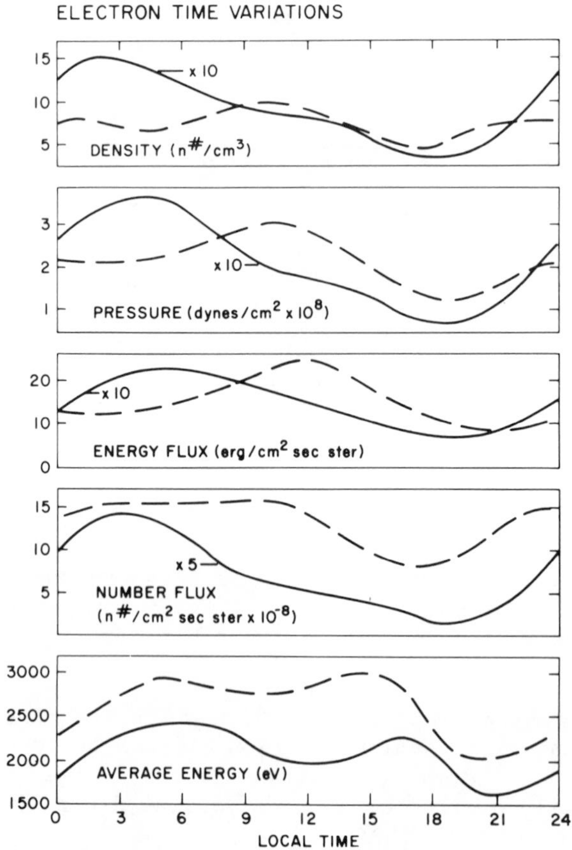

Fig. 11a The predicted local time variations in the four moments and mean energy of the electrons; from Ref. 3. Values are for Ap = 15 (solid curve) and Ap = 297 (dashed curve).

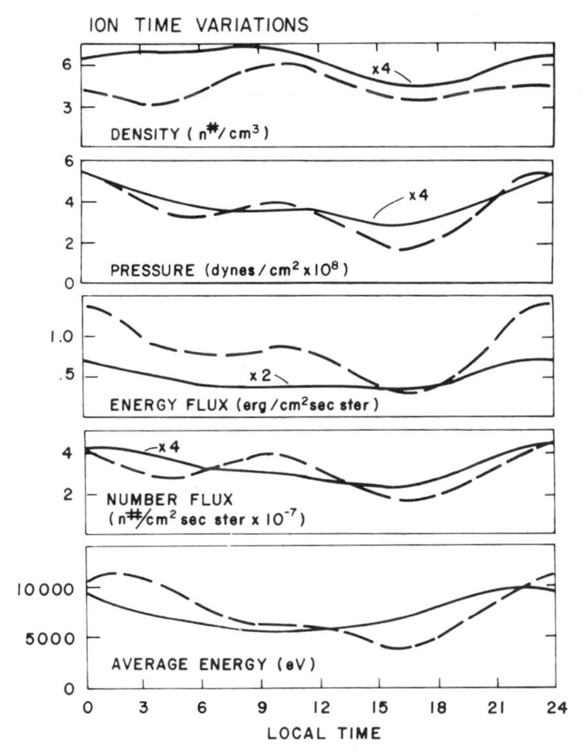

Fig. 11b The predicted local time variations in the four moments and mean energy of the ions; from Ref. 3. Values are for Ap = 15 (solid curve) and Ap = 297 (dashed curve).

The previous two model types provide most of the data that are needed to study the effects of spacecraft charging. In order to give a complete picture of the current status of such modeling efforts, however, at least some mention of the other two types of models is necessary. Further, as these more complex models ultimately will be required if an accurate understanding of the magnetospheric environment is ever to be had, we will briefly review the basic concepts (see Ref. 23 for details).

The typical example of static field models is that developed by McIlwain[44] in conjunction with the ATS-5 geosynchronous data. By careful analysis of the data, McIlwain constructed time-independent electric and magnetic field models that were capable of reproducing the observed spectra. Some results of his analysis for typical electron and ion trajectories in his model fields are given in Figs. 12a and 12b. (Note that $\mu = 1/2\ mV^2/B$.) Many variations exist on McIlwain's model.[45-49] The major difference between these models is the

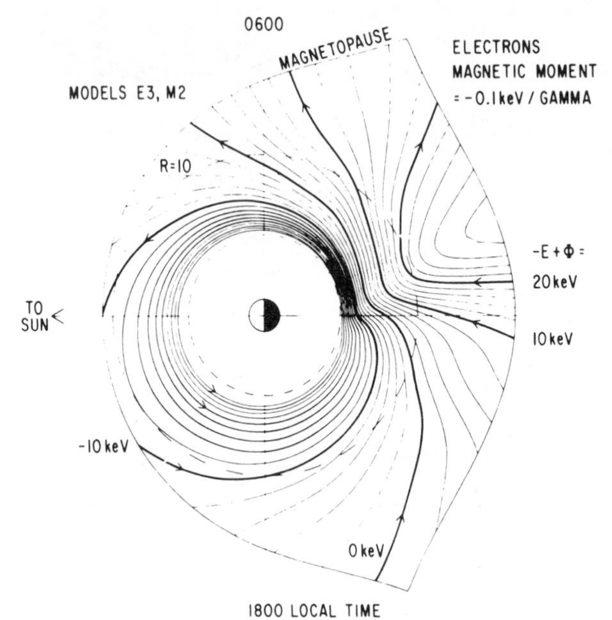

Fig. 12a Trajectories for electrons having μ = 0.1 keV/γ in the McIlwain model; from Ref. 44.

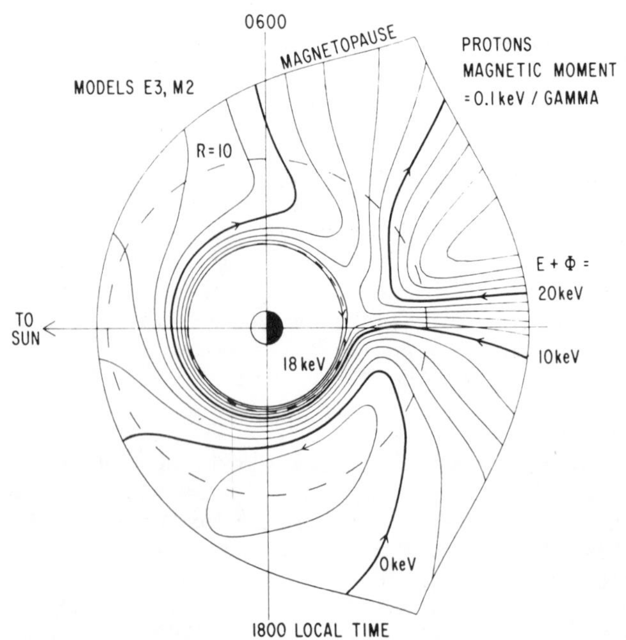

Fig. 12b Trajectories for protons having μ = 0.1 keV/γ in the McIlwain model; from Ref. 44.

exact treatment of the manner in which hot plasma reaches the near-Earth environment: the so-called injection event.[42] There is basic disagreement on whether the hot plasma that engulfs a satellite in geosynchronous orbit during a magnetic storm is local plasma that has been energized or plasma accelerated in from antisunward of the Earth.

The static models of the plasmasphere and upper ionosphere divide into two principal categories: ionospheric diffusion and $\bar{E} \times \bar{B}$ drift. Chappell et al.[35] (see Fig. 13a) and, to a lesser extent, Wolf[47] and Chen[50] model the drifts of particles in the plasmasphere in a fashion similar to McIlwain's (i.e., by $\bar{E} \times \bar{B}$ drift). Others[51-53] model the upward diffusion of ionospheric particles into the exosphere and plasmasphere. Recently, more comprehensive models[54,55] (Fig. 13b) have been produced in this latter category by using a collisionless kinetic theory to compute the diffusion along the field line of various constituents.

Although time-dependent models of the magnetosphere are not yet sufficiently developed for spacecraft charging analysis, they do promise ultimately to provide a tool capable of simulating the time-varying magnetospheric plasma parameters at any point in space to any detail desired. Specifically, Roederer and Hones[56] have successfully reproduced many of the features of the ATS-5 data by assuming a static field model upon which they superimposed a time-varying convection electric field component. In a similar manner, Smith and coworkers[57-59] have used a time-varying convection electric field potential[60]:

$$\phi = AR^2 \sin \left(2\pi \frac{LT}{24}\right) \tag{11}$$

where

LT = local time dependence (0 at midnight)
R = radial distance from Earth

A is taken to be of the form[61]

$$A = \frac{0.045}{(1 - 0.159\, K_p + 0.009\, K_p^2)^3} \tag{12}$$

Assuming a dipolar magnetic field, Smith and his co-workers[57-59] graphically track the movement of plasma at different energies from the distant Earth's tail into the vicinity of the plasmasphere.

The final models to be discussed are the three-dimensional, time-dependent models of the Rice University group under

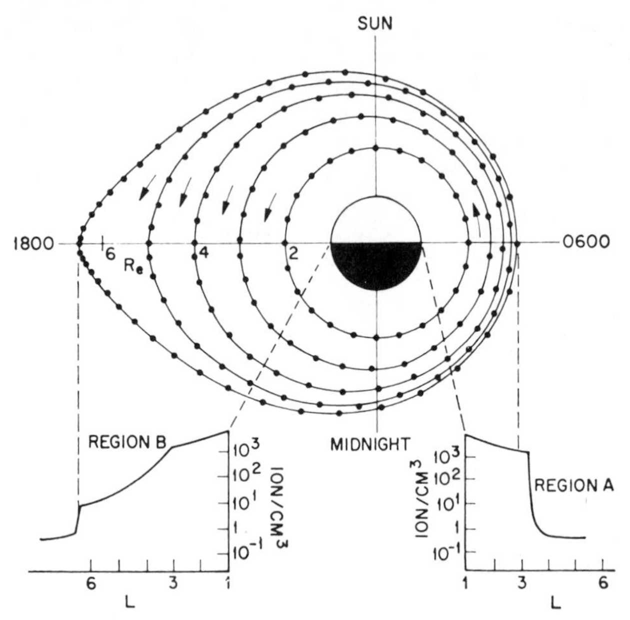

Fig. 13a Calculated motion of flux tubes in the plasmasphere; from Ref. 35. (Distance between dots gives approximate drift distance of the flux tubes in 1 hr.)

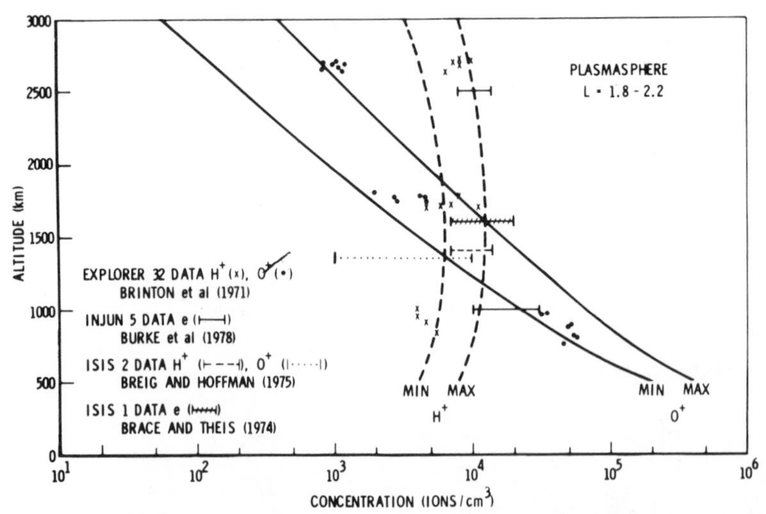

Fig. 13b Comparison between daily variations in various in-situ observations of the topside ionosphere and a plasmasphere model (see Ref. 55) having two different boundary conditions "MAX" and "MIN" at 500 km. (Particle densities at the boundary doubled between MIN and MAX; electron density is equal to the sum of H^+ and O^+ densities.)

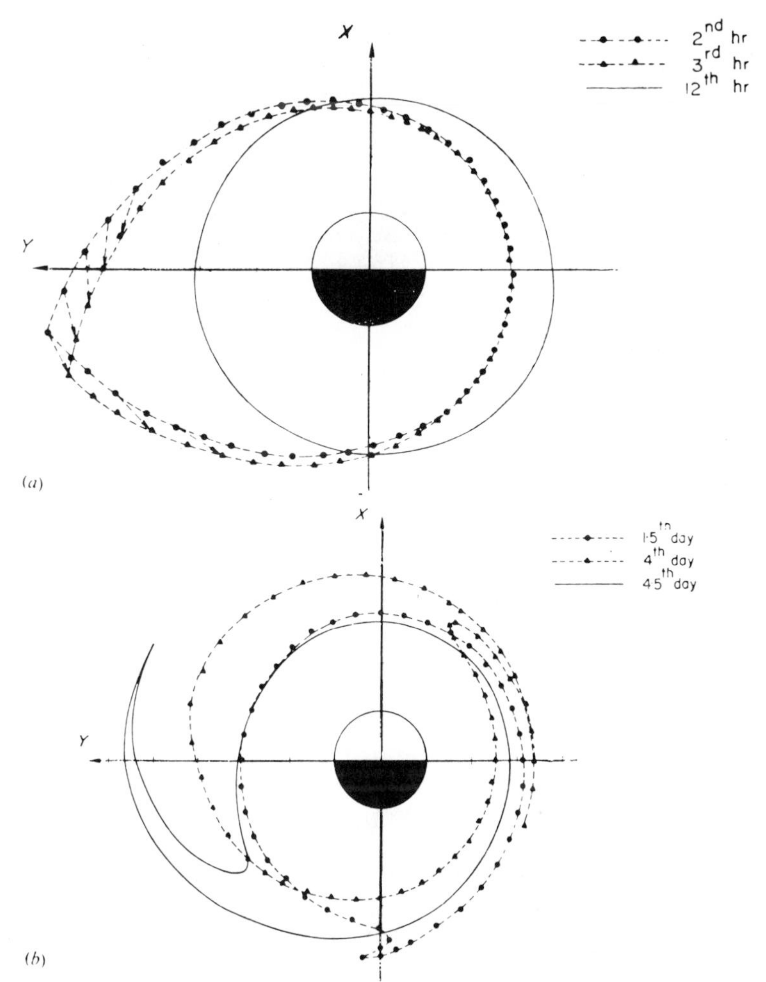

Fig. 14 Distortion of the plasmapause at various times following a factor of two decrease in the convection field at time t = 0; from Ref. 63.

R. A. Wolf.[47,50,62-66] Unique among magnetospheric models, they calculate the particle drifts and resulting effects on the electric and magnetic fields in a self-consistent fashion. In this sense, their models closely resemble spacecraft sheath models.[67-69] Some of their results for the plasmasphere[63] are presented in Fig. 14. Note in particular how the plasma "tails" off from the plasmasphere as a function of time following an increase in the terrestrial electric field. This graphically illustrates the great complexity of the magnetospheric plasma regime and the need for detailed, three-dimensional models.

Before concluding the review of the magnetosphere, it is useful to take note of means of predicting the state of the magnetosphere.[70] Current techniques[43] rely heavily on measurements of the geomagnetic index K_p or observations of the solar wind. Recently, evidence has been presented that sufficient real-time plasma data exist to generate a spacecraft charging index.[71] If this is ultimately proven true, then it will be possible to continually monitor the near-Earth environment and warn against spacecraft charging or evaluate its effects after the fact.

Recently, there has been increasing scientific interest in the plasma environments of Jupiter, Venus, and the other planets. With the Mariner, Pioneer, and Voyager missions, this interest has passed from mere speculation to active research. In Table 2 are listed some of the important spacecraft charging parameters for the solar wind, Earth, Jupiter, and Venus. The Jupiter data are from a review by Goldstein and Divine,[72] and the Venus data are from Robinson and Holman.[73] The solar wind data are from a variety of sources. The data for less than 1 AU are from Schwinn et al.[74] while the data for greater than 1 AU are estimated.[75] At best, Table 2 should be interpreted as containing order-of-magnitude estimates.

IV. Spacecraft Charging Models

The basic spacecraft-charging equation, in terms of the current densities (or number fluxes) is

$$J_E(V) - (J_I(V) + J_{SE}(V) + J_{BSE}(V) + J_{PH}(V) + J_B(V) + J_{SI}(V) + J_{RLC}(V)) = J_T(V) \tag{13}$$

where

V = satellite potential

J_E = ambient electron flux to a point on satellite surface

J_I = ambient ion flux to a point on satellite surface

J_{SE} = secondary electron flux due to J_E

J_{SI} = secondary ion flux due to J_I

J_{BSE} = backscattered electrons due to J_E

J_{PH} = photoelectron flux

J_B = active current sources such as charged particle beams or ion thrusters

J_{RLC} = resistive, capacitive, and inductive coupling currents between a point and adjoining surfaces

Table 2 Rough estimates of "typical" plasma parameters for various space environments.

Region	Altitude	N_o, cm^{-3}	Characteristic Energy, kT, eV		λ_D, m	
			Ions, H^+	Electrons	Ions, H^+	Electrons
Solar Wind	0.3 a.u.	50 (10-100)	40 (10-100)	65 (10-100)	6.7	8.5
	1.0 a.u.	5 (0.5-100)	10 (0.5-100)	50 (10-100)	10.5	23.6
	5.0 a.u.	0.2 (0.1-10)	1 (0.1-50)	20 (5-50)	16.7	74.5
Venus						
Ionosphere	200-1000 km	10^5-10^2	0.1	0.1		
Earth						
Ionosphere	150 km	$6x10^5$	0.1	0.2	.003	.004
Plasmasphere	1000 km-3 R_E	10^4-10^2	1	1	.075-.75	.075-.75
Plasmasheet	5.62 R_E	2 (1-10^3)	5000	2500	373	263
High Latitude	...	0.1 (0.01-1)	200	200	333	333
Jupiter						
Ionosphere	4000 km	$3x10^5$	0.07	0.07	.004	.004
Plasmasphere	4.5-5.5 R_J	100 (50-200)	100	4	7.45	1.49
Ring Current	8.1-9.1 R_J	12	400	4	43	4.3
High Latitude	...	0.1 (0.01-1)	400	4	471	471

The spacecraft-charging problem consists of two phases: 1) the determination of the J's as functions of ambient conditions and potential, and 2) the determination of V so that $J_T = 0$ at every point on the satellite surface in the case of a nonconducting satellite, or for the entire satellite for a conducting satellite. As in any complex physical situation, the solution of Eq. (13) can be carried out at different levels of sophistication. The purpose of this section is to review the rudiments of spacecraft-charging theory and give an overview of the methods that have been developed to solve Eq. (13). Several reviews of spacecraft charging theory and of modeling techniques already exist (see Refs. 5, 10, 14, 69, 76-81) so only a cursory description of the models will be given here.

The first part of this section will deal with the concept of Debye length. Based on the ratio of satellite size to Debye length, two examples of current calculations, one for a "thick" sheath and one for a "thin" sheath will be presented. In the context of these simple examples, models of increasing complexity will be reviewed. The section will conclude with a discussion of the most up-to-date models available.

An important concept in probe theory or spacecraft charging is that of Debye length, the distance over which a probe (satellite) significantly perturbs the ambient medium. (Note: This is a definition of the satellite sheath. In actuality, the Debye length is only one measure of this sheath. The sheath thickness is not only dependent on potential and charge but, through the so-called "presheath," can influence the plasma up to the order of the satellite radius.[60]) Specifically, electrons and ions form a cloud around any charged surface, which is a function of the particle's energy and density and the potential on the surface. By way of illustration, assume the electrons are in thermal equilibrium above a plate such that

$$n_e(X) = n_{eo} e^{\frac{qV(X)}{kT_E}} \tag{14}$$

where

n_{eo} = electron density at infinity (i.e., V = 0)
X = height above plate
T_E = electron temperature

Poisson's equation states that

$$\frac{d^2V}{dX^2} = -4\pi q\,(n_I(X) - n_e(X)) \tag{15}$$

Substituting in for n_I the approximation $n_I(X) \approx n_{eo}$, for small qV/kT_E (this is the "linearized space charge" approximation [69])

$$\frac{d^2V}{dX^2} = 4\pi q n_{eo} \left(\frac{qV}{kT_E}\right) \tag{16}$$

The solution to this is

$$V(X) = V_o e^{-X/\lambda_D} \tag{17}$$

where $\lambda_D = \left(\frac{kT_E}{4\pi q^2 n_{eo}}\right)^{\frac{1}{2}}$ the classical Debye length approximation. (The "linearized space charge" approximation can be extended to multidimensional geometries. An analytic fit to the spherical case is discussed later in this volume.[66])

The characteristic distance λ_D is the approximate distance over which the surface potential V_o is attenuated by 1/e. Values of λ_D for different plasma regimes are listed in Table 2. The terminology "thick" sheath and "thin" sheath implies that the region over which the satellite affects the ambient plasma is larger or smaller than the characteristic dimensions of the spacecraft. Usually, this reduces to determining whether λ_D is larger than the satellite radius (thick sheath) or smaller (thin sheath). From Table 2, for a 1-m-radius spacecraft with no active surfaces in low Earth orbit the term thin sheath is appropriate. For geosynchronous orbit unless the structure is 10 m or larger, the thick sheath approximation is appropriate. (Note: As outlined later in this volume,[69] if active surfaces (i.e., exposed potentials driven by the satellite systems) are present, these criteria will change.) In the following discussion, depending on which approximation applies, specific simplifying assumptions will be made which allow computations of the J's in Eq. (13).

As a first example of sheath theory, assume a large structure such that $\lambda_D \ll r_s$, where r_s is the radius of the surface. Assuming that, at the surface ($X = r_s$), the potential is V_o and that the surface is nearly planar relative to the sheath dimensions, at distance Y ($Y = X-r_s$) from the surface

$$\frac{d^2V(Y)}{dY^2} \simeq -4\pi q\, n(Y) \qquad \text{Poisson's equation} \tag{18a}$$

$$J = q\, n(Y)\, v(Y) = \text{constant} \qquad \text{current continuity} \tag{18b}$$

$$\frac{1}{2} mv(Y)^2 = q\, V(Y) \qquad \text{energy conservation} \tag{18c}$$

In order to solve these equations, it is assumed that V and $\frac{dV(Y)}{dY}$ are 0 at some distance Y = S which determines the sheath thickness (this is called the "space-charge limited" assumption). The solution becomes

$$J = \frac{1}{9\pi} \left(\frac{2}{mq}\right)^{\frac{1}{2}} V_o^{3/2}/S^2 \tag{19}$$

This is the space-charge limited "solution" or Child-Langmuir law for a plane. If J is replaced by $J_o = K^* q n_o (kT/m)^{\frac{1}{2}}$ (where $(2\pi)^{\frac{1}{2}} \leq K^* \leq 1$,[19] the lower limit corresponds to Eq.(2) and the upper limit is for monoenergetic plasma), then the sheath thickness can be estimated:

$$S = \frac{1}{3} (2/kT)^{\frac{1}{4}} (V_o/q)^{3/4} (1/\pi n_o)^{\frac{1}{2}} (1/K^*)^{\frac{1}{2}} \tag{20}$$

This sheath thickness determines the region over which charge is collected and is important, therefore, in determining the maximum current that can flow for a given V. (This solution is for a planar geometry. Parker[69] gives an analytical fit to Langmuir's result for a spherical geometry.)

In the thick-sheath limit, $\lambda_D \gg r_s$. To a good approximation, for a particle at position r approaching a satellite with its center at r = 0, by conservation of energy and momentum,

$$\frac{1}{2} m v_o^2 = \frac{1}{2} m v(r_s)^2 + q V(r_s) \tag{21a}$$

$$mRv_o = m r_s v (r_s) \tag{21b}$$

where

v_o = velocity in ambient medium
R = position where v_o is measured
v = velocity at minimum approach distance r_s (= $v_\perp$)

Solving for the impact parameter R (any particle having r ≤ R will reach r_s):

$$R^2 = r_s^2 \left(1 - \frac{2qV(r_s)}{mv_o^2}\right) \tag{22}$$

$(R - r_s)$ is equivalent to the sheath thickness S defined for a thin sheath, since it is the size of the region from which particles can be drawn. By definition, for the same body and the appropriate limiting plasma conditions, $S \ll (R-r_s)$.

The total current density striking the satellite surface for a monoenergetic beam is

$$J(V) = \frac{I}{4\pi r_s^2} = J_o \left(1 - \frac{2qV(r_s)}{mv_o^2}\right) \tag{23}$$

where I = total current to satellite, or the ambient current which would pass through an area $4\pi R^2$

J_o = ambient current density at $R = I/4\pi R^2$

This is the so-called "thick-sheath, orbit limited" current relation. For a general distribution function F, Liouville's theorem states that

$$F(v) = F'(v') \tag{24}$$

where F' and v' are the distribution function and velocity at the surface of the spacecraft. F and v are the ambient distribution and velocity at the end of the particle trajectory connected to the satellite surface where F' and v' are measured If F is assumed to be a Maxwell-Boltzmann distribution f, then by Eq. (21a):

$$f_i'(v') = e^{\pm qV/kT_i} f_i(v') \tag{25}$$

(+ for electrons, - for ions)

where

$$(v')^2 \geq \max\left(0, \pm \frac{2qV}{m_i}\right)$$

In the thick-sheath limit, if spherical symmetry is assumed, by substitution into the integral in Eq. (2) and integration [with the limits on v' imposed by Eq. (25)] :

$$J_E = J_{EO} \begin{cases} e^{qV/kT_e} & V < 0 \quad \text{repelled} \\ 1 + \dfrac{qV}{kT_e} & V > 0 \quad \text{attracted} \end{cases} \tag{26}$$

$$J_I = J_{IO} \begin{cases} e^{-qV/kT_I} & V > 0 \quad \text{repelled} \\ 1 - \dfrac{qV}{kT_I} & V < 0 \quad \text{attracted} \end{cases}$$

where

$$J_{i0} = \frac{qn_i}{2}\left(\frac{2kT_i}{\pi m_i}\right)^{\frac{1}{2}}$$

For the attracted species, the results are identical to the orbit-limited solution Eq.(23) if $1/2\ mv_o^2$ is replaced by kT_i. This is the reason the name thick-sheath, orbit limited current is used for this approximation. Prokopenko and Laframboise[82] have rederived (in more general terms) the current density to a sphere, infinite cylinder, and infinite plane (i.e., three-, two-, and one-dimensions) for a Maxwell-Boltzmann distribution for the orbit-limited solution in the manner just discussed.

(The original derivation, for a more restricted set of assumptions, was by Mott-Smith and Langmuir.[83]) Their results are:

For the attracted species

$$J_i = J_{i0} \begin{cases} (1 + Q_{is}) & \text{sphere} \\ \left[2\,(Q_{is}/\pi)^{\frac{1}{2}} + e^{Q_{is}}\ \mathrm{erfc}\,(Q_{is}^{\frac{1}{2}})\right] & \text{cylinder} \\ (1) & \text{plane} \end{cases} \tag{27}$$

For the repelled species

$$J_i = J_{i0}\, e^{Q_{is}}$$

where

$$Q_{is} = \pm \frac{qV}{kT_i} \quad \text{(+ for electrons, - for ions)}$$

For a sphere, the results are identical to Eq. (26). As the Debye length is, by definition, small relative to structural dimensions for the infinite planar case, it should be equivalent to a thin-sheath approximation, so that $J_i = J_{i0}$ for the attracted species (i.e., the approximation made earlier to estimate the thin sheath thickness). Thus, conceptually, the change in current density for the attracted species as the sheath (or geometry) is varied from a thick (spherical) to a thin (planar) sheath is the same as varying χ between 1 and 0 where[21]

$$J_i = J_{i0}\,(1 + \chi Q_{is}) \qquad 0 \leq \chi \leq 1 \tag{28}$$

Note that for the repelled species on the other hand, the results are always the same.

For either the thick sheath or thin sheath, if wake effects are neglected, the total current is obtained by multiplying the collection area of the satellite by the ambient current density J_{i0} (usually given by either Eq. (2) or $qn_o\,(kT/m)^{\frac{1}{2}}$. The ratio $J_i/_{i0}$ is then equivalent to the ratio of the collection area (either $4\pi R^2$ or $4\pi(S + r_s)^2$ for a spherical satellite) to satellite area. (For a spherical satellite it would be $4\ r_s{}^2$.) In going from a thin sheath ($S \ll r_s$) to a thick sheath (r_s) to a thick sheath ($r_s \ll R$), the ratio goes from 1 to $R^2/r_s{}^2$ for a spherical satellite i.e., χ goes from 0 ($I = 4\pi r_s{}^2 J_{i0}$) to 1 ($I = 4\pi R^2 J_{i0}$) as implied by Eq. (28).

Returning to Eq. (13), approximations are now available for $J_E(V)$ and $J_T(V)$ for either a thick or thin satellite sheath. Given these currents to a surface, Eq. (13) for a single point can be solved provided the secondary currents, J_{PH}, and the backscattered currents are known for V such that $J_T(V) = 0$ (J_B and J_{RLC} are assumed to be 0). Useful potential calculations can be made in this manner based on the assumption that (see Section II)

$$J_{SE} \sim a\ J_E \qquad (29)$$
$$J_{SI} \sim b\ J_I$$
$$J_{BSE} \sim c\ J_E$$

Eq. (13) reduces, given these approximations and assuming a thick sheath, to[84]

$$V = -\frac{kT_E}{q}\ \mathrm{Ln}\left[\frac{J_{EO}\ (1 - a - c)}{(1+b)\,J_{IO}\left(1-\frac{qV}{kT_I}\right)+J_{PH}(X_m)}\right] \quad V < 0 \qquad (30)$$

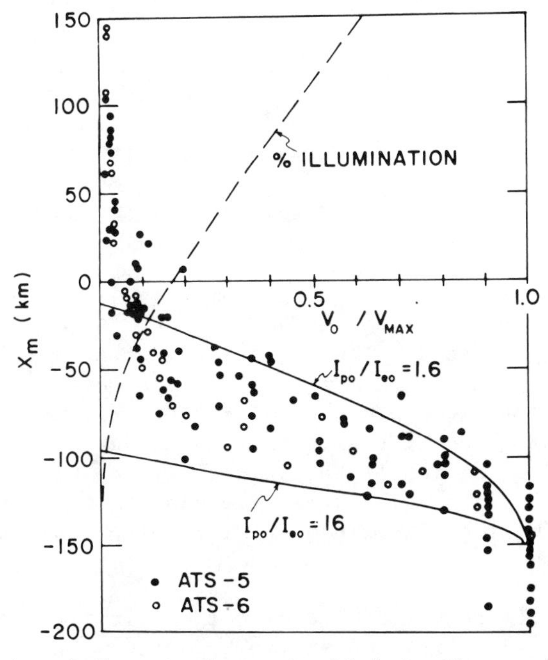

Fig. 15 Ratio of the varying potential to the maximum potential when the spacecraft was in eclipse; predicted curves for I_{PO}/I_{EO} (= $J_{PO}/4\ J_{EO}$) values of 1.6 and 16; percent illumination as a function of X_m (using same scale as that for $V(X_m)/V_o$); from Ref. 84.

where X_m = minimum altitude of ray from satellite to center of sun. This equation applies to the conditions most often considered in current spacecraft-charging codes, i.e., conditions at geosynchronous orbit during a geomagnetic storm. During such periods, qV/kT_I is usually less than 1. Further, J_{PH} either is much greater than $(1 + b)\ J_{IO}$ or is 0 (i.e., the satellite is in the Earth's shadow or the surface is shaded) so that

$$V \simeq -kT_E/q \ \mathrm{Ln}\left[J_{EO}\ (1-a-c)/J_{PH}\ (X_m)\right] \quad \text{sunlight} \tag{31a}$$

$$V_o \simeq -kT_E/q \ \mathrm{Ln}\left[J_{EO}\ (1-a-c)/J_{IO}\ (1+b)\right] \quad \text{eclipse} \tag{31b}$$

Dividing Eq. (31b) into Eq. (31a)

$$V/V_o = \frac{\mathrm{Ln}\left[J_{EO}\ (1-a-c)/J_{PH}(X_m)\right]}{\mathrm{Ln}\left[J_{EO}\ (1-a-c)/J_{IO}\ (1+b)\right]} \tag{32}$$

Eq. (21) (for various values of X_m, J_{EO}/J_{PHO}, and for $J_{EO}/J_{IO} \simeq 27$, $a \sim .4$, $b \sim 3$, and $c \sim .2$) is plotted in Fig. 15 along with actual observations from the geosynchronous satellites ATS-5 and ATS-6. The potentials in eclipse range from -300 to -10000 V and correspond to a wide range of ambient parameters. Thus, even a simple model is capable of estimating the effects of a complex charging phenomena: eclipse passage at geosynchronous orbit. Applications of this type of model are discussed later in this volume.[85]

Prokopenko and Laframboise[82] have applied their results in Eq. (27) to different ambient conditions. Some of these results are listed in Table 3. The variations in potential with geometry (or varying χ) are quite large: the potential is significantly larger for the planar (thin-sheath) geometry. More important, however, is their discovery of multiple roots to Eq. (13). The meaning of this discovery is still to be determined, but it implies that the time history of the charging process may be important in determining the final potential configuration.

The single-point model does not explicitly consider the problem that satellites consist of a variety of surfaces and that differential charging of those surfaces is possible. The currents between surfaces must be included in Eq. (13) to model this properly. Finite element or circuit models,[73,86,87] as this class of models is termed, consist of many points, each representing a surface on the satellite. Besides the ambient, secondary, backscattered, and photoelectron currents considered in the single-point model, the coupling currents to each point, J_{RLC}, are included to estimate the currents between surfaces. A circuit schematic from a Pioneer Venus simulation[73] is presented in Fig. 16. Time variations are handled to a certain extent by inductive and capacitive elements making these models applicable to a wider range of problems than the single-point models.

The probe models just discussed have advantages and disadvantages. At one extreme, such models allow a straightforward development and physical interpretation of many of the phenomena associated with spacecraft charging. The models of Parker,[14,69,88,89] Prokopenko and Laframboise,[82] and Whipple[5] have been of particular value in this. The models developed along these lines generally involve little computation time. If spacecraft had remained unsophisticated and relatively insensitive to arcing, these models would have been adequate. At the other extreme, with the advent of CMOS and VLSI circuitry, effects (e.g., particle focusing) which were once considered minor and therefore were ignored in the probe models, are now major systems problems. Intimate knowledge of plasma sheath dynamics becomes a necessity: this knowledge cannot be provided effectively by probe models. Several models which attempt to self-consistently calculate individual particle trajectories through the sheath will be discussed later. Although all depart from the probe models, they include in them important elements based on probe theory and Eq.(13) so that many of the ideas discussed previously are still applicable.

The most straightforward models conceptually, though perhaps most demanding computationally, are the so-called "particle-pushing" codes. In these models, many individual or groups of similar particles are followed simultaneously by the computer as they move through the satellite plasma sheath. The computer keeps track of surface interactions (backscattered and secondary electrons and photoelectrons), interactions between the satellite fields and the particles, and interactions among the particles themselves. Results from this type of model,[67] illustrating the time-dependence of the potential on a spherical satellite as a function of time for various ambient conditions, are presented in Fig. 17. The rapid rise time could conceivably cause circuit upsets as only $\sim$ 2V are necessary to trigger many integrated circuits. The model also has successfully modeled the effects of plasma oscillations and predicted potential minima in agreement with probe theories. The major problems with this type of model are run time and numerical instabilities. Although it has recently been extended to

Table 3 Floating potentials of shaded surfaces of synchronous-altitude spacecraft for orbit-limited ion collection in spherical, infinite cylindrical, and planar symmetries.[a]

Material	"Quiet" Spectrum Floating Potential (volts)			"Disturbed" Spectrum Floating Potential (volts)		
	3-D	2-D	1-D	3-D	2-D	1-D
Gold	-28.0	-28.1	-28.6	-3570	-6520	-15,450
Aluminum	-1370	-2060	-5180	-6610	-11,250	-21,570
Aluminum with oxide coating	+0.68	+0.68 or -723* or -1010	+0.68 or -490 * or -4210	-6200	-10,690	-20,920
Quartz	+0.55	+0.55	+0.55 0 -922 * or -3190	-5860	-10,240	-20,380
Aquadag	-1560	-2390	-5900	-7090	-12,010	-22,350
Beryllium Copper	+0.64	+0.64	+0.64 or -522 * or -3420	-5710	-9870	-19,830
Beryllium Copper Activated	+1.4	+1.4	+1.4	+0.41 or -908* or -3910	+0.54 or -843* or -7520	+0.54 or -776* or -17,580

* Unstable

[a] From Ref. 79

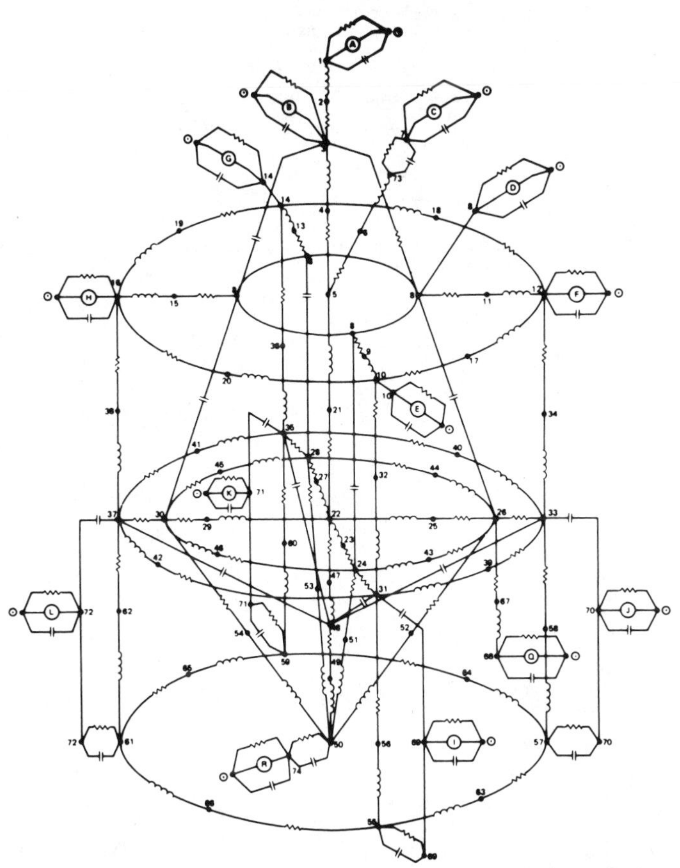

Fig. 16 Example of a finite element (circuit) model of the Pioneer Venus satellite; from Ref. 73.

model a finite cylinder, this type of model has yet to be run for the three-dimensional case because of the severe demands that it places on run time.

As with models of the low energy plasma, consideration of physical principles allows a reduction in the number of parameters that are necessary to adequately specify the satellite sheath. Models by Parker [14,69] are based on the simultaneous, self-consistent solution of two equations in the satellite sheath:

Poisson's equation $$\nabla^2 V = 4\pi q\ (n_s + n_E - n_I) \qquad (33a)$$

Vlasov equation $$\mathbf{v} \cdot \vec{\nabla} f_i - \frac{q_i}{m_i} \vec{\nabla}\ V(r) \cdot \vec{\nabla}_V\ f_i = 0 \qquad (33b)$$

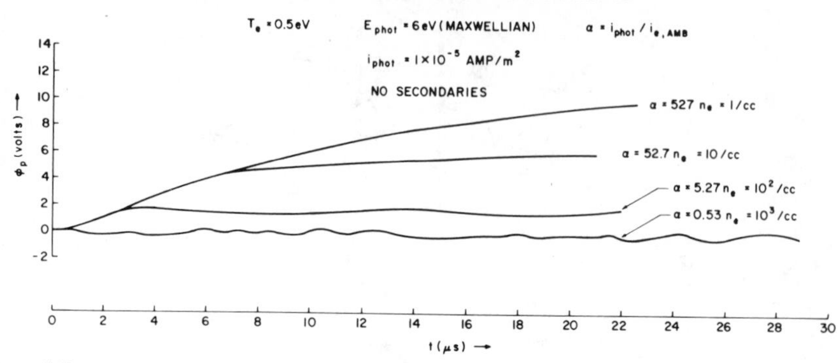

Fig. 17 Spacecraft voltage transients following the "turn-on" of the plasma at t = 0. The results are for a sphere (no secondary emission). α is the ratio of photoelectron to ambient electron temperature, and E_{phot} is the mean energy of the photoelectrons (assumed to be Maxwellian).

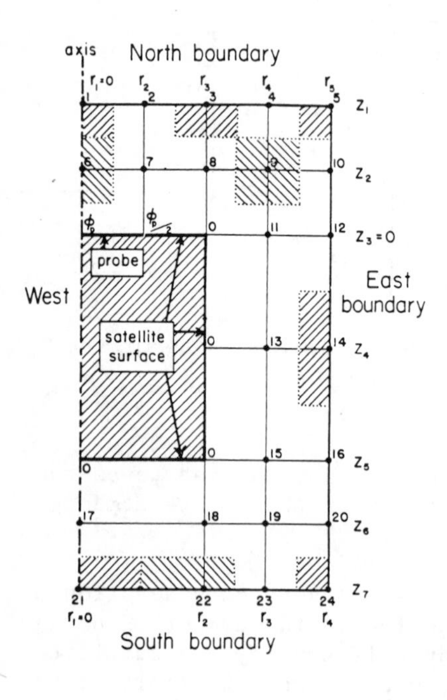

Fig. 18 Example of grid used by Parker to estimate the potential drop and plasma sheath near a cylindrical satellite; from Ref. 14. (The Poisson and Vlasov equations are solved at each indicated point).

where

n_E = ambient electron density

n_I = ambient ion density

n_S = surface-emitted electron density

$\vec{\nabla}, \vec{\nabla}_V$ = gradient operators with respect to position and velocity space, respectively

As outlined by Parker, the first step is the calculation of the number densities as a function of spatial position $\vec{r}$:

$$n_i(\vec{r}) = \iiint f_i(\vec{r},\vec{v})\, dv_X\, dv_Y\, dv_Z \qquad (34)$$

Parker[14] has developed an efficient method, called the "inside-out" method, for carrying out this integration by carefully determining the types of orbits that particles passing through the sheath are on. Equation (33b) states that f is constant along a particle trajectory [see Eqs. (24) and (25)]. If $V(\vec{r})$ is known, then, the particle trajectories can be traced backwards from the point in question to their origin in the ambient medium (or on the satellite surface) where f is known, hence, the name "inside-out." Equation (34) can then be integrated at different grid points on the satellite surface or in the satellite sheath (see Fig. 18). Once the n_i are known as a function of position $\vec{r}$, Eq. (33a) can be converted to finite difference form and solved by straightforward numerical techniques to obtain $V(\vec{r})$. In practice, this procedure is iterated until the $n_i(\vec{r})$ and $V(\vec{r})$ are simultaneous solutions of Eqs. (33) and (34).

Parker's method and related techniques[91] have been extended to a variety of situations. Parker[14] has employed his method to estimate the sheath and wake surrounding low-altitude satellites. Based on Parker's techniques, Reiff et al.[92] have calculated the currents to the solar power satellite. Their method has some disadvantages since it cannot be used to estimate time dependency except as a series of quasi-equilibrium states.

The most ambitious model currently available for computing satellite potentials in the thick-sheath limit is the NASCAP (NASA Charging Analyzer Program) computer code.[68,93-96] This code combines a solution of the Poisson equation and a probe-charging model along with a complex graphics package to compute the detailed time behavior of charge deposition on spacecraft surfaces. An approximate circuit model of the satellite is used to estimate voltage changes during each time step for an implicit potential solver. The propagation of particle beams through the satellite sheath is computed by orbit-tracing in the sheath field. The code has several options available, ranging from a less-detailed code capable of calculating the differential potential on a simple laboratory material sample, to a detailed code capable of modeling individual booms and sensor surfaces on a complex satellite such as the P78-2 SCATHA satellite.[68] The primary intent of the model is to compute the effects of satellite geometry on the satellite photosheath. The first step in the NASA Charging Analyzer Program model is to insert the satellite geometry and material content. The code has provisions for modeling (using a variety of simple three-dimensional building blocks) surfaces ranging from cubes and planes to complete satellites with their booms and solar panels (Fig. 19a). In the second stage the currents to the satellite surfaces are computed by treating each surface element as a current-collecting probe. The inside-out technique is then used to model the photosheath. Time dependency is included in the analysis (Fig.19b).

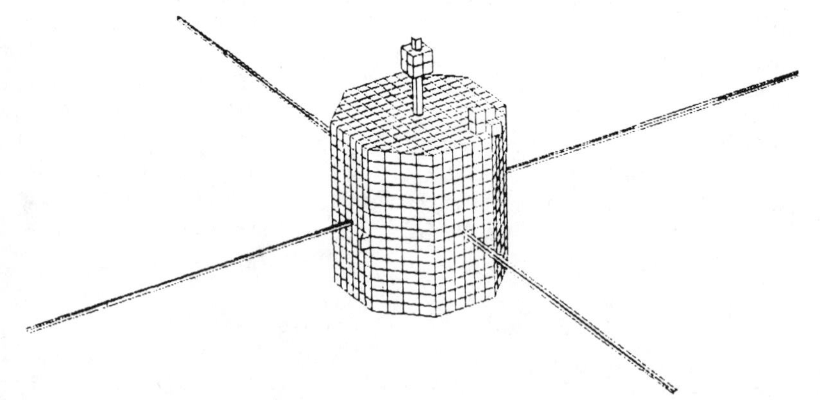

Fig. 19a NASCAP/AFGL computer simulation of the SCATHA (P78-2) satellite; from Ref. 95.

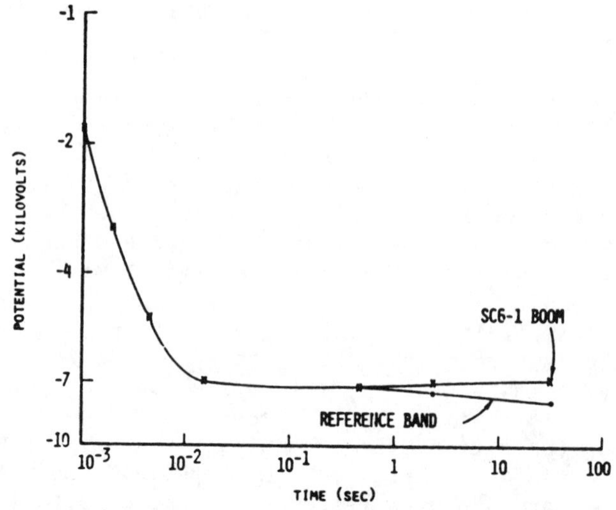

Fig. 19b NASCAP/AFGL computer simulation of time-dependence of potential for two surfaces on the SCATHA (P78-2) satellite; from Ref. 95.

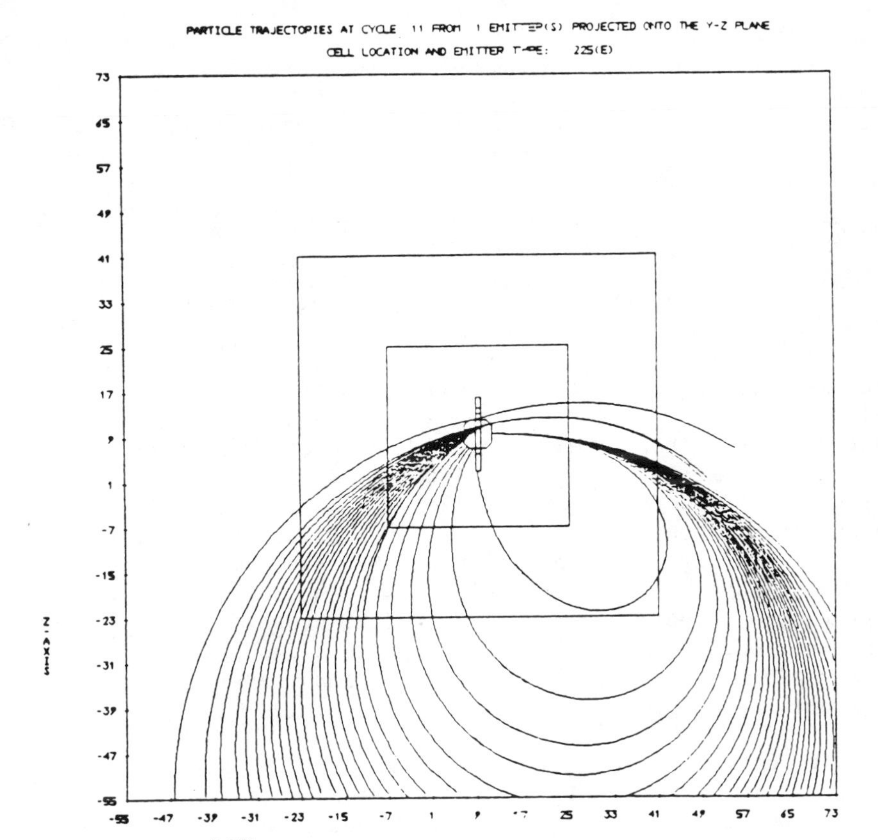

Conducting object potential = +0.98 keV.
Uniform magnetic field = 0.1 gauss along -X axis.
Plot boundary = grid 4.
Mean beam energy = 1.1 keV.
Beam energy spread = 0.1 keV (25 energies).
Beam angular spread = 0°.
Electron emission current = 1 ma.
Highest grid which particles entered = 6.

Fig. 19c Particle trajectories from an electron emitter as computed by the NASCAP/AFGL code; from Ref. 68.

The analysis is carried out on successively larger, coarser nested grids,[68,94,95] allowing the trajectories of particles emitted from the satellite to be traced in the space surrounding the satellite (Fig. 19c).

The NASCAP program combines the best of the circuit element model, the inside-out technique, and Langmuir probe theory in one model along with the added advantage of detailed graphical results. Unfortunately, with this gain in capability the code has become quite large and requires hours of computing time for the more detailed models. Two steps currently are being taken to alleviate this problem. First, efficient versions have been developed for specific computational tasks such as the SCATHA version discussed in Rubin et al.[68] Secondly, in conjunction with a number of laboratory and in-situ satellite experiments, an attempt is being made to verify all aspects of the code (see Refs. 96 and 68).

V. Spacecraft Charging Interactions

In previous sections the theory of spacecraft charging and methods for estimating satellite potential variations and plasma sheaths have been presented. Once a sheath forms and potential differences are created, a number of effects are possible, most of which are deleterious. Typical is the plot of satellite operational anomalies as a function of local time at geosynchronous orbit presented in Fig. 20.[97] The cause of this clustering near local midnight is spacecraft charging: intense fluxes of energetic electrons associated with injection events are encountered near local midnight which lead to arcing and hence to control circuit upsets and operational anomalies. In this section, effects such as these will be described and examples presented where possible.

To date the most impressive spacecraft-charging interactions have been charge buildup and the subsequent arcing. Charge buildup has been clearly observed in a number of recent

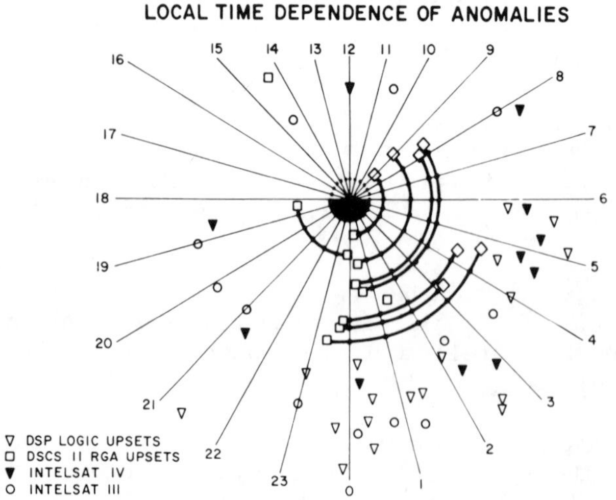

LOCAL TIME DEPENDENCE OF ANOMALIES

▽ DSP LOGIC UPSETS
□ DSCS II RGA UPSETS
▼ INTELSAT IV
○ INTELSAT III

Fig. 20 Local time plot of satellite operational anomalies (radial distance has no meaning); from Ref. 97.

experiments. Nowhere are the data quite so well documented, however, as for the ATS-5 and ATS-6 geosynchronous satellites.[98] A typical sample of data in spectrogram format for day 59, 1976, for ATS-6 is presented in Fig. 21a. In this type of

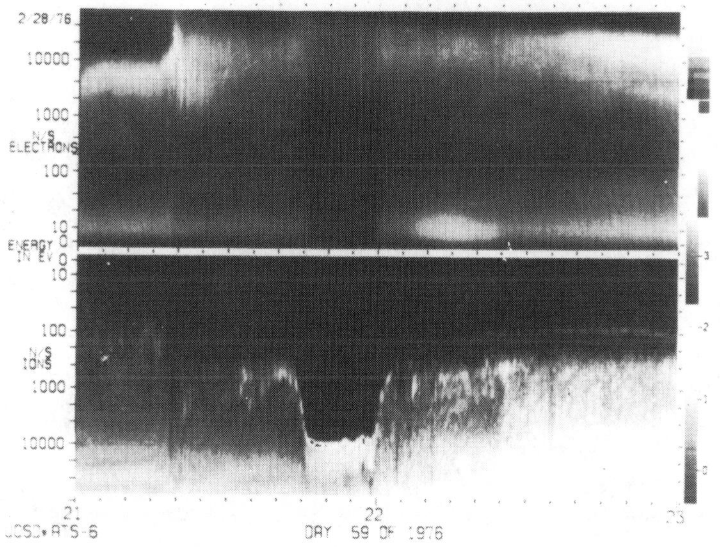

Fig. 21a Spectrogram of ATS-6 count rates as a function of time and energy for day 59, 1976. The spacecraft potential is -10,000 V at 2200 UT.

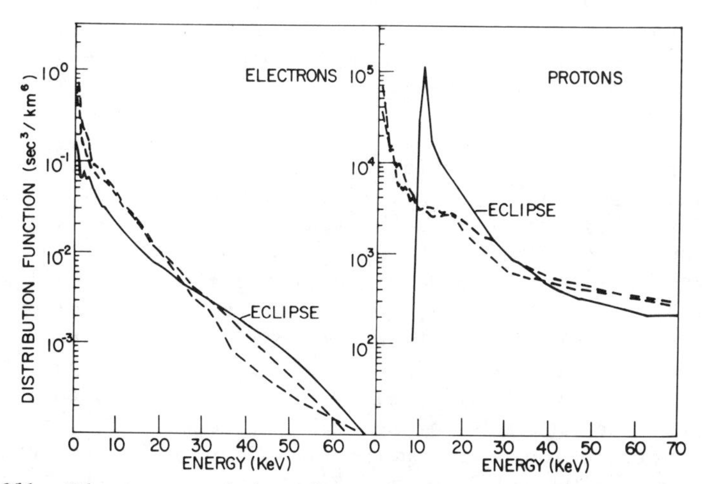

Fig. 21b Electron and ion distribution functions vs energy for day 59, 1976. (Dashed lines represent the spectra before and after eclipse; solid lines represent eclipse spectra.)

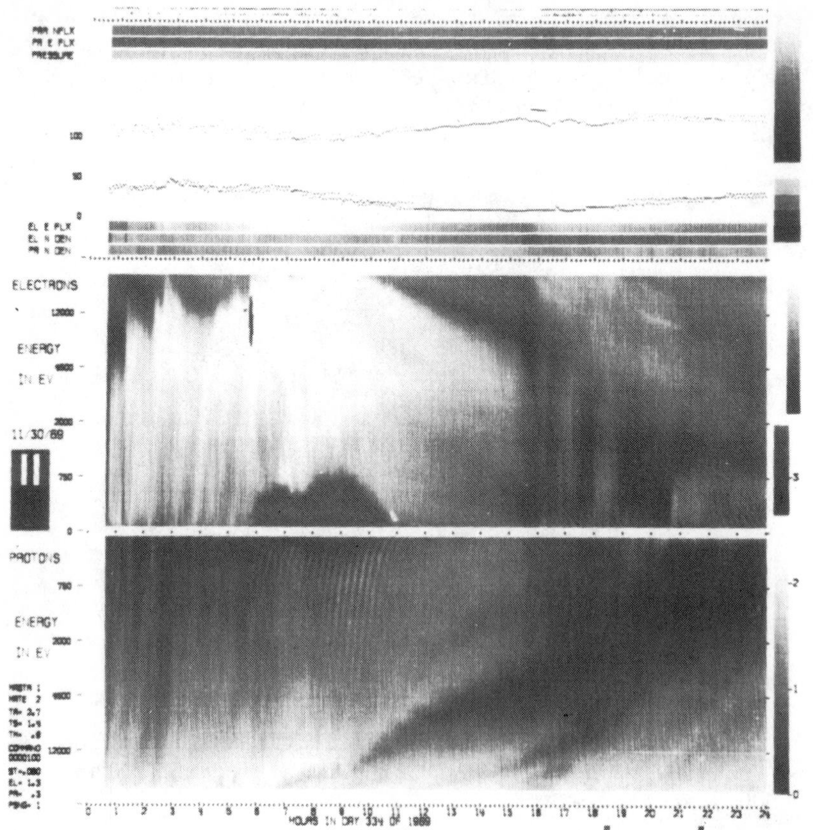

Fig. 22 Day 334, 1969, spectrogram from ATS-5; from Ref. 98. The feathered pattern in the ions between 0700 and 1100 UT and the corresponding loss of low energy electrons is real and the result of differential charging.

plot,[29] the y-axis is particle energy, the x-axis is time, and the z-axis (shading) is particle count rates. The interpretation of this figure is that the spacecraft obtained a potential of -10,000 V relative to the space plasma between 2100 and 2200 universal time when the satellite passed into eclipse. As exemplified by Eq. (31), the satellite entered eclipse and the photoelectron flux went to zero shifting the already negative (-100 V) potential even more negative. The negative potential accelerates all positive ions as they approach the satellite, adding energy qV to each particle. Thus, zero energy positive ions gain energy qV and appear as the bright band between 2100 and 2200 UT. Electrons, in contrast, are decelerated, giving the drop-out in the electron spectra between 2100 and 2200 UT. Actual spectra for before and after the eclipse are presented in Fig. 21b demonstrating this effect. Potentials as high as -2000 V in sunlight and -20,000 V in eclipse have been observed on ATS-6.

Another effect of potential variations, namely differential charging, is also visible in the ATS-5 and ATS-6 spectrograms. In Fig. 22, a spectrogram from day 334, 1969, for ATS-5 is presented for the detector looking parallel to satellite spin axis. Between 0500 and 1100 UT, a feathered pattern is visible in the ions, and a dropout is observed in the electrons below 750 eV. These patterns are not visible in the detectors looking perpendicular to the satellite spin axis. The explanation[98] is that a satellite surface near the detectors has become differentially charged relative to the detectors, resulting in a preferential focusing of the ion fluxes and a deficiency in the electron fluxes to the parallel detectors.

Given that differential charging can take place, whether through potential differences on adjoining surfaces or through charge deposition in dielectrics, arcing can occur. Arcing, defined as the rapid (~ n sec) rearrangement of charge by punch-through (internal breakdown of dielectrics), by flashover (dielectric breakdown), between surfaces, or between surfaces and space, is not well understood. In this volume, Balmain[99] discusses the details of surface discharges while Frederickson[15] discusses the source of dielectric breakdown. A typical arc-discharge pulse[100] is plotted in Fig. 23. Balmain[99] finds that surface discharges display characteristics which scale with variations in specimen area according to well-defined power laws: peak current scales as the 0.50 power of the area, released charge as 1.00, energy dissipation as 1.50, and pulse duration as 0.53.

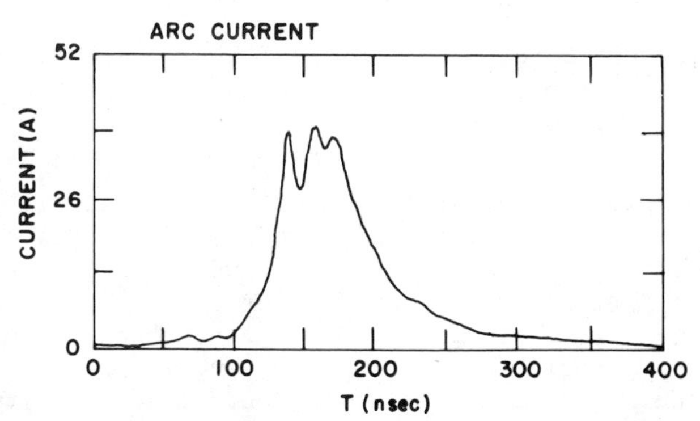

Fig. 23 Oscilloscope tracing of discharge currents into a conductor supporting a mylar specimen; from Ref. 100.

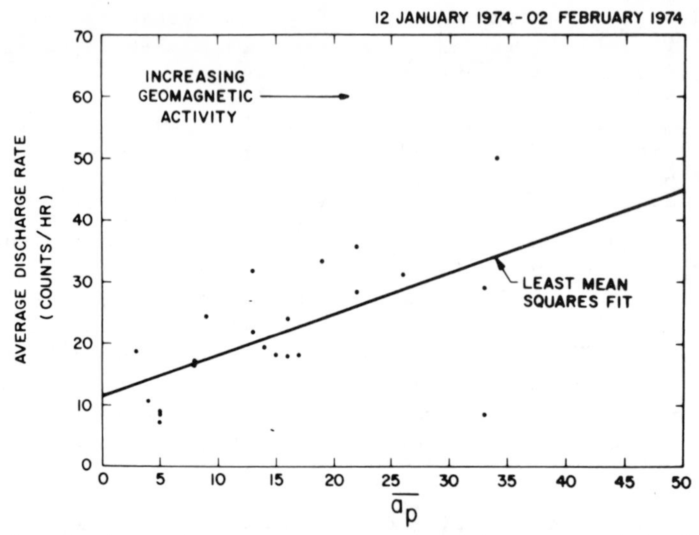

Fig. 24 Number of arcs per hour as a function of daily average Ap for a geosynchronous satellite; from Ref. 102.

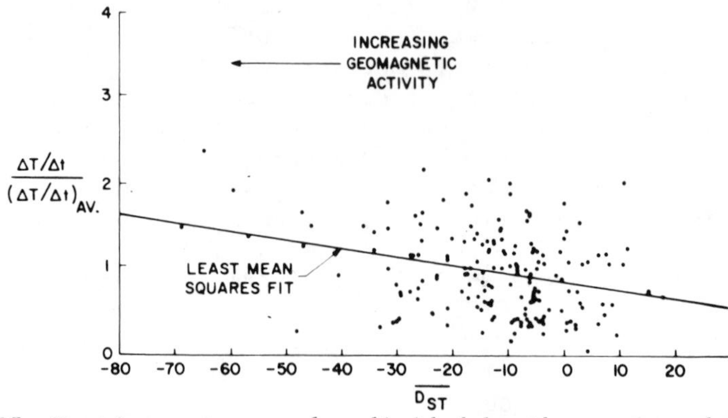

Fig. 25 Heating rate per day divided by the average heating rate for the entire time period as a function of the daily D_{st} index; from Ref. 101.

Another result for surface arcs[18,101] in the laboratory is that the breakdown potential on a negative surface varies from -500 V at low earth orbit to -10,000 V at geosynchronous orbit, implying that arcing should not be a common occurrence. In Fig. 24,[102] the arcing rate on a geosynchronous satellite shows a steady increase with the geomagnetic index Ap. Since arcing is common at low levels of geomagnetic activity, a discrepancy exists between laboratory and in-situ measurements, which underscores the need for further analysis. (The P78-2 satellite also has demonstrated arcing and high potential whereas the GEOS II satellite has not; this inconsistency further points up the lack of understanding in this key area.)

The effects of arcing are somewhat better understood than the process itself. The current pulse (Fig. 23) that results generates an electrical pulse in spacecraft systems either by direct current injection or by induced currents due to the associated electromagnetic wave.[103] Balmain[99] and Nanevicz and Adamo[101] have analyzed the effects of arc discharges on material surfaces. Balmain[99] gives several examples of holes and channels of micron size in dielectric surfaces. Nanevicz and Adamo[101] find large-scale physical damage to solar cells such as fracturing of the cover glass.

Contaminant ions, caused by thrusters (ionic or chemical) or outgassing of satellite materials, can be trapped within the satellite sheath and preferentially deposited on negatively-charged spacecraft surfaces. Cauffman[104] and others[105] have estimated that as much as 50 Å of material can be deposited on optical surfaces in as little as 100 days. In Fig. 25,[101] the heating rate of sensors on a geosynchronous satellite apparently rose with geomagnetic activity. This is believed to be due to increased contaminant deposition during periods of geomagnetic activity and, therefore, increased charging (see Fig. 8b). Such deposition also may alter secondary emission and photoelectron properties.

Another problem considered to be of possible importance is parasitic power loss at low orbital altitudes due to interactions between the ambient plasma and exposed high voltage sources such as solar cells. McCoy et al.[19] and Stevens[18] have estimated this effect at low earth orbit to result in a 10% power loss for voltages in excess of 5000 V. As discussed previously, at -100 V arcing begins at low earth orbit and is a significant problem for potentials of -1000 V at geosynchronous orbit. Apparently, then, the power loss and damage due to arcing is much more serious on high-voltage arrays than is parasitic power loss.

Several miscellaneous effects related indirectly to charging also have been observed. As discussed in a previous section, if the velocity of a satellite is comparable to the ambient ion velocities, a plasma wake[12] and associated plasma turbulence are generated which can measurably affect scientific measurements near the satellite (Fig. 4). The $\bar{v} \times \bar{B}$ electric field likewise distorts sensitive electric field measurements and leads to ohmic heating and drag (both low for most small satellites). A further effect is "multipacting," a process whereby the varying electric field on a transmitter antenna or similar device accelerates the local plasma in such a way that an electron avalanche can occur. That is, an electron initially accelerated by the transmitter field generates secondaries. If the transmitter is in resonance with the electron travel time between transmitter surfaces, then the secondaries will generate still more secondaries until an electron avalanche occurs.

VI. Prevention of Spacecraft Charging

Although varying the satellite-to-space potential allows the measurement of very low energy plasma,[38] charge buildup on satellite surfaces is not, in general, a desired phenomenon. In order to eliminate or at least limit the worst effects of spacecraft charging, several techniques have been developed. Although the obvious solution is to develop systems which can withstand the worst effects, this is not always a feasible or desirable method. Alternatives to this "brute force" method will be described in this section.

The simplest method for preventing spacecraft charging effects is simply to employ sound design techniques: use conducting surfaces where possible and proper grounding techniques. Although a large satellite-to-space potential can occur, differential charging, the major spacecraft-charging problem, is significantly reduced by these procedures. Several different methods have been developed to assure an adequately conducting surface. Nonconducting surfaces on the GEOS series of geosynchronous satellites, for example, were coated with indium oxide. Since solar cells are the primary nonconducting surfaces on GEOS, and since indium oxide is sufficiently transparent to sunlight that it does not degrade their operation, this technique has been quite successful[106] at keeping satellite potentials between 0 and -150 V, even in eclipse. Such coating techniques, however, can be expensive and difficult if large surfaces are involved.

A technique that may be applicable to large surfaces involves the use of electron and ion emitters. Grard[107] and Gonfalone et al.[108] discuss the application of such systems to actual satellite systems. The latter paper describes the successful application of a low current (mA) electron emitter on the ISEE-1 satellite. The ISEE-1 is in a highly elliptical (300 km to 23 R_E) orbit so that it spends a long time in the solar wind. The ISEE-1 results indicate that the electron cloud emitted by the experiment successfully clamped the potential of the satellite at a few volts positive to the ambient plasma. Purvis and Bartlett[17] report results from ion and electron emitters on the geosynchronous ATS-5 and ATS-6 satellites. These results indicate that electron emission alone reduces the potential from several thousand to several hundred volts negative. Use of an ion emitter and neutralizer successfully clamps the satellite-to-space potential near zero. Similar success has been demonstrated by the beam experiments on the SCATHA P78-2 satellite.[109] Since there may be some difficulties with these techniques if, as Frederickson[15] implies, grounding the surface enhances dielectric breakdown, further research is still needed.

Careful selection of satellite materials can reduce spacecraft charging. Although thermal control surfaces, which are necessary on many satellites, generally consist of dielectric materials, selection of the materials according to their secondary emission properties can substantially reduce charge buildup. Rubin et al.[110] have demonstrated that for materials with a secondary emission greater than one, the plasma temperature must be several times the energy at which this occurs if a satellite is to charge up. Garrett and Rubin[84] present observational data from ATS-5 in support of this and imply that the ambient electron temperature must be above 2 keV before the satellite begins to charge.

Several other techniques have been proposed. (For example, see Beattie and Goldstein[111] for methods of protecting the Jupiter probe.) Recent results even indicate that dielectric materials may be altered by the arcing process in a manner which greatly reduces future arcing.[112] It is currently thought, however, that the techniques previously described are adequate for charge prevention. Basically, spacecraft-charge prevention is a matter of good design technique: ground well and avoid cavities in which charge can be deposited.

VII. Conclusions

Spacecraft charging is a multi-faceted, complex problem which has been the subject of several conferences and books. A review such as is presented here can only give the reader an overall view of the problem and of methods to compute its effects and eliminate them. Starting with a description of the major current sources responsible for charge deposition, many key facets of the charging problem were discussed. These ranged from a review of the space plasma environment to an outline of spacecraft sheath theory. Several techniques for charge control were proposed. Since other papers in this volume are intended to put the details of these charging issues into perspective, this review has, hopefully, fit those details into the context of the overall spacecraft-charging problem.

References

[1]Langmuir, I., General Electric Review, Vol. 27, July 1924, pp. 449-458, and Physical Review B - Solid State, Vol. 28, March 1926, pp. 727-735, reprinted in Collected Works of Irving Langmuir, Vol. 4, edited by G. Suits, MacMillan, New York, 1961.

[2]Garrett, H. B., Mullen, E. G., Ziemba, E., and DeForest, S. E., "Modeling of the Geosynchronous Plasma Environment - Part 2, ATS-5 and ATS-6 Statistical Atlas," AFGL-TR-78-0304, 1978.

[3]Garrett, H. B., "Modeling of the Geosynchronous Orbit Plasma Environment - Part I," AFGL-TR-77-0258, 1977.

[4]Grard, R.J.L., "Properties of the Satellite Photoelectron Sheath Derived from Photoemission Laboratory Measurements," Journal of Geophysical Research, Vol. 78, June 1973, pp. 2885-2906.

[5]Whipple, E. C., "The Equilibrium Electric Potential of a Body in the Upper Atmosphere," NASA X-615-65-296, 1965.

[6]Garrett, H. B. and DeForest, S. E., "Time-Varying Photoelectron Flux Effects on Spacecraft Potential at Geosynchronous Orbit," Journal of Geophysical Research, Vol. 84, May 1979, pp. 2083-2088.

[7]Baragiola, R. A., Alonso, E. V., and Florio, A. O., "Electron Emission from Clean Metal Surfaces Induced by Low-Energy Light Ions," Physical Review B - Solid State, Vol. 19, January 1979, pp. 121-129.

[8]Sternglass, E. J., "Backscattering of Kilovolt Electrons from Solids," Physical Review B - Solid State, Vol. 95, July 1954, pp. 345-358.

[9]Garrett, H.B., "Spacecraft Potential Calculations - A Model," AFGL-TR-78-0116, 1978.

[10]Kasha, M. A., The Ionosphere and Its Interaction with Satellites, Gordon and Brush, New York, 1969.

[11]Wildman, P.J.L., "The Dynamics of a Rigid Body in the Space Plasma," published elsewhere in this volume.

[12]Parker, L. W., "Differential Charging and Sheath Asymmetry of Nonconducting Spacecraft due to Plasma Flows," Journal of Geophysical Research, Vol. 83, October 1978, pp. 4873-4876.

[13] Henderson, C. L. and Samir, U., "Observations of the Disturbed Region Around an Ionospheric Spacecraft," Planetary Space Science, Vol. 15, October 1967, pp. 1499-1513.

[14] Parker, L. W., "Calculation of Sheath and Wake Structure About a Pillbox Shaped Spacecraft in a Flowing Plasma," Proceedings of the Spacecraft Charging Technology Conference, edited by C. P. Pike and R. R. Lovell, AFGL-TR-77-0051/NASA TMX-73537, 1977, pp. 331-366.

[15] Frederickson, A. R., "Radiation Induced Dielectric Charging," published elsewhere in this volume.

[16] Liemohn, H. B., Holze, D. H., Copeland, R. C., and Leavens, W. M., "Ion Thruster Plasma Dynamics Near High Voltage Surfaces on Spacecraft," published elsewhere in this volume.

[17] Purvis, C. K. and Bartlett, R. O., "Active Control of Spacecraft Charging," published elsewhere in this volume.

[18] Stevens, N. J., "Space Environmental Interactions with Biased Spacecraft Surfaces," published elsewhere in this volume.

[19] McCoy, J. E., Konradi, A. and Garriott, O. K., "Current Leakage for Low Altitude Satellites," published elsewhere in this volume.

[20] Parker, L. W., "Plasma Sheath Effects and Equilibrium Voltage Distributions of Large High-Power Solar Arrays," Spacecraft Charging Technology - 1978, edited by R. C. Finke and C. P. Pike, NASA CP-2071/AFGL-TR-79-0082, 1979, pp. 341-357.

[21] Parker, L. W., "Effects of Plasma Sheath on Solar Power Satellite Array," AIAA Paper 79-1507, 1979.

[22] Jemiola, J., "Spacecraft Contamination: A Review," published elsehwere in this volume.

[23] Garrett, H. B., "Review of Quantitative Models of the 0 to 100 keV Near-Earth Plasma,"Reviews of Geophysics and Space Physics, Vol. 17, May 1979, pp. 397-417.

[24] Vasyliunas, V. M., "Magnetospheric Plasma," Solar Terrestrial Physics/1970, edited by E. R. Dyer, Reidel Publ. Co., Dordrecht, Netherlands, 1972, pp. 192-211.

[25] Young, D. T., "Ion Composition Measurements in Magnetospheric Modeling," Quantitative Modeling of the Magnetospheric Processes, Geophysical Monograph, Ser. 21, edited by W. P. Olson, American Geophysical Union, Washington, D.C., 1979.

[26] DeForest, S. E., "The Plasma Environment at Geosynchronous Orbit," Proceedings of the Spacecraft Charging Technology Conference, edited by C. P. Pike and R. R. Lovell, AFGL-TR-77-0051 /NASA TMX-73537, 1977, pp. 37-51.

[27] Chan, K. W., Sawyer, D. M., and Vette, J. I., "A Model of the Near-Earth Plasma Environment and Application to the ISEE-A and -B Orbit," NSSDC/WDC-A R&S 77-01, July 1977.

[28] Vasyliunas, V. M., "A Survey of Low-Energy Electrons in the Evening Sector of the Magnetosphere with OGO 1 and OGO 3," Journal of Geophysical Research, Vol. 73, May 1968, pp. 2839-2884.

[29] DeForest, S. E. and McIlwain, C. E., "Plasma Clouds in the Magnetosphere," Journal of Geophysical Research, Vol. 76, June 1971, pp. 3587-3611.

[30] Su, S.-Y. and Konradi, A., "Average Plasma Environment at Geosynchronous Orbit," Spacecraft Charging Technology - 1978, edited by R. C. Finke and C. P. Pike, NASA CP-2071/AFGL-TR-79-0082, 1979, pp. 23-37.

[31] Garrett, H. B., Rubin, A. G. and Pike, C. P., "Prediction of Spacecraft Potentials at Geosynchronous Orbit," in Solar-Terrestrial Prediction Proceedings, Vol II, edited by R. F. Donnelly, 1979.

[32] Frank, L. A., "On the Extra-Terrestrial Ring Current During Geomagnetic Storms," Journal of Geophysical Research, Vol. 72, August 1967, p. 3753-3762.

[33] Craven, J. D., "A Survey of Low-Energy (E > 5 keV) Electron Energy Fluxes Over the Northern Auroral Regions with Satellite Injun 4," Journal of Geophysical Research, Vol. 75, May 1970, pp. 2468-2474.

[34] Carpenter, D. L.,"Whistler Studies of the Plasmapause in the Magnetosphere, 1: Temporal Variations in the Position of the Knee and Some Evidence on Plasma Motions Near the Knee," Journal of Geophysical Research, Vol. 72, February 1966, pp. 693-709.

[35] Chappell, C. R., Harris, K. K., and Sharp, G. W., "The Morphology of the Bulge of the Plasmasphere," Journal of Geophysical Research, Vol. 75, July 1970, pp. 3848-3856.

[36] Chappell, C. R., "Recent Satellite Measurements of the Morphology and Dynamics of the Plasmasphere," Reviews of Geophysics and Space Physics, Vol. 10, November 1972, pp. 951-979.

[37] Carpenter, D. L. and Park, C. G., "On What Ionospheric Workers Should Know About the Plasmapause - Plasmasphere," Review of Geophysics and Space Physics, Vol. 11, February 1973, pp. 133-154.

[38] Lennartsson, W. and Reasoner, D. L., "Low-Energy Plasma Observations at Synchronous Orbit," Journal of Geophysical Research, Vol. 83, May 1978, pp. 2145-2156.

[39] DeForest, S. E. and Wilson, A. R., "A Preliminary Specification of the Geosynchronous Plasma Environment," DNA 3951T, 1976.

[40] Mauk, B. H. and McIlwain, C. E., "Correlation of Kp with the Substorm Plasma Sheet Boundary," Journal of Geophysical Research, Vol. 79, August 1974, pp. 3193-3196.

[41] Freeman, J. W., "Kp Dependence of the Plasma Sheet Boundary," Journal of Geophysical Research, Vol. 79, October 1974, pp. 4315-4317.

[42] Kivelson, M. G., Kaye, S. M., and Southwood, D. J., "The Physics of Plasma Injection Events," AGU Chapman Conference on Magnetospheric Substorms, UCLA Institute of Geophysics and Planetary Physics, Pub. No. 1878, 1978.

[43] Garrett, H. B. and DeForest, S. E., "An Analytical Simulation of the Geosynchronous Plasma Environment," Planetary and Space Science, Vol. 27, August 1979, pp. 1101-1109.

[44] McIlwain, C. E., "Plasma Convection in the Vicinity of the Geosynchronous Orbit," Earth's Magnetospheric Processes, edited by B. M. McCormac, Reidel Pub. Co., Hingham, MA, 1972, pp. 268-279.

[45] Kavanagh, L. D., Jr., Freeman, J. W., Jr. and Chen, A. J., "Plasma Flow in the Magnetosphere," Journal of Geophysical Research, Vol. 73, September 1968, pp. 5511-5519.

[46] Roederer, J. G. and Hones, E. W., Jr., "Electric Field in the Magnetosphere as Defined from Asymmetries in the Trapped Particle Flux," Journal of Geophysical Research, Vol. 75, July 1970, pp. 3923-3926.

[47] Wolf, R. A., "Effects of Ionospheric Conductivity on Convective Flow of Plasma in the Magnetosphere," Journal of Geophysical Research, Vol. 75, September 1970, pp. 4677-4698.

[48] Konradi, A., Semar, C. L., and Fritz, T. A., "Substorm-Injected Protons and Electrons and the Injection Boundary Model," Journal of Geophysical Research, Vol. 80, February 1975, pp. 543-552.

[49] Walker, R. J. and Kivelson, M. G., "Energization of Electrons at Synchronous Orbit by Substorm-Associated Cross-Magnetospheric Electric Fields," Journal of Geophysical Research, Vol. 80, June 1975, pp. 2074-2082.

[50] Chen, A. J., "Penetration of Low-Energy Protons Deep Into the Magnetosphere," Journal of Geophysical Research, Vol. 75, May 1970, pp. 2458-2467.

[51] Angerami, J. J. and Thomas, J. O., "Studies of Planetary Atmospheres, 1: the Distribution of Electrons and Ions in the Earth's Exosphere," Journal of Geophysical Research, Vol. 69, November 1964, pp. 4537-4559.

[52] Schunk, R. and Walker, J.C.G., "Thermal Diffusion in the Topside Ionosphere for Mixtures which Include Multiply Charged Ions," Planetary and Space Science, Vol. 17, May 1969, pp. 853-868.

[53] Mayr, H. G., Fontheim, E. G., Brace, L. H., Brinton, H. C., and Taylor, H. A., Jr., "A Theoretical Model of the Ionosphere Dynamics with Interhemisphere Coupling," Journal of Atmospheric and Terrestrial Physics, Vol. 34, October 1972, pp. 1659-1680.

[54] Lemaire, J. and Scherer, M., "Exospheric Models of the Topside Ionosphere," Space Science Reviews, Vol. 15, March 1974, pp. 591-640.

[55] Chiu, Y. T., Luhmann, J. G., Ching, B. K., and Boucher, D. J., Jr., "An Equilibrium Model of Plasmaspheric Composition and Density," Journal of Geophysical Research, Vol. 84, March 1979, pp. 909-916.

[56]Roederer, J. G. and Hones, E. W., Jr., "Motion of Magnetospheric Particle Clouds in a Time-Dependent Electric Field Model," Journal of Geophysical Research, Vol. 79, April 1974, pp. 1432-1438.

[57] Smith, P. H., Bewtra, N. K., and Hoffman, R. A., "Inference of the Ring Current Ion Composition by Means of Charge Exchange Decay," NASA TM-79611, 1978.

[58] Smith, P. H., Hoffman, R. A., and Bewtra, N. K., "A Visual Description of the Dynamical Nature of Magnetospheric Particle Convection in a Time-Varying Electric Field," Transactions of the American Geophysical Union, Vol. 59, April 1978, pp. 361-368.

[59] Smith, R. H., Bewtra, N. K., and Hoffman, R. A., "Motions of Charged Particles in the Magnetosphere Under the Influence of a Time-Varying Large Scale Convection Electric Field," Quantitative Modeling of the Magnetospheric Processes, Geophysical Monograph, Ser. 21, edited by W. P. Olson, American Geophysical Union, Washington, DC, 1979.

[60]Ejiri, M., Hoffman, R. A., and Smith, P. H., "The Convection Electric Field Model for the Magnetosphere Based on Explorer 45 Observations," NASA/GSFC, X-625-77-108, 1977, also in Journal of Geophysical Research, Vol. 83, October 1978, pp. 4811-4815.

[61] Grebowsky, J. M. and Chen, A. J., "Effects of Convection Electric Field on the Distribution of Ring Current Type Protons," Planetary and Space Science, Vol. 23, July 1975, pp. 1045-1052.

[62] Jaggi, R. K. and Wolf, R. A., "Self-Consistent Calculation of the Motion of a Sheet of Ions in the Magnetosphere," Journal of Geophysical Research, Vol. 78, June 1973, pp. 2852-2866.

[63] Chen, A. J. and Wolf, R. A., "Effects on the Plasmasphere of a Time-Varying Convection Electric Field," Planetary and Space Science, Vol. 20, April 1972, pp. 483-509.

[64] Wolf, R. A., "Calculation of Magnetospheric Electric Fields," Magnetospheric Physics, edited by B. M. McCormac, Reidel Pub. Co., Dordrecht, Netherlands, 1974, pp. 167-177.

65
Southwood, D. J., "The Role of Hot Plasma in Magnetospheric Convection," Journal of Geophysical Research, Vol. 35, December, 1977, pp. 5512-5520.

66
Harel, M., Wolf, R. A., and Reiff, P. H., "Preliminary Report of the First Computer Run Simulating the Substorm-Type Event of 19 September 1976," Appendix A, Annual NSF Report for Grants ATM-74-21185 and ATM-74-21185., 1978.

67
Rothwell, P. L., Rubin, A. G., and Yates, G. K., "A Simulation Model of Time-Dependent Plasma-Spacecraft Interactions," Proceedings of the Spacecraft Charging Technology Conference, edited by C. P. Pike and R. R. Lovell, 1977, pp. 389-412.

68
Rubin, A. G., Katz, I., Mandell, M., Schnuelle, G., Steen, P., and Roche, I., "A 3-D Spacecraft Charging Computer Code," published elsewhere in this volume.

69
Parker, L. W., "Plasmasheath-Photosheath Theory for Large High-Voltage Structures," published elsehwere in this volume.

70
Garrett, H. B., "Low Energy Magnetospheric Plasma Interactions with Space-Systems - The Role of Predictions," Proceedings of 1979 Solar-Terrestrial Prediction Workshop, 1979.

71
Garrett, H. B., Schwank, D. C., Higbe, P. R., and Baker, D. N., "Comparison Between the 30-80 keV Electron Channels on ATS-6 and 1976-059A During Conjunction and Application to Spacecraft Charging Prediction," Journal of Geophysical Research, (accepted for publication).

72
Goldstein, R. and Divine, N., "Plasma Distribution and Spacecraft Charging Modeling Near Jupiter," Proceedings of the Spacecraft Charging Technology Conference, edited by C. P. Pike and R. R. Lovell, AFGL-TR-77-0051/NASA TMX-73537, 1977, pp. 131-242.

73
Robinson, P. A., Jr. and Holman, A. B., "Pioneer Venus Spacecraft Charging Model," Proceedings of the Spacecraft Charging Technology Conference, edited by C. P. Pike and R. R. Lovell, AFGL-TR-77-0051/NASA TMX-73537, 1977, pp. 297-308.

74
Schwenn, R., Rosenbauer, H., and Muhlhauser, K. H., "The Solar Wind During STIP II Interval: Stream Structures, Boun-

daries, Shocks, and Other Features as Observed by the Plasma Instruments on Helios-1 and Helios-2," Contributed Papers to the Study of Travelling Interplanetary Phenomena/1977, edited by M. A. Shea, D. F. Smart, and S.T. Wu, AFGL-TR-77-0309, 1977.

[75] Smith, E. J. and Wolfe, J. H., "Pioneer 10, 11 Observations of Evolving Solar Wind Streams and Shocks Beyond 1 AU," Study of Travelling Interplanetary Phenomena, edited by M. A. Shea et al., Reidel Co., Hingham, MA, 1977, pp. 227-257.

[76] Chen, F. F., "Electric Probes," Plasma Diagnostic Techniques, edited by R. H. Huddlestone and S. L. Leonard, Academic Press, Inc., New York, 1965, pp. 113-200.

[77] Whipple, E. C., Jr., "Modeling of Spacecraft Charging," Proceedings of the Spacecraft Charging Technology Conference, edited by C. P. Pike and R. R. Lovell, AFGL-TR-77-0051/NASA TMX-73537, 1977, pp. 225-235.

[78] Knott, K., "The Equilibrium Potential of a Magnetospheric Satellite in an Eclipse Situation," Planetary and Space Science, Vol. 20, August 1972, pp. 1137-1146.

[79] Rosen, A., (ed.) Spacecraft Charging by Magnetospheric Plasma, Progress in Astronautics and Aeronautics, Vol. 47, AIAA, New York, 1976.

[80] Pike, C. P. and Lovell, R. R., (ed.), Proceedings of the Spacecraft Charging Technology Conference, AFGL-TR-77-0051/ NASA TMX-73537, 1977.

[81] Finke, R. C. and Pike, C. P., (ed.), Spacecraft Charging Technology-1978, NASA CP-2071/AFGL-TR-79-0082, 1979.

[82] Prokopenko, S.M.L. and Laframboise, J.G., "Prediction of Large Negative Shaded-Side Spacecraft Potentials," Proceedings of the Spacecraft Charging Technology Conference, AFGL-TR-77-0051/NASA TMX-73537, 1977, pp. 369-387.

[83] Mott-Smith, H., Jr., and Langmuir, I., "The Theory of Collectors in Gaseous Discharges," Physical Review B - Solid State, Vol. 28, May 1926, pp. 727-760.

[84] Garrett, H. B. and Rubin, A. G., "Spacecraft Charging at Geosynchronous Orbit-Generalized Solutions for Eclipse Passage," Geophysical Research Letters, Vol. 5, December 1978, pp. 865-868.

[85] Garrett, H. B. and Gauntt, D., "Spacecraft Charging During Eclipse Passage," published elsewhere in this volume.

[86] Inouye, G. T., "Spacecraft Potentials in a Substorm Environment," Progress in Astronautics and Aeroanutics: Spacecraft Charging by Magnetospheric Plasma, Vol. 47, edited by A. Rosen, AIAA, New York, 1976, pp. 103-120.

[87] Massaro, M. J., Green, T., and Ling, D., "A Charging Model for Three-Axis Stabilized Spacecraft," Proceedings of the Spacecraft Charging Technology Conference, AFGL-TR-77-0051/NASA TMX-73537, 1977.

[88] Parker, L. W., "Computer Methods for Satellite Plasma Sheath in Steady-State Spherical Symmetry," Lee W. Parker, Inc., 1975, AFCRL-TR-75-0410.

[89] Parker, L. W., "Theory of Electron Emission Effects in Symmetric Probe and Spacecraft Sheaths," Lee W. Parker, Inc., 1976, AFGL-TR-76-0294.

[90] Parker, L. W., "Time-Dependent Sheath Model in Cylindrical Geometry for SCATHA Spacecraft," Lee W. Parker, Inc., 1979.

[91] Laframboise, J. G., "Theory of Spherical and Cylindrical Langmuir Probes in a Collisionless Maxwellian Plasma at Rest," University of Toronto, Institute of Aerospace Studies, June 1966, Report No. 100.

[92] Reiff, P. H., Freeman, J. W., and Cooke, D. L., " Environmental Protection of the Solar Power Satellite," published elsewhere in this volume.

[93] Katz, I., Cassidy, J. J., Mandell, M. J., Schnuelle, G. W., Steen, P. G. and Roche, J. C., "The Capabilities of the NASA Charging Analyzer Program," Spacecraft Charging Technology - 1978, edited by R. C. Finke and C. P. Pike, NASA CP-2071/AFGL-TR-79-0082, 1979, pp. 101-122.

[94] Katz, I., Parks, E. E., Wang, S. and Wilson, A., "Dynamic Modeling of Spacecraft in a Collisionless Plasma," Proceedings of the Spacecraft Charging Technology Conference, edited by C. P. Pike and R. R. Lovell, AFGL-TR-77-0051/NASA TMX-73537, 1977, pp. 319-333.

[95] Schnuelle, G. W., Parks, D. E., Katz, I., Mandell, M. J., Steen, P. G., Cassidy, J. J., and Rubin, A., "Charging Analysis of the SCATHA Satellite," Spacecraft Charging Technology-1978, edited by R. C. Finke and C. P. Pike, NASA CP-2071/AFGL-TR-79-0087, 1979, pp. 123-143.

[96] Roche, J. C. and Purvis, C. P., "Comparison of NASCAP Predictions with Experimental Data," Spacecraft Charging Technology - 1978, edited by R. C. Finke and C. P. Pike, NASA CP-2071/AFGL-TR-79-0082, 1979, pp. 144-157.

[97] McPherson, D. A. and Schober, W. R., "Spacecraft Charging at High Altitudes: The SCATHA Satellite Program," Progress in Astronautics and Aeronautics: Spacecraft Charging by Magnetospheric Plasmas, Vol. 47, edited by A. Rosen, AIAA, New York, 1976.

[98] DeForest, S. E., "Spacecraft Charging at Synchronous Orbit," Journal of Geophysical Research, Vol. 77, February 1972, pp. 651-659.

[99] Balmain, K. G., "Surface Discharge Effects," published elsewhere in this volume.

[100] Balmain, K. G., Cochanski, M., and Kremer, D. C., "Surface Micro-Discharges on Spacecraft Detectors," edited by C. P. Pike and R. R. Lovell, Proceedings of the Spacecraft Charging Technology Conference, AFGL-TR-77-0051/NASA TMX-73537, 1977, pp. 519-526.

[101] Nanevicz, J. E. and Adamo, R. C., "Arcing Occurrence and Its Effects on Space Systems," published elsewhere in this volume.

[102] Shaw, R. R., Nanevicz, J. E., and Adamo, R. C., "Observations of Electrical Discharges Caused by Differential Satellite Charging," Progress in Astronautics and Aeronautics: Spacecraft Charging by Magnetospheric Plasmas, Vol. 47, edited by A. Rosen, AIAA, New York, 1976, pp. 61-76.

[103] Mindel, I. N., "DNA EMP Awareness Course Notes," 3rd ed., DNA 2722T, 1977.

[104] Cauffman, D. P., "Ionization and Attraction of Neutral Molecules to a Charged Spacecraft," SAMSO-TR-73-263, 1973.

105
Jemiola, J. M., Proceedings of the USAFA/NASA International Spacecraft Contamination Conference, AFML-TR-78-190/NASA-CP-2039, 1978.

106
Wrenn, G., personal communication, November 1978.

107
Grard, R. J. L., "The Multiple Applications of Electron/Emitters in Space," Proceedings of the Spacecraft Charging Technology Conference, edited by C. P. Pike and R. R. Lovell, AFGL-TR-77-0051/NASA TMX-73537, 1977, pp. 203-221.

108
Gonfalone, A., Pedersen, Fahleson, U. V., Falthammar, C-G., Mozer, F. S. and Tobert, R. B., "Spacecraft Potential Control on ISEE-1,"Spacecraft Charging Technology-1978, NASA CP-2071/AFGL-TR-79-0082, 1979, pp. 256-267.

109
Cohen, H., personal communication, June 1979.

110
Rubin, A. G., Rothwell, P. L. and Yates, G. K., "Reduction of Spacecraft Charging Using Highly Emissive Surface Materials," Proceedings of the 1978 Symposium on the Effects of the Ionosphere on Space and Terrestrial Systems, Naval Research Lab., Washington, D.C., 1978.

111
Beattie, J. R. and Goldstein, R., "Active Spacecraft Potential Control System for the Jupiter Orbiter with Probe Mission," Proceedings of the Spacecraft Charging Technology Conference, edited by C. P. Pike and R. R. Lovell, AFGL-TR-77-0051/NASA TMX-73537, 1977, pp. 143-165.

112
Frederickson, A. R., personal communication, November 1979.

SPACECRAFT CHARGING DURING ECLIPSE PASSAGE

H. B. Garrett* and D. M. Gauntt+
Air Force Geophysics Laboratory, Hanscom Air Force Base, Mass.

Abstract

The passage of a space structure through the Earth's (or moon's) shadow is attended by a change in the photoelectron flux from the surface of the spacecraft. If, as is often observed in and near geosynchronous orbit, the ambient electron flux is sufficient, spacecraft charging will result. In this paper, the detailed variation of the photoelectron flux will be modeled. Using this and other simple models of the spacecraft charging phenomena, the changing potential on a typical geosynchronous satellite will be estimated. The model will then be extended to encompass the case of a large (10-km diam) passive circular structure (the space-based radar) and of a large (100 km^2) passive square structure (the solar power satellite). Depending on the material, significant potential gradients are possible across such objects. Although little danger is expected from eclipse passage if proper design criteria are followed, the results do indicate the need for caution in the design of any spacecraft expected to spend time in the geosynchronous (or similar) plasma environment.

I. Introduction

The problem of spacecraft charging[1,2] is generally recognized and at least qualitatively understood by the space physics community (see books by Rosen,[3] Pike and Lovell,[4] Finke and Pike[5]). An appreciation of the possible effects of charge buildup on spacecraft surfaces has resulted in the development of very sophisticated codes capable of modeling in detail the various spacecraft charging processes.[6-9] Several less sophisticated but more efficient models have been advanced by various groups [10-13] for the analysis of specific effects and the calculation of these effects on actual systems. The need has arisen recently to apply these models to large

* Research Physicist.

+ Student Aide.

space structures.[14,15] As yet, however, only a few[16-19] of these modeling efforts have considered the effects of temporally varying solar illumination and the associated variation of photoelectron flux. The purpose of this paper is to review current knowledge in this area and to analyze the effects on space structures.

In the first section of this article, we will review the theory of spacecraft charging as it applies to the problem of entry and exit of a passive satellite into and out of the Earth's (or moon's) shadow. Passive means that no potentials other than those due to the ambient environment are considered. A simple model capable of explaining the phenomenon will be reviewed and compared with actual observations from the ATS-5 and ATS-6 geosynchronous satellites. The model will then be modified in order to estimate the effects of a time-varying photoelectron flux on two large space structures, the space-based radar (SBR) and the solar power satellite (SPS), when they are not operating. The possible impact on passive space structures of the resulting potential will be discussed briefly.

II. General Problem

A. Charge Balance

The basic problem in spacecraft charging is to determine the potential of a point relative to ground (taken to be the ambient space plasma in this study) such that all the possible currents to that point balance to zero. Expressed algebraically:

$$I_E - (I_I + I_{SE} + I_{SI} + I_{BSE} + I_P + I_L) + I_T = 0 \qquad (1)$$

where
- I_E = ambient electron current
- I_I = ambient ion current
- I_{SE} = secondary electron current due to ambient electrons
- I_{SI} = secondary electron current due to ambient ions
- I_{BSE} = backscattered electron current due to ambient electrons
- I_P = photoelectron current
- I_L = leakage current to other satellite surfaces
- I_T = current due to ion (or electron) thrusters

All of these currents are time varying, and at any given instant in time, Eq. (1) may not be satisfied. The system will, however, respond in order to satisfy Eq. (1). In this paper the effects of I_T will be ignored and I_L will be assumed to be due only to resistances. (Capacitance, inductance, and imposed potentials, present when the SPS is generating power, are ignored). Under these circumstances, the time constant for Eq. (1) to be satisfied is on the order of milliseconds,[6] while the characteristic decay time of solar illumination during eclipse passage is on the order of seconds. Thus, Eq. (1) is an adequate starting point for the discussion of the effects of eclipse passage on space structures.

All the currents in Eq. (1) are complex functions of vehicle potential, shape, angle of illumination, material, and in particular, the details of the resulting plasma sheath and wake surrounding the structure. The sophisticated models described in the introduction are capable, at great cost in computing time, of taking all these effects into account. Fortunately (as will be discussed shortly), efficient algorithms do exist so that the satellite-to-space potential can be found using simple approximations. In subsequent sections various means of estimating the currents in Eq. (1) using these approximations will be presented.

B. Photoelectron Emission

In this section a detailed model of the photoelectron flux is presented. First, the geometry of the sun-Earth-satellite system is reviewed. A simple model of the atmospheric attenuation[19] is next described. Then, since the exact atmospheric attenuation with wavelength and the photoemission properties of satellite materials are not known accurately, an iterative procedure using actual data is described. This procedure allows an absolute determination of the photoelectron flux escaping from a satellite and its sheath.

The fundamental coordinate system employed in calculating satellite eclipse passage is presented in Fig. 1. Once the satellite position in space relative to the Earth and the direction vector to the center of the sun are determined, the quantity X_m (the minimum altitude above the surface of the Earth of a ray to the center of the sun from the satellite) can be found.[17] As will be seen, the computation of the solar illumination as a function of X_m allows the removal of variations in time and position between various eclipse passages at geosynchronous orbit, greatly simplifying the computations.

Following the procedure outlined in Refs. 17 and 19, once X_m is known the atmospheric attenuation is calculated in a straightforward manner. The satellite illumination by the sun, as a function of height X and wavelength λ for grazing

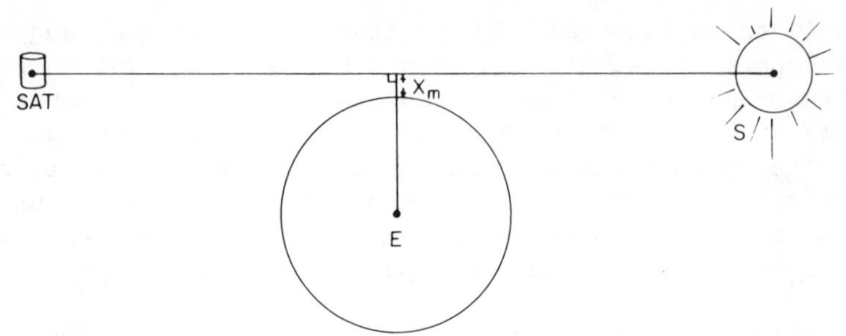

Fig. 1 Illustration of the meaning of X_m, the altitude of the minimum ray path from the satellite to the center of the sun.

incidence[20] is approximately

$$\phi(\lambda, X) = \phi_\infty(\lambda)\, e^{-\tau} \tag{2}$$

where

τ = $n(X)\, \sigma_a(\lambda)\, (2\pi R_e H)^{1/2}$
R_e = Earth's radius
$n(X)$ = number density of atmospheric constituent
H = scale height of constituent at X
X = minimum ray path altitude from arbitrary point on solar disk to satellite
$\phi_\infty(\lambda)$ = unattenuated solar flux at wavelength λ
$\sigma_a(\lambda)$ = absorption cross section at wavelength λ

Considering the lack of knowledge about the exact wavelengths involved, there is no justification at present for assuming anything more complex than a single-constituent exponential atmosphere. Thus

$$n(X) = n_o \exp\left[-(X-X_o)/\delta X\right] \tag{3}$$

where n_o is density at reference altitude X_o, and δX is scale height. Then the unattenuated sunlight as a function of X (where variations in λ are ignored) is

$$\phi(X) = \phi_\infty \exp\left(-\exp\left[-(X-Z_o)/\delta Z\right]\right) \tag{4}$$

where Z_o is reference altitude (to be determined), δZ is scale height (to be determined), and ϕ_∞ is the total flux at $X = \infty$ (i.e., no attenuation). The percentage of illumination, $F(X_m)$, reaching the satellite is obtained by integrating Eq. (4) over all X's appropriate to that X_m and dividing by the result for no attenuation ($X_m = \infty$).

Injun 5, a highly inclined (81 deg), low altitude (2543x677 km) satellite, had a spherical electrostatic analyzer which gave direct measurements of the photoelectron current as the satellite passed in and out of eclipse,[21] allowing a calibration of the attenuation model. The percent of photoelectron current as a function of X_m is given in Fig. 2. Also plotted are model predictions of the percent of illumination for values of Z_o and δZ corresponding to no atmosphere ($Z_o = 0$, $\delta Z = 0$), an ozone atmosphere ($Z_o = 76.4$ km, $\delta Z = 4.67$ km),[22] and an empirical fit ($Z_o = 90$ km, $\delta Z = 40$ km) to the data. This latter fit (henceforth called the Injun 5 model) will allow an estimate of the photoelectron current as a function of X_m.

Returning to Eq. (1), given all the relevant currents, except for the photoelectron current, to a single point

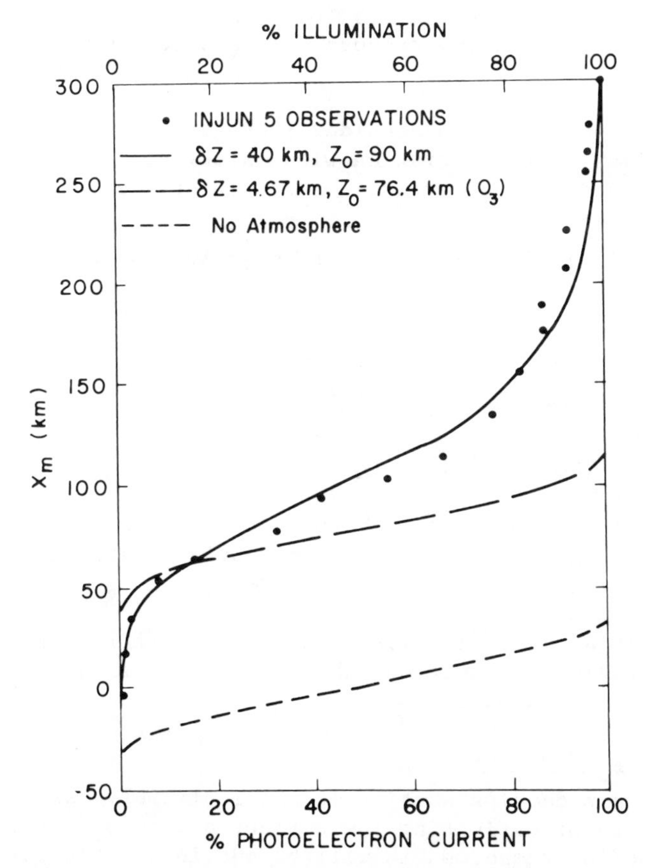

Fig. 2 Percent of solar illumination reaching the spacecraft and percent of photoelectron current measured by Injun 5 as functions of X_m, the minimum ray path altitude.

(I_L = 0 in this case) and the potential as a satellite passes into and out of eclipse, it is possible to calculate the residual current necessary to give the observed potential as a function of X_m. Since the percent of photoelectron current by the procedure just outlined is known as a function of X_m, Z_o, and δZ, the absolute photoelectron current can be determined by empirically fitting the residual current. This is done in Fig. 3 for 21 eclipse passages of the geosynchronous satellite ATS-5 for two cases: no atmosphere and the Injun 5 model. Knowing the shape of ATS-5, the absolute current density J_{po} can be calculated from the number flux and is found to be .11 nA/cm^2 and .55 nA/cm^2, respectively. Based on the theoretical estimates[23] listed in Table 1, the Injun 5 model gives the more realistic value, a value near that of graphite. Since, however, the value determined by this method is the total flux escaping from ATS-5 (not the emitted flux), it should be taken as a lower bound on the actual emission rate at the satellite surface.

The final point to consider in regards to the photoelectron flux is its variation with satellite potential, V. Based on theoretical results,[23] DeForest has derived the following formula:

$$I_p\,(X_m, V) = I_{po} \cdot F\,(X_m) \qquad V < 0$$

$$= \frac{I_{po} \cdot F\,(X_m)}{\left(1 + \frac{V}{0.7}\right)^2} \qquad V > 0$$

where

$$F\,(X_m) = \phi(X_m)\,/\,\phi_\infty$$

I_{po} = total photoelectron current for $X_m = \infty$ and the appropriate satellite cross-sectional area

This is only an approximation, but since we will consider almost exclusively conditions where $V < 0$, it is quite adequate for modeling purposes.

C. Potential Modeling

In the preceding section it was implicitly assumed that models exist for estimating the ambient and secondary currents. In this section these models will be outlined. Methods for estimating the satellite-ambient potential for a single point and a simple extension to large structures will be discussed. As much of the theory was taken from Whipple,[1] the reader is referred to that text for greater detail.

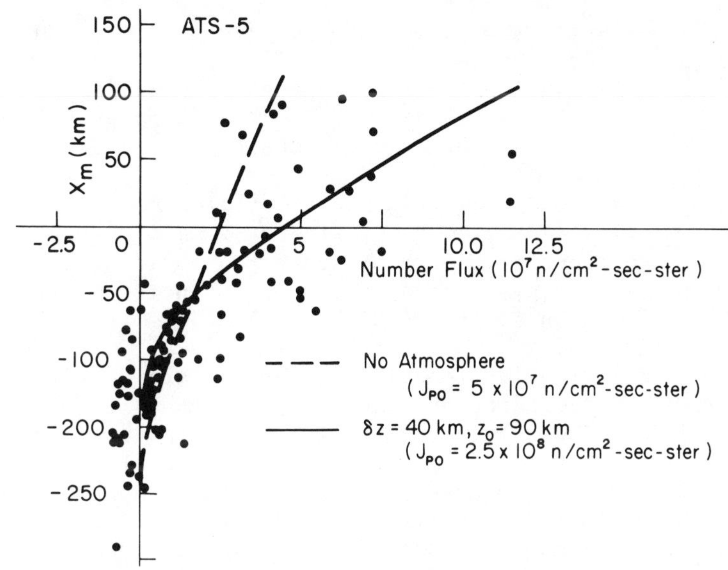

Fig. 3 Residual (photoelectron) current estimated for 21 ATS-5 eclipse passages. (Two atmospheric attenuation profiles have been fitted to the data.)

As presented in Whipple[1] and expanded in DeForest[2] and Garrett,[13] the currents to a single point in the thick sheath approximation are:

$$I_E = \frac{A2\pi q}{m_E^2} \int_0^\infty f_E(E)\ E\ dE \tag{5a}$$

$$I_I = \frac{A2\pi q}{m_I^2} \int_0^\infty f_I(E)\ E\ dE \tag{5b}$$

$$I_{SE} = \frac{A2\pi q}{m_E^2} \int_0^\infty g(E')\ dE' \int_0^\infty \delta_E(E)\ f_E(E)\ E\ dE \tag{5c}$$

$$I_{SI} = \frac{A2\pi q}{m_I^2} \int_0^\infty g(E')\ dE' \int_0^\infty \delta_I(E)\ f_I(E)\ E\ dE \tag{5d}$$

$$I_{BSE} = \frac{A2\pi q}{m_E^2} \int_0^\infty dE' \int_{E'}^\infty \frac{G}{E}\left(\frac{E'}{E}\right) f_E(E)\ E\ dE \tag{5e}$$

where
A = surface area of structure
q = electronic charge
m_E, m_I = mass of electrons and ions
E = energy of incident particle
E' = energy of emitted particle

f_E, f_I = distribution functions for incident electrons and ions

g = emission spectrum for secondary electrons (integral is assumed to be 1 if potential is less than 0, or 0 if potential is greater than 0)

δ_E = secondary electron yield for electrons

δ_I = secondary electron yield for ions

$G(E'/E)$ = percent of electrons scattered at a given energy E' as a result of an incident electron at energy E

If the distribution functions f_E and f_I are known, then Eq. (5) can be integrated numerically provided δ_E, δ_I, and G are specified. These latter variables are material-dependent and, currently, not well defined for most materials. In order to solve Eq. (5), functions for δ_E, δ_I, and G appropriate to aluminum were assumed. To make the model general, a calibration procedure using ATS-5 data was employed in order to correct for the effects of other materials.

Specifically, the actual distribution functions between 50 eV and 50 keV from the University of California experiment[25] on ATS-5 were inserted in Eq. (5), and the detailed integrations were carried out numerically. The spectra immediately before eclipse entry or after eclipse exit (termed "sunlit"), and the spectra immediately after eclipse entry or before eclipse exit (termed "eclipsed"), were integrated to give I_E, I_I, I_{SE}, I_{SI}, and I_{BSE}. V was found by inserting trial-values in Eq. (1) until a value was found to satisfy the equation.

Table 1 Photoelectron emission characteristics[a]

Material	Work function, eV	Saturation Flux, 10^{12} n_e/sec-m^2	Saturation Current Density, $\mu A/m^2$
Aluminum oxide	3.9	260	42
Indium oxide	4.8	190	30
Gold	4.8	180	29
Stainless steel	4.4	120	20
Aquadag	4.6	110	18
LiF on Au	4.4	90	15
Vitreous carbon	4.8	80	13
Graphite	4.7	25	4

[a]From Ref. 23

I_{SE}, I_{SI}, and I_{BSE} were each multiplied by a calibration constant which was varied until the potentials predicted by this procedure agreed, in a least squares sense, with the observed potentials. The results, for calibration constants of 1.3, .55, and .4 for I_{SE}, I_{SI}, and I_{BSE}, respectively, are plotted in Fig. 4. The agreement is quite good ($\pm$500 V for potentials between 0 and -10,000 V), indicating the validity of the procedure.

Considerable experimentation[25] has indicated that the sum of two Maxwell-Boltzmann distribution functions (or, in many cases, a single distribution function) well approximates the geosynchronous plasma population, either electrons or ions. Thus, at the risk of some loss in quantitative accuracy but a significant gain in computation efficiency, f_E and f_I can be assumed to be Maxwell-Boltzmann distributions of the form

$$f(E) = n \left(\frac{m}{2\pi T}\right)^{3/2} e^{-E/T} \tag{6}$$

where

- f = distribution function for electrons (f_E) or ions (f_I)
- T = Maxwell-Boltzmann temperature for electrons (T_E) or ions (T_I) (T is assumed to have the same units as E (eV))
- n = number density, for electrons (n_E) or ions (n_I)

We can approximate the integrals in Eq. (5) by (assuming a thick sheath)[26]

$$I_E = \frac{Aqn_E}{2} \left(\frac{2T_E}{\pi m_E}\right)^{\frac{1}{2}} \chi_E \tag{7a}$$

$$I_I = \frac{Aqn_I}{2} \left(\frac{2T_I}{\pi m_I}\right)^{\frac{1}{2}} \chi_I \tag{7b}$$

$$I_{SE} = I_E \cdot \begin{cases} 0 & V > 0 \\ 0.843 - 0.842\, e^{-0.0286 T_E} & T_E \leq 200 \text{ eV} \quad V \leq 0 \\ 0.143 + 0.740\, e^{-0.0003 T_E} & T_E \geq 200 \text{ eV} \quad V \leq 0 \end{cases} \tag{7c}$$

$$I_{SI} = I_I \cdot \begin{cases} 0 & V > 0 \\ (4.78 - 0.653\, e^{0.974\times10^{-4} V}) - (0.53 + 3.78\, e^{0.00018 V}) \cdot \\ \cdot\, e^{-(0.572\times10^{-4} + 0.00018\, e^{0.00014 V}) T_I} & V \leq 0 \;; \end{cases} \tag{7d}$$

$$I_{BSE} = I_E^{\bullet} \begin{cases} \frac{2}{9} & V \leq 0 \\ (0.253 - 0.0027\, e^{0.072V}) + (0.0867 + 0.657\, e^{-0.184V}) \cdot \\ \cdot\, e^{-(0.00017 + 0.0098\, e^{-0.231V})\, T_E} & V > 0 \end{cases} \tag{7e}$$

where it was assumed that

$$\delta_E \simeq \begin{cases} 2.55\,(\log E)^{9.8}\, e^{-(\log E)/0.25} & V < 0;\ E > 1 \text{ eV} \\ 0 & \text{otherwise} \end{cases}$$

$$\delta_I \simeq \begin{cases} 5e^{-4060/(E+300)} & V < 0,\ E \geq 700 \text{ eV} \\ .086 & V < 0,\ E \leq 700 \text{ eV} \\ 0 & V > 0 \end{cases}$$

$$G \simeq \frac{4}{3} \left(\frac{E'}{E}\right) \left(1 - \frac{E'}{E}\right)$$

The functions χ_E and χ_I are (8a)

$$\chi_E = \begin{cases} 1 + \frac{qV}{T_E} & V > 0 \\ \exp(qV/T_E) & V < 0 \end{cases}$$

$$\chi_I = \begin{cases} \exp(-qV/T_I) & V > 0 \\ 1 - \frac{qV}{T_I} & V < 0 \end{cases} \tag{8b}$$

These approximations are appropriate for a thick sheath, spherical symmetry, and isotropic flux. Equations (7c)-(7e), δ_E, and δ_I are for aluminum and must be corrected as previously described. Results for the simple two Maxwell-Boltzmann model and Eq. (7) are plotted in Fig. 4. The error is, as would be expected, larger (about $\pm$1200 V) than when actual spectra are used. Such accuracy, considering the present status of spacecraft charging models, is quite adequate.

The equations for I_E, I_I, I_{SE}, I_{SI}, and I_{BSE} have been derived for actual spectra and for the assumption of a Maxwell-Boltzmann distribution. Observational data indicate these equations are adequate for computing the potential befor a small satellite like ATS-5 at geosynchronous orbit where the size of the satellite ($\sim$1-10m) is smaller than the

Debye length (~250m) when the satellite is in eclipse. The extension to large structures is not straightforward and only recently have attempts been made to model the detailed sheath surrounding a large structure and the deviation from a thick-sheath approximation.[14,15] We will make the assumption that, to first order, the thick sheath approximation is extendable to large structures if the structure can be approximated by segments having a width less than the Debye length (~100 m). Further, this assumption will be tested by comparing the results with the one-dimensional (planar), thin-sheath model.[8]

That is, for a Maxwell-Boltzmann distribution

$$\chi_I = 1 \quad V < 0 \qquad (9)$$
$$\chi_E = 1 \quad V > 0$$

The other terms remain the same.

Given that a large structure can be approximated by an electrical analog consisting of segments (or nodes) connected

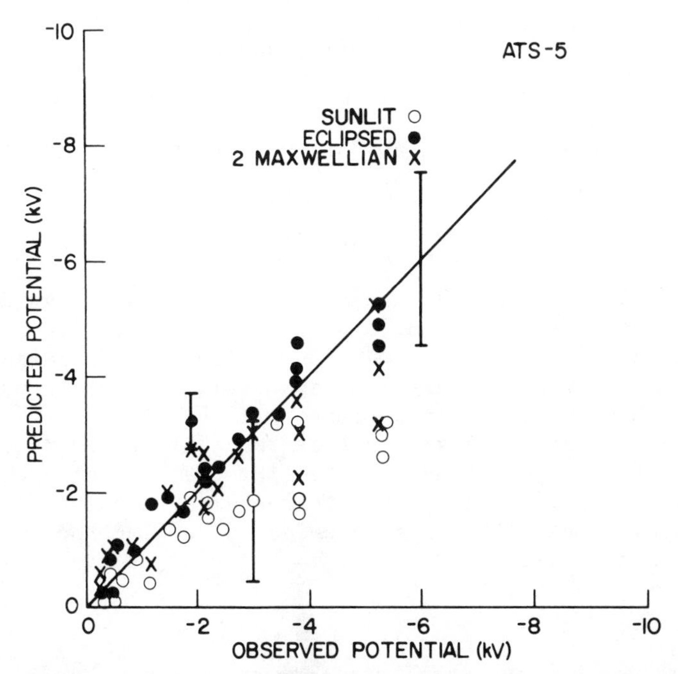

Fig. 4 Predicted and observed potentials for ATS-5 for total eclipse of the sun. (Closed symbols are for values calculated using the eclipsed spectra; open symbols are for values calculated using the sunlit spectra; crosses are estimated potentials using a two-Maxwellian approximation.)

by resistors, Eq. (1) can be expanded so that the large structure is approximated by

$$I_E(V_o) - (I_I(V_o) + I_{SE}(V_o) + I_{SI}(V_o) + I_{BSE}(V_o) + I_p(X_o, V_o)) = \frac{V_o - V_1}{R_1} \tag{10}$$

$$I_E(V_1) - (I_I(V_1) + I_{SE}(V_1) + I_{SI}(V_1) + I_{BSE}(V_1) + I_p(X_1, V_1)) = \frac{V_1 - V_o}{R_1} + \frac{V_1 - V_2}{R_2}$$

$$\vdots$$

$$I_E(V_n) - (I_I(V_n) + I_{SE}(V_n) + I_{SI}(V_n) + I_{BSE}(V_n) + I_p(X_n, V_n)) = \frac{V_n - V_{n-1}}{R_n}$$

where X_i = point on satellite along direction of movement
V_i = potential at point X_i, $i = 0, 1, \ldots . n$
R_i = resistance between X_{i-1} and X_i

The solution of this set of equations by iterative techniques is straightforward, regardless of whether the thick or thin sheath approximation is used.

III. Results

In order to test the accuracy of the models just developed, the results for a single point (small) satellite, the ATS-5, are compared with observation during eclipse passage. A generalized result for eclipse passage will be described. Based on these results, an estimate will be made of the potential across a large, passive structure as it enters into eclipse. The calculations are made for the case of a thick-sheath, n-point model and for a thin-sheath, n-point model. The effects of varying resistances and photoelectron emission rates also are analyzed.

A. Single Point

The eclipse potential variations assuming a single-point, thick-sheath model have been compared with observations.[18,19] The results from the latter paper for two typical eclipse passages of ATS-5 are presented in Fig. 5. These results are for a detailed calculation with the actual spectra. Added to

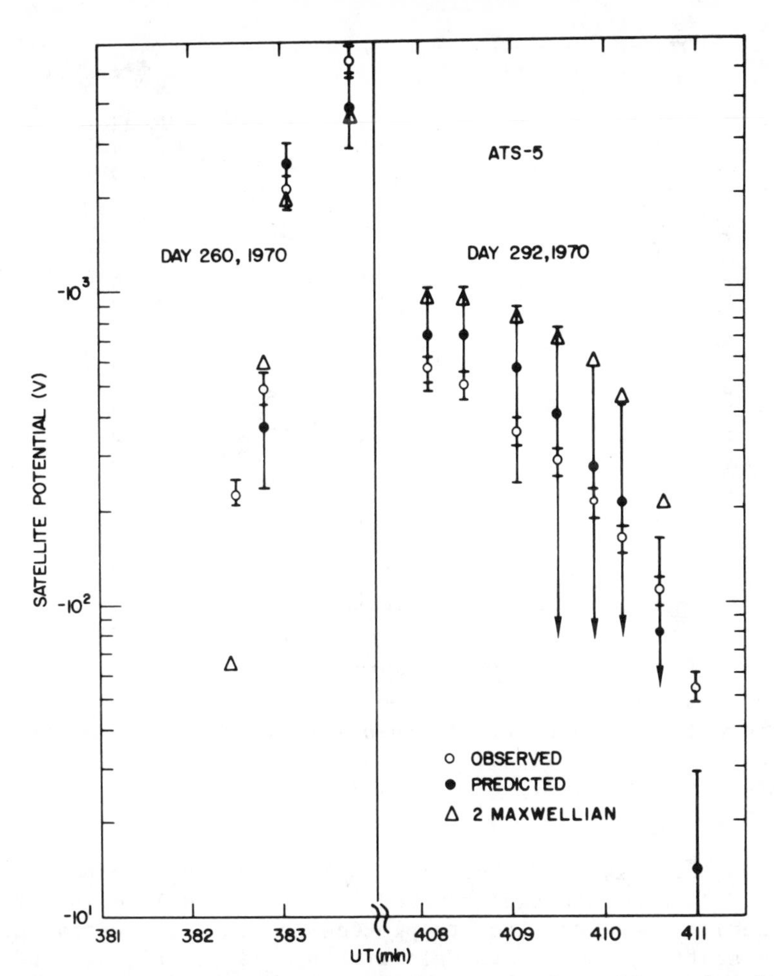

Fig. 5 Observed and predicted (average of sunlit and eclipsed spectra) potentials for the eclipse entry of ATS-5 on Day 260, 1970, and the eclipse exit on Day 292, 1970. (Triangles are estimates using a two-Maxwellian approximation.)

the figure are the results for the analytic model using the approximations represented by Eq. (7). Although clearly not as accurate, the results do indicate agreement over the range of observations. The results for all eclipse passages studied are presented in Fig. 6 for the detailed spectra calculations. Although the log-log scale tends to be misleading, the results are still impressive since the error is only $\pm$700 V.

An interesting result is obtained if the ATS-5 (and ATS-6) potentials during eclipse passage are plotted as a

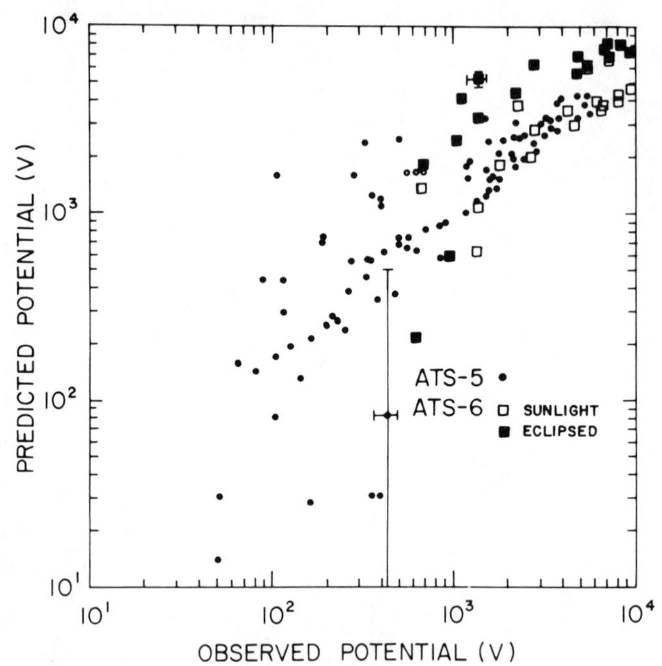

Fig. 6 Observed and predicted potentials for ATS-5 and ATS-6 for the entire data base. (Values for ATS-5 are the averages of the sunlit and eclipsed spectra; for ATS-6 the different spectra are indicated.)

function not only of X_m but of $V(X_m)$ divided by the potential V_o (the maximum potential reached when the satellite is totally eclipsed). These results are plotted in Fig. 7. Instead of a random scatter, the points fall within a fairly narrow band. This band is given by the following simple theory: Assume that the approximations for I_{SE}, I_{SI}, and I_{BSE} are given by aI_E, bI_I, and cI_E and that a, b, and c are constant (equal to approximately .4, 3, and .2). Then, upon substitution of Eqs. (7a) and (7b) for I_E and I_I,[18] Eq. (1) becomes

$$\frac{V(X_m)}{V_o} \approx -\text{Ln}\ [.4 + 2.5\ (I_{PO}/I_{EO})\ F\ (X_m)] \tag{11}$$

I_{EO} is the ambient electron current for 0 potential, $|V|/T_I \ll 1$, and the ion current is assumed to be 1/25 I_{EO}. This is the equivalent to assuming a thin sheath. Results for two values of the ratio I_{PO}/I_{EO} are shown. The curve represented by Eq. (11) is a general result for a wide range of actual ambient conditions at geosynchronous orbit near midnight.

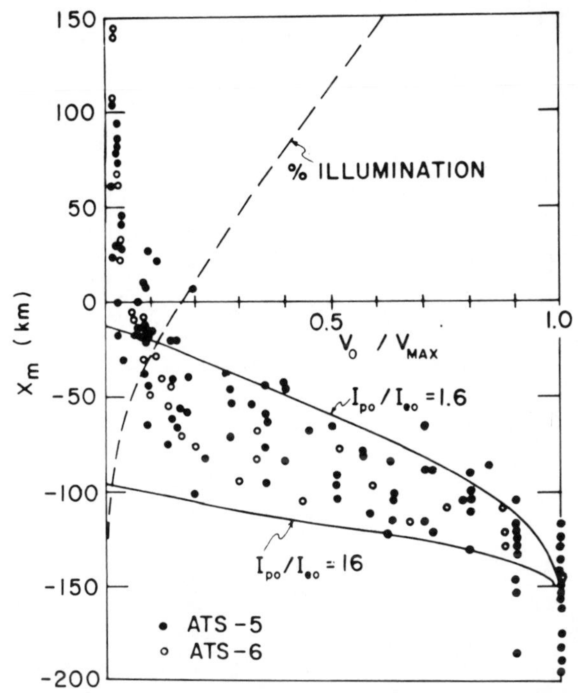

Fig. 7 Ratio of the varying potential to the maximum potential when the spacecraft ATS-5 and ATS-6 were in eclipse. (Also shown are the predicted Eq. (11) curves for I_{PO}/I_{EO} values of 1.6 and 16 and the percent illumination as a function of X_m.)

As will be seen shortly, it approximates the average satellite-to-space potential curve for a large space structure and also can be used to estimate the maximum potential to be expected between the ends of such a structure.

B. Results for Large Space Structures

Assuming that a one-or two-component Maxwell-Boltzmann distribution represents the ambient environment and that the assumption of aluminum-like properties (calibrated to ATS-5 to account for composite materials) is adequate for analysis purposes, a qualitative (if not quantitative) estimate of the effects of a varying photoelectron flux on a large space structure can be made. For example, given the simple model just outlined, the effects of a change from a thick-sheath, single-point satellite to a thin-sheath, plane or disk structure can be easily studied by employing the three-dimensional and one-dimensional approximations discussed previously. The effects of varying satellite shape may be studied by comparing the

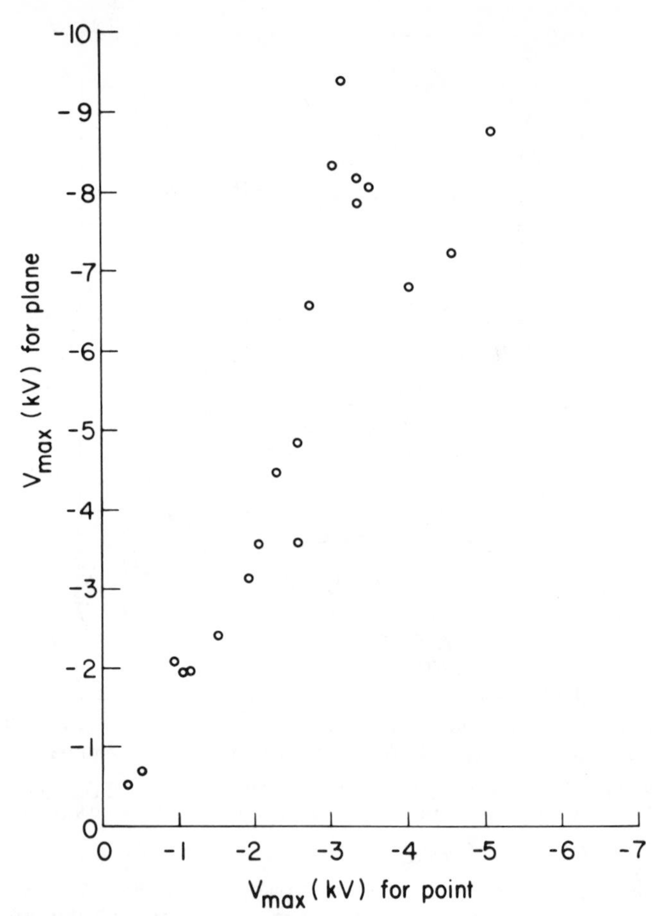

Fig. 8 Plasma data from the 21 ATS-5 eclipse passages studied used to estimate the eclipse potential based on the assumption of a plane or point.

SPS results to the SBR. The effects of varying resistance and photoemission also are easily evaluated by our model. (Note that although the photoelectron flux is, to a certain extent, attracted to or trapped in the vicinity of the spacecraft, this effect is not explicitly modeled here; it can be studied easily by assuming it results in a varying resistance across the structure.) Variations in the resistance and photoemission and their effects for the thick- and thin-sheath approximations and for a 100 km^2 (SPS) and a 10-km diam disk (SBR) are discussed here and thus readily evaluated.

As an initial example, in Fig. 8 the differences between thin-sheath and thick-sheath approximation are compared.[28]

Table 2 Space plasma parameters from ATS-5

	Day 289, 1969		Day 291, 1970	
	N, cm^{-3}	T, eV	N, cm^{-3}	T, eV
Electrons	.292	1060	.835	998
	.550	6070	.227	3910
Ions	.057	103	.276	319
	.890	8090	.911	9530

Specifically, the plasma data from the 21 ATS-5 eclipse events studied earlier were used to compute the satellite potential in eclipse, V_{max}, assuming either a planar or point approximation. That is, for the attracted species, χ is assumed to be given by

$$\chi_I = 1 - Z \frac{qV}{T_I} \quad V < 0 \tag{12}$$
$$\chi_E = 1 + Z \frac{qV}{T_E} \quad V > 0$$

Where

Z = 1 for point or thick sheath

0 for planar or thin sheath

Assuming a planar approximation gives potentials nearly twice those of the point (or thick-sheath) approximation. As the planar case yields "worst case" values, this assumption will be used throughout unless noted otherwise.

In Fig. 9, the change in potential as a function of position along the SPS for five different resistivities is plotted. A "no atmosphere" model was assumed for simplicity in this and subsequent plots.[28] The ambient conditions observed by ATS-5 on day 289, 1969 (the highest charging event observed, see Table 2), were assumed to be the baseline conditions. The satellite center was located at X_C = 150 km, where the potential at the center of the array was approximately -5990 V with respect to the plasma. This is exactly the same potential obtained if the value of the solar illumination, averaged over X_i (position relative to the center of the array), was substituted into Eq. (11) for a single point assuming a thin sheath. Also plotted in Fig. 9 is a curve for the SBR. There is little difference between the SBR and SPS curves. The main difference is a reduction in the absolute potential drop. This is readily attributed to the difference in areas between the two structures.

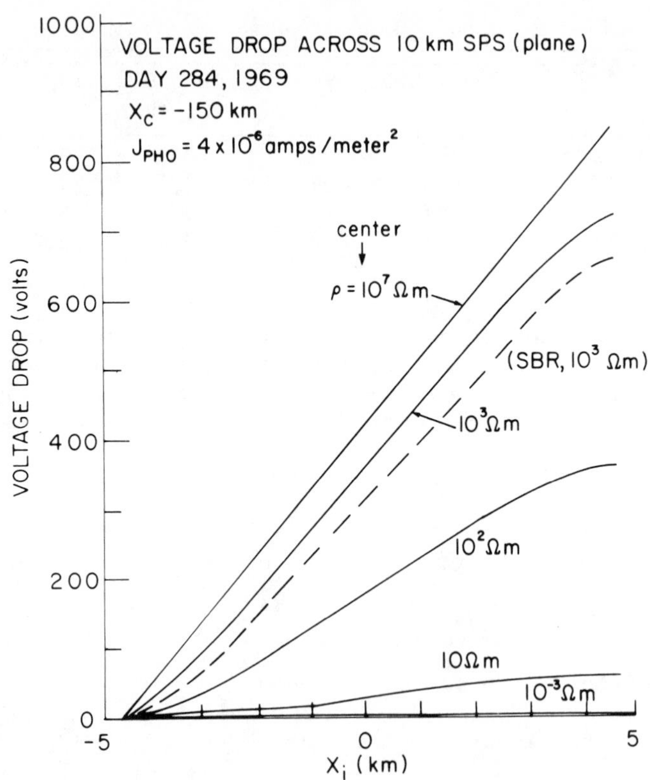

Fig. 9 Voltage drop across the SPS (and SBR) for varying resistivities on Day 289, 1969. (Satellite center is at X_m = 150 km.)

Gauntt[29] has shown that, to a high degree of accuracy, the variations with X along the array in Fig. 9 are given by the differential equation

$$-a_1^2 \, (V_c + a_2 \, X) = \left(\frac{\partial^2}{\partial X^2} + g \, (X) \frac{\partial}{\partial X} - a_1^2 \right) V \tag{13}$$

where $a_1 = \frac{\rho}{t} \frac{\partial J}{\partial V} T \Big|_{V = V_c}$

$$a_2 = \frac{\partial V}{\partial X} \Big|_{X=0}$$

ρ = resistivity (assumed constant)
t = thickness
V_c = potential at center of array (X = 0)
X = position along array
J_T = total current (density) into a point on array (= 0, but $\frac{\partial J_T}{\partial V} \neq 0$)

$$g = \frac{1}{L}\frac{dL}{dX}$$

L = width of array

For the SPS, g = 0, and Eq. (13) can be solved to give V as a function of X:

$$V = a_4 \sinh (a_1 X) + a_2 X + a_3 \tag{14}$$

where

$$a_3 = V_c$$

$$a_4 = \frac{-a_2}{a_1 \cosh \left(\frac{-sa_1}{2}\right)}$$

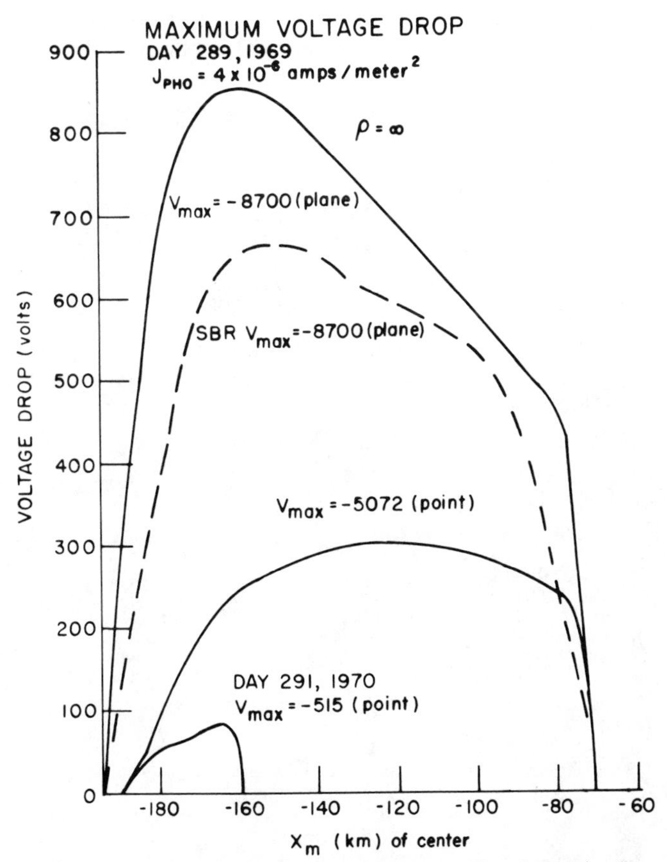

Fig. 10 Maximum voltage drop across SPS (and SBR) for different conditions and charging assumptions as a function of the X_m position of the center of the structure.

s = length of structure

(On day 289, 1969, typical values of $\frac{\partial J}{\partial V}\Big|_{V=V_c}$ and $\frac{\partial V}{\partial X}\Big|_{X=0}$ were 10^{-10} A - V^{-1}- m^{-2} and 10^{-1} V - m^{-1} respectively.)

The maximum potential drop across a structure occurs for infinite resistance. This means that the ends of the array are electrically isolated and can be treated as single points. Equation (11), which gives the potential on a single point as a function of X_m, may then be used to estimate the potential on both ends as the array is eclipsed; their difference being the maximum potential drop. In Fig. 10, using the detailed model, we have by this process calculated the maximum potential drops to be expected on the SPS and SBR as they are eclipsed for several extreme cases. (Note that day 291, 1970, was the weakest charging event observed.) Data for day 289, 1969, are employed to estimate the SPS potential drop for the thick- and

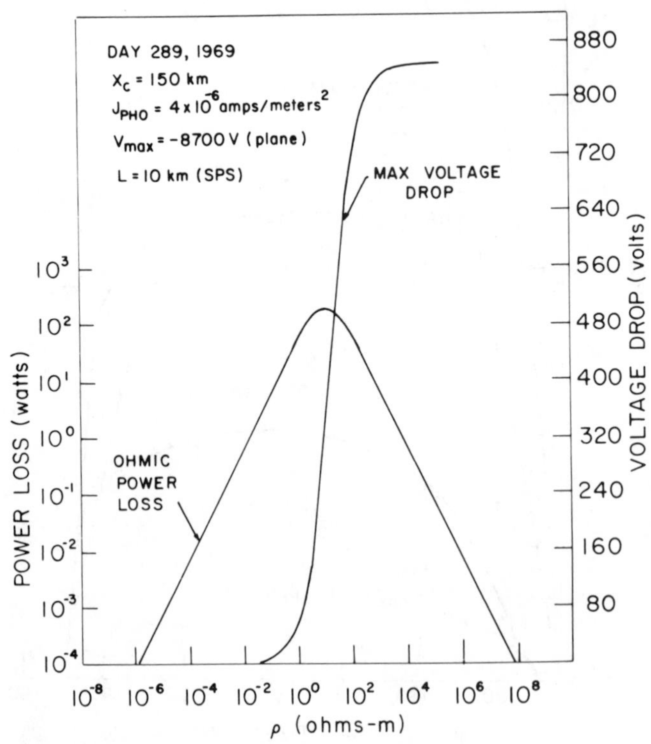

Fig. 11 Total power loss and voltage drop across SPS as functions of the resistivity ρ. (Satellite center is at X_m= 150 km.)

thin-sheath approximations. The potential drops are equal near the sunlit end of the diagram as qV/T_I goes to zero (see Eq. (11)), making the point and planar models equivalent. Strong asymmetries are exhibited on the eclipse side of the diagram for the planar and point models, however, since qV/T_I now becomes the dominant term. Even larger asymmetries are seen when the day 291, 1970, data are compared with the day 289, 1969, data. This is an obvious result of the two different environments. In contrast, the SBR results show only a slight deviation from the SPS results, as observed earlier.

Given estimates of the average potential as a function of the position of the center of the array(Eq. 11) and the variation in potential across the array as a function of resistance, Eq. (14), one can estimate many of the parameters of interest to the designer. As an example, in Fig. 11 the maximum potential drop ΔV at 150 km is plotted as a function of resistivity. Given these two quantities, ΔV and ρ, the power loss in the array can be calculated as a function of resistivity.[29] This function, also plotted in Fig. 11, shows a maximum between 10^2-10^4 ohm-m (roughly the resistivity of silicon or germanium). Gauntt[29] has shown that the value of the resistivity for which the power is maximum can be empirically fit by

$$\rho_{max} = L^{-2} \, (J_{PH})^{-\frac{1}{2}} \, 10^{10.3} \tag{15}$$

where L = length of structure km
J_{PH} = photoelectron flux at structure, $A\text{-}m^{-2}$
ρ_{max} = maximum resistivity, ohm-m

Although very small when averaged over large arrays like the SPS or SBR, the heating represented in Fig. 11 may be a consideration in the design of extremely thin, long structures since it may result in a periodic stress. (It would, however, be in phase with other larger thermal effects associated with eclipse passages.)

IV. Conclusions

To summarize, a simple though effective model for the calculation of spacecraft-to-space potential was presented. An algorithmic version of this model was developed. The results for both models were compared to actual observations and found to be in good agreement during eclipse passage. In the second half of the report, the single-point model was extended to include large, passive structures. The effects of assuming a thick sheath or thin sheath during eclipse passage were estimated. Similarly, the effects of varying shape, photoelectron emission, and conductance were analyzed. Simple, first-order approximations were advanced for estimating some of these variations (such as the maximum potential drop and the power loss across an array).

To conclude, simple point models can explain a) the changes in potential on spacecraft passing in and out of eclipse, and b) the effects of these changes on large space structures. Since the problem of eclipse passage has not yet been treated accurately by the detailed, three dimensional models, and these models have not been calibrated in detail with real data, the models presented here currently provide the only estimates of the effects of eclipse passage on space vehicles. Finally, any model of eclipse passage, to be truly useful, will need to be extended to include active structures such as an operational SPS. The SBR will most likely be a passive structure (in the spacecraft charging sense) so that the analysis presented here is readily applicable. Whether or not the SPS will be turned off (i.e., passive) during eclipse passage in order to avoid large voltage surges as the solar cells are extinguished is presently unknown. An active structure model would allow an evaluation of this effect. Even so, the single-point, passive model predicts many interesting effects due to the voltage surges associated with eclipse passage which deserve consideration.

References

[1]Whipple, E. C., "The Equilibrium Electric Potential of a Body in the Upper Atmosphere," NASA X-615-65-296, 1965.

[2]DeForest, S. E., "Spacecraft Charging at Synchronous Orbit," Journal of Geophysical Research, Vol. 77, February 1972, pp. 651-659.

[3]Rosen, A. (editor), Spacecraft Charging by Magnetospheric Plasma, Progress in Astronautics and Aeronautics: Vol 47, AIAA, New York, 1976.

[4]Pike, C. P. and Lovell, R. R., (Ed.), Proceedings of the Spacecraft Charging Technology Conference, AFGL-TR-77-0051/NASA TMX-73537, 1977.

[5]Finke, R. C. and Pike, C. P., (Ed.), Spacecraft Charging Technology - 1978, NASA CP-2071/AFGL-TR-79-0082, 1979.

[6]Rothwell, P. L., Rubin, A. G., and Yates, G. K., "A Simulation Model of Time-Dependent Plasma-Spacecraft Interactions," Proceedings of the Spacecraft Charging Conference, edited by C. P. Pike and R. R. Lovell, AFGL-TR-77-0051/NASA TMX-73537, 1977, pp. 389-412.

[7]Parker, L. W., "Calculation of Sheath and Wake Structure About a Pill Box-Shaped Spacecraft in a Flowing Plasma," Proceedings of the Spacecraft Charging Technology Conference, edited by C. P. Pike and R. R. Lovell, AFGL-TR-77-0051/NASA TMX-73537, 1977, pp. 331-366.

[8]Prokopenko, S.M.L., and Laframboise, J. G., "Prediction of Large Negative Shaded-Side Spacecraft Potentials," Proceedings of the Spacecraft Charging Technology Conference, edited by C. P. Pike and R. R. Lovell, AFGL-TR-77-0051/NASA TMX-73537, 1977, pp. 369-387.

[9]Katz, I., Cassidy, J. J., Mandell, M. J., Schnuelle, G. W., Stein, P. G., and Roche, J. C., "The Capabilities of the NASA Charging Analytic Program," Spacecraft Charging Technology - 1978, edited by R. C. Finke and C. P. Pike, 1979, NASA CP-2071/AFGL-TR-79-0082, 1979, pp. 101-122.

[10]Massaro, M. J., Green, T., and Ling, D., "A Charging Model for Three-Axis Stabilized Spacecraft," Proceedings of the Spacecraft Charging Technology Conference, edited by C. P. Pike and R. R. Lovell, AFGL-TR-77-0051/NASA TMX-73537, 1977, pp. 237-269.

[11]Robinson, P. A., Jr., and Holman, A. B., "Pioneer Venus Spacecraft Charging Model," Proceedings of the Spacecraft Charging Technology Conference, edited by C. P. Pike and R. R. Lovell, AFGL-TR-77-0051 and NASA TMX-73537, 1977, pp. 297-308.

[12]Purvis, C. R., Stevens, N. J., and Oglebay, J. C., Charging Characteristics of Materials: Comparison of Experimental Results with Simple Analytical Models," Proceedings of the Spacecraft Charging Technology Conference, edited by C. P. Pike and R. R. Lovell, AFGL-TR-77-0051/NASA TMX-73537, 1977, pp. 459-486.

[13]Garrett, H. B., Spacecraft Potential Calculations - A Model, AFGL-TR-78-0116, 1978.

[14]Parker, L. W., "Plasmasheath-Photosheath Theory for Large High-Voltage Structures," published elsewhere in this volume.

[15]Reiff, P. H., Cooke, D. L., and Freeman, J. W., "Environmental Protection of the Solar Power Satellite," published elsewhere in this volume.

[16]Inouye, G. T., "Spacecraft Potentials in a Substorm Environment," Progress in Astronautics and Aeronautics: Spacecraft Charging by Magnetospheric Plasma, Vol. 47, edited by A. Rosen, AIAA, New York, 1976.

[17]Garrett, H. B., "Effects of a Time-Varying Photoelectron Flux on Spacecraft Potential," AFGL-TR-78-0119, 1978.

[18]Garrett, H. B., and Rubin, A. G., "Spacecraft Charging at Geosynchronous Orbit-Generalized Solutions for Eclipse Passage," Geophysical Research Letters, Vol. 5, October 1978, pp. 865-868.

[19]Garrett, H. B., and DeForest, S. E., "Time-Varying Photoelectron Flux Effects on Spacecraft Potential at Geosynchronous Orbit," Journal of Geophysical Research, Vol. 84, No. A5, May 1, 1979a, pp. 2083-2088.

[20]Swider, W., "The Determination of the Optical Depth at Large Solar Zenith Distance," Planetary Space Science, Vol. 12, August 1964, pp. 761-782.

[21]Burke, W. J., Donatelli, D. E., and Sagalyn, R. C., "Injun 5 Observations of Low-Energy Plasma in the High-Latitude Topside Ionosphere," Journal of Geophysical Research, Vol. 83, No. A5, May 1978, pp. 2047-2056.

[22]Weeks, L. H., Cuikay, R. S., and Corbin, J. R., "Ozone Measurements in the Mesosphere During the Solar Proton Event of 2 November 1969, "Journal of Atmospheric Science, Vol. 29, No. 6, September 1972, pp. 1138-1142.

[23]Grard, R.J.L., "Properties of the Satellite Photoelectron Sheath Derived from Photoemission Laboratory Measurements," Journal of Geophysical Research, Vol. 78, No. 16, June 1973, pp. 2885-2906.

[24]DeForest, S. E., and McIlwain, C. E., "Plasma Clouds in the Magnetosphere," Journal of Geophysical Research, Vol. 76, No. 16, June 1971, pp. 3587-3611.

[25]Garrett, H. B., and DeForest, S. E., "An Analytical Simulation of the Geosynchronous Plasma Environment," Planetary Space Science, Vol. 27, No. 8, August 1979, pp. 1101-1109.

26
Tsipouras, P., and Garrett, H. B., "Spacecraft Charging Model - 2 Maxwellian Approximation," AFGL-TR-79-0153, 1979.

27
McCoy, J. E., Konradi, A., and Garriott, D. K., "Current Leakage for Low Altitude Satellites," published elsewhere in this volume.

28
Gauntt, D. M., "Large Space Structure Charging During Eclipse Passage," AFGL In-House Report, 1979 (submitted for publication).

OCCURRENCE OF ARCING AND ITS EFFECTS ON SPACE SYSTEMS

Joseph E. Nanevicz* and Richard C. Adamo+
SRI International, Menlo Park, Calif.

Abstract

One of the consequences of spacecraft charging is the occurrence of electrical discharges or arcs on the affected satellite. The evidence for the occurrence of such arcs and their effects on spacecraft are reviewed both in the light of the author's experience and based on the published results of other works in this area.

Early evidence for the occurrence of arcs was generated in the course of laboratory simulations of charging of satellite thermal control materials. Piggy back instrumentation carried on a non-NASA satellite confirmed the occurrence of discharges in synchronous orbit, and related them to periods of electron injection during magnetic substorms.

The effects of arcing on space systems have also been studied in the laboratory and on spacecraft. Laboratory experience demonstrates that arcing leads to physical damage to thermal control surfaces, and that it generates electromagnetic noise pulses which can affect spacecraft systems. Orbital data are presented to corroborate the laboratory results.

I Introduction

As synchronous orbit satellites were developed and became more widely used in a variety of military, scientific and commercial applications, it became apparent that many displayed certain types of unanticipated behavior. The undersirable consequences of these unexpected events ranged from the relatively innocuous, such as uncommanded circuit switching,[1] to more serious temperature rises[2] and included major vehicle-system failures.[3]

Reports that satellites could charge to potentials of thousands of volts during geomagnetic substorms[4] suggested

* Associate Director, Electromangetic Sciences Laboratory.
+ Program Manager, Electromagnetic Sciences Laboratory.

static charging and subsequent arc discharges as a likely mechanism to explain many of the anomalies observed on operational spacecraft. Laboratory experiments, analytical programs, and orbital measurements were, therefore, undertaken to explore and quantify the charging processes, the arcing, and the resulting effects on spacecraft systems.

At the time of this writing, the areas of spacecraft charging and discharge characterization are very active. Laboratory programs are underway and the Spacecraft Charging at High Altitudes (SCATHA) program P78-2 satellite has been successfully launched and is beginning to generate data. Thus, it is certain that substantial refinements will occur in our understanding of the various aspects of the occurrence of arcing and its effects. The need for information about the various facets of spacecraft charging is sufficiently immediate, however, that it is appropriate to present information at the current level of understanding, recognizing that this information will be verified or modified as this technical area matures. In this section, the authors will endeavor to concentrate on general results that appear to be well established. The primary emphasis will be on presenting broad concepts and profitable ways of viewing the arc discharge process and its effects. The approach will be engineering oriented, anticipating that the reader is interested in the arcing process insofar as it is likely to affect the operation or lifetime of satellites or satellite subsystems. Use will be made of the authors' experiences as well as the work reported by others active in this field.

II Arcing Occurrence

General

The actual occurrence of arcs on space vehicles has been demonstrated by a variety of means with varying degrees of directness. Historically, interest in this area began with observations of anomalous satellite behavior. These observations along with vehicle potential and spacecraft environment measurements, suggested the possible usefulness of laboratory simulations of surface charging and arcing. Piggyback instrumentation was later installed and flown on an operational satellite to detect transient electromagnetic signals associated with arcs on the surface of the satellite. Continuing experience with operational satellites has provided additional confirmation of undesirable system behavior consistent with the occurrence of arcing. Gradually, better physical models for the charge storage process and for the discharge mechanisms are evolving. In view of the degree of activity in these areas, a substantial improvement in understanding of the causes and effects of arc occurrence is expected.

Arcs in Laboratory Simulations

In 1973, in connection with the development of instrumentation for arc detection on a non-NASA geosynchronous satellite, SRI conducted a number of laboratory experiments to simulate the charging of insulating satellite surfaces and to study the properties of the resulting discharges.[5] In the first set of these experiments[6] it was found that typical spacecraft thermal control materials Optical Solar Reflectors (OSR), Teflon, and Kapton exposed to an electron beam with a 10-keV energy and a current density of 1-20 nA/cm^2 produced discharges from the surfaces of these materials to the substrate on which they were mounted.[2] These arcs were visible as flashes of light over the surface of the material, with greatest intensity where the discharge terminated on the substrate (in the gaps between the rows of OSR and at the edges of the Teflon and Kapton). Reducing the beam energy below roughly 6 keV eliminated these arcs. The rate of occurrence of arcing increased with increasing beam current density. The simple laboratory set up used for these experiments is shown in Fig. 1.

Orbital data from the satellite of Refs. 1 and 5 showed evidence of discharges even during geomagnetically quiet times when energetic electrons were not present. This motivated an effort to identify discharge mechanisms which would occur under these conditions. Previous experience with dielectrics in plasmas indicate that breakdowns can occur when charge extracted from the plasma is deposited on thin films of oxide and other insulating materials on metal electrodes within the plasma region. Charge from the ionized gas accumulates on the film surface and develops very high dc field strengths in the film. When the dielectric strength of the film is exceeded, the film breaks down, and a visible flash or scintillation is observed. This mechanism was investigated in an experiment in which an Iridite-treated magnesium surface (a material commonly employed in satellite construction) was installed in a plasma-filled chamber and biased 50 volts with respect to the test chamber walls. Flashes of the sort reported by Malter[8] were observed accompanied by electro-magnetic noise pulses.

Meulenberg, in 1975, suggested that charge storage and subsequent discharges could occur in the bilayer formed near

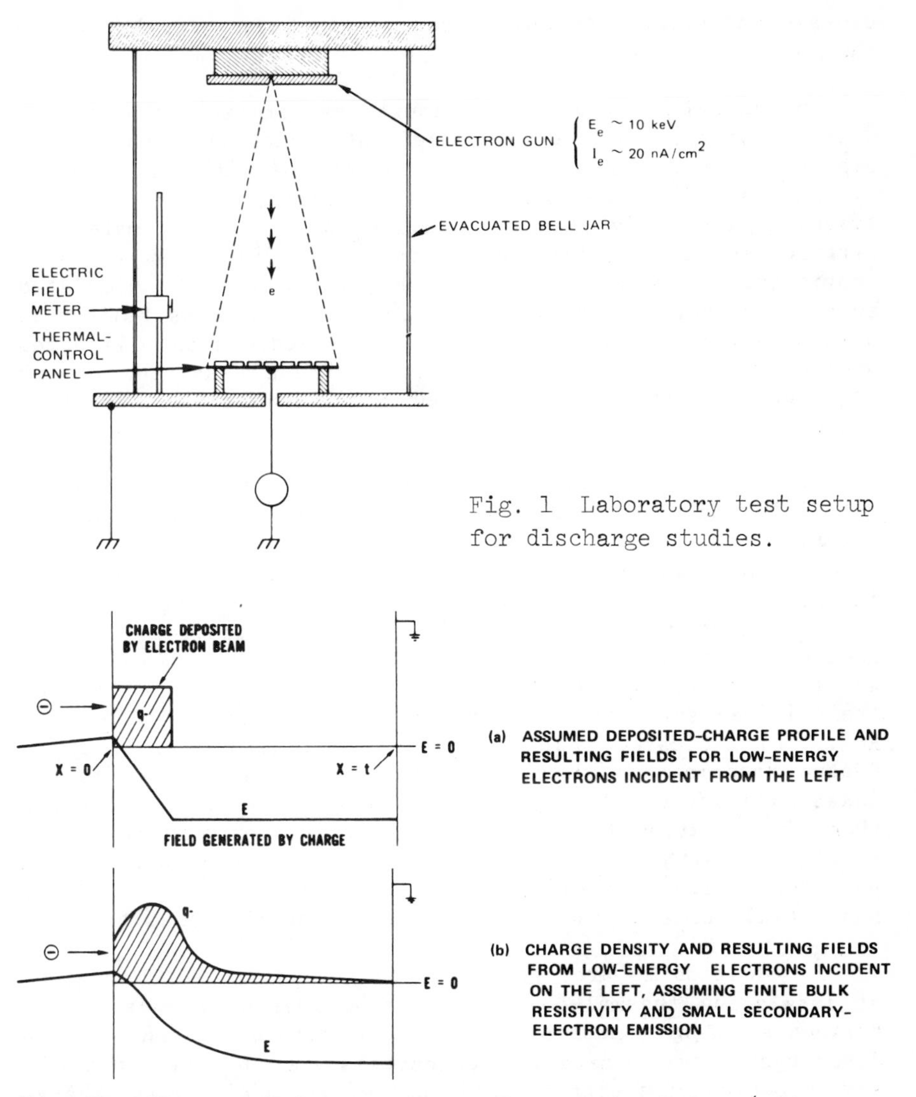

Fig. 1 Laboratory test setup for discharge studies.

Fig. 2 Bilayer of charge near surface of insulator (after Meulenberg).

the surface of an insulator when incident electrons penetrate to their stopping distance in the insulating material,[7] as shown in Fig. 2. Since his experimental work was done in a windowless chamber, the visual appearance of these breakdowns was unknown. They were electrically detected on a number of typical spacecraft thermal-control materials, including aluminized Kapton and Teflon. Meulenberg's results were highly significant in that they offered a mechanism by which charge storage and breakdown could occur without puncture through to the substrate.

In general, virtually all laboratory simulations of spacecraft-insulator charging and subsequent arcing have employed an electron gun to illuminate the target material in a vacuum system free of ambient plasma. N. J. Stevens has used such a system in the study of a wide variety of spacecraft materials arranged in diverse structural forms.[8,9] His system includes provisions for simultaneous exposure of samples to simulated solar illumination and to an electron beam. Stevens finds that solar illumination of certain materials reduces or eliminates arcing with incident electron beam current densities having magnitudes normally associated with magnetic substorm injections.[10]

Studies of the bulk conductivities of various insulating materials indicate that bulk conductivity behavior is highly material sensitive and, for Kapton, it can increase by five orders of magnitude under one solar con-stant of illumination.[11] Solar illumination also produces photoemission currents that can reduce negative surface potentials. Thus, whether or not arcs occur on a satellite is highly dependent on the types of materials used in its fabrication and on whether the critical surfaces are illuminated by the sun. The work of Ref. 13 indicated that the increased bulk conductivity of Kapton associated with solar illumination can persist for at least 36 h after illumination has ended. This is much longer than the duration of a night-time eclipse. Accordingly, a satellite with an outer surface covered with such photoconductive materials could be expected to be less susceptible to arc discharges even under eclipse conditions.

In addition to arcs generated by charge stored on regions of insulating thermal-control material, discharges can occur between adjacent conductors charged to different potentials. A discharge between metallic conductors can be an extremely severe source of electromagnetic noise, since all of the charge stored on a conductor can flow freely and is available to participate in the discharge. In the case of a discharge on a dielectric material, the charge must propagate from the region of storage through the discharge region by means of a series of avalanching processes.

Discharges from metallic conductors have been observed in the laboratory in connection with simulation tests of thermal blanket material.[9,12] These blankets are made up of a number of layers of insulating material, each coated on one side with a conducting film. Unless great care is taken to provide an electrical connection between each of these conducting layers, it is possible for differential charging and subsequent arcing to occur between adjacent layers or between a layer and the substrate.

Direct Orbital Data on Arc Occurrence

Direct evidence confirming the occurrence of arcs on satellite surfaces has been obtained from the piggyback instrumentation described in Refs. 1 and 5. The results of these measurements provide significant insights in a number of areas. For example, Fig. 3 (reproduced from Ref. 1) demonstrates that the rate of discharge occurrence is affected by substorm current injection, but that discharges can occur independently of satellite eclipse. This implies that at least one arcing mechanism can operate even in the presence of photoelectrons generated when the satellite is sunlit. The data of Fig. 3 also indicate that, on this satellite, the bulk conductivity of the surface materials is sufficiently low, even under sunlit conditions, that arcs are produced by the charge deposited by the substorm electron injection. In this regard, it should be noted that the discharge pulse sensor on this satellite was located in a region of second-surface quartz OSR. The data of

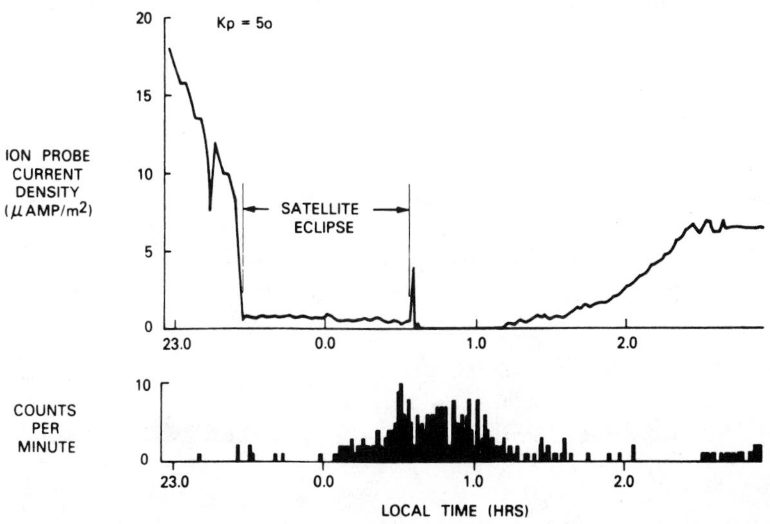

Fig. 3 Static-charge experiment data recorded during disturbed geomagnetic field conditions through satellite eclipse. (An energetic electron injection event and discharge rates as high as ten per minute are observed during this pass. The discharges are associated with the injection event (note depressed ion probe current) rather than with satellite eclipse.)

Ref. 11 indicate that, of the materials typically used for spacecraft thermal control, quartz is one of the best insulators and that its conductivity change due to solar illumination is one of the smallest observed.

Figure 4 (also reproduced from Ref. 1) indicates that, although the rate of discharge occurrence is increased by the electron injection near midnight, discharges occur throughout the entire 24-h period. Additional data from Ref. 1 indicate that, even on days during which no substorm activity was observed, discharge pulses were detected throughout each 24-h period. These results indicate that at least one of the charge-storage and discharge processes can occur in the absence of measurable energetic electron injections. For example, if the surface material is highly insulating, the few energetic electrons striking the thermal control surfaces can accumulate at their electron-stopping distances below the surface until the internal fields become sufficiently high to initiate a discharge. These electrons can accumulate within the material independent of the fact that a substantially larger photoelectron current may be flowing outward from the surface of the dielectric region.

The data from Ref. 1 indicate that there is a fundamental difference between the arcs occurring during a geomagnetic substorm and those occurring during undisturbed conditions. Fig.

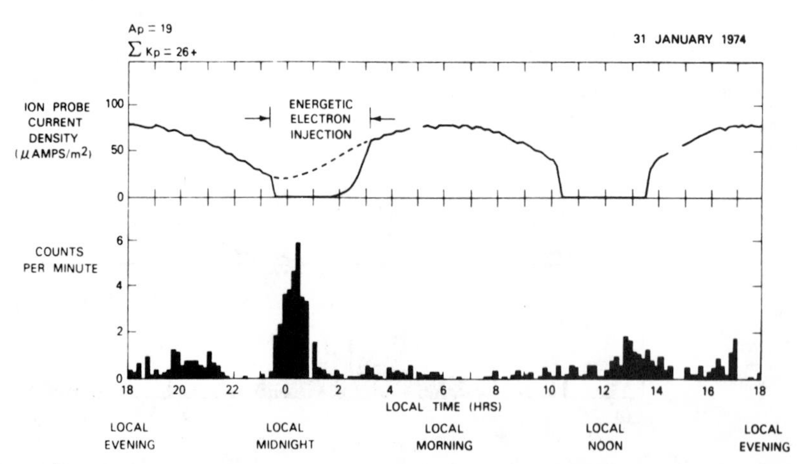

Fig. 4 Static-charge experiment data from an orbit for which substorm activity was observed. (Depressions of the ion-probe current near local midnight are caused by the injection of energetic electrons to synchronous altitudes. Discharge rates as high as six per minute are measured by the pulse counter during the electron injection event.)

5 shows that, during normal undisturbed conditions, arcing occurs just as the region of the spacecraft in which the discharge sensor is located becomes sunlit. During a substorm, on the other hand, the conditions shown in Fig. 6 apply, and there is no relationship between sun position and the occurrence of arcs.

Indirect Orbital Evidence of Arc Occurrences

A substantial portion of the evidence for arc discharge occurrence on operational satellites has been generated from observations of the behavior of systems intended to be immune to transient interference. That spacecraft anomalies are related to the occurrence of discharge pulses can be inferred from the results of Ref. 1. There it is shown that each satellite system anomaly is virtually always accompanied by the occurrence of a discharge pulse within the same 1 sec. telemetry frame.

Numerous satellite systems reportedly display anomalous behavior in orbit--particularly during geomagnetic substorm disturbances. The correlation of anomalous system behavior with substorm occurrence strongly suggests that electromagnetic pulses generated by arcing affect the functioning of satellite electronics systems.

In this regard, it should be observed that the primary motivation for the performance of the piggyback experiments described in Refs. 1 and 5 was the fact that the thermal behavior of the satellite did not agree with that projected for the system. It was suggested that this unanticipated behavior could be the result of thermal-control-material deterioration caused by spacecraft charging and subsequent arc discharges.

Present spacecraft do not normally carry instrumentation to detect the occurrence of substorms and arcing. Although indi-

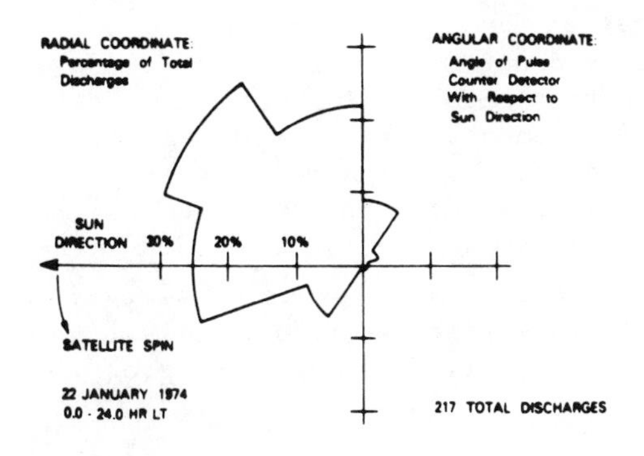

Fig. 5 Spin modulation of discharges observed during a quiet geomagnetic period.

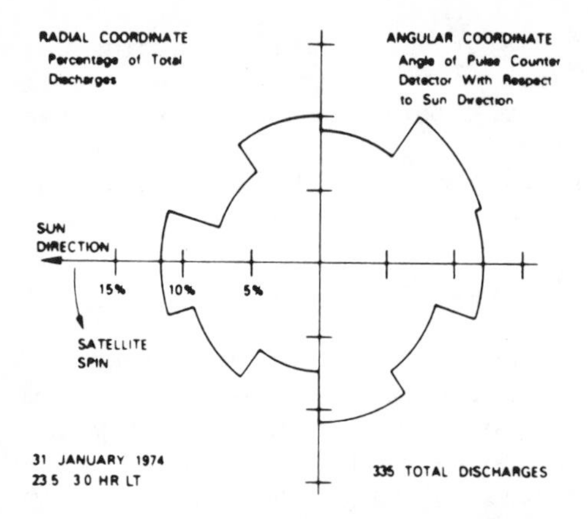

Fig. 6 Spin modulation of substorm-produced discharges near local midnight.

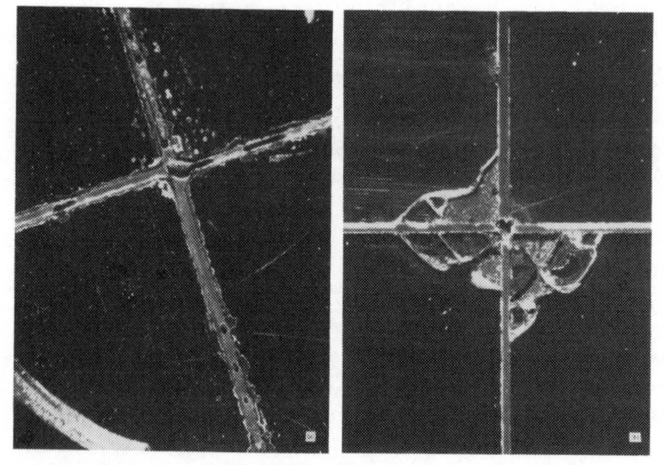

Fig. 7 Discharge damage to OSR panel.

rect orbital data cannot be used to state unequivocally that the anomalous behavior of a particular system can be ascribed to arcing, there are a sufficient number of anomalous events compatible with this explanation that it is prudent to seriously consider this indirect evidence. A number of studies are of particular significance in this regard. Fredricks and Scarf present a concise summary of arguments for associating satellite system anomalies with spacecraft charging.[13] A somewhat more detailed report was prepared by Coge and Pearlston.[14] A thorough study of satellite anomalies was conducted by the Aerospace Corporation[15] before there was a substantial awareness of spacecraft charging and its effects. Thus, many anomalous events are described but not attributed to arcing on the satellite.

III Arcing Effects on Space Systems

General

Satellite charging and the subsequent discharges can have deleterious effects both on the physical structure of the satellite and on its electromagnetic environment. The basis for these effects is the rapid release of the electrostatically-stored charge deposited on or within insulating surfaces. The electric arc breakdown produces localized heating and evolution of material at the site of the arc. Even relatively minor material damage at the discharge site may be of concern, as in the case of critical thermal control panels. Alternatively, it is possible for the discharge effluents to degrade system behavior by being redeposited on critical optical surfaces.

Since the arc discharge involves the rapid motion of charged particles, a transient electromagnetic signal is generated by the breakdown. If the electromagnetic coupling and system susceptibility are appropriate, this transient signal can affect the functioning of spacecraft electronic systems.

Physical Effects of Arcing

In connection with laboratory studies of spacecraft charging, various examples of material damage have been observed. Perhaps the earliest of these is described in Ref. 2, which was written before the satellite of Refs. 1 and 5 was launched. During the development of the satellite instrumentation, an OSR panel was exposed to a 10-keV electron beam with current densities ranging from 1 to 10 nA/cm^2. Following the charging and discharging, it was found that damage to the OSR cells had occurred. A photograph of this OSR damage was shown in Refs. 5 and 6.

More recently, tests have been conducted on OSR panels in connection with the characterization of electromagnetic signals generated by spacecraft discharges.[16] The panels used in these tests were exposed to a beam of 20-keV electrons with a current density 10 nA/cm^2 for extended periods of time while a number of electromagnetic transient-signal measurements were made. The damage to two of these test panels is shown in Fig. 7. In the panel shown in Fig. 7a, the damage was confined primarily to the gap regions between the individual cells. The adhesive between the cells was blown out in the damaged regions and the silver was etched away from the edges of the cells. The effluent blown out by the discharges was deposited as a region of discoloration along each of the gaps. The discoloration deposit was most intense in the regions showing the greatest damage to the silver or the adhesive. There appeared to be minimal damage to the quartz itself, even at the edges of the cells.

The test panel in Fig. 7b shows a greater degree of damage. The adhesive is driven out entirely from the regions of the gaps, and there are areas of discoloration along these gaps where the effluent has redeposited. The damage to the silver is substantial, as evidenced by the large dark spots at several places along the gaps. Finally, there is appreciable damage to the quartz itself in the large, dark spot at the junction of the four cells near the center of the photograph.

The damage to a sample of silvered teflon tape is shown in Fig. 8. Substantial regions of silver have been blown away from the back of the sheet. With this material, it was found that the damage tends to be concentrated along lines running diagonally in the figure. This is the direction that discharges over the surface are directed and corresponds to scratches in the tape along which the tape backing was slit during manufacture to facilitate its removal.

From an examination of the sample of Fig. 8 under an optical microscope, it appeared that the outer surface of the tape had not been damaged or punctured (except for one or two of the larger isolated regions which had a volcano-like appearance). However, melting and subsequent healing may have occurred.

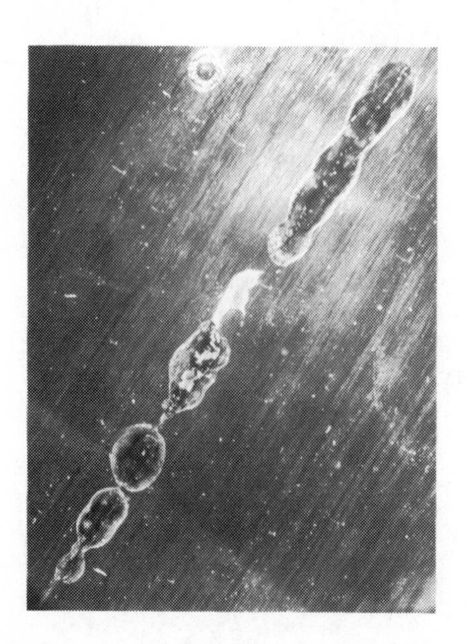

Fig. 8 Discharge damage to silvered teflon tape sample.

Since there were no regions of discoloration around the damaged regions in Fig. 8, it appears reasonable to conclude that no effluent was emitted and that puncturing or other damage of the outer surface had not occurred.

In the Malter discharge[17] experiments described in Refs. 5 and 18, it was found that material was blown out by the discharge and deposited in a ring around the actual puncture site, as shown in Fig. 9, which is reproduced from these references. Although the greatest density of deposited material was in the immediate vicinity of the discharge, it is certain that the effluents can be propelled to substantial distances from the site. Such material deposited on a critical optical surface will gradually degrade its performance.

Laboratory tests demonstrating damage to thermal blankets have been conducted by several investigators. Inouye has shown that the high-peak currents associated with arcing in these materials can gradually destroy connections to ground straps installed to prevent potential differences between the various

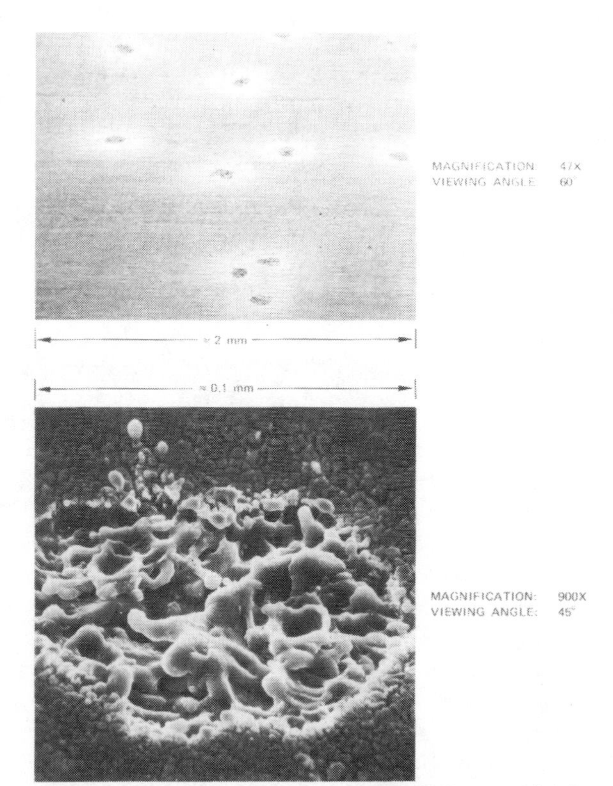

Fig. 9 Damage caused by low-voltage malter discharges.

thermal blanket layers and the satellite frame.[12] Stevens has reported similar results from his laboratory work.[9]

Orbital Experience

Orbital experience tending to demonstrate physical effects of arcing on spacecraft performance is predominantly associated with thermal behavior of satellites. On certain satellites it has been found that the temperature rises more rapidly than predicted from outgassing and redeposition calculations based upon the materials used in construction. Various studies have been made of satellite temperature rates of change as a function of magnetic substorm activity. An example of such a correlation study is shown in Fig. 10 (reproduced from Fig. 9 of Ref. 1). Other studies have compared satellite thermal performance immediately before and after a major substorm. In general, it appears that the performance qualities of the thermal control materials are degraded by the occurrence of substorms.

It is not clear whether the deterioration is caused by loss of silvering as illustrated in Figs. 8 and 9 or whether it stems from redeposition of the discharge effluents generated by the arcing. If effluent deposition were the responsible mechanism, one would expect reports of corresponding optical system deterioration following substorm occurrence. The authors are not aware of reports of such deterioration. Although this may be interpreted to mean that such redeposition is not occurring, it

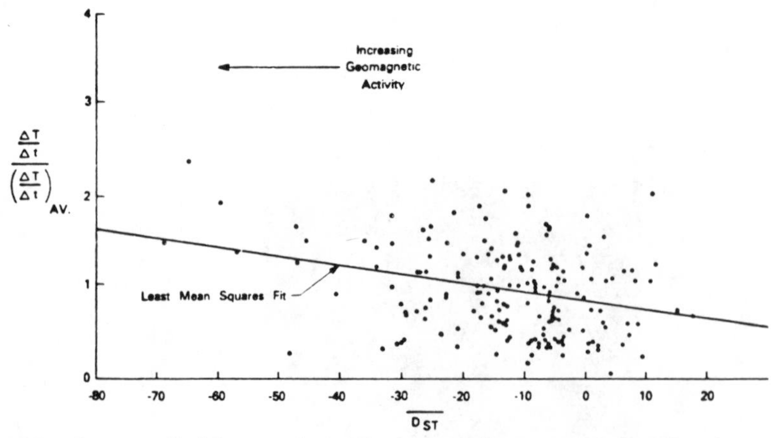

Fig. 10 A correlation exists between increases in the temperature of sensors on these satellites and increases in magnetospheric substorm activity as determined by D_{ST}. (This suggests the possibility that part of this temperature rise could be caused by phenomena associated with satellite charging.)

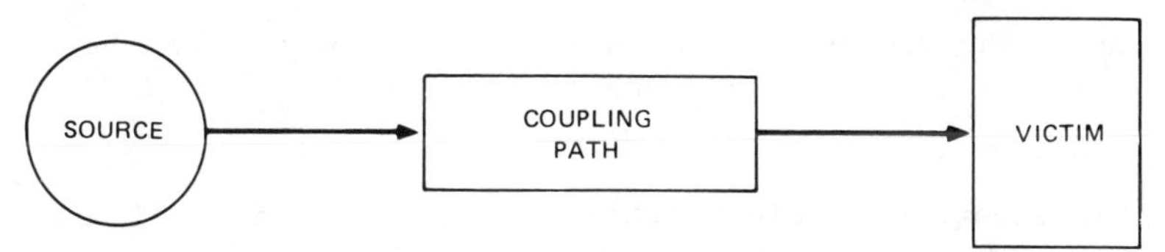

Fig. 11 Schematic characterization of the interference couplin process.

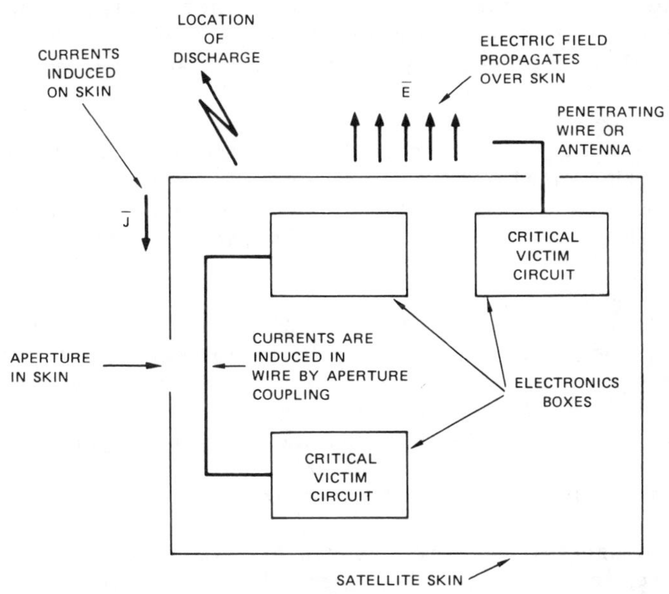

Fig. 12 Illustration of the electromagnetic processes associated with arc-coupling on a satellite.

may also indicate that optical system degradation is not as readily detected as a change in the thermal properties of a satellite.

Electromagnetic Effects of Arcing

The rapid motion of electric charge in the arcing process generates transient electromagnetic signals that can couple into electronic systems on the satellite. This situation is illustrated schematically in Fig. 11. The arc generates an electromagnetic signal source that excites various coupling paths, ultimately inducing a signal in the victim circuit.

To estimate the magnitude and nature of the signals induced in the victim circuit, it is necessary to be able to define the characteristics of the source and the coupling to the victim. Whether or not it is necessary to be concerned about the signals induced in the victim circuit depends upon the victim sensitivity to damage or upset and upon the consequences of the victim upset.

The electromagnetic situation existing on a satellite during an arc discharge on the surface is illustrated in Fig. 12. The arc discharge processes induce a transient electric field normal to the outside surface of the skin on the satellite. This field excites currents in wires penetrating through the skin of the vehicle as shown at the right side of the figure. These signals, in turn, may affect electronic systems inside the vehicle.

The discharge also excites skin currents on the outside surface of the satellite. These currents excite apertures and joints in the skin which induces currents in wires entirely on the interior of the satellite, as shown at the left side of the figure.

From Fig. 12, it is evident that the occurrence of electromagnetic signals associated with arcing is detected most easily using sensors designed to couple strongly to the electromagnetic signals produced on the exterior of the satellite. It is also evident that systems on the interior of the satellite may be affected by electromagnetic pulses coupled through inadvertent paths of the sort shown in Fig. 12.

Electromagnetic effects associated with discharges on OSR panels were investigated during the development of the satellite instrumentation of Ref. 5. A record of a transient signal measured during these tests was presented in Ref. 6. More recently, SRI has been involved in the characterization of the electromagnetic signals generated by the electrical breakdown

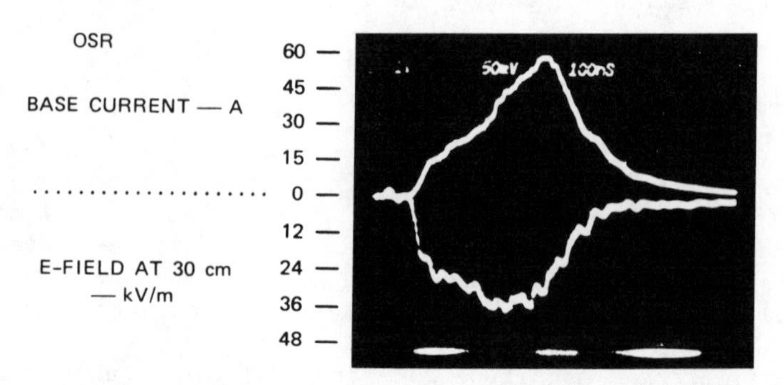

Fig. 13 Transient electromagnetic signals generated by a discharge on an OSR panel.

of spacecraft insulating materials, using a test setup similar to that shown in Fig. 1. Preliminary results of this work were presented in Ref. 16. Additional measurements have been made with the test set-up slightly modified to eliminate some of the shortcomings discussed in Ref. 16 and to permit measurements of skin current, sample-base current, and electric field in the vicinity of the discharge.[19]

Fig. 13 shows the electromagnetic record of a typical breakdown occurring on the surface of an OSR panel. A positive unipolar pulse is generated in the test sample base, return-current circuit, indicating that negative charge is driven away from the sample surface during a breakdown. During the first 370 ns, the current rises to its maximum value of nearly 60 A, then decays monotonically. The negative-going electric-field (E-field) pulse, roughly a mirror image of the base current, is created by electrons in the breakdown plasma being driven upward. Their dipole moment increases as they leave the surface, generating the electric-field pulse.

The results of subsidiary experiments using a four-grid structure of the sort described in Ref. 6 indicated that most of the electrons are expelled to at least 10 cm from the surface of the OSR panel. These same experiments indicated that the electron velocity is greater than 5.8×10^6 m/s. At this velocity, they would require only 65 ns to travel the 38-cm distance to the grounded electron source above the sample. Thus, it is likely that during the course of the discharge illustrated--lasting over 400 ns--a column of electrons extending from the test sample to the electron source was established.

It is interesting to estimate the electric-field intensity that such a column of charge would produce and to compare this result with the measured value of electric field. From the upper trace in Fig. 13, during the first 50 ns, the average current is 7.5 A. Thus, approximately 3.5×10^{-7} coulomb (C) of charge has been evolved. Assuming that the electron velocity is 5.8×10^6 m/s, this charge, distributed over a column 30-cm high, will produce an electric field of 57 kV/m at a distance of 30 cm. This is in reasonable agreement with the measured value of 24 kV/m; particularly since the electron velocity was probably substantially greater than 5.8×10^6 m/s, so that the charge density in the electron column would actually be lower than that used in this calculation.

To obtain an understanding of how the induced E-field depends on the radial distance from the discharge, an experiment involving two E-field sensors was performed. The reference sensor was permanently mounted 30 cm from the discharge, while a second sensor monitored electric field at various positions farther from the source. The data from this experiment indicate that the field intensity decreases at least as fast as $1/r^3$, where r is the distance from the discharge to the sensor location.

Experiments were also conducted using aluminized Kapton and silver-coated Teflon test samples. Similar results, differing primarily in pulse risetime and pulse amplitude, were obtained.

It is important to recognize that a discharge generated in a laboratory setup, such as that illustrated in Fig. 1, will not duplicate all of the features of a discharge from an in-orbit satellite. For example, in the top trace of Fig. 13, the current flows for 400 ns until it reaches a peak of 60 A. This means that the charge evolved from the base and collected by the electron gun was 1.2×10^{-5} C. On the other hand, a typical satellite with a self-capacitance of 200 pF would be charged to a potential of 60 kV by the removal of this quantity of charge. Limiting processes on the satellite would, therefore, restrict the blow-off current substantially before 400 ns. Thus, the results of the laboratory measurements can best be used to define the general characteristics of the arc discharge source, while recognizing that proper adjustments will have to be made to apply these results to the case of an orbiting satellite.

It should be noted that the removal of 3.5×10^{-7} C of change in the first 50 ns of the discharge would result in charging the same satellite to only 1.75 kV. Peak electric fields of tens of kilovolts as calculated above and as measured in Fig. 13 are, therefore, entirely possible on a satellite.

The magnitudes of the signals generated are highly significant. Transient-field charges of tens of kilovolts per meter occurring in a period of several tens of nanoseconds have been measured roughly 30 cm from the center of the test sample. Such transient fields are comparable to those normally associated with nuclear electromagnetic pulse (EMP) events or nearby lightning. It has long been recognized that lightning and EMP can seriously affect unprotected electronic systems, and that deliberate measures must be taken to harden systems against these electromagnetic threats.

3. Orbital Evidence of Arcing

The most direct evidence of the occurrence of arcs on satellite surfaces was obtained with the electromagnetic pulse-counting instrumentation described in Refs. 1 and 5. Selected results from this program have herein been presented in Figs. 3-6.

The instrumentation system incorporated an electric dipole sensor installed on the exterior of the satellite in a region of quartz second-surface OSR. Laboratory measurements, with the flight system prior to launch, indicated that its sensitivity allowed the detection of all discharges occurring within approximatley 1 m of the sensor.

On the test satellite, provisions were made to monitor orbital anomalies of the sort that were typical of this satellite. These included an anomalous detector response and an uncommanded reset of a satellite command circuit. Historically, these anomalies occurred on several other satellites in this series, in addition to the satellite on which this experiment was flown.

During the period of operation of the experiment of Ref. 5, several occurrences of both types of these anomalies were reported. For the anomalous detector response, a total of 35 anomalies occurred. For 31 of these cases, a discharge was observed by the experiment coincidental with the anomalous detector response. These anomalous responses were reported on several different days, and the probability that the coincidences occurred by chance is essentially zero. There were also a total of six uncommanded circuit resets, and five of these resets occurred coincident with the detection of discharges by the experiment. Thus, data from this experiment have confirmed the hypothesis that these types of orbital engineering anomalies can be caused by interference from electrical discharges.

In January 1979, the P78-2 spacecraft, sponsored by the joint NASA/Air Force Spacecraft Charging at High Altitudes (SCATHA) Program[20,21] was successfully launched. Experiments aboard this vehicle are designed to provide a detailed characterization of the charged-particle environment at geosynchronous altitudes as well as to monitor the effects of this environment (1) on the potentials and currents of sample thermal control materials, (2) on the nature of electromagnetic transients generated by the occurrence of arcs, and (3) on the rate of deposition of contaminants on selected materials.

At the time of this writing, preliminary analyses of only limited quantities of data from P78-2 have been accomplished.[22]

The P78-2 payload includes a Transient Pulse Monitor (TPM)[23,24] as part of the on-board "housekeeping" instrumentation. The TPM is designed to detect and characterize electromagnetic transients occurring at four locations internal to the spacecraft. These include: two signal cables that are part of the main vehicle wiring harness (one having low-impedance ter-

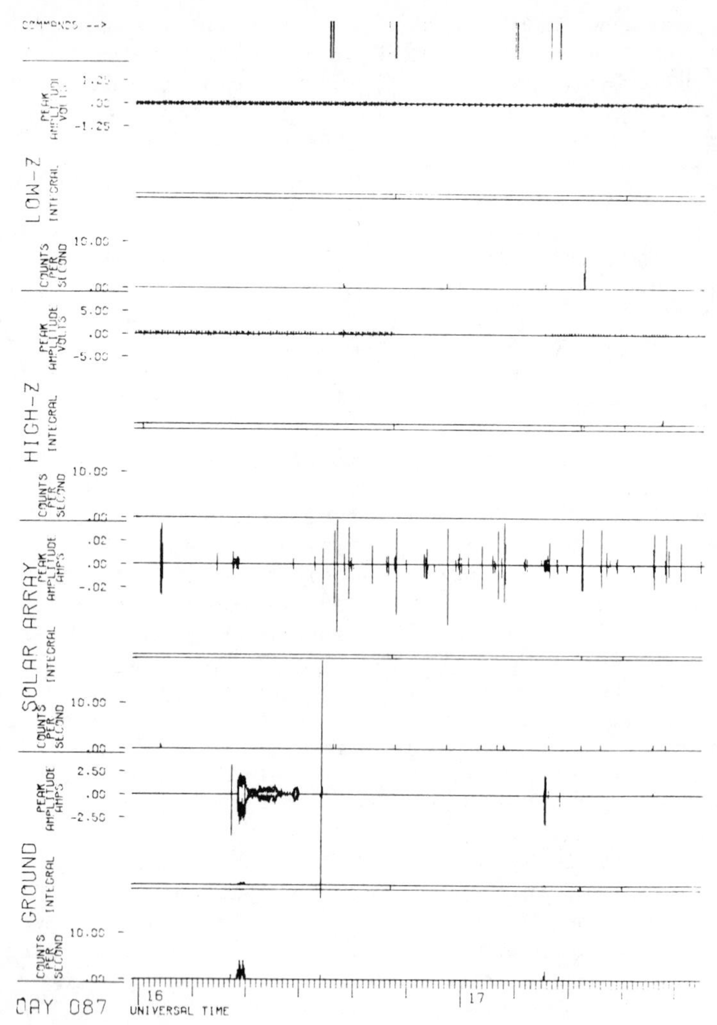

Fig. 14 P78-2 transient pulse monitoring data; 28 March 1979 --1559 to 1745 UT.

minations, the other having high-impedance terminations at each end), a power lead from the upper solar array to the spacecraft power-conditioning unit (PCU), and a ground lead from the PCU to the "single-point" ground.

Fig. 14 summarizes data obtained by the TPM on March 28, 1979, from 15:59 to 17:45 Universal Time (UT). During the period the spacecraft passed through Earth eclipse, several commands were issued to onboard systems (as indicated at the top of the figure), and the on-board SC9 charged-particle experiment, provided by the University of California at San Diego (UCSD), detected the occurrence of a -8 kV charging event.

Many of the transient signals detected by the TPM during this period are the result of normal systems operations and are not attributable to spacecraft charging events. These include the low-level transients on the Low-Z and High-Z sensors caused by internal clock signals distributed to many of the experiments through the main wiring harness, many of the solar array sensor transients caused by PCU switching, several command-related transients, and the large ground-sensor transient at 16:34 UT caused by turn on of the vehicle transmitter carrier signal.

The spacecraft entered eclipse through the Earth's penumbra from 16:15-16:17 UT. Throughout this and the next 10-to-15 minute period, numerous bursts of activity were detected by the TPM on both the solar array and ground sensors. Similar activity, although of shorter duration, was detected upon eclipse exit from 17:13 to 17:15 UT. At present, activity of this type, frequently observed during eclipse entry and exit, is believed to be caused by thermal control system and power regulator switching, by redistribution of accumulated charge due to changes in photoemission, or by a combination of these effects.

At 16:37:31 UT, the TPM High-Z and Low-Z sensors, as well as the SCI-8B Aerospace Corp. pulse-analyzer sensor detected an electrical transient that cannot be attributed to normal vehicle system operations. Just prior to this time, the SC9 payload indicated that the vehicle charged rapidly to approximately -8 kV.

At 17:14:48 UT, the TPM High-Z, Low-Z and solar array sensors and the SC1-8B sensor again detected an unexplained transient. The SC9 payload indicated that the spacecraft potential dropped rapidly from -2 kV 0 at this time. The available data strongly suggest that the events at both 16:37:31 and 17:14:48 UT are the result of spacecraft-charging induced arc discharges.

Based upon the preliminary analysis of a limited quantity of data from P78-2, both the frequency of occurrence and the amplitudes of arc-discharge-induced transients are small as compared to those produced by the normal operation of on board systems. It should be noted, however, that the characteristics of each transient are determined not only by the characteristics of the discharge source, but also by the exact nature of the coupling path between the location of the source and the locations of the transient sensors. It is, therefore, extremely likely that significantly higher amplitude transients are induced at other unmonitored locations on the P78-2 vehicle. It is also true, as discussed in Section III C1, that the characteristics of the coupling paths on any spacecraft are critically dependent upon the exact details of the vehicle's design and that the P78-2 vehicle incorporates many electromagnetic shielding and design features that differ from those of most other operational spacecraft. It is anticipated that detailed analysis of the data from the P78-2 vehicle will provide an improved overall picture of the causes and effects of spacecraft-charging processes and will allow more accurate laboratory simulations of both charging and arcing phenomena to be performed and applied to other current and future spacecraft.

IV. Summary

A combination of laboratory and orbital data indicates with a high degree of confidence that, under certain orbital conditions, arc discharges can occur on materials normally used in the fabrication of the outer surfaces of satellites. These same measurements indicate that physical damage caused by arcing can degrade the performance of thermal- control systems and that the intense electromagnetic transients associated with arcing can upset or damage spacecraft electronic systems.

Ongoing laboratory and orbital measurement programs are providing much needed information for a more thorough understanding of these processes.

Based on the present state of knowledge in this area, it appears that, in the design of future satellites and space systems, serious consideration must be given to design techniques to both minimize the occurrence of arcing caused by spacecraft-charging phenomena and to reduce the possibility of undesirable or catastrophic failure of systems or subsystems should unexpected arcing occur.

References

1 Shaw, R. R., Nanevicz J. E., and Adamo, R. C., "Observations of Electrical Discharges Caused by Differential Satellite-Charging," Published as Paper SA41, Spacecraft Charging By Magnetospheric Plasmas, Progress in Astronautics and Aeronautics, Vol. 47, for the American Geophysical Union Spring Annual Meeting, Wash., D.C., June 1975, MIT Press, 1976.

2 Moore, D. W., "Electrical Discharge Damage of Second Surface Mirrors," Internal Rept., The Aerospace Corporation, El Segundo, Calif., July 27, 1972.

3 Inouye, G. T., "Spacecraft Charging Model," TRW Systems Group, Redondo Beach, Calif., for AIAA 13th Aerospace Sciences Meeting, Pasadena, Calif., January 20-22, 1975.

4 DeForest, E., "Spacecraft Charging at Synchronous Orbit," Journal of Geophysical Research, Vol. 77, Feb. 1972.

5 Nanevicz, J. E., Adamo, R. C., and Scharfman, W. E., "Satellite-Lifetime Monitoring," Final Rept. for Aerojet Electrosystems Company, Azusa, Calif., Contract F04701-71-C-0130, SRI Project 2611, March 1974.

6 Adamo, R. C., and Nanevicz, J. E., "Spacecraft-Charging Studies of Voltage Breakdown Processes on Spacecraft Thermal Control Mirrors," published as Paper SA42, Spacecraft Charging by Magnetospheric Plasmas, Progress in Astronautics and Aeronautics, Vol. 47, for the American Geophysical Union Spring Annual Meeting, Wash., D. C., June 1975, MIT Press, 1976.

7 Meulenberg, A. Jr., "Evidence for a New Discharge Mechanism for Dielectrics in a Plasma," Published as Paper SA70, Spacecraft Charging by Magnetospheric Plasmas, Progress in Astronautics and Aeronautics, Vol. 47, for the American Geophysical Union Spring Annual Meeting, Wash., D. C., June 1975, MIT Press, 1976.

8 Stevens, N., Lovell, R., and Gore, V., "Spacecraft-Charging Investigation for the CTS Project," published as Paper SA43, Spacecraft Charging by Magnetospheric Plasmas, Vol.47, Progress in Astronautics and Aeronautics, for the American Geophysical Union Spring Annual Meeting, Wash., D. C., June 1975, MIT Press, 1976.

9 Stevens, N., et al., "Testing of Typical Spacecraft Materials in a Simulated Substorm Environment," Proceedings of the Spacecraft Charging Technology Conference, edited by C. P. Pike and R. Lovell, AFGL-TR-77-0051, NASA TM X-73537, pp. 431-457 (1977).

10 Berkopec, D., Stevens, N., and Stuman, C., "The Lewis Research Center Geomagnetic Substorm Simulation Facility," Proceedings of the Spacecraft Charging Technology Conference, AFGL-TR-77-0051, No. 364, NASA TMX-73537, Feb. 1977.

11 Adamo, R. C., and Nanevicz, J. E., "Effects of Illumination on the Conductivity Properties of Spacecraft Insulating Materials," NASA Contract NAS3-20080, SRI Project 4904, SRI International, Menlo Park, Calif., July 1977.

12 Hoffmaster, D. K., Inouye, G. T., and Sellen, J. M. Jr., "Surge Current and Electron Swarm Tunnel Tests of Thermal Blanket and Ground Strap Materials," Proceedings of the Spacecraft Charging Technology Conference, AFGL-TR-77-0051, No. 364, NASA TMX-73537, Project ILIR 7661, Feb. 1977.

13 Fredricks, R. W., and Scarf, F. L., "Observations of Spacecraft Charging Effects in Energetic Plasma Regions," presented at the Sixth ESLAB Symposium on Photon and Particle Interactions with Surfaces in Space, Noordwijk, Holland, 26-29 September 1972, (Rept. No. 09670-7021-R0-00, August 1972).

14 Coge, J. R., and Pearlston, C. B., "Final Report on High Altitude Charging on Communication Satellites," Prepared for European Space Research and Technology Center, Domeinweg, Noordwijk, the Netherlands, Aerospace Rept. ATR-76(7568)-2, The Aerospace Corporation, El Segundo, Calif., Jan. 1977.

15 Buehl, F. W., and Hammersand, R. E., "A Review of Communications Satellites and Related Spacecraft for Factors Influencing Mission Success," Vol. I (Analyses) TOR-0076(6792)-1 and Volume II (Appendices A, B, C), TOR-0076(6792)-1, The Aerospace Corporation, El Segundo, Calif., Nov. 1975.

16 Nanevicz, J. E., Adamo, R. C., and Beers, B. L., "Characterization of Electromagnetic Signals Generated by Electrical Breakdown of Spacecraft Insulating Materials," Presented at Spacecraft Charging Technology -1978, NASA Conference Pub. 2071, AFGL-TR-79-0082, November 1978.

17 Malter, L., "Thin Film Field Emission," The Physical Review, Vol. 50, pp. 48-58, July 1936.

18 Nanevicz, J. E., and Adamo, R. C., "Malter Discharges as a Possible Mechanism Responsible for Noise Pulses Observed on Synchronous-Orbit Satellites," Published as Paper SA71, Spacecraft Charging by Magnetospheric Plasmas, Progress in Astronautics and Aeronautics, Volume 47, for the American

Geophysical Union Spring Annual Meeting, Wash., D. C., June 1975, MIT Press, 1976.

[19] B. Milligan, et al.,"Spacecraft Discharge Characterization - Test Setup, Quick-Look Experiments, and Preliminary Model Development," Final Rept., Phase I, Contract F04-701-78-C-0165, SRI Project 8071, SRI International, Menlo Park, Calif., (to be published).

[20] McPherson, D. A., Cauffman, D. P., and Schober, W., "Spacecraft Charging at High Altitudes -- The SCATHA Satellite Program," AIAA Paper 75-92, AIAA 13th Aerospace Sciences Meeting, Pasadena, Calif., Jan. 1975.

[21] Durrett, J. C., and Stevens, J. R., "Description of the Space Test Program P78-2 Spacecraft and Payloads," Presented at Spacecraft Charging Technology - 1978, NASA Conference Pub. 2071, AFGL-TR-79-0082, November 1978.

[22] Fong, J. C., et al., "Preliminary Engineering Analysis Results From the P78-2 Satellite (SCATHA)," AIAA 18th Aerospace Sciences Meeting, AIAA-80-0331, Pasadena, Calif., Jan. 1980.

[23] Nanevicz, J. E., and Adamo, R. C., "Transient Response Measurements on a Satellite System," Proceedings of the Spacecraft Charging Technology Conference, AFGL-TR-77-0051, No. 364, NASA TMX-73537, Project ILIR 7661, Feb. 1977.

[24] Adamo, R. C., Nanevicz, J. E., and Hilbers, G. R., "Development of the Transient Pulse Monitor (TPM) for SCATHA/P78-2," presented at Spacecraft Charging Technology - 1978, NASA Conference Pub. 2071, AFGL-TR-79-0082, Nov. 1978.

SURFACE DISCHARGE EFFECTS

K.G. Balmain*
University of Toronto, Toronto, Canada

Abstract

Specimens of Mylar sheet were exposed to a 20-kV electron beam. The resulting surface discharge arcs were photographed, and the discharge current into a metal backing plate was measured as a function of time. Discharge damage tracks were photographed using a scanning electron microscope and a light microscope. The area of the Mylar sheet was defined by a round aperture in a close-fitting metal mask, and the current pulse characteristics were plotted against area on log-log paper. The plots appear as straight lines (due to power-law behavior) with slopes of 0.50 for the peak current, 1.00 for the charge released, 1.50 for the energy dissipated, and 0.53 for the pulse duration.

Introduction

Arc discharges on thin sheets of spacecraft dielectric material have been discussed at length in the published proceedings of various conferences[1-4] and in review papers.[5-7] These discharges have been given names such as "punchthrough," "blowoff," and "flashover," all very vivid in keeping with their strength, dramatic appearance, and destructiveness. To date, direct measurements in space of discharge properties have been few,[8,9] so almost all of the direct observations and measurements have been done in laboratory vacuum chambers in which 10- to 30-keV electrons at current densities of the order of 10 nA/cm^2 have been directed at the insulating materials under test. Exposure to these electrons produced sporadic lightning-like arcs across the dielectric surface,[10] the arcs having a spider-web or feathery appearance, sometimes seeming to emanate from a very bright spot. These bright focal points often occurred near the edge of the material or were associated with

---

Presented in part at the 1978 USAF/NASA Spacecraft Charging Technology Conference, Colorado Springs, Colo., Oct. 31-Nov. 2, 1978.

*Professor, Department of Electrical Engineering.

punctures through the material. In general, each discharge arc produced a strong, short pulse of current into the metal surface underlying the dielectric sheet.

Earlier work by Gross[11,12] showed that a dielectric block irradiated with 2-MeV electrons could retain its charge for a long period, releasing it in an arc discharge stimulated by the insertion of a grounded needle. The resulting damage tracks were concentrated at the stopping distance of the 2-MeV electrons. Meulenberg[13] showed that breakdown-level electric fields could develop between the embedded layer of electrons and the dielectric surface which could be somewhat positive as a result of secondary emission. Therefore, breakdown could occur from the embedded electrons either to the surface (blowoff) or to the underlying metal substrate (punchthrough) depending on the circumstances. Such initial breakdown at a specific point could then be followed by an arc discharge propagating through the layer of embedded charge (flashover), as suggested by the damage tracks observed by Gross. Discharge propagation thus could provide the mechanism for mobilizing charge in sufficient quantity to produce the phenomenon of charge "cleanoff" observed by many laboratory experimenters.

Most laboratory experiments to date have involved dielectric areas much smaller than the exposed dielectric areas existing on operational synchronous-orbit satellites, so the question of area scaling of charge/discharge phenomena arises naturally. Certainly it is easier and faster to carry out

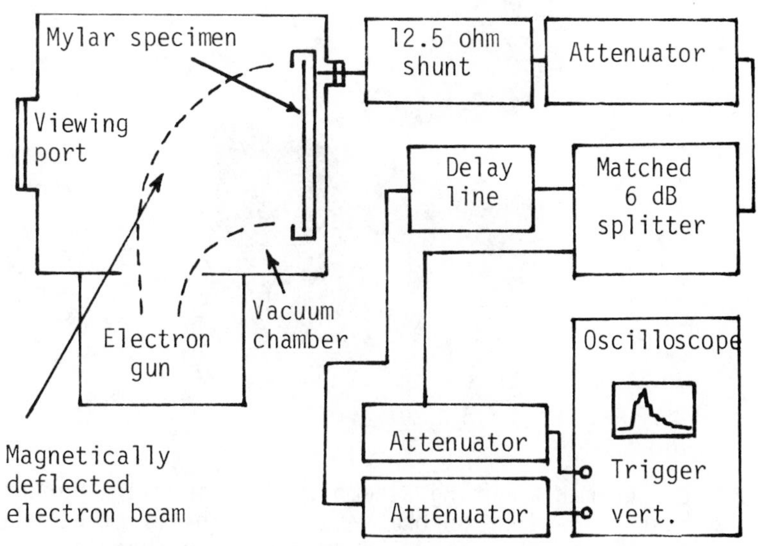

Fig. 1 The experimental apparatus.

small-scale experiments, compared with large-scale experiments in vacuum chambers large enough to hold spacecraft components or even an entire spacecraft.

It has been shown[14] that, for surface discharges on metal-backed polymer dielectrics, the peak discharge current is proportional to the surface area raised to a power "p" lying between 0.5 and 0.8 for the range of areas between 0.2 cm^2 and 20 cm^2. The most consistent results shown were for Teflon with a value of p = 0.58, whereas the results from Mylar gave p = 0.76 with more scatter in the plotted points. In the foregoing experiments the dielectric surface area was determined by cutting the test specimen to size in the shape of either a square or a rectangle with a 2:1 side length ratio. It is also possible to define the effective charged area by controlling the size of the incident electron beam,[14] a procedure which can give charged areas in the range of 10^{-5} cm^2 to 10^{-3} cm^2, along with peak discharge currents which for Teflon lie approximately on the line with slope p = 0.58 referred to above, extrapolated downward from much larger areas.

Cutting a dielectric specimen to size means that the stressed and perhaps damaged edges are exposed to the electron beam. Furthermore, the edges adjacent to the grounded metal substrate in most experiments, receive a progressively larger

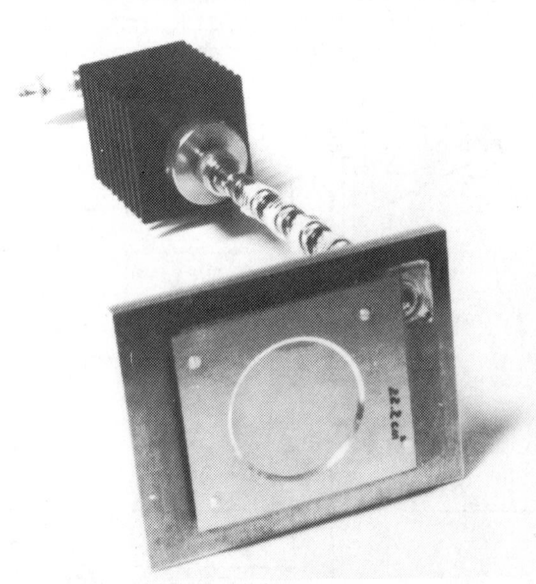

Fig. 2 The specimen mounting assembly with circular-aperture aluminum mask. The four 50-ohm shunts and the first attenuator are attached to the coaxial feed-through receptacle.

electron flux as the incident electron beam is deflected away from the accumulating charge in the mid-region of the specimen. Clearly a better means of defining the effective charged area is required, one such means being the metal mask technique described below.

Experiment Design

The vacuum chamber and electrical instrumentation are shown in Fig. 1. A vacuum-sealed bulkhead receptacle carries the discharge pulse signal to a 10-ohm termination consisting of four 50-ohm coaxial feed-through terminations (comprising a 12.5-ohm shunt) plus the 50-ohm input to the attenuator and measurement cable. The attenuators, splitter, and delay line are all matched and calibrated up to 4 GHz. All cables are matched at both ends to prevent ringing. The oscilloscope has a 400-MHz bandwidth. Precautions were taken to insure adequate shielding of operating personnel from X-radiation.

Figures 2 and 3 show the specimen mounting plate and a typical circular-aperture metal mask used to define the charged area on the specimen surface. For each such area the experimental procedure was to set the incident 20-keV electron beam to a current density which produced approximately one discharge every 15 sec. For the largest areas this resulted in a current density of about 100 nA/cm^2 and for the smallest areas about 10 μA/cm^2; variation of the current density over this range

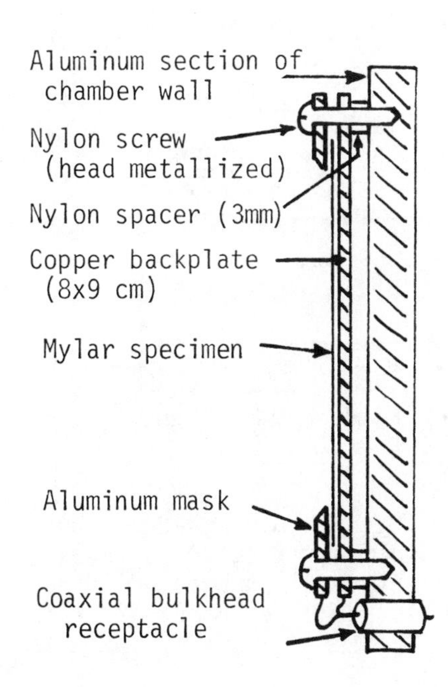

Fig. 3 Details of the specimen mounting assembly.

affected only the discharge rate and had no noticeable effect on the amplitude or shape of the discharge current pulses for the 125-μm thick Mylar specimens under test.

Phenomena Expected

Figure 4 shows the expected effects of exposing a dielectric sheet to a monenergetic beam of electrons. First, the 20-keV electrons penetrate to the stopping depth, estimated to be 7 to 8 μm. As the embedded charge builds up, the electron energy upon impact is reduced, resulting in somewhat shallower deposition.[15] The accumulated electrons induce positive charge in the underlying metal substrate and in the nearby edge of the metal mask, and the incident beam is deflected by electrostatic forces to deposit the highest charge concentration close to the mask edge.

Relatively thin specimens or specimens with edges shielded against the incident electron beam have a tendency to discharge from the embedded electron layer through to the substrate via the "punchthrough" arc[16] indicated as "a" in Fig. 4. In cases

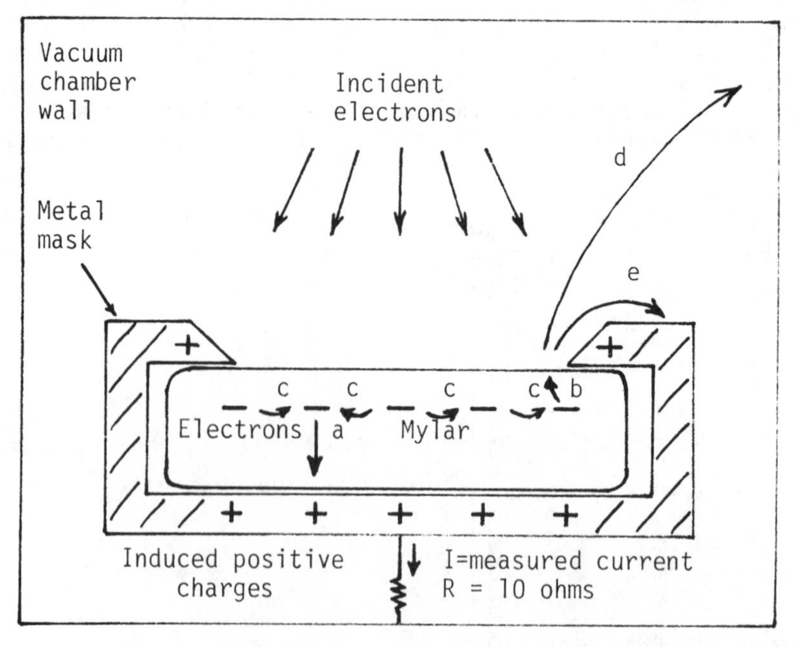

Fig. 4 Illustration of possible charging and discharging phenomena. a) punchthrough arc, b) blowoff arc (arc to surface), c) flashover arc (propagating subsurface discharge), d) ejected electrons going to chamber wall, and e) ejected electrons going to mask.

where there is a high near-surface field and especially close to an edge, an arc through to the surface ("b") or "blowoff" arc will occur, accompanied by a strong ejection of electrons. This will leave behind a high-field region inside the dielectric, resulting in a propagating "flashover" arc "c" through the layer of embedded electrons. Depending on the circumstances, the flashover may produce sporadic or continuous blowoff of material as it propagates.

The punchthrough arc is sufficiently energetic when it occurs to leave a puncture all the way through the dielectric specimen. The incident electrons would see this hole as a grounded spot and would tend to focus on it so that the punchthrough hole would then become the focal point for subsequent discharge activity in the form of flashover arcs and blowoff arcs.

The discharge current measurement consists of measuring the voltage across the resistor R in Fig. 4. It is very important to note that the only discharge currents measured by this method are those arising from the ejected-electron orbits "d" which terminate on the chamber walls. Electron orbits "e" terminating on the mask or punchthrough arcs will not result directly in the flow of current through R.

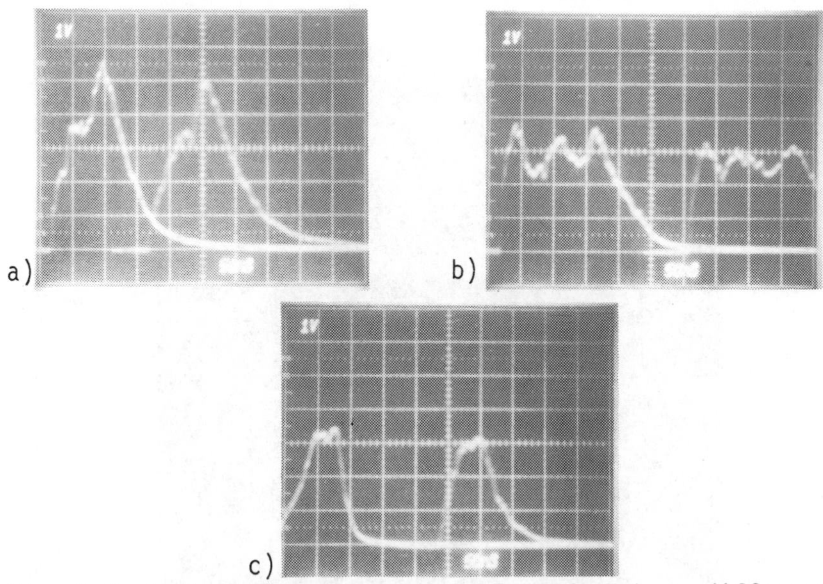

Fig. 5 Typical discharge current pulses for three different Mylar specimens. a) Area = 22.2 cm^2, I_m = 44.5 A and 39.7 A. b) Area = 22.2 cm^2, I_m = 30.2 A (see Fig. 7 for photograph of arc), and I_m 26.2A. c) Area = 47.6 cm^2, I_m = 110.7 A and 101.2 A.

The Arc Discharges

The measured discharge current pulses of Fig. 5 provide some idea of the normal range of pulse shapes encountered. For each setup (all experimental conditions fixed) it was quite common to measure 5 to 10 current pulses of nearly identical shapes, as suggested by the nearly identical pairs in Fig. 5. The shortest pulses measured (for small dielectric areas, not shown) exhibited a small degree of negative overshoot and ringing, possibly because their rise and fall times were close to the limit for the oscilloscope used.

Visual inspection of the discharge arcs revealed that no two arc patterns are identical although there may be a degree of similarity and a few common bright-spot focal points in a given setup, especially near the mask edge. Typical discharge arcs are shown in Figs. 6-9.

Discharge Damage

In view of the maximum electron penetration distance of about 8 μm, one might expect to see damage down to approximately this depth. In the search for this damage, a specimen was selected for which progressively increasing mask diameters had been used and on which approximately 15 discharges per mask had

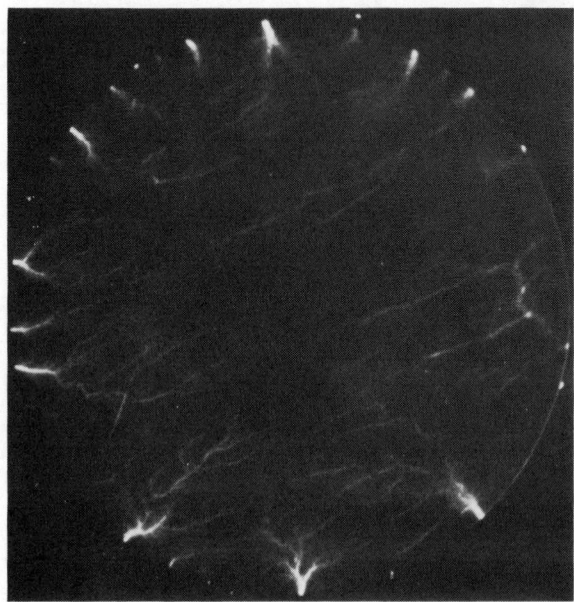

Fig. 6 Arc showing simultaneous edge streamers. Mylar area = 22.2 cm^2.

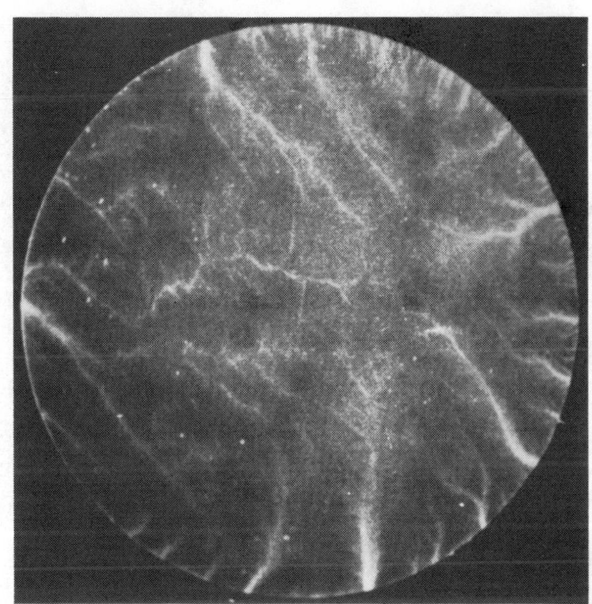

Fig. 7 Arc corresponding to pulse with I_m = 30.2 A in Fig. 5. Mylar area = 22.2 cm^2.

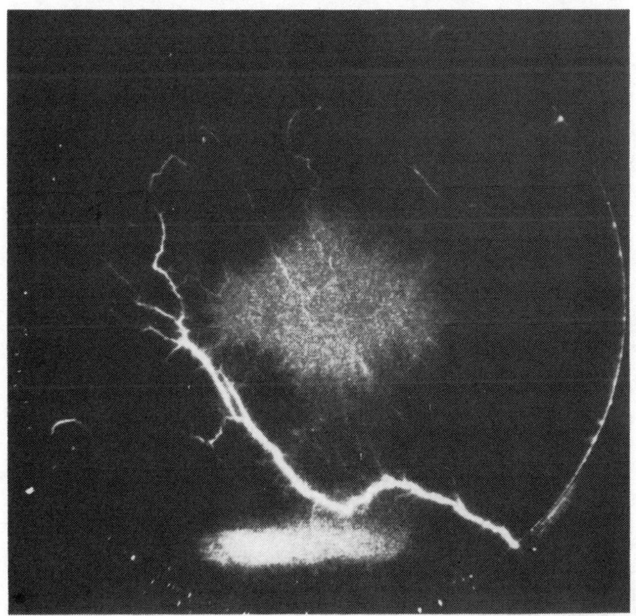

Fig. 8 Arc with single primary streamer. Mylar area = 47.6 cm^2. Two light shaded regions are the result of electron-impact luminescence during an unusually long open-shutter period before the discharge.

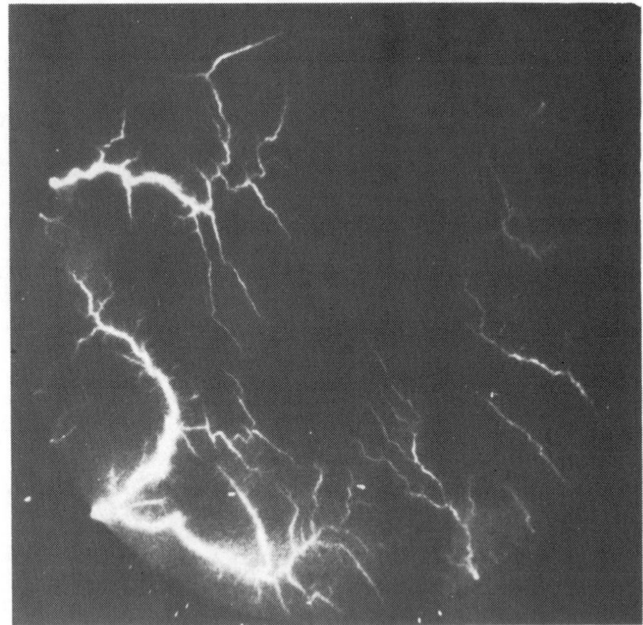

Fig. 9 Arc with multiple primary streamers. Mylar area = 47.6 cm^2.

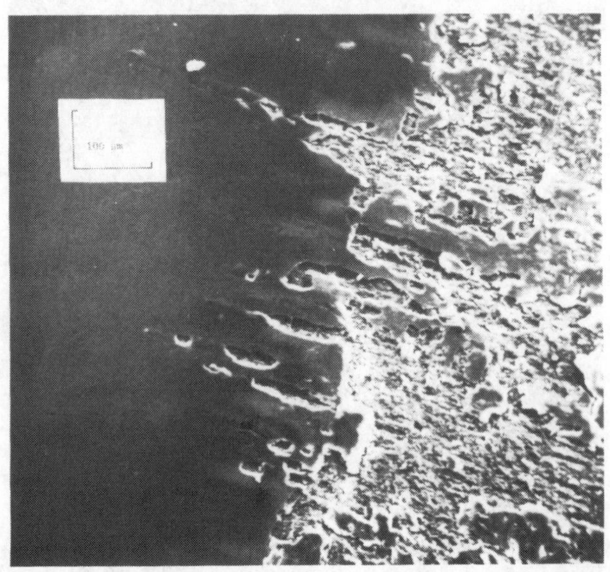

Fig. 10 Scanning electron microscope image of damage along one of the mask-edge lines on a single specimen for which a sequence of progressively larger mask apertures had been used.

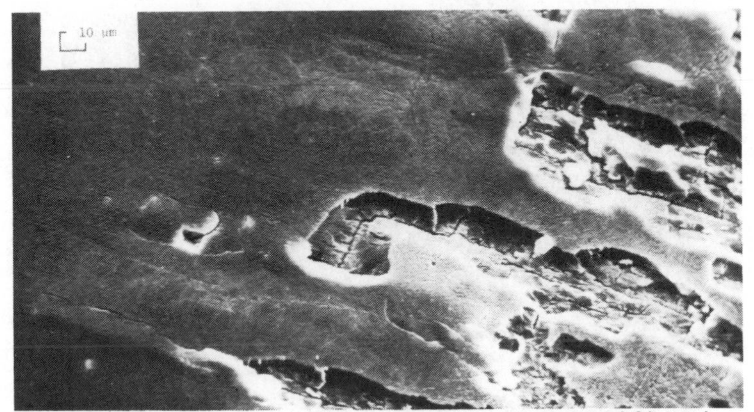

Fig. 11 Scanning electron microscope image: magnification of Fig. 10.

Fig. 12 Scanning electron microscope image: magnification of Fig. 11.

occurred. The specimen was vacuum-coated with gold and viewed in a scanning electron microscope. The resulting micrograph (Fig. 10) clearly shows two regions of greatly differing damage separated by one of the lines along which the mask edge had been positioned.

Micrographs of increasing magnification are shown in Figs. 11-14, in which it is clear that the most characteristic damage consists of a deep channel of less than 1 µm in width at the bottom of a much wider depression whose surface apparently was blown off by the force of the discharge arcs. Also visible, especially near the top of Fig. 14, are two apparent holes in the side of the depression.

Fig. 13 Scanning electron microscope image: magnification of Fig. 12.

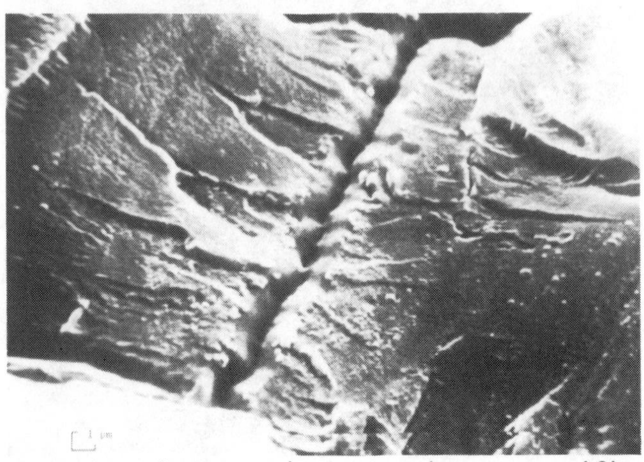

Fig. 14 Scanning electron microscope image: magnification of Fig. 13. Note two tunnel openings near top of page.

Transmitted-light microscope photographs also were taken and are shown in Figs. 15-17. These reveal a network of damage tracks underlying the relatively undamaged portion of the surface. Two of these damage tracks open out into the two holes already mentioned with respect to Fig. 14. It therefore appears that the subsurface damage tracks are tunnels along which the discharge arcs are channeled, sometimes blowing off the top few microns of the material surface. The tunnel radii appear to be approximately 0.3 μm, and the tunnel depths appear to be mostly 4 to 8 μm below the surface.

It is worth noting that complete puncture of the specimen (or punchthrough) was extremely rare in these experiments on

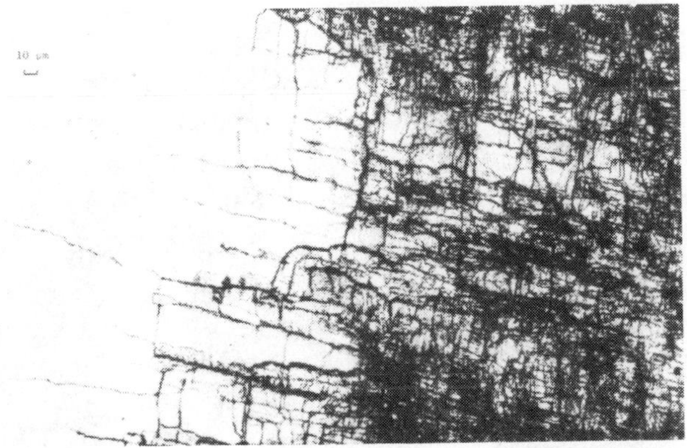

Fig. 15 Transmitted-light photograph corresponding to Fig. 11.

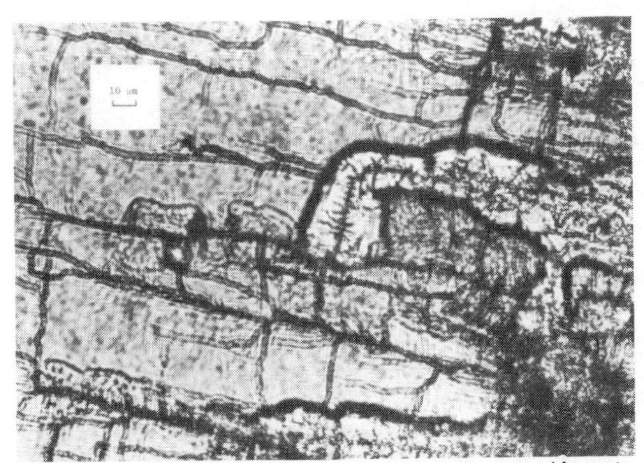

Fig. 16 Transmitted-light photograph corresponding to Fig. 12.

125-μm thick Mylar. In fact, it occurred in only one sequence of experiments involving the use of progressively larger masks on a single specimen.

Area Scaling

A sequence of experiments was carried out using the full range of mask diameters available and using a new, unexposed specimen for each mask size. The peak currents I_m are plotted in Fig. 18 against specimen area, and each averaged point is accompanied by a vertical bar indicating the full range of currents used to calculate the average. The straight line drawn through the larger-area points has a slope of 0.50, indicating that the peak currents are proportional to the square root of the area - that is, proportional to the linear dimensions of

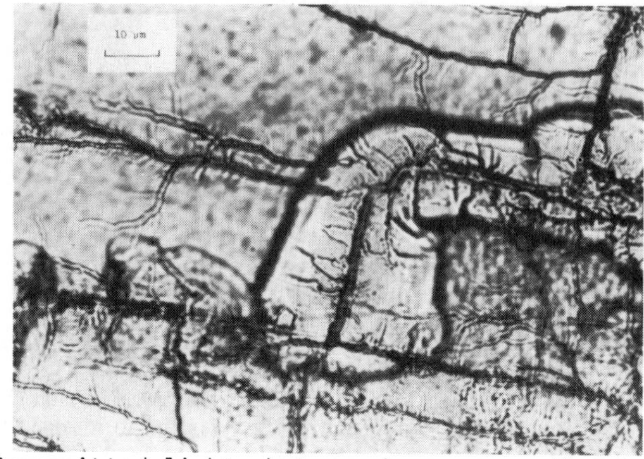

Fig. 17 Transmitted-light photograph corresponding to Fig. 13.

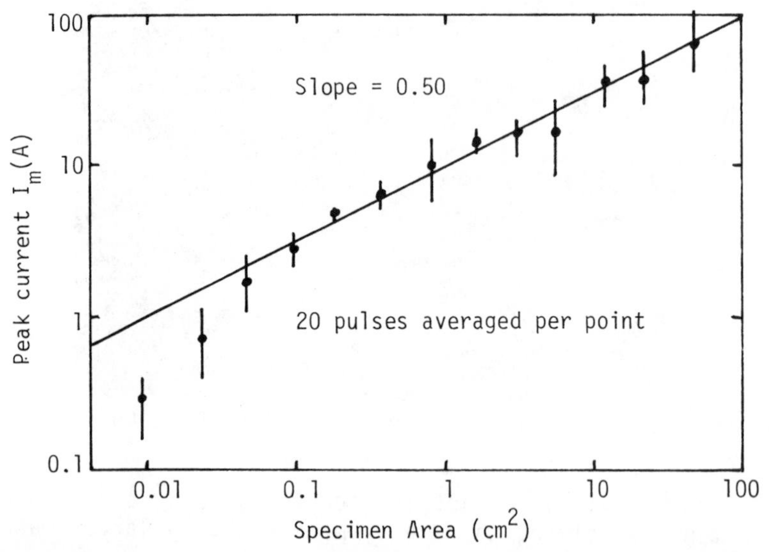

Fig. 18 Variation of peak current I_m with Mylar specimen area, a different unexposed specimen being used with each mask aperture size.

the aperture. It is worth noting that this straight line extrapolates to $I_m \approx 1000$ A at an area of 1 m^2.

The charge Q passing through the measurement system is given by

$$Q = \int I \, dt$$

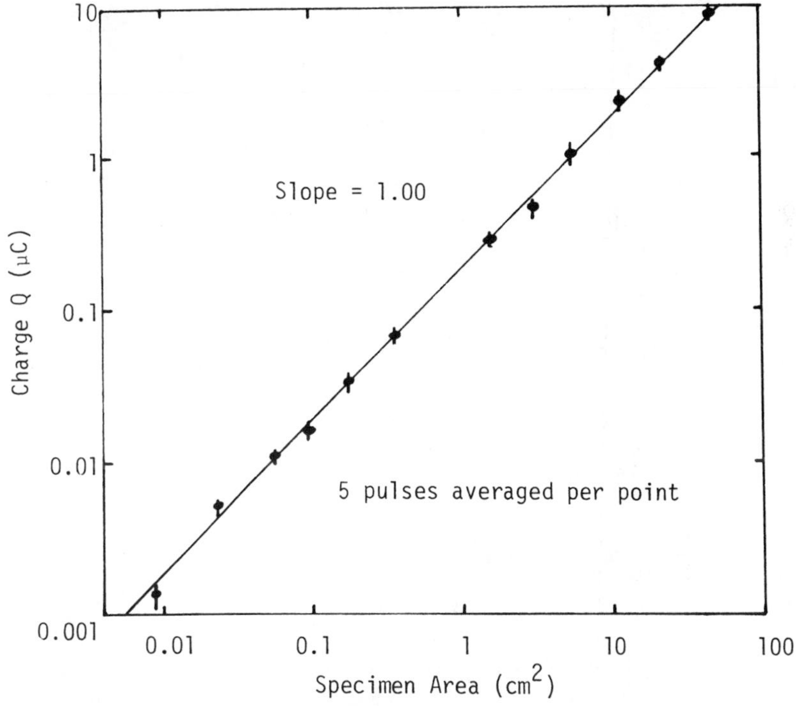

Fig. 19 Variation of released charge $Q = \int I\, dt$ with Mylar specimen area for the same conditions as in Fig. 18.

an integration which was carried out using data from oscilloscope photographs. The results of these computations are shown in Fig. 19 which includes a straight-line approximation having a slope of 1.00 indicating charge proportionality to specimen area.

The energy dissipated in the load resistor R is given by

$$E = R \int I^2\, dt$$

and the resulting graph of energy against area is shown in Fig. 20. The highest energy point plotted on the graph is 5 mJ; this is sufficient to burn out a typical 50-ohm coaxial microwave attenuator rated at 20 W average, a lesson which was learned by experience during the early stages of the research program!

The pulse duration was calculated from the relation

$$T = \frac{1}{I_m} \int I\, dt$$

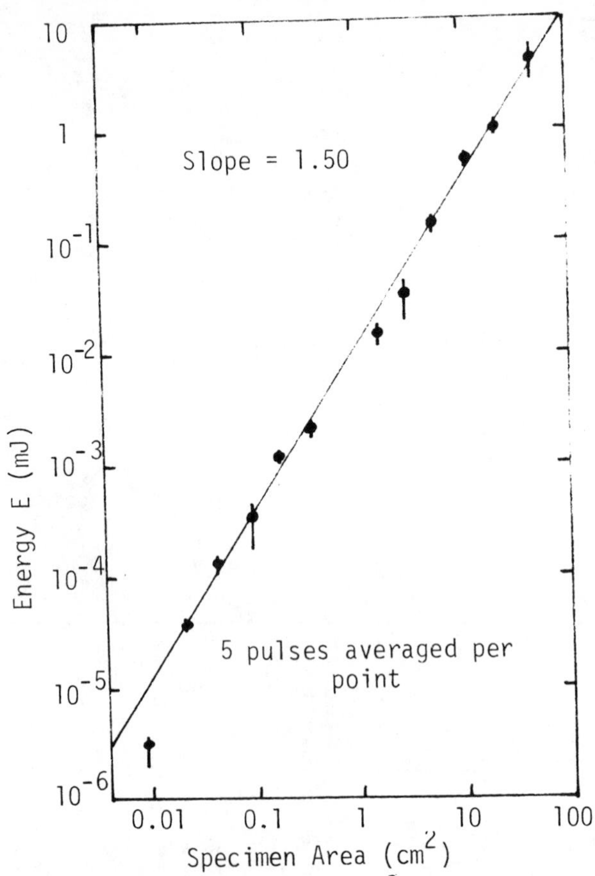

Fig. 20 Variation of energy $E = R \int I^2 dt$ with mylar specimen area. The energy is dissipated in a load resistor R = 10 ohms. The conditions are the same as in Fig.18.

and the resulting graph of duration against area is shown in Fig. 21. The straight-line approximation has a slope of 0.53 although one would have expected a slope of 0.50 because the duration calculated in this way is just the ratio of released charge to peak current. Departure from the straight-line approximation is noticeable for small areas and thus for short pulses, this departure taking the form of extended pulse durations and reduced peak currents as well. The probable cause of this effect is the 400-MHz bandwidth of the oscilloscope.

Figure 21 also can be used to estimate the propagation velocity of the flashover arc, assuming that the pulse duration is determined by the propagation time of an arc across the

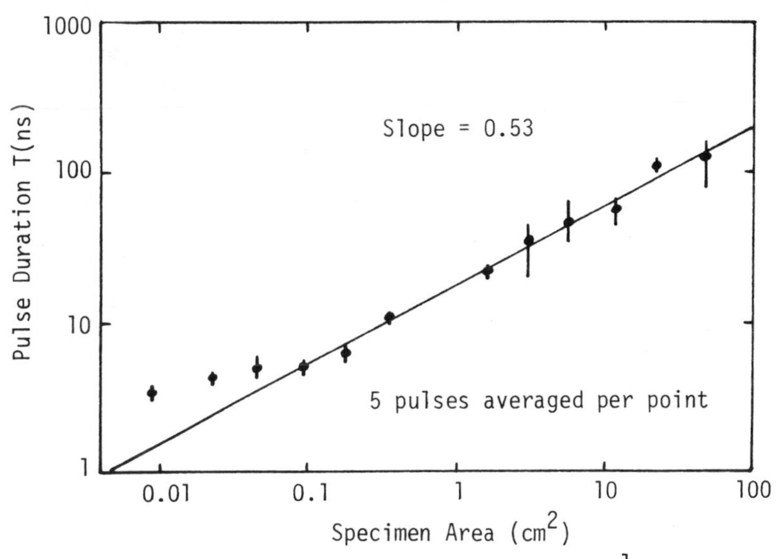

Fig. 21 The variation of pulse duration $T = I_m^{-1} \int I\,dt$ with Mylar specimen area for the same conditions as in Fig. 18.

specimen. From the graph, an aperture radius of 1 cm corresponds to a pulse duration of 33 nsec, the ratio giving a velocity of 3×10^5 m/sec.

In order to assess cumulative arc effects, an experimental sequence was carried out using progressively increasing mask sizes on the same Mylar specimen. As already indicated, this experiment produced occasional punchthroughs, and also this specimen was used for the damage photographs already discussed. In spite of the differences in procedure and damage, the resulting peak current graph of Fig. 22 is essentially indistinguishable from the graph of Fig. 18 for experiments employing unexposed specimens for each area. This similarity may be due to the fact that the results of Fig. 22 are for specimens having an unexposed annular ring of dielectric adjacent to the mask edge for each new, larger area in the experimental sequence.

The reverse sequence also was tried, involving progressively decreasing mask areas. The peak current, released charge, energy dissipated, and pulse duration are graphed as functions of area in Figs. 23-26. Material fatigue was most evident at the smallest areas. In these cases one or two discharges occurred immediately after the electron beam was turned on, followed by a complete absence of discharges, so that no data was recorded. Apart from this there is little difference

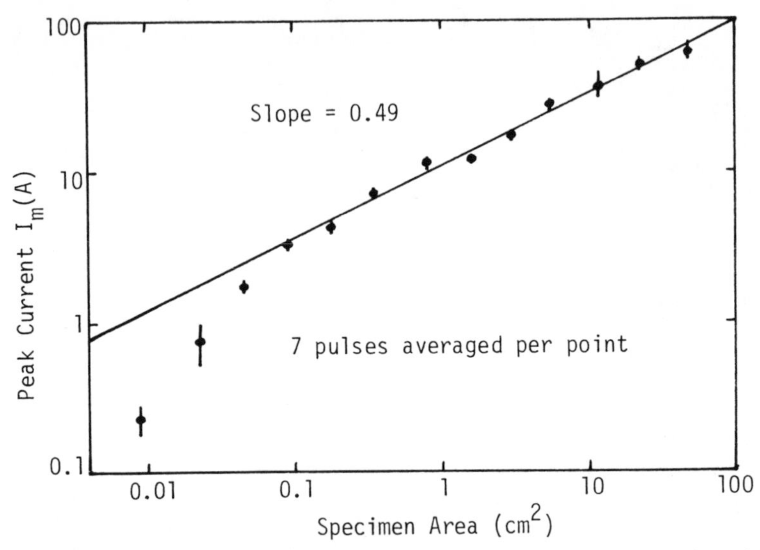

Fig. 22 Peak current I_m for an experimental sequence in which a single specimen was used with progressively increasing mask aperture diameters.

in comparison with the cases discussed previously, the difference amounting to a slight decrease in peak current and a slight increase in pulse duration.

Interpretation

The proportionality of the released charge to the specimen area is demonstrated in Fig. 19 in which the plotted points exhibit little scatter and little departure from exact proportionality even for the smallest areas tested. This suggests that a constant fraction of the charge incident on any given area discharges via electron orbits going to the chamber wall. Now suppose that a fraction of the charge discharges via electron orbits going directly to the metal mask ("e" in Fig. 4). However, the outside dimensions of the mask are constant, so the exposed mask area increases as the exposed dielectric area decreases. Therefore, one might expect that an increasing fraction of the ejected electrons would go to the mask as the dielectric specimen area is decreased, leaving a smaller fraction for the current to the walls. This apparently does not happen, according to Fig. 19, so that the discharge current going to the mask is most probably negligible.

If the discharge were similar to the discharge of a capacitor through a resistor connected across the capacitor terminals, then the peak current would be proportional to the incident beam energy (capacitor voltage), which is constant, but independent of specimen area (capacitance). Also, the energy dissipated in the resistor would be proportional to area. Neither of these proportionalities applies to the surface discharges, so the discharge mechanism is not similar to that of a capacitor.

The appearance of the discharge arcs and the arc damage suggests that the discharge must propagate at a finite velocity. One such finite-velocity mechanism is the electron-hopping mechanism suggested by Inouye and Sellen.[17] Leadon and Wilkenfeld[18] also have estimated the inductance and capacitance of the discharge arc to calculate the discharge time duration. Regardless of the precise mechanism involved, suppose that a discharge initiates at a point and expands outward uniformly at a constant velocity. If the ejected electrons emerge from the wave front, then the discharge current would be proportional to the length of the wave front. The maximum wave front length would be controlled by the linear dimensions of the exposed dielectric, so the peak current would be proportional to the square root of the specimen area, exactly as observed. The pulse duration also would have the same proportionality, approx-

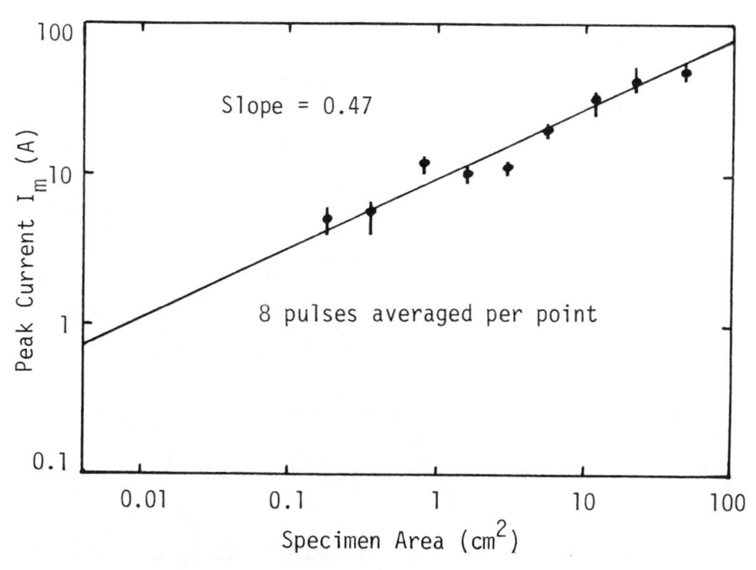

Fig. 23 Peak current I_m for an experimental sequence in which a single specimen was used with progressively decreasing mask aperture diameters.

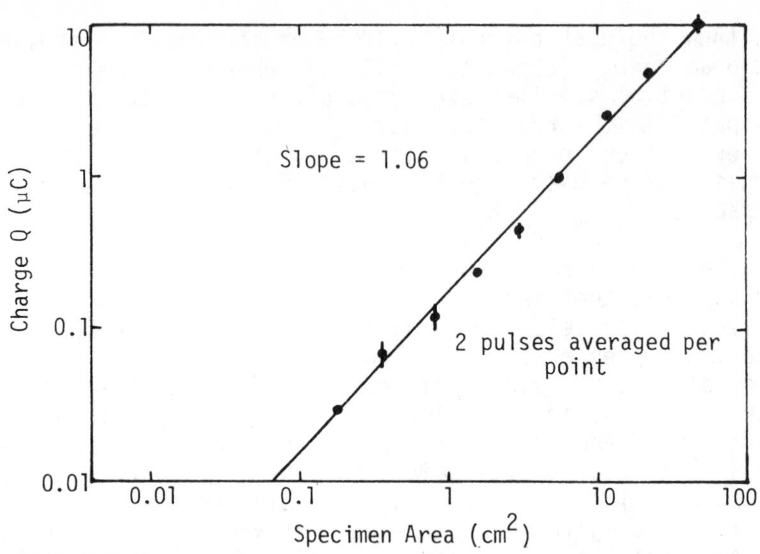

Fig. 24 Released charge Q for the same conditions as in Fig.23.

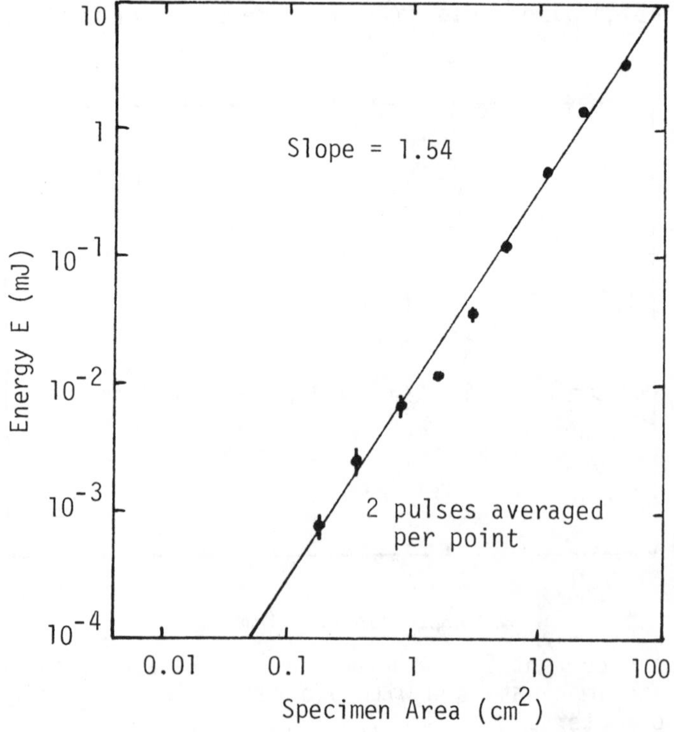

Fig. 25 Energy dissipated E for the same conditions as in Fig.23.

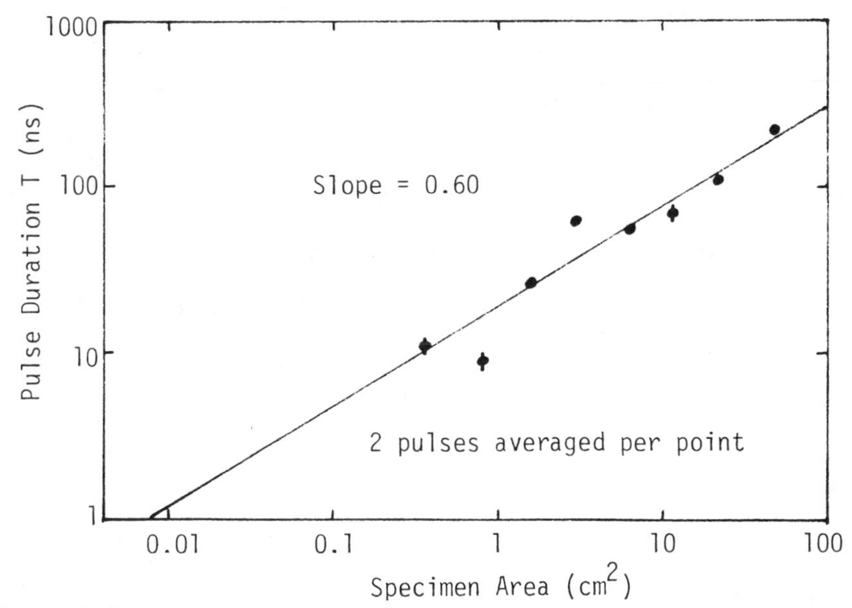

Fig 26 Pulse duration T for the same conditions as in Fig. 23.

imately as observed. A sequence of linear, branching discharge paths should produce the same result overall.

If the discharge propagation were entirely subsurface and all electron ejection from the point of initiation, then the process would involve two velocities. These would be the velocity of the discharge wave front and the velocity of subsurface electron propulsion toward the point of ejection. Thus the apparent velocity deduced from measured currents would in fact be the sum of these two velocities.

A vivid analogy is that of a "flash flood" caused by a sudden, heavy rainstorm over a specified area. If the water runoff velocity were constant, then the peak runoff water flow would be proportional to the square root of the storm area, and the runoff duration would have the same proportionality.

Conclusions

Surface discharge arcs on metal-backed Mylar exhibit electrical characteristics which scale with variations in specimen area according to well-defined power laws. The characteristics identified are peak current, released charge, energy dissipated, and pulse duration, and the respective powers are 0.50, 1.00, 1.50, and 0.53.

The discharge arcs appear to be channeled into hair-like branching tunnels at the penetration depth, with occasional blowoffs of surface material. These tunnels do not cover the entire surface, however, so there remains an open question as to the discharge mechanism between the tunnels. There are signs of material fatigue in the electrical measurements of discharge properties, but these indications of fatigue seem small in view of the very extensive surface damage visible in the scanning electron microscope, especially near the mask edge.

The surface discharges appear to propagate at a well-defined velocity of about 3×10^5 m/sec. Propagation at a constant velocity is consistent with the experimental observations of the area scaling laws for peak current and pulse duration.

In considering the application of these results to materials on synchronous-orbit spacecraft, there are several differences between the laboratory conditions and the conditions in space which should be kept in mind. In the laboratory, the incident electron fluxes used were much higher than those encountered in space, the higher fluxes having been employed in order to have a high enough discharge rate for reasonably rapid accumulation of data and in order to have reasonably predictable discharge occurrence times for high-quality arc photography. Even though electrical independence of electron flux was noted in the flux range employed, there does exist the possibility of different behaviour at much lower fluxes. In addition, unlike the situation in space, the laboratory experiments involved monoenergetic electrons normally incident on the dielectric, with no incident ions and no ambient plasma. As for the various types of materials used on spacecraft, previous work[14] has suggested that different types of polymers have much in common with respect to discharge properties, so it is possible that the results of this study will apply approximately to other highly insulating polymers.

Overall, the most remarkable result of this experimental study is that, in spite of the variability in discharge arc appearance from one arc to the next, in spite of extensive damage inflicted on the dielectric surface, and in spite of the obvious complexity of the detailed discharge process, there still exists a high degree of order in the electrical properties of the current pulses observed. This degree of order strongly suggests the existence of a set of well-defined physical principles governing the behavior of surface discharges and underlines the importance of seeking out these principles.

Acknowledgments

The research was supported by the Natural Sciences and Engineering Research Council of Canada under Grant No. A-4140. The experiments and data reduction were carried out by G.R. Dubois, whose contributions are gratefully acknowledged.

References

[1]Rosen, A., ed., Progress in Astronautics and Aeronautics: Spacecraft Charging by Magnetospheric Plasmas, Vol. 47, AIAA, New York, 1976.

[2]Pike, C.P. and Lovell, R.R., eds., Proceedings of the Spacecraft Charging Technology Conference, Rept. AFGL-TR-77-0051/ NASA TMX-73537, Feb. 24, 1977.

[3]Goodman, J.M, ed., Effect of the Ionosphere on Space and Terrestrial Systems, Proceedings of an NRL/ONR - sponsored conference, Arlington, Va., January 24-26, 1978, U.S. Government Printing Office Stock No. 008-051-00069-1.

[4]Proceedings of the 1978 USAF/NASA Spacecraft Charging Technology Conference, Colorado Springs, Colo., Oct. 31-Nov. 2, 1978.

[5]Rosen, A., "Large Discharges and Arcs on Spacecraft," Astronautics & Aeronautics, Vol. 13, June 1975, pp. 36-44.

[6]Rosen, A., "Spacecraft Charging: Environment - Induced Anomalies," Journal of Spacecraft and Rockets, Vol. 13, March 1976, pp. 129-136.

[7]Rosen, A., "Spacecraft Charging by Magnetospheric Plasmas," IEEE Transactions of Nuclear Science, Vol. NS-23, Dec. 1976, pp. 1762-1768.

[8]Nanevicz, J.E., and Adamo, R.C., "Transient Response Measurements on a Satellite System," Ref. 2, pp. 723-734.

[9]Shaw, R.R., "Electrical Interference to Satellite Subsystems Resulting from Spacecraft Charging," Ref. 3, pp. 337-345.

[10]Stevens, N.J., Berkopek, F.D., Staskus, J.V., Blech, R.A., and Narciso, S.J., "Testing of Typical Spacecraft Materials in a Simulated Substorm Environment," Ref. 2, pp. 431-457.

[11]Gross, B., "Irradiation Effects in Borosilicate Glass," The Physical Review, Vol. 107, July 15, 1957, pp. 368-373.

[12]Gross, B., "Irradiation Effects in Plexiglas," Journal of Polymer Science, Vol. 27, 1958, pp. 135-143.

[13]Meulenberg, A., "Evidence for a New Discharge Mechanism for Dielectrics in a Plasma," Ref. 1, pp. 237-246.

[14]Balmain, K.G., Kremer, P.C., and Cuchanski, M., "Charged-Area Effects on Spacecraft Dielectric Arc Discharges," Ref. 3, pp. 302-308.

[15]Beers, B.L., Hwang, H.-C., Lin, D.L., and Pine, V.W., "Electron Transport Model of Dielectric Charging," Ref. 4, pp. 209-238.

[16]Yadlowsky, E.J., Hazelton, R.C., and Churchill, R.J., "Characterization of Electrical Discharges on Teflon Dielectrics Used as Spacecraft Thermal Control Surfaces," Ref. 4, pp. 632-645.

[17]Inouye, G.T. and Sellen, J.M., "A Proposed Mechanism for the Initiation and Propagation of Dielectric Surface Discharges," Ref. 3, pp. 309-312.

[18]Leadon, R. and Wilkenfeld, J., "Model for Breakdown Process in Dielectric Discharges," Ref. 4, pp. 704-710.

ACTIVE CONTROL OF SPACECRAFT CHARGING

Carolyn K. Purvis*
NASA Lewis Research Center, Cleveland, Ohio

and

Robert O. Bartlett+
NASA Goddard Space Flight Center, Greenbelt, Md.

Abstract

The concept of active control of spacecraft charging by charged particle emission is described. Active potential control experiments using the ATS-5 and ATS-6 geostationary spacecraft are discussed, and results of these experiments are presented. Previously reported results are summarized, and a guide to reports on these data are provided. Experimental evidence presented indicates that emission of electrons only is not effective in maintaining spacecraft potential near plasma potential for spacecraft with electrically insulating surfaces. Emission of a low energy plasma, however, is effective for this purpose.

Introduction

A satellite immersed in a plasma environment will come into equilibrium with the environment by acquiring surface charges such that the net current to it is zero. It will thus acquire a nonzero potential relative to the space plasma potential. This potential will vary as the currents to the spacecraft change. For geosynchronous spacecraft, dramatic changes in potential occur in response to changes in these currents.

In particular, when energetic substorm plasmas are present, the removal and reinstatement of photoelectron current caused by eclipse passage are observed to result in dramatic shifts in the potentials of the ATS-5[1] and ATS-6[2] spacecraft. Both spacecraft charge negatively during substorm activity; this is consistent with the observation that the ambient elec-

---

*Physicist.

+Aerospace technologist.

tron fluxes are larger than the ambient ion fluxes. No large positive excursions have been observed. In sunlight, the most negative potentials reported have been -300 V for ATS-5[1] and -2200 V for ATS-6.[3] In eclipse, the comparable potentials are -10 kV for ATS-5 and -20 kV for ATS-6.[4] The difference in most negative potentials observed in eclipse is believed to be due primarily to an increase in geomagnetic activity between the ATS-5 and ATS-6 missions.[5] In 1974, when both spacecraft were eclipsed simultaneously, they charged to similar potentials.[6] Clearly, the large excursions in spacecraft potential during eclipse passage could be eliminated by finding a suitable replacement for the photoelectron current, which could be turned on during eclipse. Ideally, one could go a step further and devise a method to emit appropriate currents to control spacecraft potential at all times.

The potentials discussed here are those of the spacecraft structures (electrical reference) with respect to plasma potential; these are "absolute" or overall potentials. For a spacecraft with electrically isolated surfaces (i.e., insulating surfaces), differential charging can occur because of either differences in the properties of surface materials (e.g., secondary electron yield) or differences in the "environment" between surfaces (e.g., one surface sunlit, and another shadowed). The latter effect is expected to create larger differential potentials than the former. Differential charging effects must be considered in devising strategies for active control of spacecraft potential by charged particle emission.

The ATS-5 and ATS-6 spacecraft each carried a University of California, San Diego (UCSD) Auroral Particles Experiment (which provides the information on spacecraft potential) and a Cesium Ion Engine Experiment. These instruments have been used jointly to investigate active control of spacecraft potentials by charged particle emission.[7] Neither of these instruments was designed with active control of spacecraft potential as an objective, but analysis of the plasma data taken during their operation has yielded much valuable information on the subject.

In what follows, the ATS-5 and ATS-6 spacecraft and relevant instrumentation are described briefly, an overview of the experiments conducted is presented, published results are identified, experiments results are summarized, and conclusions are drawn based on these results. It is anticipated that more definitive results will be forthcoming from the active control experiments planned for the P78-2 (SCATHA) spacecraft which was launched in January 1979. SCATHA carries a modified version of the ATS-6 UCSD detector system, a charge ejection system, and other diagnostics specifically designed to study active spacecraft potential control.[8]

ATS-5 and ATS-6 Systems

ATS-5 Spacecraft and Instrumentation

The ATS-5 spacecraft was launched in Aug. 1969 into geosynchronous orbit. Its configuration is shown in Fig. 1. The spacecraft has a cylindrical geometry, 1.3 m in diameter and 2 m in length. It is divided into three cylindrical sections of approximately equal length. Most of the experiments and spacecraft systems are contained in the center section. The two outermost sections are open-ended shells to which solar cells have been mounted and which have an outer surface primarily of quartz cover glass. The center section is covered with a fiberglass skin to which a nonconductive thermal control paint has been applied. Therefore, the outermost surface of ATS-5 is an electrical insulator.

Two contact ion engine systems are on board ATS-5. The location of the No. 2 engine relative to the ATS-5 UCSD

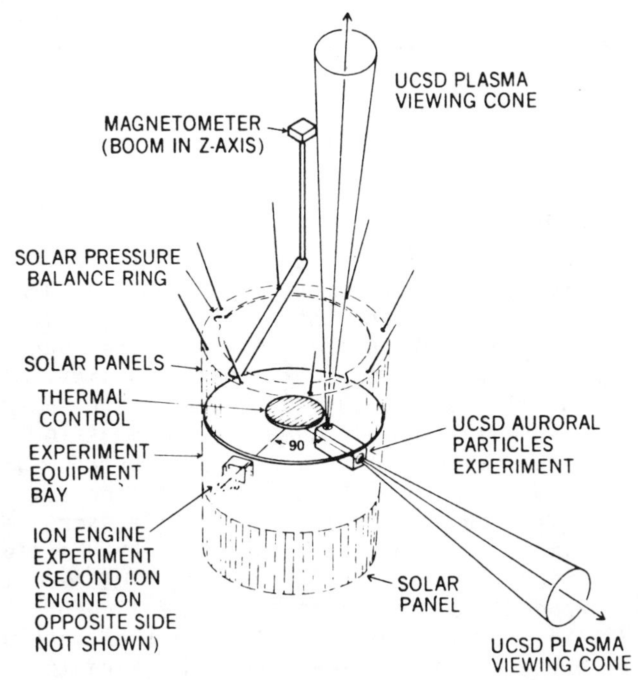

Fig. 1 ATS-5 orbital configuration.

Auroral Particles Experiment is shown in Fig. 1. The No. 1 engine is on the opposite side of the spacecraft. Due to a design fault in the ATS-5, the spacecraft could not be despun and spins at 76 rpm around the cylinder axis.

The ATS-5 ion engine system has been described in detail by Worlock et al.,[9] and the constraints on its use during active control tests are summarized by Purvis et al.[6] Its contact ion source was designed to deliver 1 mA of singly charged cesium ions which are neutralized by a hot filament electron source. The ion engine exhaust aperture in the spacecraft skin is 5 cm in diameter. The neutralizing filament is recessed 2.5 cm within the spacecraft and operates at spacecraft potential. It consists of a resistively heated filament of yttrium doped tantalum, 0.18 mm in diameter, operated at a temperature of 1700° C. At this temperature, the neutralizer is emission limited at about 3 mA. The minimum resolvable neutralizer emission current is 6 μA. No neutralizer current has been observed during active control operations using the neutralizer alone.

The UCSD Auroral Particles Experiment on ATS-5 consists of two pairs of plasma detectors. These are mounted to the body of the spacecraft (see Fig. 1) so that one pair looks parallel to the spacecraft spin axis and the other pair looks perpendicular to it. Each pair of detectors is comprised of an electron detector and an ion detector which cover the energy range from 50 eV to 50 keV. These detectors have been described in more detail by DeForest and McIlwain.[10]

ATS-6 Spacecraft and Instrumentation

The ATS-6 spacecraft was launched in May 1974 into geosynchronous orbit. Its orbital configuration is shown in Fig. 2. The distance between the ends of the two solar arrays is 16.5 m. The near-cubical module at the focus of the 9.1-m parabolic reflector is about 1.6 m on a side. The outer surface of most of the structure is covered with Kapton thermal insulation. However, all conductive elements of the structure and the vapor-deposited aluminum surfaces of the thermal blankets are bonded to the common spacecraft ground. The parabolic reflector is formed utilizing a dacron mesh with a copper coating. The copper is covered with a noncontinuous coating of silicon rubber. While the copper mesh of the reflector is grounded to the structure, the reflector's surface conductivity is dominated by the silicon rubber insulator. Thus, because the solar cells are covered by quartz glass, the majority of the outer surface of ATS-6 is nonconducting.

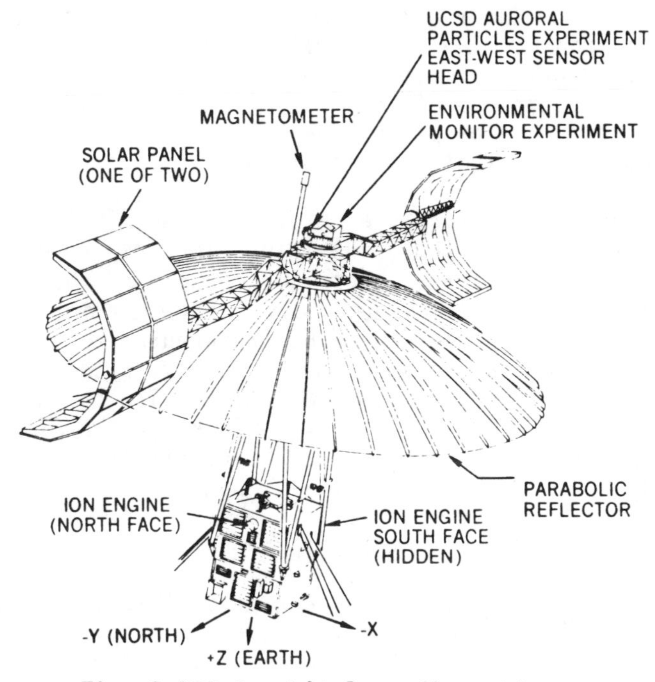

Fig. 2 ATS-6 orbital configuration.

There are two cesium bombardment ion engine systems on ATS-6. They are located on the north and south faces of the Earth viewing module, as shown in Fig. 2. The ATS-6 ion engine system has been described in detail by James et al.[11] Each system produces a 115 mA beam of cesium ions which is neutralized by electrons extracted from a second cesium plasma. This electron source, known as a plasma bridge neutralizer, is of interest when studying active control of a geosynchronous spacecraft because it can serve as a source of ions as well as electrons. Its operation has been discussed by Bartlett et al.[12]

For the discussion of active control tests, it is important to note that this plasma bridge neutralizer has two modes of operation, known as plume and spot modes. These represent a low-current, high-voltage plasma discharge (plume mode) and a high-current, low-voltage plasma discharge (spot mode). They occur in sequence after the neutralizer is commanded on. The plume mode discharge ignites about 6 to 12 min after the "neutralizer on" command is given; the neutralizer shifts to spot mode about 10-15 min after plume mode. In spot mode, which is used for normal ion engine operations, the neutralizer is emission limited at 3 A of electrons and a few milliamperes of ions. The minimum resolvable emission current is 1 mA of electrons only. No neutralizer emission current has been observed during operations using the neutralizer alone. Although the occurrence of spot mode is evident in the telemetry, the occurrence of plume mode is not and must be estimated based on preflight testing of the system.

The UCSD Auroral Particles Experiment on ATS-6 is an outgrowth of the ATS-5 system. The main detectors are arranged in two electron-ion pairs. These are mounted on the Environmental Monitor Experiment (see Fig. 2), one pair in the N-S plane and one pair in the E-W plane (referenced to the spacecraft axes). They can be swept mechanically in their respective planes to obtain angular information. The energy range covered by these detectors is 1 eV to 80 keV. The ATS-6 detectors are described in more detail by Bartlett et al.[11] and by McIlwain.[13]

Active Control Experiments

A few operations of the ion engine systems (with the ion beams on) on ATS-5 and ATS-6 have occurred in full sunlight. Because of the high spin rate of ATS-5, its ion engine could not be operated in the manner for which it was designed. Brief operations were conducted in 1972 and 1973, and some data were collected on spacecraft potential.[14] Each of the ATS-6 ion engines was operated, one of them continuously for 92 hr. These operations occurred in 1974.

Most of the active control experiments have been performed using the neutralizers alone and were conducted during eclipses, when the probability of environmental charging was high. Some daylight neutralizer operations also have been conducted with each spacecraft.

Neutralizer only operations were conducted in eclipse with the ATS-5 hot-wire filament during each eclipse season from the fall of 1974 through the spring of 1978. Most of these were done with the neutralizer of the No. 2 engine. The neutralizer was commanded on after the spacecraft had entered eclipse (usually $\sim$10 min after eclipse entry), left on for 4 or 5 min, and then commanded off. During 1977 and 1978, some longer (10-15 min) operations were conducted using each neutralizer, and some tests using both neutralizers in a "staggered" pattern (No. 1 on, No. 2 on, No. 1 off, No. 2 off) were conducted.

Neutralizer operations in eclipse were conducted with the ATS-6 plasma bridge neutralizer during the fall of 1976 and the spring and fall of 1977 eclipse seasons. Two types of operation occurred. In the first type, the neutralizer was brought into full operation (spot mode) before entry into eclipse, remained on for 10 min after eclipse entry, and was commanded off. In the second type of experiment the neutralizer operation was accomplished entirely during eclipse.

As noted previously, the ATS-5 and ATS-6 spacecraft charge to similar potentials in similar environments in eclipse, despite the great differences in configuration.[9] Therefore, data from eclipse operations of the ion engine neutralizers on these two spacecraft can be used to compare the effectiveness of the ATS-5 hot wire filament and the ATS-6 plasma bridge neutralizer as active control devices for spacecraft charging.

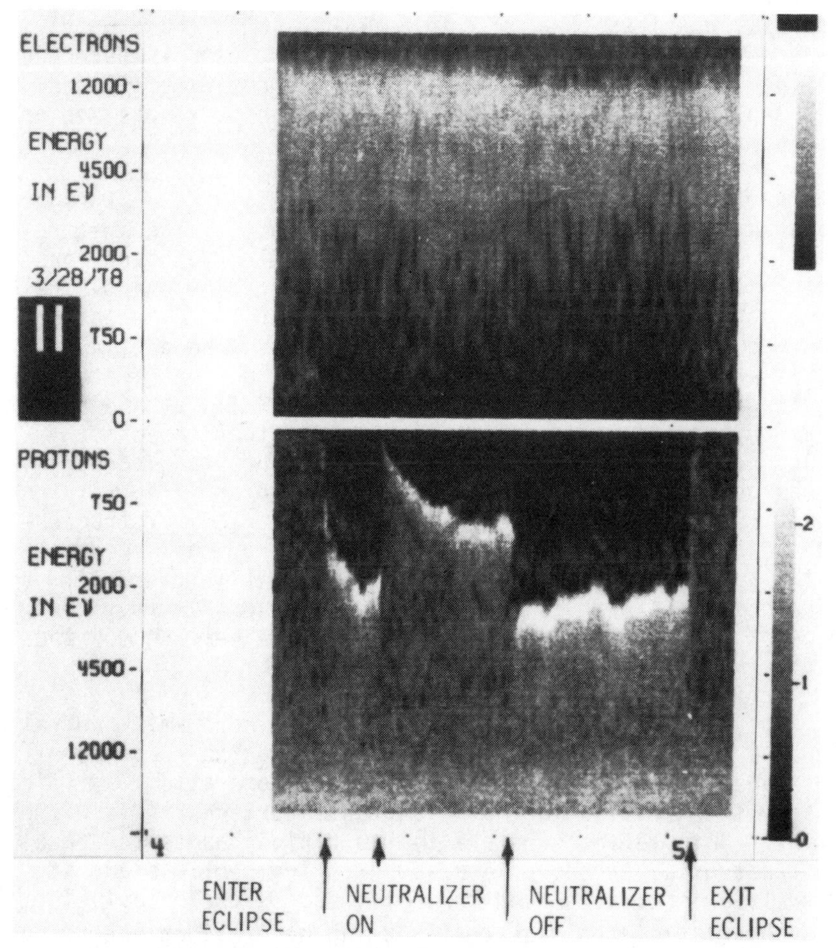

Fig. 3 ATS-5 spectrogram (March 28, 1978, day 87).

Details of the active experiments, including command logs and eclipse times, are given by Olsen and Whipple.[15] A number of workers have studied the data from these operations.[2,5,9,11,15-20] Table 1 presents a chronological listing of reports on active control of ATS-5 and ATS-6 and indicates the types of operations of each spacecraft discussed.

ATS-5 Results

The limited data from the ATS-5 ion engine operations as a thruster indicated no change in the spacecraft potential, which was less than 50 V in magnitude (the lower limit of ATS-5 potential measurements). In a brief unneutralized operation during turnoff, however, the spacecraft charged to about -3000 V.[14] Similarly, operations of the hot wire filament alone in sunlight produced no measurable effect. Spacecraft potential was less than 50 V in magnitude when these operations were conducted. The most interesting results have been obtained during eclipse operations of the hot wire filament when the spacecraft potential was more than 50 V negative and thus apparent in the particle data.

Particle data for a neutralizer operation in eclipse are shown in an energy-time spectrogram in Fig. 3. The spectrogram is a plot of the energy of arriving electrons and ions versus time [given in Universal Time (UT)]. The energy scale for electrons increases upward and that for ions increases downward so that they have a common origin. The particle fluxes are indicated by a gray scale (at the right of the spectrogram). Dark areas correspond to low fluxes and bright areas to high fluxes.[10] The potential of the spacecraft structure is reflected in the ion data. When the spacecraft is charged negatively, no ions are observed arriving with energies less than the spacecraft potential. Ions of true energy near 0 eV arrive with an apparent energy corresponding to the spacecraft potential through which they have fallen and appear as a peak in the flux at this energy. Thus the spacecraft's potential at various times is indicated by the bright band of ions in the spectrogram.

The data in Fig. 3 were taken during the 15-min neutralizer operation on March 28, 1978 (day 87). The response of the spacecraft to eclipse and neutralizer operation is typical of the ATS-5 results for both long and short neutralizer operations. Notable features are 1) the rapid changes in spacecraft potential at entry into and exit from eclipse and at neutralizer activation and turn off, and 2) the relatively slow charging of the spacecraft during neutralizer operation.

Table 1 Reports on ATS-5 and ATS-6 Active Control (in chronological order)

Report	Ref.	ATS-5 Ion engine with neutralizer	ATS-5 Neutralizer only: Sun	ATS-5 Neutralizer only: Eclipse	ATS-6 Ion engine with neutralizer	ATS-6 Neutralizer only: Sun	ATS-6 Neutralizer only: Eclipse
DeForest, Goldstein (1973)	14	√[a]	...	...	...	...	...
Bartlett, Goldstein, DeForest (1975	12	...	...	√	√	...	...
Goldstein, DeForest (1976)	2	...	...	√	√	...	...
Purvis, Bartlett, DeForest (1977)	6	...	...	√	...	...	√
Goldstein (1977)	16	...	...	...[b]	√	...	...
Olsen, Whipple (1977)	17	...	√	√	√	√	√
Olsen, Whipple, Purvis (1978)	18	...	...	√	√	√	√
Olsen (1978)	19	...	...	...	√	√	√
Bartlett, Purvis (1979)	7	...	...	√	√	...	√
Olsen, Whipple (1979)	20	...	...	...	√	√	√
Olsen, Whipple (1979)	15	...	√	√	√	√	√

[a]Includes one brief unneutralized ion beam operation during takeoff.

[b]Laboratory study.

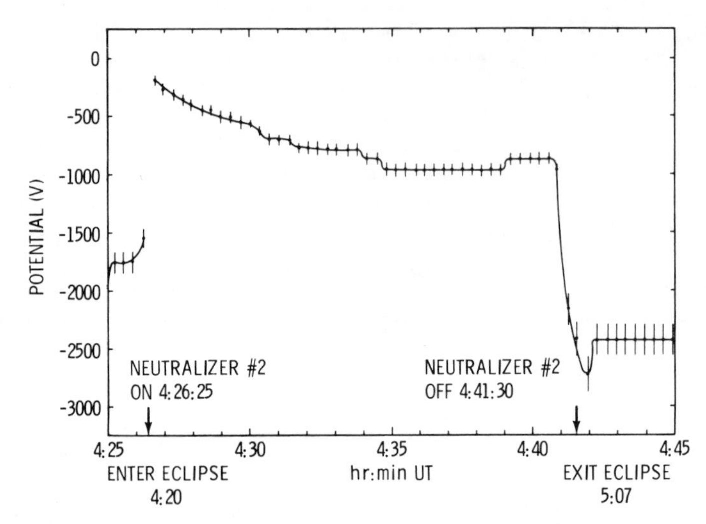

Fig. 4 ATS-5 neutralizer/eclipse operation (March 28, 1978, day 87).

The effect of hot wire filament neutralizer operation on spacecraft potential is more quantitatively indicated in Fig. 4, which is a linear plot of spacecraft potential vs time. Prior to neutralizer activation the spacecraft was charged to about -1700 V. When the neutralizer was turned on, the spacecraft initially discharged to a potential less negative than -200 V (a complete energy scan requires about 20 sec) and subsequently charged to about -1000 V during the 15 min operation. When the neutralizer filament was turned off, the spacecraft charged more negatively and reached a steady potential of about -2500 V in about 2 min.

The initial rapid discharging at neutralizer activation is similar to the discharging observed at exit from eclipse. It is believed to represent a change in potential of the entire spacecraft: both the structure and insulating surfaces change potential simultaneously while any potential difference between structure and insulation remains unchanged. The subsequent slow charging of the structure is believed to result from differential charging of the surface insulation near the neutralizer aperture forming a potential barrier which suppresses electron emission from the filament. The filament on ATS-5 is mounted about 2.5 cm inboard of the spacecraft skin, which has thermal insulation in its vicinity. Suppression of electron emission from such a filament in this geometry by buildup of potential on the "skin" has been measured in laboratory tests by Goldstein.[16]

The belief that absolute charging or discharging (i.e., of the structure and insulation together) can occur quickly, while differential charging of insulating surfaces relative to the underlying structure occurs slowly (in minutes or tens of minutes), is based on the observation that the capacitance of the spacecraft to the exterior plasma is much less than that of the insulating surfaces to the structure. This difference in rates of absolute and differential charging has been observed by DeForest in his studies of ATS-5 charging[21] and by Purvis et al.[22] in laboratory investigations of charging rates. Observed rates of differential charging are consistent with the rate at which the ATS-5 structure charges during neutralizer operation.

The potential overshoot observed at neutralizer turnoff (see Fig. 4) also is assumed to be a differential charging effect.[15] It appears to be a more prominent feature of long neutralizer operations than of short ones. This is consistent with the differential charging hypothesis because more differential charging is expected to occur during long operations.

Dual ATS-5 hot wire filament neutralizer operations were conducted during the fall of 1977 and spring of 1978 eclipse seasons. These consisted of operating the two hot filaments in a "staggered" pattern so that a single neutralizer was on at the beginning and end of the test sequence, and both were on in the middle. Three data sets were obtained in which

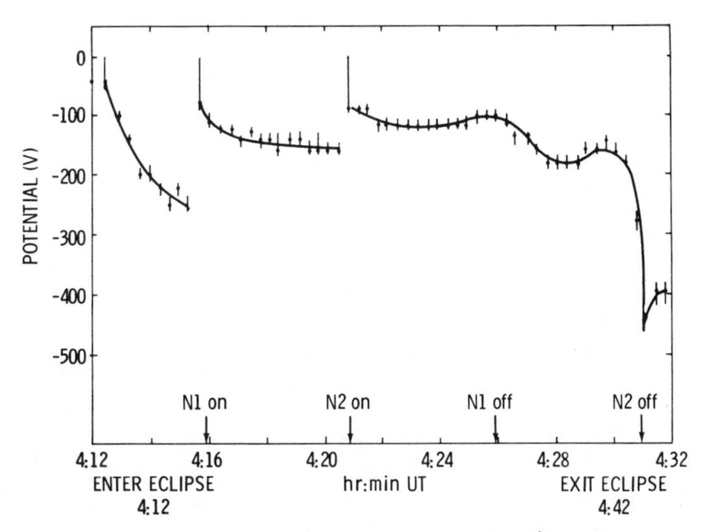

Fig. 5 ATS-5 neutralizer/eclipse operation (October 4, 1977, day 277).

there was observable charging of the spacecraft. Data from the operation on October 4, 1977 (day 277), is shown in the line plot in Fig. 5. Activation of the first neutralizer had the expected result; the spacecraft potential was reduced (in magnitude) abruptly to near zero and then slowly became more negative. When the second neutralizer was turned on, the spacecraft's potential repeated the sudden-shift/slow-recovery pattern, leveling out at a somewhat less negative potential than that with the single neutralizer operation. This behavior is reasonable based on the differential-charging/emission-suppression arguments given above. When neutralizer No. 1 was turned off, the spacecraft potential became more negative again, albeit in a rather uncertain and therefore puzzling manner. Turnoff of the second neutralizer resulted in behavior similar to that seen in single neutralizer tests.

Figure 6 shows data from another two-neutralizer operation on February 28, 1978 (day 59). Again, activation of the first neutralizer and extinction of the second resulted in the expected single-neutralizer type response. In this case, activation of the second neutralizer had essentially no effect on the spacecraft potential. The structure potential was about -1200 V at activation of the second neutralizer on day 59 of 1978, compared to a potential of only -150 V at this point on day 277 of 1977. It is reasonable to assume that the differential charging also was more severe during the operation on day 59 of 1978. In this case, emission from the sec-

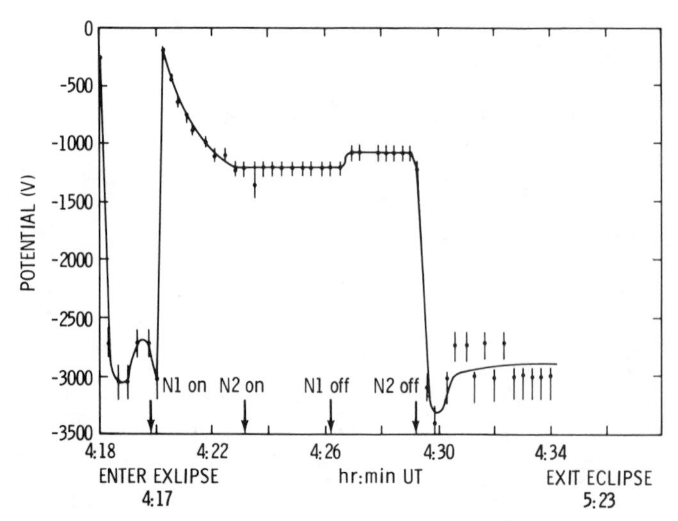

Fig. 6 ATS-5 neutralizer/eclipse operation (February 28, 1978, day 59).

ond neutralizer may have been immediately suppressed and the spacecraft potential thus unaltered by its activation. When neutralizer No. 1 was turned off, the spacecraft potential was reduced in magnitude. This response, which also was observed in the third dual neutralizer operation,[15] was unexpected and is not presently understood.

ATS-6 Results

The effects of operating the ATS-6 ion engines as thrusters (i.e., both ion beam and neutralizer together) have been reported by a number of investigators (see Table 1). During ion engine operation, the spacecraft potential was maintained at about -4 or -5 V, even during magnetospheric substorm activity, which would have been expected to cause a charging of the spacecraft without engine operation.[2,20] Furthermore, there are indications that differential charging was suppressed during thruster operation and that the measurement of environmental data was enhanced by a constant spacecraft potential.[19]

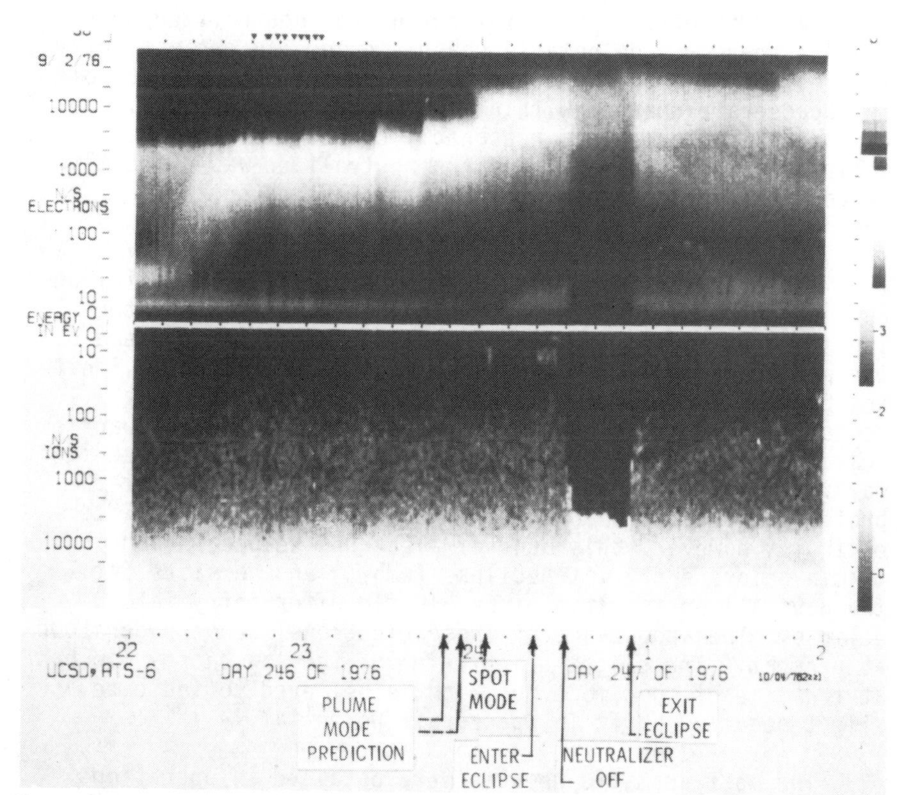

Fig. 7 ATS-6 neutralizer/eclipse operation (Sept. 2-3, 1976, days 246-247).

Operations of the ATS-5 plasma bridge neutralizer alone in sunlight were conducted during 1976. Data from operations in Aug. and Nov. have been presented by Olsen.[19] During the Nov. operations the spacecraft potential was taken from about -100 V to near zero on two occasions. In other daylight operations (when the spacecraft potential was near zero), a band of low energy ions sometimes appeared. These are believed to represent a slight negative shift in potential. Changes in the low energy electron spectrum which accompany neutralizer operation in sunlight are believed to represent reduction of differential charging.

Of particular interest to the problem of active spacecraft charging control are the operations of the ATS-6 plasma bridge neutralizer during eclipses. As noted previously, two types of eclipse operation occurred. In the first type, the neutralizer was brought into full operation (spot mode) before entry into eclipse. Data from such a test on Sept. 2-3, 1976 (days 246-247) are shown in spectrogram form in Fig. 7. It is clear from the ion data that the spacecraft remained within about 10 V of plasma potential during the neutralizer operation as there was no notable change in the ion data at entry into eclipse. When the power to the neutralizer was cut off, the spacecraft charged within minutes to about -4000 V. Because the environment is reasonably constant during the eclipse passage, operation of the neutralizer was evidently responsible for maintaining the spacecraft potential near zero during the initial 10 min of eclipse.

While no significant changes in spacecraft potential occurred during neutralizer operation, there are some interesting features in the low energy particle data. At the end of the period during which the plume mode is predicted to ignite, the band of low energy electrons usually associated with photoelectrons and secondaries trapped by a potential barrier near the spacecraft[23] fades. This effect has been noted in daylight operations of the neutralizer and has been interpreted as a reduction of differential charging.[18-20] Interestingly, however, this band of electrons reappears about 12 min before entry into eclipse (umbra) and persists until the neutralizer is extinguished 10 min after entry into eclipse. The appearance of these electrons is not understood at present. The slight brightening of the low energy ion band at the time of spot mode ignition is believed to indicate a slight negative shift in spacecraft potential.[15,19]

The most dramatic results were obtained in operations conducted entirely during eclipse. Figure 8 shows data from

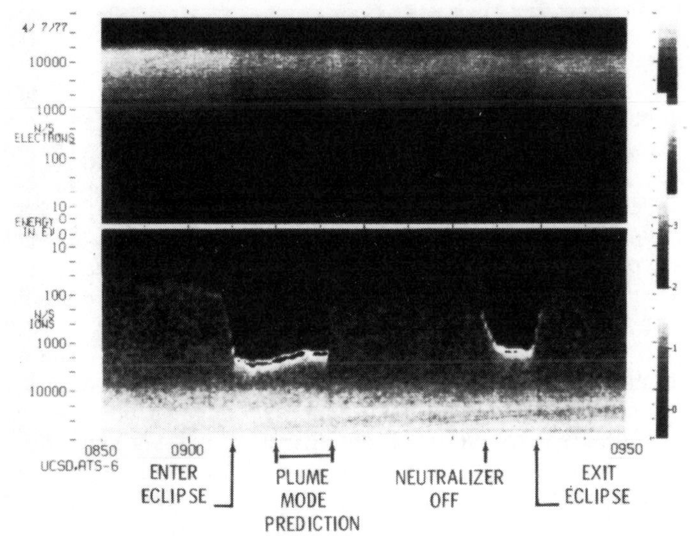

Fig. 8 ATS-6 spectrogram (day 97 of 1977).

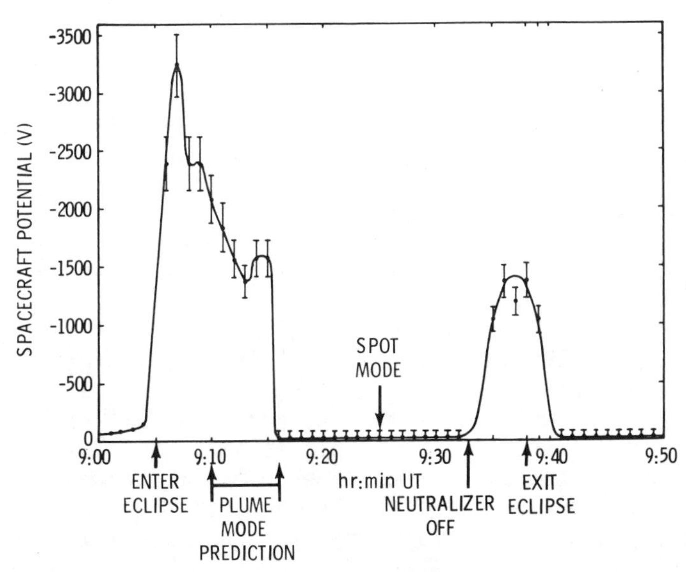

Fig. 9 ATS-6 neutralizer/eclipse operation (day 97 of 1977).

this type of test on April 7, 1977 (day 97), in spectrogram form; a plot of the spacecraft potential vs time during this operation is shown in Fig. 9. The absence of natural low energy ions during this test makes it impossible to determine the spacecraft's potential accurately; however, it is clearly near zero. The spacecraft was charged to about -100 V before entry into eclipse. In eclipse it charged to several thousand volts negative and was at -1500 V when the plume mode apparently occurred at the end of the predicted period, consistent with the behavior shown in Fig. 7 and with expectations based on its power supply.[7] At this time, the spacecraft potential was reduced to near zero in a time too short to be measured with the 16-sec energy range scan of the UCSD instrument. The spacecraft potential was maintained near zero until the neutralizer power was turned off. At that time, the spacecraft again charged to nearly -1500 V and remained charged until exit from eclipse. The time constants for these changes in potential are similar to those shown in Fig. 7. The rapidity with which the neutralizer reduced the spacecraft potential, compared to the rate at which the spacecraft changed in potential in response to natural charging and discharging, indicates that the current output of the neutralizer dominated the plasma currents during these operations. The largest negative potential observed during these active control tests with ATS-6 was -8000 V. Spacecraft response to neutralizer activation was similar to that shown in Figs. 8 and 9 for the -1500 V case.

Summary and Conclusions

Operations of the ATS-5 hot filament neutralizer in eclipse have shown that emission of electrons can reduce the potential of a charged spacecraft to near zero. However, differential charging of the insulating surfaces of the spacecraft resulted in suppression of electron emission and consequent charging of the spacecraft during filament operation.

It follows from this interpretation of the ATS-5 active control data that a device which can emit only electrons is not suitable for maintaining a spacecraft near plasma potential during eclipse unless the spacecraft's surfaces all are conducting. It should be possible to emit enough electrons to keep the spacecraft structure near plasma ground if the emitting device were sufficiently far from the spacecraft body and/or could accelerate the electrons sufficiently. Because insulating surfaces would be charged by the ambient plasma to negative kilovolt potentials, a body-mounted electron emitter would have to accelerate electrons to kilovolt potentials before they could overcome local potential barriers. A boom-mounted emitter would have to accelerate electrons to an energy greater than the potential at its location due to the charged spacecraft surfaces. Because the screening distances at geosynchronous altitude are large, very long booms or large accelerating potentials are needed. In either case, if the attempt to maintain the spacecraft structure near plasma potential were successful, kilovolt potentials would be developed across surface insulation.

Operation of the ATS-6 ion engine clamped the spacecraft at a fixed potential of about -4 or -5 V. It also suppressed differential charging. Operation of the ATS-6 plasma bridge neutralizer maintained the spacecraft potential within a few volts of the ambient plasma potential for all observed plasma conditions, both in sunlight and in eclipse. It successfully reduced potentials as large as -8000 V to near zero within the 16 sec time resolution of the plasma detector's energy scan. Spacecraft potential was consistently brought to near zero by the occurrence of plume mode operation and with an emitted electron current of less than 1 mA. Based on this evidence, it is believed that a low energy plasma discharge provides an effective means for controlling spacecraft potential.

Acknowledgments

The authors wish to acknowledge the contributions of C. E. McIlwain, S. E. DeForest, E. C. Whipple, Jr., and R. C. Olsen of the University of California, San Diego, and of R. Goldstein of the Jet Propulsion Laboratory.

References

[1]DeForest, S. E., "Spacecraft Charging at Synchronous Orbit," Journal of Geophysical Research, Vol. 77, Feb. 1972, pp. 651-659.

[2]Goldstein, R. and DeForest, S. E., "Active Control of Spacecraft Potentials at Geosynchronous Orbit," Progress in Astronautics and Aeronautics: Spacecraft Charging By Magnetospheric Plasmas, Vol. 47, edited by A. Rosen, AIAA, New York, 1976, pp. 169-181.

[3]Reasoner, D. L., Lennartsson, W., and Chappell, C. R., "Relationship Between ATS-6 Spacecraft-Charging Occurrences and Warm Plasma Encounters," Progress in Astronautics and Aeronautics: Spacecraft Charging By Magnetospheric Plasmas, Vol. 47, edited by A. Rosen, AIAA, New York, 1976, pp. 89-101.

[4]Rubin, A. G. and Garrett, H. B., "ATS-5 and ATS-6 Potentials During Eclipse," Spacecraft Charging Technology-1978, NASA CP-2071, 1979, pp. 38-43.

[5]Garrett, H. B., Mullen, E. G., Ziemba, E., and DeForest, S. E., "Modeling of the Geosynchronous Orbit Plasma Environ-

ment - Part 2. ATS-5 and ATS-6 Statistical Atlas," Air Force Geophysics Laboratory TR-78-0304, 1978.

[6]Purvis, C. K., Bartlett, R. O., and DeForest, S. E., "Active Control of Spacecraft Charging on ATS-5 and ATS-6," Proceedings of the Spacecraft Charging Technology Conference, edited by C. P. Pike and R. R. Lovell, AFGL-TR-77-0051 & NASA TM X-73537, 1977, pp. 107-120.

[7]Bartlett, R. O. and Purvis, C. K., "Summary of the Two Year NASA Program for Active Control of ATS-5/6 Environmental Charging. Spacecraft Charging Technology-1978, C. P. Pike, eds., NASA CP-2071, 1979, pp. 44-58.

[8]Cohen, H. A., Sherman, C., Mullen, E. G., Huber, W. B., Masek, T. D., Sluder, R. B., Mizera, P. F., Schnauss, R., Adamo, R. C., Nanovich, J. E., and Delorey, D. E., "Design, Development, and Flight of a Spacecraft Charging Sounding Rocket Payload," Spacecraft Charging Technology-1978, NASA CP-2071, 1979, pp. 80-90.

[9]Worlock, R., Davis, J. J., James, E. L., Ramirez, P., and Wood, O., "An Advanced Contact Ion Microthruster System," AIAA Paper 68-552, June 1968.

[10]DeForest, S. E. and McIlwain, C. E., "Plasma Clouds in the Magnetosphere," Journal of Geophysical Research, Vol. 76, June 1971, pp. 3587-3611.

[11]James, E. L., Worlock, R. M., Dillon, T., Gant, G., Jan, L., and Trump, G., "A One Millipound Cesium Ion Thruster System," AIAA Paper 70-1149, Aug. 1970.

[12]Bartlett, R. O., DeForest, S. E., and Goldstein, R., "Spacecraft Charging Control Demonstration at Geosynchronous Altitude," AIAA Paper 75-359, Mar. 1975.

[13]McIlwain, C. E., "Auroral Electron Beams Near the Magnetic Equator," Physics of the Hot Plasma in the Magnetosphere, edited by B. Hultquist and L. Stenflo, Plenum Publishing Corp., New York, 1975, pp. 91-112.

[14]DeForest, S. E. and Goldstein, R., "A Study of Electrostatic Charging of ATS-5 Satellite During Ion Thruster Operation," (California University and Jet Propulsion Laboratory, California Institute of Technology; NASA Contract NAS7-100 and JPL-953675.) NASA CR-145910, 1973.

[15]Olsen, R. C. and Whipple, E. C., "Active Experiments in Modifying Spacecraft Potential: Results from ATS-5 and ATS-6," University of California at San Diego SP-79-01, 1979.

[16]Goldstein, R., "Active Control of Potential of the Geosynchronous Satellite ATS-5 and ATS-6," Proceedings of the Spacecraft Charging Technology Conference, edited by C. P. Pike and R. R. Lovell, AFGL-TR-77-00S1. NASA TM X-73537, 1977, pp. 121-129.

[17]Olsen, R. C. and Whipple, E. C., "Active Experiments in Modifying Spacecraft Potential: Results from ATS-5 and ATS-6," University of California (UCSD-SP-77-01, California at San Diego SP-77-01 & NASA CR-152607, 1977.

[18]Olsen, R. C., Whipple, E. C., and Purvis, C. K., "Active Modification of ATS-5 and ATS-6 Spacecraft Potentials," Presented at the Symposium on the Effect of the Ionosphere on Space and Terrestrial Systems, Naval Research Laboratory and Office of Naval Research, Washington, D.C., Jan. 1978, Paper 4-8.

[19]Olsen, R. C., "Operation of the ATS-6 Ion Engine and Plasma Bridge Neutralizer at Geosynchronous Altitude," AIAA Paper 78-663, Apr. 1978.

[20]Olsen, R. C. and Whipple, E. C., "Operations of the ATS-6 Ion Engine," Spacecraft Charging Technology-1978, NASA CP-2071, 1979, pp. 59-68.

[21]DeForest, S. E., "Electrostatic Potentials Developed by ATS-5," Photon and Particle Interactions with Surfaces in Space, edited by R. J. L. Grard, D. Reidel Publishing Co., Dordrecht-Holland, 1973, pp. 263-276.

[22]Purvis, C. K., Staskus, J. V., Roche, J. C., and Berkopec, F. D., "Charging Rates of Metal-Dielectric Structures," Spacecraft Charging Technology-1978, NASA CP-2071, 1979, pp. 507-523.

[23]Whipple, E. C., Jr., "Observation of Photoelectrons and Secondary Electrons Reflected from a Potential Barrier in the Vicinity of ATS-6," Journal of Geophysical Research, Vol. 81, Feb. 1976, pp. 715-719.

A THREE-DIMENSIONAL SPACECRAFT-CHARGING COMPUTER CODE

Allen G. Rubin*
Air Force Geophysics Laboratory, Hanscom Air Force Base, Mass.

Ira Katz,+ Myron Mandell,* Gary Schnuelle,* Paul Steen,*
Don Parks,** and Jack Cassidy+
Systems, Science and Software, Inc., La Jolla, Calif.

and

James Roche*
NASA Lewis Research Center, Cleveland, Ohio

Abstract

A computer code is described which simulates the interaction of the space environment with a satellite at geosynchronous. Employing finite elements, a three-dimensional satellite model has been constructed with more than 1000 surface cells and 15 different surface materials. Free space around the satellite is modeled by nesting grids within grids. Applications of this NASA/ Spacecraft Charging Analyzer Program (NASCAP), code to the study of a satellite photosheath and the differential charging of the SCATHA (Satellite Charging at High Altitudes) satellite in eclipse and in sunlight are discussed. In order to understand detector response when the satellite is charged, the code is used to trace the trajectories of particles reaching the SCATHA detectors. Particle trajectories from positive and negative emitters on SCATHA also are traced to determine the location of returning particles, to estimate the escaping flux, and to simulate active control of satellite potentials. The extensive graphical capability of the code enables one to unravel complex three-dimensional effects.

---

*Research Scientist.
+Program Manager.
**Senior Research Scientist.

I. Introduction

The SCATHA satellite, which is at present in orbit, is instrumented to study the problem of spacecraft charging in and near geosynchronous orbit. The spacecraft charging problem originally was discovered on the ATS-5 geosynchronous satellite in 1972[1,2] when hot plasma injections during substorms charged the satellite on the order of -10 kV. This high potential, as well as potential gradients between various portions of a spacecraft, sometimes causes arcing. These arcs couple rf voltages into the spacecraft electronics and disrupt satellite operations.

Since spacecraft charging is a troublesome problem for satellite operation, the Air Force and NASA are engaged in a joint program to investigate, understand, and control spacecraft charging. As part of this program, a computer code has been developed to model spacecraft charging.[3,4,5] This code, called NASA/ Spacecraft Charging Analyzer Program, (NASCAP) has been adapted to modeling the details of SCATHA.

NASCAP computations are carried out in a series of nested grids where each successive grid has twice the spacing of the previous grid; this system permits the modeling of a space of virtually any size desired. Figure 1 shows the nested grids with an object in the inner grid. The code employs simple shapes as building blocks to model complex objects: the basic shapes are cubes or slices of cubes. In addition, thin plates and thin cylinders are used to more accurately model booms and solar cell arrays (see Figure 2). Particle fluxes in the ambient environment may be input as single or double Maxwellian distributions or as actual observed particle energy spectra. Particle fluxes accounted for in the code are the incident electrons and protons, backscattered and secondary electrons, secondary electrons due to protons, and photoelectrons. Species other than hydrogen may be included. The solar flux incident upon a spacecraft is simulated using three-dimensional shadowing routines. The computational methods and many of the results from these computations will be discussed in this paper. Running times for SCATHA models are between 0.5 and 2 hr on the CDC 6600.

II. The Computational Cycle

The NASCAP computation is based on alternate solutions to the Poisson equation (in order to determine the electric field on the satellite surfaces and in the surrounding space) and calculations of particle motion and charge deposition. These computations are carried out in a series of time steps chosen to be appropriate to the specific problem. During each time step a field solution, charge deposition, or particle motion calculation is carried out. The computation assumes that charging proceeds through a series of equilibrium states (the so-called quasistatic assumption). The general techniques for solving for these equilibrium states will be outlined here. We will follow this with a discussion of the specific methods (either reverse trajectory or spherical probe) employed in the analysis.

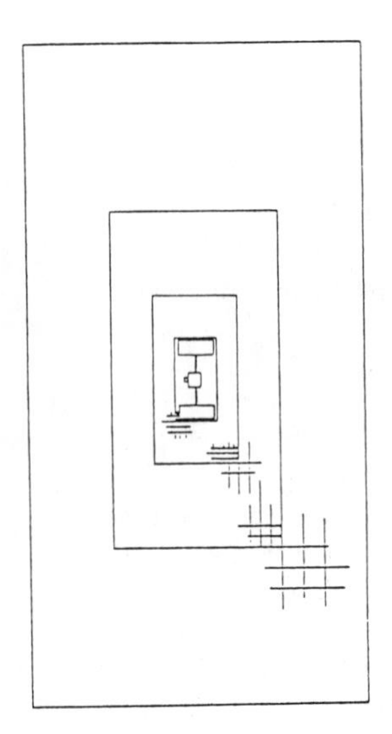

Fig. 1 Nesting of successive grids.

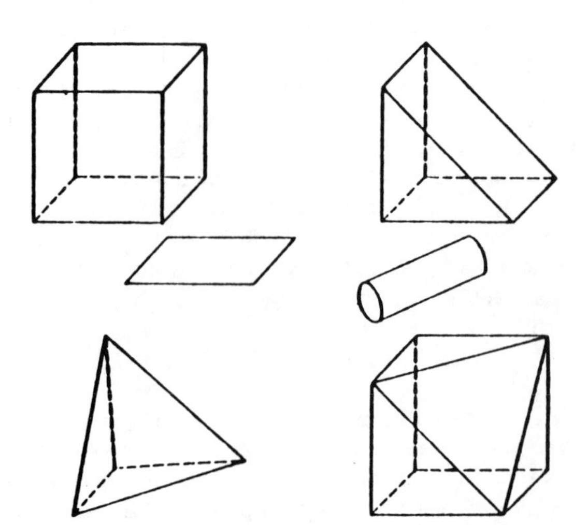

Fig. 2 Object building blocks.

General Techniques

The Poisson or Laplace equation (depending on the problem treated) is solved by the scaled conjugate gradient method. The scaled conjugate gradient method is applied to an array of linear equations resulting from the finite element formulation of the field problem. In finite difference methods, differential equations that provide an exact representation are approximated over finite spatial intervals. In the finite element method, however, the continuous body is first represented by an approximating body of finite elements. The equations describing the potential about the approximating body are then evaluated exactly. In the former case, exact equations are solved approximately; in the latter, approximating equations are solved exactly.

The scaled conjugate gradient method is an iterative method of solving the 10,000 and more linear equations which are generated at each time step. Because of the large number of equations to be solved, a fast, efficient method must be used. The iterative methods used take advantage of the sparseness of the coefficient matrix, require a minimum of storage, and (for the conjugate gradient method) converge in 10-40 iterations.

In the ordinary conjugate gradient method,[6,7] we are faced with solving a system of linear equations

$$\vec{\vec{M}}\vec{x} = \vec{y}$$

where y is a vector representing the charge at every point on a three-dimensional grid. The vector x is the solution, the potential at every point, M is a large square matrix. We choose an initial estimate of the solution, x^o, and define a residual vector

$$\vec{r}^o = \vec{y} - \vec{\vec{M}}\vec{x}^o$$

$$p^o = r^o$$

Then we iterate:

$$a_i = \left[\vec{r}^i\right]^T \vec{r}^i \Big/ \left[\vec{p}^i\right]^T \vec{\vec{M}}\, \vec{p}^i$$

$$\vec{x}^{(i+1)} = \vec{x}^{(i)} + a^{(i)}\, \vec{p}^{(i)}$$

$$\vec{r}^{(i+1)} = \vec{r}^{(i)} - a^{(i)}\, \vec{\vec{M}}\, \vec{p}^{(i)}$$

$$b^{(i)} = \left[\vec{r}^{(i+1)}\right]^T \vec{r}^{(i+1)} \Big/ \left[r^{(i)}\right]^T \vec{r}^{(i)}$$

$$\vec{p}^{(i+1)} = \vec{r}^{(i+1)} + b^{(i)}\, \vec{p}^{(i)}$$

The scaled conjugate gradient method improves convergence in cases of large zone size and very thin dielectric skins by scaling the large coproduct matrix so that all diagonal elements are of order unity.

Charge Deposition

Two code options exist for calculating the charge deposited on a cell. These are the reverse trajectory sampling method and the spherical probe approximation. Considerations of computer time and the desired accuracy determine which method is preferable.

In the reverse trajectory sampling method,[8] the space charge densities and the incident flux on any surface element can be determined by tracking time-reversed orbits starting from an object surface element and invoking the principle of phase space volume invariance along particle trajectories in a collisionless plasma. If a particle starting at (r_o, v_o) on an object surface reaches (r,v), then:

$$f'(\vec{r}_o, \vec{v}_o) = f(\vec{r}, \vec{v})$$

where f is the phase space density. Thus, if $r = r(r_o, v_o, t)$ is a point remote from the object, the orbit is one which a particle incident from remote distances can follow to the object: therefore, the flux of plasma particles incident on the surface can be determined by

$$f'(\vec{r}_o, \vec{v}_o) = f(\vec{V})$$

where f(V) is the distribution function of particles in the undisturbed plasma.

The procedure to be followed in connecting the spectrum of particles in a given element at r_o to the unperturbed plasma spectrum is:

1) Track particles with specified velocities v_o from their point of origin r_o along time-reversed trajectories until they reach the outer boundary of the computational mesh at r_p, or until it is clear that the particles will never reach the outer boundary.

2) Identify the incident particle spectrum at r_o as

$$f'(\vec{r}_o, \vec{v}_o) = f'(\vec{r}_p, \vec{v})$$

where v is the velocity with which particles on their time-reversed trajectories reach the outer boundary at r_p.

If v_t is the terminal velocity at the outer boundary due to the backward trajectory tracing of a particle emitted with an initial velocity v_i at the surface, then

$$f'(\vec{r}_o, \vec{v}_i) = f(\vec{r}_p, \vec{v}_t)$$

is the distribution function value for v_i. By scanning these sampling points throughout the velocity space and completing their backward trajectory calculations, we can construct the whole distribution function $f'(r_o, v_o)$.

For each species, we have to integrate the distribution function to obtain the charging current density, that is:

$$\vec{j}_e = \hat{n} \int_0^{2\pi} \int_0^1 \int_{v_m(\mu,\phi)}^{\infty} q v_o^3 f_o'(v_o, \mu, \phi)\, dv_o\, \mu\, d\mu\, d\phi$$

where $\mu = \cos\theta$. $v_m(\mu, \emptyset)$ is the minimum escape speed of an electron emitted from the surface element with an angle $(\theta, \emptyset)$ with respect to n, and $f_o'(v_o, \mu, \emptyset)$ will include the allowable ranges of θ and $\emptyset$ for the incident flux (which are determined experimentally by the backward trajectory calculations).

The second option used in NASCAP, the spherical probe formulation, is, in principle, exact for a sphere in an isotropic Maxwellian plasma. The differential particle flux per unit area to the satellite is

$$\frac{d^2 f}{dE\, d\Omega} = p \left(\frac{kT}{2\pi m}\right)^{\frac{1}{2}} E \exp\left[-(E+qV)/kT\right] \frac{\cos\theta}{\pi}$$

where E is the energy of incidence, q the particle charge, and V the surface voltage. (The formula requires $E + qV \geq 0$.) For this option, the incident flux is given by

$$p \left(\frac{kT}{2\pi m}\right)^{\frac{1}{2}} \exp(-qV/kT) \qquad qV > 0$$

$$p \left(\frac{kT}{2\pi m}\right)^{\frac{1}{2}} \left(1 - \frac{qV}{kT}\right) \qquad qV < 0$$

The secondary emission and backscatter are calculated as

$$\Gamma = 2\pi \int_{E_{min}}^{\infty} dE \int_0^1 d(\cos\theta)\, \gamma(E,\theta) \frac{d^2 f}{dE\, d\Omega}$$

where γ is the relevant coefficient. This second option is much faster than the reverse-trajectory method but does not allow non-isotropic local fuel to modify collection currents.

III. Code Applications in Spacecraft Design

In order to find the points of highest voltage stress on a spacecraft in the design stage, a model of the spacecraft is first constructed. A fairly accurate geometrical, electrical, and materials model of the spacecraft can be obtained using the program. The charging code is then run for several environments to study the variation in potentials at various points on the surface. In this way, weak points in the design can be pinpointed. Similarly, the code is used to simulate the plasma and solar radiance encountered in a variety of orbits in order to compute spacecraft-charging effects. (The environment about planets such as Jupiter are simulated just as easily as that of the Earth.)

The sheath field has obvious effects on electric field measurements, low energy plasma measurements, and particle energy spectrum measurements up to about 100 keV. In designing experiments, therefore, it is desirable to have a computation in which the effects of spacecraft sheath fields can be accounted for. Experiments involving emission of charged particles require a computation of orbits in the sheath field. Such computations are greatly facilitated by the NASCAP code. This code has emitter and detector routines which trace particle trajectories, predict return currents during active control, and plot detector response. We will discuss the results of these computations as they apply to an operational satellite - the SCATHA satellite.

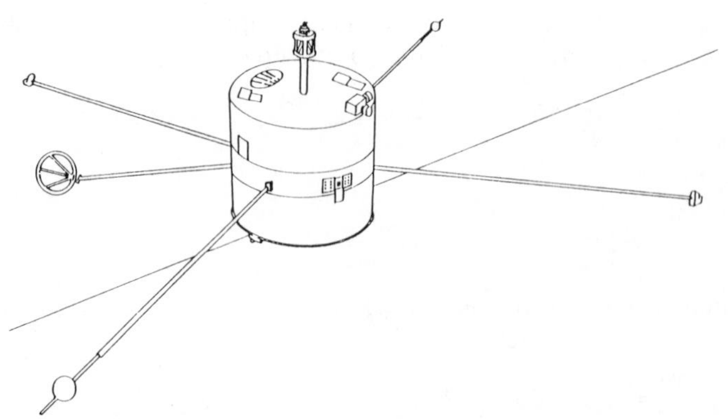

Fig. 3 Drawing of the SCATHA satellite.

SCATHA Model

Figure 3 gives the configuration of the SCATHA spacecraft which is instrumented to study spacecraft charging and is the first spacecraft ever to be modeled by a code before flight. A SCATHA model is shown in Figs. 4 and 5; Figure 6 gives a perspective view. A comparison of the actual SCATHA geometrical features with the gridded SCATHA model is given in Table 1. It should be noted that the model reproduces most of the features of SCATHA quite faithfully. Table 2 lists the fifteen surface materials included in the SCATHA model. Figure 7 shows their placement on SCATHA. Where the materials properties were not well known, the best available values have been employed.

Potential Contours

As an example of the versatility of the NASCAP code, the potential sheath around SCATHA (employing the model just described) in the plane of the booms in eclipse is shown in Fig. 8. The particle temperatures were chosen to simulate a severe substorm. As demonstrated in this diagram, the spacecraft potential is drawn out along the booms as a result of coupling of the boom surfaces to the underlying cable shields. Figure 9 shows the potential 1 m below the plane of the booms, where the booms still have a substantial effect.

Figure 10 presents an example of extreme differential charging on SCATHA. After initial charging during eclipse, the sunlight was turned on. The photoemission resulted in about 3 kV of differential charging along the booms.

Table 1 Comparison of actual SCATHA geometrical features to gridded NASCAP model[a]

	SCATHA	Model
Radius	33.6 in	32.0 in
Height	68.7 in	68.0 in
Solar array height	29 in	27.2 in
Belly b and height	11.3 in	13.6 in
SC9-1 experiment	9.2x6x8	9.1x4.5x9.1 in
SC6-1 boom	1.7 in (radius)	1.7 in (radius)
	118 in (length)	113.2 in (length)
Surface area	2.16×10^4 in	2.11×10^4 in^2
Solar array area	1.23×10^4	1.15×10^4 in^2
Forward surface area	0.36×10^4	0.34×10^4 in^2

[a]Zone size = 4.53 in (11.5 cm).

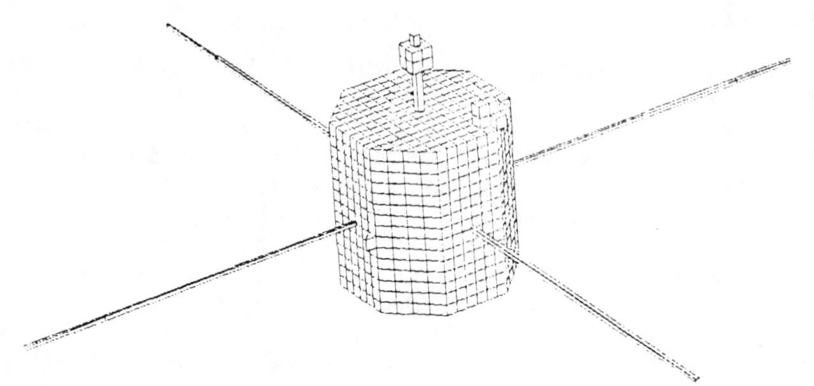

Fig. 4 Perspective view of a 4-boom SCATHA model.

Table 2 Exposed surface materials

Designation	Material
GOLD:	gold plate
SOLAR:	solar cells, coated fused silica
WHITEN:	nonconducting white paint (STM K792)
SCREEN:	SC5 screen material
YELOWC:	conducting yellow paint
GOLDPD:	88% gold plate with 12% conductive black paint in a polka dot pattern
BLACKC:	conductive black paint (STM K748)
KAPTON:	kapton
SI02:	SiO_2 fabric
TEFLON:	teflon
INDOX:	indium oxide
YGOLDC:	conducting yellow paint (50%) gold (50%)
ML12:	ML12-3 and ML12-4 surface
ALUM:	aluminum plate
BOOMAT:	platinum banded kapton

Eclipse Charging of the SCATHA Satellite

The SCATHA model has been employed for a preliminary study of charging in eclipse. Figure 11 shows the potential as a function of time for a substorm environment of 1 particle/cm^3 electron temperature of 20 keV, and ion temperature of 40 keV. Charging takes place in three stages. In 10^{-2} sec, the vehicle charges to -7 kV relative to the ambient plasma, close to saturation. This saturation potential is maintained for 50 sec at which time the vehicle starts to charge differentially.

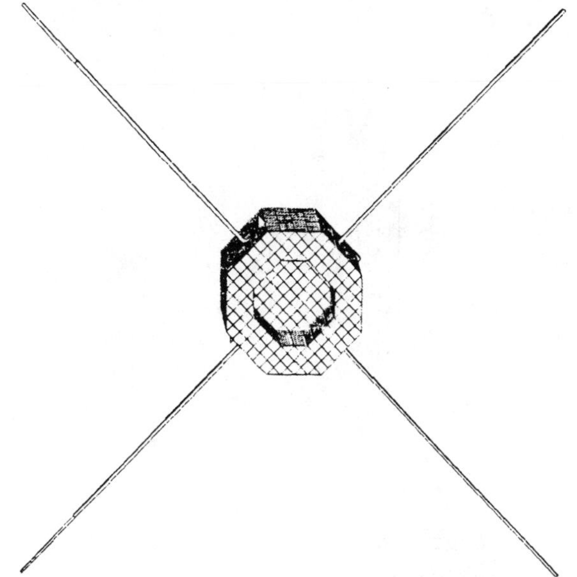

Fig. 5 View showing the modeling of the bottom cavity of SCATHA.

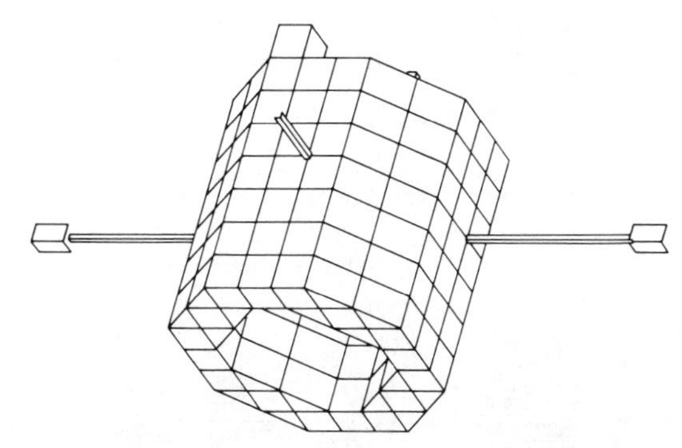

Fig. 6 Perspective view one-grid SCATHA model (zone size = 23 cm).

Differential charging continues out past 100 sec, - steady state is not reached in any relevant time. Thus, differential charging will be a continuous, dynamic process since the external environment will change much more rapidly.

Code Uses in Flight Data Analysis

The Emitter-Detector routine traces the orbits of ions and electrons in the spacecraft sheath field and the external fields. The Detector subroutine does the following: given

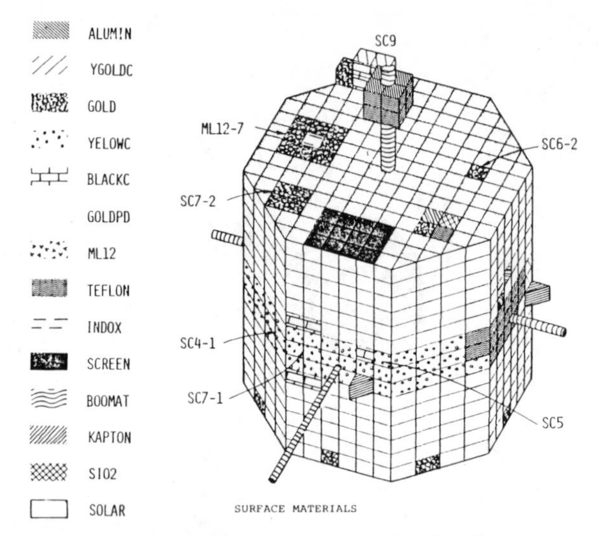

Fig. 7 Illustration of accuracy with which materials placement on the satellite surface can be treated.

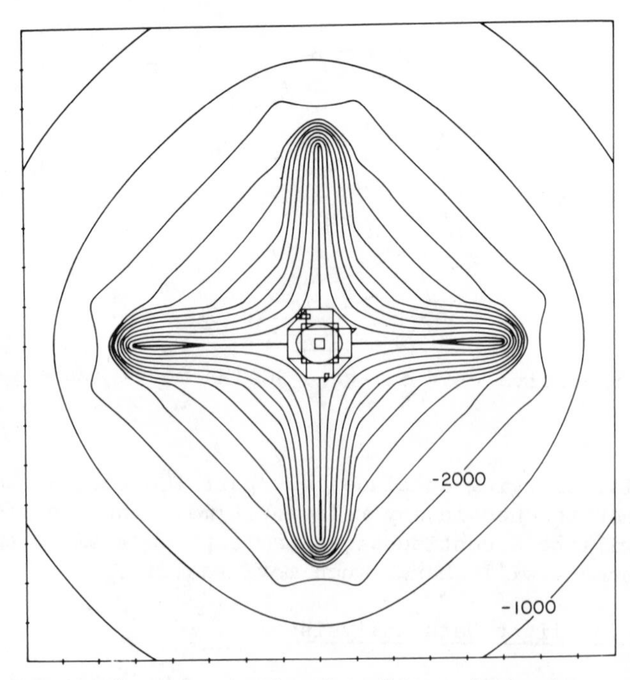

Fig. 8 Potential contours in the plane of the booms about SCATHA in eclipse for a severe charging condition (along x-y plane of z = 17).

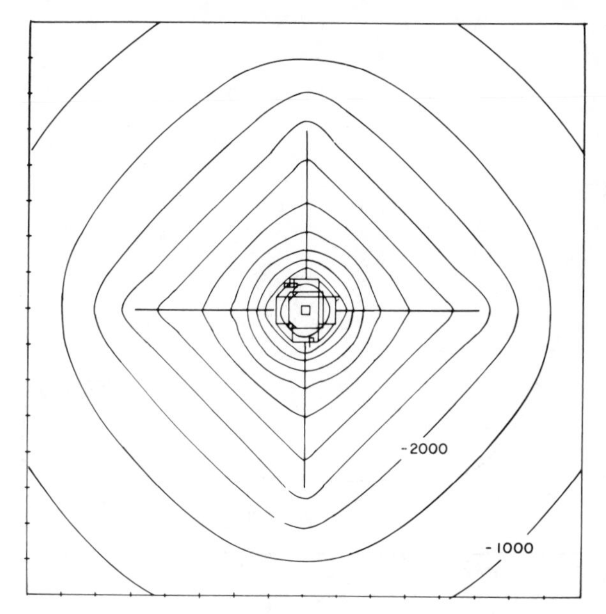

Fig. 9 Potential contours 1 m below the SCATHA center in eclipse.

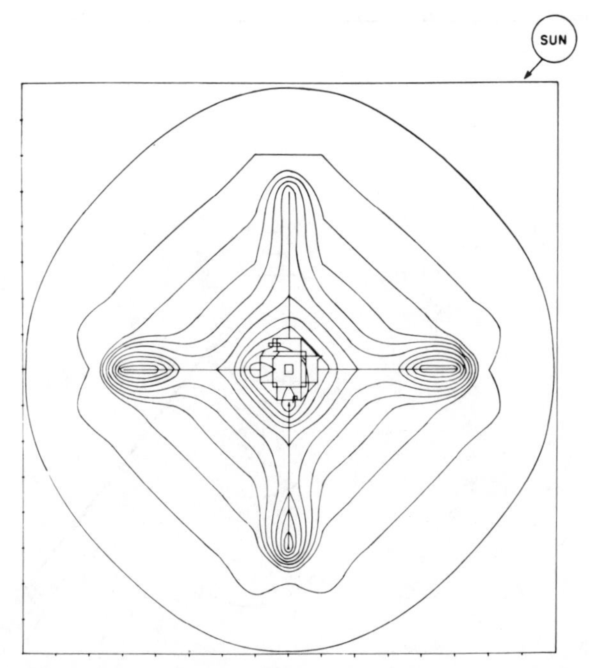

Fig. 10 A case of severe differential charging for the sunlit case (t = 38 sec).

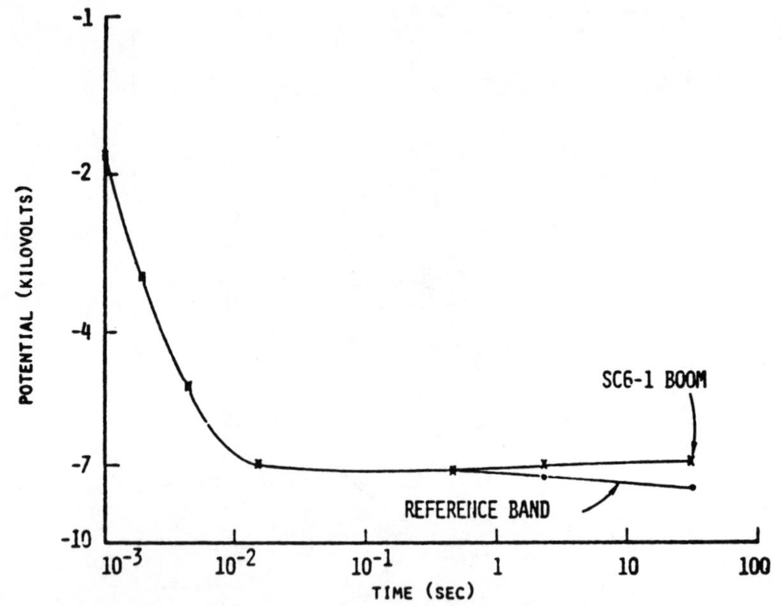

Fig. 11 Spacecraft potential vs time for two points on SCATHA satellite.

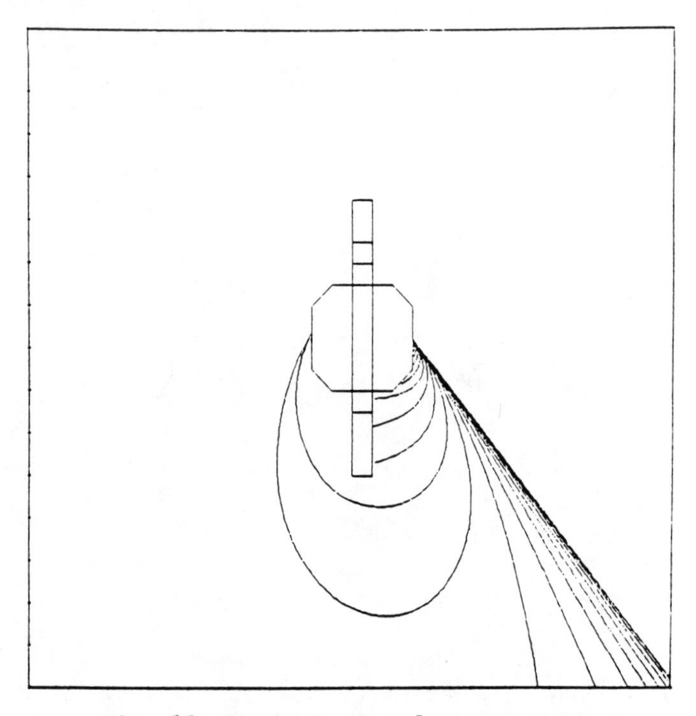

Fig. 12 Trajectories for protons.

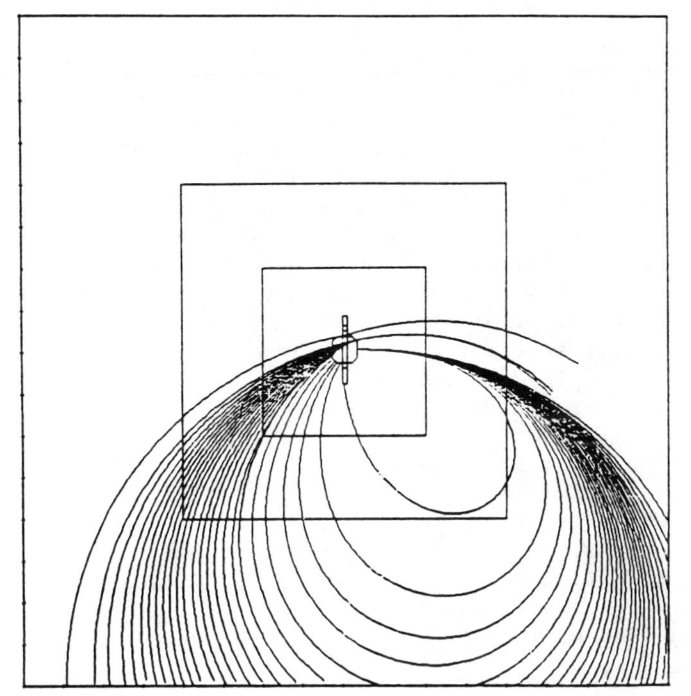

Fig. 13 Particle trajectories from electron emitter.

the characteristics of a detector (such as sensitivity, solid angle, view angle, and energy range), particle trajectories reaching the detector are traced outward from the detector through the sheath and external fields. In this way, by looking at the origin of a particle, one can tell whether it came from space or from another portion of the spacecraft. An example of a detector calculation is shown in Fig. 12.

Figure 12 shows trajectories for protons received at 25 different detector energies logarithmically spaced from 10 eV to 50 keV for a spacecraft charged to -4.5 keV with no magnetic field. This figure is typical of the graphical output generated by the code. The straight line trajectory is for the highest energy particles. Notice that at some of the lower energies, particles may originate on the spacecraft surface. Note also that the detector collects particles from a larger solid angle for lower energy particles when the spacecraft is charged, in effect increasing the geometrical factor.

Emitters

SCATHA contains both electron and ion guns for studies of active charge control. The currents emitted by the guns can charge the spacecraft both positively and negatively. In the absence of natural charging, ions can be emitted to produce charging to desired negative potentials, simulating substorm effects. Alternatively, when a natural charging event occurs, the electron gun can be activated to study the process of reduction of surface charge. In order to analyze these processes, the NASCAP code has provisions for one or more low-density particle emitters at arbitrary surface cell locations on the satellite. Particle trajectories from these emitters are calculated and displayed graphically. In addition, currents escaping from and returning to the satellite are calculated.

As an example of these calculations, particle trajectories from emitters are shown in Fig. 13. In order to trace trajectories over large distances, the computation is carried out in embedded meshes. In the figure, three meshes are used with a monopole potential assumed outside of the largest mesh. Each successive mesh has twice the spacing of the preceding mesh. A closeup view of the returning particle trajectories is shown in Fig. 14.

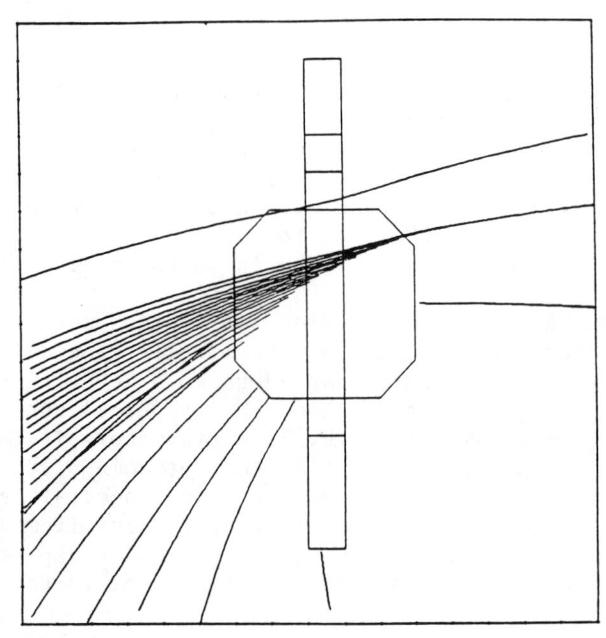

Fig. 14 Closeup view of returning particles in the inner grid.

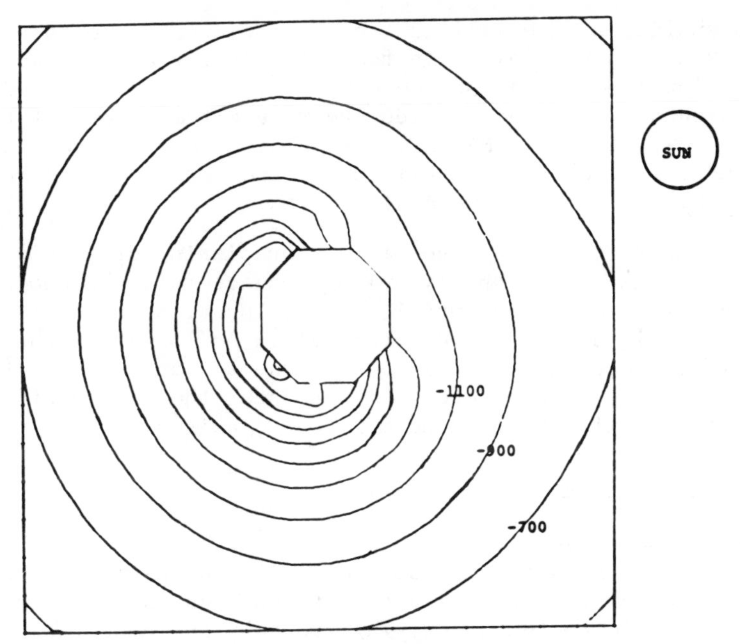

Fig. 15 Potential contours about an illuminated cylinder showing that a highly negative potential exists close to the positive sunlit surface because of the influence of the shaded side.

Photosheaths

NASCAP has a sheath routine which computes the charge density due to photoelectrons as well as the resulting potential contours. Figure 15 shows the potential contours around a teflon cylinder in sunlight, and Fig. 16 the charge density. Note that the potential on the sunlit surface is -1100 V for a moderate substorm with $T_e = T_i = 1.3$ keV. NASCAP sometimes predicts high negative potentials on sunlit dielectric surfaces. This daylight charging comes about because a saddle point in the potential,[9] which builds up due to the photosheath, initially blocks photoelectrons from escaping. The sunlit side charges negatively until the saddle becomes low enough for a portion of the photoelectrons to escape. The photosheath saddle point exists for a slowly spinning or nonspinning spacecraft.

The ubiquitous existence of daylight charging predicted by the NASCAP code is easy to understand once one knows of its existence. It does, however, introduce a radical change in the way people think about charging. Previously, one assumed that sunlit surfaces would remain nearly uncharged. Now it is clear that a sunlit surface will charge to thousands of volts when the dark side potential is in the kilovolt region. In addition, one previously expected photoelectrons from the sunlit side to drift around to the shaded side and reduce the potential there. We now know, thanks to NASCAP, that photoelectrons are constrained close to the region in which they are produced and affect the potential only locally.

In contrast to the case of the nonspinning or slowly spinning satellite, on a rapidly spinning spacecraft the photoelectrons reduce the potentials at all locations on the satellite. The highest negative potential is no longer at the equatorial shaded side but is now at the top and bottom of the satellite. The lowest potentials are near the center of the satellite for illumination normal to the axis of rotation.

Discharge Analysis

NASCAP has facilities for simulating the effects of arc discharges on the satellite surface. Discharges may be of two types: a punch-through discharge (which arcs through a surface dielectric to the underlying ground), and a surface blowoff arc (which removes charge from a dielectric surface). It

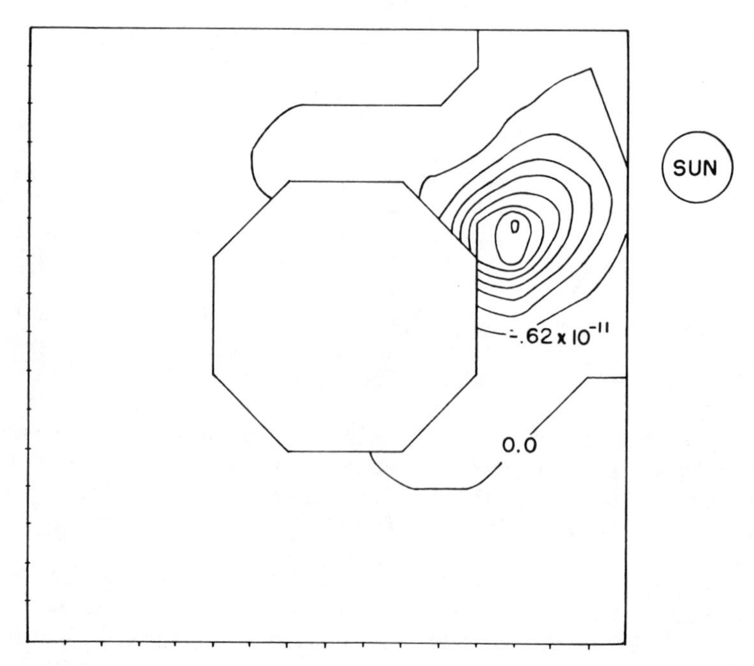

Fig. 16 Charge density contours about an illuminated cylinder.

has been shown that the majority of spacecraft arcing takes place by the surface blowoff mechanism. NASCAP simulates the effects of arcs by allowing an arc current to flow when the potential gradient in a given region exceeds a predetermined value. The potential distribution before and after arcing, as a function of time, are then calculated employing NASCAP.

Antenna Design for Electric Field Measurements

A final application for the NASCAP code is in the design of antennas to measure electric fields. If one wishes to measure an electric field between the tips of antennas which are covered with dielectric materials, it is important to know the electric field produced around the antenna by the ambient plasma in the absence of an external electric field. NASCAP may be used to calculate this field and to choose the materials for the electric field measuring antenna so that the electric field due to the ambient plasma is minimized. Although this has not been done to date, it is another potential application of the NASCAP code in designing satellites.

IV. Summary

The purpose of this paper has been to describe a tool for studying spacecraft charging: the NASCAP code. We have outlined the procedures the program follows in first defining the satellite and its materials in a consistent fashion for the computer and then estimating the potential on the satellite surfaces and in the surrounding space. We have described how it treats photosheaths, charging and differential charging vs time, shadowing, trajectories from particle emitters, the response of detectors, and the effects of spacecraft spin. We have, using the SCATHA satellite (the first satellite ever designed specifically to study the effects of spacecraft charging), presented these results in terms of an operational satellite. As SCATHA is currently returning a wealth of data on spacecraft charging, we hope shortly to be able to validate the detailed predictions of the code. Even without this validation procedure, we believe that the NASCAP code currently is capable of providing significant information to the satellite designers so that steps can be taken early in the design phase to alleviate spacecraft charging, with a resultant major savings in cost.

References

[1]DeForest, S. E., "Spacecraft Charging at Synchronous Orbit," Journal of Geophysical Research, Vol. 77, February 1972, pp. 651-658.

[2]Knott, K., "The Equilibrium Potential of a Magnetospheric Satellite in an Eclipse Situation, Planetary and Space Science, Vol. 20, May 1972, pp. 1137-1146.

[3]Katz, I., Parks, D. E., Mandell, M. J., Harvey, J. M., and Wang, S. S., "NASCAP, a Three-Dimensional Charging Analyzer Program for Complex Spacecraft," IEEE Transactions on Nuclear Science, Vol. 6, December 1977, pp. 2276-2280.

[4]Katz, I., Cassidy, J. J., Mandell, M. J., Schnuelle, G. W., Steen, P. G., and Roche, J. C., "The Capabilities of the NASA Charging Analyzer Program," Spacecraft Charging Technology - 1978, edited by R. C. Finke and C. P. Pike, NASA CP-2071/AFGL-TR-79-0082, 1979, pp. 101-122.

[5]Mandell, M. J., Katz, I., Schnuelle, G. W., Steen, P. G., and Roche, J. C., "The Decrease in Effective Photocurrents Due to Saddle Points in Electrostatic Potentials Near Differentially Charged Spacecraft," IEEE Transactions on Nuclear Science, Vol. 25, December 1978, pp. 1-13.

[6]Hestenes, M. R. and Stiefel, E., "Method of Conjugate Gradients for Solving Linear Systems," Journal of Research of the National Bureau of Standards, Vol. 49, December 1952, pp. 409-415.

[7]Jennings, A., "Matrix Computations for Engineers and Scientists," John Wiley and Sons, New York, 1977.

[8]Parker, L. W., and Whipple, E. C., "Theory of a Satellite Electrostatic Probe," Annals of Physics, Vol. 44, January 1967, pp. 126-161.

[9]Fahleson, U., "Plasma-Vehicle Interactions in Space; Some Aspects of Present Knowledge and Future Development," Photon and Particle Interactions with Surfaces in Space, R. J. L. Grard (ed.), D. Reidel Publishing Co., Dordrecht, Holland, 1963, pp. 563-569.

Chapter III—Radiation Effects on Space Systems

RADIATION EFFECTS ON SPACE SYSTEMS AND THEIR MODELING

A. L. Vampola*
The Aerospace Corporation, Los Angeles, Calif.

Abstract

Space systems are subject to degradation of performance and damage by the charged particle populations trapped within the Earth's magnetic field. Spacecraft encounter electrons, protons, and ions with energies from a few eV to many MeV in various regions of the magnetosphere. As a result, components suffer radiation damage, logic upsets occur, sensors experience elevated background levels, and, near synchronous altitudes where hot tenuous plasmas occur, differential charging with subsequent arcing may be experienced. Past efforts have produced satisfactory models of the trapped energetic charged particle population in most regions of the magnetosphere. Efforts are continuing in such diverse areas as the interaction of spacecraft with hot plasmas and damage mechanisms in microcircuitry.

Introduction

The energetic particle population trapped in the Earth's magnetosphere has been a significant factor in every space flight since the beginning of the artificial satellite era. Explorer I, the U. S.'s first satellite, contained a Geiger-Mueller (G-M) tube to measure cosmic rays. The energetic protons trapped in the Earth's magnetic field penetrated the shielding and, together with bremsstrahlung from energetic electrons impinging upon the satellite, saturated the count-rate circuitry and G-M tube for significant portions of each orbit. Thus, James Van Allen discovered the trapped radiation zones named after him and also became the first to experience deleterious effects in his equipment due to the trapped radiation. In the early years, the trapped radiation was only a curiosity to be studied and to be considered in the design of sensors. However, the Starfish nuclear event in 1962 injected vast numbers of energetic electrons into the magnetosphere. During the first few days, the dose rate from these electrons was sufficient to disable several satellites in orbit. The radiation environment

---

*Senior Scientist.

immediately became a significant factor in the design of spacecraft subsystems, especially the solar power arrays. However, the subsequent decay of the Starfish electrons (typical lifetime of 300 days in the inner zone) brought a certain amount of complacency back into the design of satellite systems. The radiation environment was benign as far as any hardware, other than solar cells and sensors, was concerned.

The evolution of solid state circuitry, with ever-diminishing dimensions in elements, eventually brought the radiation environment back into play as a concern in satellite design. Complimentary metal oxide silicon (CMOS) technology, with its low power requirements and small geometry, promised to be the spacecraft designer's solution to a need for very complex spacecraft subsystems; its very low tolerance for radiation, however, reopened the problem of radiation dose in orbit. It also precipitated research into defining better models of the radiation environment (Vampola et al., 1977; Vette et al., 1979). In this review, we shall first discuss the radiation environment and its effects on systems in space and then summarize the status of modeling efforts in this area.

The Radiation Environment

The radiation environment can be considered from two points of view: spatial distribution and type of particle. In the immediate vicinity of the Earth, external to the magnetosphere, the particle environment consists of the solar wind, cosmic rays, and occasionally energetic ions from the sun and energetic electrons from the sun and Jupiter. The solar wind is basically a high-speed (≈ 400 to 700 km/sec) low-temperature, low-density plasma. The major ionic species is H^+, with an admixture of a few percent He^{++}. The solar wind has a major effect on the transport of the solar magnetic field embedded within it, which in turn has a controlling role in geomagnetic activity. The solar wind and the interplanetary magnetic field (IMF) embedded in it cause magnetic substorms and storms. With the exception of this role, the solar wind does not significantly affect spacecraft systems or sensors. (Low-energy plasma particle detectors and electric/magnetic field sensors do respond to it, but usually that is their purpose.)

Cosmic rays interact with spacecraft primarily through activation of sensitive sensor systems; i.e., they constitute a background in optical sensors, particle detectors, etc. The energy contribution from cosmic rays is so low as to be only a minor consideration for manned missions and negligible for hardware. The sole exception is an effect in low-level logic elements which will be discussed briefly later. Since cosmic rays have access to the magnetosphere along Störmer orbits, sensors which are affected by cosmic rays may be subject to their interference anywhere within the magnetosphere, and appropriate design techniques must be utilized to negate these effects.

The energetic particles from the sun, primarily protons and electrons, are a different matter. Whereas the flux and dose rate from cosmic rays is the order of $1/cm^2$ sec and less than 10^{-4} rad/hr, a large solar flare can produce a flux of energetic protons (≥ 5 MeV) of $10^5/cm^2$-sec-ster and a dose rate in excess of 100 rads/hr. Many solar proton events are accompanied by significant fluxes of energetic electrons (Lin and Anderson, 1967). The electron energy spectra are usually soft (Vampola, 1969) and of relatively low intensity. The largest electron event measured had a peak flux of 1.4×10^5 e/cm^2-ster for $E_e >$ 300 keV (West and Vampola, 1971). Typical electron events are lower by an order of magnitude. Since solar electron events, when they occur, always accompany proton events and most of the energy is carried by the protons because of their higher average energy and greater fluxes, the electron component can be ignored except for situations in which sensors are insensitive to the protons but can respond to the electrons. Energetic electrons at the Earth's orbit which have Jupiter's magnetosphere as their origin (Teegarden et al., 1974) are sufficiently low in intensity that they are of no concern except for electron sensors.

Some relatively minor solar flares produce significant fluxes of high-Z nuclei (Zwickl et al., 1978). Such nuclei are also present in typical proton flares, but they constitute a very small portion of the total flux. These energetic ions and electrons have direct access to the Earth's magnetosphere over the polar caps and, for protons with $E_p \geq 5$ MeV, to synchronous altitudes. (Presumably, energetic electrons also gain access to synchronous altitudes but are not detectable in the presence of the energetic electron fluxes normally there.)

Having discussed the particle environment outside the magnetosphere, we will now briefly discuss the particle environment starting near the Earth and progressing back out to the magnetopause. For these regions of space, the magnetic field geometry is the structure by which the particle populations are organized. It is therefore more convenient to use McIlwain's parameter L, which was derived for just that purpose (McIlwain, 1961). The L value of a field line is essentially the radial distance from the center of the Earth (in units of Earth radii) to the minimum B on that line (the magnetic equator).

The region up to about L = 2.5 is considered the inner-zone and the region beyond about L = 3 to the magnetopause is considered the outer-zone. Between the two zones is the slot, a region which during magnetically quiet periods has relatively few electrons. The inner-zone and outer-zone terminology really applies to electron populations but is also convenient in discussing the proton populations. The electron flux is soft, i.e., it has a steep energy spectrum with few electrons above 1 MeV. However, the inner zone also contains a high-energy proton population, with fluxes in excess of 10^3 p/cm^2-sec-ster above 50 MeV. These proton fluxes peak at about L=1.45. Electron fluxes peak a little further out, at about L = 1.5 for E_e = 1 MeV.

The slot region is the result of a diminishing radial diffusion rate with decreasing L and an increasing loss rate due to wave-particle interactions. During magnetic storms, the slot is filled by electrons from the outer zone which are under an enhanced diffusion rate because of the storm. The particles then decay away after the storm, with some diffusing into the inner zone (Vampola, 1971).

The outer zone, which includes the synchronous altitude orbits, is a more interesting region because of its complexity and variability. Both electron and proton fluxes respond to magnetic storms. For the protons, no slot occurs such as that exhibited by the electrons. The outer zone is just a continuation of the inner zone proton fluxes, with a systematic increase in the L value at which the proton flux peaks with diminishing energy per particle. Protons with E_p = 10 MeV peak at about L = 2.5 while the peak intensity is at about L = 3 for 1 MeV protons. The peak flux for 10 MeV protons is about 10^4/cm^2-sec (omnidirectional), while the 1 MeV protons peak at about 10^8 in the same units, decreasing to 10^5 at L = 5 and 10^3 at synchronous. These values are typical. Short time variations of up to a factor of 50 have been reported low on the field line. Equatorial variations are typically much smaller.

The electrons exhibit even greater variability. Several orders of magnitude increases occur following large magnetic storms. After a major magnetic storm, the flux of 3 - 5 MeV electrons in the outer zone can increase from quiet-time levels that are near the cosmic ray background rate to hundreds of electrons/cm^2-sec-ster-keV, with an energy spectrum of E^{-3}. The integrated fluence above 1 MeV can be as high as 10^7 e/cm^2-sec. These energetic electrons have a relatively long residence time (tens of days) because the strength of their source of replenishment (radial diffusion of lower energy electrons from farther out in the magnetosphere which are energized during their inward transport) is relatively equivalent to the loss rate (radial diffusion inward and pitch-angle scattering down the field lines into the atmosphere).

At the higher altitudes, synchronous and above, a quasi-equilibrium condition exists for electrons in the .1 - 1 MeV range because of the frequent magnetic disturbances which occur there. The more energetic electron fluxes, $E_e \geq 1$ MeV, are more variable and appear to be controlled by the solar wind speed, with major effects produced by the high speed streams from coronal "holes" (Paulikas and Blake, 1978).

At altitudes corresponding to $L \geq 5$, magnetic substorms and storms inject quantities of hot plasma. The understanding of the dynamics of injection and subsequent transport within the inner magnetosphere are still under intensive investigation, and no definitive picture is yet available. For a more thorough discussion of the general morpholoby of the radiation belts, the NASA monograph SP-8116 in the Space Vehicle Design Criteria series (NASA, 1975) is a useful reference.

Radiation Effects

Some radiation effects in spacecraft systems are covered in detail in sections following this overview. The effects on man in space are covered extensively in the proceedings of a workshop held in Berkeley in 1978 (Lawrence Berkeley Laboratory, 1978) and will not be addressed here. The primary concern to spacecraft subsystems is the radiation dose or dose-rate. Direct damage to electronic components can occur at quite low doses. As an example, CMOS memory elements have been observed to fail at accumulated doses as low as 2000 rads. In the outer electron zone ($L \simeq 3.5$ after large magnetic storms), dose rates as high as 1000 rads/hr beneath .35 g/cm^2 of shielding have been observed. Beneath thicker shields (1.7 g/cm^2), dose rates in excess of 100 rads/hr also have been observed. At the synchronous altitude, accumulated doses on the order of 10^5 rads/yr beneath 0.5 g/cm^2 are to be expected. Typical "soft" metal oxide silicon (MOS) devices have a threshold for damage of 10 - 20 kilorads. Obviously, the radiation environment and the shielding have to be considered very carefully when such devices are to be used in a space mission.

In addition to direct damage to electronic components, direct damage to solar cells is always considered. A typical design philosophy estimates the degradation to be expected in orbit during the mission lifetime and sizes the power system to meet the requirements at the end of mission after the expected degradation has occurred. For very long missions in severe radiation environments, direct damage to optical components (discoloration and reduced transparency) also may have to be considered.

A second type of effect which is directly related to dose is the "soft" failure. For a high-Z (typically iron) nucleus with moderate energy (≈ 2 MeV per nucleon), the specific ionization rate is such that the energy deposited in a microjunction in semiconductor logic can cause a change-of-state of the device (a bit-flip) (Sivo et al., 1979). The high-Z ion can be either a galactic cosmic ray at the low energy end of the spectrum or a solar cosmic ray emitted during a flare. In some cases the result of the energy deposition is a 'latch-up' which destroys the device. Usually, the only result is a bit error which proper design of memory devices can automatically correct with high probability of success. However, in nonself-correcting devices, the bit error rate (up to 4 bit errors/hr-megabit, Kolasinski et al., 1979) can inconvenience or damage a mission. If the microjunction is in a logic element other than memory, the resulting logic state could be a disallowed state (not achievable through normal operation) and could either cause failure of some part of a circuit or "latch up" in a disallowed state. Furthermore, a "sneak" or "phantom" transistor could be turned on by such energy deposition. Certain devices contain these potential transistors in their substrates by virtue of their geometry of doping and etching. Such a transistor is not under the control of the circuit. Usually it appears as an internal short within a chip and can be turned off only by removing power from the chip. Damage to the chip may occur when such a "sneak" transistor is activated.

Another deleterious effect associated with energetic particles is the interference or background they produce in sensitive sensors. The Cerenkov radiation produced in glass elements in low-level star sensors by cosmic rays or by energetic protons in the inner zone can rival the signal from the target star, causing erroneous results. The logic in the system has to be designed with these effects in mind. In an optical or x-ray spectrometer, the element which detects the photons usually is sensitive to energy deposited directly within it by penetrating particles, and proper design must include identifying and rejecting such events. Other senors, designed to measure infrared radiation or trapped particles, are also subject to such interference. In general, any element of a spacecraft which is sensitive to small amounts of energy, of the order of 10^{-6} ergs or 10^{6} hole-electron pairs deposited in a fraction of a microsecond, will be subject to background effects.

A final example of an effect which has been ascribed to energetic particles, and which seems to be a probable cause of some anomalous behavior of spacecraft in orbit, is cable-dielectric charging. The proposed mechanism involves irradiation by large fluxes of energetic electrons (such as are present in the outer zone after large magnetic storms) and embedding charge in insulators. Numerical calculations show that charge can build up to the breakdown point in cable dielectrics. The resulting spark between the dielectric and ground sheath or center conductor will inject a fast signal into the system that may affect the system logic. If the signal looks like that produced by breakdown between insulating materials and ground on the exterior of a spacecraft because of plasma charging (discussed later), the pulse is very short. The typical rise time is less than 10 nsec, and it decays to less than 10% of its initial height by 1 μsec.

At high altitudes, typically near synchronous orbit or higher, magnetic storms and substorms inject hot plasma from the tail of the magnetosphere. In the presence of the hot plasma, spacecraft can charge to significant negative potentials (> 10 kV) because of the much greater mobility of the electrons compared to the ionic constituent of the plasma. The resulting charge on the spacecraft can lead to a number of deleterious effects, some of them potentially lethal to the hardware. The negative potential may enhance the accumulation of contaminants by spacecraft surfaces through the mechanism of outgassing, photoionization of the outgassed molecule, and reattraction by the potential on the spacecraft. Thermal control surfaces, especially second-surface mirrors, optical devices, and certain sensors, are particularly vulnerable to this type of degradation. If the spacecraft is in sunlight, photoemission of electrons from the sunlit portions of the vehicle may neutralize the charge buildup due to the plasma. The self-shadowed portions of the vehicle, unless they are electrically conducting and connected to a sunlit conductor, will remain at high negative potential. The potential difference between these differentially charged surfaces may result in an arc which could interfere with the operation of the vehicle or could even destroy a critical component. The charging phenomenon is not now well understood, but it is under investigation by the Spacecraft Charging At High Altitude (SCATHA) program. The program involves both laboratory work and a satellite, STP 78-2, in a near-synchronous orbit for purposes of data accumulation.

Modeling the Radiation Environment

The standard models of the radiation belts in general use are those developed at the National Space Science Data Center at the NASA Goddard Space Flight Center. The first ones were issued by J. Vette while at Aerospace Corporation in 1965, and they utilized virtually all of the rocket and satellite data available for the proton model, AP1, and all of the satellite data available for the electron model, AE1. The models were produced in an iterative fashion by assuming a spectrum and calculating the response of various satellite instruments to this spectrum. The result was then used to select a better spectrum, and calculations were continued. The inconsistencies between data sets (several orders of magnitude in some cases) were handled by assigning a "confidence factor" to each data set and producing a weighted-average of all the sets. The modeling effort was transferred to NASA Goddard Space Flight Center under the direction of J. Vette, and it continues today.

The task of producing a reliable model is a formidable one because of the difficulty of getting the large data bases into the appropriate form for modeling purposes. The task usually takes several years for a single data base. The problem of reliability of the data continues. Current techniques involve much better calibration and understanding of instrument responses prior to launch than in the early years of space research, but inconsistencies between data sets still are a fact of life. Some of the problems arise from the differing detection techniques used by various investigators. The great variability of the outer zone electron fluxes is an additional complicating factor. As a result, new models are issued whenever sufficient new reliable data become available to enable a significant improvement over the old model. Models covering energetic protons in the inner zone, electrons in both the inner and outer zones, low energy protons in the outer zone, and an electron model for the synchronous orbit have been issued. The decay of the Starfish electrons in the inner zone, local time variations in the outer zone, and solar-cycle variations in the inner and outer zones have been included in various models. Storm effects in the outer zone are now being studied for inclusion in the next generation of models. For a thorough discussion of the history of the individual models, AP1 through AP8 and AE1 through AE7, see the review paper by Vette et al., 1979.

The current version of the electron model of the magnetosphere utilizes several individual models. AE5 covers the region $1.2 \leq L \leq 2.8$. It was issued in 1972 and incorporated provisions for the decay of the Starfish flux, a solar-cycle dependent quiet-time flux, and enhancements due to magnetic storms. A variant, calculated for solar-maximum using the basic AE5 model, was issued as AE6. These models appear to be entirely adequate as a description of the inner zone (errors less than a factor of two) and will continue to be useful for the foreseeable future. The outer zone electrons had been represented by AE4 until recently, but the model was found deficient in high energy fluxes during and following magnetic disturbances (Vampola et al., 1977). An interim model, AEI-7, has been issued for use until a better model incorporating magnetic storm effects can be constructed. Two versions of AEI-7 were actually issued: AEI-7 Hi utilized data from the satellite OV1-19, and AEI-7 Low utilized data from the AZUR satellite, both of which made measurements above 4 MeV during solar maximum. AEI-7 Hi predicts doses under thick shielding about an order of magnitude higher than the AE4 model and half an order of magnitude higher than AEI-7 Low. Recent measurements by dosimeters in the outer zone prior to solar maximum (and which should be representative of a solar cycle average as addressed by AEI-7 Hi) indicate that even AEI-7 Hi may be low by a factor of two in predictions of dose (Blake, 1980). One of the drawbacks in modeling the outer zone, the lack of measurements of energetic electrons near the equator, is being remedied with instrumentation on the STP 78-2 spacecraft.

The situation with regard to proton models is much better. Accurate measurements of proton fluxes and energies are much easier to obtain and there has been much less discrepancy between data sets. AP8 (Sawyer and Vette, 1976) covers the energy range .1 - 400 MeV. Solar cycle variations below 1000 km (due to changes in the scale height of the atmosphere) for $L < 3$ are included by splitting AP8 into two versions, AP-8 MIN and AP-8 MAX.

Future plans for modeling include: a) fine-resolution temporal effects in the low energy protons due to magnetic storms, b) a storm-time representation of the energetic electron fluxes, and c) a description of substorm effects on the plasma population in the outer magnetosphere. An environmental "atlas" is being written at the Air Force Geophysics Laboratory and will be issued in the near future.

References

Blake, J. B., "GPS Dosimeter Results," Air Force Systems Command/Space Division Tech. Rep., 1980.

Kolasinski, W. A., Blake, J. B., Price, W., and Smith, E., "Simulation of Cosmic Ray Induced Soft Errors and Latchup in Integrated-Circuit Computer Memories," in Proceedings of the IEEE 1979 Conference on Nuclear and Space Radiation Effects, 1980, in press.

Lawrence Berkeley Laboratory, Proceedings of the Workshop on the Radiation Environment of the Satellite Power System, Lawrence Berkeley Laboratory, Berkeley, Calif., LBL-8581, 1978.

Lin, R. P., and Anderson, K. A., "Electrons $\geq$ 40 keV and Protons $\geq$ 500 keV of Solar Origin, Solar Physics, 1,1967, pp. 446-464.

McIlwain, C. E., "Coordinates for Mapping the Distribution of Magnetically Trapped Particles," Journal of Geophysical Research, Vol. 66, 1961, pp. 3681-3692.

NASA, "The Earth's Trapped Radiation Belts," NASA SP-8116, 1975.

Paulikas, G. A. and Blake, J. B. "Energetic Electrons at Synchronous Altitude 1967-1975," Air Force Systems Command/Space Division, Los Angeles, Calif., SAMSO-TR-78-75, 1978.

Sawyer, D. M. and Vette, J. I., "AP-8 Trapped Proton Environment for Solar Maximum and Solar Minimum," National Space Science Data Center, Greenbelt, Md., NSSDC/WDC-A-R&S 70-76, 1976.

Sivo, L. L., Brettschneider, M., Price, W., and Pentecost, P., "Cosmic Ray Induced Soft Error in Static MOS Memory Cells, in Proceedings of the IEEE 1979 Conference on Nuclear and Space Radiation Effects, 1980, in press.

Teegarden, B. J., McDonald, F. B., Trainor, J. H., Webber, W. R., and Roelof, E. C., "Interplanetary MeV Electrons of Jovian Origin, Journal of Geophysical Research, Vol. 79, 1974, pp. 3615-3622.

Vampola, A. L., "Energetic Electrons at Latitudes Above the Outer-Zone Cutoff, Journal of Geophysical Research, Vol. 74, 1969, pp. 1254-1269.

Vampola, A. L., "Natural Variations in the Geomagnetically Trapped Electron Population," Proceedings of the National Symposium on Natural and Manmade Radiation in Space, edited by E. A. Warman, NASA TM X-2440, 1972, p. 539-547.

Vampola, A. L., Blake, J. B., and Paulikas, G. A., "A New Study of the Magnetospheric Electron Environment," Journal of Spacecraft and Rockets, Vol. 14, 1977, pp. 690-695.

Vette, J. I., Teague, M. J., Sawyer, D. M., and Chan, K. W., "Modeling the Earth's Radiation Belts," Proceedings of the International Solar-Terrestrial Predictions Workshop, edited by R.Donnelly, NOAA/ERL, Boulder, Colo., 1979.

West, H. I., Jr. and Vampola, A. L., "Simultaneous Observations of Solar-Flare Electron Spectra in Interplanetary Space and Within the Earth's Magnetosphere," Physical Review Letters, Vol. 26, 1971, pp. 458-463.

Zwickl, R. D., Roelof, E. C., Gold, R. E., and Krimigis, S. M., "Z-Rich Solar Particle Event Characteristics 1972-1976," Astrophysical Journal, Vol. 225, 1978, pp. 281-303.

COSMIC RAY EFFECTS IN VERY LARGE SCALE INTEGRATION

John N. Bradford*
Rome Air Development Center, Hanscom Air Force Base, Mass.

Abstract

The reduced size and operational energy associated with microelectronic circuitry has created a situation wherein a cosmic ray, in a single pass, can generate a false signal or otherwise damage a circuit element. In this work an analytic expression for the event rate of false signals (a.k.a. soft errors) is derived, based upon the geometric probability of a cosmic-ray track occurring in the sensitive volume of a micro-circuit device. The analyses show that substantial reduction in the soft error rates from heavy cosmic rays can be achieved by device designs which incorporate large operational energies and small depletion collection volumes. The permanent damage to oxide layers in metal-oxide-semiconductor (MOS) devices is estimated. The special role of protons as initiators of soft errors and permanent damage is explored. These processes are then described in the context of very large scale integration (VLSI) electronics in Earth satellite environment. Proton non-elastic scatter events which produce energetic recoiling Silicon nuclei are identified as the chief cause of proton initiated soft errors.

Introduction

Microelectronic circuits (VLSI) subjected to cosmic-ray or other heavy ion bombardment undergo false switching-error writing response. The requirement of the circuit to produce such a response is the deposition of a threshold amount of energy in the form of ionizing reactions in a small fairly well-defined volume of the circuit. The requirements for the ion or cosmic ray are that the energy be sufficient to penetrate any shielding, with kinetic energy in excess of the circuit threshold, ΔE, and linear energy transfer great enough to deposit ΔE in a track whose length is only several microns.

This work describes the dose distribution surrounding a cosmic-ray track and the impact of that energy deposition in

*Deputy for Electronic Technology.

the form of hole-electron pairs in VLSI circuitry, specifically as it applies to dynamic Random Access Memories. An event rate for errors in satellite borne electronics exposed to the cosmic flux is derived. The possibilities of permanent damage to the oxide layers in MOS devices are described, and finally the role of protons in these interactions is developed.

Size and Dose Distributions near Heavy Ion Tracks

When a fast heavy ion penetrates condensed matter, some of the orbital electrons are stripped from the ion, and a beam of such ions then has a mean ionic charge, $z \leq Z$; z has been measured widely as a function of energy and velocity of the ion and is given by Barkas[1]

$$z = Z(1 - e^{-15\beta Z^{-2/3}}) \qquad (1)$$

The ion loses energy through electron collisions and generates electron flux at a rate given by Mott[2]

$$\frac{dn}{d\omega} = \frac{2\pi N z^2 e^4}{mv^2} \cdot \frac{1}{\omega^2} \quad 1 - \beta^2 \, \omega/\omega_m + \frac{\pi\beta z}{137} \left(\frac{\omega}{\omega_m}\right)^{\frac{1}{2}}\left(1-\frac{\omega}{\omega_m}\right) \qquad (2)$$

where $dn/d\omega$ = the number of electrons between energy ω, $\omega + d\omega$ generated per cm of ion track in matter containing N electrons/cm^3 and $\omega_m = 2mc^2\beta^2\gamma^2$; e, m are the electron charge and mass and $\beta = v/c$.

The above electron generation mechanisms can be coupled with electron transport models to produce a relatively detailed picture of the dose and electron flux near a heavy ion track. Kobetich and Katz[3] and Hamm[4] have made extensive calculations of the energy deposited near an ion track with the following result:

$$D = kz^2/r^2 \qquad (3)$$

That is, the energy deposition, D, in ergs/g at radius r from a cylindrical track falls off as $1/r^2$ and is proportional to z^2, where, from Eq. 1 above, z is the mean ionic charge of the ion or cosmic ray. Both z and k are functions of β. Recent measurements at Brookhaven[5] have verified the $1/r^2$ dependence over a range of 1 to 300 nm (10 Å), as shown in Fig. 1.

The column of ionization which a fast heavy ion leaves in its wake contains central core energy densities of 8×10^{20} eV/cm^3 for protons to 10^3 times greater for heavy cosmic rays (e.g., Fe^{56}). The track can be viewed as a cylinder whose radius is approximately 1μ and within which the dose falls off from the center as r^{-2}. For this work it is the size of the cylinder in relation to the size of electronic devices, as shown in Fig. 2, and the energy deposited therein that is of interest.

Track Length Distribution in 3-Dimensions

Recent increases in the density of planar technology electronic microcircuits have led to devices which are capable of interruption by the passage of a single cosmic ray or other highly ionizing particle. One of the more prominent of these is the dynamic random access memory cell of the MOS type. The interaction of fast heavy ions with MOS memory cells can create writing errors. Two sources have been identified: 1) cosmic rays[6-8] and 2) uranium and/or thorium α decay[9] from ceramic packaging materials adjacent to the memory cell. The underlying requirement for memory cell upset is the deposition of an amount of energy, greater than some threshold energy, which

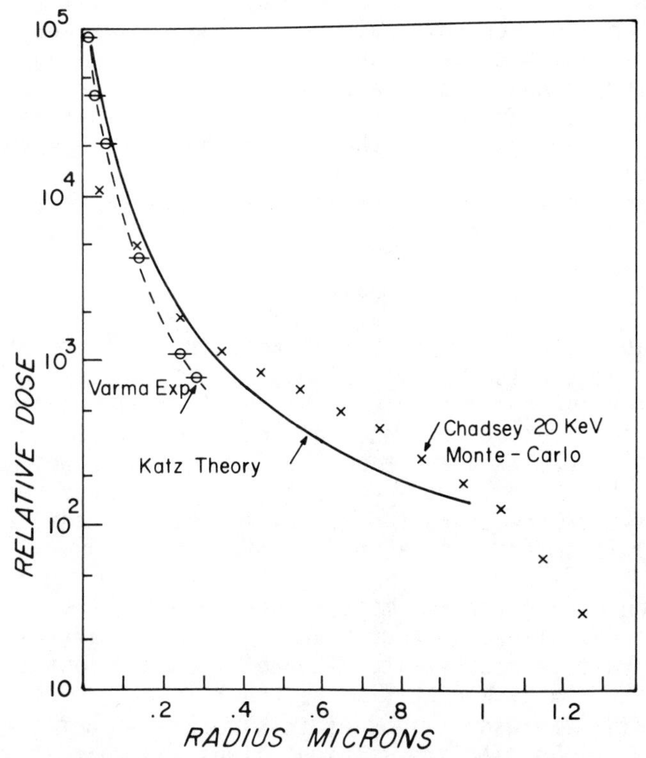

Fig. 1 Comparison of radial distributions for heavy ions, SEM electron beams, theory, and experiment.

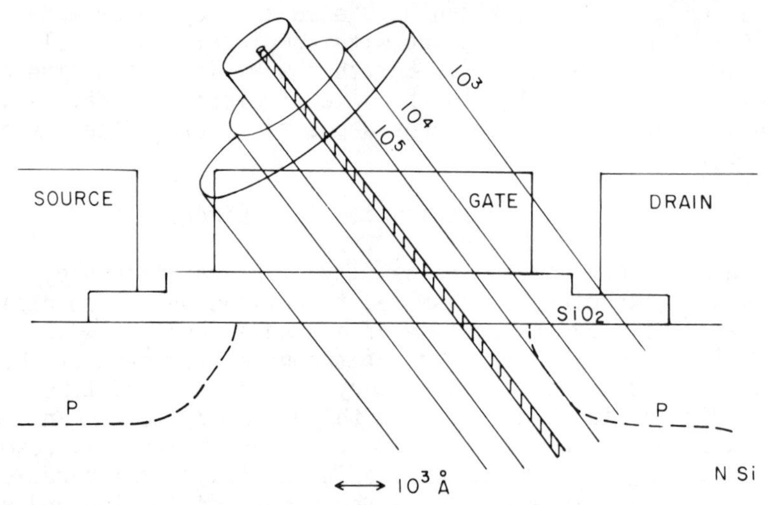

Fig. 2 Heavy ion track in perspective of an MOS device of VLSI size. Concentric circles are dose in rads in silicon.

generates an electron-hole pair flux. The subsequent collection of the charges causes a writing error. The energy must be deposited in the sensitive region of the device, i.e., where the charges are collectable. In general that region is near the gate and source of the MOS device. The possibility of large diffusion lengths enlarges that region. The dose surrounding a cosmic-ray track as it impinges normally on an MOS structure is shown in Fig. 3.

Analysis

The approach to the problem taken here is to treat the collection volume (sensitive volume) of the electronic device as a right rectangular volume. The cosmic rays traverse this volume without deflection, and the tracks are thus straight lines. This permits a solution based on the geometric probability of the chord lengths generated when a convex volume (surface everywhere characterized by a positive radius of curvature) is immersed in a uniform isotropic flux.

An ion moving through matter loses energy at a rate called its LET (linear energy transfer). For small distances this loss rate is a constant. The total energy lost by the ion is the LET x ℓ, where ℓ is the track length in the region of interest. The assumptions of this work are that the heavy ion tracks in question are straight lines and that the volume of interest is a rectangular parallelpiped, $h \times a \times b$, $h < a$

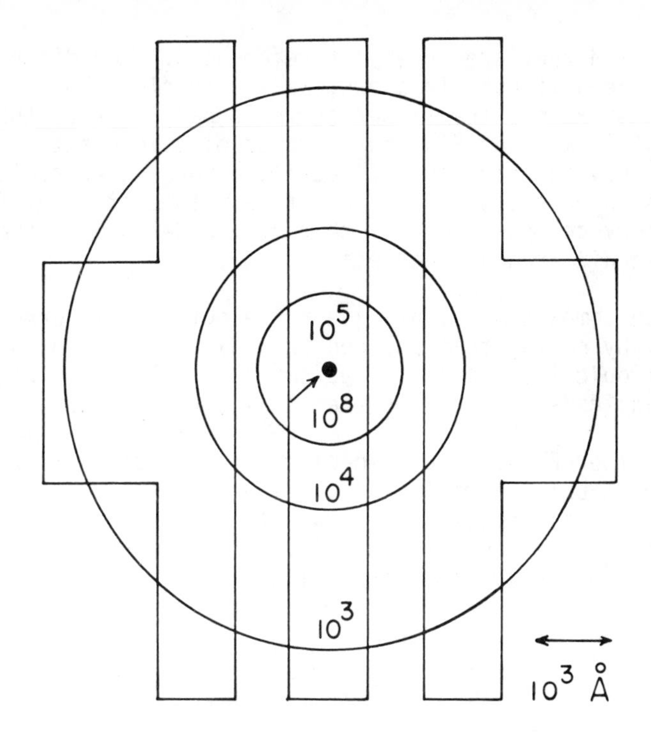

Fig. 3 Dose distribution surrounding a heavy ion track normal to VLSI device.

< b. Actual typical dimensions are $h \sim 2\mu$; $a,b \sim 20\mu$ in large scale integration (LSI) and h, a, $b < 1\mu$ in VLSI.

What is required is the probability that a cosmic-ray track length, within the collection volume, is greater than or equal to the minimum length needed for the electronic response to occur. The probability distribution which answers that requirement is called the sum distribution. The sum distribution for chord lengths in a rectangular volume, $C(\ell)$, was determined exactly from the work of Coleman[10] who solved the problem for the 2-dimensional rectangle and unit cube, of Kellerer[11] who developed the formalism which leads to the 3-dimensional sum distribution for right cylinders of arbitrary cross section, and of Bradford[12] who applied the formalism to a rectangular parallelpiped.

The resulting analytic functions $C(\ell)$ are plotted in Figs. 4 and 5 for several cases. The symmetry of the infinite plane slab and the sphere permits easier evaluation of the integrals, and they are shown for references. The distribution for the unit cube is shown, as is the distribution for a volume h x 4h x 6h. The latter curve is pertinent to the attempts of Pickel and Blandford[8] to predict an event rate for cosmic-ray interruption of satellite borne electronics. The volume in question is the sensitive volume of each of 24 4K RAM memory cells, generally that volume defined by the source-to-drain area and depletion depth.

We can compute an event rate for electronic interrupt using $C(\ell)$ by noting that any convex body immersed in a uniform, isotropic flux, φ particles/cm^2/sec, will experience $(\varphi S/4)$ transits/s, where S is the total surface area.

$\Phi(x)$, the LET spectrum, specifies the particle flux for each value of LET. A minimum track length along which sufficient energy is deposited to cause the electronic response to occur is determined by each LET value. The relationship which leads to that minimum track length is

$$\ell = \frac{\Delta E_{threshold}}{LET \cdot \rho}$$

where ρ is the density and the ΔE threshold is the minimum energy deposition required for electronic response. $C(\ell)$ then gives the fraction of track lengths generated by the

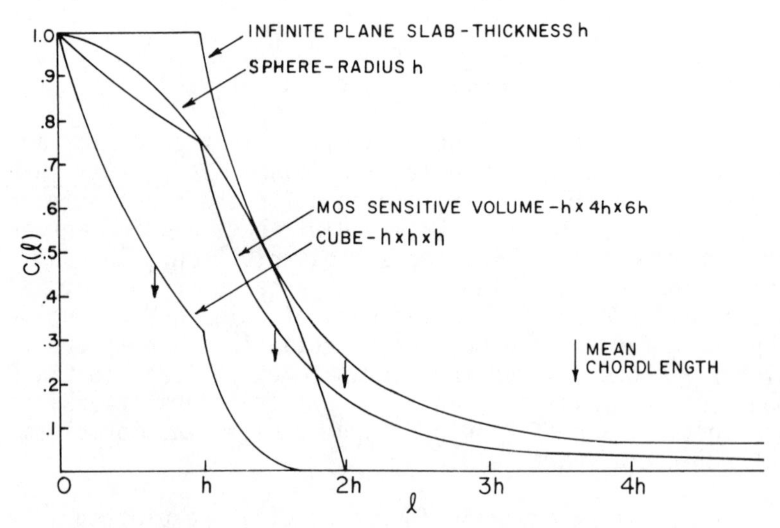

Fig. 4 Sum distribution functions for chord lengths in various geometries.

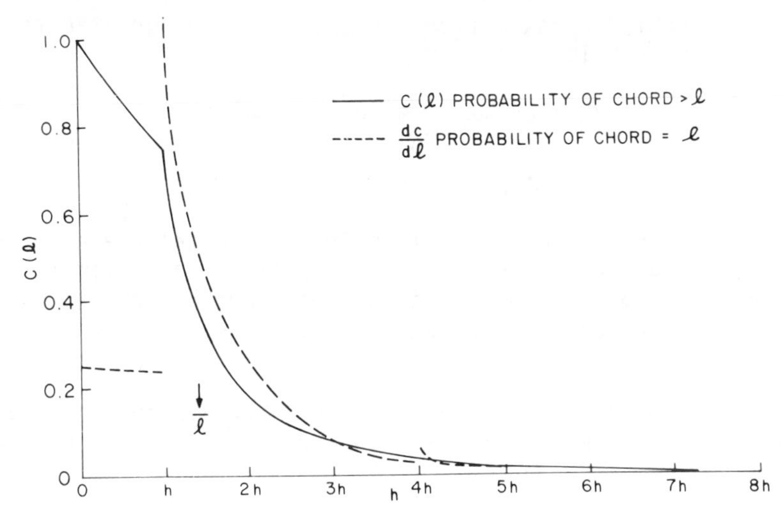

Fig. 5 Differential and sum distributions for track lengths in VLSI device. Volume is of rectangular shape, h x 4h x 6h.

particle flux with that LET value which can cause the electronic response. Integration over the LET spectrum then yields the total event producing combinations of LET and sufficient track length. The event rate thus is given by

$$n = S/4 \int_{x_0}^{x_1} \Phi(x)\ C(\Delta E/\rho x)dx \text{ events per sec per device} \quad (4)$$

assuming 100% efficiency for the process (e.g., charge collection), $\Phi(x)$ is the LET flux in particles/[$cm^2 \cdot s \cdot$(MeV cm^2/g)], x is LET in MeV cm^2/g and $\varphi = \int \Phi(x)dx$

Since $C(\ell)$ is a reasonably complicated function, the event rate integral usually requires machine solution. An approximation can be obtained by noting that the dominant terms in $C(\ell)$ fall off like $1/\ell^r$, $r > 2$ and that $\Phi(x) \sim \Phi_0 (x_0/x)^P \cdot P = 3.8.$[13] Equation (4) is thus reducible to simple form by taking the following definitions for the integrand:

$$\Phi(x)=(x_0)(x_0/x)^P=16\text{x}10^{13}(10^3/x)3.8$$

$$\text{particles/}[m^2 \cdot s \cdot (\text{MeV } cm^2/g)] \quad (5a)$$

with the values for x_0, $\Phi(x_0)$, P taken from Ref. 13. For $C(\ell)$ take

$$C(\ell) = C(h)(h/\ell)^{2.2} = 0.75\ (h\rho x/\Delta E)^{2.2} \quad (5b)$$

as a reasonable approximation for chord lengths greater than h in rectangular volumes.[12]

$$x_0 = \Delta E/\rho \ell_{max} = 10^3 \text{ MeV cm}^2/\text{g} \quad (6a)$$

$$x_1 = \Delta E/\rho \ell_{min} = 7\times10^3 \text{ Mev cm}^2/\text{g} \quad (6b)$$

are the limits in the integral of Eq. (4), where the numbers are for the 4K RAM of Ref. 8. The effect of these limits in determining the portion of two species C and Fe of the cosmic spectrum, which can participate in that case, is shown in Figs. 6 and 7.

The general solution is obtained by retaining the definitions of the limits following integration of Eq. 4 and the event rate, n, is given by

$$n = 7.5\times10^{-7} NS(h\rho/\Delta E)^{2.2}[(x_0)^{-.6}-(x_1)^{-.6}]$$

$$= 7.5\times10^{-7} NS(h\rho/\Delta E)^{2.2}[(\rho\ell_{max}/\Delta E)^{.6}-(\rho\ell_{min}/\Delta E)^{.6}]$$

$$= 7.5\times10^{-7} NS(h\rho/\Delta E)^{2.8}[(\ell_{max}/h)^{.6} -1]$$

$$= 7.5\times10^{-7} NS(h\rho/\Delta E)^{2.8}[(10\Delta E/h\rho)^{.6}-1] \text{ events/}$$

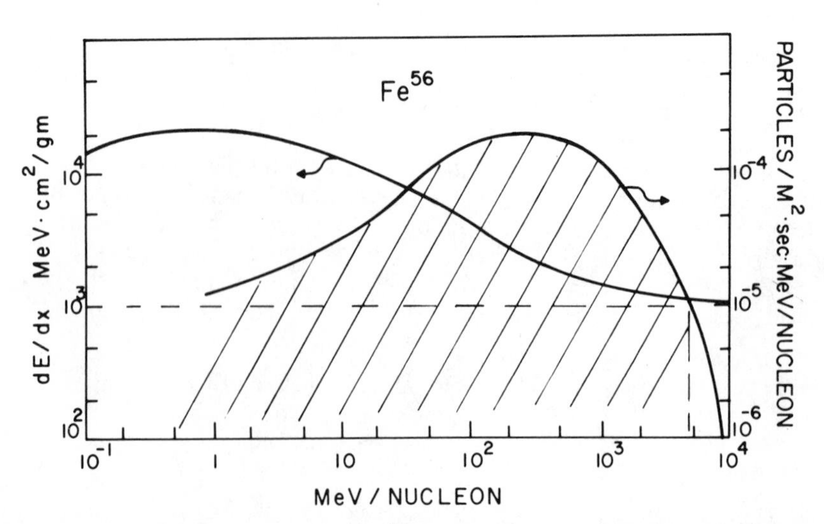

Fig. 6 Cosmic flux for Fe^{56} with LET values for the same energy range. Shaded area shows population which can cause electronic interrupt of Ref. 8.

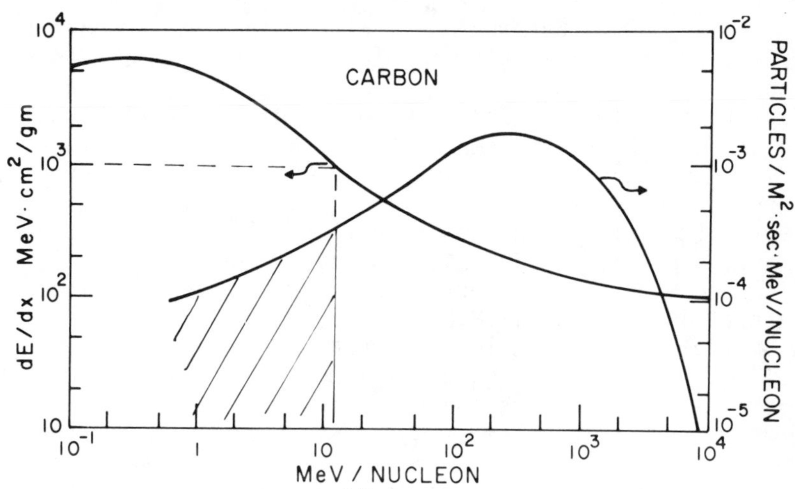

Fig. 7 Cosmic flux for C^{12} with LET values for the same energy range. Shaded area shows population which can cause electronic interrupt of Ref. 8.

Since ℓ_{min} = h in this approximation, ℓ_{max} is the major diagonal of the volume. This result is valid for cosmic ray flux with LET > 10^3 MeV cm^2/g, where

S = surface area of device sensitive volume in μ^2
h = device sensitive volume minimum dimension (presumably depletion depth) in μ
ΔE = switching energy in MeV
N = number of devices
ρ = density in g/cm^3

This result is based on the unshielded LET spectrum of Heinrich[13] for $6 \leq z \leq 26$. To take shielding into account one need only multiply these results by an attenuation factor from Heinrich's paper, $\sim e^{-3.91(t/50)}$, where t is shield thickness in g/cm^2. This factor applies only to the total LET spectrum > 10^3, not to any individual species.

The event rate, Eq. (7), can be used by electronics designers or system analysts to anticipate soft error rates in satellite-borne electronics devices. The basic requirement for such utilization is to identify those memory cells or switches which can be characterized as having a rectangular collection volume and within which a threshold energy deposition, ΔE, can be specified. By way of example, when the numbers from the Pickel and Blandford[8] analysis of the 4K RAMs in satellites are used in Eq. (11), one gets 203 events/year for the 98,309 cells involved. This compares reasonably well with the rate calculated in Ref. 8.

Equation (7) has further merit in that it demonstrates the event rate sensitivity to physical parameters. Clearly emphasized is the necessity to scale the minimum dimension, h (depletion depth), faster than the threshold energy, ΔE, as one designs the higher density VLSI technology. Failure to utilize an analysis which includes the track length distribution can be seen in Ref. 8 where the mean chord length is used instead of the full distribution. This leads to the erroneous conclusion in that paper that the minimum LET rate which can produce the electronic error is 4.6×10^3 MeV cm^2/g, whereas, in fact, it is 9.6×10^2 MeV cm^2/g. Finally, as is intuitively clear, the longer track lengths can come only from ions with larger angles of incidence to one of the planes of the device. This response to angle of incidence, which agrees with the behavior of $C(\ell)$, has been measured experimentally.[14,15] Thus, establishing a threshold energy for the device error sets a limit on the angle of incidence for each LET value.

Permanent Damage (Oxide)

Permanent damage in the oxides of MOS devices occurs when charges are trapped in the oxide. That charge presence causes a shift in the threshold voltage and, with sufficient shift, a disabling of the device. Normally, this effect is of concern to nuclear survivability and is caused by the γ and x-ray fluxes of that environment. Nonetheless, a cosmic ray in a single pass can create enough charge pairs, which are subsequently trapped, to cause threshold shifts. In VLSI the voltage shifts required to disable the device are smaller, and hence one can anticipate that a permanent damage rate will ensue.

Equation (7) can be used to some value in this case. ΔE is set as the energy required to achieve some average dose in the oxide; then

$$\Delta E(\text{MeV}) = 1.6 \times 10^8 D(\text{rads}) V(\text{cm}^3) \tag{8}$$

If the area is shrunk to VLSI dimension with $V = 0.1 \times 1 \times 1 \times 10^{-12}$ cm^3, for 10^4 rads or greater, $\Delta E = 0.16$ MeV and n = 1 events/year/10^5 devices; for 10^3 rads or greater, $\Delta E = .016$ MeV, but $E/\ell_{min}\rho = 606$ MeV cm^2/g, which violates the validity criterion (LET $\geq 10^3$) for Eq. (7). Therefore, one must resort to the formal solution of Eq. (4) over the entire LET range for evaluation. Many times as many ions qualify as a result of this lower LET requirement.

One should keep in mind that, because of the dramatic nonuniformity of the dose distribution in the track, even though the average dose is greater than 10^3 rads, significant regions of the oxide have received > 10^5 rads. The response of devices to such nonuniform irradiation is not well established.

Expected Contributions to LSI/VLSI Soft Error Due to Solar and Cosmic Protons

The interaction of heavy ions with microelectronics devices was examined in the preceding section. The nature of the interaction is such as to produce error signals or switching, and the chief means of avoiding those errors seems to be 1) one very small dimension associated with the sensitive volume of the device, 2) large ΔE for writing or switching, and 3) redundant programming or circuits. In all the foregoing analyses the mode of energy deposition is the LET due to ionizing collisions. The foregoing analyses also recognize a minimum in the LET rate necessary to produce the undesirable effects. That minimum is currently about 10^3 MeV cm^2/g, and the cosmic flux greater than that value has been characterized[13] by a simple expotential [e.g., $\Phi(LET) = \Phi_0(x_0/x)^P \cdot \Phi_0 = 4x10^{-3} m^{-2}$ srd^{-1} sec^{-1} MeV^{-1} cm^{-2} g].

The vast majority of the cosmic-ray population has LET rates less than 10^{-3} MeV cm^2/g, however. Protons, for example, have maximum rates of only ~10^2 MeV cm^2/g at energies of a few MeV. Since the preponderant portion of the proton cosmic flux has LET rates of 2 MeV cm^2/g (480 ev/μ in Si) for a 1μ size device, the threshold energy, (ΔE_{th}), would need to be as small as 500 eV for protons to qualify for error production via ionization loss. Protons can participate even when ΔE_{th} is large, however, via nuclear reactions and by elastic scattering.

Thus, the role protons play must be examined in two views: 1) in trapped belt or solar flare exposures, in spite of the small LET rates, the flux can be large enough to cause permanent oxide damage through accumulated total dose; 2) cosmic and/or trapped belt plus solar flare exposure with protons acting singly to produce soft errors. The permanent oxide damage is not different from that encountered in x-ray fluxes from nuclear weapons and, since studies of that effect are plentiful in literature, it will not be dealt with here.

Nuclear Interactions

The proton-induced nuclear reactions of interest are p, n; p, 2n; p, pn; p, α,; and p, spall. The mechanism by which these reactions can trigger soft errors is the ionization energy loss of the recoil residual nucleus and, in the case of the p, α, and p, spall reactions, through the ionization losses of the reaction products. The total proton inelastic scatter cross section rises from a few hundred millibarns at 4 MeV to 1200 mb at 30 MeV and then declines to ~100 mb over several hundred MeV, averaging 350 mb from 100 to 400 MeV.[16,17] The p, p'; p, n; p, 2n; p, n reactions contain the majority of this cross section. For these reactions to trigger soft error, the recoiling target nucleus must be the agent for ionization energy deposition. Unfortunately, the recoil kinetic energy distribution has not been information of acute interest to nuclear researchers and often is lumped into the residual excitation energy. Thus, the fraction of the inelastic reactions which lead to a recoiling nucleus of sufficient energy to create soft error is not readily available.

Low-energy α particles already have been identified as a cause of soft errors in RAM cells. Thus one is bound to investigate reactions which produce α or larger fragments. The Si (p, α) reactions, as with all light nuclei, have relatively flat cross sections for $E_p \gtrsim 100$ MeV at about 12 mb after peaking at 60 mb at 25 MeV.[17,18] The Q value for He^3 production is 16.9 MeV and for He^4 is 7.7 MeV. At proton energies several times these, the p, α is a direct reaction, e.g., little residual nucleus excitation or recoil and, as a result, the α particles are produced in the forward direction at high energies (E_p - Q). In this energy range particles will have LET rates of the order of 5 keV/μ and, in the dimensions of LSI (20μ), cannot deposit the required energy (~5 MeV). It would require moderation (i.e., through shielding) of the incident proton flux to energies in the tens of MeV range to produce an abundance of α particles with low energies and high LET rates.

Spallation reactions have been measured[19] in Al^{27} and reflect total cross sections of about 6 mb for fragment production (i.e., F^{20}, O^{19}, Be^7). Although the fragments may differ with a Si^{28} target, the strength of the cross section will not change much from Al^{27}. Thus, the combined p, α and p spall cross section (~17 mb for E_p > 100 MeV) leads to the following estimated reaction rates:

$$n \text{ reactions/year} = \phi_{proton/yr \cdot cm^2} x \rho N_A V \qquad (9)$$

In the LSI case analyzed by Pickel and Blandford[8] $V = 1030\mu^3$ and $n = 2.3 \times 10^{-4}$/(year·device) = 23/(year·10^5 device) due to p, α and p spall.

In a solar flare, proton flux levels increase dramatically by factors of as high as 10^5 for periods of the order of 24 hr. In such a flare (10^5, 24-hr duration) the total number of p, α and p spall reactions in a 10^5 RAM would be ~6300. By using the total average inelastic cross section (350 mb), one would see 460 inelastic scatters per year per 10^5 devices for the cosmic-ray flux and 140,000 reactions in a solar flare of 10^5 x cosmic flux lasting for 24 hr. The number of these which lead to soft errors is not easily known and requires the energy spectrum of the reaction products for solution. Although the event rate is small for each VLSI element (since the volume is ~10^{-3} that of LSI), this is generally compensated for by the increased number of VLSI elements per chip.

Elastic Scattering

Consider a proton of kinetic energy 100 MeV incident on a Si nucleus, mass number 28. Assume that a transfer of greater than 100 keV is necessary by elastic collision for soft error to occur. The cross section is obtained from the integrated Mott-Rutherford relation for energy transfer greater than E_D.

$$\sigma_e = (\pi b^2/4\gamma^2)\{(\varepsilon-1) - \beta^2 \ln\varepsilon + \pi\alpha\beta \, [2(\varepsilon^{1/2}-1) - \ln\varepsilon]\} \qquad (10)$$

where

$$\gamma = (1 - \beta^2)^{-1/2} \qquad \beta = v/c$$

$$\varepsilon = E_m/E_D \qquad \alpha = Z_2 e^2/\hbar c$$

$$E_m = \frac{2E(E + 2mc^2)}{(1 + m/M)^2 \, mC^2 + 2E}$$

$$b = 2Z_2 e^2/Mc^2\beta^2$$

The m, v, β, γ, and E are proton parameters, Z_2 and M are for Si^{28}, and E_m is the maximum energy transfer possible. For an energy transfer, E_D, of 100 keV or greater resulting from 100-MeV proton elastic scatter, σ_e equals 208 mb. This is a value much larger than the p, α and p spall cross sections and implies that, for soft error thresholds of 100 keV, elastic scattering will contribute to soft errors. It should be noted that no devices currently available can be triggered by so little energy, but 100 keV is within the range of projection for VLSI.

Summary

1) The interaction of the heavy cosmic rays with satellite-borne microelectronics can be viewed along two lines: Soft errors (transient) and permanent damage. Soft errors can be avoided or reduced in number by several techniques; a) design in accordance with Eq. (7), b) fault tolerant circuitry, c) redundant programming. No consensus on the event rate at which soft errors become a "problem" exists at this time. Permanent damage in general cannot be shielded against but should not become a serious problem until device sizes reach VLSI. Failure levels can only be approximated until an experimental data base can be obtained from actual circuits.

2) Generation of soft errors from cosmic proton primaries by ionization loss along single tracks is not seen as a possibility for LSI/VLSI.

3) Proton generation of soft errors via elastic scattering reactions in Si^{28} become increasingly evident when protons of energy of a few tens of MeV are present and especially when the threshold energy for soft error reaches ~100KeV. Special notice should be taken of the vulnerability during solar flare time.

4) The number of nuclear reactions which will occur in volumes as minute as those characteristic of VLSI is very small. This is compensated by the increased number of elements per chip in VLSI. Thus, upsets caused by p, α; p spall etc., reactions can still be anticipated. If the population of recoiling residual nuclei which result from inelastic scatter and which have energies greater than a few hundred keV is large, then nuclear inelastic scatter can contribute an event rate of concern. (During the publishing time of this article experimental verification of the proton initiated upsets has occurred.[20] The evidence suggests that inelastic scatter can produce events at LSI dimensions as well as at VLSI.)

5) The electronics package vulnerability assessment will require an analysis based on proton inelastic scatter products coupled with determination of the appropriate track distributions. Experimental and theoretical studies of these processes were undertaken in the past[21,22] when semiconductors were much larger. The process needs to be repeated using current and projected LSI/VLSI circuitry to assess susceptibility to solar flare or trapped belt fluxes.

Conclusion

The response of VLSI circuitry to the particle fluxes encountered in Earth satellites can be partially anticipated from experience in nuclear survivability. Specifically, the knowledge of device behavior when exposed to fluxes of weakly ionizing particles which cause charge trapping in the oxides of MOS devices applies here. Soft error rates can be anticipated based on present experience with LSI memories in satellites or exposed to accelerator fluxes in the laboratory and on analyses of the kind presented here. The possibility of surprises exists because of permanent damage by heavy ion cosmic rays and the dramatically nonuniform dose distribution surrounding those ion tracks. Additional experimental work also is needed in the area of proton-nuclear reactions as they pertain to the creation of highly ionizing reaction products in electronics materials.

References

[1]Barkas, W. H., Nuclear Research Emulsions, Vol. 1, Academic Press, New York, 1963, p. 371.

[2]Mott, N., "The Scattering of Fast Electrons by Atomic Nuclei," Proceedings of the Royal Society of London, Vol. 124, 1929,p.425.

[3]Kobetich, E. and Katz, R. "Energy Deposition by Electron and δ Rays", The Physical Review, Vol. 170, No. 2, June 1968.

[4]Hamm, R. N., ORNL, private communication, 17 Oct. 1978.

[5]Varma, M. N., Baum, J. W. and Kuehner, A. V., "Radial Dose, LET, and W for O^{16} Ions in N_2 and Tissue Equivalent Gases," Radiation Research, Vol. 70, Mar. 1977, pp. 511-518.

[6]Binder, D., Smith, E. and Holman, A., "Satellite Anomalies from Galactic Cosmic Rays," IEEE Transactions on Nuclear Science, Vol. NS-22, Dec. 1975, pp. 2675-2680.

[7]Bradford, J. N., "Heavy Ion Radiation Effects in VLSI," Air Force Tech. Rept., RADC-TR-78-109, May 1978.

[8]Pickel, J. C. and Blandford, J. T., "Cosmic Ray Induced Errors in MOS Memory Cells," IEEE Transactions on Nuclear Science, Vol. NS-25, Dec. 1978, p. 1166.

[9]May, T. C. and Woods, M., Electronic Design, Vol. 11, June 1978, p. 37.

[10]Coleman, R., "Random Paths Through Convex Bodies," Journal of Applied Probability, Vol. 6, Dec. 1969, pp. 430-441.

[11]Kellerer, A. M., "Considerations on the Random Traversal of Convex Bodies and Solutions for General Cylinders," Radiation Research, Vol. 47, Feb. 1971, pp. 359-376.

[12]Bradford, J. N., "A Distribution Function for Ion Track Lengths in Rectangular Volumes," Journal of Applied Physics, Vol. 50, June, 1979, pp. 3799-3801.

[13]Heinrich, W., "Calculation of LET-Spectra of Heavy Cosmic Ray Nuclei at Various Absorber Depths," Radiation Effects, Vol. 34, Mar. 1977, pp. 143-148.

[14]May, T. C. and Woods, M., "Alpha-Particle-Induced Soft Errors in Dynamic Memories," IEEE Transactions on Electron Devices, Vol. ED-26, Jan. 1979, p. 8.

[15]Yaney, D. S., Nelson, J. T., and Vanskike, L., "Alpha-Particle Tracks in Silicon and their Effect on Dynamic MOS RAM Reliability," IEEE Transactions on Electron Devices, Vol. ED-26, Jan. 1979, p. 3.

[16]Miller, J. M. and Hudis, J., "High-Energy Nuclear Reactions," Annual Review of Nuclear Science, Vol. 9, 1959, p. 172.

[17]Barbier, M., Induced Radioactivity, Wiley, New York, 1969, p. 175.

[18]Preston, M., Physics of the Nucleus, Addison-Wesley, Reading, Mass., 1962, p. 593.

[19]Cline, J. and Nieschmidt, E., "Measurements of Spallation Cross Sections for 590 MeV Protons on Thin Targets of Copper, Nickel, Iron and Aluminum," Nuclear Physics, Vol. A169, 1971, p. 443.

[20]McNulty, P. J., Farrell, G. E., Wyatt, R. C., Filz, R. C. and Rothwell, P. L., "Soft Memory Errors in LSI Devices," IEEE Annual Conference on Nuclear and Space Radiation Effects, University of California at Santa Cruz, Santa Cruz, Calif., July 1979, pp. 215-216.

[21]Arnold, D. M., Baicker, J. A., Flicker, H., Gondolfo, D. A., Parker, J. R., Vilms, J., and Vollmer, J., Proton Damage in Semiconductor Devices, Final Rept., Contract NASI-1654, RCA Applied Research Division, Camden, N.J., Nov. 1, 1962.

[22]Simon, G. W., Denny, J. M., and Downing, R. G., "Energy Dependence of Proton Damage in Silicon," The Physical Review, Vol. 129, Mar. 1963, pp. 2454-2459.

RADIATION EFFECTS ON SOLAR CELLS

W. Patrick Rahilly*
USAF Aero Propulsion Laboratory,
Wright-Patterson Air Force Base, Ohio

Abstract

The predominant survivability problem for a photovoltaic power system in Earth orbit application is the effect of the trapped particle radiation environment. High energy electrons and protons will degrade the solar cells to a point where the satellite mission must be terminated. Several innovations have been developed to reduce the radiation effects. The change from P on N to N on P silicon and the addition of glass covers made a considerable impact on the life of solar cell space power systems in the early 1960's. Recent research has led to technological advancements that address the radiation damage problem using modifications of cell geometry and semiconductor material properties. Improvements in material purity and fabricability of thin and vertical junction structures has demonstrated considerable improvement in silicon cell performance in a radiation environment. Gallium arsenide technology for application to solar cells is rapidly evolving with high conversion efficiencies and radiation resistance demonstrated. Along with the advancement of the gallium arsenide is the advancement of the technology to fabricate multiple bandgap cascaded solar cell structures which promise conversion efficiencies nearly twice those for silicon or gallium arsenide solar cells with attendant high radiation resistance.

Introduction

Photovoltaic-battery power system technology has been employed in space for over twenty years. The presence of high energy protons and electrons trapped by the Earth's magnetic field has posed continual problems in the maintenance of conversion efficiency of the solar cells during mission life. As a result, near-Earth applications have required costly solar

---

*Physicist, Air Force Wright Aeronautical Laboratories.

cell array overdesign to ensure availability of power at the end of the mission. Although degradation of photovoltaic output in space continues to be a problem, recent advances in solar cell technology show considerable promise in reducing the deleterious effects of proton and electron radiation damage and in increasing conversion efficiency. This paper will discuss 1) the processes by which high-energy particles damage solar cells, 2) the effects on actual solar cells (using the silicon solar cell as an example) and 3) various techniques for improving solar cell lifetime and efficiency.

Particle Radiation Damage Effects on Solar Cells

Solar cells used for space applications are referred to as minority charge carrier devices. The performance of solar cells is very strongly tied to the lifetime of the minority carriers. As high-energy particles pass into the cells, damage centers are formed through disruptions of the semiconductor lattice. The number of damage centers generated grows with time in orbit and causes the minority carrier lifetime to decrease, leading to a corresponding loss in solar array output power. These effects will be discussed in more detail in this section and semi-analytic models of the various processes will be provided.

Depending on the orbit parameters (altitude, inclination, and eccentricity), the spacecraft will experience bombardment by either high energy protons or high-energy electrons or a combination of both.[1-3] For example, the worst-case orbit is one which is circular and lies in the Earth's equatorial plane with an altitude of approximately 2300 n. mi. above the Earth's surface. In this orbit, the spacecraft is constantly in the highest energy and flux region of the proton belt. As the orbit inclination is changed away from the equatorial plane, both the energy and flux of protons decrease. At geosynchronous orbit (in the equatorial plane and $\sim$18,360 n. mi. above Earth's surface) the particle radiation environment is relatively benign. In this orbit, the proton flux is essentially negligible, and the electron flux is low, but not negligible. In the worst-case orbit, a typical photovoltaic system would be useful for a few weeks. At synchronous altitude, the same power system would last several years. Synchronous altitude is not without its problems however. The sun's 11-yr cycle of solar storm activity influences power system design.[4,5] During periods when the sun is most active, solar proton flare events occur frequently, and designs must provide for the extra losses that will be induced by these protons.

Protons and electrons create different types of damage in semiconductor material.[6-10] The proton interacts more strongly with the semiconductor lattice because it is considerably more massive. As the proton enters the material, it dissipates kinetic energy through coulombic interactions with the orbital electrons of the lattice atoms. As the proton loses energy, it begins to undergo ever increasing momentum exchanges with the host atoms (per unit path length of the proton). At some point the proton will transfer a large amount of energy to a lattice atom, causing the atom to be violently dislodged from its position. This atom is called the primary knock-on, and it can, in turn, cause secondary knock-ons. The net result is a cluster of interstitial lattice atoms and an equivalent number of vacancies at the end of the proton track. The location of the heaviest damage caused by protons in the semiconductor material depends strongly on the initial proton energies. In contrast to the protons, the electrons entering the lattice dissipate most of their energy via inelastic interactions with orbital electrons, since the entering electrons and the orbital electrons are essentially of the same mass. Occasionally, sufficiently large momentum transfers occur to displace a host atom. The electron-displaced atoms, for the most part, have much lower energy transferred to them and, as a result, the primary knock-ons do not always cause secondary knock-ons. Electron damage to the lattice is characterized by point defects which are considerably more uniform (per unit volume) throughout the lattice than are proton damage effects. Fig 1 shows the ranges of protons and electrons in silicon as a function of particle energy.[11-12] The proton ranges are nearly two orders of magnitude less than those of the electron.

The presence of these damage centers alter the minority carrier lifetime in the semiconductor material. The rate at which the minority carriers are recombined with majority carriers via the damage centers is given by

$$\frac{1}{\tau} = \frac{1}{\tau_0} + \langle v \rangle \left(\Sigma \sigma_i N_i\right)_e + \langle v \rangle \left(\Sigma \sigma_k N_k\right)_p \tag{1}$$

where τ_0 is the minority carrier lifetime prior to radiation damage, $\langle v \rangle$ is the average thermal velocity of the minority carriers, and the summations subscripted with "e" and "p" represent the additive effects on the recombination rate by the various types of recombination centers caused by electron and proton radiation. The σ and N factors within the summations correspond to each of the recombination cross sections and densities of the various types of centers created.

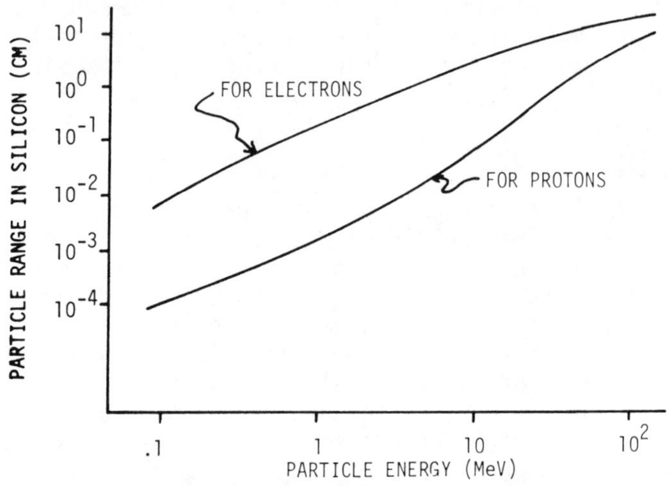

Fig. 1 Particle range vs particle energy for electrons and protons passing through silicon.

Equation (1) can be simplified by the relations

$$K_{\tau e}\Phi_e = \langle v \rangle \, (\Sigma\sigma_i N_i)_e \text{ and } K_{\tau p}\Phi_p = \langle v \rangle \, (\Sigma\sigma_k N_k)_p$$

which connect cross section α and damage center concentration N with radiation fluence Φ and a parameter K_τ defined as the carrier lifetime damage coefficient. Inserting the $K_\tau\Phi$'s into Eq. (1) yields

$$\frac{1}{\tau} = \frac{1}{\tau_0} + K_{\tau_e}\Phi_e + K_{\tau_p}\Phi_p \tag{2}$$

where the subscripts "e" and "p" have the same meaning as in Eq. (1). The value of $K\tau_e$ is strongly dependent on semiconductor type (silicon, gallium arsenide, etc) quality, and resistivity. Parameter $K_{\tau p}$ is primarily dependent on material type because of the way protons interact with the lattice atoms; it becomes more dependent on resistivity when the electrical dopant concentration in the material becomes large. Eq. (2) shows that as the particle fluences increase, the value for lifetime decreases. This decrease in lifetime reduces the average distance minority carriers can diffuse in their journey to the photovoltaic P/N junction. Fig. 2 illustrates the effect of electron damage on the solar cell current voltage characteristic.[13] The primary loss in this case is in the current. A similar figure can be shown in the case of high-energy (>10 MeV) proton damage where the cell is not protected by a coverglass. If the proton energies are low (.5-2 MeV), much damage is created near the P/N junction. This damage location affects both the current and, more seriously, the voltage of the solar cell, as indicated in Fig. 3.[14] The damage produced near the P/N junction causes shunting currents (leaky diode) which reduce the ability of the solar cell P/N junction to separate charge.

A simplified electrical model of the solar cell is given in Fig. 4. The current-voltage relationship derivable from Fig. 4 is

$$I = I_g - I_{do}\left(e^{(V+IR_s)/V_T} - 1\right) - I_{so}\left(e^{(V+IR_s)/nV_T} - 1\right) \quad (3)$$

where I_g is the light generated current, I_{do} is the reverse saturation current, I_{so} is a saturation leakage current induced by proton damage near the P/N junction, and n is a parameter associated with the quality of the diode characteristic. Eq. (3) accounts for the voltage loss effects due to low energy protons through the parameters I_{so} and n. For electron damage, I_g and I_{do} are affected, and I_{so} is negligible. For proton damage I_g and I_{do} are affected, but if the proton damage is very near the P/N junction, then I_{so} dominates changes in I_{do}. Proton-induced damage near the P/N junction does not strongly alter the value of I_g since it is dependent on minority carriers that are generated deep in the solar cell where the lifetime has not been affected.

In order to limit damage from protons, solar cells often have a clear coverglass applied which prevents the lower

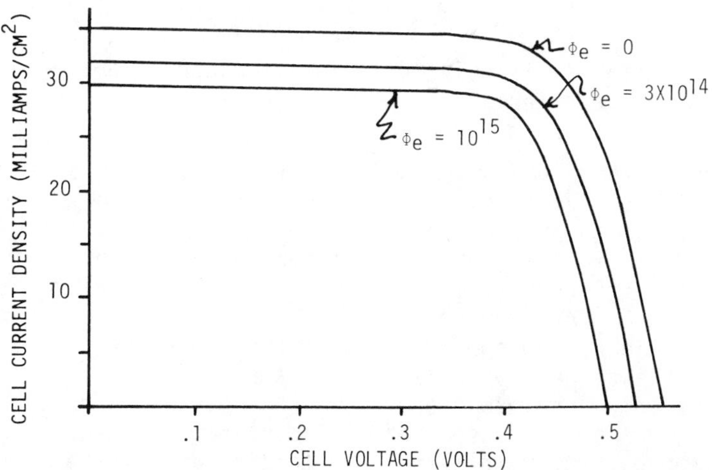

Fig. 2 Effect of 1 MeV electrons on typical silicon solar cells.

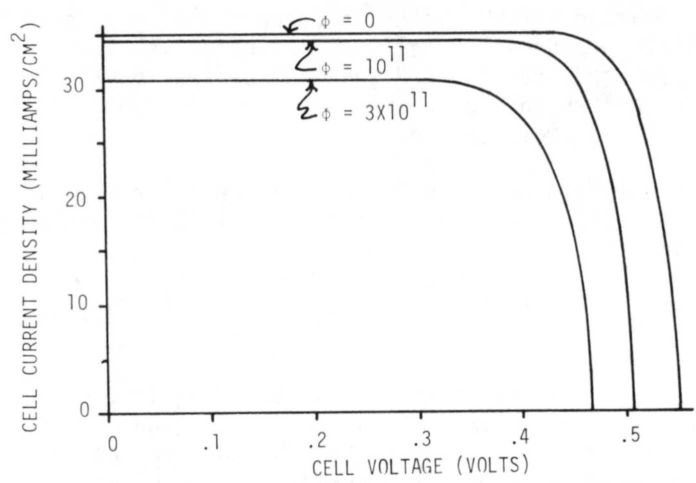

Fig. 3 Effect of .3 MeV protons on a typical silicon solar cell with no cover glass.

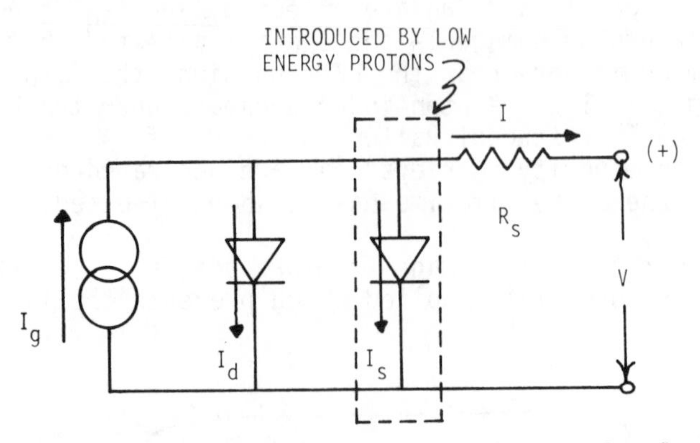

Fig. 4 Solar cell equivalent circuit diagram for low energy proton damage effect.

energy protons from entering the solar cell. The data in Fig. 1 are indicative of electron and proton ranges in coverslide material commonly used for this purpose. Most missions avoid the high-proton flux orbits in order to avoid the requirement for thick coverglasses. Even with coverglass applied, however, the higher-energy protons will still penetrate into the solar cells as low-energy protons since they lose most of their initial energy while passing through the glass. These protons entering into the cell top do not, however, cause severe problems because their incident flux is low.[15] For electrons, the coverglasses have considerably less effectiveness. Little (although measurable) gain is achieved by using coverglasses for a space environment which is predominantly composed of electrons unless the coverglasses are very thick (>20 mils). Those missions at higher orbits, such as geosynchronous altitude, employ glass covers primarily to prevent problems caused by solar proton flares. Particle radiation damage in the coverglass results in the formation of visible color centers (coverglass darkening). This causes a decrease in the amount of light reaching the solar cell for conversion to electrical power. Thus, coverglass material must be given consideration in any space photovoltaic power system design. A popular coverglass material is Cerium-doped microsheet. This particular material does not darken in the visible portion of its transmission characteristic (as do other materials) and is cost effective. Fused silica also is quite useful as a coverglass, but it is expensive.

Particle radiation entering the cells from the backside also must be given consideration in array design. This is true primarily for the electrons, as since essentially, all the protons are stopped by the array substrate materials. The success with which electrons are reduced or stopped from reaching the solar cells from the backside is very dependent on the materials used to form the array substrate.

Influence of Semiconductor Materials and Solar Cell Design

The workhorse for space power has been the silicon solar cell. Recent advances have shown considerable improvements in silicon solar cell efficiency and radiation resistance. Further, recent improvements in the fabrication of solar cells using III-V semiconductor materials show considerable promise for very high efficiency with attendant resistance to radiation damage effects. Some of these improvements are described here.

Silicon Solar Cells

The most significant improvements in increasing the useful life of silicon solar cells came in the early 1960's with the application of glass covers and the change from N-type silicon to P-type silicon as the base material.[16] The beneficial effects of adding coverglasses have been discussed. The change to P-type silicon was a result of the discovery that it was more resistant to electron damage than N-type. Briefly, minority carrier lifetime in N-type silicon of a given resistivity is more sensitive to electorn damage than the equivalent resistivity P-type silicon because the minority carrier recombination cross sections are much larger for holes in N-type than for electrons in P-type. Another factor of importance is that in N-type silicon, most of the types of damage centers induced by electron radiation are minority carrier recombination centers. In P-type silicon, a few of the species of centers are associated with recombination and the rest are traps which do not affect carrier lifetime.[17,18]

P-type silicon material has been shown to have decreasing resistance to radiation damage effects for decreasing resistivity (increasing electrical dopant concentration). No clearly definitive experiments have been performed showing why this is observed. It is known that the electrical dopant itself forms recombination sites through interaction with radiation-generated vacancies and that divacancies are also present. Plausible explanations are that the undesirable impurities are already present in the polycrystalline silicon prior to ingot growth or that they are introduced into the silicon during ingot growth (carbon, oxygen, etc). Czochralski (Cz) silicon has been found to be less radiation resistant when compared to silicon ingots obtained from the floating zone (FZ) growth technique. In the case of Cz material, oxygen and carbon are known to be present in amounts of 10^{16} to 10^{17} atoms/cm^3. In FZ material, these constituents are typically near or below 5×10^{15} atoms/cm^3. The Cz process cannot avoid oxygen contamination since the most cost effective container for the silicon melt is quartz. The cleanliness of the FZ process, to a large extent, depends on the purity of the doping gas (usually diborane gas) and the purity of the seed crystal.

Past and present silicon solar cell developments have repeatedly shown that the best silicon material to use is Cz with the 10 ohm-cm resistivity. FZ material, although more radiation resistant than Cz silicon, is roughly four to five times as expensive; the extra cost does not justify the small gain in radiation resistance. However, recent work in silicon material improvement may change these factors.[19] Research is underway on ultra-pure material with the idea that, if the impurities or undesirable constitutents are not present prior to cell fabrication (and care is taken not to introduce any during cell fabrication), the resulting cells should show more resistance to radiation damage. Further, the cells should be of high efficiency. If these anticipated gains occur, FZ material (or specially processed Cz) will be cost effective in terms of silicon solar cell performance.

Tied to the research involving high-purity silicon is research directed toward understanding the role of the electrical dopants. Investigations have indicated that gallium is superior to boron as a P-type dopant. The experiments explored the performance of solar cells fabricated from gallium, boron, and aluminum-doped FZ silicon material in relation to the phenomenon of photon degradation.[20-22] After being irradiated with electrons to some specified fluence and subsequently exposed to simulated sunlight for approximately 5 hr, boron-doped solar cells degraded in performance beyond that caused by electron damage. It also was observed that the amount of photon-induced degradation increased with increased electrical dopant concentration (lower resistivity) and was sensitive to the method of ingot growth (FZ or Cz). It was further determined that electrical dopant species was important. Data shown in Fig. 5 summarize the photon degradation effect for FZ material for the boron, aluminum, and gallium dopants relative to dopant concentration. The data clearly show that gallium is superior to boron and aluminum. The photon degradation phenomenon effects the light-generated current which is strongly dependent on minority carrier lifetime. The photon effect is not clearly understood, but it is believed to be associated with recombination-enhanced defect reactions which further reduce the minority carrier diffusion length. The second area of study was a comparison of high-purity FZ gallium-doped silicon solar cells with high-purity FZ boron-doped and conventional Cz boron-doped silicon solar cells irradiated with 1 MeV electrons. The 2 ohm-cm FZ gallium-doped cells show the same percentage drop in maximum power as do the 50 ohm-cm Cz boron-doped cells. The gallium-doped cells show excellent conversion efficiencies.

The research on material purity and electrical dopant effects will continue, with the objective of providing high end-of-life silicon solar cell efficiencies. Should it be

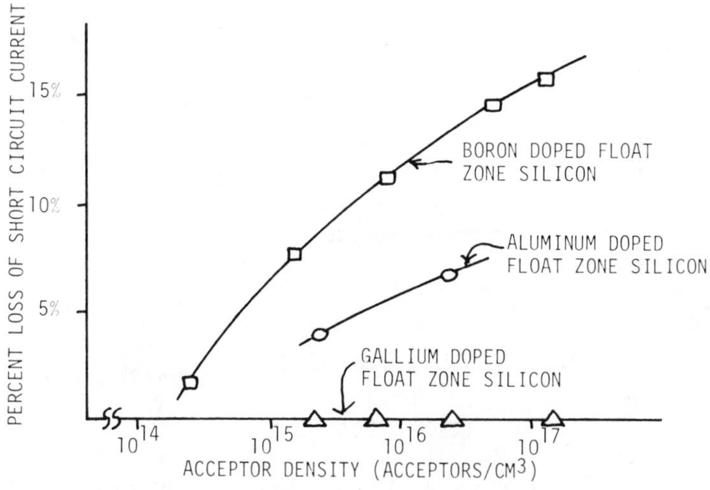

Fig. 5 Photon degradation of short circuit current.

clearly established that gallium is a superior electrical dopant , then a switch to that dopant will be emphasized.

Probably the most significant impact on resistance to radiation damage for the silicon cell is to modify the light-generated carrier collection volume geometry. It has been known for years that if silicon cells could be made sufficiently thin from high-quality material and the P/N junction could be made shallow, electron radiation damage effects would be reduced, since changes in the red portion of the spectral response would be reduced and more blue response would be obtained. For explanation, Eq. (2) is used to develop an expression relating minority carrier diffusion length L to radiation fluence (electrons only). The definition of L is given by

$$L^2 = [\mu kT/q]\tau = D\tau \tag{4}$$

where k is Boltzman's constant, T is absolute temperature, q is the charge of an electron, μ is the mobility of the minority carriers in the material, and τ is given by Eq. (2). The parameter D is defined as the minority carrier diffusion coefficient. Eq. (2) is inserted into Eq. (4) and rearranged to get

$$L = L_o / \sqrt{1 + KL_o^2\Phi} \tag{5}$$

where $L_0^2 = D\tau_0$ and $K = K\tau/D$. Next we plot L versus Φ for various L_0 and two values of K. This is shown in Fig. 6. We assume that the value of L is an indicator of the depth from which light-generated minority carriers can, on the average,

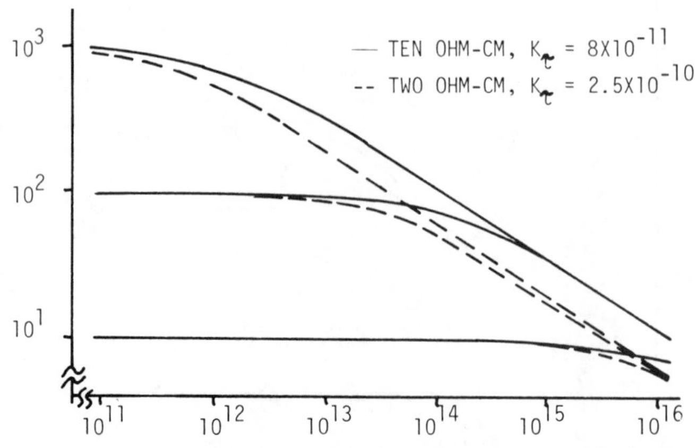

Fig. 6 Minority carrier diffusion length sensitivity to 1 MeV electron damage.

reach the P/N junction. The probability of carriers reaching the P/N junction over distance X is exp(-X/L). For a planar cell of thickness W<<L (see Fig. 7), we see that the generated carriers have a high probability of reaching the P/N junction until L has been sufficiently degraded to the point where W/L becomes large. The smaller W is, the larger Φ must be in order to affect the collection of the light-generated carriers. It must be pointed out, however, that as W decreases, the initial light-generated current decreases for indirect semiconductors such as silicon. The tradeoff for silicon is about 2 mils thickness when power at end of mission life and thermal characteristics are considered. The gain in blue response obtained from forming a shallow P/N junction is insensitive to the radiation damage since the carriers generated by the shorter wavelength light are very close to the P/N junction.

Another approach to improve radiation resistance is to modify the collection volume geometry as depicted in Fig. 6.[23-29] This structure is the vertical junction solar cell which employs orientational dependent etch processing to obtain the geometry. The majority of light-generated carriers are created very near a P/N junction. The fluence must be quite large in order to have an effect on the light-generated current. An advantage of the vertical junction geometry over the planar cell geometry of Fig. 7 is that the initial light-generated current can be as high as that for a thick planar cell, which means that excellent radiation resistance is possible without sacrificing conversion efficiency.

To illustrate the effectiveness of the geometry modifications on radiation resistance, data for a conventional cell,

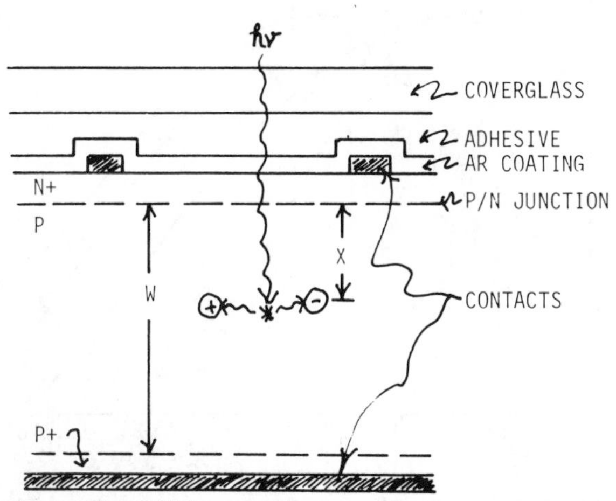

Fig. 7 Silicon solar cell cross section.

thin cell (50 μm thick) and vertical junction cell are shown in Fig. 9. The data we normalized maximum load power vs 1 MeV electron fluence. The cell data shown are for nominal 2 ohm-cm silicon material.

We see in Fig. 6 that gains are possible if the material is radiation resistant (lower values of K). The value of K is dependent on the concentration of both desirable and undesirable impurities. Significant gains in solar cell performance in a radiation environment are possible if the effects of purity, electrical dopant type, and cell geometry can be brought together into one cell type.

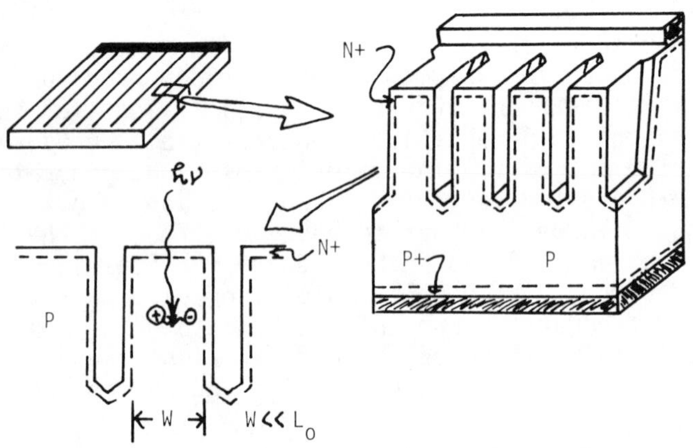

Fig. 8 Vertical junction solar cell.

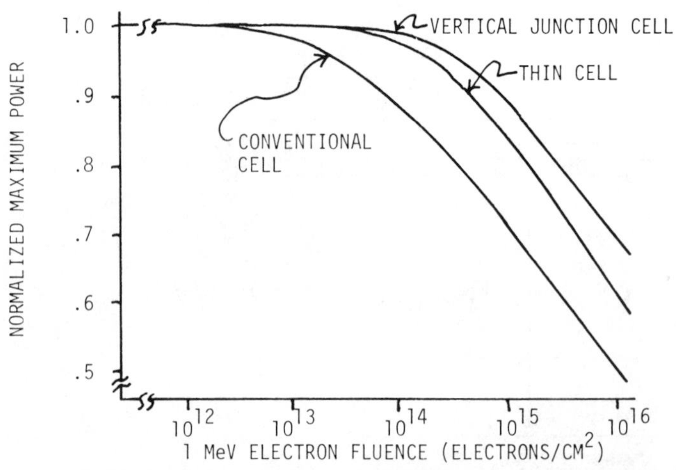

Fig. 9 Solar cell geometry effects vs electron damage.

Improvement on conversion efficiency can be made for either the planar or vertical junction geometries using the back surface field (BSF) effect.[30-33] The BSF is a highly doped P^+ region in front of the back contact, as shown in Figs. 7 and 8. The high-field region reflects minority carriers away from the back-contact, allowing the carriers an improved chance of reaching the P/N junction. The BSF can raise the generated current by more than 5% and raise the open circuit voltage by 5-10%. Another area of improvement on conversion efficiency is to provide a back contact that reflects (back surface reflector or BSR) the longer wavelength light upward toward the cell top, which in effect doubles the cell thickness.[33] The BSF and BSR approaches have more effect in the thin planar cell than on the vertical junction solar cell. To aid in the improvement of the light-generated current for the planar cells, a fabrication step called texturizing can be used.[33-36] The top surface is selectively etched (prior to N^+ diffusion) to achieve a surface made up of millions of pyramidal structures. This geometry modification significantly improves the absorption of light, thereby increasing the light-generated current. Increases in generated current of more than 20% are typical using the texturing step. However, the texturing does have one serious drawback. Since the light absorption is increased, the non-useful long wavelength light also is absorbed, resulting in an elevated operational temperature of these cells compared to nontextured cells operating under identical conditions. The absorptivity of textured cells ranges from .9 to .94 compared to .78 to .82 for planar cells and .83 to .86 for vertical junction cells.[33,37] Another method of raising the light-generated current is through the use of multi-layer antireflective coatings. This type of coating improves the current by better matching the change in the index of refraction between space and the silicon. The advantage of such coatings is low absorptivity with 8-12% increase in current.

Gallium Arsenide Solar Cells

Recent advances in gallium arsenide material growth technology have led to the development of large area (4 cm^2) AlGaAs/GaAs heteroface solar cells with very impressive space sunlight conversion efficiencies.[38-40] While silicon cells are typically in the 14-16% range, the GaAs cells are typically in the range of 16-18%.

The most dramatic improvement has come through reproducible and reliable liquid phase epitaxial growth of an aluminum-gallium arsenide window layer on gallium arsenide. Fig. 10 shows a cross-sectional view of the AlGaAs/GaAs

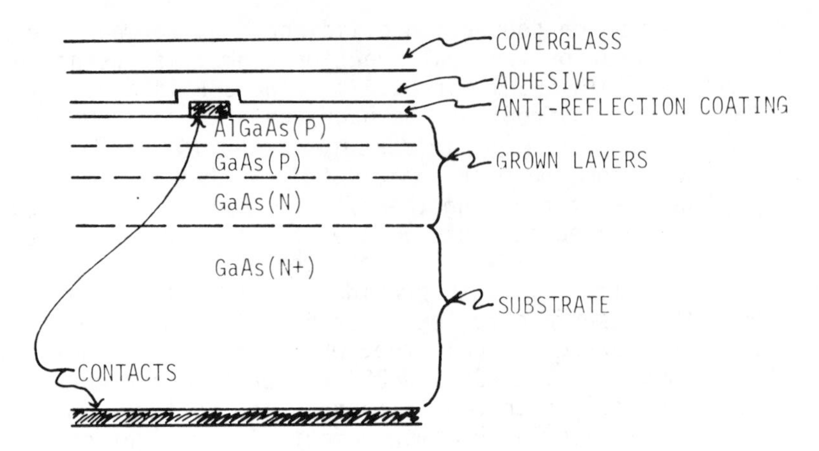

Fig. 10 AlGaAs/GaAs heteroface solar cell.

heteroface cell. The AlGaAs layer effectively shields the minority carriers from recombination at the top surface, resulting in increased light-generated current and significantly improved voltage performance as compared to GaAs cells without the window layer. The P and N layers are quite thin because of the direct bandgap nature of GaAs.

Not only are the efficiencies of these devices superior to present silicon cells, but with proper top layer designs, AlGaAs/GaAs cells have demonstrated very attractive radiation resistance characteristics. However, the radiation resistance is very sensitive to the P-GaAs layer thickness. If the P layer is more than .5μm thick, the cells become quite sensitive to 1 MeV electron damage.[39] Also, as the P layer is thinned, the sheet resistance increases thus reducing the conversion efficiency. A tradeoff between radiation resistance and electrical losses is necessary to obtain the best end of life efficiency possible within present technology.

The most important research area to address for the AlGaAs/GaAs cell is the nature and behavior of the damage centers produced by particle radiation. As for silicon, it is important to improve the semiconductor purity, thereby affecting (lowering) the value of K in Eq. (5). If K can be lowered, then the P-AlGaAs and P-GaAs layers can be made thicker, and the efficiency of these cells can exceed 20% under space sunlight conditions at 25°C. It may be found that the present purity of the material is adequate and the fabrication process is the problem area. A growing body of information is available on cells made using the liquid phase epitaxy material, but little is known about the chemical vapor deposition process in relation to radiation damage. These remarks are made with the idea that there is a potential for the activation of process-induced defects during particle irradiation. It is also necessary to keep in mind that GaAs is more complex than silicon since the number of possible defects (both process-induced and particle damage) that can form in GaAs is higher. There can be Ga vacancies and interstitials, As vacancies and interstitials, Ga divacancies, As divacancies, Ga-As divacancies, etc.

In spite of the myraid of possible types of damage centers in GaAs, the AlGaAs/GaAs cells do show promising radiation resistance. Fig. 11 illustrates the change in maximum power output vs 1 MeV electron fluence for the GaAs cell and those cells of Fig. 9. The attractiveness of the GaAs cell is clear in terms of the actual power output at each fluence. At 1 x 10^{15} electrons/cm^2 (1MeV), the GaAs device has nearly 40% higher power than the conventional silicon cells. This result has considerable impact on solar array sizing (area reduction) for meeting required end-of-life power. Higher power output at a given fluence can avoid an entire satellite redesign, resulting in a savings of many millions of dollars. Approaches to further improve the efficiency and radiation resistance are limited to material improvement and/or process improvement.

The GaAs cell has not been fabricated in a production environment and its "flight qualification" has not yet been

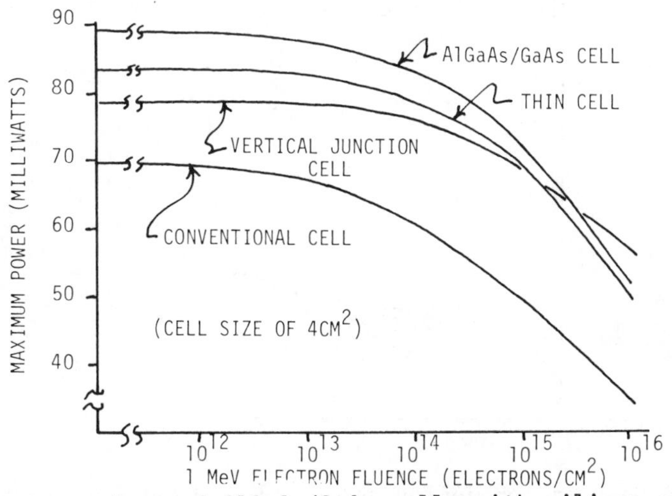

Fig. 11 Comparison of AlGaAs/GaAs cells with silicon cells of various geometries after electron irradiation.

demonstrated. These areas of concern are now being addressed with the anticipation that the technology for processing large numbers of flight-qualified GaAs cells will be available in 1982. A particular problem area is the cost of these cells. In 1978 dollars, these cells are projected to cost $20-30 apiece for a 4 cm^2 area. The projection is based on the assumption that processing and substrate cost breakthroughs will occur. With improved efficiency and radiation resistance design, however, the GaAs cell can provide lower cost arrays for end-of-life design since fewer cells will be needed and power array substrate cost reduction will result as the array areas are reduced. Research and development of large area, low-cost processing along with efforts to reduce GaAs substrate costs are underway, and significant cost reductions are anticipated by the early 1980's.

Near Term and Far Term Research and Development Emphasis

For the near term (1979 - 1983), space power system research and development programs will emphasize improvement in silicon solar cells to achieve radiation resistance and efficiencies of production cells of 16-18%. At the same time, emphasis will be placed on demonstration of flight-qualified GaAs cells in a production environment. Included in this effort will be the assembly of a flight-qualified GaAs solar cell array panel. Effort will be directed at reduced costs and improved efficiencies.

For the longer term, research and development emphasis will be placed on a derivative of the GaAs solar cell, namely the cascaded multiple bandgap solar cell.[41-43] This type of

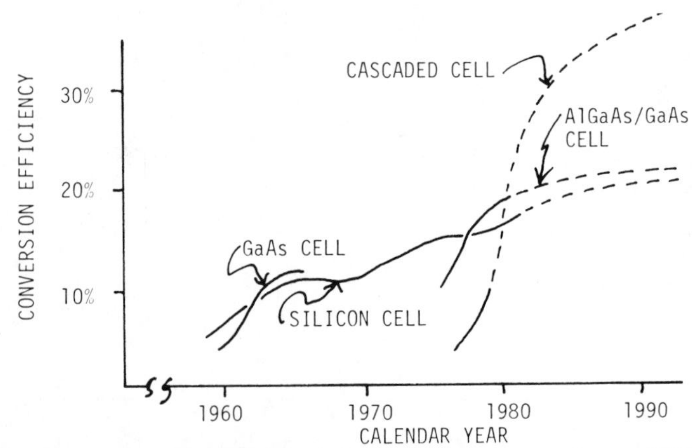

Fig. 12 Past and projected solar cell conversion efficiencies in air mass zero sunlight of unity concentration and at 25°C.

cell will be made up of binary, ternary, and possible quarternary III-V semiconductor combinations assembled together in a monolithic electrical series stack of two or more solar cells. The purpose of this type of device is to achieve 25% or greater conversion efficiency by utilizing more of the sun's spectrum. The series connections provide for higher-voltage cells. These cell types are potentially consistent with the trend toward higher-power and higher-voltage satellite power subsystems. Research and development programs are underway to demonstrate greater than 25% conversion efficiency in space sunlight at 25°C. It is anticipated that these cell types will be as resistant to radiation damage as the GaAs cell.

Summary

The development of high-efficiency and radiation-resistant solar cells for space application is summarized in Fig. 12. The data shown (beyond 1979) are only projections, since the information has yet to be generated by research and development. The trend will be away from silicon, toward GaAs, and then to multiple bandgap cells. However, the trend will be strongly dependent on the results of the GaAs cell and array panel flight qualification efforts. Also, if costs associated with the semiconductor-related technologies do not have a corresponding downward trend, silicon cells may be in use far beyond 1990.

References

[1]Lavine, J.P. and Vette, J.I., "Models of the Trapped Radiation Environment, Vol. V, Inner Belt Protons," NASA SP-3024, 1969.

[2]Lavine, J.P. and Vette, J.I., "Model of the Trapped Radiation Environment, Vol. VI, High Energy Protons," NASA SP-3024, 1970.

[3]Vette, J.I. and Lucero, A.B., "Models of Trapped Radiation Environment, Vol. III, Electrons at Synchronous Altitudes," NASA SP-3024, 1967.

[4]Webber, W.R., "An Evaluation of Radiation Hazard Due to Solar Particle Events," The Boeing Co., Rept. D2-90469, Dec. 1963.

[5]Weidner, D.K., "Natural Space Environment Criteria for 1975 - 1985 NASA Space Stations,"NASA TM X-53865, Aug. 1969; 2nd ed., Aug. 1970.

[6]Kinchar, G.W. and Pease, R.S., "The Displacement of Atoms in Solids by Radiation," Report Progress in Physics, Vol. 18, 1955.

[7]Seitz, F. and Koehler, J.S., "Displacement of Atoms During Irradiation," Solid State Physics, Vol. 2, Academic Press, 1956.

[8]Saito, H. and Hirata, M., "Nature of Radiation Defects in Silicon Single Crystals," Japanese Journal of Applied Physics, Vol. 2, No. 11, 1963, p. 678.

[9]Baiker, J.A., Flicker, H. and Vilms, J. "Proton Induced Lattice Displacement in Silicon," Applied Physics Letters, Vol. 2, No. 5, 1963, p. 104.

[10]Bulgakov, V. and Kumakhov, M.A., "Spatial Distribution of Radiation Defects in Materials Irradiated with Beams of Monoenergetic Particles," Soviet Physics - Semiconductors, Vol. 5, No. 11, 1969, p. 1334.

[11]Berger, M.J. and Seltzer, S.M., "Additional Stopping Power and Range Tables for Protons, Mesons and Electrons," NASA SP-3036, 1966.

[12]Janni, J.E., "Calculations of Energy Loss, Range Pathlength, Straggling, Multiple Scattering and the Probability of Inelastic Nuclear Collissions for 0.1 to 1000 Mev Protons," AFWL-TR-65-150, Sept. 1966.

[13]Tada, H.Y. and Carter, J.R., Solar Cell Radiation Handbook, Jet Propulsion Laboratory, Pasadena CA, 77-56, Nov. 1977, p. 3-69 to 3-78.

[14]Dye, D.L., "Current Status of Space Radiation Effects on Materials and Components," Proceedings of the Special Sessions on Protection Against Space Radiation, NASA SP-169, June 1967, p. 29.

[15]Tada, H.Y., and Carter, J.R., Solar Cell Radiation Handbook, Jet Propulsion Laboratory, Pasadena CA, 77-56, Nov. 1977, Ch. 3.

[16]Mandelkorn, J., McAfee, C., Kesparis, J., Schwartz, L., and Pharo, W., "A New Radiation-Resistant High Efficiency Solar Cell," USARDL Tech. Rept. No. 2162, Oct. 1960.

[17]Wertheim, G.K., "Energy Levels in Electron-Bombarded Silicon," Physical Review, Vol. 105, 1957, p. 1730.

[18]Wertheim, G.K., "Electron Bombardment Damage in Silicon," Physical Review, Vol. 110, 1958, p. 1272.

[19]Stella, P.M. and Opjorden, R.W., "Low Resistivity High Minority Carrier Lifetime Single Crystal Silicon Investigation," Spectrolab, Sylmar CA, AFAPL-TR-79-2031, Apr. 1979.

[20]Crabb, R.L., "Photon Induced Degradation of Electron Irradiation Silicon Solar Cells," Proceedings of the Ninth IEEE Photovoltaic Specialists Conference, May 1972.

[21]Crabb, R.L., "Photon Induced Degradation of Electron and Proton Irradiation Silicon Solar Cells," Proceedings of Tenth IEEE Photovoltaic Specialists Conference, Nov. 1973.

[22]Rahilly, W.P, Scott-Monk, J., Anspaugh, B. and Locker, D., "Electron and Photon Degradation in Aluminum, Gallium and Boron Doped Float Zone Silicon Solar Cells," Proceedings of the Twelfth IEEE Photovoltaic Specialists Conference, Nov. 1976.

[23]Wohlgemuth, J., Lingmayer, J. and Scheinine, A., "Nonreflecting Vertical Junction Silicon Solar Cell Optimization," Solarex, Rockville MD, AFAPL-TR-77-38, July 1977.

[24]Wise, J.F., "Vertical Junction Hardened Solar Cell," U.S. Patent 3,690,953.

[25]Rahilly, W.P., "Rib and Channel Vertical Junction Solar Cell," U.S. Patent 3,985,579.

[26]Rahilly, W.P., "Electron and Neutron Irradiation of Advanced Silicon Solar Cells," Proceedings of the Eleventh IEEE Photovoltaic Specialists Conference, May 1975.

[27]Lloyd, W.W., "Fabrication of an Improved Vertical Multijunction Solar Cell," Proceedings of the Eleventh IEEE Photovoltaic Specialists Conference, May 1975.

[28]Lindmayer, J., "New Developments in Vertical Junction Solar Cells," Proceedings of the Twelfth IEEE Photovoltaic Specialists Conference, Nov. 1976.

[29]Wise, J.F., "Solar Photovoltaic Research and Development Program of the Air Force Aero Propulsion Laboratory," Proceedings of the Third Solar Cell High Efficiency and Radiation Damage Conference, NASA Conf. Pub. 2097, June 1979.

[30]Wolf, M., "Drift Fields in Photovoltaic Solar Energy Converter Cells," Proceedings of the IEEE, Vol. 51, May 1963, p. 674.

[31]Godlewski, M.P., Barona, C.R. and Brandhorst, H.W., "Low-High Junction Theory Applied to Solar Cells," Proceedings of the Tenth IEEE Photovoltaic Specialists Conference, Nov. 1973.

[32]Mandelkorn, J. and Lamneck, J.H., "Simplified Fabrication of Back Surface Electric Field Silicon Cells and Novel Characteristics of Such Cells," Proceedings of the Ninth IEEE Photovoltaic Specialists Conference, May 1972.

[33]Stella, P.M., Uno, F.M. and Thornhill, J.W., "High Efficiency Solar Panel - Phase II," Spectrolab, Sylmar CA, AFAPL-TR-78-60, Aug. 1978.

[34]Haynos, J., Allison, J., Arndt, R. and Muelenberg, A., "The Comsat Non-Reflective Silicon Solar Cell: A Second Generation Improved Cell," Proceedings of the International Conference on Photovoltaic Power Generation, Hamburg, Germany, Sep. 1974, p. 487.

[35]Arndt, R.A., Allison J., and Muelenberg, A., "Optical Properties of the COMSAT Non-Reflecting Cell," Proceedings of the Eleventh IEEE Photovoltaics Specialists Conference, May 1975.

[36]Barona, C.R. and Brandhorst, H.W., "V-Grooved Silicon Solar Cell," Proceedings of the Eleventh IEEE Photovoltaic Specialists Conference, May 1975.

[37]W. Luft, TRW Systems (now at Solar Energy Research Institute, Golden CO) personal communication, Aug. 1976.

[38]Kamath, S. and Wolff, G., "High Efficiency GaAs Solar Cell Development," Hughes Aircraft Co., El Segundo CA, AFAPL-TR-78-96, Sep. 1978.

[39]Rahilly, W.P. and Anspaugh, B., "Electron, Proton and Fission Spectrum Neutron Radiation Damage in Advanced Silicon and Gallium Arsenide Solar Cells," Proceedings of the Second Solar Cell High Efficiency and Radiation Damage Conference, NASA Conf. Pub. 2020, April 1977.

[40]Moon, R.L., James, L.W., Locker, D.R., Rahilly, W.P., Meese, J.M. and Lowe, L., "Performance of AlGaAs/GaAs Solar Cells in the Space Environment," Proceedings of the Twelfth IEEE Photovoltaic Specialists Conference, Nov. 1975.

[41]Lamorte, M.F. and Abbott, D., "Two-Junction Cascade Solar Cell Characteristics under 10^3 Concentration Ratio and AMO to

AM5 Spectral Conditions," Proceedings of the Thirteenth IEEE Photovoltaic Specialists Conference, June 1978.

[42]Simmons, M., Bedair, S. and Lamorte, M.F., "Development of High Efficiency, Stack Multiple Bandgap Solar Cells," Research Triangle Institute, NC, AFAPL-TR-79-2116, Jan. 1980.

[43]Rahilly, W.P., "A Review of Air Force High Efficiency Cascaded Multiple Bandgap Solar Cell Research and Development," Proceedings of the Third Solar Cell High Efficiency and Radiation Damage Conference, NASA Conf. Pub. 2097, June 1979.

RADIATION INDUCED DIELECTRIC CHARGING

A. R. Frederickson*

Rome Air Development Center,
Hanscom Air Force Base, Bedford, Mass.

Abstract

Models for charging of dielectrics, especially polymers, by high-energy radiations $>10^3$eV are discussed. Predictions are made for controlled laboratory conditions and compared with experiment. As dielectrics with very low dark conductivity can be made to break down electrically, the coefficient of radiation-induced conductivity becomes an important material parameter. Radiations which do not fully penetrate the dielectric are found to be especially troublesome. Conduction processes within the dielectric apparently significantly enhance local electric field intensities.

I. Background

A. Purpose

Under space radiation, irradiation for material processing, electron microprobe analysis, auger analysis, ESCA analysis, electron microscopes, nuclear activation, and other irradiations, radiation quanta (>100eV) are capable of charging dielectrics to the electrostatic field breakdown level. Because electric fields of these magnitudes may radically alter the responses of the dielectric, it is necessary to predict the electric field profiles in the bulk of an irradiated dielectric. Here we discuss a method for field prediction in dielectrics irradiated by quanta above 10^3eV and present some recent results under 1 MeV electron irradiation. Applications to spacecraft are discussed since polymer dielectrics are common on external surfaces as well as on internal components of most spacecraft flown today.

B. Basic Concepts

Charged particles impacting materials are brought to rest after traveling some path length characteristic of the particle and the material.[1] For example, electrons between

---

*Deputy for Electronic Technology.

10^4 and 10^7 eV kinetic energy lose energy at "rates" between 1×10^6 eV/(g/cm^2) and 50×10^6 eV(g/cm^2) depending on the energy and on the atomic constituents of the material. (Note that if ρ, g/cm^3, is the density of the material then (g/cm^2)/ρ is a path length through the material in centimeters. The energy loss "rate" thus defined is characteristic of the material and independent of its density.) An electron with initial kinetic energy of 10^6 eV will penetrate on the order of millimeters in solids. In addition the high-energy charged particles will excite many secondary electrons in the material with energies significantly above the classical conduction band. Thus, a very complex spectrum of excited electrons in the bulk material results from the irradiation by monoenergetic charged particles.[2]

High-energy photons (>10^3 eV) are brought to rest more "slowly" than particles and create diverse spectra of secondary photons and electrons whose initial energies are significant fractions of the original photon energy. The photons ultimately lose most of their energy to electron kinetic energy, and we are back to the conditions of the previous paragraph. High-energy electrons can, in turn, create high-energy photons which can then penetrate deeper into materials. Thus, whether we initially irradiate using particles or photons we generally produce a complex spectrum of both photons and particles[3,4,5] which travel significant distances through the bulk of materials.

This motion of charged particles constitutes an electric current in the classical sense and can produce significant electromagnetic disturbances. A nuclear weapon exploded in the atmosphere can induce damaging millisecond transients in electric utility power grids for tens of miles. A radiation hot cell viewing port (very thick glass window) can suddenly fracture after years accumulation of internal charge. How large can the electric field become? Consider mythical "perfect insulators" where $\rho \approx 1$ g/cm^3 and $\sigma \approx 0$ mho/cm. Irradiation electrons are being slowed in the bulk material by atomic interactions at rates exceeding 10^6 eV/cm. Therefore, the spatial trajectories of the irradiation electrons will not be altered substantially until bulk electric fields exceed 10^6 V/cm. Since such high-field strengths will break down solid dielectrics, high-energy irradiation is, in principle, capable of destroying "perfect" dielectrics.

If the net current profile of charged particles in the bulk of the dielectric is denoted by $\bar{J}(x,y,z;t)$ C/cm^2 sec then the time rate of change of space charge q is given by

$$dq/dt\ (x,y,z;t) = -\ \text{div}\ \bar{J}(x,y,z;t)\ \text{C/cm}^3\ \text{sec} \quad (1)$$

and the space charge density is given by integration as

$$q(x,y,z;t) = q(x,y,z;t=0) - \int_0^t \text{div}\ \bar{J}(x,y,z;t^*)dt^* \quad (2)$$

We will limit further discussion to the common quasistatic case where the time-varying magnetic flux is negligible. Then the electric field is due entirely to this space charge and can be determined easily if we know $\bar{J}(x,y,z;t)$.

The net current profile is composed of two terms: a) the high-energy radiation driven current $\bar{J}_r$, which is not usually influenced by electric fields, and b) the low-energy currents $\bar{J}_c$ (such as conduction, diffusion and others), which can be influenced by electric fields, directly or indirectly.

$$\bar{J}(x,y,z;t) = \bar{J}_r(x,y,z;t) + \bar{J}_c(x,y,z;t) \quad (3)$$

The high-energy currents J_r are the driving functions which cause bulk dielectric space charge fields. The low-energy currents $\bar{J}_c$ often act to diminish the electric fields but in some cases actually enhance the fields over a restricted spatial region. The simultaneous inclusion of these two terms J_r (independent of E) and $\bar{J}_c$ (dependent on E) is what makes this problem different from the more usual solid-state space charge problems.

Since $\bar{J}_r$ is such an important term we are forced to view the solid-state dielectric from an unusual perspective, looking down from the multikilovolt electron to the conduction band rather than looking up from the fermi level. In this case the usual concept of exciting electrons from the fermi sea into the conduction bands is insufficient. Instead, to determine $\bar{J}_r$ we have to first ask: What are all the induced charge motions with energies above the conduction band? We find a significant portion of the necessary information in the literature on radiation transport and shielding. This literature appears to be adequate for quanta above 10 keV but further work is needed below 10 keV.

The radiation causes nonequilibrium charging both macroscopically and microscopically. The radiation cascade (slowing down) spectra is so rich in photon and electron spectral content[6] as well as in "solid state" excitations such as plasmons, excitons, etc., that electrons are displaced by many processes within the dielectric from their initial resting position. These displaced electrons and their parent ions can then contribute to "dielectric conductivity" as well as space charge fields. It appears that these displaced pairs (electron-ion) are a major contributor to the conductivity in irradiated dielectrics, and are much more important than thermally generated carriers in many irradiated dielectrics.[7,8]

C. Analytical Models

Perhaps the most extensive published analytical formulations for electric fields and currents in electron irradiated dielectrics and comparisons with experiments are by Bernhard Gross et al.[9,10] This and related work is motivated by practical considerations in the electret[11] field and insulator leakage currents in irradiated radiation dosimeters. The models consist essentially of two steps:

1) High energy monoenergetic electrons penetrate a uniform distance into a dielectric and form a sheet of charge at a depth of the electron "range".[12] The resulting electric field is proportional to the quantity of charge residing in the sheet (see Fig. 1).

2) Radiation-induced conductivity allows some of the charge to bleed off during the irradiation. As the electric field grows higher, the conduction currents increase until they exactly cancel the addition of incident high-energy electrons in the sheet, and equilibrium has then been attained. The electric field attained at equilibrium may or may not exceed breakdown strength, depending on the magnitude of radiation-induced conductivity.

Many modifications to the models have been discussed[9,10] (delayed conductivity, trap modulated conductivity, total dose dependent conduction, etc.), but the essential concepts remain. The difficulty with analytical models is that they are not broadly applicable because a) even monoenergetic electrons do not form a narrow sheet of charge but are distributed throughout the depth up to one range, and b) both broad spectrum and fully penetrating irradiations are not considered in the analysis. Recent work[13] has shown that improved analysis is possible but a fully general analytic solution to the electric fields and currents in irradiated dielectrics is yet to be obtained.

Some interesting experiments have been performed which relate nicely to these analytical models. Lichtenberg discharge patterns have been created in electron irradiated dielectrics, and the pattern lies near and parallel to the plane (sheet) of charge.[14,15,16] The large electric fields created in prior irradiation diminishes the depth of motion by later irradiating electrons.[17]

D. Numerical Models

Consider a one-dimensional dielectric in which currents, space charge, and electric fields reside. Let the system be quasistatic, i.e., let $\partial B/\partial t = 0$. We are left with the equations

$$\overline{\nabla}\cdot\overline{D} = 4\pi q \tag{4}$$

$$\overline{D} = \varepsilon\overline{E} \tag{5}$$

$$\partial q/\partial t = -\overline{\nabla}\cdot\overline{J} \tag{6}$$

where

q = space charge density, C/cm^3
E is in V/cm
t is in sec

As discussed earlier, $\overline{J}$ has the form

$$\overline{J} = \overline{J}_r + \overline{J}_c$$

where

$\overline{J}_r$ = function of the radiation transport
$\overline{J}_c$ = function of electric field

A numerical solution to these equations has been discussed elsewhere.[18,19,20,21,22]

Calculations of the time dependent fields and currents have been presented[18-24], using the numerical model, for X-rays and electrons incident on various dielectrics. To date, the calculations have assumed that the radiation-driven current $\overline{J}_r$ is given by high-energy radiation transport codes and that the conductivity is everywhere proportional to the local dose rate.[25] The photon and electron energies for the calculations varied from roughly 10^4 to 10^6 eV, and for nearly every one-dimensional geometrical arrangement the results indicate that peak electric fields usually occur at 10^6 V/cm in polymers and in some glasses.

The difference between the analytical and numerical techniques is essentially one of completeness. The numerical

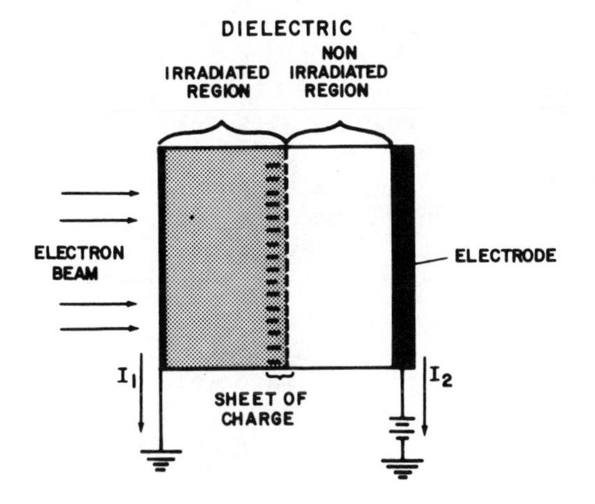

Fig. 1 Simplified model which allows for analytic solutions as discussed in Refs. 9 and 10.

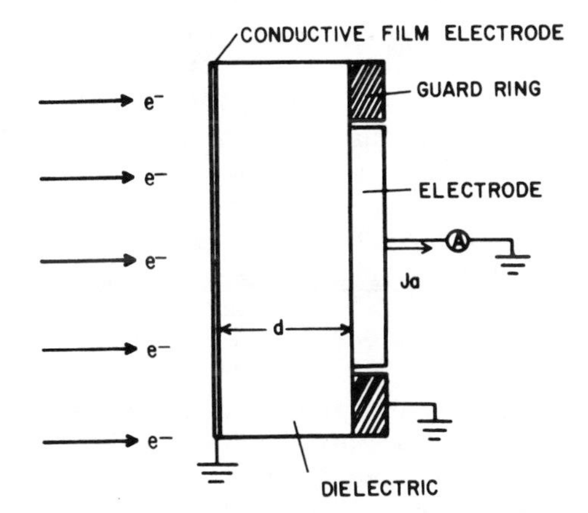

Fig. 2 Dielectric irradiation geometry for the 1 MeV electron irradiation experi ments reported here.

technique on the computer is able to include correct, typically complex functions for $\overline{J}_r$ and $\overline{J}_c$ and avoids errors introduced by the approximations required in the analytical modelling. In addition, the numerical technique is, in principle, able to encompass a great variety of functional dependencies [e.g., $\overline{J}_c = \overline{J}_c$ (x,t,E, total dose), $\overline{J}_r = \overline{J}_r$ (x,t,E), and others]. This permits us to include physical effects (such as nonohmic contacts, trap filling, photoconduction, radiation damage, and tunneling) individually or collectively. However, the numerical technique has the disadvantage of losing didactic clarity, and we must intuit the cause-and-effect relationships rather than analytically display them.

II. Experiment: Model Comparisons

We are currently in the process of experimentally determining the validity of the numerical model. The model predicts both spatial and time dependence of the currents, fields, and charge distributions, as well as the time dependence of the replacement current. The experiments are much more difficult than the model. I am unaware of a good method for measuring the time and position dependence of the internal dielectric currents during irradiation. Measurement of space charge distributions has been performed and published elsewhere[26,27] but is not an easy, precise procedure. Only the time-dependent replacement current is easily measured.

Present experiments are described in Fig. 2. The time-dependent replacement current in a series of materials is measured for comparison with the numerical model. The electron beam is provided by a Dynamitron accelerator and is scattered into a broad beam by a titanium foil in which the electrons lose approximately 70 keV energy. The beam uniformity has been measured using thermal luminescent dosimeters and its intensity is within 20% of the average intensity everywhere over the area of the dielectric. The dimension d is very much less than the width of the guard ring, so edge effects are negligible.

In this early modelling work we suppress all second-order effects and concentrate on the major process. Contact effects, field-dependent conductivity, radiation-dependent trapping, and many other effects are ignored in the initial modelling of the experiments. We assume that conductivity everywhere in the dielectric is the sum of the dark conductivity and the local radiation-induced conductivity. The induced conductivity is assumed proportional[25] to the local dose rate in analogy with semiconductors and gases in which the number of conducting ion-electron pairs produced is proportional to the energy deposited by the radiation. The local dose rate is obtained from the fortran algorithm in Ref. 28. The currents $\overline{J}_r$, produced directly by the electron irradiation, are obtained implicitly from components of the algorithm in Ref. 28. [In the fortran notation of Ref. 28, pp. 237-8, the expression used here for $\overline{J}_r$ = (1 - FB/FEB)*EXP(-ASB).] The algorithm results are in agreement with unpublished and published data, but the error bars in this data are relatively large.

In the figures we show the calculated[28] incident high-energy electron current profile $\overline{J}_r$ which reflects the scattering, slowing down, and stopping of electrons in the dielectric. We also show the calculated[28] dose depth profile. Total current profiles ($\overline{J}_r + \overline{J}_c$) are then shown at several times during the constant intensity irradiation which began at time t = 0. Charge densities and electric fields also are plotted as a function of time. The replacement current J_a, measured by meter A, is then plotted for both the calculated and experimental data: the comparison of these two curves reflects on the completeness with which we are modelling the dielectric response.

The replacement current J_a is mathematically determined from the earlier equations with the constraint that

$$\int_0^d \bar{E}(x,t)\cdot\bar{dx} = 0 \qquad (7)$$

$$J_a(t) = 1/d \int_0^d \left(\bar{J}_r(x) + \bar{J}_c(x,t) + \varepsilon\, \partial\bar{E}(x,t)/\partial t\right)\cdot d\bar{x} \qquad (8)$$

$$= J_r(x) + J_c(x,t) + \varepsilon\, \partial E(x,t)/\partial t \text{ for } 0<x<d \qquad (9)$$

These equations arise from the boundary conditions described in Fig. 2.[13]

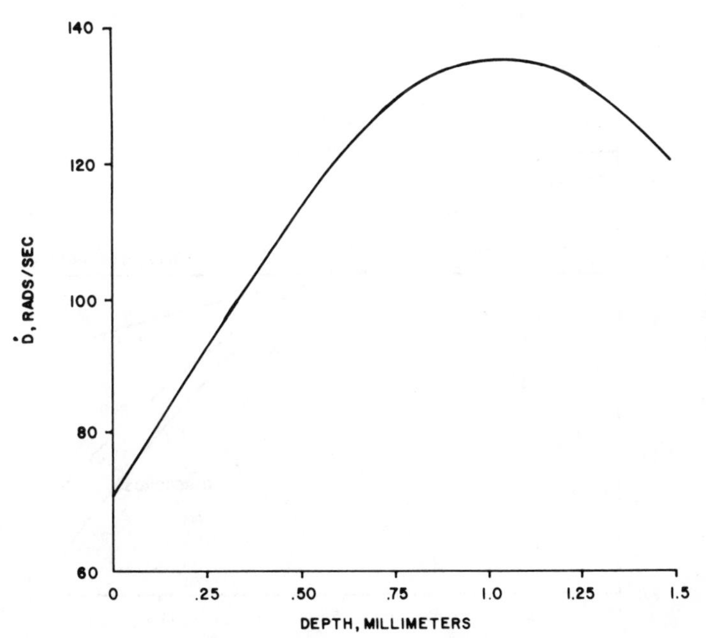

Fig. 3 Dose rate profile for polyvinylchloride (PVC) sample irradiated by 3.9×10^{-10} A/cm^2 1 MeV electrons. (Profile based on Ref. 28.)

Figures 3 through 7 describe results for a polyvinylchloride (PVC) dielectric 1.52 mm thick irradiated by 1 MeV electrons. Approximately one-third of the incident electrons are stopped in the dielectric, and the remainder are either backscattered or transmitted through to the rear electrodes. Figure 3 describes the dose depth profile[28] within the PVC dielectric; the dose profile remains constant through the irradiation. Figure 4 describes the current profile at various times.

At 0 sec the beam is turned on producing a profile appropriate to 1 MeV electrons on this sample. This current profile deposits space charge in the sample, which in turn causes conduction currents to flow. The process continues until div $\overline{J} = 0$, at which time equilibrium has been attained and no further changes occur (in this case at ≈ 5000 sec) even though the irradiation continues. Figure 5 describes the space charge distributions at several times, and Fig. 6 describes the resulting electric field profiles. It is assumed that the initial space charge density is zero. Even if this assumption is incorrect, the equilibrium profiles ($t > 5000$ sec) would be as shown in these figures; the continued irradiation will ultimately eliminate the initial con-

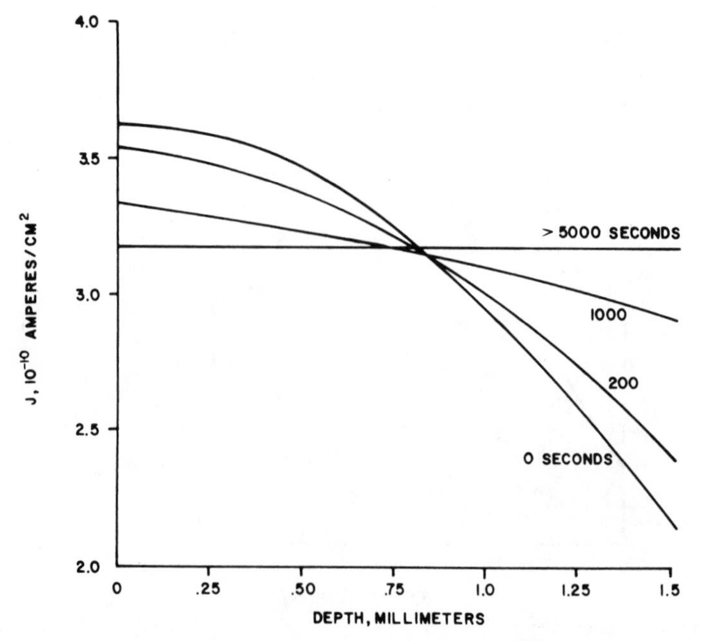

Fig. 4 Theoretical total current profiles $\overline{J}(x,t)$ in PVC at several times after commencement of irradiation.

ditions and produce the equilibrium described in Figs. 4-6 at late times.

Figure 7 compares the predicted and experimental meter current results on the PVC sample. The measurement of the incident 1 MeV electron beam current intensity may be in error by 10%. The initial current profile $\overline{J}_r(x)$ also may be in error by 10% or more, especially at deep penetration. The backscattering effects at the rear electrode (aluminum) have not been quantitatively included. Thus the measured and predicted curves agree within experimental error over most of the time. Both experiment and prediction give us a meter current which increases as the irradiation progresses.

Especially interesting is the peak in the response at 800 sec. In all three samples in which it has been seen, this peak occurred slowly enough so that the electrometer (Kiethley vacuum tube electrometer) could respond. The rear electrode in these three samples was a pressed on aluminum plate. One sample with a painted on carbon (aquadag) rear electrode did not exhibit such a peak. The cause of the peak in Fig. 7 is not clear; it could be a slow breakdown near the rear elec-

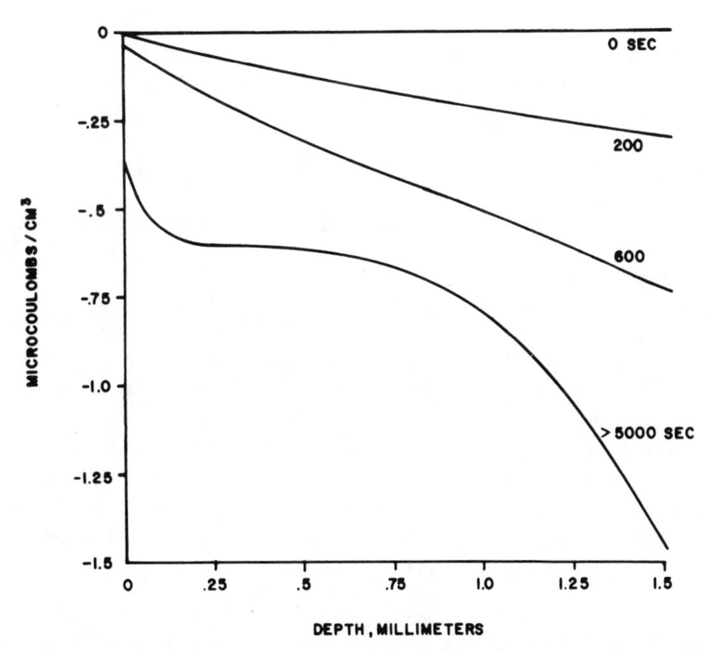

Fig. 5 Theoretical space charge density in PVC at several times after commencement of irradiation; the space charge density at 0 sec is assumed.

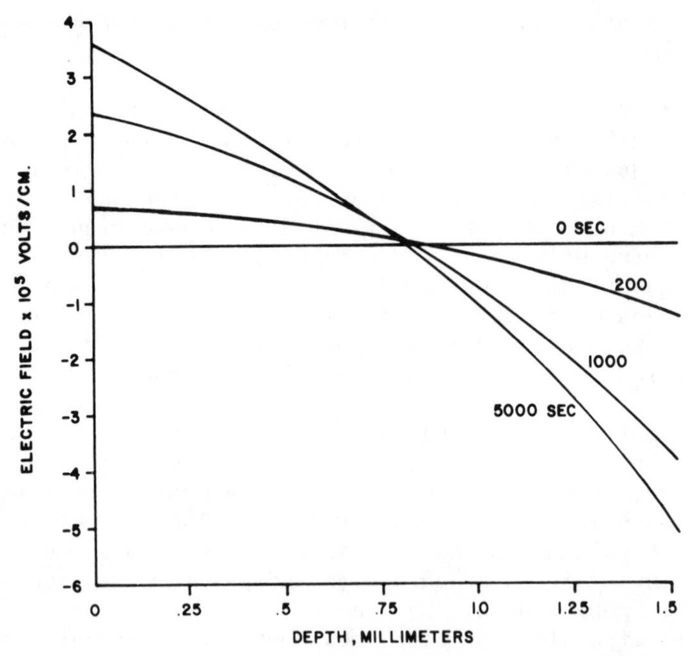

Fig. 6 Theoretical electric field profiles at several times in PVC.

trode where the field magnitude is estimated at $3x10^5$ V/cm or it could be a discharge induced by an initial internal space charge assisted by the radiation-driven fields. This peak is not likely to have occurred from a breakdown in the front half of the dielectric since it has the wrong polarity. The irradiation continued to 5000 sec, and no further peaks were observed.

Four different combinations of functional fits to dose profile and fast electron beam current profile were tried to fit the experiment of Fig. 7. They all predicted an increasing J_a meter time dependence. They all fit equally well. The prediction shown is based on the algorithm of Ref. 28 which is most completely documented. One parameter is adjustable to fit the data, the coefficient of radiation induced conductivity, k. This parameter was obtained quantitatively from the next PVC sample and seemed to fit this sample reasonably well also. K is used in the function[25]

$$\overline{J}_c(x,t) = k\dot{D}(x)\overline{E}(x,t) \quad (10)$$

and for this sample,

$k = 1.7x10^{-18}$ sec/ohm cm rad, and $\dot{D}(x)$ is given by Fig. 3.

Figures 8 through 14 describe the response of a 6.17-mm-thick polyvinylchloride sample to 1 MeV electron irradiation, 4.66×10^{-10} A/cm^2. The sample density is 1.46 g/cm^3. The incident electron beam cannot penetrate the sample, and new effects accrue. Figure 9 shows how the incident beam J(0 sec) is totally stopped in the first 5 mm of the dielectric. Figure 10 describes the growth of space charge density; notice that an incident primary electron is most likely to be stopped at about 60% of the extrapolated range.

Primary electrons and conduction electrons which are thermalized in the dielectric beyond the charge centroid (which is at about 1.8 mm) are driven further into the dielectric by conduction in the space charge field; they are then "stopped" by the nonconducting region beyond 3 mm where the dose rate is negligible. Electrons which are thermalized at depths less than 1.8 mm are "conducted" to the front electrode by the space charge field, through a region of high dose rate and high conductivity. This process continues and increases with time producing the results shown in Figs. 11-13. (This discussion is descriptive only; it is not meant to imply that only electrons are "conducting.")

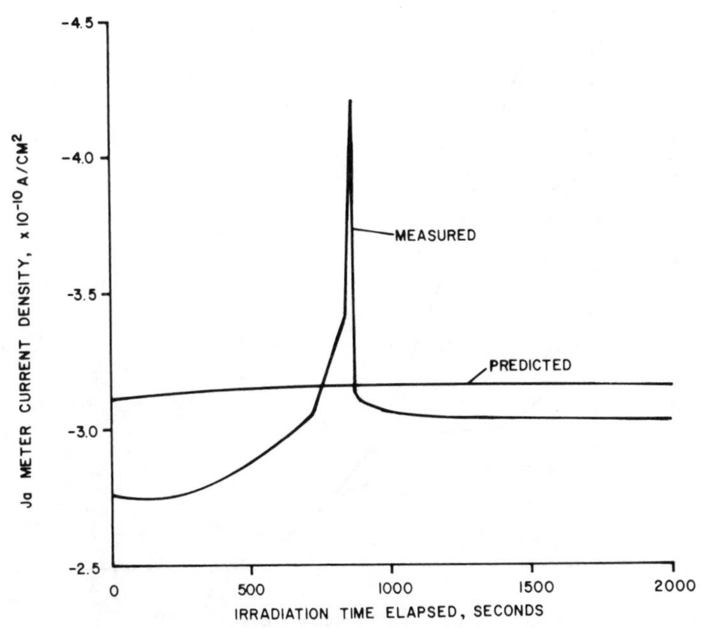

Fig. 7 Theoretical and experimental meter currents $J_a(t)$ in 1.52 mm thick PVC irradiated by 3.9×10^{-10} A/cm^2 1 MeV electrons.

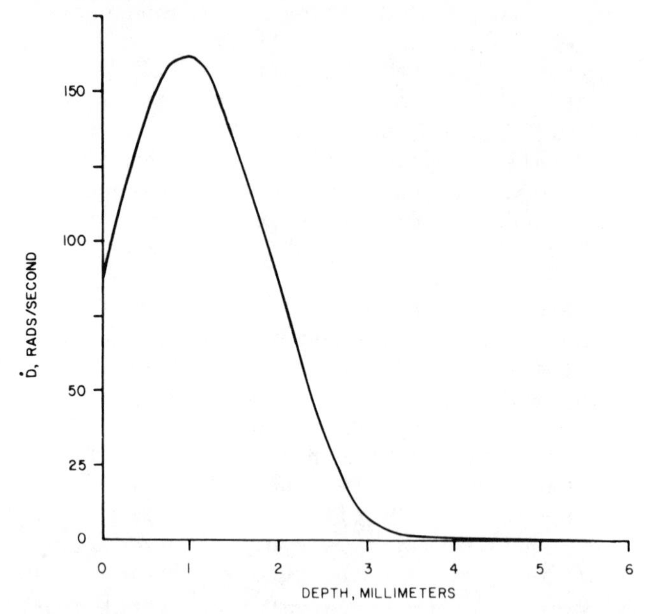

Fig. 8 Theoretical dose-depth profile in 6 mm PVC irradiated by $4.66\text{x}10^{-10}$ A/cm^2 1 MeV electrons.

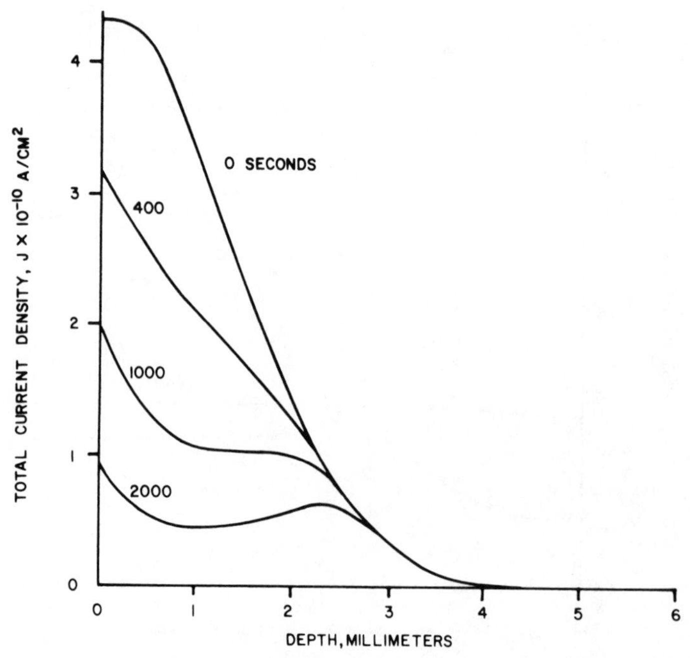

Fig. 9 Theoretical total current density in 6 mm PVC at various times after commencement of irradiation.

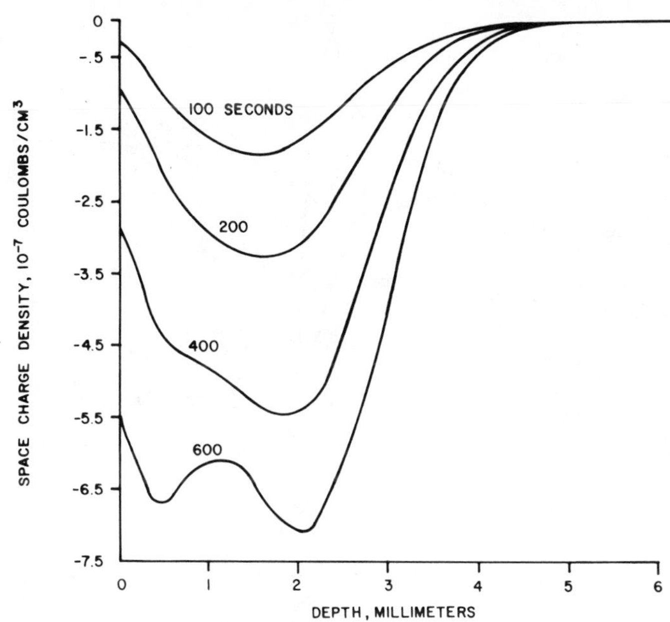

Fig. 10 Theoretical space charge density at various times in the 6 mm PVC sample.

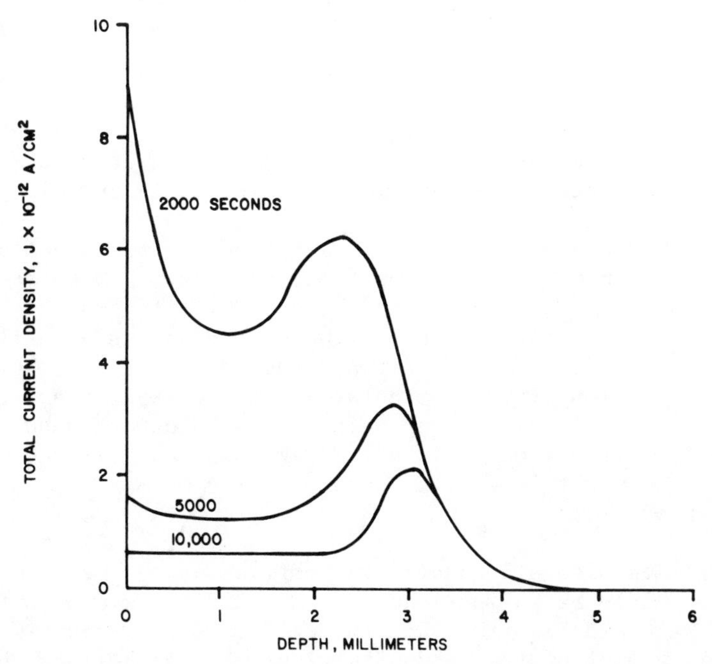

Fig. 11 Theoretical total current density in 6 mm PVC at late times in the irradiation; at 6 mm the current is close to but not exactly zero.

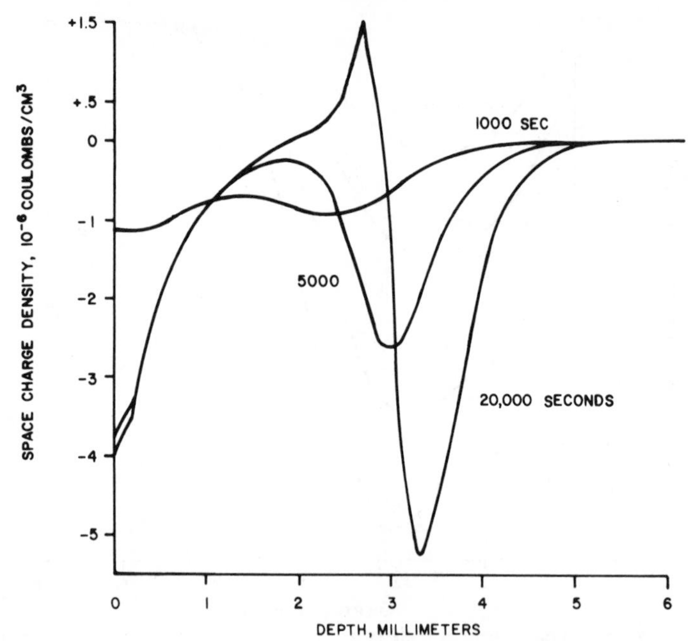

Fig. 12 Theoretical space charge density at late times in the 6 mm PVC; note the development of positive space charge.

Figures 8-13 describe the following important effects:

a) Nonpenetrating beams may produce greater electric fields than penetrating beams. (Compare Figs. 13 and 6).

b) There are additional longer time constants associated with nonpenetrating beams. The penetrating beam responses can be crudely described by a single exponential in time.[25] However, the nonpenetrating beam also has a positive space charge effect and an unirradiated region effect, both of which have long time constants. (Compare Figs. 4 and 5 with Figs. 11 and 12). In principle, if the dark conductivity were zero and the sample did not break down, equilibrium would never be fully obtained with nonpenetrating beams under continued irradiation.

c) Nonuniform conductivity profiles may be very important. Since radiation-induced conductivity is proportional to dose rate and is often much larger than dark conductivity, Fig. 8 is a plot of the conductivity profile as well as the dose rate profile. The development of positive space charge at late times indicated in Fig. 12 is due to conduction currents acting with this conductivity profile. The electric

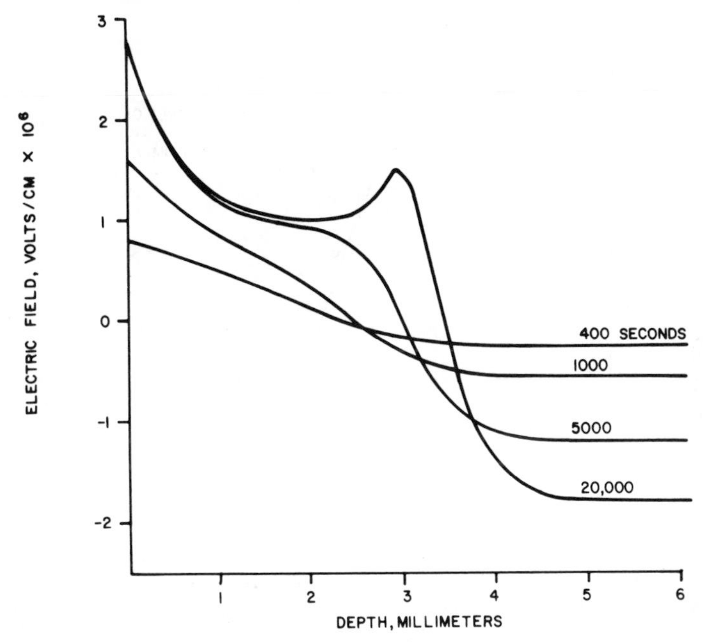

Fig. 13 Theoretical electric fields in PVC at late times.

field intensity interior to the sample near the positive space charge may exceed the fields at the front and rear electrodes at late times.[23] Any source of spatially nonuniform conductivity will cause this kind of effect. Thus, nonuniform conductivity profiles may be a cause for breakdown in irradiated dielectrics.

Figure 14 compares the predicted meter current with the experimental observation. The unknown parameter k, coefficient of radiation-induced conductivity, was adjusted so that the decay rate of J_a most closely matched the experimentally observed decay rate.[25] The resulting value of k(1.7×10^{-18} sec/ohm cm rad) was used for all the calculated curves in Figs. 3 through 14. A 10% change in k value produces an unacceptable disagreement between prediction and experiment in Fig. 14. It is encouraging to see good agreement between theory and experiment. Only very small breakdowns were observed in this sample. Some very recent experiments have been performed with more precise measurement of $\overline{J}_r(x=0)$ and even better agreement has been obtained.

Figures 15-18 describe results for a polystyrene sample 3.38-mm thick, 1.06 g/cm^3 density, and irradiated by 3.56×10^{-10} A/cm^2 of 1 MeV electrons. This is a penetrating

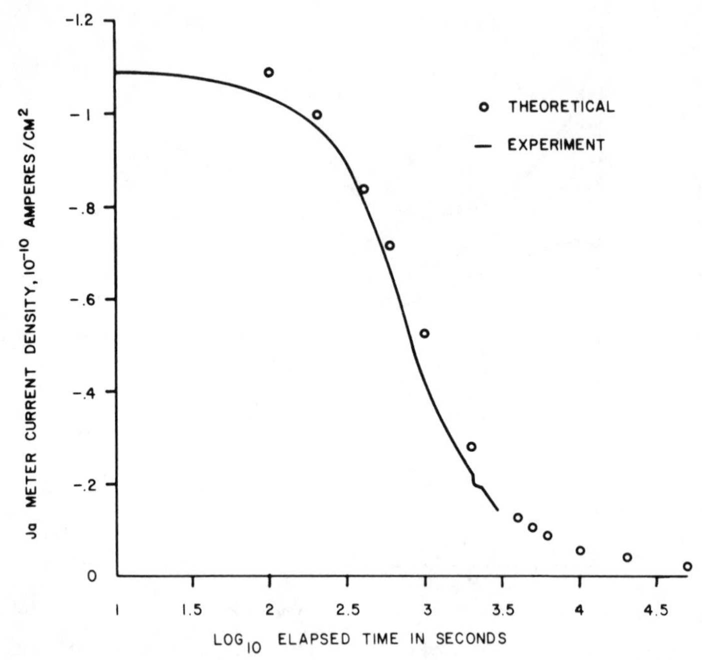

Fig. 14 Comparison between theoretical and experimental meter currents J_a in 6 mm PVC.

radiation case, and the results are somewhat similar to the penetrating radiation on PVC. The interesting differences are:

a) No anomalous pulse in the meter current was seen. (Compare Figs. 18 and 7).

b) To fit the theoretical time constant to the experimental meter current time constant, k was adjusted to $2x10^{-17}$ sec/ohm cm rad, fully an order of magnitude different than the PVC case.

c) The peak electric field is roughly an order of magnitude smaller than that for the penetrated PVC sample. If the geometries of the two samples had been identical (including dielectric constant and density), we would predict the ratio of peak fields to equal the inverse of the ratio of induced conductivity coefficients.

Other experiments have been performed under electron and gamma ray irradiations. It is difficult to obtain $\dot{D}$ and $\overline{J_r}$ profiles for gamma beams, and the few cases we have studied have given only fair agreement with the models.[25] Some electron experiments were preliminary and not well quantified.

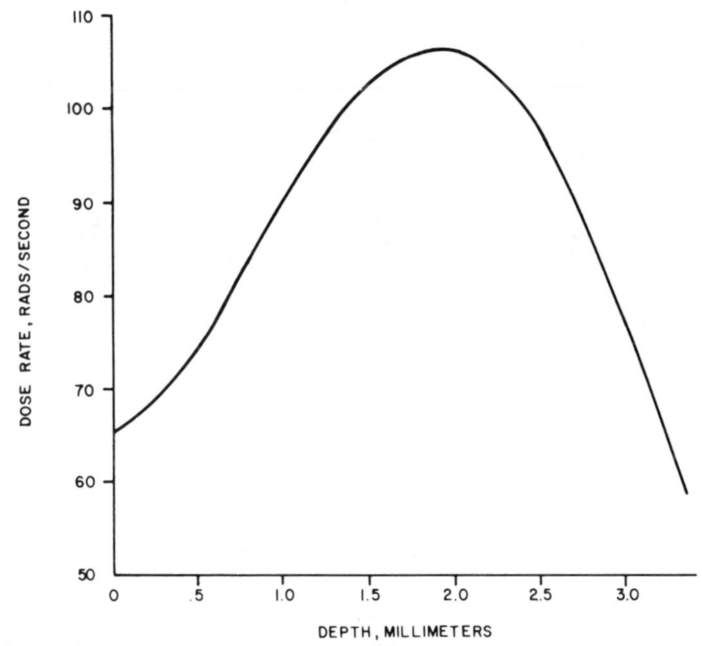

Fig. 15 Dose-depth profile in 3.38 mm thick polystyrene under irradiation by 3.56×10^{-10} A/cm^2 1 MeV electrons.

However, with both gamma rays (Cobalt 60) and 1 MeV electrons we have seen both mild and violent electrometer fluctuations characteristic of breakdowns. Lichtenberg patterns were produced in a few of the samples irradiated by electrons. More electron experiments are planned.

III. Discussion

A. Essential Concepts

High-energy radiations have been used in the past to irradiate materials under the assumptions that no electric fields were developed and that uniform depth effects would accrue. Recent developments in dosimetry and in radiation transport indicate that such assumptions are usually wrong. As we have seen in the preceding section, both photon and electron irradiations are intrinsically capable of producing fields of 10^7 V/cm. Dosimetry effects will be constant in depth only for rare circumstances where the material is homogeneous, is thin relative to the penetration of the radiation, does not significantly perturb the irradiation flux, and is surrounded by equivalent atomic number material.

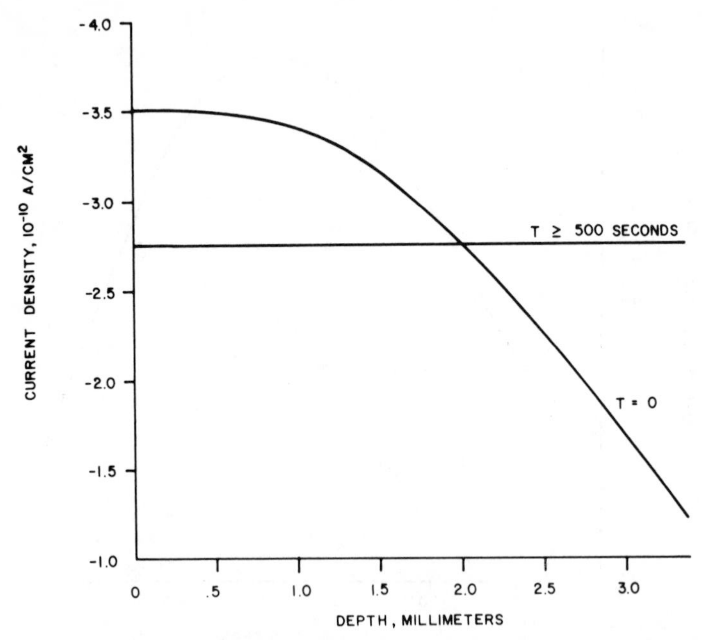

Fig. 16 Theoretical total current profile in polystyrene at various times.

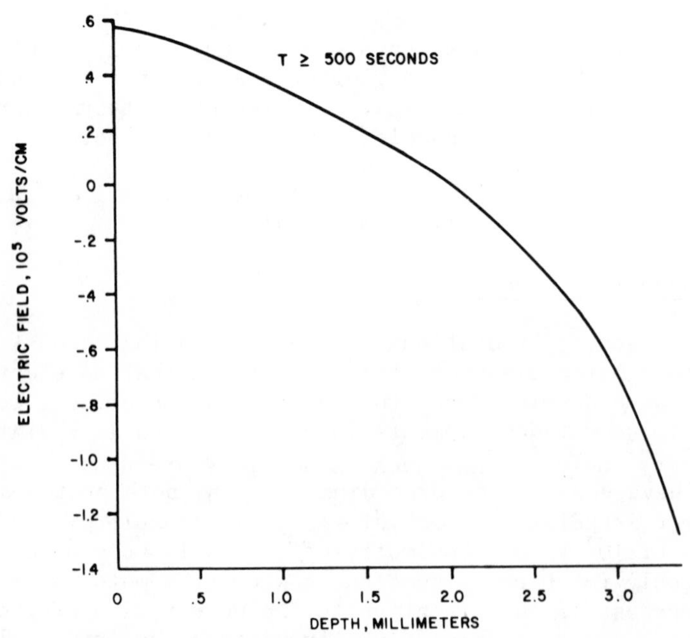

Fig. 17 Theoretical equilibrium electric field profile in the polystyrene sample.

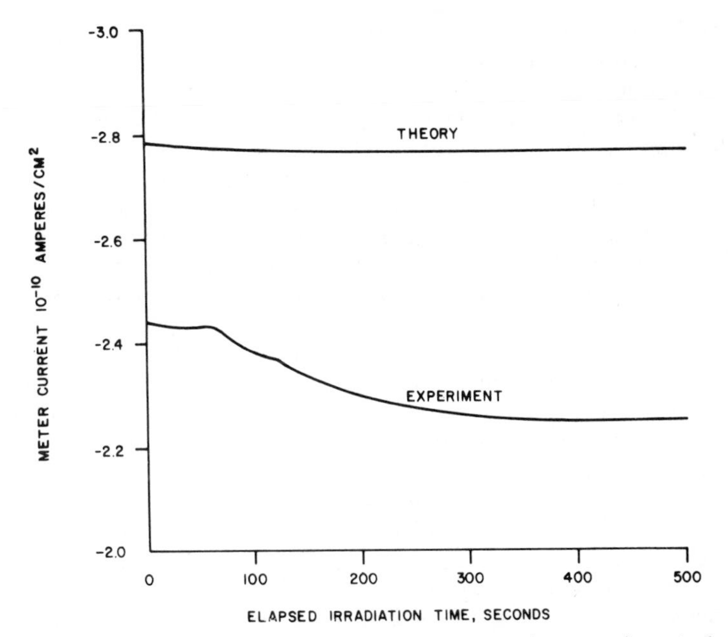

Fig. 18 Comparison between theoretical and experimental meter current J_a for the polystyrene sample. The nearly 20% disagreement is attributed mainly to incident current density calibration uncertainty.

In short, the radiation will most often create nonuniform dose depth effects and significant electric fields (in dielectrics).

The literature provides sufficient data to calculate the magnitude of the radiation-driven fields. This radiation transport is not dependent on the chemical and solid-state nature of the material. Except in the case of very high electric fields[29] ($\gtrsim 10^6$ V/cm), the radiation transport is a function primarily of the atomic number and the density of the material. The high-energy quanta interact with the material atomic constituents creating a cascade of lower energy quanta. It is not until the cascade goes below 100 eV that material-dependent effects compound the problem; well before this energy is reached, the radiation has delivered nearly all the electrostatic energy of which it is capable. For electrostatic purposes, the events below 100 eV can be lumped into conduction or quasi-conduction mechanisms.

At this time, the conduction processes are empirically determined by direct measurement under irradiation. All effects which would look like conductivity (including field-dependent perturbation of the high energy cascade flux) are empirically lumped into two terms: dark conductivity and radiation-induced conductivity. This approach is driven by the need to estimate the radiation-induced fields in materials where the appropriate conduction processes[7] are not sufficiently known. Even polarization currents are lumped into the conductivity current term in many cases. Of course some conduction processes are highly nonlinear, and one must remember that this formalism may be significantly improved with the simple inclusion, when available, of actual conduction processes.

Irradiation historically has been an important tool for investigating chemical and solid-state structures; quanta from 0.1 eV to MeV have been used for many purposes. Most often the radiation excites the process of interest; any further information about the radiation and its interactions is neglected. It is unfortunate that we cannot neglect the radiation transport in the case of dielectrics; the electrostatic fields generated by the radiation in dielectrics probably will alter the chemical and solid-state processes in which we are interested. It becomes necessary to quantify the radiation transport process in order to account for the high electric fields in irradiated dielectrics.

B. Current Problems

Several areas of endeavor seem potentially fruitful at this stage.

1) We need nondestructive methods for determining the field or charge density profiles in the dielectric, preferably during irradiation.

2) Contact effects in metal-semiconductor and metal-dielectric structures are known to be important in some problem areas. How does the high-energy radiation cascade flux alter the contact effects? Put another way, does the flux of radiation-generated low-energy excited electrons at a dielectric conductor interface render the contact ohmic?

3) We need methods to quantify contact effects, especially so that we can discuss radiation-driven electric fields in the very thin film electronic devices arriving in the electronic parts inventory.

4) With the widespread use of electron microscopes, auger spectrometers, ESCA, LEED, synchrotron uv beams, and other diagnostic tools, one can expect the growth of data on

the radiation slowing-down cascade spectrum. Application to dielectrics and space charge should follow naturally.

5) Polymers are known to have significant initial charge distributions from the moment of manufacture. Some polymers are produced with an irradiation process. It would be helpful to have this space charge quantified prior to irradiation experiments. Conceivably, the addition of a small amount of radiation-induced space charge to an initial space charge could trigger a discharge.

6) In Fig. 13 we see that a region of positive space charge internal to the dielectric is associated with a region of large electric fields and is caused by the fact that conductivity is not constant through the depth of the dielectric; the conductivity is everywhere proportional to the local dose rate. Other causes for nonconstant conductivity profiles such as layered dielectrics could similarly create breakdown strength electric fields internal to the dielectric (internal dipole or higher order fields).

C. Application to Spacecraft

The modelling described above is applicable to dielectrics with electrodes such as capacitors or metalized dielectric sheets. Dielectrics in spacecraft do not always have contacts or electrodes, but similar charging effects occur; only the electrostatic boundary conditions [Eqs. (7-9)] are changed. We require information on secondary electron emission and its field dependence for dielectrics in order to quantitatively model electrodeless dielectrics, but such data is severely limited. We need to substantiate the bulk effects before we can feel sure about the space charge densities in the dielectrics, and that is the purpose of the work reported here. After the bulk effects have been substantiated, one can incorporate the secondary emission term at the surface with its field dependence. Only then will we be able to predict the internal and external fields due to space charge in electrodeless dielectrics.

How can we model, control, or minimize the effects of the fields outside the spacecraft dielectrics as well as inside? Remembering that the space radiation environment is not constant and that some dielectrics have a relaxation time of months to years, we can perform the following estimates:

1) It is conceivable that a worst case could occur, that a space charge density throughout the dielectric could build up fields close to the dielectric breakdown point. The

fields external to such a dielectric easily could be determined and their effect on the spacecraft quantified. Where these effects and any subsequent breakdown effects are not significant, the dielectric would be satisfactory.

2) In some cases the breakdown electric field would not be approached due to a combination of dielectric-spacecraft geometries and the incident radiation spectrum. Care must be taken, in calculations of these effects, to include every region of the spectrum which is capable of introducing breakdown space charge densities within one relaxation time (months or years). Synergistic effects among regions of the spectrum make this kind of calculation difficult. Several tens of internal field calculations [18-25] in one dimension using monoenergetic electrons, gamma rays, or x rays, one at a time, from 10 keV to 1 MeV have been performed, and fields nearly always exceed 10^5 V/cm. It seems unlikely that incident spectral sensitivities could be used to advantage.

3) The conductivity of the dielectric can be used to prevent significant field buildup.

a) Dark conductivity would be the most effective field preventer. At typical space current fluence (1nA/cm^2), a dark conductivity of 10^{-12} ohm^{-1} cm^{-1} would keep fields generally below 10^3 V/cm within the dielectric. Dielectrics of this conductivity combined with the other necessary attributes (such as thermal, optical, and chemical properties) might be obtainable.

b) Radiation-induced conductivity (or photoconductivity) can be used to advantage in many cases. For penetrating radiations where all regions of the dielectric are irradiated so that the induced conductivity is substantially greater than the dark conductivity, experience with the field calculations[25] indicates that

$$E \lesssim \frac{1\text{x}10^{-12}(\text{sec v/ohm cm}^2\text{ rad})}{k(\text{sec/ohm cm rad})}$$

Notice that this is in agreement with the calculations provided earlier in this article. For nonpenetrating radiations the maximum fields will not be as severely limited by the induced conductivity; the extent of the limiting has not been determined. It would be advantageous to use or develop dielectrics with high coefficients of photoconductivity.

4) The spacecraft can, in principle, be "neutralized" by the emission of low-energy positive and negative charge

carriers into the space surrounding the craft. The geometrical complexity of a large craft would require several emitters, since the charged particle trajectories could be localized by the charged dielectric fields. This ejected space charge cloud could minimize the fields external to the dielectrics but would not solve the internal dielectric field problem. It could make the internal field problem worse.

5) Experience with irradiated dielectrics indicates that large flaws such as cracks, holes, or sharp edges significantly enhance the probability of breakdown. Impact by particulate matter (micrometeorite) could have a similar effect. Design to minimize these problems would help lower the arcing probability.

6) It must be remembered that space radiations have a temporal dependence in both intensity and energy spectrum. It is obvious from the nonpenetrating beam results that the following scenario may be important in space. A thick dielectric may be initially irradiated by 10-50 keV electrons for several hours, finally approaching equilibrium without an imminent breakdown. Then the spectrum changes, and we begin introducing 50-500 keV electrons, which penetrate deeper beyond the earlier established charge centroid, producing enhanced fields internal to the dielectric. Such a sequence of irradiations could substantially increase the field magnitudes, causing breakdowns.

Because the charge storage of dielectrics is so long lasting, the synergisms among temporally separated irradiations could be critical for space application. We need data to bound the temporal dependence for space spectra with reference to dielectric response. Spectral sequences other than the above example could be very important; synergisms with optical photoemission and photoconductivity come immediately to mind. There appears to be a large number of potentially critical sequences of spectrum changes which may need to be considered.

The use of monoenergetic laboratory irradiations does not cause effects significantly different from the effects induced by typical space radiations. However, for any particular situation, large differences might be seen between two differing monoenergetic radiations, and similarly large differences might be seen between two differing maxwellian energy distributions. We use the laboratory irradiations to quantify the material parameters. Then for any given space spectra, the dielectric response can be theoretically determined. One must know the space spectra with certainty because temporal changes and geometry/beam-energy synergisms are likely to play a pivotal role in the dielectrics used today (i.e., dielectrics with extremely low dark conductivities and low photoconductivity).

References

[1] Studies in Penetration of Charged Particles in Matter, National Academy of Sciences, National Research Council, Washington, D.C., Publication 1133, 1964.

[2] Ritchie, R. H., Tung, C. J., Anderson, V. E., and Ashley, J. C., "Electron Slowing-Down Spectra in Solids," Radiation Research, Vol 64, Oct. 1975, pp. 181-204.

[3] Evans, R. D., The Atomic Nucleus, McGraw Hill, New York, 1955, Chap. 23-25.

[4] Hubbell, J. H., "Survey of Photon Attenuation-Coefficient Measurements," Atomic Data, Vol. 3, Nov. 1971, pp. 241-297.

[5] Siegbahn, K. (ed.), Alpha-, Beta- and Gamma-Ray Spectroscopy, Vols. 1 and 2, North Holland Publishing Co., Amsterdam, 1965.

[6] Wuilleumier, F. J. (ed.), Photoionization and Other Probes of Many-Electron Interactions, Plenum Press, New York, 1976.

[7] Wintle, H. J., "Photoelectric Effects in Insulating Polymers and Their Relation to Conduction Processes," IEEE Transactions on Electrical Insulation, Vol. EI-12, April 1977, pp. 97-113.

[8] Frankevich, E. L. and Yakovlev, B. S., "Radiation Induced Conductivity in Organic Solids," International Journal Radiation Physics and Chemistry, Vol, 6, 1974, pp. 281-296.

[9] Gross, B., Sessler, G. M., and West, J. E., "Charge Dynamics for Electron-irradiated Polymer-foil Electrets," Journal of Applied Physics, Vol. 45, July 1974, pp. 2840-2851.

[10] Nunes de Oliveira, L. and Gross, B., "Space-charge-limited Currents in Electron-irradiated Dielectrics," Journal of Applied Physics, Vol. 46, July 1975, pp. 3132-3138.

[11] Perlman, M. M. (ed.), Electrets, Charge Storage and Transport in Dielectrics, The electrochemical Society, Inc., Princeton, N. J., 1973.

[12]Studies in Penetration of Charged Particles in Matter, National Academy of Sciences, National Research Council, Washington, D.C., Publication 1133, 1964, pp. 228-268.

[13]Gross, B. and Leal Ferreira, G. F., "Analytic Solution for Radiation-induced Charging and Discharging of Dielectrics," Journal of Applied Physics, Vol. 50, March 1979, pp. 1506-1511.

[14]Gross, B., "Charge Storage Effects in Dielectrics Exposed to Penetrating Radiation," Journal of Electrostatics, Vol. 1, 1975, pp. 125-140.

[15]Gross, B., "Irradiation Effects in Plexiglass," Journal of Polymer Science, Vol. XXVII, Jan. 1958, pp. 135-143.

[16]Gross, B., "Irradiation Effects in Borosilicate Glass," Physical Review, Vol. 107, July 1957, pp. 368-373.

[17]Gross, B. and Nablo, S. V., "High Potentials in Electron-Irradiated Dielectrics," Journal of Applied Physics, Vol. 38, April 1967, pp. 2272-2275.

[18]Frederickson, A. R., "Charge Deposition, Photoconduction, and Replacement Current in Irradiated Multilayer Structures," IEEE Transactions on Nuclear Science, Vol. NS-22, Dec. 1975, pp. 2556-2561.

[19]Frederickson, A. R., Radiation Induced Electrical Current and Voltage in Dielectric Structures, AFCRL-TR-74-0582 (National Technical Information Service or Defense Documentation Center), Nov. 1974.

[20]Pigneret, J. and Stroback, H., "Electrical Response of Irradiated Multilayer Structures," IEEE Transactions on Nuclear Science, Vol. NS-23, Dec. 1976, pp. 1886-1896.

[21]Matsuoka, S., Sunaga, H., Ryuichi, T., Hagiwara, M., Araki, K., "Accumulated Charge Profile in Polyethylene During Fast Electron Irradiations," IEEE Transactions on Nuclear Science, Vol. NS-23, Oct. 1976, pp. 1447-1452.

[22]Berkley, David A., "Computer Simulation of Charge Dynamics in Electron-Irradiated Polymer Foils," Journal of Applied Physics, Vol. 50, May 1979, pp. 3447-3453.

[23]Frederickson, A. R., "Electric Fields in Irradiated Dielectrics," Spacecraft Charging Technology - 1978, NASA Conf. Pub. 2071, AFGL-TR-79-0082, 1979, pp. 554-569.

[24]Beers, B. L., Hwang, H., Lin, D. L., Pine, V. W., "Electron Transport Model of Dielectric Charging," Spacecraft Charging Technology - 1978, NASA Conf. Pub. 2071, AFGL-TR-79-0082, 1979, pp. 209-238.

[25]Frederickson, A. R., "Radiation Induced Currents and Conductivity in Dielectrics," IEEE Transactions on Nuclear Science, Vol. NS-24, Dec. 1977, pp. 2532-2539.

[26]Gross, B. and Wright, K. A., "Charge Distribution and Range Effects Produced by 3-MeV Electrons in Plexiglass and Aluminum," The Physical Review, Vol. 114, May 1959, pp. 725-727.

[27]Evdokimov, O. B. and Tubalov, N. P., "Stratification of Space Charge in Dielectrics Irradiated with Fast Electrons," Soviet Physics Solid State, Vol. 15, March 1974, pp. 1869-1870, (American Institute of Physics, Copyright).

[28]Tabata, T., and Ito, R., "An Algorithm for the Energy Deposition by Fast Electrons," Nuclear Science and Engineering, Vol. 53, Feb. 1974, pp. 226-239.

[29]Evdokimov, O. B., Kononov, B. A. and Yagushkin, N. I., "Current Limitation for a Fast Electron Beam Due to Space Charge in an Insulator," Izvestiya Vysshikh Uhebnykh Zavedenii, Fizika, Vol. 9, Sep. 1975, pp. 139-141. Translated in Soviet Physics Journal, 1978, Plenum Pub. Co., New York, pp. 1339-1341.

PROTON UPSETS IN LSI MEMORIES IN SPACE

P.J. McNulty*, R.C. Wyatt+, and G.E. Farrell§
Clarkson College, Potsdam, N.Y.

and

R.C. Filz† and P.L. Rothwell†
Air Force Geophysics Laboratory Hanscom AFB, Mass.

Abstract

Two types of large scale integrated dynamic Random-Access-Memory devices were tested and found to be subject to soft errors when exposed to protons incident at energies between 18 and 130 MeV. These errors are shown to differ significantly from those induced in the same devices by alphas from an ^{241}Am source. There is considerable variation among devices in their sensitivity to proton-induced soft errors, even among devices of the same type. For protons incident at 130 MeV, the soft error cross sections measured in these experiments varied from 10^{-8} - 10^{-6} cm^2/proton. For individual devices, however, the soft error cross section consistently increased with beam energy from 18 - 130 MeV. Analysis indicates that the soft errors induced by energetic protons result from spallation interactions between the incident protons and the nuclei of the atoms comprising the device. Because energetic protons are the most numerous of both the galactic and solar cosmic rays and form the inner radiation belt, proton-induced soft errors have potentially serious implications, for many electronic systems flown in space.

I. Introduction

The soft error phenomena are becoming an important environmental problem for electronic systems flown in space. Soft errors or bit upsets are anomalous changes in the infor-

---

* Professor of Physics, Physics Department.

\+ Recent MS Graduate; presently at Pattern Analysis and Recognition Corp., New York.

§ Graduate Student, Physics Department.

† Research Physicist, Space Physics Division.

mation stored at certain locations in a semiconductor memory device without observable damage to the device itself. They represent a limitation on systems flown in space. In electronic systems the soft errors must be either avoided by the proper choice of circuit components or corrected through software. Both approaches involve significant design changes and increases in the mass and bulk of the system package.

Our present understanding of the soft errors begins with the simulation studies by Binder et al[1] that gave evidence that the soft errors previously exhibited by bipolar digital components flown in space were the result of the passage of a heavy cosmic ray nucleus with an atomic number above iron ($Z \geq 26$) through one of the sensitive circuit elements of the device.

The soft error rates observed on satellites have increased significantly since then, presumably as a result of the introduction of large scale integrated (LSI) devices into the electronic systems being flown.[2,3] The decrease in volume occupied by each element on the LSI chip results in a corresponding decrease in the increment of charge needed to differentiate between the logic states of an element. This has led to an increase in radiation sensitivity for LSI devices that was illustrated recently by May and Woods[4-6] who demonstrated soft errors in LSI devices exposed to alphas. Presumably the radiation sensitivity of the devices will increase further as devices are reduced in size (i.e., increase in the number of circuit elements per device) from the LSI regime to that of very large scale integrated (VLSI) circuits. Because of the naturally occurring alpha emitters in almost all materials, including those used to jacket memory devices, there have been a number of recent studies of the environmental implications of alpha-induced soft errors.[4-7] Guenzer et al[8] recently demonstrated neutron-induced soft errors in 16K dynamic random-access-memory (RAM) devices with 16,384 (16K) memory locations.

Since protons are more numerous than any of the heavier cosmic ray nuclei at almost all locations in space, a series of experiments were initiated to determine whether protons contribute to the soft error problem. These experiments are described in some detail in Section III. Physical mechanisms previously proposed to explain anomalous signals observed in Defense Meteorological Satellite Program (DMSP)[9] and Landsat satellite systems[10] and in the human visual system are described and considered as sources of soft errors in LSI circuits in Section II. Protons would be most likely to deposit a large amount of energy within the microscopic volume of a sensitive circuit element if they either traverse the sensitive element near the end of their range (where the rate of ionization loss is greatest) or undergo a nuclear interaction in or near the element. In Section IV the results of testing with two types of 4K dynamic RAM devices demonstrate the existence of soft errors induced by protons both at high energies and near the end of their range[11]. Analysis shows that more than one type of target element on the devices is sensitive to proton interactions.

II. Mechanisms

Any model of the soft error phenomena will almost certainly involve the interactions of energetic charged particles with some sensitive microscopic volume element or elements on the device. These sensitive regions may correspond to the memory cells themselves, the reference capacitance elements, the bit lines, the sense amplifiers, or some other circuit structures on the device. Any mechanism would then involve the creation of electron-hole pairs in or about the sensitive structure. Similar microdosimetric considerations played a role in earlier studies of another cosmic ray induced transient phenomena, the light flashes experienced by astronauts as a result of exposure of the human visual system to the cosmic rays. (The term light flashes denotes a variety of visual phenomena experienced by Apollo and Skylab astronauts with their eyes closed and adapted to darkness.[12,13]) Laboratory simulations using human subjects exposed to accelerator beams[14-22] and theoretical studies[23,27] indicate that at least three different physical mechanisms contribute and, depending on the type of visual experience or the region of space, any of the three might dominate. All three are potential mechanisms for soft errors in electronic memory systems.

The first mechanism is ionization loss along the trajectory of the primary cosmic ray nucleus. The rate at which the particle deposits energy along its trajectory is known as the linear energy transfer (LET), and some experimenters found evidence of a threshold LET for visual phenomena.[17,25] This implied that the probability that a visual sensation would be experienced depended upon the amount of energy deposited (or the number of ionizations created) within a retinal summation unit, a microscopic volume on the retina less than 300 μm in diameter and 30 μm thick.[27] These retinal summation units typically integrate signals over a thousand or more photoreceptor elements. The similarity to the standard soft error model[1-8] is striking. Figure 1 shows schematically a heavy cosmic ray nucleus traversing one of the microscopic circuit elements on an LSI

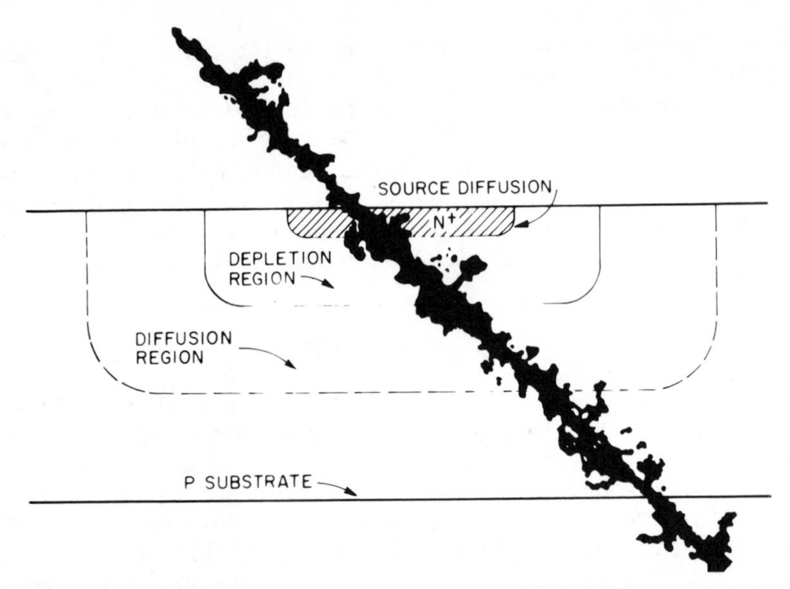

Fig. 1 Schematic representation of a high LET particle traversing an LSI circuit element.

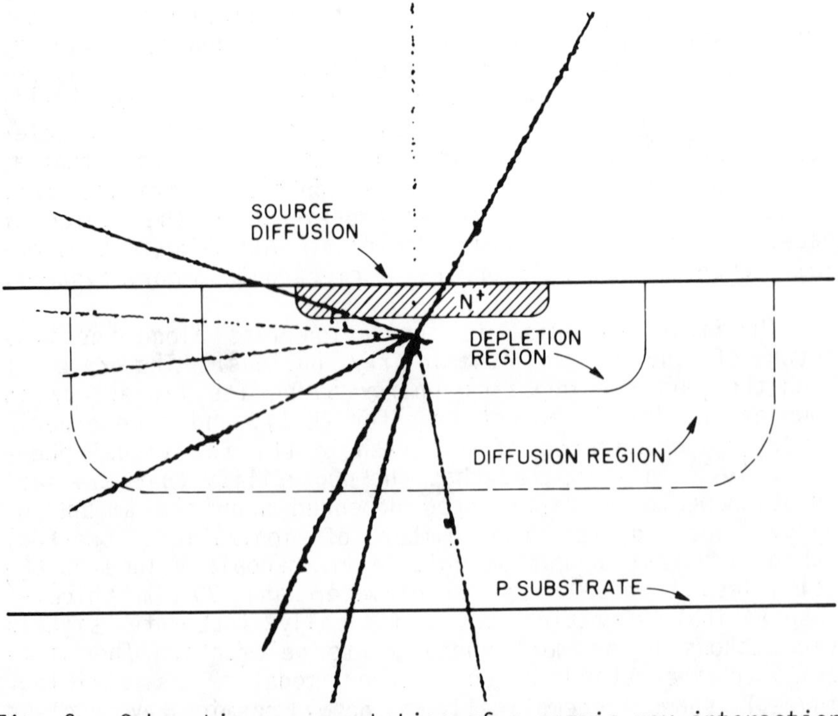

Fig. 2 Schematic representation of a cosmic-ray interaction.

device. The conventional model for soft errors requires the deposition of a threshold amount of energy in or about the depletion region of such an element.[1-7]

The second mechanism by which an energetic charged particle can deposit the same amount of energy within either a retinal summation unit[24,26,27] or a memory element[11,28,29] is by means of a nuclear interaction between the primary particle and a nucleus of the medium. Figure 2 shows a schematic representation of a cosmic-ray-induced nuclear interaction in a circuit element. The total energy deposited in the sensitive volume about the depletion region is the sum of the energies deposited by each of the secondary charged particles emerging from the event (mostly protons and alphas) and the recoiling nucleus. The relative contribution of the recoiling nucleus increases as the dimensions of the sensitive volume decrease. The trajectory of the primary particle is shown incident normal to the device, and the track is represented as that of a near minimum ionizing particle to emphasize the fact that nuclear interactions provide a mechanism by which low as well as high LET particles may induce soft errors.

Proton-induced nuclear stars provide a mechanism to allow low LET trapped protons to induce the unexpectedly high light-flash rates observed when Skylab entered the region of the inner radiation belt known as the South Atlantic Anomaly.[13] Moreover, they provide a reasonable explanation for the imperfections or blips in satellite photographs that were found to be correlated with the energetic proton flux incident on the DMSP spacecraft.[9] A major goal of the re-

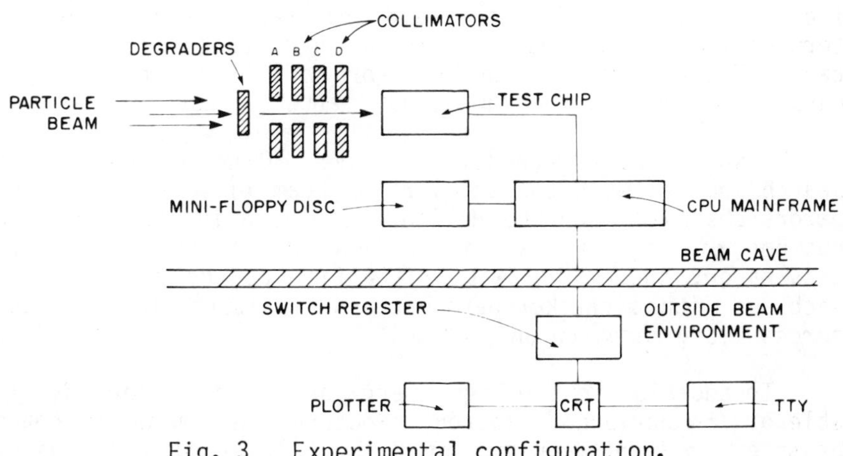

Fig. 3 Experimental configuration.

search described in this paper was to determine whether nuclear interactions contribute to soft errors in LSI devices.

A third possible loss mechanism is Cerenkov radiation. This mechanism requires an appropriate amount of transparent material in front of the sensitive region. While this condition is satisfied for the light flashes experienced by astronauts[12] it does not apply to LSI devices. Cerenkov radiation should, therefore, be a poor source of soft errors in LSI circuits.

III. Methods

The LSI devices studied were 4K dynamic RAMs, Intel's C2107B and National Semiconductor's MM-5280. The higher energy proton exposures were carried out at the Harvard Cyclotron. The configuration used for these exposures is shown schematically in Fig. 3. The 158 MeV beam from the cyclotron passed through 0.318 in. of brass, emerging at an energy of 130 MeV. The beam energy at the test device was controlled by inserting lucite degraders into the beam upstream of the chip. Exposures were carried out at beam energies, at the test device, of 18, 32, 51, 91, and 130 MeV. The beam energy was relatively well defined at the higher energies ($\pm$ 2 MeV at 130 MeV) but quite spread out at 18 MeV ($\pm$ 9 MeV). All proton irradiations to date were carried out at normal incidence. The device being irradiated was connected by about 3 ft of flat ribbon cable to a memory board of an 8080 CPU based Imsai Microcomputer. The device remained an integral part of the system's memory during irradiation. The microcomputer was connected by means of a switch register to a terminal, a printer, and a digital plotter located outside the radiation area. The operator was free to reset, reinitialize, and transfer control among the terminal, printer, and plotter without entering the beam cave. The memory was searched for soft errors either upon manual command, or automatically after preset intervals.

No soft errors were found in over 70 hr of running the search program with the accelerator beam off. Nor were any errors observed when the device was placed in the beam cave but not directly in the beam. For the data described in this paper, the device was initialized at the beginning of each run with a checkerboard pattern of alternating ones and zeroes along each row and column.

In addition to the high energy protons that were available at the Harvard Cyclotron, exposures were made to proton beams at incident energies of 0.93, 1.3, and 1.8 MeV using

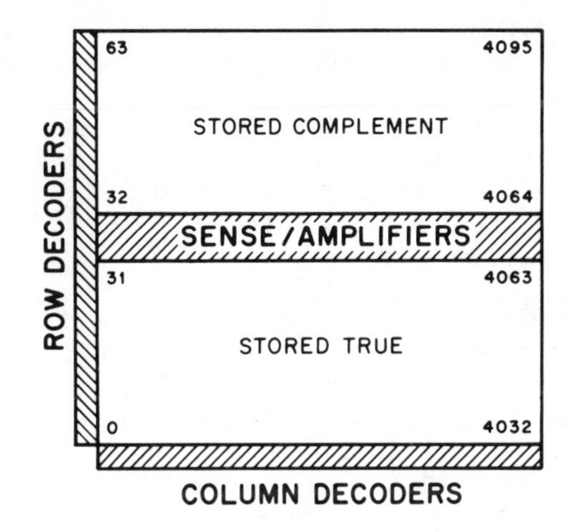

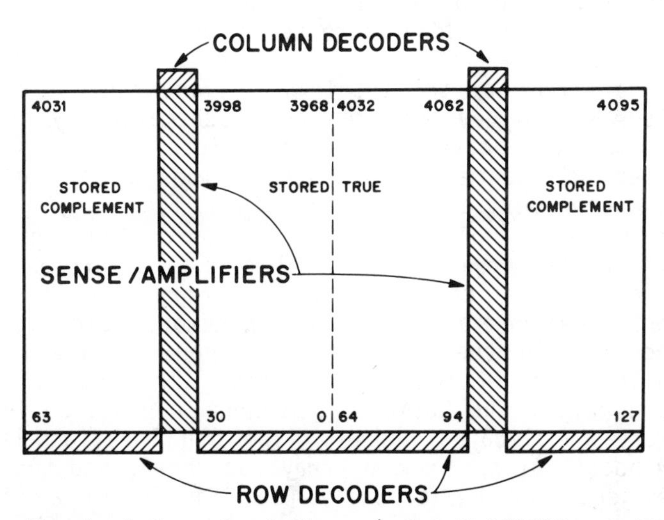

Fig. 4 Physical layout of the a) Intel C2107B and b) National Semiconductor MM5280 devices used for proton irradiations.

the RARAF Van de Graaff Facility at Brookhaven National Laboratory. These exposures were carried out to determine whether protons could induce soft errors without nuclear interactions. These beams had LET values in silicon of 42, 35, and 28 KeV/ μ m compared to 1 - 5 KeV/ μ m for the cyclotron beams. If ionization loss was a significant mechanism at the cyclotron energies (18-130 MeV), then the same devices should be even more sensitive to the low energy protons. The Van de Graaff proton beams typically had a full-width-half-maximum (FWHM) spread in energies of only a few percent but included a 5-20% contamination of lower energy protons.

For comparison irradiations of these devices also were carried out with 4.3 MeV ^{3}He nuclei at the RARAF Van de Graaff. The ^{3}He beam contained a contamination of approximately 20% protons. Exposures also were carried out to alphas from an ^{241}Am source.

At intervals during the irradiation, a device exhibiting changes was tested for hard errors (i.e., damage to the memory at one or more locations) The test consisted of reading and writing new data at each address in a sequence of four or five steps that changed the information at each location at least twice. All the data presented in this report involved devices that were, as yet, free of observable hard errors. The occurrence of a hard error was usually coincident with a sudden increase in the error rate.

The layout of the Intel C2107B and the National Semiconductor MM5280 devices are shown schematically in Figures 4a and 4b, respectively. In the Intel device the sense amplifiers divide memory into two regions while the National Semiconductor device is divided into four regions by two series of sense amplifiers. The corresponding regions where data is stored true or in complement form are shown. For the Intel device the sequential memory locations occupy adjacent locations along columns in memory starting with the 0th location in the lower left corner and ending at the 4095th location in the upper right. The memory locations are laid out in the National Semiconductor device in a somewhat more complicated manner. The 0th location is at the center bottom of the memory array, as shown, and the addresses are sequenced horizontally (see Fig.4b in a repetative pattern of the form 0, 1, 3, 2, 4, 5, 7, 6. This pattern also determines the sequence of rows also and the memory locations 3068 through 4095 would occupy the second column from the top in Fig. 4b.

IV. Results

The most striking feature of the proton-induced soft errors is the correlation between the type of error and its address in memory. Such correlations were evident in all the devices of both types tested. Typical soft-error patterns following exposure to 20, 32, 91, and 130 MeV protons are shown in Figs. 5a - 5d for an Intel C2107B device. The corresponding patterns for 32, 91 and 130 MeV proton exposures of a National Semiconductor MM5280 device are shown

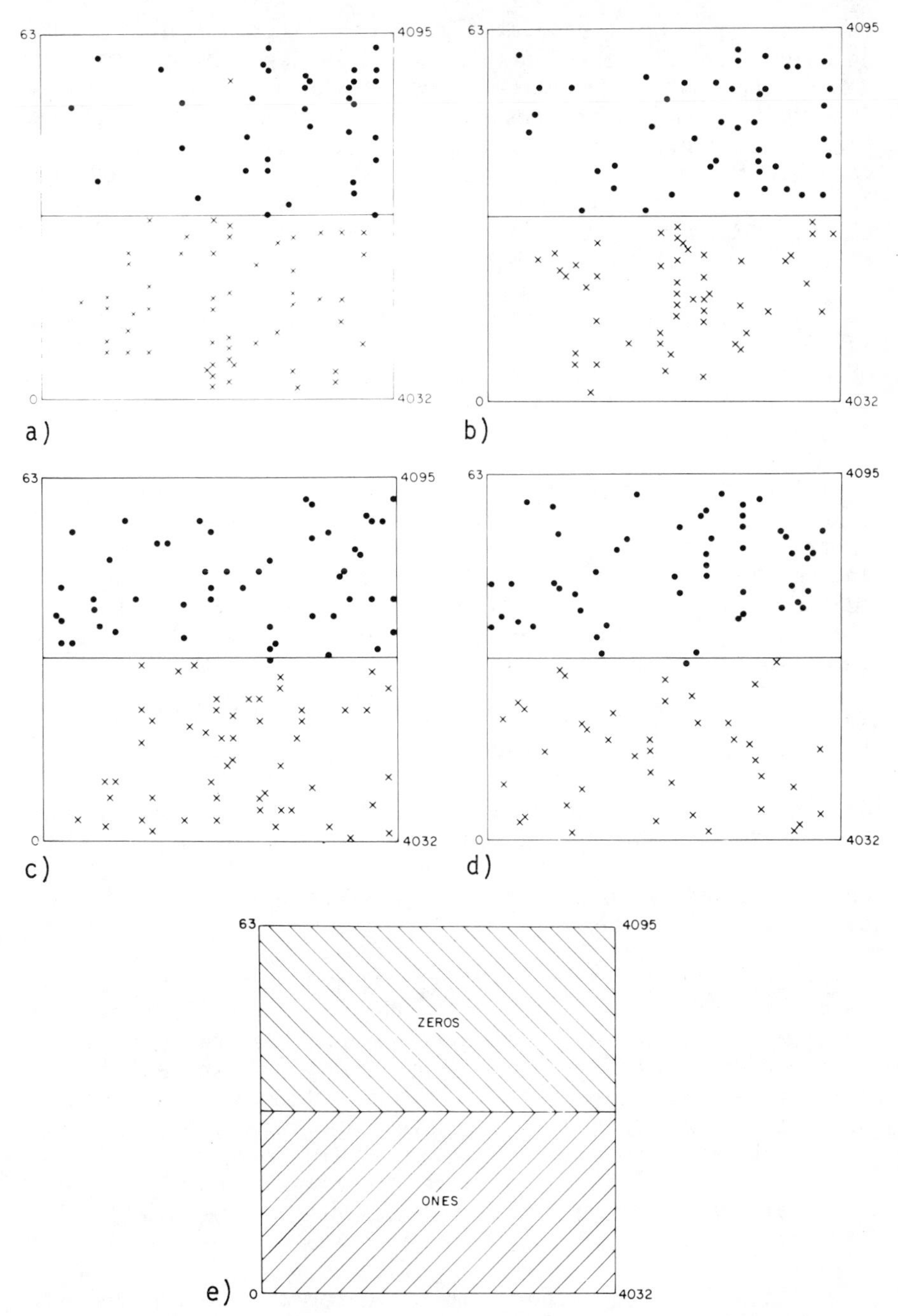

Fig. 5 Error maps at a) 20 MeV, b) 51 MeV, c) 91 MeV, and d) 130 MeV for Intel C2107B device following exposure to protons; e) Typical turn-on pattern for Intel C2107B devices.

in Figs. 6a - 6c, respectively. Both figures are bit maps on which each error's position represents its address in memory; the two-dimensional space of the figures is divided into 4096 memory addresses distributed in 64 columns of 64 locations each. The map starts with the 0th address in the lower left hand corner. Memory locations 0 - 63 occupy the left edge of each figure, 64 - 127 the next column of locations, and so on, until locations 4032 through 4095 at the extreme right of the display.

The individual memory locations are not shown in the figure except where a soft error has occurred. Errors which correspond to changes in logic state from one to zero are represented as slanted crosses and zero to one transitions as solid circles.

The address space used for the error maps of Figs. 5 and 6 corresponds to the physical sequence of the memory locations on the Intel device (as shown in Fig. 4a) and not the sequence on the National Semiconductor devices. In both maps the soft errors corresponding to zero to one changes (solid circles) and those from one-to-zero (slanted crosses) segregate into distinctly separate regions of memory. These regions are the same for devices of the same type and are delineated in the figures by solid lines. For the Intel devices but not for the National Semiconductor devices, these regions are also the regions of true and complement storage. For both device types, these regions of zero-to-one and one-to-zero errors reflect the architecture of the device in that they correspond to the regions of ones and zeros that are typically obtained when the system is first energized, the so-called turn-on patterns.

The Intel devices tested typically had turn-on patterns consisting of zeros in the upper half of memory and ones in the lower, and the National Semiconductor devices had alternating regions of eight rows each of all ones and all zeros. Turn-on patterns for the Intel and National Semiconductor devices are shown schematically in Figs. 5e and 6d, respectively. By repeated power-ups of the main frame of the microcomputer, it was often possible to transform the original National Semiconductor turn-on patterns to one nearly its complement.

The soft-error patterns in the error maps shown in Figs. 5 and 6 vary appreciably with beam energy. Soft errors in which zero logic states changed to ones were particularly sensitive to the incident particle energy. The ratio of zero-to-one and one-to-zero type upsets is plotted vs inci-

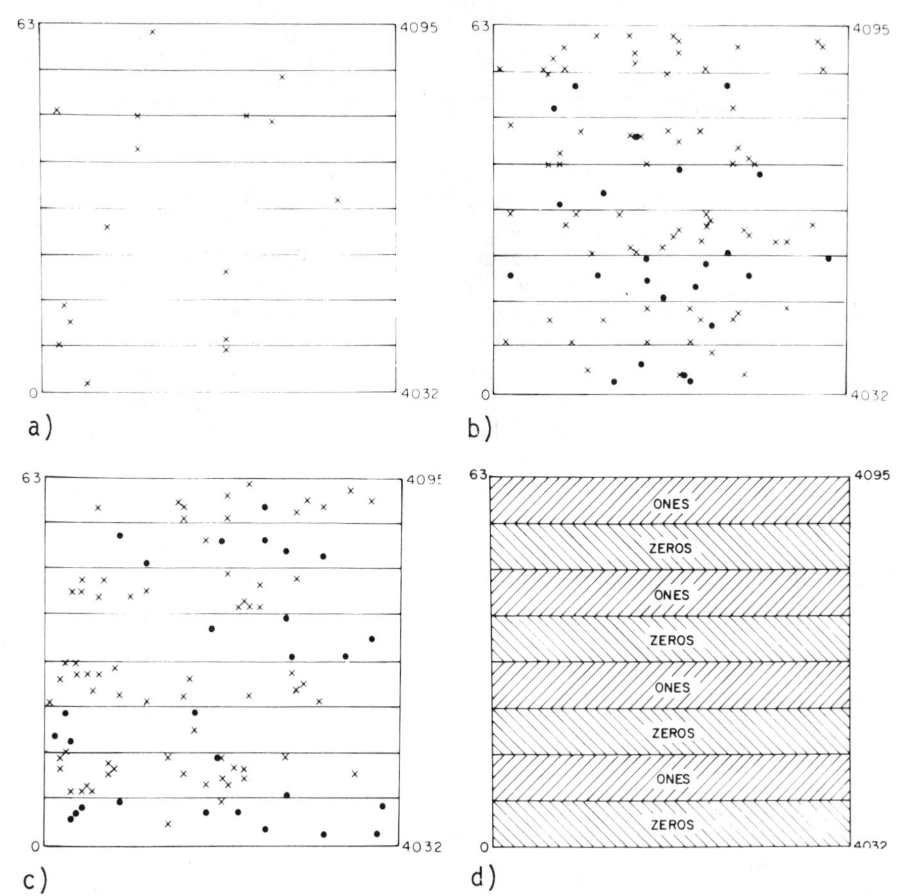

Fig. 6 Error maps for National Semiconductor device MM-5280 when exposed to protons at a) 32 MeV; b) 91 MeV; c) 130 MeV; d) Typical turn-on pattern for the National Semiconductor devices.

dent proton energy in Fig. 7 for individual Intel and National Semiconductor devices. While the ratio increased with beam energy for all the devices tested, there was considerable variation among devices, even those from the same manufacturer. The data shown in Fig. 7 are from measurements on single devices and should be interpreted as a qualitative representation of the trend to be expected from other devices of the same type.

At low proton energies (<2MeV) all the soft errors were one-to-zero. The sharp rise in the relative number of zero-to-one errors with beam energy reflects differences in either the physical mechanisms of energy deposition or the circuit elements sensitive to upsets. The fact that these differences are dependent on beam energy implies that different energy deposition thresholds are involved.

Typical error maps obtained by exposing an Intel C2107B device to 4.3 MeV ^{3}He particles at the RARAF Van de Graaff is shown in Fig. 8a, and a similar plot for a National Semiconductor device exposed to ^{4}He nuclei (alphas) from an ^{241}Am source is shown in Fig 8b. Again, the memories were initialized with a checkerboard pattern of ones and zeros. Clear differences between the helium and proton data are evident when Figs. 8a and 8b are compared with Figs. 5 and 6. First, only one zero-to-one type errors occurred in the Intel device exposed to ^{3}He and none occurred in the National Semiconductor device exposed to alphas. Besides the absence of zero-to-one soft errors, the helium data of Figs. 8a and 8b exhibit considerably more structure than is evident in Figs. 5 and 6 for protons. The errors in Figs. 8a and 8b occur almost exclusively in certain columns.

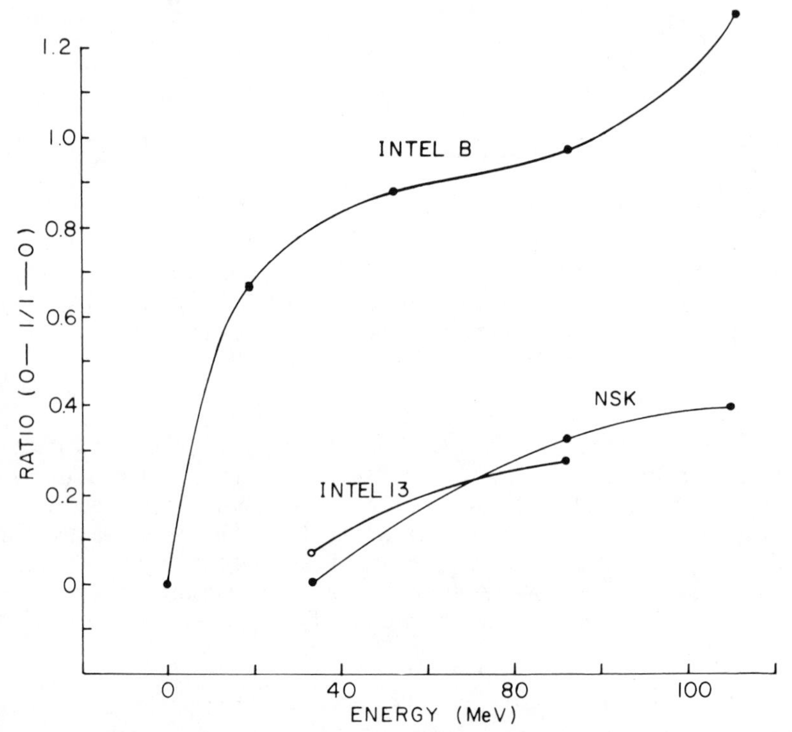

Fig. 7 Ratio of zero to one to one to zero transitions vs beam energy.

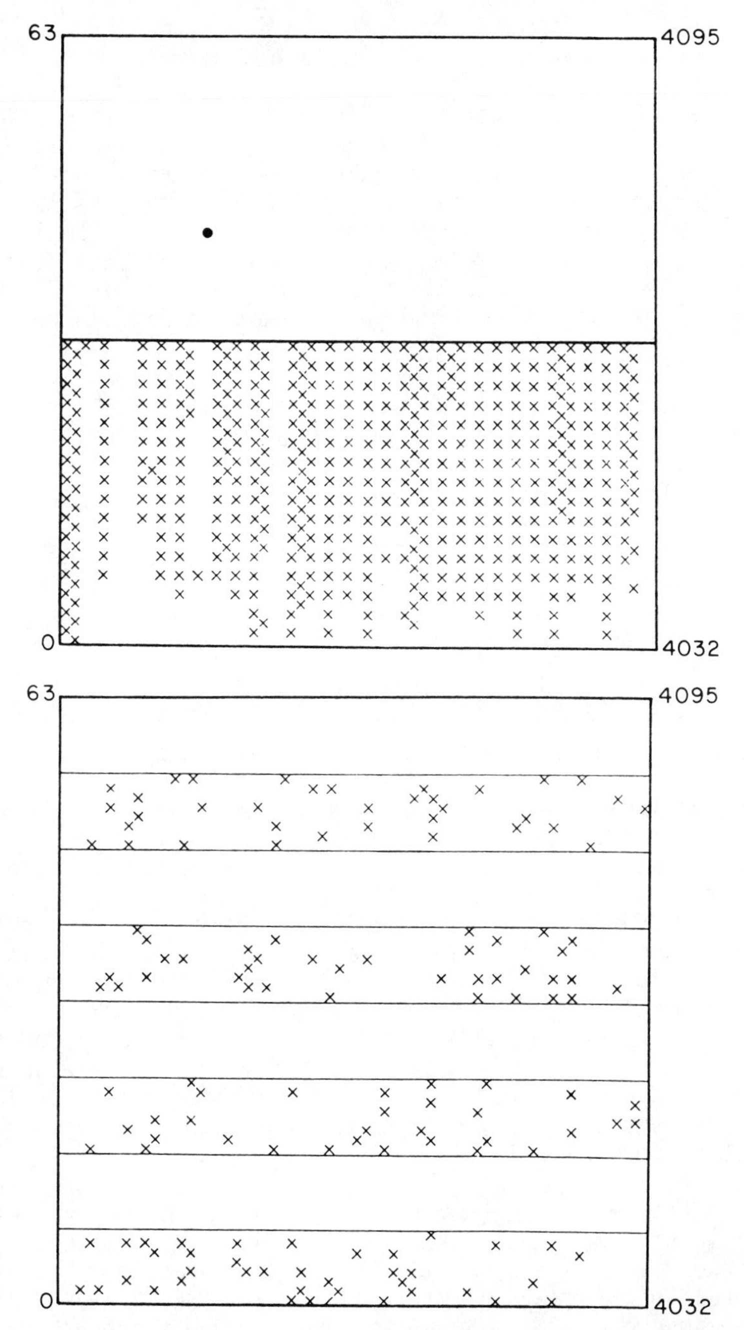

Fig. 8 a) Error map for Intel device C2107B exposed to 4.3 MeV ^{3}He. b) Error map for National Semiconductor device MM-5280 exposed to alphas from an ^{241}Am source.

Moreover, in the Intel devices these columns appeared to fill from the center line in Fig. 8a downward.

The soft error patterns of Figs. 8a and 8b appear to be in agreement with the earlier studies of May and Woods[4-6] and Yaney et al[7] who concluded that alpha-induced soft errors are not due to interactions in the memory cells themselves; they found that the bit lines and sense amplifiers were likely targets. The column structure exhibited in Figs. 8a and 8b is much less evident in the 130 MeV proton data in Figs. 5a and 5b, and the columns in Fig. 4a definitely did not fill from the center line down.

The structure evident in the error maps and the variation of the patterns with beam energy are not understood but appear inconsistent with models assuming that the memory elements are themselves the only sensitive targets. In that case the errors would be randomly distributed on an error map, as in fact they are for errors induced by stopping heavy ions[3].

Soft Error Cross Sections

The soft error cross sections were obtained by dividing the number of first soft errors by the proton fluence. The soft error cross sections measured in this study are plotted in Fig. 9 vs beam energy for the Intel and National Semiconductor devices. Data points obtained at different energies on the same devices are connected by lines. Considerable variation among devices by the same manufacturer is evident, but for the same device there seems to be a clear increase in the soft error cross section with energy. This increase occurs despite the fact that the LET and the total inelastic cross sections are decreasing with increasing energy over this interval. The explanation, presumably, lies in the fact that the average energy released in the interaction (and that transferred to the recoiling nucleus) increase with beam energy, and the cross section for events which result in large local depositions of energy probably increase with beam energy over the range 18-130 MeV.[30]

In the series of irradiations discussed here, the memory was reinitialized each time a soft error was discovered, and the records for a number of device exposures were examined to determine whether the sensitivity of the devices varied with exposure. No evidence for significant changes in the soft error cross section for first soft errors as a function of dose were found.[11]

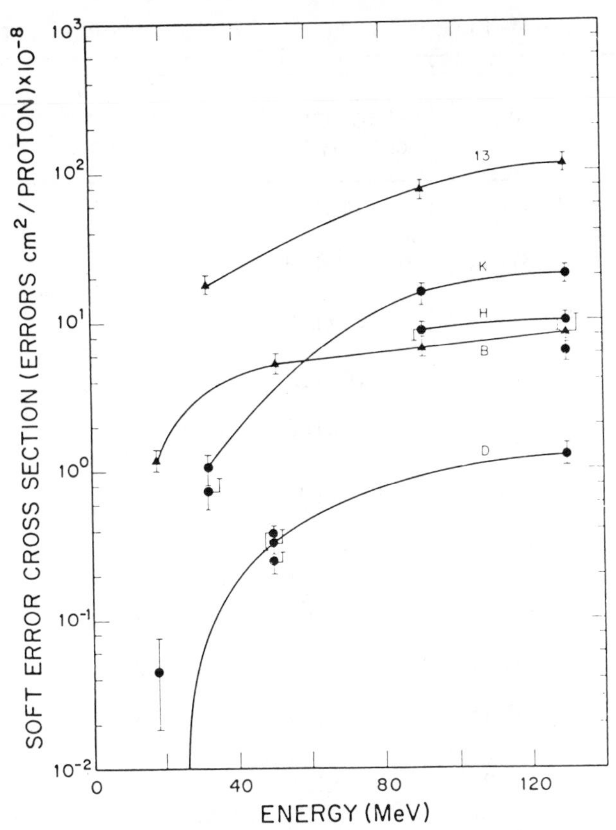

Fig. 9 Soft error cross section vs beam energy. (Lines connect data points obtained for the same devices; circles and triangles represent National Semiconductor and Intel devices, respectively.

The devices were found to be less sensitive to the low energy protons despite their much higher LET values (28-42 KeV/ μ m vs 1-5 KeV/ μ m for the cyclotron beams). Table 1 gives the proton errors observed in a number of different devices exposed to 0.93, 1.3 and 1.8 MeV protons. For most of the devices the soft error cross section was zero. The non zero cross sections in Table 1 are not necessarily evidence that the device is sensitive to individual low energy protons. Individual elements on the device would experience enough proton traversals to absorb 4 MeV or more within a refresh cycle at rates that were greater than the observed error rates. However,in the data given in Table 1, there is no evidence of dose-rate dependence of the type to be expected multiple hits were required for a soft error.

Table 1 Soft Errors induced by low energy protons.

Energy MeV	Device	Instanteous flux $cm^{-2}sec^{-1}$	Number errors	Dose rads
0.93	NS-5	$1.4x10^6$	0	25,000
0.93	Intel-14	$1.4x10^6$	0	25,000
0.93	Intel-2	$1.4x10^6$	37	3,600
0.93	Intel-2	$1.4x10^6$	0	15,000
0.93	Intel-3	$1.4x10^6$	3	465
1.3	Intel-3	$8.6x10^6$	7	200
1.3	Intel-1	$7.5x10^6$	2	15,000
1.3	Intel-1	$1.5x10^7$	0	4,000
1.3	Intel-4	$1.5x10^7$	0	22,000
1.3	Intel-4	$9.5x10^7$	0	2,000
1.8	Intel-2	$1.0x10^7$	0	1,060
1.8	Intel-3	$1.4x10^7$	8	500

Different circuit elements may become sensitive to soft errors at the higher proton energies as a result of the higher local energy deposition through interactions. This would be consistent with the changes in the type of errors evident in Figs. 5 - 7.

V. Error Rates in Spacecraft Systems

The soft error cross sections plotted in Fig. 9 range from 10^{-8} to 10^{-6} errors cm^2/proton at the higher proton energies. If, in the absence of measured values, the soft error cross section is assumed to be constant at energies above 130 MeV, then the soft-error rates to be expected in space can be estimated. A spacecraft system with 10^5 bits of memory distributed in 25 memory devices of the type tested in this study, flying in a region of deep space characterized by an energetic proton flux of 2 protons $cm^{-2}sec^{-1}$, would exhibit proton-induced soft errors of from 0.04 to 4 errors per day. This is comparable to the rates currently being reported.[2] The high proton fluence in space compensates somewhat for the relatively low proton cross sections. In the radiation belts or during solar flares, the proton-induced nuclear interactions would, of course, be greatly increased, resulting in correspondingly higher soft error rates.

VI. Conclusion

Energetic protons have been shown to induce logic upsets in two types of 4K dynamic RAM. These devices must, therefore, be soft error sensitive to the entire charge spectrum of the cosmic rays. The cross sections measured for protons are at least four orders of magnitude lower than those measured by Blandford et al.[2] and Kolasinski et al.[31] for 180 MeV stopping argon and krypton ions incident on similar devices. In the latter case the measured cross sections were comparable to the cross-sectional areas occupied by the sensitive regions of all the memory cells on the device, and those data apparently are consistent with the hypothesis that a heavy ion must deposit 8-20 MeV or more within the sensitive volume associated with a memory cell in order to change the logic state of that element.

The lower soft error cross section protons may still be a significant source of upsets in memory systems flown in space because the lower cross section is compensated by the fact that protons are considerably more abundant in space than are heavy ions capable of depositing 3 MeV or more in ionization loss along a few microns of trajectory. The proton contribution to the radiation-induced soft errors to be expected for specific devices flown in space can be determined experimentally using protons available at accelerator facilities.

The range of proton soft error cross sections measured in this experiment (see Fig 9) are clearly smaller than the physical dimensions of any LSI circuit element. The exact dimensions of the element that might be sensitive to soft errors are not known because of proprietary restrictions, and their determination was beyond the scope of this work.

The soft errors induced by protons with incident energies between 18 and 130 MeV exhibit clear differences from those induced by alphas [4-7] as shown by comparing Figs. 5 and 6 with Fig. 8. The structure evident in the error maps and the differences between zero to one and one to zero type errors distinguish proton-induced soft errors from those reported recently for stopping argon ions.[3,31] At low energies the differences from alpha-induced upsets is less pronounced. This is consistent with the findings of Guenzer et al[8] who report agreement between upsets obtained with 6-14 MeV neutrons and rough predictions based on "n, alpha" and "n, alpha n" cross sections. (During the preparation of this manuscript we have learned that Guenzer et al. have observed proton-induced soft errors at cross sections of approximately $10^{-8} cm^2$. Their data will be included in the published version of Ref. 8.) The alpha-induced soft errors and, presumably a significant fraction of the lower energy proton upsets, result primarily from events in other than the memory cells, probably in the bit lines or the sense amplifiers. As the proton energies increase the events become more randomly distributed in memory and zero to one upsets increasingly appear. This may reflect a transition from "p, alpha" events to spallation events as the dominant mechanism for soft errors. The high energies released in the latter events make additional circuit elements available for soft errors.

There is as yet no reliable single parameter which can be used to classify particles according to whether they will induce soft errors in space. The value of the particle's LET is possibly sufficient to determine whether a cosmic ray particle will deposit sufficient energy by ionization loss to induce a soft error if it traverses a sensitive volume element. However, a low value of LET does not rule out soft errors because low-LET nuclear particles can induce errors through nuclear interactions. Since only those interactions which result in a threshold localized deposition of energy can be expected to contribute, the total cross sections for inelastic nuclear scattering also are not quantitatively useful. The situation is further complicated by evidence of more than one type of sensitive circuit element on both the devices tested.

The soft error phenomena is fast becoming a serious environmental problem for systems in space. Considerable data involving a variety of device types and a broad range of energies are needed before quantitative determination of the severity of the problem can be made.

Acknowledgements

The indispensable cooperation and many helpful suggestions of A. Koehler of the Harvard Cyclotron, S. Marino and N. Rohrig of the RARAF facility of Brookhaven National Laboratory, and P.Toumbas of Clarkson College are greatfully acknowledged. This work was supported in part by Rome Air Development Center contract number F30602-78-C-0102 and NASA Contract NSG 9059.

References

[1]Binder, D., Smith, E.C., and Holman, A.B., "Satellite Anomalies from Cosmic Rays", IEEE Transactions on Nuclear Science, Vol. NS-22, December 1975, pp. 2675-2680.

[2]Pickel, J.C. and Blandford, J.T., "Cosmic Ray Induced Errors in MOS Memory Cell", IEEE Transactions on Nuclear Science, Vol. NS-25, October 1978, pp. 1166-1167.

[3]Blandford, J.T., Rockwell International, personal communication.

[4]May, T.C. and Woods, M.H., "A New Physical Mechanism for Soft Errors in Dynamic Memories", Reliability Physics 16th Annual Proceedings 1978, San Diego, Calif., April 1978, IEEE, New York, pp. 33-40.

[5]May, T.C. and Woods, M.H., "Alpha-Particle-Induced Soft Errors in Dynamic Memories", IEEE Transactions on Electron Devices, Vol. ED-26, January 1979, pp. 2-9.

[6]May, T.C., personal communication.

[7]Yaney, D.S., Nelson, J.T., Vanskike, L.L., "Alpha-Particle Tracks in Silicon and their Effects on Dynamic MOS RAM Reliability", IEEE Transactions on Electron Devices, Vol. ED-26, January 1979, pp. 10-16.

[8]Guenzer, C.S., Wolicki, E.A., and Allas, R.G., "Single Event Upset of Dynamic Rams by Neutrons and Protons", Proceedings of the IEEE Conference on Nuclear and Space Radiation Effects Santa Cruz, Calif., July 1979.

[9]Filz, R.C., and Katz, L., "An Analysis of Imperfections in DMSP Photographs Caused by High Energy Solar and Trapped Protons", Air Force Report AFCRL-TR-74-0469, September 1974.

[10]Croft, T.A., "Nocturnal Images of the Earth from Space", Stanford Research Institute Report No. 68197, March 1977.

[11]Wyatt, R.C., McNulty, P.J., Toumbas, P., Rothwell, P.L., and Filz, R.C., "Soft Errors Induced by Energetic Protons", Proceedings of the IEEE Conference on Nuclear and Space Radiation Effects, Santa Cruz, Calif., July 1979.

[12]Pinsky, L.S., Osborne, W.Z., Bailey, J.V., Benson, R.E., Thompson, L.F., "Light Flashes Observed by Astronauts on Apollo II through Apollo 17" Science, Vol. 183, March 1974, pp. 957-959.

[13]Pinsky, L.S., Osborne W.Z., Hoffman, R.A., and Bailey, J.V., "Light Flashes Observed by Astronauts on Skylab 4", Science, Vol. 188, May 1975, pp. 928-930.

[14]Fremlin, J.H., "Cosmic Ray Flashes" New Scientist, Vol. 47, July 1970, pp. 42.

[15]Tobias, C.A., Budinger, T.F., Lyman, J.T., "Radiation-induced Light Flashes observed by Human Subjects in Fast Neutron, X-ray and Positive Pion Beams", Nature, Vol. 230, April 1971, pp. 596-597.

[16]McNulty, P.J., "Light Flashes produced in the Human Eye by Extremely Relativistic Muons", Nature, Vol. 234, November 1971, pp. 110.

[17]Budinger, T.F., Bichsel, H., and Tobias, C.A., "Visual Phenomena Noted by Human Subjects on Exposure to Neutrons of Energies less than 25 Million Electron Volts", Science, Vol. 172, May 1971, pp. 808-870.

[18]McNulty, P.J., Pease, V.., Bond, V.P. "Schimmerling, W., Vosburgh, K.G., "Visual Sensations Induced by Relativistic Nitrogen Nuclei", Science, Vol. 178, October 1972, pp. 160.

[19]McNulty, P.J., Pease, V.P., and Bond, V.P., "Visual Sensations Induced by Cerenkov Radiation", Science, Vol. 189, August 1975, pp. 453-454.

[20]McNulty, P.J., Pease, V.P., Bond, V.P., "Muon-Induced Visual Sensations", Journal of the Optical Society of America, Vol. 66, January 1976, pp. 49-55.

[21]McNulty, P.J., Pease, V.P., and Bond, V.P., "Visual Sensations Induced by Relativistic Pions", Radiation Research, Vol. 66, June 1976, pp. 519-530.

[22]McNulty, P.J., Pease, V.P., and Bond, V.P., "Visual Sensations Induced by Individual Relativistic Carbon Ions: With and Without Cerenkov Radiation", Science, Vol. 201, July 1978, pp. 341-343.

[23]Madey, R., and McNulty, P.J., "Frequency of Light Flashes Induced by Cerenkov Radiation From Heavy Cosmic-Ray Nuclei",

Proceedings of the National Symposium on Natural and Manmade Radiation in Space, edited by E.A. Warman, January 1972, pp. 757-766.

[24]McNulty, P.J., and Madey, R., "Direct Stimulation of the Retina by the Method of Virtual Quanta for Heavy Cosmic-Ray Nuclei", Proceedings of the National Symposium on Natural and Manmade Radiation in Space, edited by E.A. Warman, January 1972, pp. 767-772.

[25]Tobias, C.A., Budinger, T.F., and Lyman J.T., "Human Visual Response to Nuclear Particle Exposures", Proceedings of the National Symposium on Natural and Manmade Radiation in Space, edited by E.A. Warman, January 1972, pp. 757-766.

[26]Rothwell, P.L., Filz, R.C. and McNulty, P.J., "Light Flashes Observed on Skylab 4: The Role of Nuclear Stars", Science, Vol. 193, September 1976, pp. 1002-1003.

[27]McNulty, P.J., Pease, V.P., Bond, V.P., Filz, R.C., and Rothwell, P.L., "Particle Induced Visual Phenomena in Space", Radiation Effects, Vol. 34, March 1977, pp. 153-156.

[28]Bradford, J.N., "Nonequilibruim Radiation Effects on VLSI", IEEE Transactions on Nuclear Science, Vol. NS-25, October 1978, pp. 1144-1145.

[29]Bradford, J.N., "Expected Contributions to LSI/VLSI RAM Soft Error Rates by Solar and Cosmic Protons -- Some Implications For Beam Weaponry", Air Force Report RADC-TM-79-ES-02, March 1978.

[30]Bertini, H.W.,"Spallation Reactions: Calculations", Spallation Nuclear Reactions and Their Applications, Edited by B.S. P.Shen and M.Merker, D.Reidel Publishing Co., Boston, MA, 1976, pp. 27-48.

[31]Kolasinski, W.A., Blake, J.B., Anthony, J.K., Price, W.E., and Smith, E.C., "Simulation of Cosmic-Ray-Induced Soft Errors and Latch Ups in Integrated Circuit Computer Memories", Proceedings of the IEEE Conference on Nuclear and Space Radiation Effects, Santa Cruz, CA, July 1979.

Chapter IV—Large Space Systems Interactions with the Space Environment

REVIEW OF INTERACTIONS OF LARGE SPACE STRUCTURES WITH THE ENVIRONMENT

N. John Stevens*
NASA Lewis Research Center, Cleveland, Ohio

Abstract

Environmental interactions are defined as the response of spacecraft surfaces to the charged-particle environment. These interactions are divided into two broad categories: spacecraft passive, in which the environment acts on the surface, and spacecraft active, in which the spacecraft or a system on the spacecraft causes the interaction. The principal spacecraft passive interaction of concern is the spacecraft charging phenomenon. The spacecraft active category introduces the concept of interactions with thermal plasma environments and Earth magnetic fields which are important at all altitudes. This report serves as an introduction to these active interactions. Motion induced mechanical stresses in large space structures and high voltage, large power system interactions are considered.

Introduction

Very large spacecraft are being proposed for future space missions. These spacecraft are to be used for such activities as manufacturing, scientific exploration, power generation, and habitation in locations ranging from low Earth orbits (200-400 km) to geosynchronous orbit and beyond.[1-7] Structures proposed for these missions range in size from a 10-30-m fabrication and demonstration model and a 40-200-m diam antenna, to the several kilometer dimensions of the Solar Power Satellite (SPS).[8] Because of their large sizes, structures are being designed with relatively lightweight materials to achieve the low densities required for transportation to space. As a result, the spacecraft designer faces new problems of spacecraft interaction with the environment. This report reviews these interactions.

Presented as Paper 79-0386 at the AIAA 17th Aerospace Sciences Meeting, New Orleans, La., Jan. 15-17, 1979.

*Head, Spacecraft Environment Section.

Spacecraft must function in the space environment. Anomalous behavior of geosynchronous satellite systems has shown that the environment is not completely benign. Interactions between the charged-particle environment and spacecraft exterior surfaces (i.e., spacecraft charging) can cause disruptions in spacecraft systems.[9,10] The size of the new generation of spacecraft will be on the order of the ion gyroradii at geosynchronous altitudes, which can increase interactions. The proposed spacecraft physical dimensions also are such that there is growing concern for the effect that the spacecraft can have on modifying the environment.

Proposed large, high-power systems ranging from tens of kilowatts[11] to gigawatts[8] have given rise to new environmental interactions. As power levels rise, operation at high voltages is mandatory to reduce electrical losses while maintaining reasonable weight. A Satellite Power System (SPS) design configuration calls for the generation of 10 GW at 40 kV. To

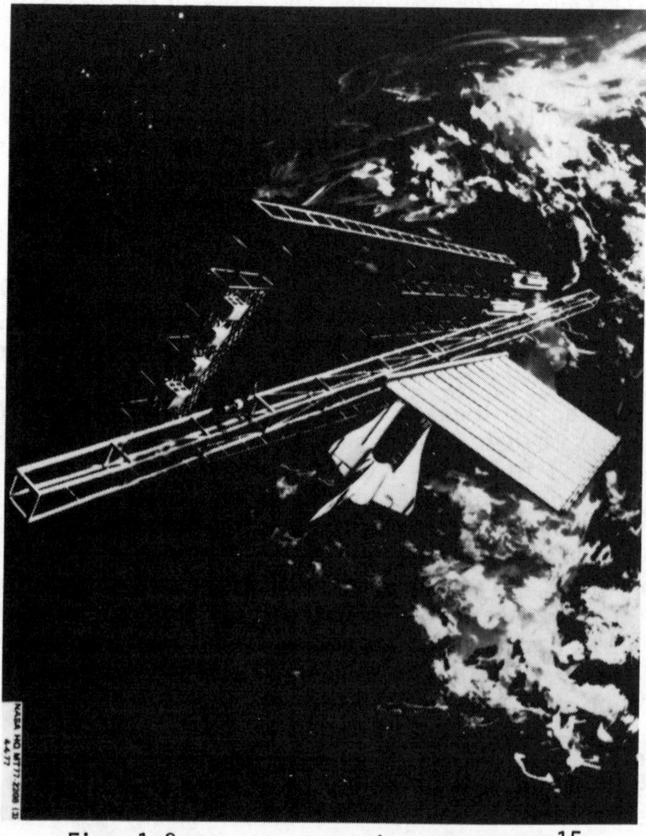

Fig. 1 Space construction facility.[15]

date, the highest operational voltage used in space is the 100-V system on Skylab.[12] At this voltage interactions with the charged particle environment are negligible.[13] Operation at higher voltages in a plasma environment, however, can influence system performance.

To illustrate the types of large structures proposed for future missions, consider the system shown in Fig. 1.[14] This system is a space construction platform with a 250-kW power array attached to provide power for space construction operations and technology demonstrations. Note the relative size of the Shuttle orbiter compared to the structure. It is the interactions of such a large structure with the charged particle environment that are of concern. Considering the cost and complexity of future large structures, these interactions must be understood, evaluated, and neutralized, if necessary, in the program design phase. This report reviews two possible interactions between active spacecraft surfaces and this charged-particle environment. The categories of interactions are defined and only briefly described; subsequent papers will deal in more detail with specific aspects.

Categories of Spacecraft-Environmental Interactions

Spacecraft-environmental interactions can be defined as the response of spacecraft surfaces to the charged-particle environment in space. Although these surfaces can be charged by this environment at all altitudes, the interactions are of concern only when they influence system performance. Interactions of concern between spacecraft and the environment are illustrated in Fig. 2. A pictorial representation of a large spacecraft configuration employing a large, high-power solar array is shown. There are two broad categories of interactions indicated: 1) spacecraft passive, where the charged-particle environment acts on spacecraft surfaces, and 2) spacecraft active, where the spacecraft or a system on the spacecraft causes the interaction.

The principal passive interaction of concern in category 1 is the spacecraft charging phenomenon. Since the spacecraft charging interaction with passive spacecraft has been presented elsewhere, it will not be discussed further here. Interactions in category 2 involve the motion-induced charging effects in large structures (i.e., due to spacecraft velocity) and electric field-induced charging effects in high-voltage, large space power system (i.e., due to spacecraft voltages accelerating particle flows to surfaces). These interactions will be discussed in the following sections.

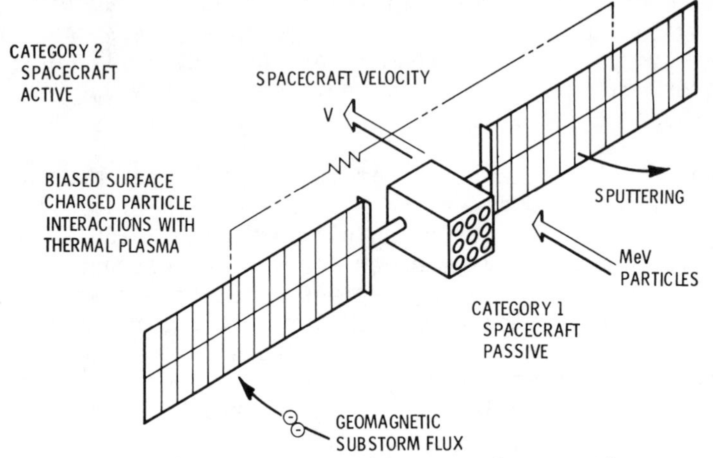

Fig. 2 Spacecraft-environment interactions.

Both categories of interactions are controlled by the charged-particle flux. The net current to the conductive surfaces and to each element of an insulator surface must be zero. This means that the surface voltages will be adjusted (relative to space plasma potential) until currents balance. Incoming currents and secondary, backscatter, and photoemitted currents must be known in order to predict surface voltages. Coupling between various parts of the spacecraft occurs not only on the surface but also through the plasma environment, thus complicating computations. Three-dimensional analytical techniques are required to predict surface voltages even for relatively simple geometries. Such analytical techniques are detailed elsewhere.

Large Space Systems Interactions

In this section interactions between charged particle environments and two classes of large active space systems will be discussed. The two classes of systems are large space structures and space power systems.

Large Space Structure Interactions

Introduction. The large structures envisioned for space have been described in journals and at conferences. These structures range from huge to momentous. To date, designers have been more concerned with devising means of building and assembling such structures than with the effects of possible interactions with the environment. These large structures will, for example, move through the weak magnetic field around

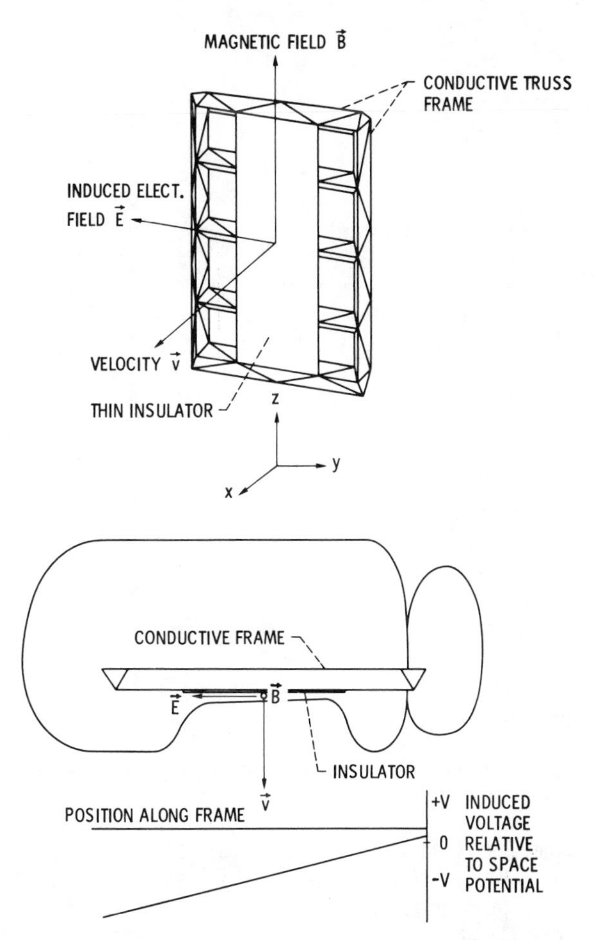

Fig. 3 Large space structure interaction: a) Induced field, and b) Possible voltage sheaths.

Earth, and this motion will induce small electric fields which can generate forces in the structure and cause particle interactions. It is the purpose of the following discussion to point out possible interactions that could influence the structure design.

Structure-environmental interactions. Consider the pictorial representation of a large structure moving in a low Earth orbit (about 400 km) across magnetic field lines as illustrated in Fig. 3a. This structure is assumed to be built with an open triangular truss network made from conductive materials. Across the central portion there is assumed to be a thin insulation cover. This idealized structure could be in

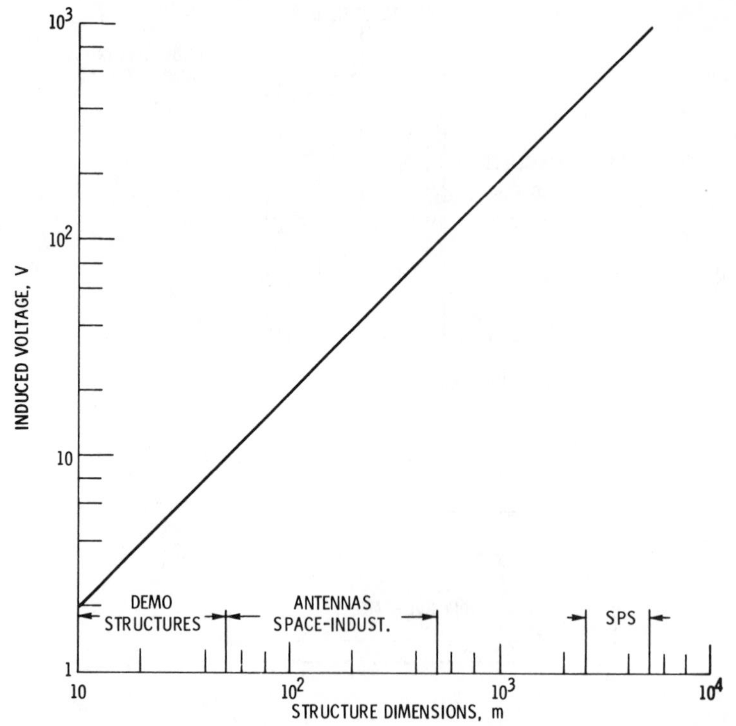

Fig. 4 Voltages induced in structure (E = 0.2 V/m).

reality a communications platform, an antenna system, or space power module. For convenience, it is assumed that the structure is moving perpendicular to the magnetic field, and solar effects are neglected.

It is known that an electric field $\vec{E}$ will be generated in a conductor moving with velocity $\vec{v}$ in a magnetic field $\vec{B}$:

$$\vec{E} = \vec{v} \times \vec{B}$$

For the coordinate system assumed in this example, the electric field will be induced in those conductors perpendicular to both the velocity and magnetic field (i.e., those conductors in the y direction). For the velocity and magnetic field at low Earth orbit, this electric field is on the order of 0.2 V/m.[15] Since any surface in space must have net zero current flows and since electrons are much more mobile than ions, this electric field will be maintained such that, relative to plasma potential, the conductive surfaces will have a small area at a slight positive potential and the rest will be negative. A situation similar to that shown in Fig. 3b could

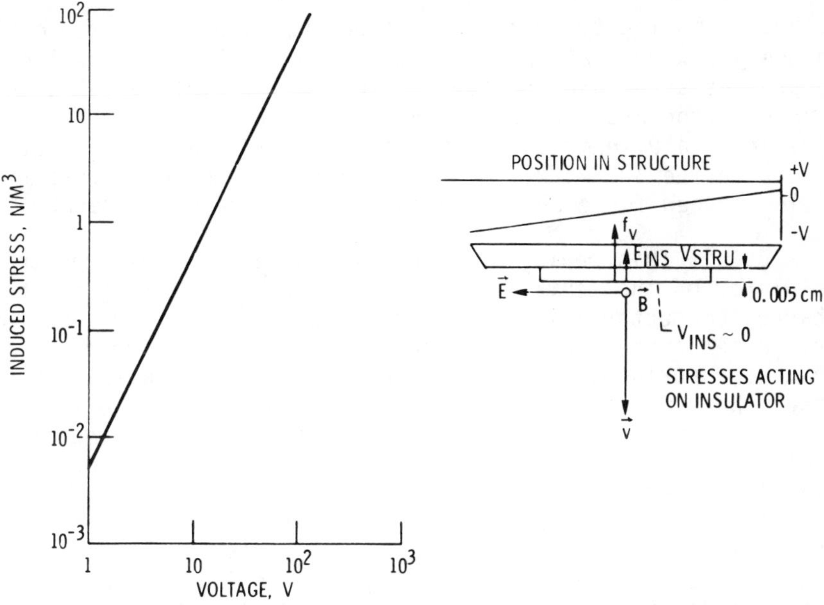

Fig. 5 Stress induced in 0.005 cm insulator by voltage across insulator.

exist. This is a view looking in the x-y plane. The insulator surface will come to a slightly negative potential depending on its current balance conditions and independent of the conductive structure. Since there can be velocity effects (wake- and ram-induced changes in density), the voltage sheaths could be distorted as shown in the figure. The expected voltages as a function of structure dimensions for this electric field are shown in Fig. 4. Until one starts considering multikilometer-sized structures, these voltages are not too large. Since the magnetic field diminishes with distance from Earth, these induced effects are negligible at geosynchronous altitudes.

The force or stress induced in the structure due to the electric field can be computed by means of Maxwell's stress tensor[16] as (for the simplified geometry):

$$T = \frac{1}{2}\vec{D} \cdot \vec{E} = \frac{1}{2}\varepsilon\varepsilon_0 E^2$$

where ε is dielectric constant and ε_0 is the permittivity of space (8.8×10^{-12} $C^2/N\ m^2$). Substituting in for the electric field and for typical values for ε results in a stress on the order of 10^{-12} N/m^3. For comparison, consider the forces involved in moving typical large structures from low Earth orbit to geosynchronous at the maximum 0.01 g acceleration desired to prevent structural damage.[17] This constraint converts into forces on the order of 100 N/m^3 for structures having densities similar to aluminum, and of the order of 10^{-3} N/m^3 for a structure like the SPS module.[18] In either case the electrostatic-field-induced force should be too small to be of concern.

But is this really true in all cases? Consider the thin insulator on the central portion of structure resting on the charged conductor. In space the surface of the insulator must have a net current of zero independent of the structure. The insulator surface, then, could be close to zero volts. Therefore, there will be a differential voltage across the insulator and an electric field through the insulator. This electric field can result in significant stress in the insulator. For example, a 2-mil (0.005-cm) thick thick insulator with a differential voltage of 20 V will have a stress induced on the order of 0.2 N/m^3. This is not a negligibly small stress. The results of a computation of the stress induced on a 2-mil-thick insulator as a result of various differential voltages is shown in Fig. 5.

Discussion. The above computations indicate that fairly large stresses can be induced in thin insulator (dielectric) surfaces by relatively small differential voltages anticipated for large structures. These computations admittedly are simplistic and have neglected several interactions that could either worsen or alleviate the stresses. The exercise is meant only to illustrate an effect associated with large structures and to point out that care must be used in judging which effects should be incorporated and which ignored in dealing with large structures.

A much more detailed analysis is required before any definite conclusions can be reached on the induced stresses. Even if this more complete analysis shows that relatively large stresses do exist, these stresses can be compensated for in the design phases. With reasonable examination of the areas where insulators, conductors, and space come together, interaction effects can be minimized, and large structures can be built for space applications.

Space Power System Interactions

Introduction. One class of large structures envisioned for future missions are space power systems utilizing solar arrays. A 25-kW system has been proposed to supplement the

Shuttle orbiter power capabilities.[11] Future plans call for systems with power generating capabilities of up to 500 kW to be launched in the late 80's and early 90's.[14] When the power levels rise, operating voltages must increase above the 30-100 V levels presently used in order to improve efficiency and reduce weight. It is the electric fields generated by higher operating voltages of these power systems that cause interactions with the space plasmas. This interaction has been studied in ground simulation facilities[19-22] and a space flight experiment.[23] Results of these studies can be used to determine the reactions involved in these phenomena.

High voltage surface-plasma interactions. This interaction is illustrated in Fig. 6, which shows a solar array system in a space environment. In the standard construction of this array, cover slides do not completely cover the metallic interconnects between the solar cells. These cell interconnects are at various voltages depending on their location in the array circuits. Because the array is exposed to space plasmas, the interconnects act as biased plasma probes attracting or repelling charged particles. At some location on the array the generated voltage is equal to space plasma potential. Cell interconnects that are at voltages V_+ above the space potential will attract an electron current which depends upon electron density and energy, as well as the voltage difference between the interconnect and space. Those interconnects that are at voltages V_- below space plasma potentials will repel electrons and attract an ion current. The voltage distribution in the interconnects relative to space potential must be such that electron and ion currents are equal. This flow of particles can be considered a current loop through the power system to

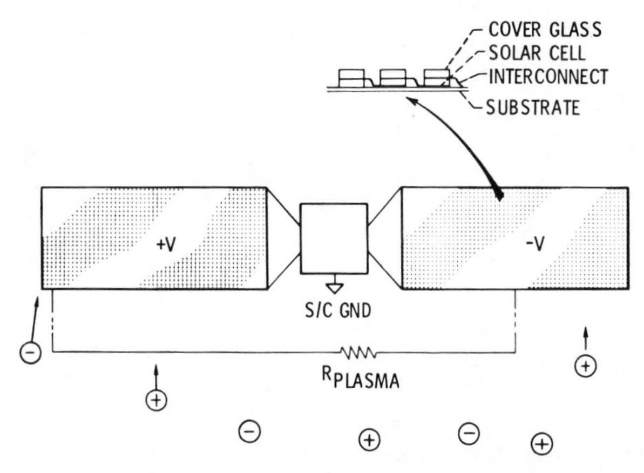

Fig. 6 Spacecraft higher voltage system-environment interactions.

space. It is a parallel electrical load with the power system and, as such, represents an additional power loss. One would expect this interaction to be more pronounced at low Earth orbits because of the high number density of low-energy plasma (see Fig. 7).

Ground simulation tests of biased solar array segments have demonstrated the behavior of these systems when exposed to plasma environments. Although discussed in great detail in Stevens[23] and McCoy et al.,[24] a short discussion of the effects on a small segment consisting of 24 2×2-cm cells connected in series (area ∿ 100 cm^2) mounted on a fiberglass board will be given here for illustration.[21] Briefly, a thermal plasma environment with densities of ∿10^3 (cm^{-3}) and ∿10^4 (cm^{-3}) and energies of about 1 eV was generated in a vacuum tank. The solar cell circuit on the test array was biased by laboratory power supplies in both positive and negative voltage steps from 0 to ±1000 V, relative to tank ground. The plasma coupling current (through the environment) and the voltage profile across the solar array surface were measured at each voltage step. These results are shown in Figs. 8a and b. The voltage profiles were similar for both plasma density tests and only one set has been reproduced.

When low positive bias voltages (≤100 V) were applied to the segment, the quartz cover slides acquired a slight negative

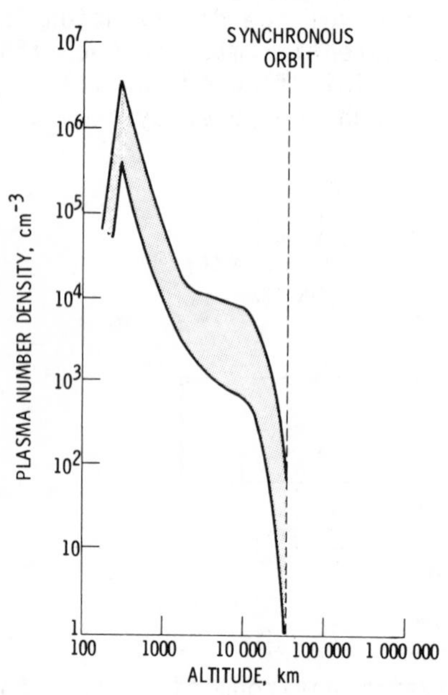

Fig. 7 Plasma number density vs altitude in equatorial orbit.[20]

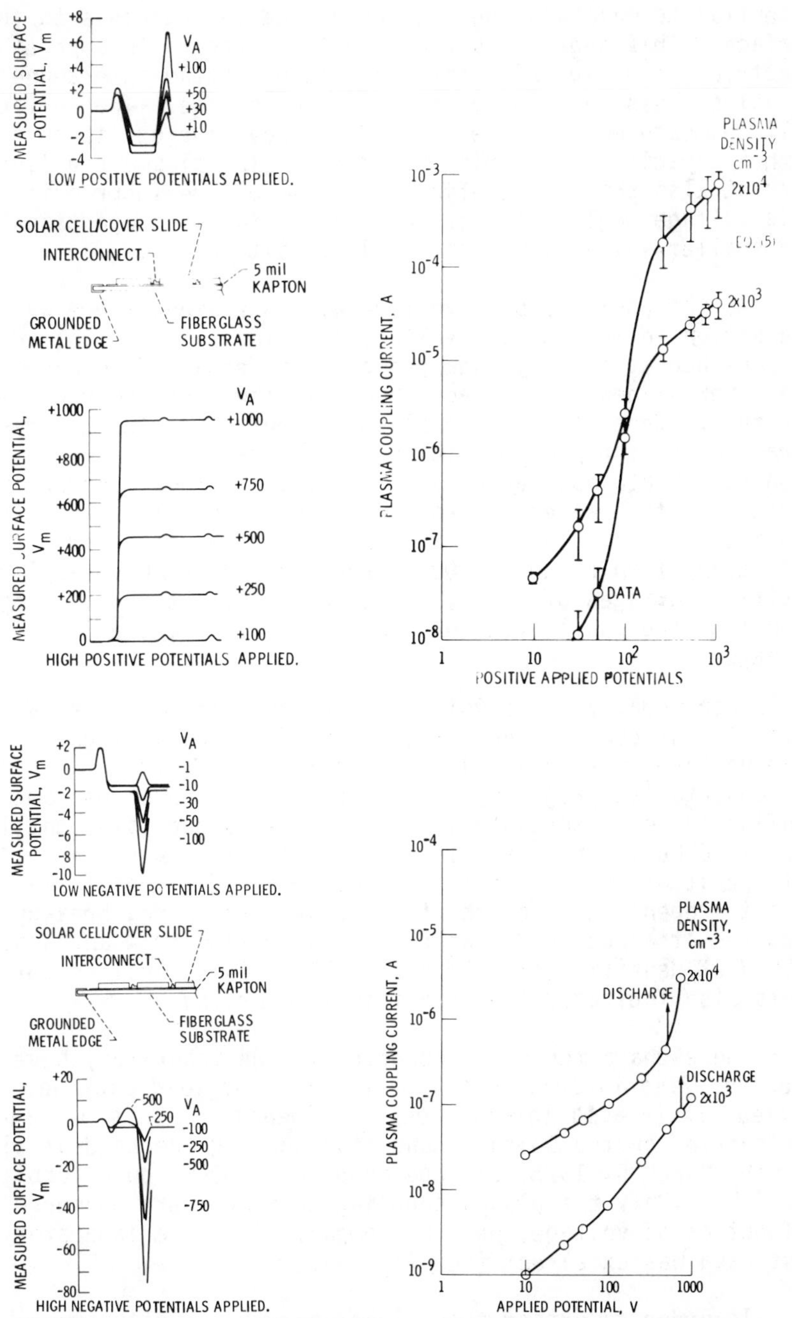

Fig. 8 Solar array surface voltage profiles and coupling currents: a) Positive applied potentials, and b) Negative applied potentials.

potential to maintain equal electron and ion currents to that surface. This negative surface voltage appears to suppress the electric field expansion into the plasma at the interconnects to values less than 10% of the applied voltage. The surface voltage measurements were taken 3 mm above the quartz surface with a capacitively-coupled voltmeter. The plasma coupling current also showed the effect of this voltage suppression; in this voltage regime, the current collection is proportional to the smaller voltage, not the applied voltage.

As the positive bias voltage was increased, there was a transition in the surface voltage profiles; surface voltage sheaths had apparently "snapped-over" or expanded to encompass the cover slides. A voltage sheath is the distance required for the voltage to decay to plasma potential due to the rearrangement of plasma particles. Snap-over seemed to occur when the sheath approached solar cell dimensions. Effective surface voltage after snap-over was about 50 V less than the applied voltage. The plasma coupling current also indicated this transition at about 100 V (see Fig. 8a). Above applied positive voltages of 100 V, the current collection was proportional to the panel area and the 0.8 power of the effective voltage.

When negative bias voltages were applied to the solar cell segment, the quartz cover slides again assumed a slightly negative voltage ($\sim$-2 to -4 V), suppressing the fields at the interconnects (see Fig. 8b). Instead of a snap-over phenomenon, confinement of interconnect electric fields persisted until the field built up to a point where discharges occurred. The voltage at which breakdown occurred appeared to be plasma-density dependent. For the tests considered here, breakdown occurred at about -600 V at densities of $\sim 10^4$ cm^{-3} and about -750 V at densities of $\sim 10^3$ cm^{-3}. The plasma coupling currents also indicated the transitions to arcing.

These characteristics observed in the laboratory have been verified in space with an auxiliary payload package called PIX (Plasma Interaction Experiment).[25] This package was carried on the Delta second stage during the Landsat III launch, March 5, 1978, and operated in a 900-km polar orbit for 4 hr. Only the plasma coupling currents were measured as a function of voltage, but the comparison to the laboratory test data was excellent (see Fig. 9).

In order to extend these laboratory results to space power system interactions, one must know the floating potential of the array relative to space. This is a complex computation which other papers[26,27] attempt to carry out theoretically.

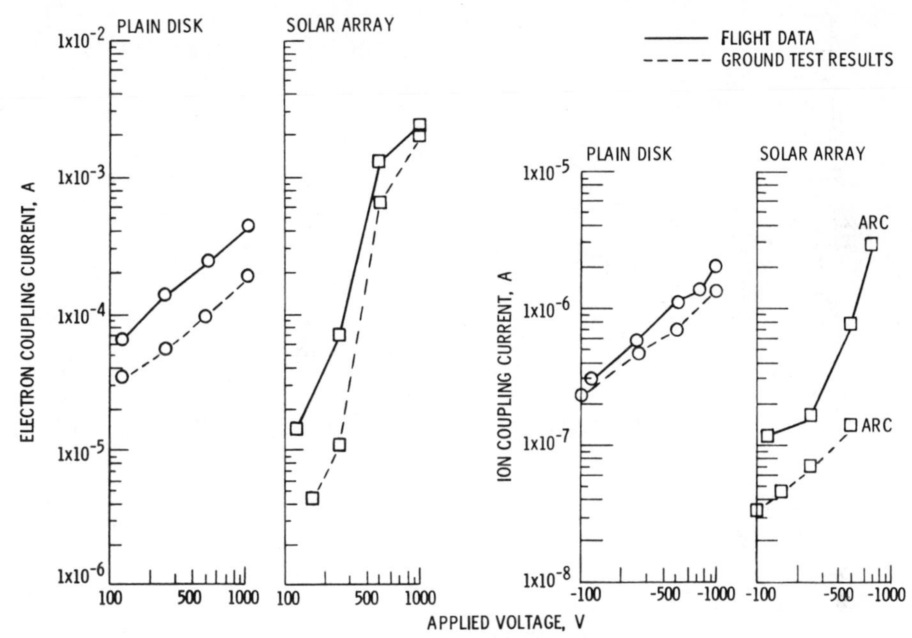

Fig. 9 Comparison of PIX flight results with ground simulation tests: a) Positive applied voltages, and b) Negative applied voltages.

Here the approximation that the array will be no more than 10% positive V_+ and 90% V_- is made so that the effect of the plasma interactions can be estimated. This split in voltage is probably conservative, and the array will float at only a few percent positive. The system operating voltage will be the sum of the absolute values of the floating potentials (i.e., $V_L = |V_+| + |V_-|$). Using the above split in voltages and the laboratory-derived characteristics, the ratio of the plasma coupling current to the operating current is shown in Fig. 10 as a function of operating voltage. This curve is typical for any power level in a 400-km orbit. It should be compared with similar results presented in this volume.

It is apparent that plasma coupling currents are negligibly small at operating voltages less than 500 V, and power systems operating at this voltage are feasible from an environmental interaction viewpoint. The limitation in operations at higher voltages appears to be arcing in the negative portions of the array. If this arcing is truly an electric field confinement effect, then a technology investigation should lead to practical methods of overcoming this limitation.

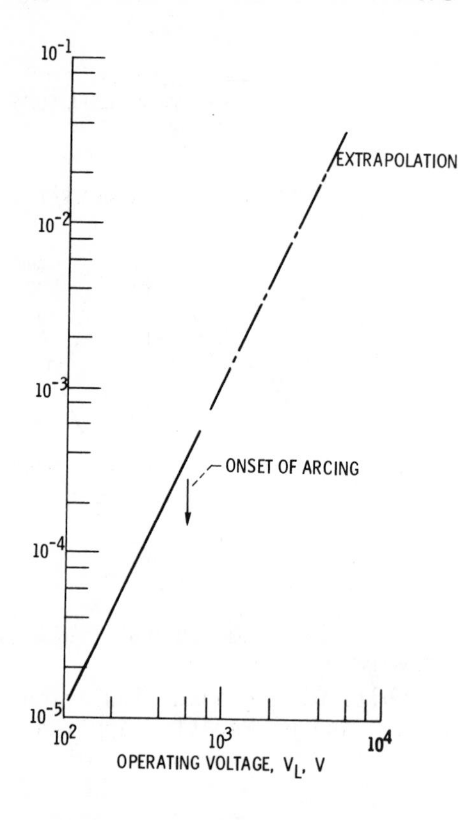

Fig. 10 Plasma coupling losses, all power levels 400 km orbit ($V_+ = 0.1\ V_{op}$ and $V_- = 0.9\ V_{op}$).

Extrapolating these results to geosynchronous orbits, coupling current losses should be even less of a concern. However, the arcing problem does exist. Laboratory data has shown that arcing could occur when operating voltages exceed 5000 V.

Discussion. Laboratory data has indicated that high voltage systems in space must be carefully engineered. There can be interactions between the space charged-particle environment and the biased conductors surrounded by insulator surfaces. While the anticipated plasma coupling current power loss probably will be negligible, a breakdown condition exists which, unless it is overcome, will limit operating voltages.

These comments are based on what were essentially short-term experiments. The space power systems have been proposed for multiyear operation (up to 30 yr for the SPS). Therefore, long term effects must be evaluated for high-voltage system space plasma interactions. Such items as the effect of long term deposition of charges in and on insulator surfaces and the influence of electrostatically enhanced contamination must be assessed before these high-voltage systems can safely and reliably function in space.

Concluding Remarks

Future space missions have requirements which cannot be satisfied by present spacecraft and necessitate the orbiting of very large deployable or erectable structures. These new spacecraft will have dimensions ranging up to kilometers and will use lightweight materials to achieve the required low density. These proposed new spacecraft must function in a charged-particle space environment, which means that the possible interactions between this environment and spacecraft surfaces must be identified and evaluated.

Two broad categories of spacecraft surface-charged particle environmental interactions have been identified: spacecraft passive, when the environment acts on the spacecraft, and spacecraft active, when a spacecraft system causes the interaction. The principal interaction in the first category is the spacecraft charging phenomenon. A detailed discussion of this phenomenon can be found elsewhere in this volume.

The second category of interactions involves motion-induced effects on large structures and the use of on-board high-voltage systems. There are indications that spacecraft motion through the Earth's magnetic field can induce additional and significant stresses across insulator-conductor interfaces. These stresses must be assessed and relieved in large structure designs. The use of high-voltage systems in space requires care similar to that employed for ground applications of high-voltage systems. In both cases, breakdowns are possible unless reasonable design guidelines are used. Both types of interactions considered here were more serious for low-Earth orbit than for geosynchronous orbit.

All interactions discussed here involved the requirement that the net currents to surfaces be zero. This applies to conductors as well as insulators. In space, surface voltages will automatically adjust so that this requirement will be satisfied. Electric fields from these charged surfaces will interact in the plasma environment, influencing particle fluxes. It is the goal of the technology investigations to understand these complex interactions and to devise means to minimize their effects on spacecraft system performance. Models of these effects and suggested means for alleviating them will be discussed elsewhere.

References

[1]"Outlook for Space," NASA SP-386, Jan. 1976.

[2]Johnson, R. D. and Holbrow, C., eds., "Space Settlements, A Design Study," NASA SP-413, 1977.

[3]Woodcock, G. R., "Solar Satellites, Space Key to Our Future," Astronautics and Aeronautics, Vol. 15, July/Aug.1977, pp.30-43.

[4]"Satellite Power System (SPS) Feasibility Study," Rockwell International Corp., Downey, Calif., SD76-SA-0239-1, 1976. (NASA CR-150439.)

[5]"Systems Definition of Space-Based Power Conversion Systems," Boeing Aerospace Co., Seattle, Wash., D180-20309-1, 1977. (NASA CR-150209.)

[6]Poeschel, R. L. and Hawthorne, E. I., "Extended Performance Solar Electric Propulsion Thrust System Study, Vol. 2, Baseline Thrust System," Hughes Research Labs., Malibu, Calif., 1977. (NASA CR-135281.)

[7]Friedman, L. D., Carrol, W., Goldstein, R., Jacobson, R., Kievit, J., Landel, R., Layman, W., Marsh, E., Ploszaj, R., and Rowe, W., "Solar Sailing - The Concept Made Realistic," AIAA Paper 78-82, Jan. 1978.

[8]Nansen, R. H., "Solar Power Satellite: Can We Afford to Ignore It? AIAA Conference on Large Space Platforms. Future Needs and Capabilities, Los Angeles, Calif., Sept. 1978.

[9]Rosen, A., ed., Progress in Astronautics & Aeronautics: Spacecraft Charging by Magnetospheric Plasmas, Vol. 47, AIAA, New York, 1976.

[10]Pike, C. P. and Lovell, R. R., eds., Proceedings of the Spacecraft Charging Technology Conference, Air Force Geophysics Lab. AFGL-TR-0051. NASA TM X-73537, Feb. 1977.

[11]Mordan, G. W., "The 25 kW Power Module - The First Step Beyond the Baseline STS," AIAA Paper 78-1693, Sept. 1978.

[12]Woosley, A. P., Smith, O. B., and Nassen, H. W., "Skylab Technology Electrical Power System," ASME Paper 74-129, Aug. 1974.

[13]Purvis, C. K., Stevens, N. J., and Berkopec, F. C., "Interaction of Large, High Power Systems with Operational Orbit Charged Particle Environments," NASA TM-73867, 1977.

[14]Savage, M. and Haughey, J. W., "Overview of Office of Space Transportation Systems Future Planning," Future Orbital Power Systems Technology Requirements, NASA CP-2058, 1978, pp. 71-92.

[15]Whipple, E. C., Jr., "The Equilibrium Electric Potential of a Body in the Upper Atmosphere and in Interplanetary Space," NASA TM X-55368, 1965.

[16]Panofsky, W. K. H., and Phillips, M., Classical Electricity and Magnetism, Addison-Wesley, Reading, Mass., 1965.

[17]Terwilliger, C. H., "Electric Propulsion for Near-Earth Space Missions," Boeing Aerospace Co., Seattle, Wash., D180-24819-1, Oct. 1978.

[18]"Solar Power Satellite-System Definition Study, Parts 1 and 2, Vol. 2: Technical Summary," Boeing Aerospace Co., Seattle, Wash., D180-22876-2, 1977. (NASA CR-151666.)

[19]Kennerud, K. L., "High Voltage Solar Array Experiments," Boeing Aerospace Co., Seattle, Wash., 1974. (NASA CR-121280.)

[20]Stevens, N. J., "Solar Array Experiments on the SPHINX Satellite," NASA TM X-71458, 1973.

[21]Stevens, N. J., Berkopec, F. D., Purvis, C. K., Grier, N. T., and Staskus, J. V., "Investigation of High Voltage Spacecraft System Interactions with Plasma Environments," AIAA Paper 78-672, Apr. 1978.

[22]McCoy, J. E. and Konradi, A., "Sheath Effects Observed on a 10-Meter High Voltage Panel in Simulated Low Earth Orbit Plasmas," Second USAF/NASA Spacecraft Charging Technology Conference, U.S. Air Force Academy, Colorado Springs, Colo., Oct. 1978.

[23]Stevens, N. J., "Space Environmental Interactions with Biased Spacecraft Surfaces," Published elsewhere in this volume.

[24]McCoy, J., Konradi, A., and Garriott, O. K., "Current Leakage for Low Altitude Satellites," Published elsewhere in this volume.

[25]Grier, N. T. and Stevens, N. J., "Plasma Interaction Experiment (PIX) Satellite Results," Second USAF/NASA Spacecraft Charging Technology Conference, U.S. Air Force Academy, Colorado Springs, Colo., Oct. 1978.

[26] Parker, L. W., "Plasmasheath-Photosheath Theory for Large High-Voltage Structures," Published elsewhere in this volume.

[27] Reiff, P. H., Freeman, J. W., and Cooke, D. L., "Environmental Protection of the Solar Power Satellite," Published elsewhere in this volume.

SPACE ENVIRONMENTAL INTERACTIONS WITH BIASED SPACECRAFT SURFACES

N. John Stevens*
NASA Lewis Research Center, Cleveland, Ohio

Abstract

Large, high-voltage space power systems are being proposed for future space missions. These systems must operate in the charged-particle environment of space, and interactions between this environment and the high-voltage surfaces are possible. Ground simulation testing has indicated that dielectric surfaces that usually surround biased conductors can influence these interactions. For positive voltages greater than 100 V, it has been found that the dielectrics contribute to the current collection area. For negative voltages greater than -500 V, the data indicate that the dielectrics contribute to discharges. Using these experimental results a large, high-voltage power system operating in geosynchronous orbit was analyzed with the NASCAP code. Results of this analysis indicated that very strong electric fields exist in these power systems. A technology investigation is required to understand the interactions and develop techniques to alleviate any impact on power system performance.

Introduction

Large space systems are being proposed for future applications such as manufacturing, technology demonstrations, communications, and beaming power for Earth usage.[1-6] These systems are proposed for operations in orbits ranging from low Earth (200-400 km) to geosynchronous. These future applications will require space power systems capable of generating from 25 kW[7] to multikilowatts[8] to gigawatts required for the Solar Power Satellite (SPS).[4,5] Since the power level is proportional to the surface area, these power systems will be large with dimensions ranging from 10's to 100's of meters.

---

*Head, Spacecraft Environment Section.

It will be necessary for these power systems to operate at elevated voltages to reduce line losses and minimize system weight.[9] At operating voltages in the kilovolt range, interactions with the charged-particle environment are possible. Since the highest operating voltage reported for satellites to date is 100 V,[10] there is limited space experience to guide the system designer in constructing a high-voltage space power system. There has been ground testing of biased surfaces in plasma environments to determine possible interactions,[11-19] as well as several analytical treatments of the impact of these interactions on performance.[20-25] Due to the limited facility size and complex insulator/conductor geometry of solar arrays, however, these experimental and analytical treatments can only serve as indicators of what might happen. While probe theory can be used to compute biased-conductor current collection, the tests have shown that the insulation surrounding the conductors does have a profound influence on the collection phenomenon. The role of the insulator in this interaction must be understood before corrective techniques can be devised to minimize the impact of the interaction on system performance.

The interactions of concern are illustrated in the high-voltage space power system shown in Fig. 1. This system consists of two large solar array wings surrounding a central body or spacecraft. Depending upon the physical size factors, this system could represent a "direct-drive" solar electric propulsion spacecraft (i.e., one in which the high voltage needed for the electric thrusters is generated directly on the solar array wings[26]) or an SPS. In either case the solar arrays are assumed to be assembled in what is called standard

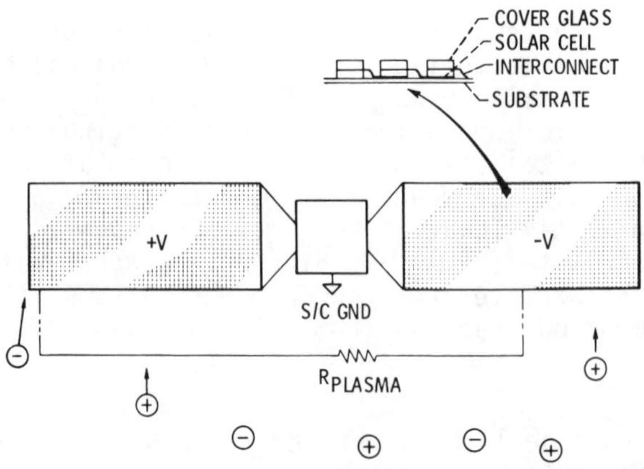

Fig. 1 Spacecraft higher voltage system-environment interactions.

construction techniques. This means that the cover slides do not completely shield the metallic interconnects from the environment. These cell interconnects are at various voltages depending on their location in the array circuits. Hence, the interconnects can act as plasma probes attracting or repelling charged particles. At some location on the array, the generated voltage will be equal to the space plasma potential. Cell interconnects at voltages above this space plasma potential will collect electrons, while those at voltages below the space potential will collect ions. The voltage distribution in the interconnects relative to space must be such that these electron and ion currents are equal (i.e., the net current is zero). This flow of particles can be considered to be a current loop through the power system to space that is in parallel with the operating system on the spacecraft and, hence, is a power loss.

The severity of this plasma-coupling current depends upon the operating voltages of the power system and the charged-particle environment. Since the proposed missions consider voltages up to a maximum of 45 kV, the electric fields from exposed surfaces at these voltages will only attract the lower energy components of the environment (i.e., the "thermal plasma"). This thermal plasma environment (see Fig. 2) has particles with temperatures of about 1 eV and densities that vary from a maximum of 10^6 cm^{-3} at about 300 km to between 1 and 10 cm^{-3} at geosynchronous altitude. Hence, one should

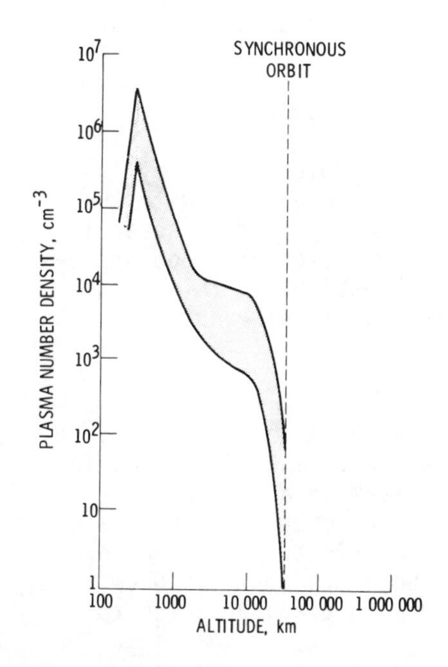

Fig. 2 Plasma number density vs altitude in equatorial orbit.

expect current coupling to be more severe at lower altitudes; only very high voltage should be a concern at geosynchronous altitudes.

Interactions with the charged-particle environment are not limited to plasma-coupling currents. The possible confinement of voltages (and electric fields) to cavities formed by the cover slides and interconnects could conceivably give rise to breakdowns in the negative voltage regions. The large, high power systems are proposed as rather flexible arrays mounted on thin, dielectric substrates. Operations at high voltages could exceed the dielectric strength of the substrate material, giving rise to additional breakdowns. In either case these breakdowns would produce transients in the power system.

Finally, the electric fields established in the dielectric substrate and cover slides can induce substantial forces in these materials. These forces must be considered in building a high-voltage space power system.

In this report the ground experimental work is reviewed to provide indicators for the interactions that could exist in the space power system. A preliminary analytical model of a large, space power system is constructed using the existing NASA Charging Analyzer Program (NASCAP), and its performance in geosynchronous orbit is evaluated. These analytical results are used to illustrate the regions where detrimental interactions could exist and to establish areas where future technology is required.

Ground Simulation Test Results

Tests have been conducted using samples biased by laboratory power supplies to determine the interactions with the plasma environment. Results have been reported by several investigators,[11-19] and the data are in reasonably good agreement. These results were summarized earlier in this volume and will be discussed in more detail in this report.

Test Procedure

Tests at the LeRC have been conducted in a 1.8 m diam by 2.5 m long vacuum facility and a 4.6 m diam by 16 m long facility. A nitrogen plasma is generated in a bombardment source and allowed to drift into the chamber. The plasma particles have a temperature of about an electron volt and densities that can be controlled from about 10^6 cm^{-3} to 10 cm^{-3}. The plasma properties are measured with diagnostic instruments (i.e., probes and Faraday cups) before and after each run. The ambient pressure in both chambers, while the sources are operating, is about 10^{-6} Torr.

The solar array segment used in these tests is shown in Fig. 3. This is a 100 cm^2 segment consisting of 24 solar cells in series mounted on a fiberglass board. During the tests in the simulation vacuum chamber, the segment is mounted such that it is electrically isolated from the chamber. Electrical leads from the segment are brought out of the chamber with high-voltage feedthroughs. These leads are connected first to an ammeter and then to a bias power supply. The power supply is referenced to the chamber which acts as the electrical ground. The circuit is completed to the segment through the plasma. The plasma coupling current collected by this segment at a given bias voltage is measured by the ammeter. The cover slide and interconnect surface voltages are measured by a capacitively-coupled probe sweeping 3 mm above the segment.[17] The test arrangement is shown schematically in Fig. 4.

Larger solar array panels with 2000 cm^2 area also have been tested in a similar manner. A series of tests also have been conducted on a solar electric propulsion array segment utilizing wrap-around interconnects. The difference between the wrap-around and standard (or conventional) interconnects is shown in Fig. 5. The results from both these test samples will be discussed in the following section of the report.

Ideally, testing should be conducted with the solar arrays themselves generating the desired voltage while floating electrically within the chamber. However, in order to study

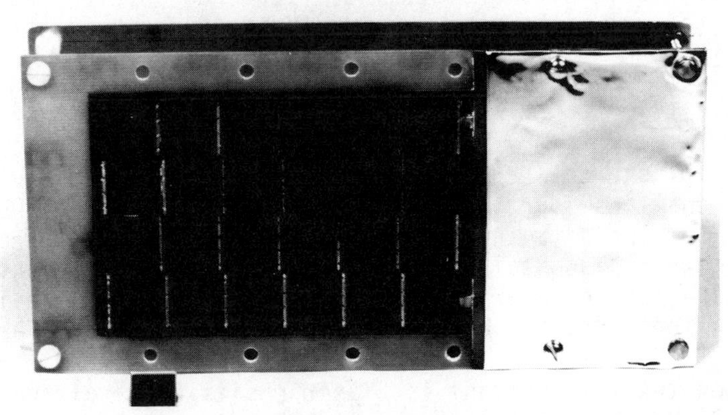

Fig. 3 Solar array segment.

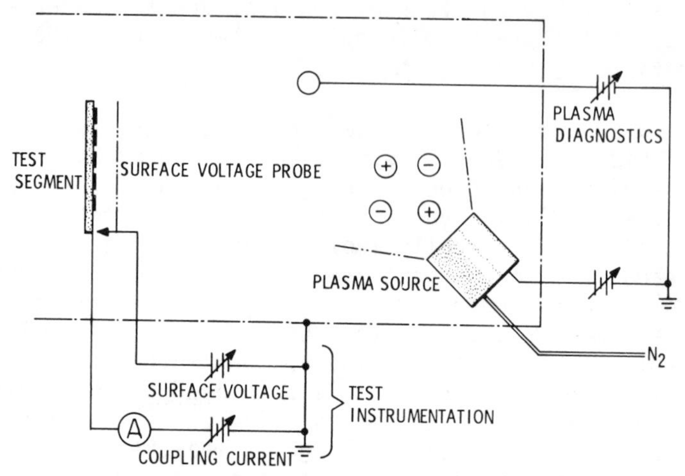

Fig. 4 Schematic diagram of test arrangement.

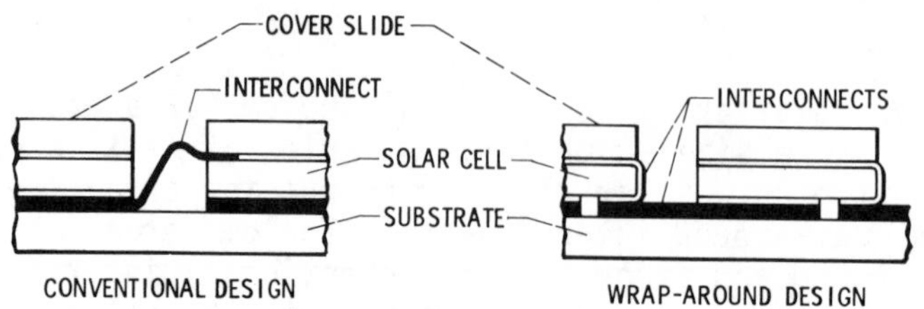

Fig. 5 Solar array interconnect configurations.

interactions in a typical series-parallel circuit solar array, a large panel would be required. Such a solar array panel would have to be tested in a correspondingly large vacuum facility to minimize the interfering effects of chamber walls. Since such facilities are often difficult to obtain, this type of testing would make a good candidate for a Shuttle experiment.

Test Results

The plasma collection currents collected over a range of plasma densities for the 100-cm^2 solar array segment is shown in Fig. 6. The data shown in Fig. 6a are for the electron coupling currents obtained when the segment is biased positive with respect to chamber ground; the data in Fig. 6b are for ion coupling currents obtained with the segment biased negative.

When the test segment is biased positive at voltages less than +100 V, the electron current collection is low. Compari-

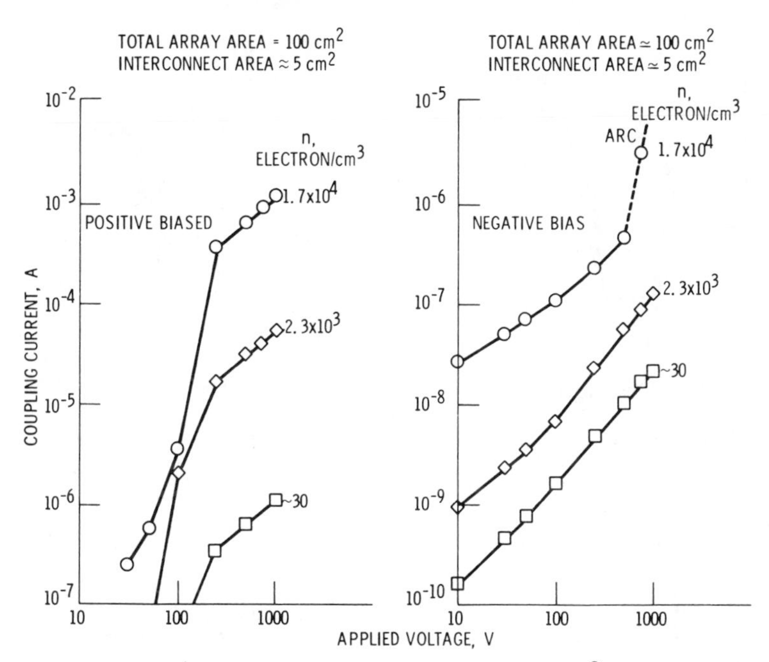

Fig. 6 Plasma coupling currents for 100 cm^2 solar array: a) Positive bias, and b) Negative bias.

sons between this data and probe theory predictions indicate that the collection at a given bias voltage and with given plasma properties is dependent only on the interconnect area. In this voltage range, the electron current collected by a segment increases uniformly with bias voltage. As the bias voltage is increased above +100 V this relationship breaks down and there is a sharp increase in the electron current collection. When the bias voltage exceeds +250 V the electron current collection again behaves as before, only with current collection (at given bias voltages and plasma properties) now being dependent on the entire panel area. This behavior occurs at all plasma densities from the simulated near geosynchronous environment (~30 cm^{-3}) to the 900 km environment (~10^4 cm^{-3}). Other tests with this segment have indicated that electron current collection with a given surface area increases linearly with bias voltage up to 20 kV.

Why does this transition in electron current collection occur? The answer can be found in the surface voltage traces obtained during these tests (see Fig. 7a). At bias voltages of less than 100 V the quartz cover slides assume the slightly negative voltage necessary to maintain a net zero current. This quartz surface voltage appears to suppress the voltage in

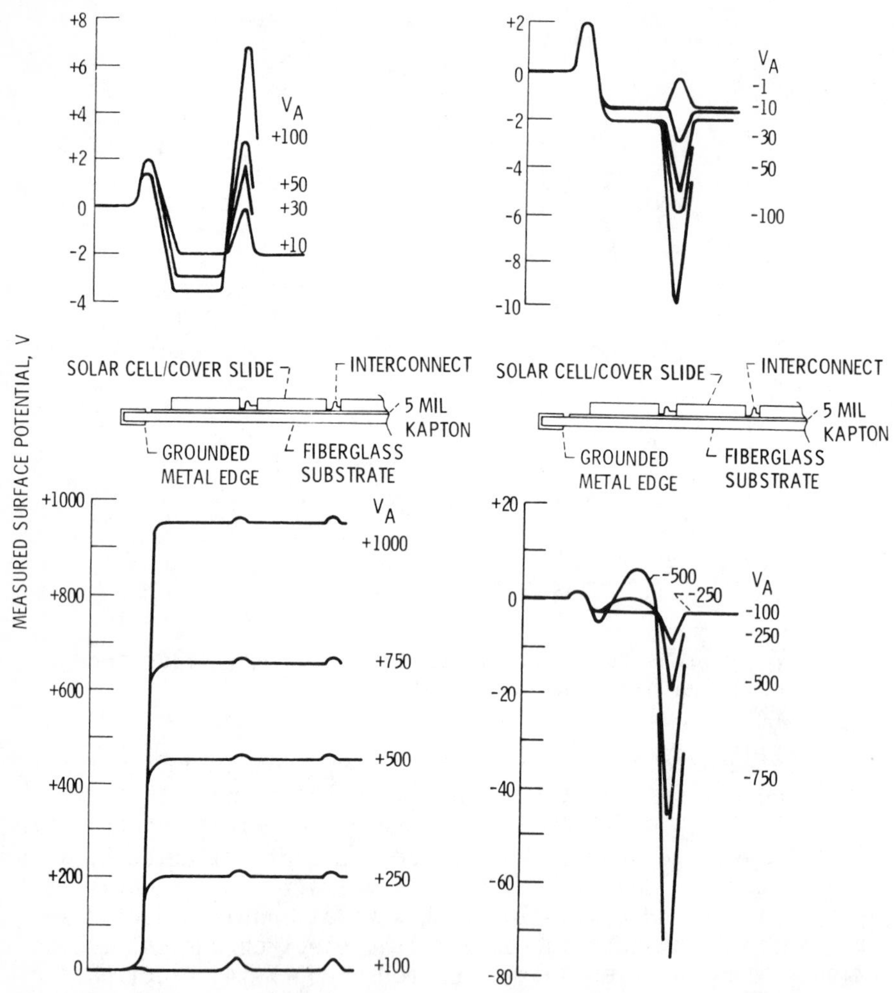

Fig. 7 Typical surface voltage profiles-solar array segment: a) Low positive potentials applied, b) Low negative potentials applied, c) High positive potentials applied, and d) High negative potentials applied.

the plasma above the interconnect to a value less than the bias voltage. Hence, the current collection area is limited to the interconnects.

Above 100 V bias there is a transition in the characteristics of the surface voltage. The quartz cover slide potential changes to a value that is about 50 V less than the bias voltage. It is as if the voltages in the interconnect region had "snapped-over," encompassing the cover slides. This oc-

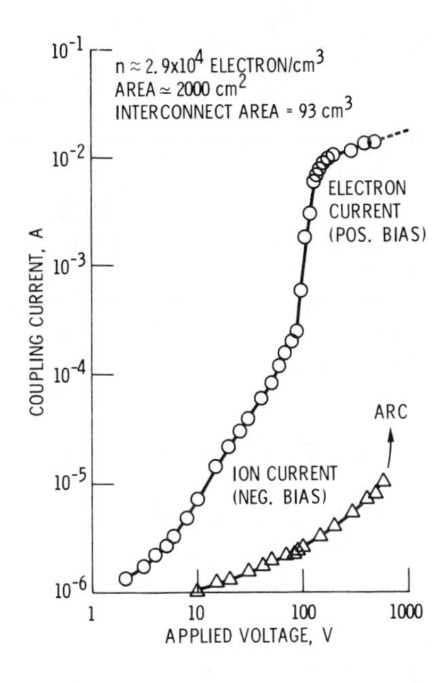

Fig. 8 Plasma coupling currents; large solar array panel.

curs due to the interconnect electric fields accelerating plasma electrons into the quartz cover slides with sufficient energy to generate secondary electrons. These secondary electrons are collected by the interconnects causing the rise in the electron current collection. The quartz cover slide surface voltage must increase to compensate for the secondary current contributions.

For the negative bias cases (see Fig. 6b) there is an abrupt transition into an arc discharge initiated at the metal interconnects. The voltage at which arcing occurs is plasma-density dependent. At densities of $\sim 10^4$ cm^{-3} (low Earth orbit) this arcing occurs above -500 V. At densities of ~ 30 cm^{-3} (geosynchronous altitudes) this arcing occurs above -5000 V. For bias voltages below these arc threshold values, the ion current collection remains relatively small, at values expected from probe theory.

The reason for the negative voltage behavior can be found in the surface voltage traces (see Fig. 7b). As negative bias voltages are applied to the segment, the quartz cover slides again assume a slight negative surface voltage to maintain the required current balance. The electric fields in the plasma due to the bias voltage are therefore confined to the region of the interconnects as before. As the bias voltage is increased, the quartz surface voltage continues to confine the interconnect voltages until the electric field existing in the cavity formed by the quartz cover slides and interconnect becomes so strong that a discharge can be triggered by field emission from the interconnect.

The characteristics of the interactions described here also have been observed in a space flight experiment called PIX, an acronym standing for Plasma Interaction Experiment.[19] This experiment was flown on the Landsat C launch on March 5, 1979, and operated in a 900 km polar orbit for over 4 hr. Bias voltages of up to ±1000 V were applied to a 100 cm^2 solar array panel. The anticipated transition in electron current collection and the breakdowns at negative voltages were recorded.

Coupling currents observed in ground simulation tests using a 2000 cm^2 solar array panel are shown in Fig. 8. For positive bias voltage cases, the transition in electron current along with the same type of area dependence discussed earlier was measured. The influence of test chamber size was observed while these tests were being conducted. The plasma in these chambers was limited by the finite volume of the chamber. When the solar array panel was biased to a value where all the available plasma electrons were collected, the coupling current saturated (i.e., remained relatively constant as the bias voltage was increased). In this test the limiting voltage was found to be about 200 V. This chamber size limitation emphasizes the need to conduct large sample, high-voltage surface tests in space. For the negative bias voltage cases, the ion current collection remained low and fairly linear with voltage until the bias exceeded -500 V at which the transition that terminates in arcing occurred.

The SEP array segment test results are shown in Figs. 9a and b. This segment had a Kapton substrate in which holes were cut to make the joints between the solar cells. Both front and back side voltage distributions are shown in Fig. 9a. The snap-over phenomenon is evident on both sides at positive bias voltages, as is the confinement at the negative bias voltage. Figure 9b shows the discharges that were photographed during this test. The light flashes correspond to the discharge points and always occur in the interconnect cavity or in the holes cut in the Kapton substrate.

Summary of Test Results

Based on the ground simulation testing one should expect that the insulator surfaces surrounding the biased conductors would influence the interactions with a charged-particle environment. For areas of a high-voltage solar array space power system that are at potentials greater than 100 V relative to

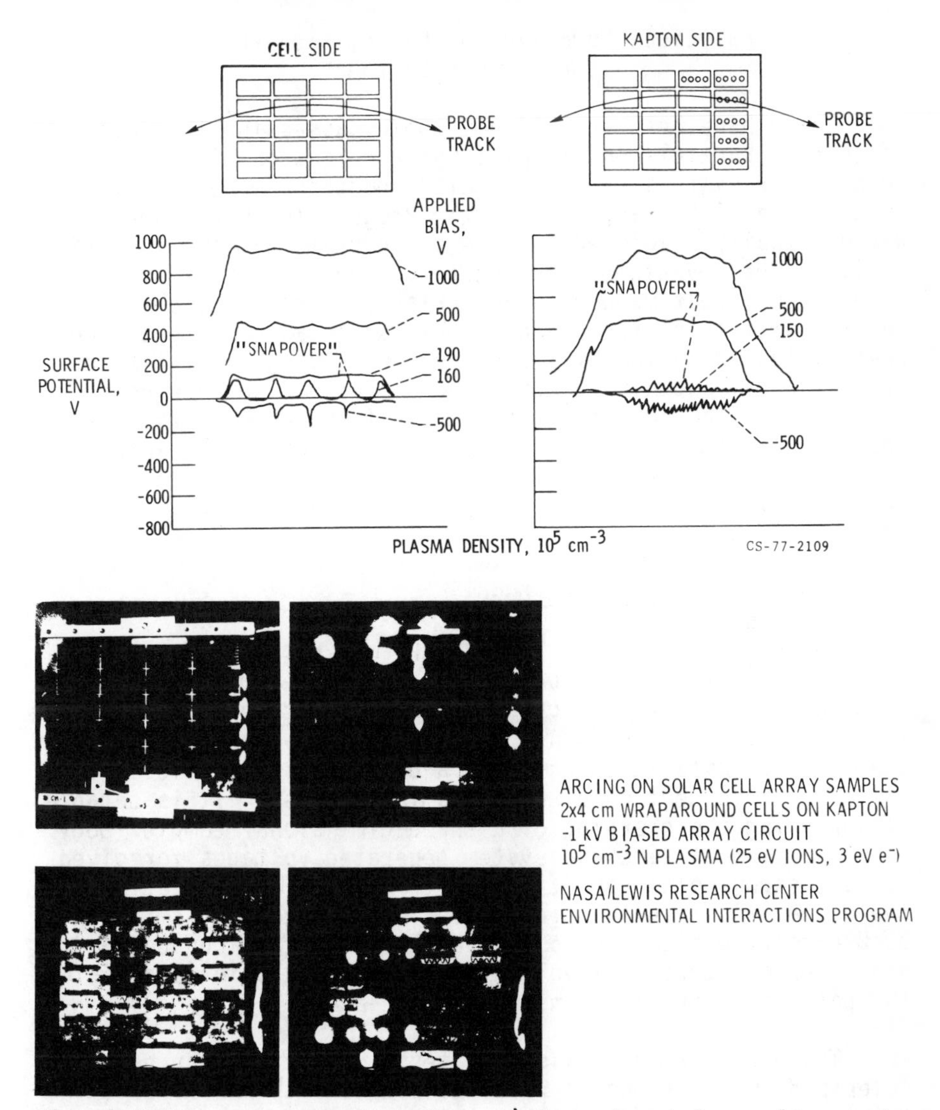

Fig. 9 SEP array segment tests: a) Experimental results, and b) Pictures of array (composite).

the space plasma potential, electron current collection should be proportional to the interconnect and cover slide area. For areas of the array that are negative with respect to the space plasma potential, current collection is limited to interconnect areas, and arcing can occur. This arcing threshold is between -500 and -1000 V in low Earth orbits and between -5000 and -10,000 V at geosynchronous.

Analytical Modelling of Large, High-Voltage Space Power Systems

There have been analytical studies assessing the impact of charged-particle environment interactions on high-voltage space power systems.[23-25] In this report the modelling is done in the NASA Charging Analyzer Program (NASCAP) code which was originally developed for the spacecraft charging investigation. The advantage of using this code is that it treats in a self-consistent manner the material's response to environmental particle fluxes. Hence, this code can compute the current collected by the biased conductors (using orbit-limited probe theory) and also compute the dielectric surface voltages. While the code can treat many of the characteristics of these high-voltage systems, it cannot yet handle all interactions. Even so, it can indicate areas where design techniques must be developed if these high-voltage space power systems are to be feasible.

NASCAP High-Voltage System Model

NASCAP description. The NASCAP code has been described previously in the literature,[27-30] and can be briefly summarized here. NASCAP is a quasistatic computational code (i.e., it assumes that currents are functions of environmental parameters, electrostatic potentials, and magnetostatic fields while not dependent on electrodynamic effects). It is capable of analyzing the charging of a three-dimensional complex body as a function of time and system generated voltages for given space environmental conditions. It includes consideration of the dielectric material properties (e.g., secondary emission, backscatter, photoemission, and bulk and surface conduction) and computes currents involving these materials in determining the potential distributions around the body.

The body must be defined in terms of rectangular parallelepipeds or sections of parallelepipeds within a 17×17×33 point grid. The body must be at least 1 grid cell thick in the version of the code used here. Only seven separate conductors can be specified in the code. The first conductor may be floating relative to space, while the other six can be biased relative to the first. The environment can be defined in terms of Maxwellian distributions by specifying the plasma densities and temperatures (in electron volts). Both normal and geomagnetic substorm environments appropriate to geosynchronous orbits can be used.

The code output includes a variety of graphical and printed data displays. Graphical output includes the material

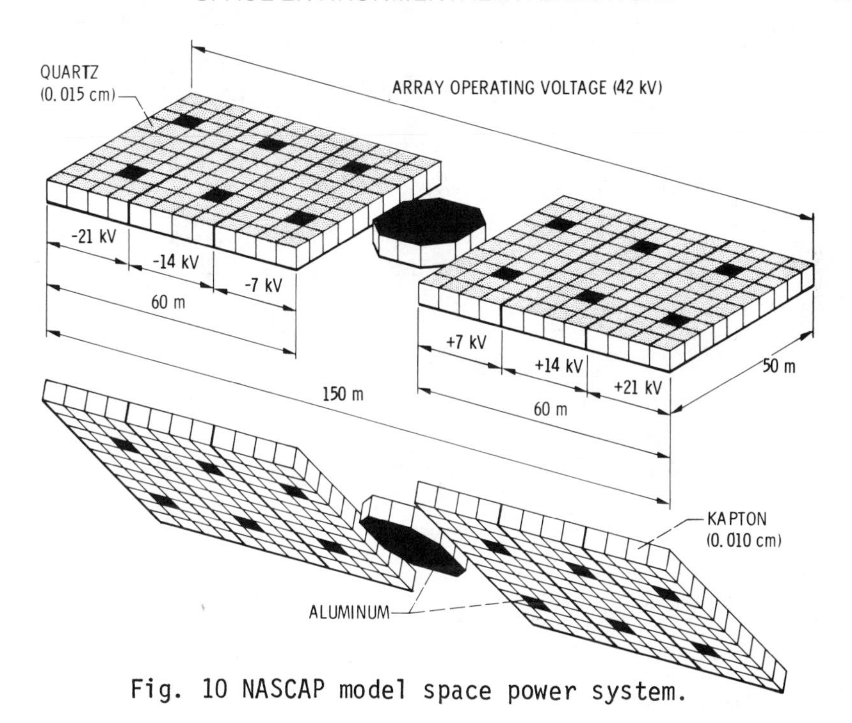

Fig. 10 NASCAP model space power system.

and perspective object definition plots, potential contour plots, and particle trajectory plots. The printed output includes a summary of all cell voltages, listings of currents to specified surfaces, and compilation of electric fields through the dielectrics, listed in decreasing order.

High-voltage system model. The NASCAP model of the high-voltage space power system considered in this report is shown in Fig. 10. This system consists of two solar array wings, each 50 by 60 m, with a central, 20-m octagonal antenna. This antenna (centerbody) is assumed to have an aluminum top and bottom while the sides are covered with 0.01-cm-thick Kapton. This body is the first conductor and is the system electrical ground, even though it is floating relative to the space plasma potential.

Each solar array wing is assumed to be divided into three sections, 20 by 50 m, with each section operating at 7000 V. The interconnects are assumed to be aluminum and exposed on the front and back of the array similar to the SEP array design. Since the NASCAP code cannot treat small gaps, the interconnects are assumed to be concentrated at two locations in each section, as shown. This exposed area represents 5% of the front and back areas of each section. This is a reason-

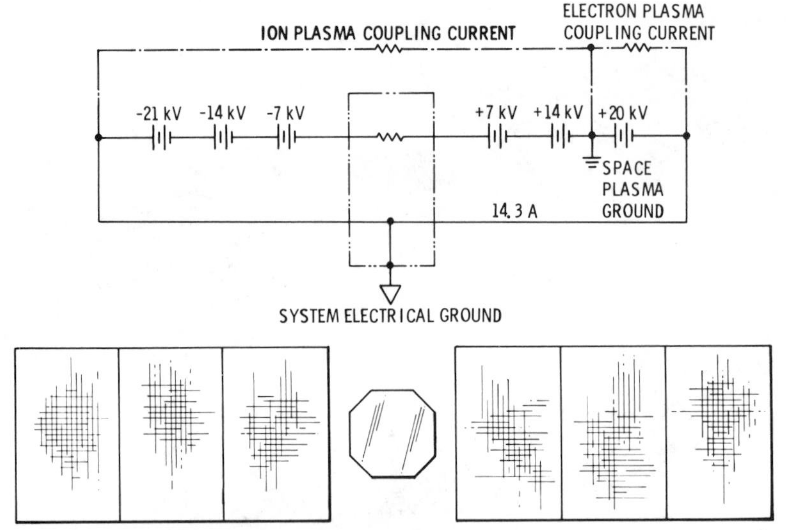

Fig. 11 Schematic diagram of high voltage power system ion plasma coupling current.

able approximation for the exposed interconnect areas. Quartz cover slides, 0.015-cm thick are used for the solar cells. The substrate and sides of the array are covered with 0.01-cm-thick Kapton.

The electrical circuit for this power system is assumed to be such that there is a 42-kV potential difference across the array (see Fig. 11). The overall power output of this system is on the order of 600 kW, with each section generating 100 kW.

Analytical Model Computations

The behavior of this system is analyzed in a quiet, plasmasphere-like, geosynchronous environment assumed to have plasma densities of 10 cm^{-3} and particle temperatures of 1 eV. The array is assumed to be in full sunlight, and only steady-state conditions are considered. The difficulties that may be encountered during eclipses and geomagnetic substorms will be left for future analysis.

Prior analytical treatment of high-voltage systems with the NASCAP code indicated that the "snap-over" phenomenon at positive voltages greater than 100 V did not automatically result from the code operation.[31] However, it could be simulated by allowing the insulator surfaces to have a low surface resistivity. A value of surface resistivity of 10^5 ohm/square

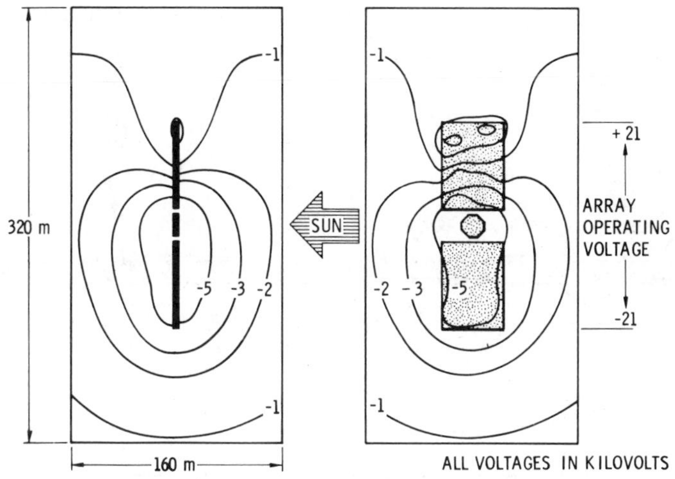

Fig. 12 Voltage distribution around power system. (System ground at -17.9 kV relative to space.)

was assumed to be appropriate and used for the quartz and Kapton substrate surfaces of the most positive-voltage section only.

The voltage distributions in the space around the system are shown in Fig. 12. These diagrams show a view of the top of the array and an edge view through the center of the array. As expected, the operating voltages used here have forced the antenna to float at a very negative value (-17.9 kV) and have established negative voltage distributions around most of the system. Large voltage concentrations are seen to exist at the simulated interconnect regions.

These voltage distributions are shown in more detail in Fig. 13. Note that only one section of the array has remained positive (+3.1 kV) and that the positive voltage distributions in space are extremely limited. This result was obtained under the assumption that only the most positive voltage wing section had low surface resistivity. Trial runs made assuming that all positive-operating voltage sections had low resistivity resulted in the +7 kV and +14 kV sections remaining negative relative to space. Hence, only the 21 kV section should be experiencing the "snap-over" phenomenon.

A view of the system sectioned through the interconnects showing the end-to-end surface voltage profile is given in Fig. 14. This voltage profile indicates that the quartz surface voltage on the negative-voltage wing remained at about -6

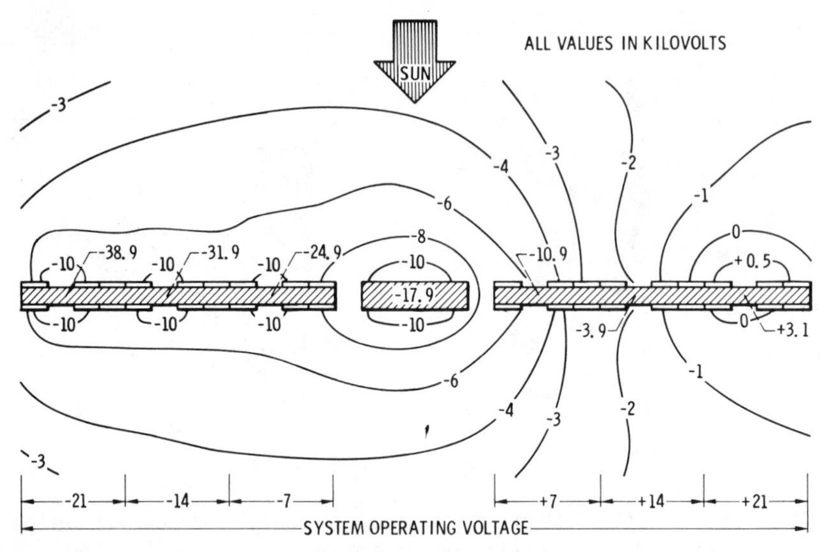

Fig. 13 Voltage distribution around space power system. (Relative to space plasma potential.)

to -7 kV while the conductive interconnects were at -25 to -39 kV, giving rise to electric fields of 1 to 2×10^6 V/cm. At these electric fields, arcing will occur. On the positive-voltage wing the fields at the interconnects are on the order of 10^5 V/cm. It is possible that arcing could occur in the first section (-10.9 kV). The absolute criteria for breakdown still has to be determined.

The simulation of the "snap-over" phenomenon also is shown in Fig. 14. The use of surface conductivity provides a reasonable approximation of this "snap-over" phenomenon, but additional work still is required to improve this simulation. The plasma coupling current collected by this system is on the order of 0.1 A, less than 1% of the bus current. Hence, this study substantiates the previous conclusions that power loss in geosynchronous orbit is not a serious problem.

The electric fields through the dielectrics are shown in Fig. 15. These fields are very strong and can produce mechanical stresses in the dielectrics. The force per unit volume induced in the dielectrics is given by

$$T = \frac{1}{2}\vec{D} \cdot \vec{E} = \frac{1}{2}\varepsilon\varepsilon_0 E^2 \text{ N/m}^3$$

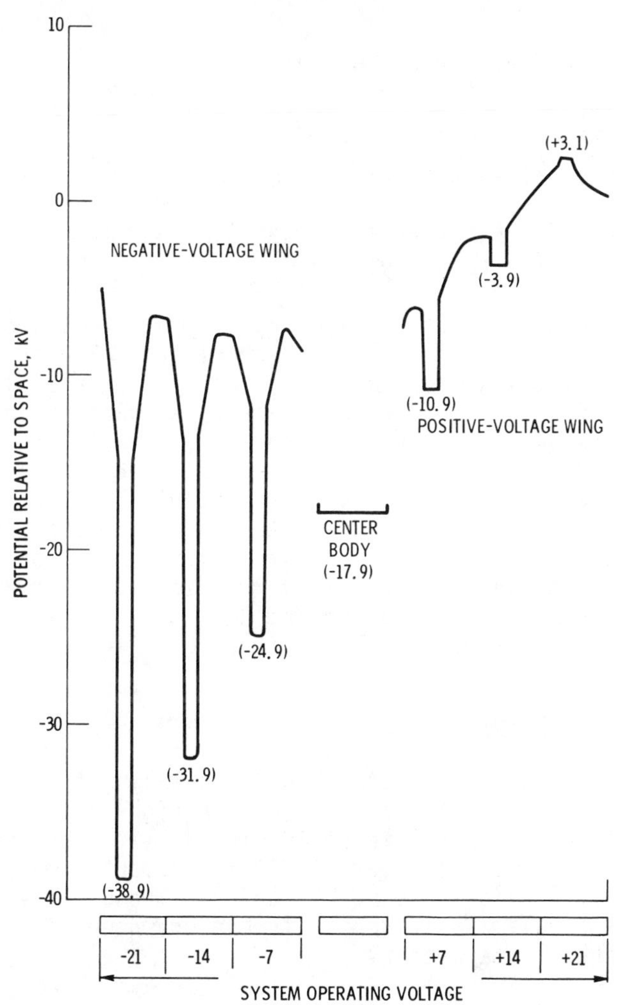

Fig. 14 Potential profile across sunlit surface. (Relative to space plasma potential.)

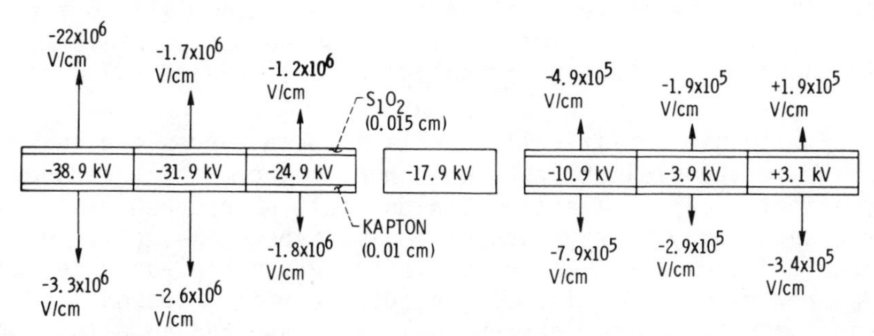

Fig. 15 Electric fields in dielectric materials; high voltage power system.

where

ε = relative dielectric constant

ε_0 = permittivity of space, 8.85×10^{-12} $C^2/N\cdot m^2$

$\vec{D}$ = electric displacement, C/m^2

$\vec{E}$ = electric field intensity, V/m

These volume forces range from 3-50 N/m^3 and must be considered in designing these systems.

Finally, the question of energy storage in these dielectrics must be considered. At the predicted voltages, the energy capacitively stored in the dielectric surfaces is on the order of megajoules. From the spacecraft charging investigation it is known that the discharge of a few joules of energy, capacitively stored in spacecraft surfaces, can disrupt electronic surfaces and degradate thermal control surfaces. A discharge in the high-voltage system of even a fraction of this stored energy could be catastrophic. Since there is insufficient data on discharge phenomena to determine if such a breakdown could actually occur and how much energy would be lost, additional studies are required to evaluate this possible threat.

Discussion of Results

In its present form, the NASCAP code is not a perfect tool for analyzing interactions between high-voltage solar arrays and charged-particle environments. It does, however, include the dielectric materials in the computations of the voltage distributions around the object. Hence, the code can be used as a valid indicator of technology needs for understanding these interactions.

This study indicates that the first barrier to overcome in the use of high voltages for space applications is the arcing in the negative-voltage wings. Since the array must float predominately negative relative to space, there will be the possibility of such arcing. This phenomenon is amenable to solution provided that a technology effort is undertaken.

The possible difficulties with induced forces and energy storage in the dielectrics also can be overcome by a technology effort. If the electric fields could be reduced, then the force and energy storage would be diminished. Reduction of electric fields could be accomplished by a change of materials or by reducing the resistivity of dielectrics proposed for these systems. These changes could increase the coupling currents so that trade-off studies must be conducted.

An improved analytical tool is required in order to assess better the interactions. Since these systems are large, ground simulation testing will continue to be limited to small samples and will be geared more toward understanding processes rather than demonstrating system feasibility. Extrapolations from test articles to complete systems and from vacuum facility simulations to space environments must be done with this analytical tool.

The above problems and the means leading to their solution are the areas in which logical concern should be expressed for high-voltage system operations in space. The interactions are not insurmountable; they are not ultimate barriers to the use of high-voltage space systems. They must, however, be faced, studied, and understood.

Concluding Remarks

Large, high-voltage space power systems are being proposed for future applications in both low Earth orbit and geosynchronous altitudes. These systems will have exposed conductive surfaces that will be at various voltage levels relative to the space plasma potential. These conductive surfaces will be surrounded by thin dielectric surfaces such as quartz or Kapton. The space charged-particle environment will interact with these complex conductor-insulator systems, influencing the system performance.

Ground simulation testing has indicated that when solar array circuits are biased to voltages greater than +100 V the electron current collection is enhanced. It is believed that the bias field accelerates plasma electrons into the cover slides, generating secondary electrons which are collected. When solar array segments are biased to large negative voltages (>-500 V), arcing occurs. This arcing appears to be confined to cavities such as those formed by cover slides and interconnects or in holes in the substrates. Both of these phenomena are plasma-density dependent.

A preliminary analytical model of a high-voltage space power system has been developed using the NASCAP code. The performance of this system operating at 42 kV is analyzed in geosynchronous environmental conditions. This analysis indicates that the system would float about 3 kV positive and 39 kV negative relative to the space plasma potential. Under these conditions the power lost through the environment is less than 1%, as expected. However, electric fields within the dielectrics (0.015 cm quartz cover slides and 0.01 cm Kapton substrates) and at the interconnects are in excess of 10^6 V/cm, giving rise to breakdowns. These electric fields also can induce strong mechanical forces within the dielectrics and store considerable energy in the dielectrics. These could cause difficulties in the proposed long lifetime of these systems.

The existing ground test data and analysis strongly indicates severe detrimental interactions between high-voltage systems and charged-particle environments. These interactions are not believed to be insurmountable. The data and studies are based on present day construction techniques and materials choices. With a technological investigation to understand the interactions, it will be possible to devise means of controlling these interactions and of guaranteeing the successful operation of high-voltage space power systems.

References

[1]"Outlook for Space," NASA SP 386, 1976.

[2]Johnson, R. D. and Holbrow, C. (eds.), "Space Settlements, A Design Study," NASA SP 413, 1977.

[3]Woodcock, G. R., "Solar Satellites, Space Key to Our Future," Astronautics and Aeronautics, Vol.15, July/Aug.1977, pp. 30-43.

[4]"Satellite Power System (SPS) Feasibility Study," Rockwell International Corp., Downey, Calif., SD76-SA-0239-1, 1976. (NASA CR-150439.)

[5]"Systems Definition of Space-Based Power Conversion System," Boeing Aerospace Co., Seattle, Wash., D180-20309-1, 1977. (NASA CR-150209.)

[6]Bekey, I., "Big COMSATS for Big Jobs at Low User Costs, Astronautics and Aeronautics, Vol. 17, Feb. 1979, pp. 42-56.

[7]Mordan, G. W., "The 25-kW Power Module - The First Step Beyond the Baseline STS," AIAA Paper 78-1693, Sept. 1978.

[8]Savage, M. and Haughey, J. W., "Overview of Office of Space Transportation Systems Future Planning," Future Orbital Power Systems Technology Requirements, NASA CP-2058, 1978, pp. 71-92.

[9]Stevens, N. J., "Spacecraft System-Charged Particle Environmental Interactions (lead)," Proceedings of the Spacecraft Charging Technology Conference, Air Force Academy, Colorado Springs, Colo., Oct. 1978. (NASA CP-2071, 1979.)

[10]Woosley, A. P., Smith, O. B., and Nassen, H. W., "Skylab Technology Electrical Power System," AAS Paper 74-129, Aug. 1974.

[11]Cole, R. W., Ogawa, H. S., and Sellen, J. M., Jr., "Operation of Solar Cell Arrays in Dilute Streaming Plasmas," TRW Systems, Redondo Beach, Calif., TRW-09357-6006-R000, Mar. 1968. (NASA CR-72376.)

[12]Herron, B. G., Bayless, J. R., and Worden, J. D., "High Voltage Solar Array Technology," AIAA Paper 72-443, Apr. 1972.

[13]Kennerud, K. L., "High Voltage Solar Array Experiments," Boeing Aerospace Co., Seattle, Wash., Mar. 1974. (NASA CR-121280.)

[14]Domitz, S., and Grier, N. T., "The Interaction of Spacecraft High Voltage Power Systems with the Space Plasma Environment," Proceedings of the Power Electronics Specialists Conference, Institute of Electrical Engineers, Inc., New Jersey, 1974, pp. 62-69.

[15]Grier, N. T. and Domitz, S., "Current from a Dilute Plasma Measured Through Holes in Insulators," NASA TN D-8111, 1975.

[16]Stevens, N. J., "Solar Array Experiments on the SPHINX Satellite," NASA TM X-71458, 1973.

[17]Stevens, N. J., Berkopec, F. D., Purvis, C. K., Grier, N. T., and Staskus, J. V., "Investigation of High Voltage Spacecraft System Interactions with Plasma Environments," AIAA Paper 78-672, Apr. 1978.

[18]McCoy, J. E. and Konradi, A., "Sheath Effects Observed on a 10-Meter High Voltage Panel in Simulated Low Earth Orbit Plasmas," Proceedings of the Spacecraft Charging Technology Conference, Air Force Academy, Colorado Springs, Colo., Oct. 1978. (NASA CP-2071, 1979.)

[19]Grier, N. T. and Stevens, N. John, "Plasma Interaction Experiment (PIX) Flight Results," Proceedings of the Spacecraft Charging Technology Conference, Air Force Academy, Colorado Springs, Colo., Oct. 1978. (NASA CP-2071, 1979.)

[20]Knauer, W., Bayless, J. R., Todd, G. T., and Ward, J. W., "High Voltage Solar Array Study," Hughes Research Labs., Malibu Calif., May 1970. (NASA CR-72675.)

[21]Springgate, W. F. and Oman, H., "High Voltage Solar Array Study," Boeing Co., Seattle, Wash., D2-121734-1, 1969. (NASA CR-72674.)

[22]Fralick, G. C., "Calculation of Current Collected in Dilute Plasma Through a Pinhole in the Insulation Covering a High Voltage Surface," NASA TN D-7957, 1975.

[23]Purvis, C. K., Stevens, N. J., and Berkopec, F. C., "Interaction of Large High Power Systems with Operational Orbit Charged-Particle Environments," (NASA TM-73867, 1977.)

[24]Parker, L. W., "Plasma Sheath Effects and Voltage Distributions of Large High-Power Satellite Solar Arrays," Proceedings of the Spacecraft Charging Technology Conference, Air Force Academy, Colorado Springs, Colo., Oct. 1978. (NASA CP-2071, 1979.)

[25]Freeman, J. W., Cooke, D., and Keiff, P., "Space Environmental Effects and the Solar Power Satellite," Proceedings of the Spacecraft Charging Technology Conference, Air Force Academy, Colorado Springs, Colo., Oct. 1978. (NASA CP-2071, 1979.)

[26]Parks, D. E. and Katz, I., "Spacecraft Generated Plasma Interactions with High Voltage Solar Array," AIAA Paper 78-673, Apr. 1978.

[27]Katz, I., Mandell, M. J., Schneulle, G. W., Cassidy, J. J., and Roche, J. C., "The Capabilities of the NASA Charging Analyzer Program," Proceedings of the Charging Technology Conference, Air Force Academy, Colorado Springs, Colo., Oct. 1978. (NASA CP-2071, 1979.)

[28]Katz, I., Parks, D. E., Mandell, M. J., Harvey, J. M., Brownell, D. H., Jr., Wang, S. S., and Rotenberg, M., "A Three Dimensional Dynamic Study of Electrostatic Charging in Materials," Systems Science and Software, La Jolla, Calif., SSS-R-77-3367, Aug. 1977. (NASA CR-135256.)

[29]Mandell, M. J., Harvey, J. M., and Katz, I., "NASCAP User's Manual," Systems Science and Software, La Jolla, Calif., SSS-R-77-3368, Aug. 1977. (NASA CR-135259.)

[30]Katz, I., Parks, D. E., Mandell, M. J., Harvey, J. M., Wang, S. S., and Roche, J. C., "NASCAP, A Three-Dimensional Charging Analyzer Program for Complex Spacecraft," IEEE Transactions, Nuclear Science, Vol. 24, Dec. 1977, pp. 2276-2280.

[31]Stevens, N. J., Roche, J. C., and Mandell, M. J., "NASCAP Modelling of High-Voltage Power System Interactions with Space Charged-Particle Environments," (NASA TM-79146, Feb. 1979.)

PLASMASHEATH-PHOTOSHEATH THEORY FOR LARGE HIGH-VOLTAGE SPACE STRUCTURES

Lee W. Parker*
Lee W. Parker, Inc., Concord, Mass.

Abstract

This work presents a new method for rigorously computing sheath structures of large spherical bodies with high-voltage surfaces and with photoelectric/secondary emission. This method, using the author's Turning-Point Formulation, is transparently simple and results in a compact computer program. Self-consistency of the Poisson and Vlasov solutions is achieved through iteration. The power and flexibility of the method is illustrated through four sample sheath solutions, including a) the sheath of a large body (radius 100 Debye lengths) with voltage 400,000 kT/e, the most extreme combination of size and voltage solved rigorously to date, and b) the "presheath" of an extremely large body, a nontrivial and heretofore unsolved problem in a warm plasma. In addition, two approximate models are considered, a) a linearized space charge model (leading to the Debye potential for spheres), and b) the Langmuir-Blodgett spherical diode. Both approximate models tend to underestimate current collection.

I. Introduction

Charge buildups on man-made space structures are of concern because they can cause troublesome effects. Some examples of these are:

a) Power losses of solar arrays[1-3] and other artificial charge-separation systems by leakage conduction in the charged-particle (plasma) medium and due to currents from ion thrusters.[4]

b) Sputtering and erosion of negative thin-layered heat-control surfaces by attracted-ion bombardment.

---

*Principal Scientist.

c) Arcing/breakdown between or through thin dielectric layers, causing false transients and tripping of logic control circuits with subsequent loss of control.[5,6]

d) Electromagnetic stresses and figure distortions of large fragile structures.

e) Interference with on-board measurements of the plasma and field environment by perturbing the environment.[7-10]

f) Interference with operation of control devices such as ion thrusters by perturbing the ion optics and diverting their currents.[4]

g) Interference with charged-particle beams emitted for various purposes.[11-16]

h) Creation of additional high-density plasma by acceleration of ions and electrons to cause impact ionization of neutral gases (residual exhaust or thruster as well as ambient), which can exacerbate some of the above effects.[15-17]

These effects depend on the sizes and surface voltages of the space bodies/structures, on the nature of the plasma, and on the sheath geometry. Any body/structure in space, natural or man-made, develops electrostatic charges on its surface (or surfaces) due to several effects, one of which is the "sheath effect" caused by its interaction with the charged particles in its environment. Because of the surface charges and the nature of the plasma, a "sheath" develops. The sheath is a layer of net space charge of opposite sign which tends to screen the bulk of the plasma from the surface charges. The sheath confines all of the significant electric field structure, and therefore its geometry and thickness relative to certain partial scale lengths on the body, or to the full body dimension itself, are of critical importance for estimating the magnitude of the various electrical interactions of interest.

The three principal ingredients involved in these estimates are a) surface voltage or voltages, b) charged-particle fluxes, including plasma ions and electrons, photoelectrons, and backscattered and secondary electrons from the surface, and ions and electrons from thruster plasmas, and c) sheath geometry and thickness. The three ingredients depend on one another and must be mutually compatible or consistent. We do not know a priori any of these ingredients (with the exception that artificial control of spacecraft potential is now feasible through charged-particle beams emitted by on-board accelerators.[11,12] Attempts to scale up laboratory measurements to proposed large-scale space systems have the difficulty of unknown scaling laws. Hence we must rely heavily on theoretical models. In general, numerical techniques and considerable sophistication are required in order to solve the mutually-consistent problem, i.e., the simultaneous solutions of the Poisson and Vlasov equations which will be described in this paper. (See also Refs. 7-9, 18-25.) This is true whether or not the surface voltage is known. Thus, the problem is analogous to that of a Langmuir electrostatic probe in a laboratory plasma, albeit geometrically more complicated. The analogy is made more complete by the possibility of artificial control of spacecraft potential as with laboratory probes, including the "floating potential" condition as a special case. Hence, we adopt the point of view and the terminology of (generalized) Langmuir probe theory in considering the relation between surface voltage, charged-particle fluxes, and sheath geometry and thickness.

In this paper we show how the three ingredients mentioned previously may be treated theoretically. Our purpose is to clarify the role played by the sheath, to describe some models for sheath calculations, and to present some new solutions. For the purposes of illustration and relative simplicity, we confine ourselves here to spherically-symmetric geometries. This dictates that the geometry of the sheath be that of a spherical shell characterized by a length scale called its "thickness" as the parameter of interest. Because of this geometrical restriction, we exclude from the scope of this paper an interesting large class of "overlapping sheath" problems associated with more complicated three-dimensional geometries. Examples of the latter, which may be described as "differential charging" problems where different insulated portions of the surface acquire different potentials, are a) those due to applied internal voltage differentials, such as in high-power solar power satellite (SPS) solar arrays (see theoretical treatments in Refs. 1-2), or two-electrode probe systems for plasma measurements in laboratory and space (see theoretical treatments in Refs. 19, 21-23, 26, 27), and b) those due to asymmetric fluxes such as photoemission or plasma flows (see theoretical treatments in Refs. 7, 8, 28).

We will be concerned mainly with a) "large" bodies or structures, and b) "high" voltages. These terms in the present contest mean respectively that a) the body is large compared with the Debye length, and b) the surface voltage relative to space is large compared with kT/e of the plasma, where T is an "average" temperature of the plasma, and k and e denote Boltzmann's constant and the electron charge, respectively. (Strictly speaking, the ion and electron temperatures T_i and T_e are generally different, and the ratio T_i/T_e can be an important parameter.)

Thus, in low Earth orbit (LEO), say at 400 km altitude, the Debye length is about 1 cm or less, and the plasma temperature is about 0.1 eV. Hence a meter-sized spacecraft in LEO would be a "large" body, and it would be at "high" voltage if its surface voltage were larger than say 10 V (either sign). The combined limits of large size and high voltage can present a computationally difficult sheath problem. For example, in a computation to be described later, an Echo experiment rocket in LEO is assumed to be charged to 40 kV due to emission of a 40-keV electron beam.[14] The object is to determine the sheath structure. This problem is modeled by a sphere of 1-m radius (i.e., 100 Debye lengths), charged to a potential of 40 kV (i.e., 400,000 times the kT/e of the plasma), and the solution is presented in Sec. V of this paper. This is the most extreme combination of values of dimensionless potential and radius for a large body at high voltage computed self-consistently to date.

By contrast, a "small" body in LEO is on the order of 1 cm or less in dimension. This could apply to a whip antenna or small probe on an LEO spacecraft. The sheath problem is computationally simpler; without space charge the field is "Laplacian" (obtained from the solution of Laplace's equation) and depends only on the surface voltage distribution (which can also be complicated by differential charging).[1,2,7,8,19,21-23,26-28]

In geosynchronous orbit (GEO), at about 36,000 km altitude, in "fair" (nonsubstorm) conditions, the plasma Debye length is about 10 m, which provides the length scale in eclipse (i.e., in darkness). In sunlight, however, photoelectrons produce a photosheath with an effective Debye length of about 1 m. Hence, a meter-sized body (e.g., SCATHA/Air Force satellite P78-2 or an INTELSAT communications satellite) is "small" in the sense that the field is nearly Laplacian. A structure more than 10-100 m in dimension (depending on whether in darkness or sunlight) is a "large" body. A ten-meter-sized structure such as the ATS-6 dish antenna is "intermediate" in size. The plasma temperature is typically 1-10 eV, so that the body voltage is "high" above, say, 100 V.

In a substorm environment in GEO, the plasma temperature can rise to the order of 10 keV. At the same time the plasma Debye length increases by at least two orders of magnitude, to the order of a kilometer. A kilometer-sized structure such as an SPS array in the dark (i.e., eclipse) is not a "large" body, and an array voltage of 40 kV is in this case not a "high" voltage. However, in sunlight a kilometer-sized array is "large" (compared with the photosheath Debye length) and the voltage of 40 kV is "high" compared with the typical 1-V equivalent energy of the photoelectrons. Interestingly, therefore, a kilometer-sized structure may undergo a transformation from a "small" body at "low" voltage to a "large" body at "high" voltage as it passes from darkness into sunlight.

In the remainder of this paper, we discuss sheath physics. In the next section we employ the analytical approximation based on the Child-Langmuir "space-charge-limited" diode model to estimate sheath thickness, applied to examples in low Earth orbit and geosynchronous orbit. Then we discuss how rigorous sheath solutions may be obtained and present sample solutions demonstrating the structure of the plasmasheath-photosheath.

Sections III and IV present a method for rigorously calculating plasmasheath-photosheath structures by Poisson-Vlasov iteration. Section III discusses the Turning-Point Formulation originated by Parker[9,20,21] for a radially-symmetric body; this is more efficient computationally than the Effective Potential Formulation.[25] Section IV discusses the Poisson equation solution and the iteration procedure for solving the nonlinear system of Poisson and Vlasov equations for the sheath.

Sections V, VI, and VII present a number of sample sheath results of the rigorous method discussed in Secs. III and IV.

In Sec. V an extremely-high-voltage large body in LEO is treated, modeling an Echo experiment where the dimensionless potential, 400,000 in units of kT/e, is the largest value treated in any self-consistent sheath calculation to date. Section V also illustrates the use of the diode (sharp-sheath-edge) model theory of Sec. II.

In Secs. VI and VII, example solutions of photosheaths and plasmasheath-photosheaths are shown. Changes occurring as the spacecraft emerges from darkness into sunlight are discussed, as are effects on sheath structure due to changes in surface potential (e.g., which can be controlled by on-board charged-particle accelerators).

Section VIII deals with the "presheath" of an extremely large body using the principle of quasineutrality. For a warm plasma this is a nontrivial task requiring the sophistication of Sec. III.

Section IX deals with the linearized space charge model, another kind of sheath approximation, and compares the approximate results from the linearized-space-charge model and the spherical-diode model with the corresponding rigorous results. The linearized space charge model provides a simple example of the use of the Turning-Point Formulation of Sec.III.

II. Sheath Geometry and Models

Sheath structures generally require self-consistent numerical solutions of the simultaneous Poisson and Vlasov equations.[9,18-25] However, in the absence of differential charging,[1,7,8,10,19,21-23,26,27] wake effects,[7,8,18] and edge effects,[1,18,21,22] and when the voltage is "high" and the sheath is thin relative to the body dimensions and consists essentially only of attracted ambient-plasma particles, an analytical approximation is available based on a "space-charge-limited" diode model, the so-called Child (or Child-Langmuir) model.[29] This is a unipolar sheath (sharp-sheath-edge) model wherein the attracted charged particles are accelerated in a parallel beam toward the collecting plate, starting with zero energy. If e and m denote the particle charge and mass, V denotes the voltage, j denotes the current density, and S denotes the diode plate separation, then the sheath thickness may be estimated from the Child law relating V, j, and S. In centimeter-gram-second system (c.g.s.) units, this is:

$$S = (2e/m)^{1/4} \cdot V^{3/4} / (9\pi j)^{1/2} \tag{1}$$

where the sheath thickness is identified with the plate separation. In sheath thickness estimations, it is customary to replace j by the random thermal current density at the sheath edge, $j_o = en_o\,(kT/2\pi m)^{1/2}$, where n_o is the plasma density, and T is the temperature of the Maxwellian distribution. If there is also a significant drift velocity v_o (as in the case of O^+ ions in LEO) toward the surface, j_o may be replaced by $[\exp(-M^2) + \sqrt{\pi}\,M(1 + \mathrm{erf}\,M)]\,j_o$, where M is the ion Mach number $M = (mv_o^2/2kT)^{1/2}$. Thus, in practical units, taking into account both thermal and drift (ion "ram") currents at the sheath edge, the planar Eq. (1) may be written

$$S(\text{meters}) = 9.33\,\frac{V^{3/4}(\text{volts})}{n_o^{1/2}(\text{cm}^{-3}) \cdot T^{1/4}(\text{ev}) \cdot (\text{RAM})^{1/2}} \tag{2}$$

where

$$RAM = \exp(-M^2) + \sqrt{\pi} M (1 + \mathrm{erf} M) \quad (3)$$

It should be noted that Eq. (2) assumes that the surface is looking into the ram direction and is invalid if the surface looks into the wake.

Equation (2) is a planar model and cannot be used when S is comparable with or exceeds the body dimensions. Corrections for three-dimensionality, treated below, are frequently made using the analogous spherical diode model where the particles move radially inward from an outer emitter to an inner spherical collector, with no angular momentum. Langmuir and Blodgett[30] give a table of factors which may be used in conjunction with Eq. (2). The implication of sheath thickness being comparable to or exceeding body size is that the current collected is larger than the product of j_0 with the surface area of the body. The current in fact is approximated by the product of j_0 with the surface area of the sheath. Hence, significant increases in current collection occur as the sheath thickens to the order of and exceeds the body dimension.

Applying the planar sheath formula Eq. (2) to LEO, assuming T=0.1 eV and $n_0=10^6/cm^3$, we see that the thickness of an electron sheath (M=0) is S=0.017 m or 1.7 cm for a 1-V body. (An ion sheath is thinner because M>0.) Thus, a whip antenna or small probe has a fairly thick sheath, as is consistent with consideration of the Debye length. On the other hand, a meter-sized body at 1 V in LEO has a thin sheath (S=0.017 m). The sheath thickness grows with voltage and becomes comparable to the body dimension (setting S=1 m) at a surface voltage "transition value" of 200 V. In this regime, the spherical correction modeling three-dimensional effects must be used. Similarly, a 10-m-radius body and a 100-m-radius body have thin sheaths until their voltages are, respectively, of the order of 5 kV and 100 kV; somewhere before this "transition" point the spherical correction should be used. The regime where the spherical correction should be used corresponds to the regime where significantly increased currents begin to be collected [Eq. (4)].

In GEO, on the other hand, assuming nonsubstorm conditions, with T=1 eV and $n_0=1/cm^3$, and M=0, Eq. (2) shows that the sheath thickness S at 1 V is 9 m, i.e., of comparable thickness for a 10-m body (e.g., ATS-6) but rather thin for a 100-m body and very thin for a 1-km body. The sheath becomes comparable with body size, and therefore significant increases in current begin to occur, when a 100-m body has a voltage of 20 V, or when a 1-km body has a voltage of 500 V.

In substorm conditions, when T=10 keV is typical, the corresponding "transition" voltages for a 10-m body, a 100-m body, and a 1-km body are, respectively, 20 V, 500 V, and 10 kV.

By using Eq. (2) we have ignored the effect of the photosheath, and thus we have underestimated the voltage required to significantly thicken the sheath. It is not clear how Eq. (2) should be modified to correct for the presence of a photosheath.

The corrections for a spherical diode model, that enable us to extend the usefulness of Eq. (2), are given in tabular

Table 1 Spherical diode model sheath radius r_s as function of S/r_o [a]

r_s/r_o	[b]	S/r_o	r_s/r_o	[b]	S/r_o	r_s/r_o	[b]	S/r_o
1.0		0.	3.0	(2.98)	4.75	14		100.8
1.05	(1.049)	0.0515	3.2		5.50	16		128.7
1.1	(1.098)	0.1078	3.4		6.28	18		159.5
1.15		0.1679	3.6		7.12	20	(20.0)	193.1
1.2	(1.206)	0.232	3.8		8.00	30		400
1.25		0.299	4.0	(3.99)	8.92	40		669
1.3		0.370	4.2		9.87	50	(50.1)	994
1.35		0.445	4.4		10.88	60		1373
1.4		0.523	4.6		11.92	70		1803
1.45		0.605	4.8		13.0	80		2280
1.5	(1.506)	0.691	5.0	(5.025)	14.12	90		2810
1.6		0.872	5.2		15.30	100	(100.0)	3380
1.7		1.067	5.4		16.49	120		4660
1.8		1.275	5.6		17.72	140		6110
1.9		1.497	5.8		19.0	160		7730
2.0	(1.998)	1.732	6.0	(6.09)	20.3	180		9510
2.1		1.980	6.5		23.7	200	(199.4)	11440
2.2		2.24	7.0		27.4	250		16920
2.3		2.51	7.5		31.3	300		23300
2.4		2.80	8.0		35.4	350		30530
2.5		3.09	8.5		39.8	400		38600
2.6		3.40	9.0		44.3	500	(495)	57000
2.7		3.72	9.5		49.1			
2.8		4.06	10	(9.94)	54.0			
2.9		4.40	12		75.9			

[a] Based on Ref. 30.

[b] From fit, Eq. (5). $r_s/r_o = F(S/r_o)$

$$x \leq 19:\ F(x)=0.5+(0.25+x)^{1/2}+0.052xH(x-0.2)$$

$$x > 19:\ F(x)=(1.0+x^{.753})^{.7524}$$

form in Table 1 and in graphical form in Fig. 1; they are obtained from the numerical data given by Langmuir and Blodgett,[30] and are used as follows: Let r_o denote the spherical collector (body) radius, and let r_s denote the sheath radius. The data represented by Table 1 or Fig. 1 allows us to find r_s/r_o as a function F of S/r_o, where S is given by Eq. (1) or Eq. (2). Thus, the current-voltage characteristic of a sphere is given, on the basis of the diode sheath model, by

$$j/j_o = r_s^2/r_o^2 = F^2(S/r_o) \tag{4}$$

Table 1 is obtained by reformulating the numerical data in Table II of Langmuir and Blodgett[30], so as to obtain a more convenient form for our sheath estimates. For the convenience of the reader we have plotted $(r_s/r_o)-1$ as a function of S/r_o in Fig. 1, using the data in Table 1. As a further convenience we have provided an analytical fit to the data as follows: (S defined by Eq. (2))

$$\text{For } \frac{S}{r_o} \leq 19: \frac{r_s}{r_o} = \frac{1}{2} + \sqrt{\frac{1}{4} + \frac{S}{r_o}} + 0.052\,\frac{S}{r_o}\,H\!\left(\frac{S}{r_o} - 0.2\right)$$

$$\text{For } \frac{S}{r_o} > 19: \frac{r_s}{r_o} = \left[1 + \left(\frac{S}{r_o}\right)^{.753}\right]^{.7524} - \left(\frac{S}{r_o}\right)^{.5666} \tag{5}$$

where H(x) is the unit step function (H=o if x<o; H=1 if x>o).

In Table 1, the figures in parentheses show how the analytically defined r_s/r_o using Eq. (5) compares with the exact value. The fit is exact at r_s/r_o=20 and 100, and the error is no greater than about 1.5% over the entire range shown. Also, at values of r_s/r_o near unity (thin sheaths), the sheath thickness r_s-r_o is in error by at most 3%. As $S\to 0$, r_s-r_o correctly approaches S. At large values of S/r_o, r_s/r_o approaches $(S/r_o)^{0.5666}$. One may compare our fit with a similar fit suggested by Kennerud,[31] which can be put in the form $r_s/r_o \sim 0.958\,(S/r_o)^{4/7}$ using our terminology. (Kennerud's formula is accurate for r_s/r_o above 100, but becomes very inaccurate in the more useful range at lower values). Our formula represents an excellent fit over the entire range of interest.

The spherical corrections for modeling three-dimensional sheath effects now may be applied to the LEO and GEO examples previously discussed within the thin-sheath approximation [Eq. (2)]. If $(r_s-r_o)/r_o$ is small, the sheath is thin and its thickness is given by S using Eq. (2). Let (r_s/r_o)=2 define

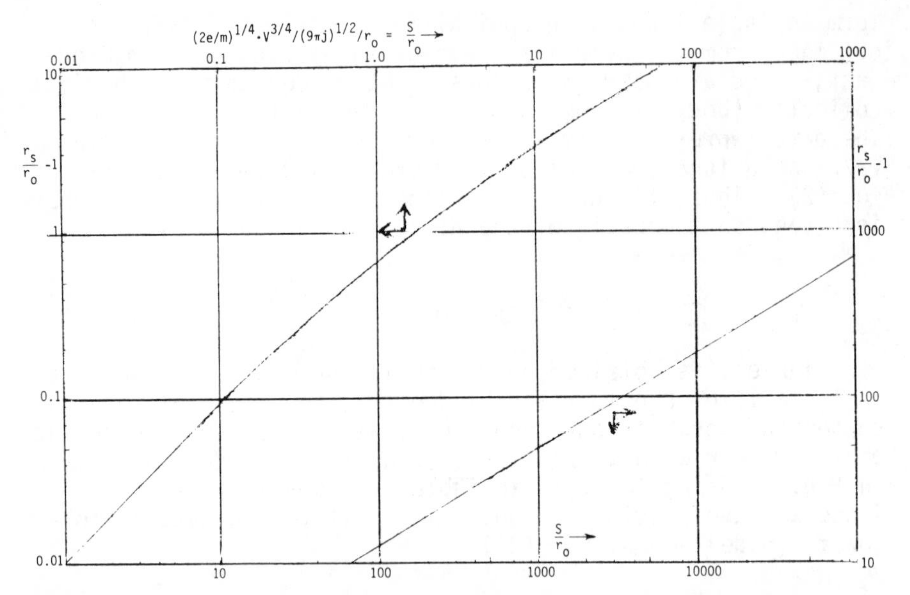

Fig. 1 Spherical sheath radius r_s vs planar sheath thickness S, Eq. (1) (Langmuir-Blodgett diode model sharp-edged sheath).

the transition point where the sheath may no longer be considered thin. From Table 1 or Fig. 1, this corresponds to $S=1.732\ r_o$.

In LEO this transition condition occurs when V=500 V, 10 kV, and 200 kV, respectively, for r_o=1 m, 10 m, and 100 m (1-m, 10-m, and 100-m bodies, respectively).

In GEO with nonsubstorm conditions (in darkness), for body radius r_o=10 m, 100 m, and 1 km, the transition voltages are, respectively, V=2 V, 50 V, and 1000 V. With substorm conditions (in darkness), the corresponding transition voltages are 50 V, 1000 V, and 20 kV.

Note that with the spherical correction the transition voltages are about twice as large as those given by the planar sheath formula alone.

Note also that the GEO transition voltages, in the case of the 10-m body in nonsubstorm conditions and in the case of all three bodies (up to 1 km) in substorm conditions, are comparable with kT/e of the plasma. These are therefore not "high" voltages, implying that the diode model is not applicable. In sunlight the photosheath should be considered. It has been neglected in the preceding approximate analysis, but its presence would tend to keep the sheath thin and to increase the transition voltage for a positive surface; a weak opposite effect may be expected for a negative surface. Photoemission effects are included later in the rigorous analysis of this paper.

Current-voltage characteristics of three-dimensional bodies may be approximated by the spherical diode model, represented by Eqs. (1)-(5). However, we defer discussion of these and of a linearized approximation until after we present a rigorous treatment of the sheath, in the next two sections. (See Sec. IX for assessments of models.)

III. Plasmasheath-Photosheath Structure by Rigorous Kinetic Theory

A three-dimensional structure/body is modeled here by a sphere while the ambient plasma is isotropic, collisionless, and stationary with respect to the structure. Photoelectrons and secondary electrons are assumed to be emitted isotropically from the surface. (In the analogous case of a laboratory Langmuir probe the emission can consist of either thermionic electrons or secondaries due to bombardment.) All charged particles that strike the surface are assumed to be neutralized. Magnetic field effects are neglected. Far from the body, the distribution functions for electrons and ions in the ambient plasma are treated in the present paper as Maxwellians. The same applies to the distribution function for emitted electrons at the surface. The problem is assumed to be independent of time, as is justified by recent time-dependent computer-simulation solutions at sufficiently long times.[32] It is desired to calculate the spatial distributions of electric potential and of electron and ion densities, as well as the currents of ions and electrons exchanged between the body and the environment. The basic problem assumes a fixed potential on the surface, but it is a natural extension to find the floating potential by solving a sequence of problems until current balance has been achieved.

In accord with the foregoing assumptions, a particle distribution function f (the density in phase space) satisfies the time-independent Vlasov equation, the first basic equation:

$$\vec{V} \cdot \nabla f - \frac{q}{m}\nabla\Phi\cdot\nabla_V f = 0 \qquad (6)$$

In Eq. (6), ∇ and ∇_V denote gradient operators with respect to position and velocity space, respectively, and q and m denote respectively the charge and mass of the particle. This equation states that f, which is a function of the vector position $\vec{R}$ and of the vector velocity $\vec{V}$, is constant along an orbit in an electric field of potential $\Phi(\vec{R})$. (The symbol V used herein to denote dimensional velocity should not be confused with V representing voltage in the previous section. Capital letters are used in this section to denote some dimensional quantities; the corresponding dimensionless quantities are denoted by lower-case letters.)

The electric potential Φ is obtained from Poisson's equation, which may be expressed in terms of the ambient-ion number density N_i, the ambient-electron number density N_e, and the surface-emitted-electron number density N_s (in the c.g.s. system of units):

$$\nabla^2\Phi = 4\pi e(N_s + N_e - N_i) \tag{7}$$

This is the second basic equation. Obtaining the numerical solution of the sheath problem requires that two fundamental subproblems be solved simultaneously on a set of radial grid points. One is the "Vlasov Problem," where the charged-particle densities are computed at the grid points, (and currents at the surface), with the potentials considered known at the same grid points. This involves solving Eq. (6). The other is the "Poisson Problem," where the potentials are computed with the charged-particle (space charge) densities known. This involves solving Eq. (7). For mutual consistency, an iteration procedure is required.

The Poisson Problem involves solving a set of linear equations representing Eq. (7) in finite-difference form and will be treated in Sec. IV. The more difficult Vlasov Problem is considered next.

The Vlasov Problem

Solution of Eq. (6) for f subject to the appropriate boundary conditions (the "Vlasov Problem") allows one to obtain the particle density $N(\vec{R})$ at a point $\vec{R}$ by integrating f over velocity space:

$$N(\vec{R}) = \iiint f(\vec{R},\vec{V})\, d^p\vec{V} \tag{8}$$

where $d^p\vec{V}$ represents a p-dimensional volume element of the velocity space. For a spherical problem p=3, while p=2 for a cylindrical problem. This integral represents the zero-th moment of the distribution. Similarly, one may obtain the flux or current density at a point $\vec{R}$ by defining V_n as the component of velocity of interest and integrating the product fV_n over velocity space (the first moment):

$$j(\vec{R}) = \iiint f(\vec{R},\vec{V})\ V_n d^p\vec{V} \tag{9}$$

On a spherical or cylindrical surface, V_n becomes V_r, the radial component of velocity.

A simple approach to solving Eq. (6) is based on its characteristics. One may determine the local value of $f(\vec{R},\vec{V})$ at an arbitrary point $\vec{R}$, by considering the orbit of a particle arriving at $\vec{R}$ with the velocity $\vec{V}$. If the orbit is traced backwards and a point is eventually reached where f is known (namely, the "source" at the surface or at infinity), then since f is constant along the particle trajectory the local value of f is identical to the known value at the source. If the orbit is found to be closed upon itself, its population can consist only of trapped particles which can have arrived there through a collisional mechanism. It is assumed here that the closed orbits, if any, are unpopulated (that is, f=0 for these). This assumption usually is made for tractability purposes, based on a plausibility argument, but its ultimate justification is difficult and requires a quantitative collisional theory. Such a theory is available,[21] but has not as yet been applied to the present sheath problem.

Since we have assumed isotropic sources at the surface and infinity, and since total energy is conserved along an orbit, the velocity distributions can depend only on the particle energy at the source. Denoting by V_∞ the particle speed at infinity, and by V_s the particle speed at the surface, we may define distinct distribution functions for ambient and emitted particles, depending on the distributions at the respective sources:

$$f_i(\infty,\vec{V}_\infty) \equiv F_\infty(MV_\infty^2/2) \tag{10}$$

$$f_e(\infty,\vec{V}_\infty) \equiv F_\infty(mV_\infty^2/2) \tag{11}$$

$$f_s(\vec{R}_s,\vec{V}_s) \equiv F_s(mV_s^2/2) \tag{12}$$

Here the subscripts i and e refer to the ambient ions (of mass M) and electrons (of mass m), respectively, while subscript s refers to the surface-emitted electrons. The arguments of F_∞

and F_s are related to the total energy which is constant along orbits. Then the distribution function for the ions at any point $\vec{R}$ can be written, specialized to a Maxwellian:

$$f_i(\vec{R},\vec{V}) = \delta_i(\vec{R},\vec{V})\ F_\infty(MV_\infty^2/2) = \delta_i(\vec{R},\vec{V})\ N_o \left[\frac{M}{2\pi kT_i}\right]^{3/2} \exp(-MV_\infty^2/2kT_i) \tag{13}$$

where N_o and T_i denote the ambient-ion density and temperature, respectively. For the electrons, there are separate distribution functions, one describing the electrons from infinity (F_∞) and the other those from the surface (F_s); these are specialized to Maxwellians:

$$f_e(\vec{R},\vec{V}) = \delta_e(\vec{R},\vec{V})\ F_\infty(mV_\infty^2/2) = \delta_e(\vec{R},\vec{V})\ N_o \left[\frac{m}{2\pi kT_e}\right]^{3/2} \exp(-mV_\infty^2/2kT_e) \tag{14}$$

$$f_s(\vec{R},\vec{V}) = \delta_s(\vec{R},\vec{V})\ F_s(mV_s^2/2) = \delta_s(\vec{R},\vec{V})(2)\ N_{so} \left[\frac{m}{2\pi kT_s}\right]^{3/2} \exp(-mV_s^2/2kT_s) \tag{15}$$

with the subscripts e and s referring, respectively, to the electrons from infinity (ambient) and those from the surface, and where T_e and T_s denote the ambient-electron and surface-electron temperatures, respectively. N_{so} denotes the surface-electron number density evaluated at the surface (with the factor of 2 in parentheses representing the doubling appropriate to a half-Maxwellian distribution at the surface). The "delta-factors" $\delta_{i,e,s}(\vec{R},\vec{V})$ contain the orbit information, that is, regarding whether the orbit connects with the appropriate source. Thus, δ_i=1 if the ion comes from infinity, while δ_i=0 if the ion comes from the probe surface. Similarly, δ_e=1 and δ_s=0 if the electron comes from infinity, while δ_e=0 and δ_s=1 if the electron comes from the surface. That is, a nonzero value for δ signifies that the orbit is "occupied." Since the δ's are step-functions, and the F-functions are known, it is only necessary to locate the boundaries of the regions in velocity space associated with the two different types of particle sources. For the symmetric problem of interest, the regions in velocity space will be found to be further subdivided according to types of orbits: those which contribute once due to a single pass through the radius of interest, and those which contribute twice because the particle reverses direction after passing the radius of interest and then recrosses the radius of interest. Thus, the δ-factors which do not vanish can take on the value 2 as well as unity.

Charged-Particle Densities

From this point onward we will use dimensionless variables, defined as follows:

r = radial coordinate R divided by the sphere radius r_o

ϕ = electric potential Φ divided by kT_i/e for ambient-plasma ions, by $-kT_e/e$ for ambient-plasma electrons, and by $-kT_s/e$ for surface-emitted electrons

v = speed V divided by $(2kT/m)^{1/2}$, where $T/m = T_i/M$ for plasma ions, $T/m = T_e/m$ for plasma electrons, and $T/m = T_s/m$ for surface-emitted electrons

v_r = radial component of velocity divided by $(2kT/m)^{1/2}$, as for v

$E = (MV_\infty^2/2 + e\Phi_\infty)/kT_i = v_\infty^2 + \phi_\infty$ for plasma ions (ϕ_∞ = plasma potential at infinity usually assumed to be zero)

$E = (mV_\infty^2/2 - e\Phi_\infty)/kT_e = v_\infty^2 + \phi_\infty$ for plasma electrons (ϕ_∞ usually assumed to be zero)

$E = (mV_s^2/2 - e\Phi_s)/kT_s = v_s^2 + \phi_s$ for surface-emitted electrons (ϕ_s = dimensionless potential at surface)

J^2 = square of angular momentum divided by $2kTmr_o^2$, with $mT = MT_i$ for plasma ions, $mT = mT_e$ for plasma electrons, and $mT = mT_s$ for surface-emitted electrons

j = surface current density divided by $j_{io} = eN_o(kT_i/2\pi M)^{1/2}$ for plasma ions, by $j_{eo} = eN_o(kT_e/2\pi m)^{1/2}$ for plasma electrons, and by $j_{so} = e(2N_{so})(kT_s/2\pi m)^{1/2}$ for surface-emitted electrons

n_i = ambient-ion density N_i divided by N_o

n_e = ambient-electron density N_e divided by N_o

n_s = surface-emitted-electron density N_s divided by N_{so}

Dimensionless Densities. The number density at a point r of charged particles (ions or electrons) having a Maxwellian distribution at its source (ambient or surface) may be written in dimensionless form as

$$n = \frac{2}{\sqrt{\pi}} \int_{v_{min}}^{\infty} \exp(-v^2 - \phi)\, v^2\, dv \int d(\cos\,\theta) \qquad (16)$$

(An expression for the current will be considered later.) Here, v denotes the local speed at the point r. The angle θ is that between the velocity vector and the radial direction at the point r. The δ-factors of Eqs. (13)-(15) are included in the limits of integration for d(cos θ) in Eq. (16). These limits will be specified later in terms of the new variables, energy and angular momentum. The quantity v_{min} is a minimum velocity, depending on the difference between source potential and local potential.

Transforming the integration over v and θ to an integration over E and J^2, where E is the total energy and J^2 is the square of the angular momentum, has the advantage that E and J^2 are constants of the motion. The pair E, J^2, then, completely characterize an orbit, independent of the variation of r along that orbit. That is, in an orthogonal r, E, J^2 space, orbits are straight lines parallel to the r axis, and the analysis of orbits which can or cannot reach a given r is thereby simplified.[20]

From the transformation

$$E = v^2 + \phi \qquad (17)$$

and

$$J^2 = r^2 v^2 \sin^2\theta \qquad (18)$$

one obtains the expression:

$$n = \frac{2}{\sqrt{\pi}} \exp(\phi_{source}) \int_{\phi_{source}}^{\infty} \exp(-E)\, dE \cdot M_n(E) \qquad (19)$$

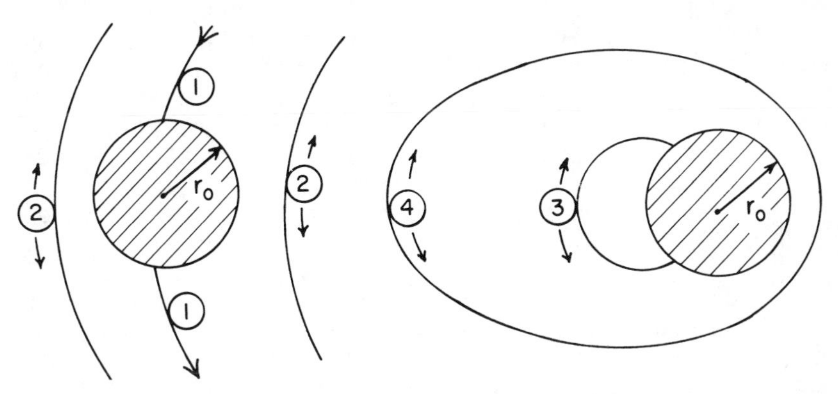

Fig. 2 The four types of orbits.

where M_n is the "monoenergetic" contribution to n, comprised of the integral over angular momenta:

$$M_n(E) = \frac{C}{2} \int \delta \cdot d[E - \phi - J^2/r^2]^{1/2} \tag{20}$$

The lower limit of the E-integral of Eq. (19), ϕ_{source}, denotes the potential at the source. For ambient ions and electrons this is $\phi_{source} = \phi_\infty$, but ϕ_∞ is usually defined to be zero. For surface-emitted electrons $\phi_{source} = \phi_s$. The factor C is unity for ambient particles, and 2 for emitted particles. (The factor 2 is associated with a half-Maxwellian of emitted particles at the surface, there being only outgoing trajectories when there is no electric field.)

There are, in general, two ranges of integration over J^2 in Eq. (20). These are associated with two of four possible types of orbits, namely, "Type 1" and "Type 2" in the case of ambient particles, and "Type 1" and "Type 3" in the case of surface-emitted particles. (The delta-factor manifests itself through the limits of integration over J^2.) The types of orbits are defined next.

The following four types of orbits, illustrated in Fig. 2, were first so-named and classified by Parker[21] (Pt. I of Ref. 21).

Type 1. The Type-1 orbit passes from infinity to the sphere, one way, or from the sphere to infinity, one way, with no turning-point. The ingoing orbit is populated by a particle from the ambient plasma, while the outgoing orbit is populated by a particle emitted from the surface. For Type-1 orbits the delta-factor is unity.

Type 2. The Type-2 orbit starts at infinity and passes inward to a minimum radius (radial turning-point) outside the sphere. It subsequently goes out again. These are two-way orbits, occupied by ambient particles going in either direction. For Type-2 orbits, the delta-factor is 2, to account for both ingoing and outgoing contributions. (It is, of course, zero if the turning-point is outside the radius of interest.)

Type 3. The Type-3 orbit starts at the sphere surface and passes outward to a maximum radius (radial turning-point), subsequently returning to the sphere. These orbits, similar and inverse to the Type-2 orbits, are also two-way orbits; they are occupied by emitted particles going in either direction. For Type-3 orbits the delta-factor is 2. (It is, of course, zero if the turning-point is inside the radius of interest.)

Type 4. This closed or "trapped" orbit circulates around the sphere indefinitely. As is usual, we assume here that trapped orbits are unpopulated and take $\delta=0$ for them, although this assumption apparently has never been justified rigorously. These orbits can be populated only by collisions. One can compute the trapped populations of these orbits rigorously by the method of Parker.[21]

Generally, the monoenergetic integrals over J^2, for flux as well as density and for emitted as well as ambient particles, all have the form:

$$M(E) = (\text{const}) \int \delta \cdot dG(J^2/r^2) \tag{21}$$

$G(X)$ is a function having various forms depending on what it represents:

$$G(X) = \sqrt{K-X} \text{ for density}$$

$$G(X) = X \text{ for flux}$$

where the total range of X is $(0,K)$. Here, X represents J^2/r^2 while K represents $E-\phi$.

The analysis of Parker[21] shows that it is possible to have Type-1 and Type-2 orbits contributing simultaneously (ambient particles), or it is possible to have Type-1 and Type-3 orbits contributing simultaneously (emitted particles); but it is not possible to have Type-2 and Type-3 orbits contributing simultaneously. Moreover, if there are any contributions at all, at least some of these come from the vicinity of $J^2=0$ in the form of Type-1 orbits. Hence, if the point r can be reached energetically, there is always a lower range of J^2 for which there are only Type-1 orbits, and there may be an upper range of J^2 for which there are either Type-2 or Type-3 orbits.

From this analysis one obtains the expression representing Eq. (20). Namely, either $M_n(E)=M_{na}(E)$ for ambient particles, or $M_n(E)=M_{ns}(E)$ for surface-emitted particles, where M_{na} and M_{ns} are defined by

$$M_{na}(E) = \underbrace{[-M_g(J_{B1}^2)+M_g(J_{A1}^2)]}_{\text{Type-1 orbits}}+2\underbrace{[-M_g(J_{B2}^2)+M_g(J_{A2}^2)]}_{\text{Type-2 orbits}} \tag{22}$$

$$M_{ns}(E) = 2\underbrace{[-M_g(J_{B1}^2)+M_g(J_{A1}^2)]}_{\text{Type-1 orbits}}+4\underbrace{[-M_g(J_{B3}^2)+M_g(J_{A3}^2)]}_{\text{Type-3 orbits}} \tag{23}$$

where

$$M_g(J^2) = \frac{1}{2r}\,[r^2(E - \phi) - J^2]^{1/2} = \frac{(g - J^2)^{1/2}}{2r} \tag{24}$$

Here the values J_{A1}^2 and J_{B1}^2 denote, respectively, the lower and upper limits of the range (or ranges) of J^2 associated with Type-1 orbits; J_{A2}^2 and J_{B2}^2 denote, respectively, the lower and upper limits of the range (or ranges) of J^2 associated with Type-2 orbits; and J_{A3}^2 and J_{B3}^2 denote, respectively, the lower and upper limits of the range (or ranges) of J^2 associated with Type-3 orbits.

We discuss next the analysis for evaluating J_{A1}^2, J_{B1}^2, J_{A2}^2, J_{B2}^2, J_{A3}^2, and J_{B3}^2.

Analysis of Limits on J^2. Writing the equation for the conservation of energy in the form

$$E = \phi + v_r^2 + \frac{J^2}{r^2} \tag{25}$$

we see that, along an orbit with fixed total energy E and angular-momentum-squared J^2, the radial velocity v_r will remain finite and not vanish or change sign as long as

$$E > \psi \equiv \phi + \frac{J^2}{r^2} \tag{26}$$

or, alternatively, as long as

$$J^2 < g \equiv r^2 (E - \phi) \tag{27}$$

Equation (26) may be analyzed to find maxima in the effective potential ψ. Alternatively, Eq. (27) may be analyzed to find minima in the turning-point function g. The analyses of Eqs. (26) and (27) to classify the orbits represent, respectively, the methods of the Effective-Potential Formulation and the Turning-Point Formulation. The Effective-Potential approach is represented by the work of Bernstein and Rabinowitz[33] and Laframboise,[25] while the Turning-Point approach is represented by the work of Bohm et al,[34] Allen et al,[35] Medicus,[36] and Parker.[20,21,9]

The two approaches are, of course, equivalent[20-21] and will be compared later. They involve different projections of the same three-dimensional phase space of r, E, and J^2. From the definition that v_r=0 defines a turning-point, Eq. (25) with v_r=0 defines a three-dimensional surface in r, E, J^2 space representing the locus of turning-points. A maximum of ψ occurring in an r-E projection (at constant J^2) corresponds to a minimum of g occurring in an r-J^2 projection (at constant E).[20]

Turning-Point Formulation. The Turning-Point Formulation is of principal concern in this paper. It is considered by the author to be much simpler and more efficient in practice than the Effective-Potential Formulation and has been shown to produce results[20] identical to those of Laframboise[25] who used the Effective-Potential Formulation. In using the Turning-Point Formulation to evaluate monoenergetic contributions (that is, contributions to the density or flux due to a given energy), one explores the g-function of Eq. (27) for least values (or minima) occurring in the radial range r=1 (the surface) to r=∞. For a fixed energy E there is only a single curve to analyze, such as that of each example in Fig. 3, which is an r-J^2 projection of the phase space.

Figure 3 shows three examples illustrating the use of the Turning-Point method, which depends on where and how least values ("negative bumps") occur in the g-function of Eq. (27). The least values may or may not be analytic minima. They correspond to "positive bumps" (representing centrifugal barriers) in the effective potential. In these examples E-ϕ is everywhere positive (necessarily), but ϕ is otherwise arbitrary and in particular may be nonmonotonic. All physically possible orbits are horizontal lines constrained to remain below the function $g=r^2(E-\phi)$. Orbits terminating on the g-function have turning-points there.

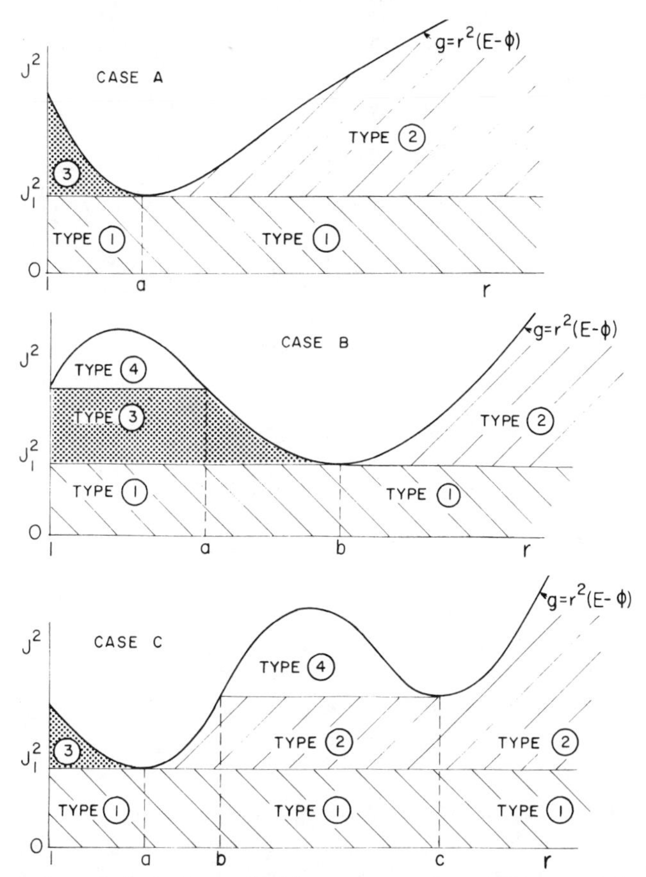

Fig. 3 Turning-Point Formulation; examples illustrating domains of four types of orbits in r,J^2 space.

In all cases (e.g., the three cases of Fig. 3) there is one principal least value (minimum) in the g-function, defined to be J_1^2. (This corresponds to the effective-potential maximum.) In Fig. 3, the least value of g occurs at r=a in cases A and C and at r=b in Case B. This value of r is the well-known "absorption radius" (defined for a given value of E).

Now we consider the limits of J^2 for any value of r. For all values of r the limits of J^2 for Type-1 orbits are $J_{A1}^2=0$ and $J_{B1}^2=J_1^2=g_{min}$. The J^2 limits of Type-2 and Type-3 orbits are as follows. $J_{A2}^2=J_1^2$ and $J_{A3}^2=J_1^2$ for all values of r. However, J_{B2}^2 and J_{B3}^2 depend on the r-position (see Fig. 3):

Case A: Two ranges of r, $J_1^2 = g(a)$.

For r in (1,a) $J_{B2}^2 = J_1^2$ (no Type-2 orbits)

$J_{B3}^2 = g(r)$

For r in (a,∞) $J_{B2}^2 = g(R)$

$J_{B3}^2 = J_1^2$ (no Type-3 orbits)

Case B: Three ranges of r, $J_1^2 = g(b)$.

For r in (1,a) $J_{B2}^2 = J_1^2$ (no Type-2 orbits)

$J_{B3}^2 = g(a)$ (secondary least value of g)

For r in (a,b) $J_{B2}^2 = J_1^2$ (no Type-2 orbits)

$J_{B3}^2 = g(r)$

For r in (b,∞) $J_{B2}^2 = g(r)$

$J_{B3}^2 = J_1^2$ (no Type-3 orbits)

Case C: Four ranges of r, $J_1^2 = g(a)$.

For r in (1,a) $J_{B2}^2 = J_1^2$ (no Type-2 orbits)

$J_{B3}^2 = g(r)$

For r in (a,b) $J_{B2}^2 = g(r)$

$J_{B3}^2 = J_1^2$ (no Type-3 orbits)

For r in (b,c) $J_{B2}^2 = g(c)$ (secondary least value of g)

$J_{B3}^2 = J_1^2$ (no Type-3 orbits)

For r in (c,∞) $J_{B2}^2 = g(r)$

$J_{B3}^2 = J_1^2$ (no Type-3 orbits)

Note that two of the limits are defined by secondary least values of g. (The radii at which these occur are "pseudo-absorption radii.") It is evident that the existence of Type-4 orbits is associated with the existence of such secondary least values. [See range r in (1,a) in Case B, and range r in (b,c) in Case C.] If the Type-4 populations were nonzero, one would integrate over the range J_{A4}^2 to J_{B4}^2 to evaluate their contribution, where J_{A4}^2 and J_{B4}^2 are evident from inspection of the g-function.

The computer logic for implementing the evaluation of the above limits on J^2 is extremely simple. We define g_i at all grid points as $r_i^2(E-\phi_i)$, where i is the grid point index and ϕ_i is the potential at the i-th point. One asks for the following three types of least values of the g_i-array:

$$\left.\begin{aligned} J_1^2 &\equiv \text{least } g_i, \text{ for all } i \\ J_{B2}^2 &\equiv \text{least } g_j, \text{ for all } j \geq i \\ J_{B3}^2 &\equiv \text{least } g_j, \text{ for all } j \leq i \end{aligned}\right\} \quad (28)$$

where i=1 corresponds to the sphere surface radius, and i=2,3, ... correspond to the 2nd, 3rd, ... etc. radial grid points away from the surface.

When g vanishes somewhere, i.e., when there are one or more negative values of g_i, as could typically occur when the potential is repulsive and E is small, we set $J_1^2=0$. In addition, for ambient particles we set $M_{na}(E)=0$ at r if g vanishes at any point between r and infinity; while for emitted particles we set $M_{ns}(E)=0$ at r if g vanishes at any point between r and the surface.

With the simple scheme given by Eq. (28), we may write down general expressions for the monoenergetic number density from Eqs. (22) and (23):

Ambient monoenergetic density

$$M_{na}(E) = \frac{1}{2r}\left[g^{1/2} - 2\,(g - J_{B2}^2)^{1/2} + (g - J_1^2)^{1/2}\right]$$

$$= \frac{1}{2} [(E - \phi)^{1/2} - 2 (E - \phi - J_{B2}^2/r^2)^{1/2} + (E - \phi - J_1^2/r^2)^{1/2}] \tag{29}$$

Emitted monoenergetic density

$$M_{ns}(E) = \frac{2}{2r} [g^{1/2} - 2 (g - J_{B3}^2)^{1/2} + (g - J_1^2)^{1/2}]$$

$$= (E - \phi)^{1/2} - 2 (E - \phi - J_{B3}^2/r^2)^{1/2} + (E - \phi - J_1^2/r^2)^{1/2} \tag{30}$$

Together with the special cases where g vanishes somewhere, the above equations are used in the general computer code for the plasmasheath-photosheath problem. It should be noted that for monoenergetic distributions of unique energy E. the density is given by Eq. (29) or Eq. (30), divided through by $E^{1/2}$. (The quantities ϕ and J^2 are then normalized by E instead of kT.)

The three limits J_1^2, J_{B2}^2, and J_{B3}^2 can be least values of g without being analytic minima in the sense that dg/dr=0. They can be equal, and there can be any number of equal least values. By this method it is possible to treat completely arbitrary variations in ϕ and therefore in g. This can be important, for example, if during the iteration procedure intermediate potential distributions appear that have peculiar shapes. There can be any number of secondary, tertiary, etc., minima. The method, therefore, is completely general and applies to nonmonotonic as well as monotonic potential functions.

This is in contrast to Laframboise's Effective-Potential Formulation method,[25] which handles no more than two maxima in the effective potential function. Moreover, ϕ is assumed to be monotonic, and electron emission is not included by Laframboise.

Dimensionless Currents. Returning to our basic integral for the current density, Eq. (9), we may write this in dimensionless form similar to Eq. (16):

$$j = 2 \int_{v_{min}}^{\infty} \exp(- v^2 - \phi)\, v^3\, dv \int d(\sin^2\theta) \qquad (31)$$

By transforming to E, J^2 coordinates through Eqs. (17) and (18), one obtains:

$$j = \exp(\phi_{source}) \int_{\phi_{source}}^{\infty} \exp(- E)\, dE \cdot J_1^2 \qquad (32)$$

Equation (32) is an expression similar to Eq. (19) for number density, except that the monoenergetic factor is now simply J_1^2 for either ambient or emitted particles. As before, ϕ_{source} is zero for ambient particles and is ϕ_s for emitted electrons.

Examples. As one application of the foregoing analysis, we consider the Langmuir probe analysis of Bernstein and Rabinowitz,[33] which is restricted to monoenergetic ambient attracted particles with no emission. They assume that the probe is large enough that the potential drops off rapidly with r, leading to a g-function (turning-point function) behavior similar to Case A of Fig. 3. For r in the range (1,a) we have $J_{B2}^2=J_1^2$ (i.e., there are only Type-1 orbits contributing in this range); for r in the range (a,∞), we have $J_{B2}^2=g$. Thus, Eq. (29) yields for the monoenergetic density in the two ranges (dividing by $E^{1/2}$ and re-normalizing ϕ and J^2):

$$\left.\begin{aligned} &r \text{ in } (1,a): \quad M_{na} = \tfrac{1}{2}\,[(1-\phi)^{1/2} - (1-\phi-J_1^2/r^2)^{1/2}] \\ &r \text{ in } (a,\infty): \quad M_{na} = \tfrac{1}{2}\,[(1-\phi)^{1/2} + (1-\phi-J_1^2/r^2)^{1/2}] \end{aligned}\right\} \qquad (33)$$

Moreover, the monoenergetic current is given simply by J_1^2.

These equations are identical except for notation to those derived by Bernstein and Rabinowitz on the basis of the Effective-Potential Formulation.

Another example leads to the ideal Langmuir expression for "orbit-limited" current, which occurs when the probe is small and the potential falls off more slowly than $1/r^2$. In this limit, g rises monotonically for attracted particles (ϕ negative) from its least value $J_1^2=g(1)=E-\phi(1)$ at the surface (r=1). The ambient-particle flux, given by Eq. (32) with $\phi_{source}=0$ and $J_1^2=E-\phi(1)$ becomes simply the well-known formula $1-\phi(1)$.

Energy Quadratures

In order to evaluate numerically the integrals over energy, one may employ quadrature formulas in the form:

$$\int_{E_{min}}^{\infty} \exp(-E)dE \cdot M(E) = \sum_{k=1}^{K} C_k \cdot M(E_k) \tag{34}$$

where M(E) denotes the monoenergetic factor dependent upon energy E, and the lower limit E_{min} may be negative. In the sum there are K coefficients C_k and abscissas (energies) E_k, with k=1,2,...,K. The coefficients and abscissas depend on the order K and nature of the quadrature scheme chosen. The coefficients C_k and abscissas E_k are evaluated as follows.

Given a potential distribution defined at a set of radial grid points r_i, there is an associated set of discrete potentials $\phi_i=\phi(r_i)$. Arrange the set of ϕ_i-values in order of increasing value. Then, if E_{min} of Eq. (34) is identified with the least value of the set of ϕ_i-values, and if E_{max} is identified with the greatest value of this set, we may split the integral and sum of Eq. (34) into two parts, one corresponding to the finite range (E_{min}, E_{max}), and the other to the semi-infinite range (E_{max}, ∞):

$$\int_{E_{min}}^{\infty} \exp(-E)dE \cdot M(E) = \int_{E_{min}}^{E_{max}} + \int_{E_{max}}^{\infty} \exp(-E)dE \cdot M(E) \tag{35}$$

First we consider the semi-infinite range (second integral), and then the finite range (first integral).

Semi-Infinite Range. For the semi-infinite range we may employ quadrature formulas developed by Steen et al.[37] or by Laframboise and Stauffer.[38] These are designed for the Maxwellian case where the integrand contains a Gaussian function as a weighting function. Let a_k, H_k denote an abscissa-coefficient pair from the data of Steen et al. Then one transforms the semi-infinite range integral to

$$\int_{E_{max}}^{\infty} \exp(-E)dE \cdot M(E) = \exp(-E_{max}) \int_{0}^{\infty} \exp(-U)dU \cdot M(U+E_{max})$$
$$= \sum_{k=1}^{K} 2H_k a_k \cdot M(a_k^2 + E_{max}) \tag{36}$$

Here we have formed the k-th coefficient as $C_k=2H_k a_k$ and the associated k-th energy abscissa as $E_k=a_k^2+E_{max}$. The coefficients H_k and abscissas a_k also may be obtained from the "One-Dimensional" table of Laframboise and Stauffer if one multiplies their coefficients by $\sqrt{\pi}$ and their abscissas by unity. (There is also the option of using the Laframboise-Stauffer "Two-Dimensional" or "Three-Dimensional" abscissas and coefficients as they suggest for a cylindrical or spherical problem, respectively. However, it is the author's opinion based on experience with both methods that this yields no significant gain in practice.)

Finite Range. The finite range of energies (E_{min}, E_{max}) defined by the set of ordered values of ϕ_i consists of a number of unequal energy intervals. The number of such intervals is equal to the number of grid points minus one. Consider one of these intervals, and assume its energy range is (A,B). Then its contribution to the energy integral may be written for second order as:

$$\int_{A}^{B} \exp(-E)dE \cdot M(E) = C_1 M(E_1) + C_2 M(E_2) \tag{37}$$

where

$$\begin{cases} u_1 = b - \sqrt{b^2 - a} \\ u_2 = b + \sqrt{b^2 - a} \end{cases} \tag{38a}$$

$$\begin{cases} b = 0.5(I_0 I_3 - I_1 I_2)/(I_0 I_2 - I_1^2) \\ a = (I_2^2 - I_1 I_3)/(I_1^2 - I_0 I_2) \end{cases} \tag{38b}$$

$$\begin{cases} C_1 = 2u_1\exp(-A)(I_1-I_0u_2)/(u_1-u_2) \\ C_2 = 2u_2\exp(-A)(I_1-I_0u_1)/(u_2-u_1) \end{cases} \tag{38c}$$

and where

$$I_0 = \frac{\sqrt{\pi}}{2}\,\mathrm{erf}(\sqrt{B-A})$$

$$I_1 = 0.5(1-\exp(A-B))$$

$$I_2 = 0.5(I_0-\sqrt{B-A}\,\exp(A-B))$$

$$I_3 = I_1-0.5(B-A)\,\exp(A-B) \tag{38d}$$

It may be noted that replacing B by infinity and A by zero in the above equations yields the abscissas and coefficients of Steen et al. for order 2.

For small (B-A), one may expand the foregoing equations to obtain to lowest order:

$$C_1 = C_2 = C \cong 0.5\sqrt{B-A}$$

$$u_1 \cong (1-1/\sqrt{3})C$$

$$u_2 \cong (1+1/\sqrt{3})C \tag{38e}$$

After computing the u's and C's as above, one forms the energies by:

$$E_1 = u_1^2 + A$$

$$E_2 = u_2^2 + A \tag{39}$$

This completes the procedure for solving the Vlasov Problem, that is, evaluation of densities and currents when the potential function is given on a set of grid points. In the next section we treat the solution of Poisson's equation and nonlinear numerical iteration to achieve a consistent solution of the Poisson-Vlasov system.

IV. Poisson's Equation (Calculation of Potentials) and Iteration Procedure for Poisson-Vlasov Solution

Poisson's equation is given by Eq. (7). It is convenient to write this equation in dimensionless form. In nondimensionalizing, one may choose the ambient parameters N_o and T_e as basic. We have also defined N_{so} as the surface-electron number density evaluated at the surface. Then Poisson's equation may be expressed in terms of the following dimensionless variables:

r = radial coordinate divided by r_o

ϕ = electric potential divided by kT_e/e

n_i = ambient-ion density divided by N_o

n_e = ambient-electron density divided by N_o

n_s = surface-electron density divided by N_{so}

λ_D = ambient-electron Debye length $(kT_e/4\pi N_o e^2)^{1/2}$ divided by r_o (that is, the "Debye number")

The resulting equation may be written:

$$\frac{d^2(r\phi)}{dr^2} = \frac{r}{\lambda_D^2}\left[n_s \frac{N_{so}}{N_o} + n_e - n_i\right] \qquad (40)$$

The form of the left-hand side implies that this equation is to be solved for $r\phi$, from which ϕ is obtained by division by r. Alternatively, one may nondimensionalize by choosing as basic the emission parameters N_{so} and T_s, where T_s is the effective temperature of an emitted Maxwellian distribution. Then the equation may be expressed in terms of the same dimensionless variables except the following:

ϕ = electrical potential divided by kT_s/e

λ_{DS} = surface-electron Debye length $(kT_s/4\pi N_{so} e^2)^{1/2}$ divided by r_o (that is, the "Debye number" based on the surface emission)

Then the equation may be written:

$$\frac{d^2(r\phi)}{dr^2} = \frac{r}{\lambda_{DS}^2}\left[n_s + \frac{N_o}{N_{so}} n_e - \frac{N_o}{N_{so}} n_i\right] \qquad (41)$$

It is important to have this possible alternate formulation in cases where photoemission is possibly dominant.

Equation (40) or (41) may be solved by differencing on a set of grid points in r and solving the difference equations. The reason for choosing the forms of Eqs. (40)-(41) is that a simple second-difference operation may be employed, namely,

$$(r\phi)_{i-1} - 2(r\phi)_i + (r\phi)_{i+1} = (\Delta r)^2(\mathrm{RHS})_i \tag{42}$$

centered at the i-th grid point r_i, where $(\mathrm{RHS})_i$ denotes the value of the right-hand side of (40) or (41) evaluated at r_i, and Δr denotes the interval of a uniform grid. Nonuniform intervals are treated in Ref. 20 (App. A).

The "double-sweep" method of solution for such tridiagonal systems of linear equations is well known. Floating boundary conditions to represent the boundary condition at infinity, appropriate for electrostatic probe and satellite problems, and the solution for nonuniform grids, are treated by Parker et al.[20,24]

Given the method for evaluating the ion and electron densities (Sec. III), the numerical solution of the sheath problem requires the solution of two fundamental subproblems, the "Poisson Problem" and the "Vlasov Problem". The Poisson Problem is solved to yield the potentials at the grid points when the charged-particle densities are given; conversely, the Vlasov Problem is solved to give the charged-particle densities at the grid points when the potentials are given. For mutual consistency an iterative process is required.

A suitable procedure for the "Poisson-Vlasov" iteration is given and analyzed by Parker and Sullivan[24] and Parker.[19,20] In the procedure, one mixes successive iterates of the charge density (or alternatively the potential iterates). This means that one mixes a fraction (α) of the most recently computed charge-density distribution [RHS in Eq. (42)] with the complementary fraction ($1-\alpha$) of the previously used charge-density distribution. For small Debye numbers, the fraction α must be small. There is an upper limit for α, and correspondingly a minimum number of iterations, such that convergence is assured. These depend on the type of boundary condition used to represent infinity, and on the number of Debye lengths between the surface and the outer grid boundary (as well as on the properties of RHS). It can be shown that a) for small Debye numbers the number of iterations required is proportional to the square of the inverse Debye number, and b) no mixing is required for convergence if the boundary of the grid is less than about π Debye lengths from the surface.[19,24]

The boundary condition where the potential floats in accord with an assumed inverse-square-law behavior is physically appropriate[25] and is found to be computationally efficient for the purely-ambient-plasma problem.[20,24] This condition also is used in the present calculations.

V. Sample Results: Plasmasheath of Large Body at Extremely High Voltage

A computer program (PARKSS = Parker Spherical Sheath) has been developed to implement the analysis of this paper (Secs. III and IV). This and the following sections present a number of sample solutions. Typical numerical parameters are discussed in Refs. 9, 20, and 24.

In this section we consider a body in the ionosphere F-region, where the body is 100 times the Debye length in radius and has a surface voltage of 40 kV. This models a rocket experiment in which a 40-keV electron beam is emitted,[14] where we assume that the vehicle charges up to 40 kV. It is desired to predict the sheath structure and the return current collected from the ionosphere. The geomagnetic field is neglected in order to allow a spherically symmetric model to be used to estimate the sheath structure. Secondary and photoemission also are neglected in this calculation, but they are not expected to be important.

Since kT is about 0.1 ev in the ionosphere, the dimensionless potential is 400,000, an extremely high voltage and the largest value treated in any self-consistent sheath calculation to date. Also, the sphere modeling the rocket is assumed to have an effective radius of 1 m. Hence, the radius is about 100 times the Debye length, assumed to be of the order of 1 cm. The body is therefore a large body at extremely high voltage.

Figure 4 shows the dimensionless potential distribution in the sheath. The numerical parameters used to obtain this solution are as follows. The outer grid radius was taken at 21 r_0; there are 100 grid zones between the surface and the outer radius; the fraction α used for the Poisson-Vlasov iteration was 10^{-5}. A large outer-boundary radius and a large number of iterations were required, as dictated by the extremely high potential.

According to Fig. 4, the potential falls off to 1/10 of its surface value at r=6 (i.e., at 6 body radii), and to 1/400,000th (i.e., to the order of kT/e) at r about 20 (not shown). The collected current density is computed to be 490

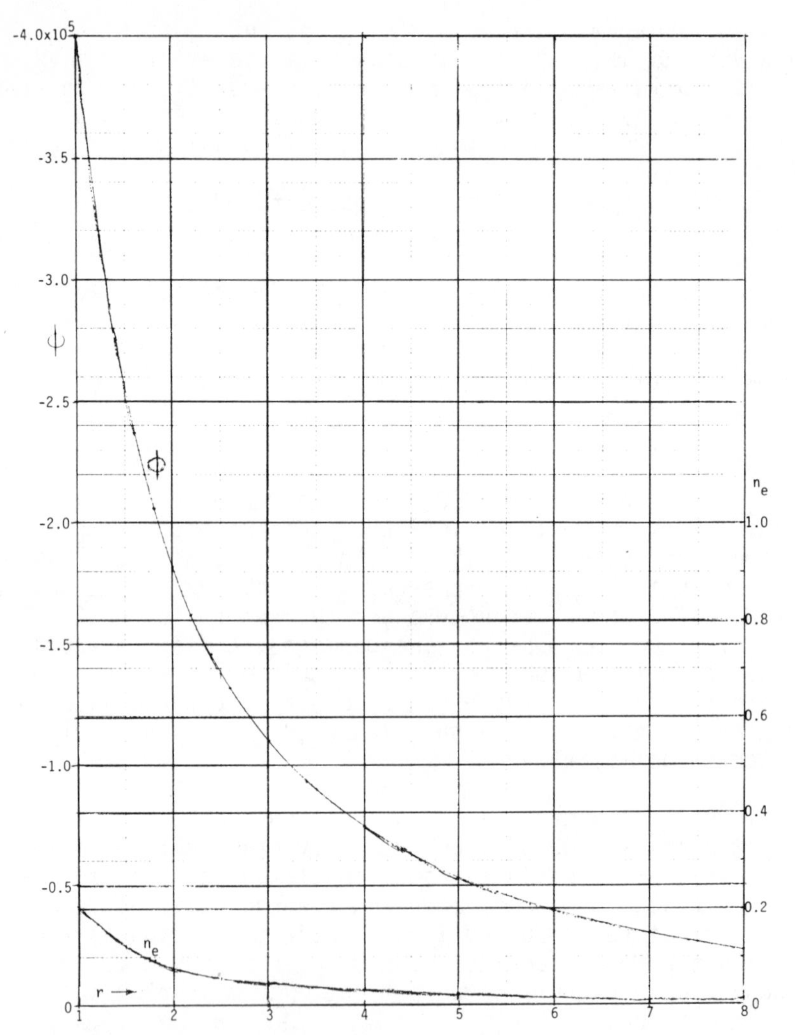

Fig. 4 Sheath of large body at extremely high voltage (eV/kT= 400,000, λ_D/r_o=0.01, j/j_o=490).

times the ambient random-thermal value. This implies a sheath radius of about 22 times the sphere radius. It is interesting to compare this value with that predicted by the diode sheath model of Sec. II, which will serve as an example of the use of the theory of Sec. II. According to Eq. (2), the planar sheath thickness is S=200 m, where V=40,000 V, T=0.1 eV, RAM=1, and n_o=55,000 cm^{-3} were used in Eq. (2) to obtain S. The value 55,000 cm^{-3} was chosen to yield a Debye length of 1 cm and is a typical value found in the ionosphere. Obviously, the spherical correction is required. Hence, from Eq. (5), Fig. 1, or Table 1, we obtain (since r_o=1 m and S/r_o=200), r_s/r_o=20.4 as the ratio of sheath to body radius. This value is quite close to the rigorous self-consistent solution value of 22. Hence, we see that under some circumstances the diode sheath theory with spherical correction may possibly be a good approximation. However, this particular example may be a fortuitous agreement. See Sec. IX for a more detailed assessment.

Also shown in Fig. 4 is the dimensionless electron density n_e (attracted particles). The ion density (repelled particles) is negligible within the range of radii in the figure. The electron density falls off, levels off at a low value, then ultimately rises to near unity (as does also the ion density) in the vicinity of r=20.

VI. Sample Results: Moderately Large Body at Moderately High Voltage with Photoemission

A spacecraft several meters in radius (perhaps a 10-m-diameter array or antenna) in the plasmasphere can be modeled by a sphere with radius of the order of 10 times the Debye length (50 cm if $N_o \sim 200$ cm^{-3}, and T ~ 1 eV). Assuming -10 V on the body, we have a dimensionless potential of ϕ_o=-10 for (attracted) ions, and a Debye number λ_D (Debye length divided by body radius) of 0.1.

Figure 5 shows ϕ, n_i, n_e, and n_s as functions of radius, both in darkness (curves labelled "A" for ambient-only with no photoemission) and in sunlight (unlabelled curves). The emission density and temperature are assumed to be the same as the ambient values (equal emission and ambient electron fluxes). This situation also can occur in GEO but may involve an on-board ion accelerator to attain ϕ_o=-10. (The curves labelled "A" have been presented in Ref. 20.)

Of interest here are the changes, due to emission, in the spacecraft sheath as it emerges from darkness into sunlight. The changes are as follows:

a) The ϕ-distribution is shifted upward slightly (more negative).

b) The ion density is raised about 10% near the surface, but the effect on the ions is insignificant beyond about 1.4 sphere radii.

c) The plasma-electron density is reduced at all radii.

d) The quasineutrality point beyond which the positive and negative charges balance ($n_i = n_e + n_s$) to within 1% is still at about the same radius (about 2 radii) as in the ambient-alone case.

e) The emitted-electron density n_s drops off very rapidly at first, then asymptotically with the inverse square of the radius. This behavior

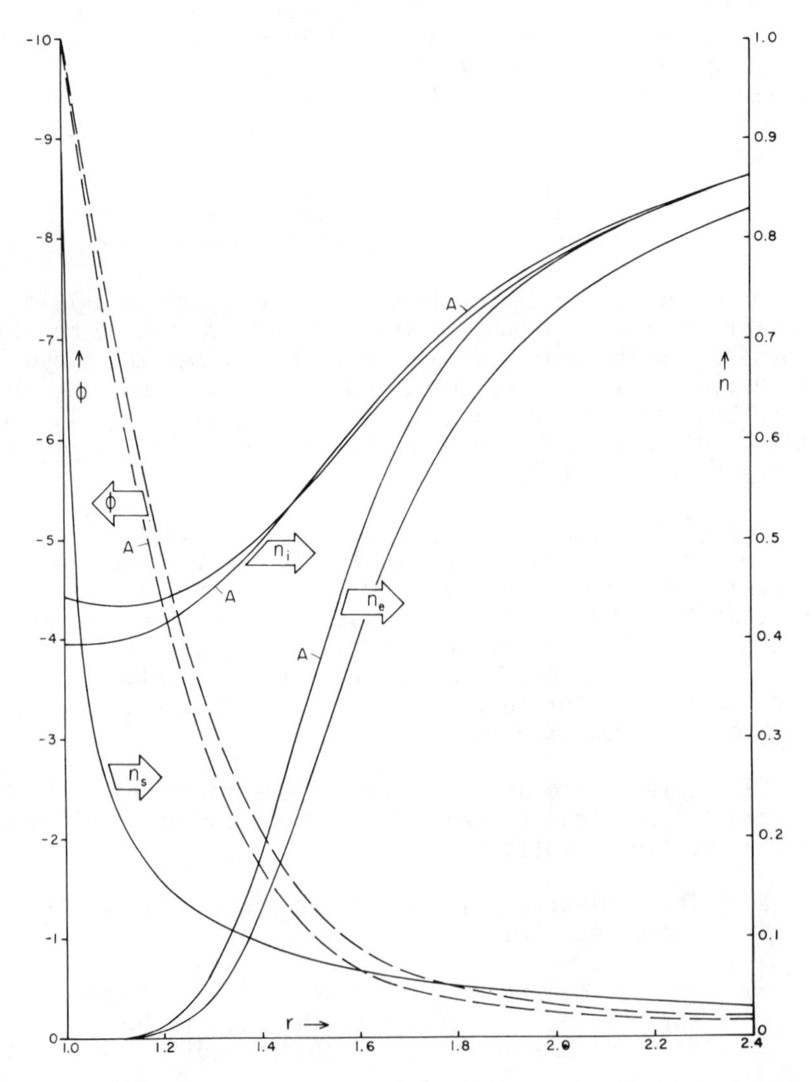

Fig. 5 Plasmasheath-photosheath of moderately large body at moderately high voltage.

is associated with electrons rolling downhill and is characteristic of "cold emission" treated in Ref. 9, where $|e\Phi_0| >> kT_S$.

VII. Sample Results: Photosheath of Intermediate-Size Body

An intermediate-size body in the plasmasphere (N_0 ~ 400 cm^{-3}, T ~ 1 eV) is of radius about 1 m, vs a Debye length of 37 cm (so that the Debye number λ_D is 0.37). Assume that we have an electron and an ion gun to control the spacecraft potential. We consider five values of surface potential, +0.8 V, +0.4 V, 0 V, -0.4 V, and -0.8 V, in order to show the effects on the photosheath-plasmasheath structure.

Figure 6 shows three sets of dimensionless potential ϕ profiles: ambient plasma alone with no emission (curves labelled "A" on the left side of the figure), emitted electrons alone with no ambient plasma (curves labelled "E" on the

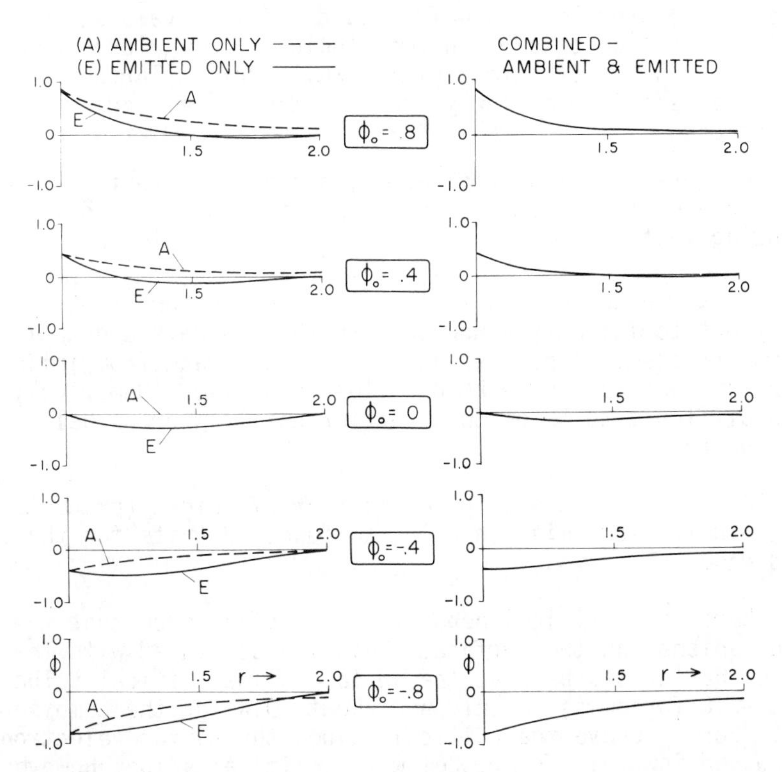

Fig. 6 Plasmasheath-photosheath structure vs body voltage (active control).

left side of the figure), and both ambient and emitted (unlabelled curves on the right side of the figure). The profiles in Fig. 6 are associated with the five values of the surface potential ϕ_0, namely, 0.8, 0.4, 0., -0.4, and -0.8, corresponding to different vertical positions in the figure as indicated by the encircled numbers. Both the ambient and emitted thermal fluxes (Maxwellian) are assumed equal, corresponding to equal ambient and photoelectron fluxes, a condition which can occur in the plasmasphere. In particular, the temperatures of the plasma ions and electrons T_i, T_e and of the emitted electrons T_s, are assumed to be all equal to one electron-volt; the density of plasma ions and electrons N_0 is equal to the emission density of photoelectrons N_{so}, and both are taken equal to 400/cm^3. Thus, the ambient and emitted thermal currents are both about 2×10^{-9} amp/cm^2, approximately the photoemission current.

Referring to the pure-ambient profiles (A), we see that ϕ is always monotone, and the profile for surface potential ϕ_0 is the negative of the profile for $-\phi_0$. Adding emission pushes the ϕ-profiles downward toward negative values, and the profiles are then possibly nonmonotonic. For ϕ_0 large and positive ϕ falls off more rapidly, while for ϕ_0 large and negative ϕ falls off more slowly, than in the corresponding pure-ambient case.

The pure-emission profiles (E) are similar to the combined-case profiles (ambient plus emitted), and both have the following features:

For ϕ_0 large and positive, ϕ is monotone downward, essentially out to infinity. However, there is always a negative minimum ("potential barrier"), that moves inward from infinity and becomes deeper as ϕ_0 becomes less positive. Numerically, it is difficult to "see" this barrier until ϕ_0 drops below about unity.

For ϕ_0=0, we have a "pure" barrier of height (or depth) about 3 units for emission only, and under 2 units for the combined case.

There is a critical negative value of ϕ_0 such that the field vanishes at the surface. This is associated with the barrier having its peak at the surface. The critical value is about -0.8 for emission only and about -0.4 for the combined case. For ϕ_0 above the critical value, the surface electrons are pushed inward. For ϕ_0 below the critical value, however, the surface electrons are pushed outward; in this case, the potential is monotone negative and electrons emitted from the surface "roll downhill." In a sense, one may think of the barrier as a virtual one whose peak has moved to within the surface. A barrier near the surface represents "space-charge-limited emission."

Comparing the ambient-alone (A) and emitted-alone (E) profiles with the combined-case profiles, the latter appears roughly to be approximable by simply averaging the A and E profiles.

Schröder[39] has presented a set of computed potential profiles similar to those shown for positive ϕ_0 in Fig. 6. Since Ref. 39 refers to interplanetary space where the photoelectric fluxes are dominant, the profiles of Ref. 39 should correspond most closely to the pure-emission (E) profiles of Fig. 6.

VIII. Sample Results: Presheath of Extremely Large Body

An extremely large body is one whose dimensions are so large that the sheath thickness can be neglected for a given body voltage. Such a body would be a Solar Power Satellite (SPS) of the order of dimension 10 km in quiescent GEO (no substorm), or a Space-Based Radar (SBR) of dimension 100 m at 500 km altitude or lower. Although the sheath thickness may be negligible relative to body size, there is a region of disturbed plasma, comparable in size to the body dimension, called the "presheath." In this region, in which the potential drop is of order kT/e, the ion and electron densities are essentially equal to each other but less than the ambient undisturbed value. Hence, the presheath is important because its effects can extend over large distances, for example, 10 km in the case of a 10-km-sized body.

The material of this section is based on work done by the author in collaboration with three other authors on a paper in preparation entitled "The Theory of Cylindrical and Spherical Langmuir Probes in The Limit of Vanishing Debye Number." The other authors are J. G. Laframboise (York University, Toronto), and M. J. M. Parrot and L. R. O. Storey (CRPE/CNRS, Orléans, France). This work is applicable to extremely large bodies and deals specifically with the presheath. The joint paper deals with cylindrical as well as spherical bodies; the cylindrical geometry is not considered here.

A calculation of the presheath structure involves a different kind of "Poisson-Vlasov" iteration from that described in Sec. IV of this paper. Instead of solving the Poisson equation as in Sec. IV, one solves the equation of quasineutrality, namely,

$$n_i(\phi) = n_e(\phi) \tag{43}$$

to obtain the potential distribution ϕ, where the symbol $n(\phi)$ means that n_i or n_e is obtained as a function of position when the ϕ-distribution is prescribed (i.e., the "Vlasov" problem). This is a nontrivial task if T>0.

The iteration method is analogous to the ordinary Poisson-Vlasov process, with a relaxation parameter α smaller than unity to avoid instability. The relaxation scheme used is

$$\phi^{n+1} = \phi^n + \alpha \cdot \ln\left(\frac{n_a}{n_r}\right) \tag{44}$$

where ϕ^n denotes the n-th iterate, and n_a and n_r denote, respectively, the attracted and repelled particle densities.

Within the assumptions of this model, the sheath (as opposed to the presheath) is so thin on the scale of the presheath that it is modeled by a discontinuity in potential at the surface. Hence, the potential at the sheath edge, ϕ_s, is part of the solution. It should be noted that ϕ_s is a function

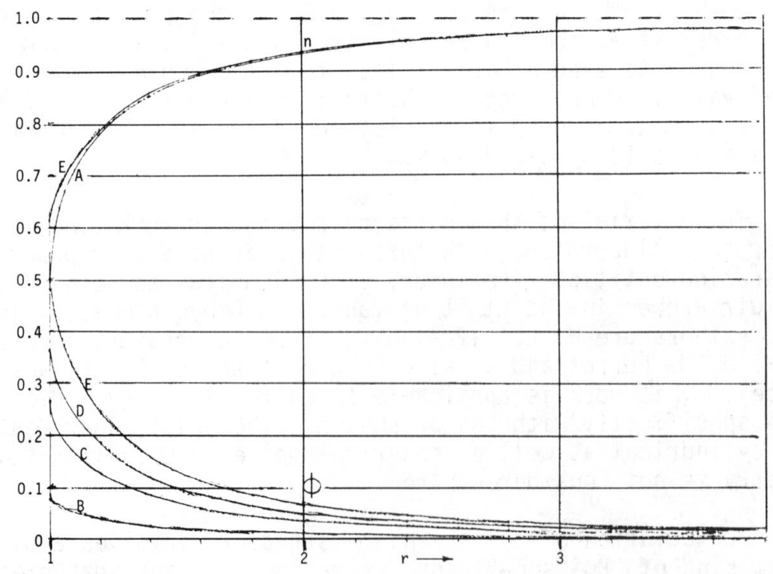

Fig. 7 Presheath structure for extremely large body; potential and density profiles vs body voltage (active control), A(ϕ_o=0), B(ϕ_o=0.1), C(ϕ_o=0.5), D(ϕ_o=1.0), E(ϕ_o=10.0).

of surface potential ϕ_0 and approaches an asymptotic finite value (0.46) as ϕ_0 becomes infinite.

Figure 7 shows how the potential in the presheath varies as a function of radius for different values of body potential ϕ_0. Since $T_i=T_e$ by assumption, the distribution of potential is identical except for sign for a positive or negative body. Figure 7 also shows plasma density (n_i or n_e) as a function of radius at two extreme values of body potential, $\phi_0=0$ and $\phi_0=10$, where $\phi_0=10$ is already in the asymptotic domain. This figure indicates that the presheath density does not vary much as ϕ_0 increases, whereas of course the potential profile ϕ does. The potential and density profiles appear to have infinite slope at the surface.

The current of attracted particles collected by the large body increases continuously from its random thermal value j_0

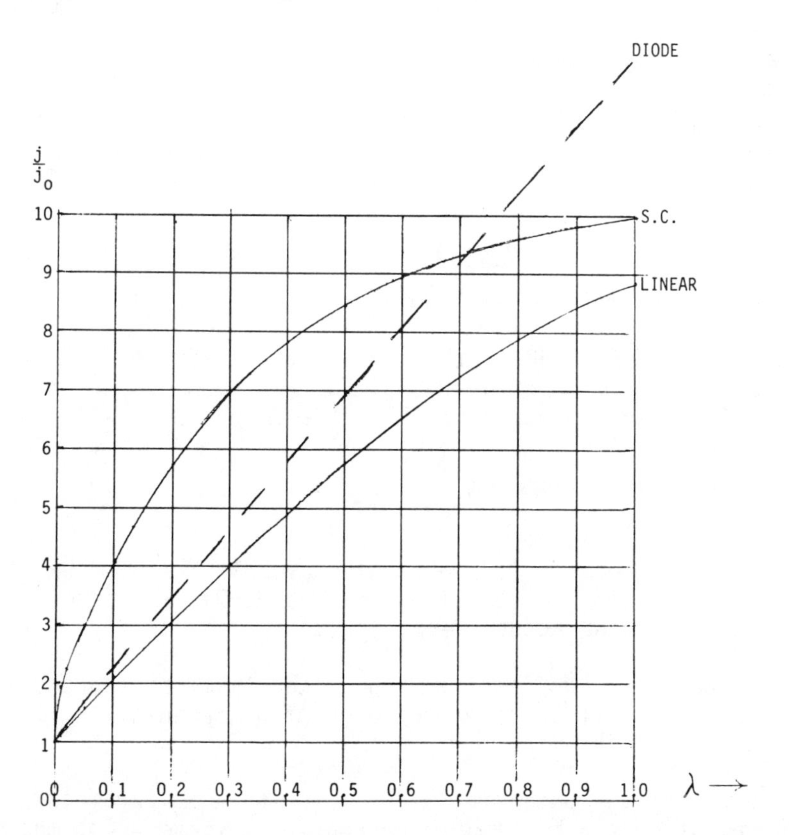

Fig. 8 Approximate models (linear and diode) vs rigorous (self-consistent) solutions, current vs Debye number, at $\phi_0=-10$.

at ϕ_o=0, to about 1.45 j_o as ϕ_o approaches infinity. This is at variance with what one may predict on the basis of one-dimensional theory, namely, a constant value of attracted current $j=j_o$ at all body potentials. It is of interest to note that the cylindrical body also collects asymptotically about 1.45 times its random-thermal current, suggesting that the current is independent of body shape (while the density and potential profiles are not).

IX. Linearized Space Charge Model, and Comparisons Among Models

It is of interest, because of the complexity of rigorous sheath calculations, to explore various approximate methods for this type of calculation, especially for large bodies. In Sec. II we presented the sharp-sheath-edge diode model. In this section we present the linearized model leading to the well-known Debye potential (or "screened Coulomb" potential). The advantage of this model is that the charge density is a known (assumed) function of potential.

A simple example of an analytically defined space charge is that in which the ions and electrons are in thermal equilibrium. Thus, Poisson's equation becomes

$$\frac{d^2(r\phi)}{dr^2} = \frac{r}{\lambda_D^2}\left(e^{\phi} - e^{-\phi}\right) \tag{45}$$

in accord with Eq. (40) with n_s=0. Now consider the limit as ϕ becomes small. Then the RHS of Eq. (45) becomes $K^2 r\phi/\lambda_D^2$, where K^2=2. The equation then has the simple Debye-potential solution

$$\phi = \frac{\phi_o}{r}\exp[(1 - r)/K\lambda] \tag{46}$$

such that $\phi=\phi_o$ when r=1. We use the symbol λ rather than λ_D henceforth in order to distinguish it as a Debye-length-like parameter from the actual Debye length.

It can be shown that other physical examples lead to the same form with different values of K of order unity.[19] In the following, we take K=1.

Now that the potential is known, with λ as parameter, it is straightforward to compute current vs voltage using the Turning-Point method of Sec. III. That is, j/j_o is given for attracted monoenergetic particles by the least value of g, namely, J_1^2/E_0, where:

$$J_1^2 = \text{least value of } \{r^2 E_0 + r|\phi_0|\exp[(1-r)/\lambda]\} \qquad (47)$$

where E_0 is the energy of the particles. For a Maxwellian, one integrates J_1^2 over energies, by quadratures as in Sec. III. (However, in a high-voltage sheath, the assumption of mono-energeticity is a good approximation.)

Figure 8 shows a comparison between results using the linearized-space-charge model, the diode model of Sec. II, and results based on the rigorous self-consistent Poisson-Vlasov solutions described in Secs. III and IV, and applied to Maxwellian distributions. We show in Fig. 8 the dimensionless current as a function of λ, for a fixed body dimensionless potential of ϕ_0=-10 (attracted particles). The solid curves labelled "LINEAR" and "S.C." denote the Debye potential (linearized-space-charge) model, and the self-consistent rigorous model with $\lambda=\lambda_D$, respectively. The dashed curve, labelled "DIODE," denotes the diode model. The Debye-potential currents lie consistently below the corresponding

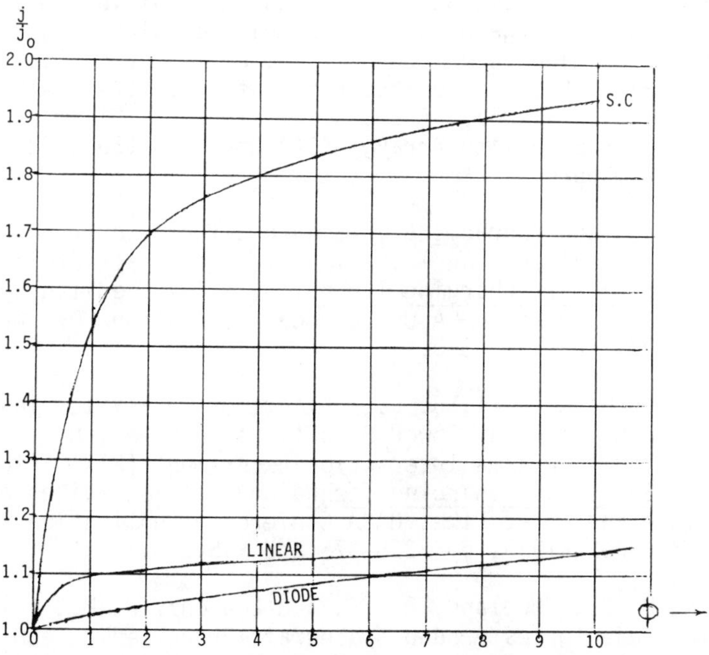

Fig. 9 Approximate models (linear and diode) vs rigorous (self-consistent) solutions, current vs potential, at Debye number = 1/100.

self-consistent currents (which also agree with Laframboise's results[25]). Moreover, the discrepancy becomes worse as λ_D becomes small. Hence, it is not at all clear that the Debye potential is a good approximation, even for an extremely large body. The diode model gives a linear characteristic below λ=1, and is apparently not much more accurate than the Debye-potential model, apparently underestimating the current at small λ, and overestimating at large λ.

Finally, in Fig. 9 we compare the current-voltage characteristics of a large body (with radius r_0 equal to 100 Debye lengths), based on the linearized model, the diode model, and the rigorous model. The two approximate models give currents well below the self-consistent results and are of the same order of magnitude, within the range of ϕ shown. At extremely large values of ϕ, however, the diode model current may become comparable with the rigorous current, as was found to be the case in the example of body voltage 400,000 kT/e.

References

[1]Parker, L. W., "Plasma Sheath Effects and Equilibrium Voltage Distributions of Large High-Power Satellite Solar Arrays," Spacecraft Charging Technology - 1978, edited by R. C. Finke and C. P. Pike, NASA Conference Publication 2071/AFGL-TR-79-0082, 1979, pp. 341-357. Also, see "Effects of Plasma Sheath on Solar Power Satellite Array," AIAA Paper 79-1507, July 1979, Williamsburg.

[2]McCoy, J. E. and Konradi, A., "Sheath Effects Observed in a 10-Meter High-Voltage Panel in Simulated Low-Earth-Orbit Plasma," Spacecraft Charging Technology - 1978, edited by R. C. Finke and C. P. Pike, NASA Conference Publication 2071/AFGL-TR-79-0082, 1979, pp. 315-340.

[3]Domitz, S. and Kolecki, J. C., "Effect of Parasitic Plasma Currents on Solar-Array Power Output," also Grier, N. T. and Stevens, N. J., "Plasma Interaction Experiment (PIX) Flight Results," Spacecraft Charging Technology - 1978, edited by R. C. Finke and C. P. Pike, NASA Conference Publication 2071/ AFGL-TR-79-0082, 1979, pp. 358-375 and 295-314.

[4]Liemohn, H. B., Copeland, R. L., and Leavens, W. M., "Plasma Particle Trajectories Around Spacecraft Propelled by Ion Thrusters," Spacecraft Charging Technology - 1978, edited by R. C. Finke and C. P. Pike, NASA Conference Publication 2071/ AFGL-TR-79-0082, pp. 419-436.

[5]Rosen, A., ed., Progress in Astronautics and Aeronautics: Spacecraft Charging by Magnetospheric Plasmas, Vol. 47, AIAA, New York, 1976.

[6]Pike, C. P. and Lovell, R. R., eds., Proceedings of the Spacecraft Charging Technology Conference, AFGL-TR-77-0051/NASA TMX-73537, Feb. 1977.

[7]Parker, L. W., "Differential Charging and Sheath Asymmetry of Nonconducting Spacecraft Due to Plasma Flows," Journal of Geophysical Research, Vol. 83, Oct. 1978, pp. 4873-4876.

[8]Parker, L. W., "Potential Barriers and Asymmetric Sheaths Due to Differential Charging of Nonconducting Spacecraft," Lee W. Parker, Inc., Concord, Mass., AFGL-TR-78-0045, Jan. 1978.

[9]Parker, L. W., "Theory of Electron Emission Effects in Symmetric Probe and Spacecraft Sheaths," Lee W. Parker, Inc., Concord, Mass., AFGL-TR-76-0294, Sept. 1976.

[10]Whipple, E. C., Jr., "Observation of Photoelectrons and Secondary Electrons Reflected from a Potential Barrier in the Vicinity of ATS-6," also Whipple, E. C., Jr., "Theory of Spherically Symmetric Photoelectron Sheath," Journal of Geophysical Research, Vol. 81, Feb. 1976, pp. 715-719 and 601-607.

[11]Purvis, C. K., Bartlett, R. O., and DeForest, S. E., "Active Control of Spacecraft Charging on ATS-5 and ATS-6," Proceedings of the Spacecraft Charging Technology Conference, edited by C. P. Pike and R. R. Lovell, AFGL-TR-77-0051/NASA TMX-73537, Feb. 1977, pp. 107-120.

[12]Cohen, H. A., Sherman, C., Mullen, E. G., Huber, W. B., Masek, T. D., Sluder, R. B., Mizera, P. F., Schnauss, E. R., Adamo, R. C., Nanevicz, J. E., and Delorey, D. E., "Design, Development and Flight of a Spacecraft Charging Sounding Rocket Payload," Spacecraft Charging Technology - 1978, edited by R. C. Finke and C. P. Pike, NASA Conference Publication 2071/AFGL-TR-79-0082, 1979, pp. 80-90.

[13]Hess, W. N., Trichel, M. C., Davis, T. N., Beggs, W. C., Kraft, G. E., Stassinopoulos, E., and Maier, E. J. R., "Artificial Aurora Experiment: Experiment and Principal Results," Journal of Geophysical Research, Vol. 76, Sept. 1971, pp. 6067-6081.

[14]Winckler, J. R., Arnoldy, R. L., and Henderson, R. A., "Echo 2: A Study of Electron Beams Injected into the High-Latitude Ionosphere from a Large Sounding Rocket," Journal of Geophysical Research, Vol. 80, Jun. 1975, pp. 2083-2088, also Israelson, G. A. and Winckler, J. R., "Effect of a Neutral N_2 Cloud on the Electrical Charging of an Electron-Beam-Emitting Rocket in the Ionosphere: Echo IV," Journal of Geophysical Research, Vol. 84, Apr. 1979, pp. 1442-1452.

[15]Managadze, G. G., "Processes of the Charge Neutralization and the Determination of the Rocket Potential Under the Conditions of the Injection of a Powerful Electron Beam during the 'ARAKS' Experiment," also Lyachov, S. B. and Managadze, G. G., "Beam-Plasma Discharge near the Rocket (Zarnitza-II Experiment)," USSR Academy of Sciences Space Research Institute Reports 309 and 310 (Eng. trans.), 1977.

[16]Cartwright, D. G., Monson, S. J. and Kellogg, P. J., "Heating of the Ambient Ionosphere by an Artificially Injected Electron Beam," Journal of Geophysical Research, Vol. 83, Jan. 1978, pp. 16-24.

[17]Bernstein, W., Leinbach, H., Kellogg, P., Monson, S., Hallinan, T., Garriott, O. K., Konradi, A., McCoy, J., Daly, P., Baker, B., and Anderson, H. R., "Electron Beam Injection Experiments: The Beam-Plasma Discharge at Low Pressures and Magnetic Field Strengths," Geophysical Research Letters, Vol. 5, Feb. 1978, pp. 127-130.

[18]Parker, L. W., "Calculation of the Sheath and Wake Structure about a Pillbox-Shaped Spacecraft in a Flowing Plasma," Proceedings of the Spacecraft Charging Technology Conference, edited by C. P. Pike and R. R. Lovell, AFGL-TR-77-0051/NASA TMX-73537, Feb. 1977, pp. 331-366.

[19]Parker, L. W., "Theory of the External Sheath Structure and Ion Collection Characteristics of a Rocket-Borne Mass Spectrometer," Mt. Auburn Research Associates, Inc., Newton, Mass., AFCRL-71-0105, Nov. 1970.

[20]Parker, L. W., "Computer Method for Satellite Plasma Sheath in Steady-State Spherical Symmetry," Lee W. Parker, Inc., Concord, Mass., AFCRL-TR-75-0410, July 1975.

[21]Parker, L. W., "Computer Solutions in Electrostatic Probe Theory," Mt. Auburn Research Associates, Inc., Newton, Mass., AFAL-TR-72-222, Apr. 1973, also EOS Transactions of the American Geophysical Union, Vol. 54, Apr. 1973, p. 281.

[22]Parker, L. W., "Computation of Ion Collection by a Large Rocket-Mounted Mass-Spectrometer Plate at a Large Drawing-in Potential," Mt. Auburn Research Associates, Inc., Newton, Mass., AFCRL-72-0524, Sept. 1972.

[23]Parker, L. W. and Whipple, E. C., Jr., "Theory of a Satellite Electrostatic Probe," Annals of Physics, Vol. 44, Aug. 1967, pp. 126-161.

[24]Parker, L. W. and Sullivan, E. C., "Iterative Methods for Plasma-Sheath Calculations - Application to Spherical Probe," NASA TN D-7409, March 1974, also "Boundary Conditions and Iterative Procedures for Plasma Sheath Problems," Rarefied Gas Dynamics, 6th Symposium, Vol. 2, edited by L. Trilling and H. Y. Wachman, Academic Press, New York, 1969, pp. 1637-1641.

[25]Laframboise, J. G., "Theory of Spherical and Cylindrical Langmuir Probes in a Collisionless Maxwellian Plasma at Rest," University of Toronto Institute of Aerospace Studies, Rept. 100, June 1966.

[26]Parker, L. W. and Whipple, E. C., Jr., "Theory of Spacecraft Sheath Structure, Potential, and Velocity Effects on Ion Measurements by Traps and Mass Spectrometers," Journal of Geophysical Research, Vol. 75, Sept. 1970, pp. 4720-4733.

[27]Parker, L. W. and Laframboise, J. G., "Multi-Electrode Plasma Probe for Orbit-Limited-Current Measurements, II. Numerical Verification," Physics of Fluids, Vol. 21, Apr. 1978, pp. 588-591.

[28]Soop, M., "Numerical Calculations of the Perturbation of an Electric Field around a Spacecraft," Photon and Particle Interactions with Surfaces in Space, edited by R. J. L. Grard, D. Reidel, Hingham, Mass., 1973, pp. 127-136.

[29]Child, C. D., "Discharge from Hot CaO," Physical Review, Vol. 32, May 1911, pp. 492-511, also Langmuir, I., "The Effect of Space Charge and Residual Gases on Thermionic Currents in High Vacuum," Physical Review, Vol. 2, Jun. 1913, pp. 450-486.

[30]Langmuir, I. and Blodgett, K. B., "Currents Limited by Space Charge between Concentric Spheres," Physical Review, Vol. 24, Jul. 1924, pp. 49-59.

[31]Kennerud, K. L., "High Voltage Solar Array Experiments," The Boeing Co., Seattle, Wash., NASA CR-121280, Mar. 1974.

[32]Rothwell, P. L., Rubin, A. G., and Yates, G. K., "A Simulation Model of Time-Dependent Plasma-Spacecraft Interactions," Proceedings of the Spacecraft Charging Technology Conference, edited by C. P. Pike and R. R. Lovell, AFGL-TR-77-0051/NASA TMX-73537, Feb. 1977, pp. 389-411.

[33]Bernstein, I. B. and Rabinowitz, I. N., "Theory of Electrostatic Probes in a Low-Density Plasma," Physics of Fluids, Vol. 2, Mar. 1959, pp. 112-121.

[34]Bohm, D., Burhop, H. S., and Massey, H. S. W., "The Use of Probes for Plasma Exploration in Strong Magnetic Fields," Characteristics of Electrical Discharges in Magnetic Fields, edited by A. Guthrie and R. K. Wakerling, McGraw-Hill, New York, 1949, Chap. 2.

[35]Allen, J. E., Boyd, R. L. F., and Reynolds, P., "The Collection of Positive Ions by a Probe Immersed in a Plasma," Proceedings of the Physical Society (London), Vol. B70, Mar. 1957, pp. 297-304.

[36]Medicus, G., "Theory of Electron Collection of Spherical Probes," Journal of Applied Physics, Vol. 32, Dec. 1961, pp. 2512-2520.

[37]Steen, N. M., Byrne, G. D., and Gelbard, E. M., "Gaussian Quadratures for the Integrals $\int_0^\infty \exp(-x^2)f(x)dx$ and $\int_0^b \exp(-x^2)f(x)dx$," Mathematics of Computation, Vol. 23, Jul. 1969, pp. 661-671.

[38]Laframboise, J. G. and Stauffer, A. D., "Optimum Discrete Approximation of the Maxwell Distribution," AIAA Journal, Vol. 7, Mar. 1969, pp. 520-523.

[39]Schröder, H., "Spherically Symmetric Model of the Photoelectron Sheath for Moderately Large Plasma Debye Lengths," Photon and Particle Interactions with Surfaces in Space, edited by R. J. L. Grard, D. Reidel, Hingham, Mass., 1973, pp. 51-58.

CURRENT LEAKAGE FOR LOW ALTITUDE SATELLITES

J. E. McCoy,* A. Konradi,* and O. K. Garriott†
NASA Johnson Space Center, Houston, Texas

Abstract

Ionospheric plasma densities exceeding $10^{12}/m^3$ exist around satellites in low Earth orbit. Operation of large solar arrays at high voltage may drive substantial leakage currents through this surrounding plasma. Power losses which would exceed solar cell output have been observed for small test objects biased above +2,000 V in the laboratory. Estimates of these effects for very large power systems are developed. Recent large scale (10 m) lab tests are reported. Estimates based on calculations of space charge limited sheath dimensions are identified as a good working model, leading to projected power losses for large arrays increasing much more slowly than for small arrays.

Introduction

Although space is a very good vacuum, it is not an absolute one, and the ionized component of the very thin residual "gases" present is capable of causing significant electrical interactions. This has been noticed particularly by various satellites in geosynchronous Earth orbit (GEO). These satellites were observed to charge up to surface potentials of several kilovolts under solar eclipse/geomagnetic "storm" conditions. The ambient (low energy thermal ions and electrons) plasma is too thin to effectively discharge the charge acquired from high energy "storm" radiation (ions or electrons) absorbed by the satellites.

In low Earth orbit (LEO) the denser plasma of the ionosphere/plasmasphere environment easily overcomes any radiation charge accumulation. This generally eliminates any concern about unwanted charge buildup such as observed in GEO. A significant new problem for high-power space systems results,

---

Presented as Paper 79-0387 at the AIAA 17th Aerospace Sciences Meeting, New Orleans, La., Jan. 15-17, 1979.

*Physicist.
†Physicist/Astronaut.

however, because of the inverse of the GEO charging problem. In LEO, even necessary high operating voltages may be dissipated by the dense ambient plasma.

This potential problem is especially significant to the design of systems attempting to operate large solar arrays at high voltages in LEO. Normal orbits used by spacecraft in this region of operations lie within the ionospheric F2 region, where the ambient plasma densities (primarily monatomic oxygen ions) range from 10^{11} to $10^{12}/m^3$. Operation of a large solar array in this environment, with exposed surfaces at operating voltages exceeding 10,000 V, can reasonably be expected to drive substantial parasitic currents between the array and the surrounding plasma.

In this paper we will examine various approaches to the problem of equilibrium plasma currents induced by a large high-voltage solar panel in LEO. We adapt methods from plasma probe theory, used in the limit of large attracting voltages ($eV >> kT$). Unless otherwise stated, the collecting surface is considered entirely conductive, nonreflective, and nonemitting. The effects of detailed surface structure and areas of insulation are deferred for later treatment as (possibly important) perturbations confined to the inner portions of the primary current-collecting sheath, or as the sum of overlapping sheaths.

Particular attention is paid to the effects of scale size in comparing results of laboratory measurements using small (1-10 cm) test objects, in order to improve projections of expected results for large objects in space. We propose use of a model which assumes the formation of a space-charge limited sheath around any high-voltage object. This sheath provides for an equilibrium adjustment between the (undisturbed to first order) ambient plasma at "ground" potential outside the sheath and the high voltage on the surface of the object. Calculation of the total "leakage" current collected by any object (such as a solar array on a LEO satellite) then becomes a matter of calculating the outer surface area of its surrounding sheath. The total current collected will be the undisturbed plasma currents crossing the effective collecting area of this sheath surface, relatively independent of the size or surface details of the sheath-forming object.

Limitations on Current

Calculations of plasma currents to a high-voltage panel in LEO depend primarily upon identifying the applicable limitations on current flow. If nothing limited the movement of 10^{12} electrons/m^3, acceleration of a column of these electrons into a panel at +10,000 V would result in a current density

$$j = Ne\sqrt{\frac{2eV}{m}} = 10 \text{ amp/m}^2 \tag{1}$$

For a panel at -10,000 V, the heavier O^+ ions would move more slowly but still provide a current density of 56 mA/m^2. The power lost into the panel would be 100 kW/m^2 at +10kV or 560 W/m^2 at -10 kV. Comparison with a typical solar array power output of perhaps 100 W/m^2, indicates the impossibility of operating an array at 10 kV in LEO under these circumstances. Fortunately for orbital applications of large solar power arrays, a few factors have been neglected which will tend to limit the currents induced.

An attempt to limit leakage currents by the insulation of cell interconnects and other exposed conductors does not seem to be feasible. Several studies have found that even a few small pinholes in a thin layer of insulation over such conductors result in collection of as much current as a comparable array left uninsulated. Although application of thicker insulation, with careful quality control, may be acceptable for small arrays, this approach is prohibitively heavy and expensive for very large arrays.

Ionospheric Conductivity

Although for many purposes the space plasma is considered to be a perfect conductor, this assumption is not usually applicable. Finite conductivity of the ionosphere is one effect that can limit the rate that charge can move to support Eq. (1).

Ionospheric conductivity is controlled by collisions and the Earth's magnetic field. The specific conductivity, that which applies parallel to the magnetic field, is given by the equation

$$\sigma_0 = ne^2 \left[\frac{1}{m_e(\nu_e - i\omega)} + \frac{1}{m_i(\nu_i - i\omega)}\right] \tag{2}$$

Typical values in the (LEO) ionosphere range from 10-100 mho/m. This will not have much effect on the currents calculated in Eq. (1) for electric fields aligned with the magnetic field. A potential gradient of as little as 1 V/m along the magnetic field would be sufficient to drive currents in excess of 10 amp/m^2. Some other process must be identified to limit field-aligned plasma currents to a high-voltage panel.

For currents across the field, however, the greatly reduced Pederson conductivity applies. Typical values at LEO are less than 10^{-5} mho/m. A potential gradient as large as 1,000 V/m (10kV over 10m) could drive no more than 10 mA/m^2, resulting in an almost tolerable power loss of 100 W/m^2. At higher altitudes, or as sheath effects reduce collisions, the Pederson conductivity rapidly drops toward zero.

The finite ionospheric bulk conductivity is therefore of little value for reducing the plasma currents parallel to $\underline{B}$. The Pederson conductivity actually increases the current flow permitted across magnetic field lines. This current flow otherwise would be greatly limited by the radius of gyration of the charged particles about the field lines. Typical values for thermal particles are

$$R = \frac{mv_{\perp}}{eB} \approx \begin{cases} 1\text{-}10 \text{ cm for electrons} \\ 10\text{-}20 \text{ m for ions} \end{cases} \qquad (3)$$

This cross-field current could become even larger if some process were to increase the collision frequency near a high voltage panel beyond that in the normal ionosphere. This might occur due to outgassing from the satellite.

The Hall conductivity (current flow orthogonal to both $\underline{E}$ and $\underline{B}$) also exists in a collisional plasma, but it is generally negligible above 200 km and should not produce current flow to a conductor.

Orbit-limited Current Collection

The limitation to motion of electrons (and ions) across magnetic field lines in a collisionless plasma is an example of a sort of "orbit-limited" motion. Some electrons (or ions) are prevented from reaching an attracting surface even though they pass within, and are accelerated by, its electric potential well. This can cause a very significant reduction in current, as the object can only collect charge from along the volume of space contained within a "flux tube" defined by the magnetic field lines passing within 1-2 R_L.

Expressions involving (particle angular momentum about the collecting probe) orbit-limited currents are frequently encountered in treatments of current collection by a spherical probe in a low density plasma. The probe is assumed surrounded by an infinite isotropic plasma of low density, so that charge separation fields are negligible. The electron (or ion) current collected is given by

$$I = 4 \pi p^2 j_0 \qquad (4)$$

where j_o is the electron (or ion) random thermal current in the undisturbed plasma, and p is the impact parameter (the effective "collision" radius for attracted particles to be deflected to hit the probe). Then

$$\bar{j}_o = 1/4 \; N \, e\bar{v} \tag{5}$$

where $\bar{v}$ is the average electron (ion) velocity. The impact parameter can be calculated from conservation of angular momentum. For a particle with initial energy E coming from infinity, on a trajectory that would pass a distance p from the center of the probe if not deflected, the angular momentum is

$$P_\theta = p\sqrt{2 \, m \, E} \tag{6}$$

The actual momentum of a particle deflected just sufficiently to strike the surface of the spherical probe is

$$P_\theta = a\sqrt{2 \, m \, (E + eV_p)} \tag{7}$$

where a is the radius of the probe, and V_p is its surface voltage with respect to the distant plasma. Combining Eqs. (6) and (7), we obtain

$$p = a\sqrt{1 + \frac{eV_p}{E}} \tag{8}$$

Assuming a Maxwell velocity distribution for V, we obtain the usual expression

$$I = 4\pi a^2 \, j_o \quad (1 + \frac{V_p}{E}) = A_p Ne\sqrt{\frac{kT}{2\pi m}} \quad (1 + \frac{eV_p}{kT}) \tag{9}$$

The argument employed to obtain Eq. (9) was quite general. Unless prevented by some additional effect, such as space-charge or magnetic deflection, any spherical object of radius a and surface potential V_p will collect every charged particle of initial energy E that would have crossed the surface of a virtual sphere of radius p in the undisturbed plasma (undeflected by the electric field of the actual object). Any charged particle whose undisturbed trajectory would have passed outside of p cannot be collected. Even though the particle may be strongly accelerated toward the attractive potential, it cannot reach the radius of the probe, a, unless some other force allows it to lose angular momentum. The current collected is limited to that which would have crossed the spherical surface area:

$$4\pi p^2 = A_p \quad (1 + \frac{eV_p}{E}); \quad A_p = 4\pi a^2 \tag{10}$$

A similar argument for a cylindrical collector of radius a and length L leads to the equation for an attracting orbit-limited cylindrical probe

$$\begin{aligned} I &= (2\pi a L)\frac{2}{\sqrt{\pi}}\left(1 + \frac{eV_p}{kT}\right)^{1/2} j_o \\ &= A_p N e \sqrt{\frac{2kT}{\pi^2 m}}\left(1 + \frac{eV_p}{E}\right)^{1/2} \end{aligned} \tag{11}$$

Notice that current increases with voltage for a cylindrical collector more slowly than for a sphere. (For an infinite plane collector, there is no increase with voltage.)

For large collecting voltages, $V_p >> kT/e$, we obtain the current density at the collecting surface

orbit-limited

$$j = I/A_p = Ne\sqrt{\frac{kT}{2\pi m}}\left(\frac{eV_p}{kT}\right) \qquad \text{(sphere)} \tag{12a}$$

$$j = I/A_p = Ne\sqrt{\frac{2e}{\pi^2 m}}\left(V_p\right)^{1/2} \qquad \text{(cylinder)} \tag{12b}$$

$$j = Ne\sqrt{\frac{kT}{2\pi m}} \qquad \text{(plane)} \tag{12c}$$

The comparison of these "limited" collection cases with Eq. (1) yields rather surprising results. While the cylindrical collector does "limit" its surface current density to a factor of π less than the "unlimited" disaster calculated in Eq. (1), the spherical collector actually achieves a higher current density for any voltage over 4π kT/e, even though it is unable to draw charge from any distance greater than p.

The spherical collector is able to do this because of the geometry of the situation, which focuses the current into an ever smaller surface area as it approaches the collector. This forces N to increase as $r^{-3/2}$, as can be obtained from conservation of current:

$$\begin{aligned} I(r) &= 4\pi r^2 j(r) = 4\pi r^2 N(r)\, e\sqrt{\frac{2eV(r)}{m}}\,; \quad V(r) = \frac{V_p a}{r} \\ &= \text{"}I(p)\text{"} = 4\pi p^2 N_o\, e\sqrt{\frac{kT}{2\pi m}} \\ N(r) &= N_o \frac{p^2}{r^2}\sqrt{\frac{kT}{2\pi}}\sqrt{\frac{(r/a)}{2eV_p}} = N_o\ (\text{const})\ r^{-3/2} \end{aligned} \tag{13}$$

Since most tests of current leakage to small probes or arrays have been performed under basically orbit-limited collection conditions, it is not surprising that unacceptably high current densities are frequently observed even at plasma densities well under the LEO maximum of $10^{12}/m^3$ and/or using collectors with less than 10% of their surface area conducting.

Magnetic Fields and Orbit Limitation

The limitations on current collection area due to angular momentum orbit-limits are proportional to collector size and independent of particle mass. Both electrons and ions of the same initial energy are collected from the same effective surface area. The characteristic dimensions for magnetic flux-tube confinement to become important (Eq. 3) are dependent on particle mass, becoming important for electrons before ions. This is illustrated in Fig. 1 for a spherical collector (or a cylinder oriented perpendicular to the magnetic field). In the absence of a magnetic field, current is, in effect, collected from a "virtual" surface located at distance p in the (hypothetical) undisturbed plasma.

In the actual plasma, significant electric field from the collecting surface reaches well beyond the distance p and has already deflected the plasma particles from what would have been their (undeflected) trajectories. This is illustrated in Fig. 1 by the trajectories A and B, which represent the actual trajectories of two ions (assumed too massive for significant deflection by the magnetic field) whose "virtual" trajectories would have been A' and B' in the undisturbed plasma. A represents the limiting trajectory able to hit the surface at radius a and be collected. A is assymptotic at large distance to A', the "virtual" trajectory with closest approach at distance p. Ions whose "virtual" trajectories B' would pass just beyond the impact parameter p (although strongly deflected to an actual trajectory B passing well inside p), can never reach a to be collected and therefore follow an open "orbit" back to the distant plasma.

The correct current collected therefore can be determined by considering the impact parameter p as defining a "collecting surface" beyond which the plasma is undisturbed. All attracted plasma particles that cross the "collecting surface" are drawn to the collector; all particles outside the surface are unaffected by the collector's presence; and all repelled particles are reflected from the surface. This is a computational convenience of good utility. Simply calculate the relevant thermal or drift velocities of the attracted ions (or electrons) and multiply by the surface area or cross section of the "collecting surface" just as if it really

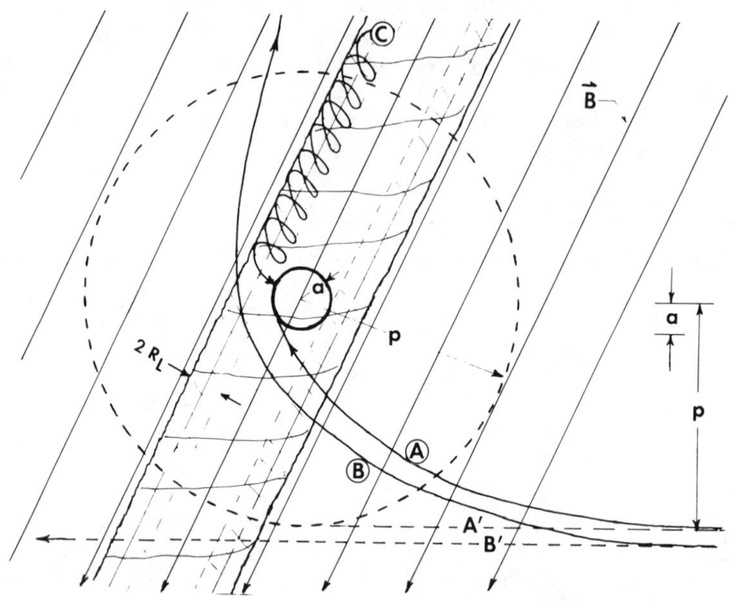

Fig. 1. Orbit-limited current collection with a magnetic field, attracting probe radius a.

existed. (Do check that $R_L > p$ and space-charge effects are negligible.)

For much lighter particles, or for larger dimensions such that p (or even a) becomes large compared to the gyroradius R_L about the magnetic field lines, the situation changes to that illustrated by trajectory C. Now only those particles spiraling along field lines within a "flux tube" of dimension roughly (4R + 2a) are able to reach the collector surface. For $a > R_L$, the effective collection area is reduced to just the projected cross section of the collector perpendicular to the field. This can reduce the current collected substantially.

Current collection along such a flux tube, assuming no charge crosses the walls, is obtained from conservation of current

$$j = Ne\sqrt{\frac{2eV}{m}} = j_o = N_o e\sqrt{\frac{kT}{2\pi m}} \tag{14}$$

where we assumed that (a) equilibrium exists in the plasma far from the collector (where V = 0 and even the most limited transport across the magnetic field can accumulate to restore N against collector depletion) and (b) within the distance along the flux tube where the particles are accelerated by the collecting potential from $\overline{v}\cdot\overline{B}/B = \sqrt{kT/2\pi m}$ to $v = \sqrt{2\ eV/m}$ there is no significant current flow across the magnetic field lines.

Therefore

$$N = N_o \sqrt{\frac{kT}{4\pi eV}} \tag{15}$$

Equation (14) has the very good feature, for reducing leakage currents, that it is independent of the collecting voltage. To the extent that ions and electrons are unable to cross field lines within the acceleration region, the total leakage current to a very large solar array of area A oriented at an angle α from perpendicular to the magnetic field will collect a total thermal leakage current (for arbitrary operating voltage):

$$I_{th} = A \cos \alpha\ N_o e \sqrt{\frac{kT}{2\pi m}} \tag{16}$$

If N and kT are the undisturbed F2 ionospheric values, for example $N_o = 10^{12}/m^3$, $kT_e = 0.2$ e, $kT_i = 0.1$ eV, $M_i = 16$ amu:

$$I_{th} = (1.2 \times 10^{-2})\ A \cos \alpha \qquad (V>0)$$

$$I_{th} = (5.0 \times 10^{-5})\ A \cos \alpha \qquad (V<0)$$

These currents are for thermal fluxes only. "Ram" currents due to particles intercepted by the motion of the array through the plasma must be added. If any plasma drift motion, $\overline{E} \times \overline{B}/B^2$, is neglected and β is taken as the angle between orbital velocity (v_s) and the sun ($\beta < 90$ deg.),

$$I_{ram} = Nev_s\ a^* A \cos \beta = (1.2 \times 10^{-3})\ A^* \cos \beta$$

For $\beta > 90$ deg., the currents may be further reduced by plasma wake effects. $A^* = a^*A$ represents an effective intercept cross section intermediate between A and πp^2, resulting again from limited mobility across magnetic field lines during acceleration of intercepted particles.

Therefore, complete magnetic gyro-orbit domination may reduce the total orbit-limited current collected to

$$I = AN_o e \left(\sqrt{\frac{kT}{2\pi m}} \cos \alpha + a^* V_s \cos \beta\right) \tag{17}$$

The minimum currents estimated in Eq. (17) may be increased by several effects. Pederson conductivity may contribute directly, or indirectly by means of increasing the effective value of kT in the source region. Second order terms involving $\nabla\cdot\overline{E}$ were shown by Parker and Murphy[5] to increase the effective cross section of the flux tube. Therefore, magnetic gyro-orbit limitations will reduce the collected current, but only to some value intermediate between Eq. (17) and the no-magnetic-field values given by the commonly quoted Eqs. (12a-c).

Space-charge Limited Current Collection

Neither the ionospheric bulk conductivity nor (angular momentum) orbit-limitations on current collection in simple geometries seem capable of providing a useful upper limit to leakage currents from the LEO ionosphere to a solar array operating in the 10 kV range. Confinement of currents to magnetic flux tubes may result in substantial limitations to the maximum leakage currents obtainable. However, calculation or experimental determination of the lower limit to efficiency of such reduction appears difficult. We will show that an even stronger limit on current collection may result from space charge effects, independent of magnetic confinement. The space-charge limited sheath behavior also exhibits characteristics which make it readily amenable to analytic calculation of total current leakage estimates and detailed computation by solution of Poisson's equation within limited regions with determinate boundary conditions.

Sheath Formation

In all analyses so far, we have assumed plasma density low enough that charge separation (space charge) effects are negligible. While this assumption generally is valid for centimeter size objects, for larger scale objects charge separation fields should dominate to form distinct space-charge-limited sheaths of limited dimension around any current collecting object.

As outlined by Langmuir in 1923,[6] when a large collecting surface is at a high negative potential with respect to the plasma, it repels electrons from its vicinity and attracts ions, thereby becoming surrounded by a positive ion "sheath" region containing a positive space charge with no negative ions or electrons. (The polarity of potential and charges is invertable; an identical behavior is obtained for a negative electron "sheath" region formed around a collector at positive voltage.) The entire difference in potential between collector and surrounding plasma becomes concentrated within this sheath; the positive space charge of the sheath adjusts itself to exactly cancel the electric field due to the negative charge on the collector, so that the field of the collector does not extend beyond the outer surface of the sheath. Neglecting terms of $kT << eV$, the total ion current collected is limited to the ions that cross the outer boundary of the sheath as a result of their existing thermal or drift motions in the undisturbed plasma. The thickness and therefore the area of the outer surface of the sheath can be calculated from space-charge equations.

For this calculation, the outer boundary of the sheath is treated as a virtual anode emitting ions (from the plasma) and the collector as a (non-emitting) cathode. It does not matter whether an anode surface is physically present (as a metal plate) or there is simply the mathematically equivalent surface that remains at constant potential and location where all plasma electrons are reflected, and the ions continue their motion into the sheath region.

To illustrate, we take the well-known Child-Langmuir Law expression for planar diode space-charge limited current

$$j = \frac{4\varepsilon_o}{9}\sqrt{\frac{2q}{m}}\,\frac{V^{3/2}}{d^2} \tag{18}$$

In this case,[7] the free variable must be the "electrode" separation (sheath thickness) d, since j already is fixed by the current density crossing the virtual "electrode" formed by the outer sheath surface. Assuming drift and satellite orbital velocities zero, this will be

$$j_o = k^* N_o q \sqrt{\frac{kT}{m}} \tag{19}$$

where

$$\sqrt{\frac{1}{2\pi}} \leqq k^* \leqq 1$$

Solving Eqs. (18) and (19) for sheath thickness and expressing kT/e as E in eV, we obtain

$$d = \frac{5.89 \times 10^3}{\sqrt{k^*}}\,\frac{|V|^{3/4}}{N^{1/2}\,E^{1/4}} \tag{20}$$

Similar expressions can be obtained for cylindrical and spherical geometry respectively, using

$$\text{cylindrical} \quad j_{oc} = \frac{4\varepsilon_o}{9}\sqrt{\frac{2q}{m}}\,\frac{V^{3/2}}{r a \beta^2} \tag{21}$$

$$\text{spherical} \quad j_{os} = \frac{4\varepsilon_o}{9}\sqrt{\frac{2q}{m}}\,\frac{V^{3/2}}{r^2 \alpha^2} \tag{22}$$

where β^2 and α^2 are quantities tabulated by Langmuir,[8,9] or obtainable from numerical solution of Poisson's Equation in appropriate coordinates. In the limit of d small compared to collector dimensions, both solutions converge to the planar diode solution.

Sheath Limitation of Leakage Current

The space-charge limited sheaths act to limit leakage values in a manner similar to the effects of magnetic confinement, i.e., by restricting growth of the ratio of collection radius to object size. Applying Eq. (20) to LEO plasma parameters, and approximating rather freely for purposes of simplifying the example, we see that distinct outer boundaries to the current-collecting sheaths surrounding a high-voltage surface should be expected to reach a limiting size of the order of 1 m/kV, nearly independent of the size of the high-voltage object.

When the assumptions in this analysis are valid, total current collected by the object depends only on the outer surface area of sheath available to intercept ambient (thermal, drift, etc.) currents existing in the undisturbed plasma outside the sheath surface. The resulting current

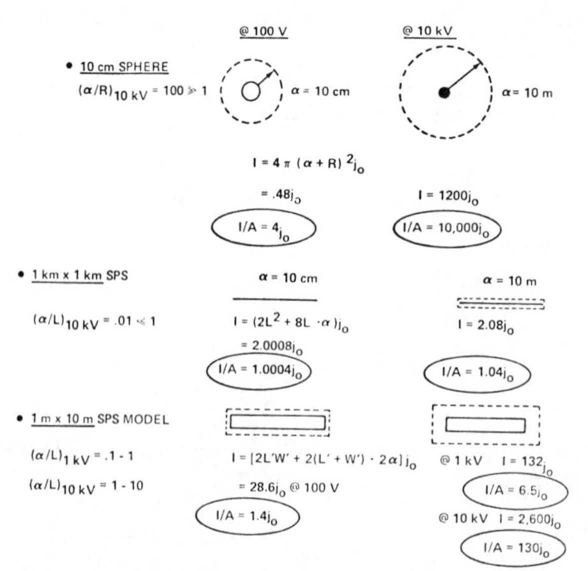

Fig. 2. Current vs. voltage vs. α/L.

- α/L = ratio of plasma sheath thickness to object size
- j_0 = current/unit area (I/A) collected at 0 V

multiplication factor at any voltage will be the ratio of sheath outer surface area to object surface area, as illustrated by Fig. 2. (In effect, sheath "conductance" per unit area becomes a function of both voltage and size, rather than a constant.) For illustration, we have assumed plasma parameters such that the resulting sheath thickness grows from 10 cm at ±100 V to 10 m at ±10 kV. This sheath becomes very large compared to the 10 cm sphere; the total current collected increases by nearly 120 (a very high "conductance" sheath). The same plasma sheath around a 1 km "SPS" array has a very small ratio of sheath-to-object size. The total current collected should increase by only a few percent, a very low sheath "conductance" which becomes even lower with increased voltage.

For experimental investigation of the validity of this sheath model, working either in LEO or in the large space simulation chamber at JSC, it is possible to observe the growth of these sheaths around a 1-10 m object with 0.1-10 kV applied. These dimensions permit a test of their behavior in "free space" without the inevitable wall effects due to sheath growth exceeding the size of small chambers. At the same time, observation can be made of the predicted transition in behavior from the small α/L situation (whose predicted current vs voltage behavior is very similar to the orbit-limited case) to the large α/L ratio situation used to project the expected values of leakage current for large arrays shown in Fig. 3.

The importance of tests at the 1-10 m size range, both in the large chamber and with Shuttle flight experiments, is readily seen as a relevant size for verification of this sheath model. (This size is applicable to design concepts for the next generation of high-power solar arrays such as the Microwave Transmission Test Article (TA-1), proposed as a key development test step in the evolution of a workable Solar Power Satellite design).

Solar Array Sheaths

The application of the sheath model to solar arrays is illustrated by the use of the TA-1 design concept shown in Fig. 4, as an example, to discuss the development of sheaths around a very large high-voltage (HV) solar array in LEO space plasma. The TA-1 is conceived as a space-fabricated 240 x 21-m solar array, developing 500 kW at up to 40 kV, and driving an array of microwave tubes for orbital tests of antenna/microwave power transmission and reception. At least part of its projected test activity would be conducted from LEO. One early

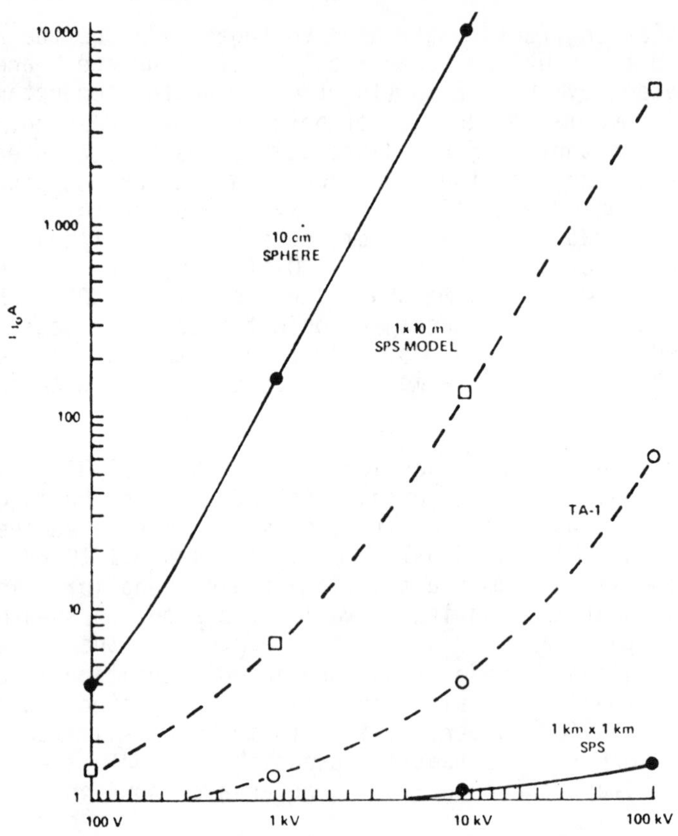

Fig. 3. Estimated sheath limited leakage currents vs. voltage for various size panels, normalized to unity at zero volts.

design decision to be made for TA-1 is whether to attempt direct operation of the entire array in series connection to obtain 40 kV, or to operate the array at some lower voltage considered safe from harmful plasma interactions and use power converters to obtain the 40 kV drive voltages for the microwave tubes.

Figure 5 shows the plasma sheath configuration expected to develop around the TA-1 array during operation in LEO with 40 kV developed in simplest series connection end-to-end. Estimated sheath dimensions are sketched around both side and plane views. The entire array will float to an equilibrium voltage, with respect to the plasma, such that total current collected is zero (i.e., total electron current exactly cancels total ion current collected by the negative portion of the array). The resulting surface voltage (with respect to the plasma) and leakage current density along the array is plotted above the plan view. The sheath thickness illustrated at each point along the face of the array is based on solution of Eq. (20) for the voltage on the array at that point. The sheaths sketched at the ends are estimates assuming the edges act as cylindrical probes without charge buildup on the back panels of the TA-1.

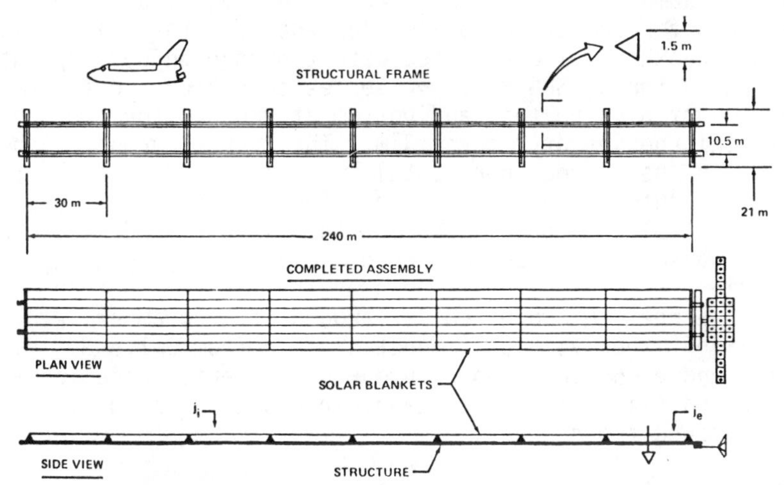

Fig. 4. Microwave Transmission Test Article: configuration concept-1.

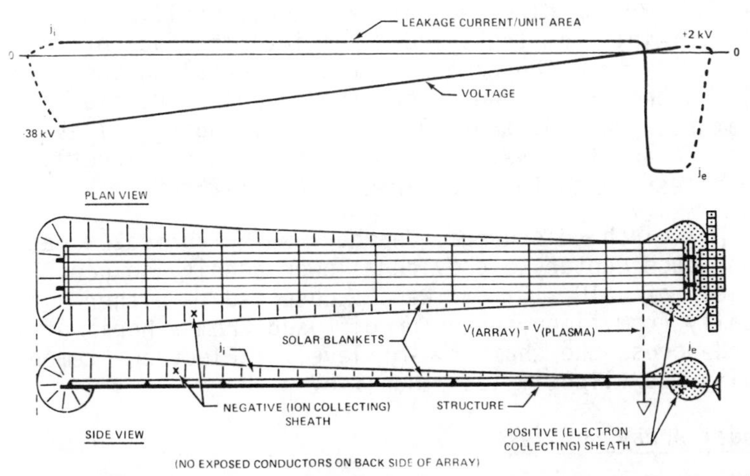

Fig. 5. Plasma sheath configuration, leakage current and voltage with regard to plasma potential. Working model for preliminary analysis of TA-1 array behavior in LEO.

1977 Chamber Test for Sheath Study

The performance of an actual test on the scale of 10 m available in the large chamber was needed to determine which scaling relations are applicable to large solar arrays. Figure 6 shows the layout of the basic configuration used for most tests. The high voltage panel (SPS) was hung near the center of the chamber, with 7-10 m of free space available in all directions for unobstructed development of the high-voltage plasma sheaths. The expected extent of sheath development is illustrated for an SPS model in series connected configuration (the top at high voltage and bottom at ground) for two typical sheath thicknesses of 1 m and 3 m. The three probes labeled 22-24 can be moved horizontally from outside the sheath to locate the outer sheath boundary (point of first observed change in plasma conditions). The sheath and associated effects also could be observed visually using low light television (LLTV) cameras shown at the first- and third-floor levels. Large solenoid coils around the chamber provided control of the vertical magnetic field from 0-1.5 G. Plasma density and electron temperature measurements were obtained from 15 half-inch spherical Langmuir probes located at various points around the chamber.

Plasma Generation

A 30-cm Kaufman thruster borrowed from NASA Lewis Research Center was used with argon gas to generate a flowing plasma of density 10^4-10^6/cm^3 directed either horizontally (across the magnetic field) from the third level into the face of the panel or vertically from the center of the floor (along the magnetic field) along the length of the panel. Plasma electron temperatures varied from 0.5-2 eV, averaging 1 eV. Ion temperature and flow velocity were not measured directly; flow energy is estimated to have varied from 15-25 eV.

A six-inch discharge chamber was fabricated at NASA Johnson Space Center to provide a lower density source. Plasma densities of 10^2-10^4 cm^3 were observed, either flowing vertically from floor center or diffused from a horizontal flow directed across the chamber 1-m level. Electron temperature typically was slightly less than 1 eV.

SPS Model Array for Test

Figure 7 shows the physical dimensions of the SPS model tested, as well as location and identification of available test connections to the copper contact strips. The simulated solar array test surface was fabricated by securing a 5 mil

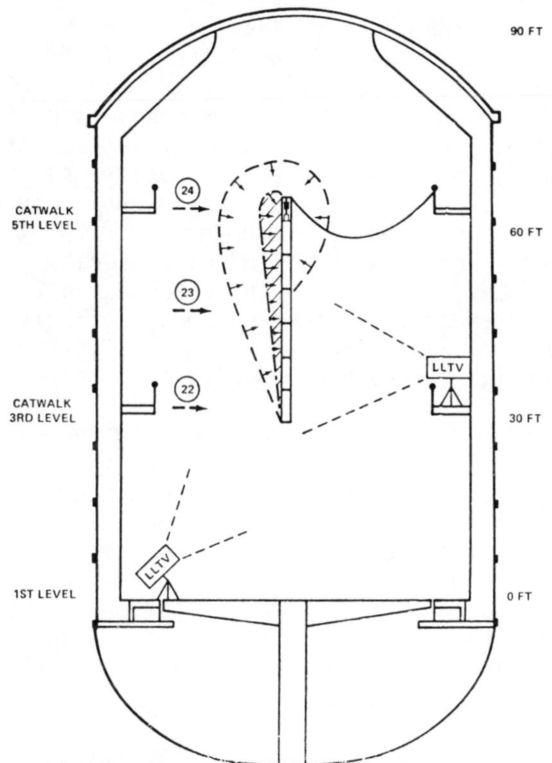

Fig. 6. 1x10 m"SPS" high voltage array test set-up in Chamber A. Plasma source located at center of first level, or above third level catwalk.

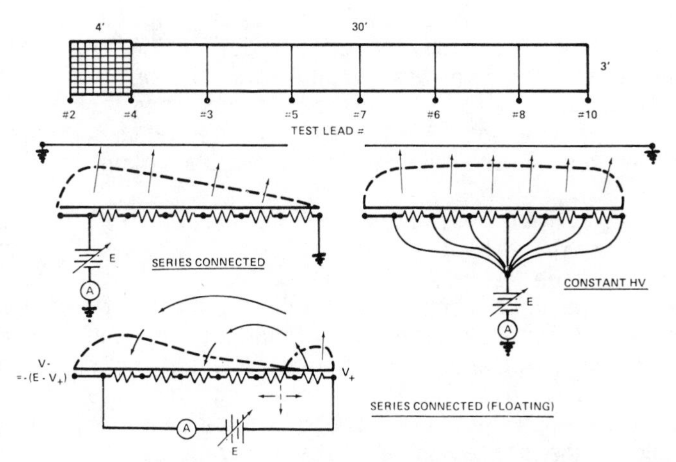

Fig. 7. "SPS" model array for test. Physical layout and available electrical test connections showing resulting sheath and plasma current configurations.

sheet of Velostat (conductive plastic) with 5 mil Kapton backing to a rigid panel of dielectric (1/3 cm of Lexan), leaving a (10 cm) exposed dielectric border along all edges. The exposed Velostat provided an electrically conducting front surface for plasma current collection, with sufficient internal resistance (140 kΩ) to provide approximately uniform voltage gradient along the panel when biased end-to-end. The panel can thus be operated to produce a monotonic voltage gradient to simulate a large array with all cells series connected end-to-end, or the whole panel can be connected to be at the same potential for specialized testing of particular voltage areas. The actual dimensions differed slightly from the nominal 1x10 m design for ease of fabrication.

For test purposes, the array was operated in one of the three electrical configurations shown. The "series connected" (floating) configuration is the actual case which would apply in space; currents close from the positive voltage (V_+) end of the array, through the conducting plasma, to the negative portion of the array. The chamber walls and lab ground are not involved in the circuit at all. The location of the point along the array which is at "plasma ground" potential will be inversely proportional to the relative ambient current densities of ions and electrons in the free plasma. For typical conditions, this will result in the array floating 97-99% negative with respect to plasma potential.

Since operation in the fully floating configuration was physically awkward, most "series connected" testing was done with the power supply and one end of the array grounded to the chamber walls. This was equivalent to testing the negative or positive portions of a floating array individually, with the return current path closing through the chamber wall (via the plasma). In either case, all voltage drops from array surface potential to plasma potential are contained within the sheath. The outer surface of the sheath is at plasma potential.

This test object was designed to produce the extreme values of current leakage possible from a large solar array or other high-voltage surface. To eliminate confusion from attempting a correct treatment of the effect of relative surface area and configuration of conductive and insulating portions of the surface of an array, the entire front surface (except the actual solar cell section and exposed dielectric border) was made conductive. The SPS model should, therefore, generate the large scale (outer) sheath configuration believed to be of primary importance in determining its equilibrium interaction with an ambient plasma. The currents thus collected will not be reduced by any surface insulation factor.

Test Objectives

In order to test the validity of the proposed approach to scaling calculations of plasma current leakage based on relative sheath-to-object size, with the sheath size limited by space-charge effects, three primary test questions were identified:

1) Existence, sharpness and size of the expected outer sheath surface.

2) Equilibrium floating potential of a "series connected" (floating) array.

3) Magnitude of leakage currents as a function of voltage (actually, sheath size).

A secondary question was the possible existence and behavior of transient current pulses (electrical breakdown or "arcs" to the plasma) reported to occur in smaller scale experiments.

Experimental Results

The fundamental result achieved was direct observation of the existence, form, and dimensions of the plasma sheaths formed about the high-voltage panel. Leakage currents between the panel and the surrounding plasma, through the observed sheaths, were recorded for comparison with the theoretically expected

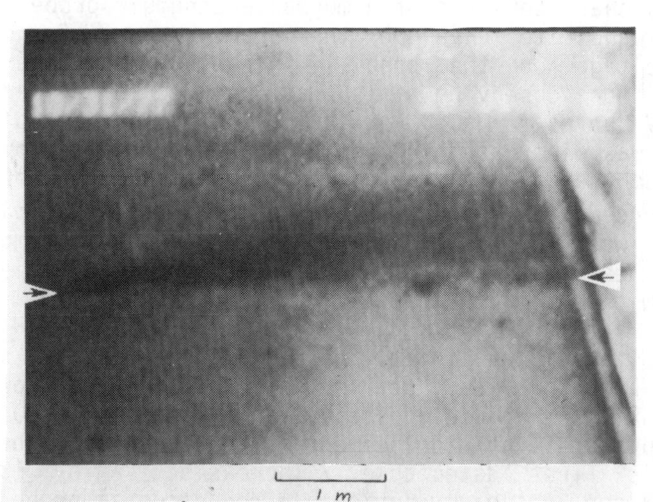

Fig. 8. LLTV image of high voltage plasma sneath. Arrows indicate location of "SPS" panel.

current transmission capacities of the sheaths. The following is a summary of results discussed in more detail by McCoy and Konradi.[7]

The existence and form of the sheaths was observed by two independent means, both of which detect the location and sharpness of the outer boundary with minimum disturbance of its configuration by physical intrusion of hardware.

Sheath Observation by LLTV

When viewed under conditions of sufficiently high plasma density against a dark background, the sheath is usually visible on LLTV, as a dark region in front of the panel which expands or contracts as a function of voltage on the panel face. Figure 8 shows a typical LLTV image of the series-connected sheath, with surface potential on the SPS increasing from 0 V at the bottom end to 1 kV near the top (about the center of the panel, viewed exactly edge-on) of the picture. The sheath is the wedge-shaped dark area, seen to increase in thickness nearly linearly with voltage to perhaps 1-2 m at 1 kV. The outer boundary generally is rather sharply defined in the LLTV image, as expected from the space-charge limited thickness hypothesis.

Sheath Detection by Probes

The second method of observing the sheaths involved watching for an alteration in the observed I vs V current collection characteristic of a moveable Langmuir probe, as the sheath expands to envelope the probe because of increasing surface voltage on the panel. A satisfactory operational technique for observing this was to record a series of log I vs V curves for electron collection as surface voltage on the SPS was increased in steps from zero until the probe was deep inside the panel's sheath. A representative set of curves thus recorded is shown in Fig. 9. The undisturbed plasma at this time was about 10^6/cc, with an electron temperature slightly less than 1 eV (as deduced from the intial curve recorded with 0 V on the panel).

As voltage is applied to the panel, no effect is seen at the probe location (i.e., at 1 m, still outside the growing SPS sheath) until the applied voltage, V_{op}, reaches -800 V. At this point, a slight displacement of the curve at higher probe voltages is first detectable. Increasing V_{op} by 100 V to -900 V causes a clearly noticeable reduction in probe current at +100 V bias, more than resulted from the entire 800 V previous change. There is yet no change at bias voltages in the retarding portion of the curve. We interpret this as indicating the probe is still (just) outside the panel's sheath boundary but near enough for a partial "shadowing" of the probe from attracted electrons.

The sheath probably has just passed the location of the probe when -1,000 V is applied to the panel. The current zero-crossing voltage has shifted. As the panel voltage is increased further, moving the location of the sheath edge further beyond the probe location, even greater bias voltages are required on the probe before its electron-attracting field is strong enough to reach beyond the electron-depleted sheath's boundary to an undisturbed plasma region containing electrons.

A set of curves similar in appearance is obtained for positive (electron-collecting) sheaths. The causes are

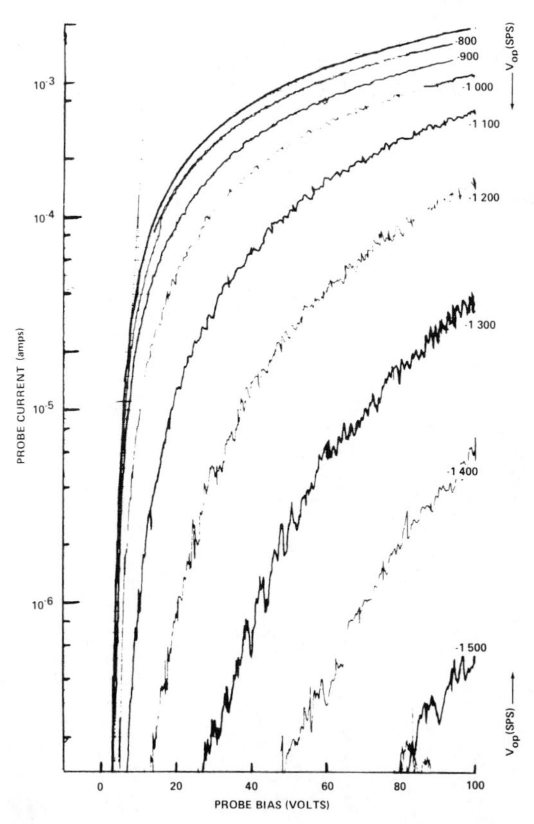

Fig. 9. Change in recorded (ln I vs. V) spherical test probe "characteristic" with SPS panel voltage (V_{op}) as sheath expands past probe.

probably quite different, since there are electrons available for collection inside this sheath.

Sheath Size

Within present limits of experimental error, the observed sheath thicknesses are proportional to $(V^{3/4}/N^{1/2})$, as expected for space-charge limited current flow with d (sheath thickness) the free variable, Eq. (20). Figure 10 shows the applied voltage required at various plasma densities for the outer sheath surface to reach a probe (probe 23 in Fig. 6) located 1 m from the surface of the array. The reference line is the theoretical thickness calculated for a planar diode with an effective electron or ion "temperature" of 1 eV. Notice that the electron sheath (shown as ⊗) is about the same size as the ion sheaths ⊙.

Sheath Current Leakage vs Voltage

The resultant current multiplication factor was observed to be much lower than observed on previous small-scale

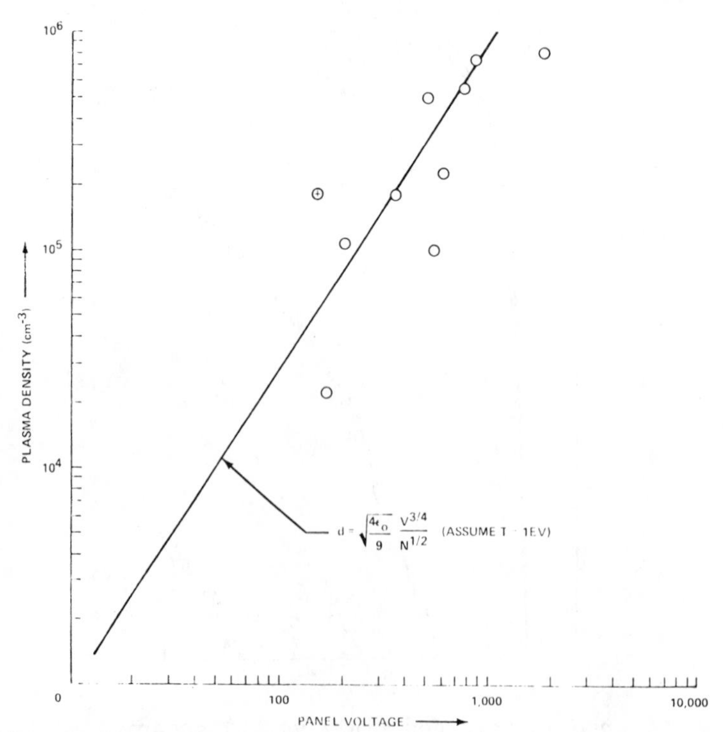

Fig. 10. Applied voltage (V_{op}) required for outer sheath boundary to reach test probe located 1 m from surface of "SPS".

(orbit-limited) tests. Figure 11 shows current leakage from the SPS to the plasma observed with V_{op} from -10 V to -3,000 V in four ambient plasma densities ranging from 10^4-10^6/cm³. The observed rate of increase in leakage current with V_{op} is seen to increase as the resulting sheaths become large compared to the panel width, as expected. The regions of sheath size are rough estimates based on calculations by Eq. (20) normalized to an actual measurement for each data set.

Floating Potential

Recalling the requirement that total current flowing to an electrically isolated panel in series-connected floating configuration must be zero for voltage equilibrium with the plasma to exist, we expect values of V_- and V_+ relative to the plasma to shift so that

$$j_{oi} A_- = -j_{oe} A_+ \tag{23}$$

where j_{oi} and j_{oe} are the ambient ion and electron current densities across the outer sheath boundary, and A_- and A_+ are

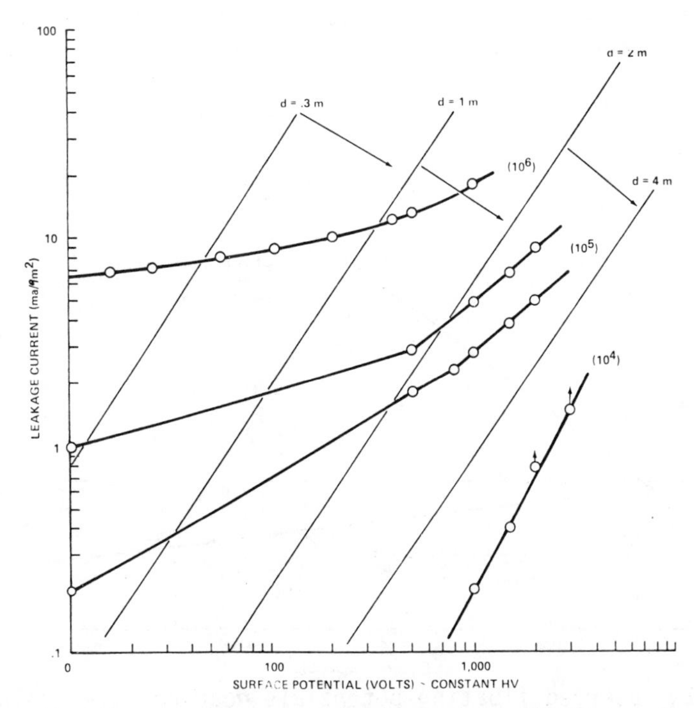

Fig. 11. Measured current leakage from "SPS" to plasma.

the effective surface areas of the negative and positive potential sheaths. We neglect current contributions from other sources, such as secondaries, and the area immediately around the $V = 0$ point along the panel. For reasonably thin sheaths, relative to panel dimensions, we can use the approximation

$$A_-/A_+ = L_-/L_+ = V_-/V_+ \tag{24}$$

where L_-, L_+ are the lengths of the sections of the panel floating negative and positive, respectively, with respect to the plasma potential. Along the panel, dV/dL is assumed constant.

Under normal plasma conditions during our test

$$-j_{oe}/j_{oi} = 43$$

Using the sheath model, we expect the panel to float about 2.3% positive, the remainder negative, under normal plasma conditions during our test, with ion flow perpendicular to the panel. This was verified experimentally. The 30-cm thruster was operated from the third level catwalk, aimed horizontal directly into the face of the panel. The experiment configuration was series-connected (floating) as shown in

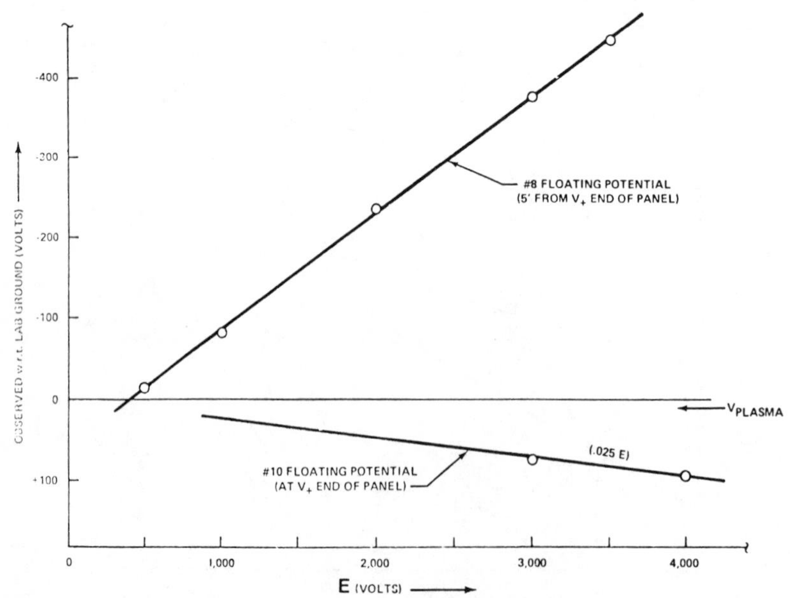

Fig. 12. Observed floating potentials near positive voltage end of simulated high voltage solar array.

Fig. 7. At -3 to 4 kV, 2.6-2.3% of the panel was observed to be at a positive potential (see Fig. 12).

Arcing to Plasma

Arcing, defined here as any sharp and transient increase in current drain to the plasma, was frequently observed. Many measurements of current loss vs voltage were limited to voltages less than 2-3 kV because arc-induced transients became so severe that useful meter readings could not be made. The arcs were observed to occur at positive voltages greater than +400 V and negative voltages greater than -1,000 V. There appears to be no repeatable dependence between plasma density and the minimum voltage for the onset of arcing. Occurrence of arcing at a particular voltage did not appear to be repeatable under apparently identical plasma conditions. The appearance of the arcs as observed on LLTV also varied greatly.

Every arc observed by the LLTV system occurred from an insulator surface. We have not observed any instance of an arc occurring from the conductive surface area of the panel. Although some arcs were small and did not affect the rest of the panel except for small pulses in the total panel current, many resulted in a sudden discharge of the entire panel voltage, which required 1-5 sec to rebuild. This was observed both as a collapse of the entire sheath to much smaller dimensions and as a drop in panel voltage, as indicated by the power supply meter.

The collapse of the entire sheath was observed by LLTV to occur even in cases where the visible discharge came from an insulator surface which was located 1-2 m out into the sheath from the panel and which had no contact with the conductive panel surface other than through the plasma. It would appear that the arcs we observed are the result of local charge buildups due to sheath currents impinging on a nonconductor in their path; this process appears to be similar to that believed to occur on satellites in GEO during substorms.

At the time of a transient arc, a much larger charge transfer is required to discharge the power supply output capacitor than the total amount contained in the sheath or in the vicinity of the visible discharge point on the insulator. Most of the observed current drain from the power supply must be due to large-scale currents developed throughout the collapsing sheath, currents substantially in excess of the equilibrium space-charge limited values. The small area of a visible flash region ($<1\ cm^2$) would not seem capable of carrying the 0.01 C discharged (in less than 1/30 sec), even if all ambient neutrals were ionized to provide bipolar charge carriers.

Further Development of Sheath Model

Basic Sheath Model

The equations for a space-charge limited sheath with unipolar current flow (no secondary emission currents from the collector) have well-known cold plasma solutions for planar, cylindrical and spherical geometries which were given by Eqs. (18), (21), and (22).

The planar case Eq. (18) was readily solved for the sheath thickness Eq. (20). Solution of the cylindrical or spherical cases requires an iterative matching of values of thickness and voltage with corresponding values of α^2 or β^2. For situations where the sheath thickness is smaller than the object size, this correct solution is generally within 50% of the more readily calculated planar solution. Therefore, for most problems which do not demand high precision, we can use a pseudo-planar model shown in Fig. 13 to obtain a sheath thickness estimate at each point of interest along the panel.

Examination of this model shows the areas which need additional development. The primary effect of the sheath is to confine the collecting potentials within a reduced region. The sheath acts to separate all charge of one (collected) polarity from the plasma and concentrate a sufficient amount of it near the collector (space charge in the sheath) to exactly cancel the electric field due to the collector voltage. The effective current collection volume is therefore reduced by confinement of the fields within the much smaller volume of the sheath. Within this volume, all previous considerations of orbit-limited or magnetic field restricted collection will still apply. Only the relevant electric field configurations have been modified.

Potential Distribution in the Sheath

The space charge is distributed within the sheath so as to maintain equilibrium at all points between the changing values of V, N and local ion velocity. For the planar case, $j = j_0$ is constant, while other geometries must conserve total current $I = jA$ across any (constant current density) surface. Non-symmetric geometries can be seen to require either computer techniques or some sort of symmetric geometry aproximation for solution.

For the situation in Fig. 13, Eq. (18) can be solved for V as a function of distance z inside the sheath boundary:

$$V(z) = \left(\frac{9}{4\varepsilon_0}\right)^{2/3} \left(\frac{m}{2e}\right)^{1/3} j_0^{\,2/3} z^{4/3} \tag{25a}$$

For a flowing plasma (k* = 1) with E = 1 eV, this becomes

$$V(z) = \frac{N_0}{10.6\times10^4} z^{4/3} \tag{25b}$$

At a plasma density $N_0 = 10^{12}$, this becomes very nearly

$$V(N = 10^{12}, E = 1) = z^{4/3} \quad (kV) \tag{25c}$$

Since current density is constant, most of the space charge is concentrated near the outer boundary of the planar sheath, decreasing with distance toward the collector as

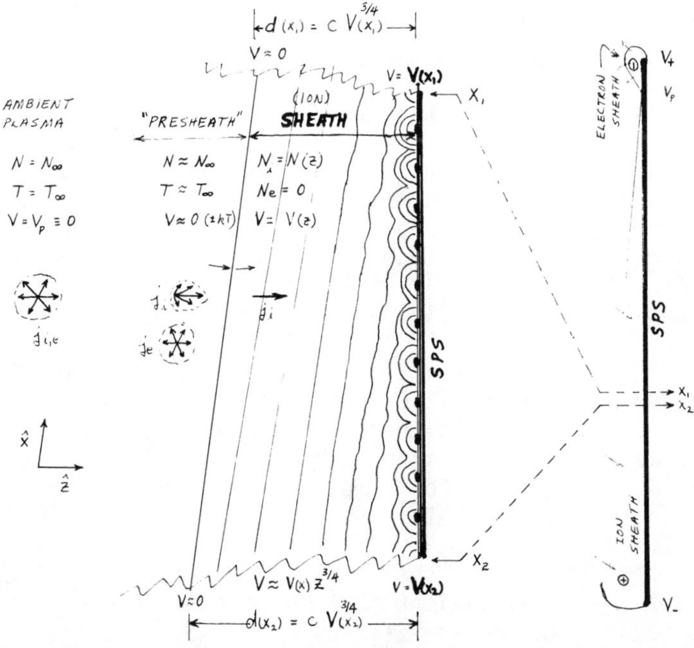

Fig. 13. Pseudo-planar model of space charge limited sheath configuration. Sheath dimensions, small compared to overall SPS, large compared to dimensions of exposed conductor/dielectric configuration on face of SPS.

$$N = N_o \sqrt{\frac{E}{eV}} = (\text{const.})\ z^{-2/3} \tag{26}$$

Bipolar Space Charge Sheath

The emission of photo-electrons or other repelled secondary charged particles from the collector surface will alter the sheath by partially canceling the existing space charge. Since the secondary particles are at rest at the collector and accelerate from there, their space density is maximum at the collector and decreases with potential towards the outer boundary in a manner similar to Eq. (26). Solution of Poisson s equation for the resulting equilibrium requires a numerical integration (even for the one-dimensional case) and is suitable for computer application.

Averaging Effect on Collector Structure

Total current collected is entirely a function of outer surface area of the sheath, which depends on sheath thickness relative to object size. Thus current collected should not be a function of the detailed surface structure of the collecting array except to the extent that this causes a change in the relative sheath thickness. The sheath tends to average out the effects of small scale voltage and current variations at the collector surface. For this reason, it would appear to be generally true that the exact area and location of exposed conductor (bare interconnects, pinholes, etc.) surface on a large solar array will not appreciably affect the total current collected by the outer surface of the surrounding sheath. This seems consistent with the results for pinholes and various arrays reported in Ref. (3) and others.

Although complete solution of the problem for actual detailed solar array surface configurations obviously requires the use of computer models that can handle both space charge and surface charge (on insulators) effects, the sheath model does provide an approach for simplifying the situation in order to make first-order estimates.

For this purpose, the actual collector surface is represented by an equivalent surface at some reduced potential V_o^* that results in the same fields at some distance z^* into the surrounding sheath as the real surface. This distance must be large compared to the scale of the surface structure (for example, cell size, 1-10 cm), but small compared to sheath thickness. The appropriate reduced potential can be estimated by application of Gauss Law to fields across the surface at z*, using total charge value inside z* based on estimated surface charge buildup on nonconductor surfaces reduced by estimated additional "image" charge induced on conductor surfaces by charging of adjacent nonconductors. The degree of charge buildup on cover slides, etc., can be estimated by considering the degree to which they are screened by the effects of orbit-limited (or even a secondary space charge sheath-within-a-sheath) charge collection by adjacent conductors within the high-velocity inflowing "environment" of the primary sheath. In this respect electrons are more easily diverted because of their lower mass/momentum. Also, the electrons provide secondaries of the same charge to cause near-surface charge enhancement and migration, leading to a general expectation of more rapid screening and lower surface charge accumulation on insulated areas of an electron-collecting array.

Solution for current to any particular small structure, such as a pinhole or cover slide, can be done as an individual problem using existing orbit-limited solutions considering the primary sheath parameters, rather than external plasma, as the relevant input quantities (so long as the local solution does not itself become strongly space-charge limited).

Presheath

Throughout the present paper we have considered the plasma immediately outside the sheath surface to be undisturbed. This is true only neglecting terms on the order of kT. It is obvious that some change must occur, at least sufficient to alter the velocity distribution function for ions (ion sheath) from omnidirectional to hemispherical. An early task in the application of this model, therefore, is a satisfactory treatment of the presheath region of unknown extent outside the sheath surface, where some effects must occur to match velocity distributions from truly undisturbed plasma to those actually existing immediately outside the primary high-voltage sheath. This is probably the same problem encountered in any finite plasma, i.e., "wall" effects on the interior plasma volume.

Conclusions

We conclude that calculations based on space-charge limited sheaths provide a promising working model for calculating design estimates of high-voltage plasma current leakage from large solar arrays and similar objects. It would appear necessary that such estimates be verified by a coordinated sequence of plasma-vacuum tests, progressing from small lab chambers to full scale flight tests, because of the large differences in applicable scaling relations which may result from subtle differences in assumed conditions. Large scale tests of the sort described here, together with adequate math models to provide detailed solutions and needed continuity between different design or test configurations, will be an important element in any development test sequence for systems involving large surfaces and high voltages.

The present results are preliminary, based on exploratory calculations and measurements intended to determine the feasibility of this type of procedure and the order of magnitude of the quantities involved. Detailed verification and extension of these results is needed; this will require extension of underlying computer models to include the space-charge effects. Cross-checking of the predictions of such models with actual measurements within 1-5 m sheaths observed during tests in the large chamber and space should be very useful to further development of both models and tests.

The present results indicate that equilibrium high-voltage leakage currents to the plasma should be much less than some earlier predictions had indicated, particularly for the very large solar arrays. The power loss and other effects caused by the transient arcing phenomena threaten to be much more significant unless adequate means are developed to understand and control them. Prospects for this are good, as the arcs do not appear to be a direct consequence of the interaction of high-voltage surfaces with an ambient plasma, as is the case for (equilibrium) current leakage.

More detailed and complete study of the large-scale high-voltage sheaths around a solar array appears basic to an adequate treatment of both problems. The plasma sheath formed around any high voltage surface envelopes all surrounding structure in an environment very similar to that of GEO during intense storm conditions. It is this environment, not the ambient LEO plasma outside the sheaths, within which all small elements of the surface structure operate and interact.

References

[1]Rosen, A. (ed.), Progress in Astronautics and Aeronautics, Vol. 47, Spacecraft Charging by Magnetospheric Plasmas, AIAA, New York, 1976.

[2]Pike, C. P. and Lovell, R. R., eds., Proceedings of the Spacecraft Charging Technology Conference, NASA TM X-73537, 1977.

[3]Kennerud, K. L., "High Voltage Solar Array Experiments," NASA CR-121,280, 1974.

[4]Johnson, F. S. (ed.), Satellite Environment Handbook, Stanford University Press, Stanford, Calif., 1961.

[5]Parker, L. W. and Murphy B. L., "Potential Buildup on an Electron-Emitting Ionospheric Satellite," Journal of Geophysical Research, Vol. 72, March, 1967, pp. 1631-1636.

[6]Langmuir, I. and Mott-Smith, "Studies of Electric Discharges in Gases at Low Pressures," General Electric Review, Vol. 27, 1927, Part I, pp. 449-455; Part II, pp. 538-548; Part III, pp. 616-623; Part IV, pp. 762-771; Part V, pp. 810-820.

[7]McCoy, J. E. and Konradi, A., "Sheath Effects Observed on a 10-Meter High Voltage Panel in Simulated Low Earth Orbit Plasma," In Proceedings of the Spacecraft Charging Technology Conference, AFGL-TR-79-0082, pp. 315-340, 1978.

[8]Langmuir, I., and Blodgett, K.,"Currents Limited by Space Charge Between Coaxial Cylinders," Physical Review, Vol. 22, October, 1923, pp. 347-357.

[9]Langmuir, I. and Blodgett, K., "Currents Limited by Space Charge Between Concentric Spheres," Physical Review, Vol. 23, July, 1924, pp. 49-60.

[10]Stevens, N. J., et al., "Investigation of High Voltage Spacecraft System Interaction with Plasma Environments," AIAA Paper 78-672, April, 1978.

ENVIRONMENTAL PROTECTION OF THE
SOLAR POWER SATELLITE

Patricia H. Reiff*, John W. Freeman+,
and David L. Cooke‡
Rice University, Houston, Texas

Abstract

This paper examines theoretically several features of the interactions of the Solar Power Satellite (SPS) with its space environment. We calculate the leakage currents through the kapton and sapphire solar cell blankets. At geosynchronous orbit (GEO), this parasitic power loss is only 0.7%, and is easily compensated by oversizing. At low-earth orbit, (LEO), the power loss is potentially much larger (3 %), and anomalous arcing is expected for the high voltage negative surfaces. Preliminary results of a three-dimensional self-consistent plasma and electric field computer program are presented, confirming the validity of the predictions made from the one-dimensional models. Lastly, the paper proposes magnetic shielding of the satellite, to reduce the power drain and to protect the solar cells from energetic electron and plasma ion bombardment. We conclude that minor modifications from the baseline SPS design can allow the SPS to operate safely and efficiently in its space environment.

Introduction

Space is by no means empty. It contains light, magnetic fields and both neutral and charged particles. The light energy is the raison d'etre for space power generation; but it can also eject photoelectrons from satellite surfaces, giving the surface a positive charge and giving it an effective conductivity (Pelizzari and Criswell, 1978).

---

* Assistant Chairman, Department of Space Physics and Astronomy.

+ Professor, Department of Space Physics and Astronomy.

‡ Graduate Fellow, Department of Space Physics and Astronomy.

Magnetic field strengths in the earth's vicinity range from 6×10^{-5}T (0.6 Gauss) at the earth's poles to 2×10^{-9} T (2γ) in the neutral sheet in the magnetotail (1γ = 10^{-5} Gauss). At the geosynchronous orbit, the magnetic field strength is roughly 1×10^{-7} T (100 γ). A magnetic field of this strength causes no threat per se to spacecraft operations; however, it plays a fundamental role in trapping energetic particles. These trapped particles respond not only to the Earth's magnetic field, but spacecraft fields as well, especially for spacecraft large in comparison to particle gyroradii (Reiff, 1976; Reiff and Burke, 1976).

Neutral particles have little effect on spacecraft operations above ∿ 600 km; at Low-Earth Orbit, however, neutrals can charge-exchange in the EOTV thruster beam (see below).

Charged particle populations at synchronous orbit are of several types and are illustrated in Figure 1. The innermost region is the plasmasphere, a torus-shaped locus of relatively dense (∿ $100/cm^3$), cool (kT ∿ 1 eV) plasma that has evaporated from the ionosphere. Because of the low energies of the plasmaspheric ions, they are considered harmless (Reasoner et al., 1976); however, they can be accelerated by spacecraft electric fields to energies high enough to do damage (tens of kilovolts). Imbedded in the plasmasphere are the radiation belts, regions of very low density but quite high energy (tens to hundreds of kilovolts) trapped radiation. This radiation can cause hazards to men and solar cells.

The remaining plasma population that can penetrate to geosynchronous orbit is the plasma sheet (Fig. 1). This tenuous plasma ($0.1\text{-}1/cm^3$) is considerably warmer (kT on the order of kiloelectron volts) than the plasmasphere (Garrett and DeForest, 1979). In addition, its presence at geosynchronous orbit is associated with substorm activity, when both the fluxes and energies are higher. It is this kind of plasma that contributes most strongly to spacecraft charging and its concomitant disruption of satellite systems (Inouye, 1976).

This paper concentrates on spacecraft charging and its effects on solar power satellite (SPS) systems, in particular the NASA/Marshall Space Flight Center (MSFC) baseline design (Hanley, 1978). "Worst case" plasma environments are used to determine possible charging hazards. Spacecraft charging is the principal focus of this paper since its effects can be severe: arc generation from exceeding break-

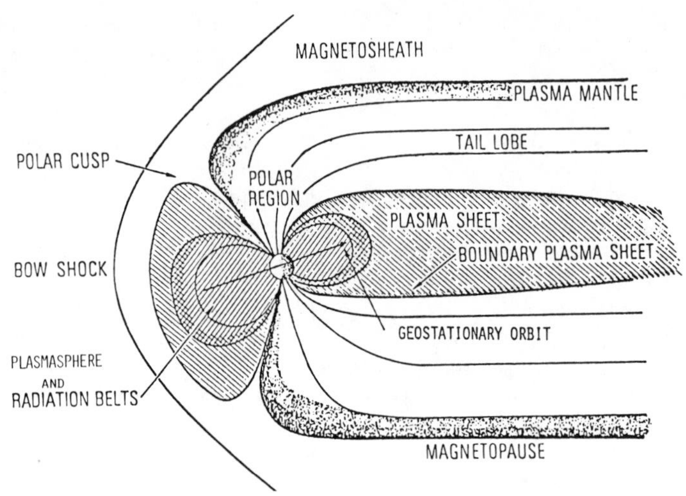

Fig. 1 Sketch of the Earth's magnetosphere (from Mizera and Fennell, 1978).

down voltages, direct electrical component damage from transients, disruption of logic and switching circuits from electromagnetic interference, change of reflective or thermal control surfaces due to the attraction of outgassed contaminants or pitting, and shock hazards for extravehicular and docking activities (see DeForest, 1972; Pike and Bunn, 1976; Shaw et al., 1976).

We will show that, under substorm conditions, the kapton substrate contemplated for use as a support blanket for the reflectors and solar cells will be subjected to near-breakdown voltage. Additional kapton insulation seems unfeasible because of weight considerations. The alternatives, higher conductivity substrates or conducting leads to the surfaces, seem more reasonable since the resulting parasitic currents are not excessive. The paper also will discuss the optimum point for grounding the spacecraft to the solar panels and outlines a method of using judicious routing of bus-bar currents to shield the satellite from particle bombardment. Although it is possible to use a similar method to magnetically align the satellite with the Earth's magnetic field (counteracting gravity-gradient torques), the fields required seem unreasonably large.

Spacecraft Charging

A body immersed in a plasma will acquire a net charge from unequal fluxes of plasma particles. For most plasmas, the electron and ion densities N_e and N_i are roughly equal, and the electron and ion temperatures T_e and T_i are comparable. Thus the electron flux J_e (proportional to $N_e \sqrt{kT_e/M_e}$) is generally much larger than the ion flux J_i , and the body acquires a negative charge sufficient to bring the currents into balance. For stationary, isothermal, singly-charged plasmas, the equilibrium unlit body potential is roughly $(kT_e/e)Ln(J_e/aJ_i)$ (Whipple, 1965), where a is a parameter (of order unity) depending on the thickness of the sheath. Exposing the body to sunlight causes photoelectrons to be ejected. For most substances, the photoelectron current is on the order of one to four nanoamps per square centimeter. Since this is comparable to or larger than most space plasma electron currents, the surface will tend to acquire a small positive charge. The actual equilibrium potential will depend on the details of the ion and electron distribution function, however (Whipple, 1976). The fluxes to a sunlit plate immersed in a plasma are shown schematically in Fig. 2. The lit side will tend to charge slightly positive, and the dark side negative.

The NASA MSFC baseline design (Hanley, 1978) is shown in Fig. 3. The surfaces on the satellite are divided into two types: active and passive, depending on whether or not voltages appear on the surface as a result of the satellite's own power supply. Passive surfaces include the solar reflectors and structural members. Active surfaces include the solar cells, interconnects, and bus bars. Active sur-

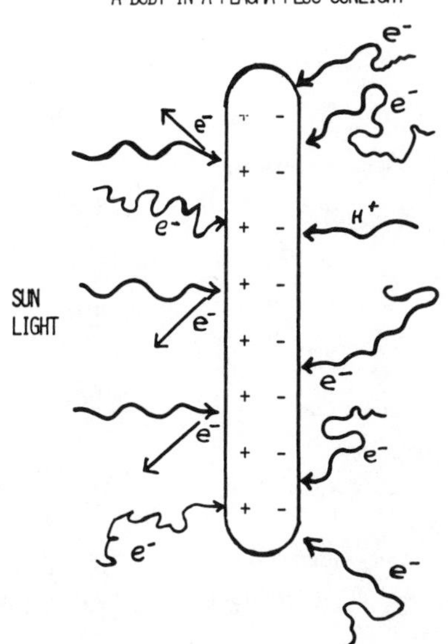

Fig. 2 Schematic of plasma and photoelectron currents.

faces may attract or repel the ambient ions or electrons depending on the polarity of the surface voltage. Currents reach the passive surfaces only by photoemission and the thermal motion of ions and electrons. (We ignore backscattered and secondary electrons.)

Calculation of Potentials

We make the simplifying assumption of a thin sheath (or 1-dimensional) approximation, i.e., the area collecting plasma is the actual geometrical area of the satellite (no focussing considered). The ambient electron and ion currents, therefore are, simply the thermal currents, given by

$$J_{i,e} = \frac{Ne}{4} \left(\frac{8kT}{\pi M}\right)^{1/2} \quad (1)$$

where N, e, T, and M are the number density, charge, temperature and mass for electrons or protons, depending on which current is calculated.

Parker (1979) has addressed the problem of a large flat-plate solar collector in space. He has found that the thin-sheath approximation is not valid at geosynchronous orbit for active structures. However, in the MSFC design, the passive, grounded reflecting panels form a trough in which the solar cells lie. Since the reflectors are conducting,

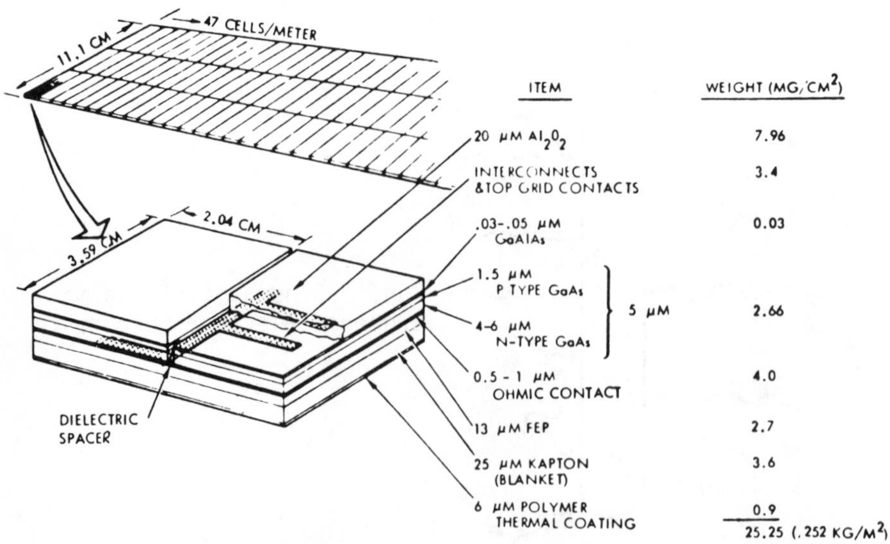

Fig. 3 Sketch of the MSFC January 25, 1978, baseline design (from Hanley, 1978).

they have a tendency to confine electric fields from the solar cells within the trough. This reduces the thick-sheath focussing effect because the electric fields do not penetrate significantly into space above the trough, and the reflectors themselves are barriers against plasma fluxes entering from the sides of the trough. Later in the paper we verify this assumption by showing results from a modified version of Parker's PANEL program for the special geometry of the MSFC design.

The analytic calculations below assume, for simplicity, an intermediate sheath approximation; i.e., no focussing of outside plasma is considered, yet the sheath is large enough that photoelectrons from the reflectors can impact the solar cell, and vice versa.

For GEO, our assumed "worst case" plasma conditions are: $N_e = N_i = 2/cm^3$, $kT_e = 5$ keV and $kT_i = 10$ keV (Inouye, 1976). This yields $J_e = 3 \times 10^{-10}$ A/cm^2 and $J_i = 1 \times 10^{-11}$ A/cm^2.

The photoelectron current density was calculated by integrating the product of the photoelectron yield function for synthetic sapphire and the solar spectrum: the resulting photocurrent density J_{pe} is 3×10^{-9} A/cm^2. A similiar calculation for aluminum yields roughly the same photoelectron current density.

It is apparent, then, that the photoelectron current will usually dominate for all sunlit surfaces at GEO. The equilibrium potential for such surfaces will be on the order of a few times the average photoelectron energy, from about 1 to 100 V positive, such as is found on the dayside of the moon (Reasoner and Burke, 1972; Freeman and Ibrahim, 1975). Passive sunlit surfaces will attain this voltage; however, for active surfaces, the finite conductivity of the cover surfaces (kapton and sapphire) will prevent this voltage from being obtained, i.e., the surface potential will more nearly follow that of the underlying solar cell.

Nightside potentials are estimated from Chopra's (1961) equation:

$$\phi \simeq - \frac{kT_e}{2e} \ln \left(\frac{M_i T_e}{M_e T_i}\right) \qquad (2)$$

For the "worst case" described above, this implies a dark-side potential of -17,000 V. Secondary electron emission or backscattering will reduce this potential somewhat. Again, passive surfaces will attain this voltage, but most active surfaces will be more nearly the potential of the underlying solar cell.

The most vulnerable active surfaces on the satellite are the solar cells because the ohmic contacts are separated from the plasma by only tens of micrometers of shielding. Figure 4 shows the dimensions and structure of the solar cell selected in the MSFC design. The GaAlAs cell is supported from below by a kapton blanket and is covered with synthetic sapphire. The sapphire coverglass is 20 μm thick and the kapton blanket is 25 μm thick.

For our study, the solar cell was idealized as a sapphire - active region - kapton sandwich as shown in Fig. 5. Plasma ions were assumed to be attracted to the negatively biased porton of the solar array and plasma electrons to the positively biased portion. Photoelectrons were assumed to leave the negative surface and be attracted to the positive surface. Secondaries were neglected. The currents used were those described previously; we assume a steady state condition. In this case the voltages across the sapphire and kapton dielectrics are the photoelectron and plasma currents multiplied by the resistance of the dielectrics. The assumed resistivity of sapphire is 10^{12} ohm-cm. Based on the measurements of Kennerud (1974) we have approximated the

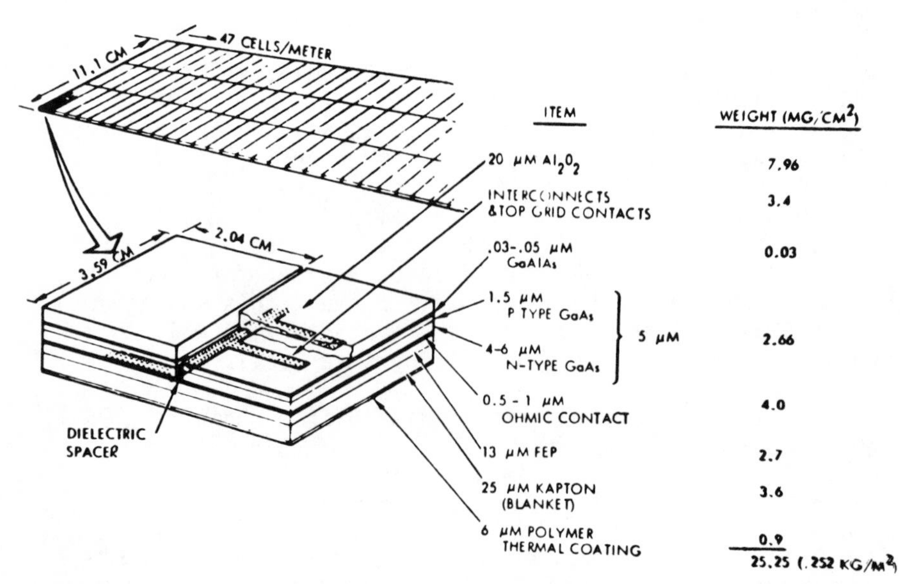

Fig. 4 Cross-section of a proposed GaAlAs solar cell (from Hanley, 1978).

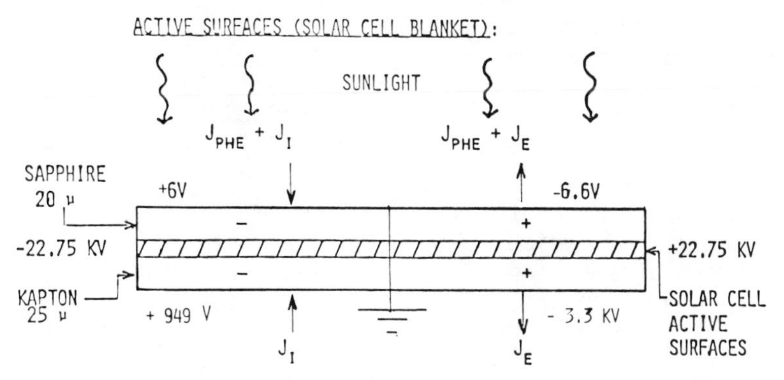

Fig. 5 Idealization of the solar cell blanket, used in calculations of electrostatic potential, for the "worst case" plasma fluxes.

resistivity of kapton by

$$\rho = 9.2 \times 10^{16} \exp -[E/1.1 \text{ KV/mil}] \text{ ohm-cm},$$

where E is the electric field across the kapton in KV/mil. The transcendental equation for the potential difference, V, through the 1-mil kapton layer is $\ln [V/K] = -V/1100$, where K is proportional to the current ($K = 9 \times 10^{16} \times$ thickness (cm) $\times$ current (A/cm^2)). This equation was solved numerically. The resulting voltages are shown on Fig. 5: a drop of 949 V through the ion-attracting side, and a drop of 3.3 KV through the electron-attracting side. In no case are the breakdown voltages exceeded; however, the voltage on the positive array is within a factor of 2 of the breakdown voltage. For an electron current ten times larger (which can certainly occur within the satellite's life-span), the voltage drop is 5.4 kV, which is near breakdown. For this reason, we recommend replacing kapton with a higher conductivity material, or else providing a current path from the solar cell to the back side. Conductive coatings will also help reduce spot arcing (McCoy and Konradi, 1979).

Kennerud (1974) and others have found anomalous arcing when solar panels are held at high voltage negative in a plasma. Typical voltages and currents required for such anomalous arcing to take place are 400 volts at 1×10^{-7} A/cm^2. Our expected ion currents to the negative portion of the solar array at GEO are 1×10^{-11} A/cm^2. Therefore, we do not anticipate anomalous arcing in the GEO environment.

The MSFC design calls for the reflectors to be constructed from 0.5 mil (12.5 μm) kapton covered with a 400 Å film of aluminum. We expect the aluminized front side potential to be fixed at 1 to 100 volts positive by photoelectron emission. Using the same analysis that was applied to the kapton solar cell blanket, we calculate the reflector back side voltage to be approximately -1.7 kV for our standard "worst case" condition, and 2.7 kV for a ten times larger electron current. The breakdown voltage for 1/2 mil kapton is 3.1 kV, which could be reached with only slightly more severe plasma conditions. Clearly, the backside must also be conducting and electrically connected to the front, or the kapton must be replaced with a higher conductivity material. A summary of the expected voltages on various surfaces during sunlit and eclipse conditions is shown on Fig. 6.

Optimizing the Grounding Point

The currents between the satellite and the plasma will adjust until the net current is zero. This means that the flow of current to the positively biased areas must equal that from the negatively biased areas. In a flat-plate collector, the balance is between plasma electron currents to the positive portions and plasma ion currents to the negative portions of the array. Since the plasma electron currents are so large, the plate will "float" substatially negative, i.e., the area of the collector with negative

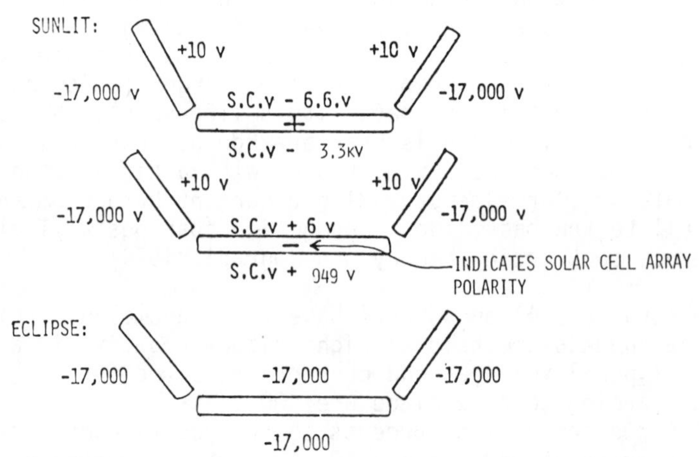

Fig. 6 Summary of voltages on the reflectors and solar cells surfaces, for solar cells at large positive voltages (top), large negative voltages (middle), and during eclipse (bottom). (Midpoint of the solar cell voltage string is assumed to be grounded to the sunlit side of the reflectors.)

potential is much larger than the corresponding positive potential area (Parker, 1979).

In the MSFC design, however, the large aluminum reflectors are also sources and sinks of photoelectrons. Photoelectrons from the reflectors will be attracted to positive portions of the solar cell array and photoelectrons from the negative portions of the solar cell array will be attracted to the neighboring reflector (Figure 7). These electrons will "hop" along the surface (Pelizarri and Criswell, 1978), adding to the power drain. Thus the photoelectron current becomes the dominant parasitic current, at least in all but the most intense substorm environments.

The large aluminum reflectors make a convenient spacecraft ground, since the sunlit sides will remain a few volts positive with respect to space. To minimize the power drain, the solar cell array should drive no new currents through the reflectors to the plasma. Thus the reflector "ground" should be tied to the solar cells in an optimium way. Accurate calculation of the 3-dimensional electric field pattern and resultant power drain including effects of the space charge and secondaries is a formidable task; an oversimplifed argument follows. If A- is the solar cell area that is negative and A+ is the solar cell area that is positive, current balance requires

$$(A-)\ (J_{pe} + 2J_i) = (A+)\ (J_{pe} + 2J_e) \qquad (3)$$

or,

$$A-/A+ = (J_{pe} + 2J_e)/(J_{pe} + 2J_i).$$

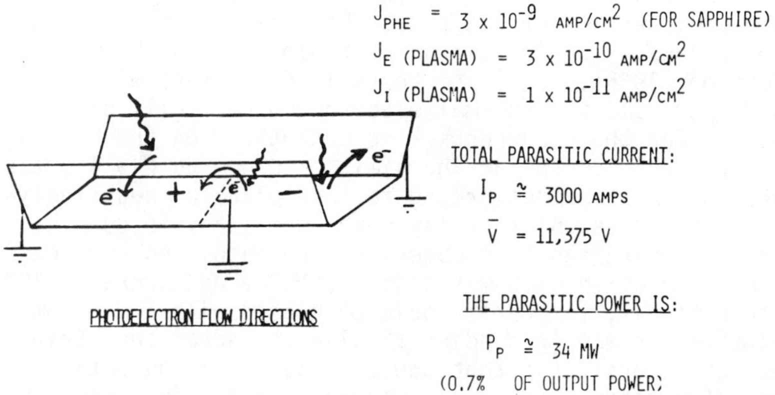

$J_{PHE} = 3 \times 10^{-9}$ AMP/CM^2 (FOR SAPPHIRE)
J_E (PLASMA) $= 3 \times 10^{-10}$ AMP/CM^2
J_I (PLASMA) $= 1 \times 10^{-11}$ AMP/CM^2

TOTAL PARASITIC CURRENT:
$I_P \cong 3000$ AMPS
$\bar{V} = 11{,}375$ V

PHOTOELECTRON FLOW DIRECTIONS

THE PARASITIC POWER IS:
$P_P \cong 34$ MW
(0.7% OF OUTPUT POWER)

Fig. 7 Summary of parasitic current densities for the SPS and the parasitic current and power loss total for one half of the Marshall satellite (5 GW system).

Here we assume that the photoelectron flux from the reflectors to the positive segments is approximately the same as the photoelectron flux from the negative segments to the reflectors. For low plasma-current regions, (e.g., the plasmasphere or the quiet plasmasheet) or for cases in which the plasma current is shielded from the surfaces magnetically, the ratio approaches unity. Even for our "worst case," the ratio is only 1.17. Therefore, we recommend grounding the midpoint of the string to the reflectors. On the other hand, at low Earth orbit plasma electron and ion ram fluxes dominate, and the grounding point must be more carefully calculated.

With the ground point determined, the parasitic load can be calculated. The principal parasitic current at GEO is from photoelectrons (Fig. 7), and is calculated to be about 3000 A . Coupled with an average potential drop of 11375 V, this implies a power loss of 34 MW, which is only 0.7% of output power, and is easily manageable by slight oversizing. This percentage power loss is comparable to that (∿ 0.1%) from a flat-plate collector (Parker, 1979). Thus optimizing the grounding point at GEO is not critical. As discussed later, however, at LEO optimization could be very important.

Currents at Low-Earth Orbit

An integral part of the SPS concept is the Earth-Orbit Transfer Vehicle (EOTV) which will transfer the SPS to GEO. It is expected to employ a high-voltage solar cell array and to operate primarily in the low-Earth orbit (LEO) environment where the plasma currents are considerably different than GEO. At 400 km altitude, the dominant ion is O^+ with a number density of $10^6/cm^3$ and a temperature of 2000 K (Johnson, 1965). Thus the thermal ion current will be 7 x 10^{-9} A/cm and the thermal electron current will be 3 x 10^{-7} A/cm . For these currents, the potentials on the EOTV will be comparable to the arcing limit of Kennerud (1974); therefore, one must expect arcing to take place on negatively-biased surfaces unless a lower-voltage array is used. Indeed, arcing has been observed from insulated surfaces in a LEO simulation vacuum tank test (McCoy and Konradi, 1979). Alternatively, the array could be biased with a minimum of negative surface (grounding the lowest end of the string to the reflectors), but that would be far from the optimum grounding scheme, and would increase parasitic losses by a factor of three.

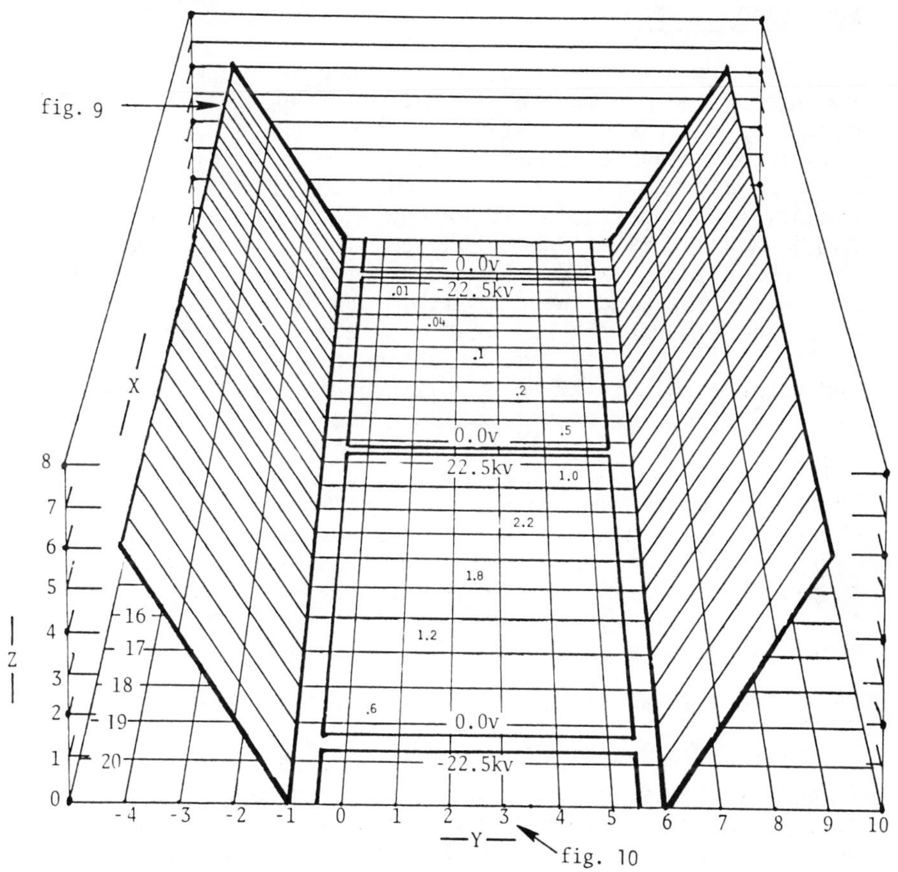

Fig. 8 Computer grid used to model 2 panels of the SPS. (Small numbers on the panel surface are the plasma electron currents normalized to random thermal currents.)

Spacecraft motion implies a substantial though varying ram flux which will cause an additional parasitic current drain of as much as 2×10^{-7} A/cm^2. Coupled with the current losses due to the thermal currents, the power loss could be as high as $\sim$3 %. As noted, however, arcing probably will occur at much lower potentials than those for which 3 % power loss would be observed. Parker (1979) has pointed out that sheath and wake effects also could substantially alter the satellite potentials and current flow.

A final source of parasitic current for the EOTV is thermal ions created by charge-exchange of neutrals with the ion thruster beam. These ions can be effectively shielded, however. For details see Freeman and Few (1979).

Non-Steady State

Until now it has been assumed that the charging currents from the plasma are steady. This approach is supported by a study of the time dependent charging of a three-axis stabilized spacecraft by Massaro et al., (1977). For all the surfaces modeled, they found that the greatest differential voltages occurred in the steady state limit, although nearly instantaneous changes in absolute potential were observed. However, in order to evaluate the effects of non-steady charging, we calculated the RC time constant or discharge time of the relevant insulators, sapphire and kapton. The RC decay time is $\rho\varepsilon$ where ρ is the resistivity and ε the permittivity. For kapton this implies a time constant of 1 hr; for sapphire, 1 sec Large magnetospheric changes can occur with 1 min - 1 hr time constants (McIlwain, 1974; Inouye, 1976). Therefore, high voltages can build up on the kapton in time intervals short compared to the discharge time. Transient charging is not expected to cause differential charging in excess of the steady state predictions; nevertheless, the large kapton time constant reinforces the previous conclusion that kapton should be replaced with a higher conductivity material.

Three-Dimensional Model

All of the foregoing analysis on parasitic loads, plasma induced voltages, etc., is based on one-dimensional plasma theory. More precise results require a three-dimensional self-consistent computer model which takes into account all plasma scurces and interactions with reflectors simultaneously. A computer program, "PANEL," written by Dr. Lee Parker (Parker, 1979), provided a convenient starting point for our model of the SPS environment. Preliminary results will be presented here. They are preliminary since we have not yet included the photoelectron current (which we showed to be important), nor have we as yet included space charge effects. Nevertheless, the results demonstrate several important features of the sheath around the SPS troughs.

PANEL utilizes a three-dimensional grid where the satellite is modeled by fixing potentials at selected grid points. Laplace's equation is then satisfied by relaxing the free space potentials until Gauss's law is satisfied for a box surrounding each point. The currents and power losses are obtained by numerically performing the integral

$$J = \int_0^\infty dv \int_0^{\pi/2} d\theta \int_0^{2\pi} d\phi \; f\,(v,\theta,\phi)\; v^3 \cos\theta \sin\theta$$

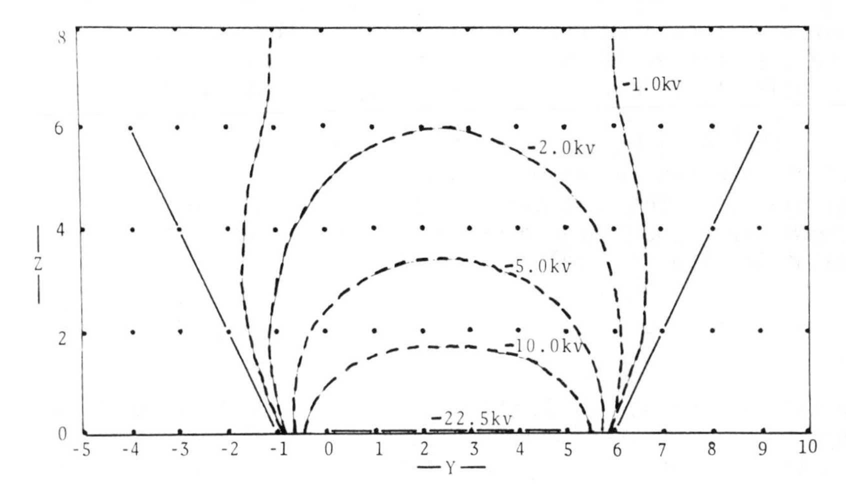

Fig. 9 Equipotential contours in the yz plane at x = 0 (indicated in Fig. 8).

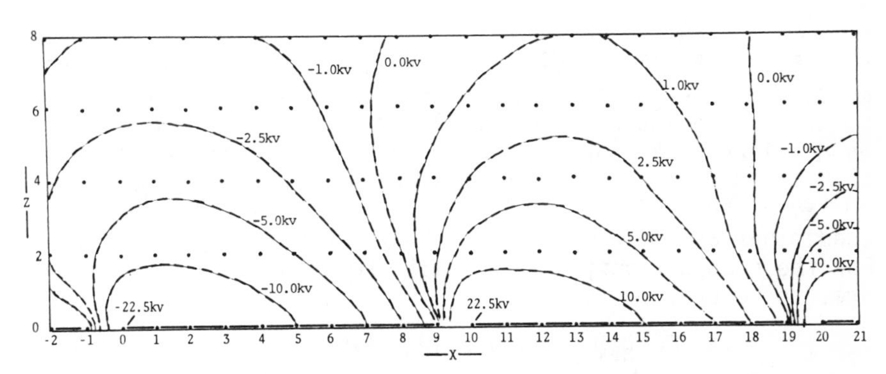

Fig. 10 Equipotential contours in the xz plane at y = 3 (indicated in Fig. 8).

where J is the current density, and f is the distribution function. The problem is then to evaluate f. For a collisionless steady state plasma, the Vlasov equation

$\vec{v} \cdot \vec{\nabla} f + \frac{1}{m} \vec{F} \cdot \vec{\nabla}_v f = 0$, states that a distribution function

is constant along a particle's path in phase space. If f is written in terms of a particle's total energy (E = T + V, the kinetic plus potential energy), f will be constant in E along the path in real space. The integral for J is then transformed into a sum using the method of gaussian gradu-atures which picks key values of E, θ, and ϕ. These values represent trajectories that are traced backwards to either source or nonsource regions to determine the value of f. Once the current is known it is multiplied by the local potential to determine the power loss at that point.

PANEL is a Laplacian calculation since space charge effects are not included in the electrostatic potential calculation. The next phase in the development of PANEL is to calculate the charge density for each point in space by evaluating the integral

$$N = \int_0^\infty dv \int_0^{\pi/2} d\theta \int_0^{2\pi} d\phi f(v, \theta, \phi) v^2 \sin\theta$$

in the same manner as described for the current calculation. Then PANEL must iterate between the potential relaxation routine and the density calculation since the density calculation depends upon the potential structure for accurate trajectories. This is known as the inside-out method (Parker, 1977) because trajectories are traced backwards in time.

Figure 8 illustrates the three-dimensional grid used to model two interior panels of a trough. Not shown are grid points at the intersection of all integer x and y values and even values of z. One unit of grid spacing corresponds to 85.0 meters, giving model dimensions of 765 m X 425 m. Fixed voltages are indicated on the figure. The assumed plasma conditions are $N_i = N_e = 2/cm^3$, $kT_i = 10$ keV, $kT_e = 5$ keV. For these conditions, the random thermal current densities are, as before:

$$J_{th,i} = 1.25 \times 10^{-7} \text{ A/m}^2$$

$$J_{th,e} = 3.79 \times 10^{-6} \text{ A/m}^2$$

The dimensionless numbers at selected points on the panels are ratios of local average electron current densities to the random electron thermal current. For the two panels modeled, PANEL traced 864 trajectories per grid square of surface. The resulting total current collected and power loss are 6.64×10^{-2} A, and 5.66×10^2 W for protons and 2.25 A and 2.72×10^4 W for electrons. Calculated potential patterns in the x = 0 plane and y = 3 plane are shown in Figs. 9 and 10, respectively. Note that potentials of only 1-2 kV extend beyond the upper limits of the trough, justifying our earlier "intermediate sheath" approximation.

Photoelectrons from the reflectors and backscattered and secondary electrons undoubtedly will be important contributor to the power loss but have not yet been modeled.

Magnetic Protection of the SPS

The SPS of necessity contains bus bars of current 10^5 A, routed between the solar panels and the microwave antennae. With judicious routing of these bus bars, the SPS can create its own protective magnetic barrier, screening out all the low energy (~100 eV) plasmaspheric plasma (which can cause power drain), and most of the energetic electrons. Parker and Oran (1979) have shown that this idea is feasible with nominal bus-bar currents. We propose modified bus-bar currents to prevent spacecraft fields from merging with the earth's magnetic field. Merging can have two harmful effects:

1) It can channel energetic particles trapped in the Earth's magnetic field towards sensitive areas of the SPS.

2) It can energize the high density plasmaspheric plasma that would otherwise be harmless.

Previous spacecraft were small in size compared to particle gyroradii, so magnetic effects were not important. The size of the SPS, however, is comparable to particle gyroradii, so magnetic effects must be taken into account. (At geosynchronous orbit, a 2 eV proton or 3 keV electron has a gyroradius of 2 km; a 50 eV proton or 80 keV electron has a gyroradius of 10 km.) In the following, in order to estimate these effects (i.e., to repel trapped particles and to minimize energy released in magnetic merging) we assume that it is important to have spacecraft magnetic fields parallel to sensitive areas (e.g., solar cells) and aligned

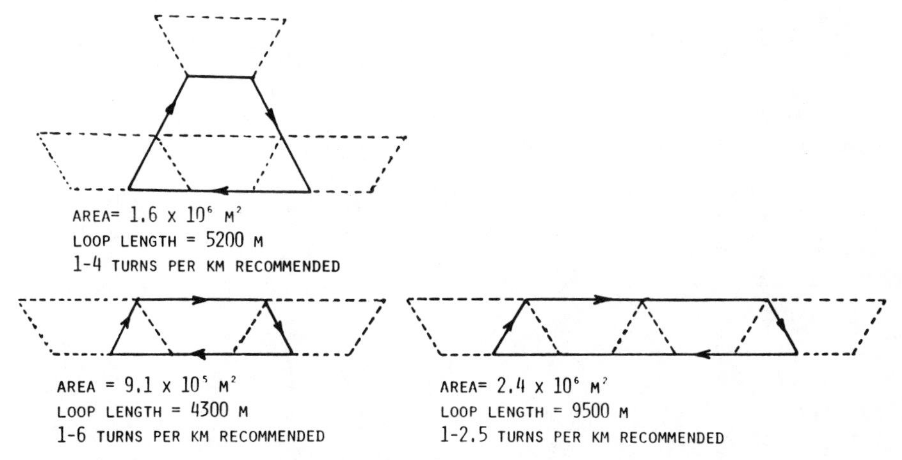

Fig. 11 Recommended current windings for several SPS configurations (view from north end).

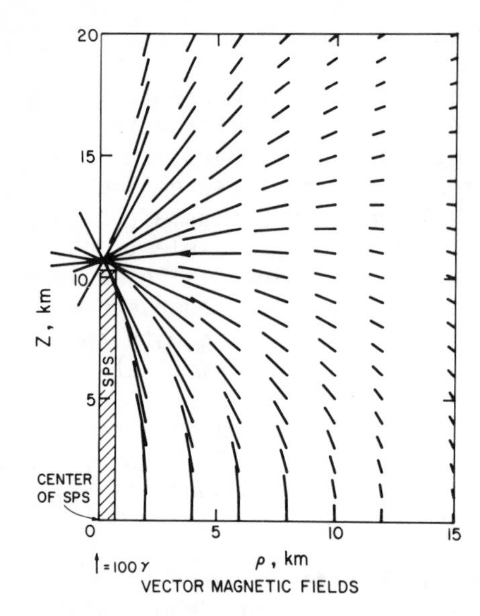

Fig. 12 Vector magnetic fields for a solenoidal current configuration, low-field case ($\mu = 10^{11}$ A-m^2 per km, 21 km total). (Z-axis is along the spacecraft (z = 0 is the center), and ρ is measured from the spacecraft axis; only one quadrant is shown, because of symmetry: $B_Z(-z) = B_Z(z)$; $B_\rho(-z) = -B_\rho(z)$.)

Table 1 Magnetic Moment Required for SPS Tasks

Task	Rigidity required	Orientation of moment	Internal field (Gauss)	Required moment, A-m^2/km†
Shielding 200 eV protons and 30 keV electrons	2×10^3	*Parallel Antiparallel	1.3 4	1×10^{11} 3×10^{11}
Shielding 3 KeV protons and 2 MeV electrons	8×10^3	*Parallel *Antiparallel	5.3 11	4×10^{11} 8×10^{11}
Shielding 30 KeV protons 10 MeV electrons	3×10^4	Parallel *Antiparallel	20 25	1.5×10^{12} 2×10^{12}
Magnetic alignment (balance gravity-gradient	N/A	*Antiparallel	92	7×10^{12}

*Recommended Orientation
†Multiply by 21 for total magnetic moment.

with the Earth's magnetic field. (Even magnetic fields perpendicular to the surface can be beneficial, however, and have been considered in Parker and Oran, 1979.)

A solenoidal bus-bar winding yields the best magnetic field configuration: at a distance, the field approaches that of a dipole, and in the vicinity of the satellite the field is parallel to the solar panels. The windings for the solenoid should enclose as much area as feasible. This will have two benefits: it will maximize the overall dipole moment while minimizing the bus bar length and thus IR losses, and will minimize the internal field. On the other hand, for spatial uniformity, one should have at least one turn per kilometer. Some possible cross-sections are shown in Fig. 11. This figure is a view from the north end of three types of trough-like SPS design and shows one turn of the helical winding each.

The field of the SPS must have sufficient rigidity to successfully deflect the species desired to be excluded. Table 1 shows magnetic moments μ required for various tasks. Two possible orientations of the SPS's dipole moment are compared: parallel or antiparallel to the Earth's dipole moment. A parallel orientation, since it adds to the local magnetic field, is more efficient at shielding the SPS from particle bombardment; however, the opposite orientation is dynamically more stable, since the SPS's moment will tend to align with the Earth's magnetic field. In fact, the moment may be used to balance gravity-gradient torques if the dipole moment is large enough. For a (uniform) body 22 km long and 4 km wide of mass 5×10^7 kg, the moment of inertia about an axis perpendicular to the length of the satellite would be 2×10^{12} kg-m^2 . The daily $\pm$ 10 deg tilt of the geosynchronous magnetic field would cause a torque on the satellite of $(\underline{\mu} \times \underline{B}) = 1.7 \times 10^4$ Nt-m, for a $\mu = 10^{12}$ A-m^2 (corresponding to 0.9 Nt of force on each end). Since the satellite is so massive, this torque will result in a daily sinusoidal tilting motion of the satellite of amplitude $\sim 10^{-5}$ degrees, completely negligible. A 10 deg tilt of the satellite toward the Earth, in contrast, will cause a gravity-gradient torque of 2.7×10^6 Nt-m, or 125 Nt at each end, requiring a magnetic moment of 1.5×10^{14} A-m^2 to balance it. Then, however, the 10 deg misalignment between the spin axis and the dipole axis of the Earth would become more important. In addition, the magnetic fields in the SPS center would be quite large (90 G.). The internal field is sensitive to the exact configuration, and can vary by a factor of two or so depending on the area and number of turns per km. The rigidity, on the other hand, is not too sensi-

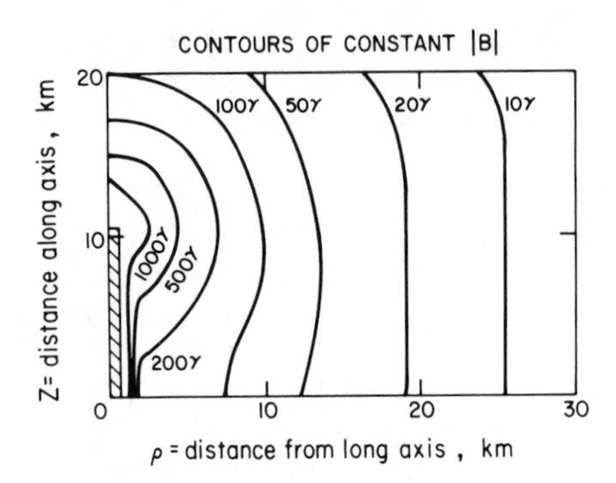

Fig. 13 Contours of constant $|B|$ for the low-field case. Only one quadrant is shown.

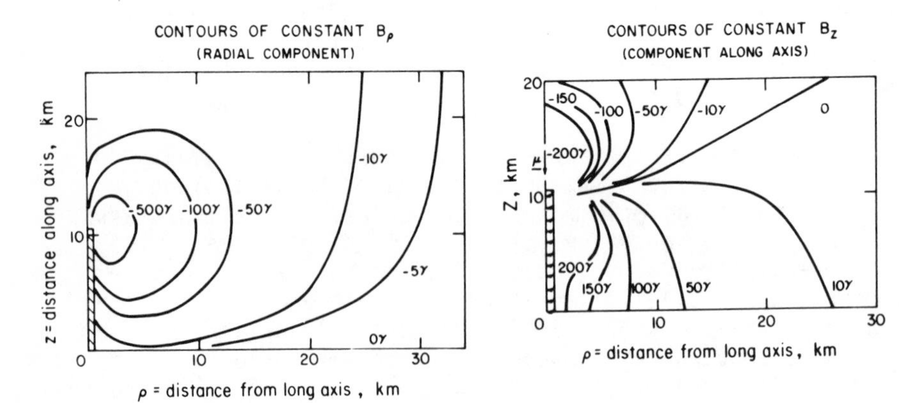

Fig. 14 Contours of constant B_ρ and B_Z, low-field case.

tive on the exact configuration, being mainly a function of overall magnetic moment.

One reasonable magnetic field configuration is shown in Figs. 12 - 14. The dipole moment assumed for these figures is the low-field case, 10^{11} A-m^2 per km, 21 km total. All components of the field are, of course, linear in the dipole moment. This model superposes 21 dipoles at 1 km intervals (simulating one turn per km). Figure 12 shows vector magnetic fields for one quadrant; Fig. 13 shows contours of constant $|B|$, and Fig. 14 shows magnetic field components. Here the z-component is measured along the long axis and the ρ component is measured from the long axis. The center of the SPS is the lower left corner ($z = 0$, $\rho = 0$). Only one quadrant is shown because of symmetry: $B_Z(z) = B_Z(-z)$; $B_\rho(z) = -B_\rho(-z)$. The field is similar to that of a solenoid and is nearly parallel to the long sides of the SPS (and therefore to the solar cells), converging at the SPS's north and south ends. (The SPS is aligned north-south to minimize the shadowing of one SPS on another in the equinox seasons.)

The field in Figs. 12 - 14 is strongest at the ends and weakest in the center; therefore, fewer wraps (or, more likely, less current per wrap) could be used at the ends and still obtain the same overall rigidity. A field of 100 extends to over 7 km from the center, and a field of 20 extends to 19 km. The overall rigidity at $\rho = 1$ km, $z = 0$ km is roughly 2000γ - km (G-cm). With a magnetic field of this orientation and strength, ions < 200 eV (including all the plasmaspheric plasma) and electrons < 30 keV (most of the plasma sheet electron fluxes) are excluded. Higher dipole moments would yield more shielding (see Table 1). Thus, it appears that magnetic protection is feasible. Because of the convergence of the field, particle fluxes will have a tendency to strike only the ends of the long axis of the SPS. Simply capping the ends of the SPS, then, will be sufficient to protect electronics and humans inside from the lower-energy particles. Such capping is also useful to prevent the plasma from the ion engines from returning to the satellite, causing a significant power drain (Freeman and Few, 1979).

Conclusions

The SPS will certainly interact with its plasma environment. It appears that, with relatively minor modifications to the NASA MSFC baseline design, these interactions will not significantly impair SPS operations. The conclusions and recommendations of this study include:

1) Arcing is likely to occur on kapton surfaces (the solar reflectors and the solar cell back surface blanket) during substorms unless the kapton is replaced by a lower resistivity material ($\rho \leq 10^{13}$ ohm-cm) or current paths from the surfaces to the solar cells are provided.

2) The SPS parasitic load under normal conditions will be about 34 MW (for a 5 GW array) at geosynchronous orbit. This 0.7 % power loss is unimportant.

3) The optimum grounding point at GEO for the SPS solar cell array is approximately the midpoint on the voltage string. At LEO, arcing considerations demand that the string be biased mostly positive, although the optimum configuration to minimize power loss would be substantially negative (see conclusion 5).

4) The solar cells may require conductive coatings. The reflector panels may require current paths linking the front and back sides.

5) Severe arcing problems are expected for negative portions of the EOTV solar cell array at LEO. Overcoming this problem by biasing the array as positive as possible will result in high parasitic loads (power losses on the order of 3 %).

6) The SPS will occasionally charge to about -20 kV during eclipses.

7) Three-dimensional computer modeling of the SPS electric field pattern and plasma currents is underway. The model shows that, for the grounding scheme used here, spacecraft electric fields extend only slightly beyond the reflectors.

8) Active magnetic plasma shielding is possible through judicious routing of bus-bars; power drain from additional lengths of bus-bars has not been calculated yet.

9) It is possible to use the internal magnetic field to align the satellite (counteracting gravity-gradient torques), but it would require an unreasonably large magnetic moment (1.5×10^{14} A-m^2).

Acknowledgements

The authors thank Dr. Lee Parker for consultation and the use of the computer program "PANEL". In addition, we have benefited from discussion with Dr. James McCoy. This work was supported by NASA under grant NAS8-33023, and by the Brown Foundation.

References

Burke, W. J., Reiff, P. H., and Reasoner, D. L., "The Effect of Local Magnetic Fields on the Lunar Photoelectron Layer While the Moon Is in the Plasma Sheet," Geochemicha Cosmochemicha Acta, Suppl. 6, Vol. 3, 1975, pp. 2985-2997.

Chopra, K. P., Reviews of Modern Physics, Vol. 33, 1961, p. 153.

DeForest, S. E., "Spacecraft Charging at Synchronous Orbit," Journal of Geophysical Research, Vol. 77, February, 1972, pp. 651-659.

Freeman, J. W. and Few, A. A., "Electrostatic Protection of the Solar Power Satellite And Rectenna," final report, NASA contract NAS8-33023, Rice University, Houston, Texas, 1979.

Freeman, J. W. and Ibrahim, M., "Lunar Electric Fields, Surface Potential And Associated Plasma Sheets, The Moon, Vol. 14, 1975, pp. 103-114.

Garrett, H. B. and DeForest S. E., "Analytic Simulation of the Geosynchronous Plasma Environment," Planetary Space Science, Vol 26, (in press), 1979.

Hanley, G. M., "Satellite Power Systems (SPS) Concept Definition Study," final report for NASA contract NAS8-32475, Rockwell International Report #5078-AP-0023, Cincinnati, 1978.

Inouye, G. T., "Spacecraft Potentials in A Substorm Environment," Progress in Astronautics and Aeronautics: Spacecraft Charging by Magnetospheric Plasmas, ed. by A. Rosen, AIAA, New York, 1976, pp. 103-120.

Johnson, F. S (ed). Satellite Environment Handbook, Stanford University Press, Stanford, California, 1965.

Kennerud, K. L., "Final Report High Voltage Solar Array Experiments," Boeing Aerospace Corp., Seattle, Wash., Rept. No. CR121280, 1974.

Massaro, M. J., Green, T., and Ling, P., "A Charging Model for Three-axis Stabilized Spacecraft," Proceedings of the Spacecraft Charging Technology Conference, edited by C. P. Pike and R. R. Lovell, Air Force Rept. AFGL-TR-77-0651, 1977, pp. 237-267.

McCoy, J. E. and Konradi, A., "Sheath Effects Observed on A 10 Meter High Voltage Panel in Simulated Low Earth Orbit Plasma," Space Craft Charging Technology-1978, edited by R. C. Finke and C. P. Pike, NASACP 2071, 1979, pp. 315-340.

McIlwain, C. E., "Substorm Injection Boundaries," Magnetospheric Physics, edited by B. M. McCormac, D. Reidel, Hingham, Mass., 1974, p. 143.

Mizera, P. F. and Fennell, J. F., "Satellite Observations of Polar, Magnetotail Lobe, and Interplanetary Electrons at Low Energies," SAMSO Rept TR-78-6, 1978.

Parker, L. W., "Plasma Sheath Effects and Voltage Distributions of Large High-power Satellite Solar Arrays," Spacecraft Charging Technology-1978, edited by R. C. Finke and C. P. Pike, NASA CP 2071, 1979, pp. 341-357.

Parker, L. W. and Oran, W. A., "Magnetic Shielding of Large High-power-satellite Solar Arrays Using Internal Currents," Spacecraft Charging Technology-1978, edited by R. C. Finke and C. P. Pike, NASA CP 2071, 1979, pp. 376-387.

Parker, L. W., "Calculation of Sheath and Wake Structure about A Pillbox-shaped Spacecraft in A Flowing Plasma," Proceedings of the Spacecraft Charging Technology Conference, edited by C. P. Pike and R. R. Lovell, Air Force Report AF6L-TR-77-0051, 1977, pp. 331-336.

Pelizzari, M. A. and Criswell, D. R., "Differential Photoelectric Charging of Nonconducting Surfaces in Space," Journal of Geophysical Research, Vol. 83, November, 1978, pp. 5233-5244.

Pike, C. P. and Brun, M. H., "A Correlation Study Relating Spacecraft Anomalies to Environmental Data," Progress in Astronautics and Aeronautics: Spacecraft Charging by Magnetospheric Plasmas, edited by A. Rosen, AIAA (New York), 1976, pp. 45-60.

Reasoner, D. L. and Burke, W. J., "Characteristics of the Lunar Photoelectron Layer," Journal of Geophysical Research, Vol. 77, December, 1972, pp. 6671.

Reasoner, D. L., Lennartsson, W., and Chappell, C. R., "Relationship Between ATS-6 Spacecraft-charging Occurrences and Warm Plasma Encounters," Progress in Astronautics and Aeronautics: Spacecraft Charging by Magnetospheric Plasmas, edited by A. Rosen, AIAA, New York, 1976, pp. 89-101.

Reiff, P. H., "Magnetic Shadowing of Charged Particles by An Extended Surface," Journal of Geophysical Research, Vol. 81, July, 1976, pp. 3423-3427.

Reiff, P. H. and Burke, W. J., "Interactions of the Plasma Sheet with the Lunar Surface at the Apollo 14 Site," Journal of Geophysical Research, Vol 81, September, 1976, pp. 4761-4764.

Shaw, R. R., Nanevicz, J. W., and Adamo, R. C., "Observations of Electrical Discharges Caused by Differential Charging," Progress in Astronautics and Aeronautics: Spacecraft Charging by Magnetospheric Plasmas, edited by A. Rosen, AIAA, New York, 1976, pp. 61-79.

Whipple, E. C., Jr., "The Equilibrium Electric Potential at A Body in the Upper Atmosphere and in Interplanetary Space," NASA T. N. X-615-65-296, 1965.

Whipple, E. C., Jr., "Theory of the Spherically Symmetric Photoelectron Sheath: A Thick Sheath Approximation and Comparison with ATS6 Observation of a Potential Barrier," Journal of Geophysical Research, Vol. 81, February, 1976, pp. 601-607.

ION THRUSTER PLASMA DYNAMICS NEAR HIGH VOLTAGE SURFACES ON SPACECRAFT

H. B. Liemohn,* D. H. Holze,+
W. M. Leavens,+ and R. L. Copeland≠

Boeing Aerospace Company, Seattle, Wash.

Abstract

Ion thrusters are being considered for propulsion of future spacecraft because of their high propulsion efficiency and easy access to solar power for long-duration missions. Operation of such thrusters requires high spacecraft potentials, and their local electric fields can draw return current from the thruster plasmas, reducing system efficiency. In this paper, the thruster plasma is assumed to be described by a collimated energetic beam ($\sim$10 keV) and a cloud of ionized thermal propellant ($\sim$10 eV) produced by charge-exchange. A simple adiabatic model is used to describe the expansion of these neutral plasmas away from the source. As the pressure falls shielding currents dissipate, and the geomagnetic field takes control of the particles. Results appropriate to proposed electric propulsion missions and the solar power satellite are presented, and operational considerations are discussed.

I. Introduction

Spacecraft requirements for long-duration missions and minimum vehicle weight have enhanced interest in ion thrusters that operate from solar energy. Although these propulsion systems have low thrust, their high propulsion efficiency during prolonged missions makes them strong candidates for interplanetary explorations to comets and asteroids and as a

---

Presented as Paper 79-2105 at the AIAA/DGLR/Princeton University 14th International Electric Propulsion Conference, Oct. 30-Nov. 1, 1979.

*Manager, Physics Technology, Boeing Aerospace Company.

+Staff Member, Physics Technology, Boeing Aerospace Company.

≠Now Staff Member, Fusion Energy Division, Oak Ridge National Laboratory.

means of station keeping for geosynchronous payloads. They are also attractive for orbit-transfer propulsion when time is not a primary limitation.

A number of small thrusters have been built and tested in space and the laboratory.[1-4] Somewhat larger propulsion modules have been conceived for future large-scale applications.[5,6] All of these thrusters generate a thermal plasma in a chamber and accelerate the ions across closely-spaced grids. The high acceleration voltage collimates the ion beam, and an adjacent electron source provides immediate neutralization. A secondary source of plasma at the outlet results from charge exchange between the beam ions and escaping neutral atoms. Approximately 15% of the beam charge is transferred to these thermals, which represent a significant source of local current. Some operational characteristics of these thrusters are summarized in Table 1.

The plasma environs around the spacecraft consist of both thruster exhaust and natural background. Some characteristics of these plasmas that are needed in the following analysis are presented in Table 2. (The 10 eV beam thermal energy in Table 2 is random motion superimposed on the 10 keV directed beam energy.) Only operations in low Earth orbit (LEO) around 400 km altitude, and geosynchronous Earth orbit (GEO) at 6.6 Earth radii geocentric are considered here to demonstrate the effects. The spacecraft is assumed to be following a circular trajectory so that its speed relative to the background magnetic field is 7.7 km/sec at LEO and zero at GEO. At LEO the natural plasma density is adequate to shield the spacecraft fields from the thruster plasma. However, the tenuous conditions at GEO allow fields to extend well beyond the vehicle dimensions.

For ion propulsion to operate efficiently from solar energy, a high-voltage solar array is required to avoid the extra weight of power converters. The local electric fields around these panels can attract the thruster plasma if Debye shielding by the ambient plasma is inadequate. Additionally, as the thruster plasma disperses, the geomagnetic field eventually takes control of the particles, and some geometries permit orbits to fold back on the vehicle surface. Such return current diminishes the effectiveness of the overall system by inhibiting propulsion and/or leaking power from the solar energy collectors.

The purpose of this paper is to investigate the interaction of the thruster plasma with its parent power source. The approach is to understand the plasma behavior qualita-

Table 1 Nominal ion thruster characteristics[5-8]

Source grid diameter, cm	30	100
Propellant gas	Hg	A
Beam current, A	2	80
Acceleration voltage, kV	1.1	10^a
Exit plane beam density, ions/cm^3	$\sim 5 \times 10^9$	
Exit plane thermal density, ions/cm^3	$\sim 10^{11}$	
Thermal ion current, A	0.2	~8
Propulsion efficiency, %	70	80
Thrust, N	0.13	6

[a]Solar power satellite station keeping operation at geosynchronous altitude.

Table 2 Plasma environment parameters[8-10]

		Natural plasma		Argon thruster plasma	
		Quiet	Substorm	Beam	Thermal
Ion density, ions/cm^3	LEO[a]	$10^5 - 10^6$	...	...	...
	GEO[b]	1 - 5	0.1 - 1	...	...
Ion thermal	LEO[a]	0.1	...	10	10
	GEO[b]	~1.0	~7000	10	10
Ion mean free path, km	LEO[a]	~10	...	$\sim 10^2$	10
	GEO[b]	$\sim 10^6$	...	$\sim 10^7$	10^3
Ion Larmor radii, km	LEO[a]	0.004	...	4	0.1
	GEO[b]	1	...	900	30

[a]LEO at 400 km.
[b]GEO at 36,000 km.

tively and make quantitative assessments where return current may be significant. The hope is to identify modes where there are negligible system losses, to determine characteristics of modes where return current levels impact the system efficiency, and to explore ways to modify inefficient configurations. In Section II a model is developed for the expansion characteristics of the energetic beam and thermal plasma. Section III describes the role of the geomagnetic field. The spacecraft field effects are treated in Section IV. Preliminary results of some of this work have been reported earlier.[11]

II. Thruster Plasma Expansion

Rather than solve the nonlinear fluid equations numerically for equilibrium conditions, we present a less complicated approach in which the trajectory equation for the average plasma column radius with time is derived analytically from conservation of energy and adiabatic constraint equations. The adiabatic fluid equations are used because they often are a valid approximation for collisionless systems as discussed in Appendix 1. The only requirement is that the third moment of the distribution function around the mean expansion speed must be small. This, along with conservation of particles, yields an equation for the average plasma density with time. It is solved only for the ion motion, assuming that there exists a charge-neutralizing background of electrons that follows the ions.

Expansion around the beam motion occurs at a speed $\overline{v}$, which is derived from the conservation of energy,

$$1/2\ M\ \overline{v}^2 = k(T_0 - T) \tag{1}$$

where M is the ion mass, $\overline{v}$ is the local mean expansion speed, k is Boltzmann's constant, T_0 is the thermal ion temperature at the source, and T is the local average ion temperature along the beam. (T decreases from its initial value T_0.) In this model T varies radially but not axially. The adiabatic equation is

$$T/N^{\gamma-1} = T_0/N_0{}^{\gamma-1} \tag{2}$$

where N is the plasma density, N_0 is the source density, and γ has the value five-thirds here.

Beam Plasma

The thruster beam ions are ejected with a mean speed much greater than the transverse expansion speed. Viewed from a reference frame fixed in the beam (the "beam frame"), the vehicle passes very quickly, moving many beam radii before the beam expands one beam radius. At any particular point on the beam, the expansion is approximately cylindrical. Over longer distances, the thruster motion is accounted for by letting the expansion time be a function of position along the beam.

Radial expansion is modelled by the adiabatic fluid equations. Rather than solve the full nonlinear equations, we assume that the beam has a density profile which expands but maintains its shape at all times. Then we solve for the characteristic beam radius R(t), where t is the time since injection of the local beam (i.e., the time since the particular section of the beam was emitted from the thruster).

Conservation of particles requires that

$$N/N_{oB} = [r_o/R(t)]^2 \tag{3}$$

where N_{oB} is the exit plane beam density and r_o and R(t) are the beam radii at the thruster and at a point downstream from the thruster where u is the local axial beam velocity. Equations (1) - (3) yield the expression

$$\overline{v} = \frac{dR}{dt} = \sqrt{\frac{2k\ T_o}{M}\left[1 - \left(\frac{r_o}{R}\right)^{4/3}\right]} \tag{4}$$

or

$$v_o\ t = \int_{r_o}^{R} \frac{dR'}{\left[1 - \left(\frac{r_o}{R'}\right)^{4/3}\right]^{1/2}} \tag{5}$$

where v_o is the rms ion source speed defined by

$$\frac{1}{2} M\ v_o^{\ 2} = k\ T_o \tag{6}$$

With the transformation

$$\sec\ \theta = (R/r_o)^{2/3} \tag{7}$$

Equation (5) can be put in the form

$$v_o\ t = \frac{3\ r_o}{2} \int_0^{\cos^{-1}(r_o/R)^{2/3}} \frac{d\theta}{\cos^{5/2}\theta} \tag{8}$$

which can be integrated to give

$$v_0\, t = \frac{r_0}{\sqrt{2}}\, F \left\{ \sin^{-1} \sqrt{2}\, \sin \left[\frac{1}{2} \cos^{-1} \left(\frac{r_0}{R(t)}\right)^{2/3} \right] , \frac{1}{\sqrt{2}} \right\} + R \sqrt{1 - \left(\frac{r_0}{R(t)}\right)^{4/3}} \tag{9}$$

Here F is the elliptic integral of the first kind. Only values of $R \gg r_0$ are of interest. For large R, Eq. (9) is closely approximated by

$$R \simeq v_0 t \tag{10}$$

For $R = 10$ m and $r_0 = 0.5$ m (one thruster), Eq. (10) gives a value of $v_0 t$ which is only 4% low.

Isodensity contours will be computed below for use in plasma pressure balance calculations. Radial spreading is caused both by thermal motion and by lack of perfect beam collimation. A Gaussian density distribution is assumed:

$$n = (r_0/R)^2\, N_{oB} e^{-(r/R)^2} \tag{11}$$

where $(r_0/R)^2$ is a normalization constant, $r = z \tan \theta_0$, z is axial distance from the thruster, and θ_0 is a characteristic angle defined by

$$\theta_0 = \alpha + 180\, v_0/\pi u \tag{12}$$

Here α is a characteristic divergence angle for the thruster itself; the second term accounts for thermal spreading with u (the beam directional velocity) = z/t. For the specific thruster design considered, $\theta_0 \approx 6.3$ deg.

Solving Eq. (11) for r determines the contour $r_n(z)$ on which the plasma density is n:

$$r_n = z \tan \theta_0 \sqrt{\ln \left[\frac{N_{oB}}{n}\left(\frac{r_0}{z \tan \theta_0}\right)^2\right]} \tag{13}$$

The axial distance from the source where density n occurs is

$$z_n = \frac{r_0}{\tan \theta_0} (N_{oB}/n)^{1/2} \tag{14}$$

An isodensity beam contour has been computed for a specific Solar Power Satellite (SPS) GEO station-keeping requirement discussed subsequently. Twenty-five 100-cm thrusters are clustered together at each corner of the large satellite. Figure 1 illustrates the beam isodensity contour, in which n = 300 ions/cm^3, produced by these 25 thrusters. An equivalent thruster radius of 25 (0.5) = 2.5 m was assumed. (Elsewhere[12] an orbital transfer vehicle example was presented in which 800 thrusters were clustered together. Although the beam plasma was treated in an entirely different way there, it is interesting that the beam of Fig. 1 reaches the length of ∿1000 km suggested earlier when the conversion is made from 25 to 800 thrusters.)

Thermal Plasma

Charge transfer within the beam, between the slow neutrals and the fast ions, produces a relatively high-density cool plasma which remains in the neighborhood of the thruster. The source characteristics of this plasma plume depend upon the charge transfer production rate and the subsequent motion of the slow ions, which are described in Appendix 2. The latter step is considerably more complicated, because the ion dynamics are dominated by the electrostatic potential structure in the charge transfer region, which in turn is determined by the beam neutralization process.

The thermal plasma[8] is emitted continually during thruster cperation. Viewed from the reference frame fixed in the vehicle, a steady-state spherical flow situation is established. Consequently, the flux across successive spherical surfaces must be constant. The flow velocity, $\overline{v}_S$ at the source, reaches a steady value v_0 within several source radii, as can be shown from the adiabatic fluid equations following exactly the same analysis as we used for the beam plasma expansion. Then the thermal plasma density must vary as the inverse square of the distance ρ from the source. Note that ρ is the independent variable, not a characteristic

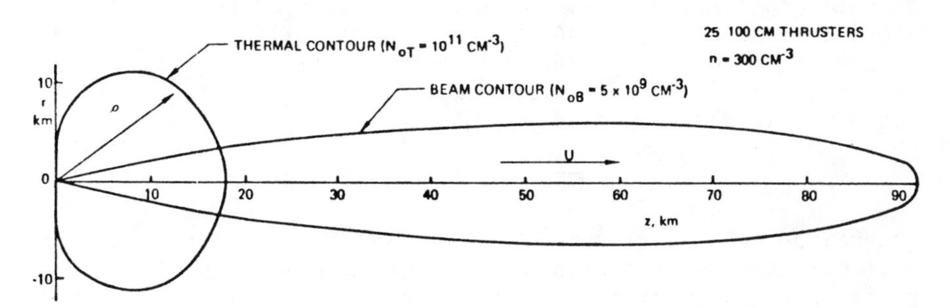

Fig. 1 Isodensity plasma contours in geosynchronous orbit.

radius as used in the beam plasma case. No approximation to the radial profile is required in this case.

The angular distribution is assumed[8] to be proportional to cos θ, where θ is the angle measured from the beam axis about the center of the source. Then the density distribution is given by

$$n = \frac{\overline{v}_s \, N_{oT}}{2 \, v_o} \left(\frac{r_o}{\rho}\right)^2 \cos\theta \tag{15}$$

where N_{oT} is the exit plane thermal density and $\overline{v}_s/2\,v_o$ is the normalization constant. Figure 1 also gives the thermal plasma isodensity contour for the thruster in geosynchronous orbit assuming $\overline{v}_s/v_o = \sqrt{1\text{ eV}/10\text{ eV}}$.

III. Geomagnetic Field Effects

The thruster plasma is sufficiently dense in the vicinity of its exit plane (see Table 1) to generate its own currents that block out the geomagnetic field. As the beam and thermal plasmas expand, however, the geomagnetic field eventually takes control of the individual particles.

One propagation criterion that has been used[13] requires that the ion energy density be greater than the geomagnetic field density, or

$$\varepsilon_m = 8\,\pi\, n\, M\, c^2/B^2 > 1 \tag{16}$$

where ε_m is the particle dielectric constant, B is the geomagnetic field, and c is the speed of light. This parameter is just the ratio of the Debye length to the Larmor radius; when $\varepsilon_m \gtrsim 1$, plasma polarization fields can be built up without interference by the magnetic field. However, a pressure balance requirement still applies and puts a more stringent limit on beam propagation [i.e., will lead to smaller beam dimensions than Eq. (16)].[14] As a first approximation, it will be assumed that most ions transfer to single particle orbits by the time the plasma dynamic pressure drops to one-tenth of the magnetic field pressure, i.e., when

$$n\, M\overline{v}^2 = (0.1)\, B^2/8\pi \tag{17}$$

The reason for the factor 0.1 in the right side of Eq. (17) is that only the outer sheath of the beam transfers to single particle behavior as the dynamic pressure drops to the value of the magnetic field pressure. The dynamic pressure must drop considerably lower before the core of the beam attains single particle behavior.

The transition contour dimensions for both LEO and GEO are given in Table 3. The densities were calculated for the criterion in Eq. (17) and used in Eqs. (13) - (15) to calculate the contours in Fig. 1. Injection velocities for ions into the geomagnetic field are assumed to be those for the free expansion, since the Larmor radii are large. These assumptions provide a reasonably consistent model for estimating geomagnetic effects.

The thermal plasma density falls much faster than that of the beam. Consequently, its transition location is very close to the vehicle as indicated by the values in Table 3. Furthermore, the thermal plasma does not function as independently of the spacecraft as the beam does; the plume stays close to the vehicle until the field takes control. Thus, these ions cause concern as a possible return current source.

The ion-beam directed velocity is substantially greater than that of the spacecraft at LEO as well as GEO. Consequently, these ions enter the geomagnetic field with their beam speed. They follow well known first-order adiabatic orbits along geomagnetic flux tubes.[15,16] Depending on their mirror altitude, they may be trapped for an extended period (possibly years above LEO).

Most applications at LEO involve accelerating spacecraft to attain higher altitude, and the ion beam is directed at a large angle to the field. The tilt of the geomagnetic axis relative to the Earth's axis and appreciable trajectory inclinations can produce angles as small as 30 deg. or less, however, allowing the beam to spiral down the field into the upper atmosphere.

At GEO the spacecraft is stationary in the geomagnetic field, and the ion beam may be directed at any angle with

Table 3 Transition location for 25 thruster plasma fluid to single particle orbits

	B, G	Transition density, ions/cm³	Beam z_n	Beam r_n	Thermal ρ_n
LEO (400 km)	0.31	1.7×10^7	400 m	26 m	76 m
GEO (36,000 km)	0.0011	3×10^2	93 km	6.2 km	18 km

respect to the field depending upon its purpose. Generally the beam ions are injected into trapped orbits of long duration. Their longitudinal drift motion and subsequent dispersion is expected to eliminate any appreciable return current to the spacecraft. However, some caution is advised to avoid those rare conditions around the geomagnetic equator that create particle mirror locations close to the vehicle which could cause bursts of return current. At the opposite extreme, beams injected nearly parallel to the field (within 2.5 deg.) follow the flux tube down to the atmosphere where they are absorbed. Due to the variety of possible operating conditions, no attempt is made to provide quantitative results here, although they are readily calculable.

At LEO it is evident from Fig. 2 that the thermal plume cannot "catch up" to the spacecraft (when local electric fields are omitted from consideration) because thermal speed is directed into the hemisphere away from the vehicle. Even those ions in the energetic Maxwellian tail of the thermal plasma move away because of a lack of collisions back toward the vehicle. A typical cluster of thermal ion orbits is shown in Fig. 3 for the case in which the spacecraft is being propelled normal to the local geomagnetic field.

The thermal ions emanating from a geosynchronous satellite thruster would follow geomagnetic orbits like those in Fig. 3, assuming no electric fields were present; however, the scale is now about 30 times larger (see Larmor radii in Table 2). Those ions injected nearly orthogonal to the local geomagnetic field can return to the spacecraft after one gyration if their velocity components satisfy the inequality

$$v_{//}/v < \ell_{//}(\text{km})/60 \tag{18}$$

where $v_{//}$ is the velocity component parallel to the field, v is

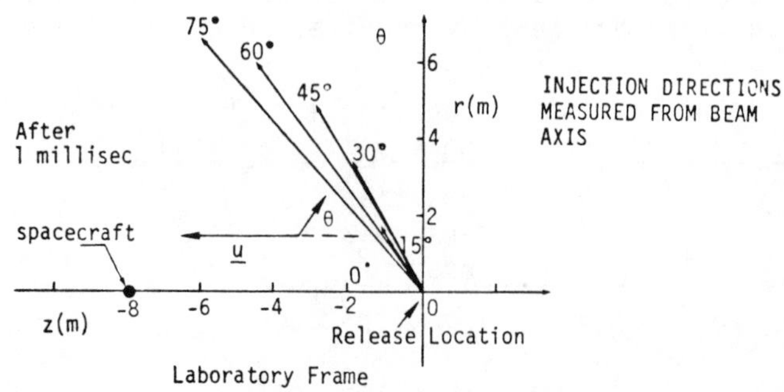

Fig. 2 Thermal ion trajectories at LEO.

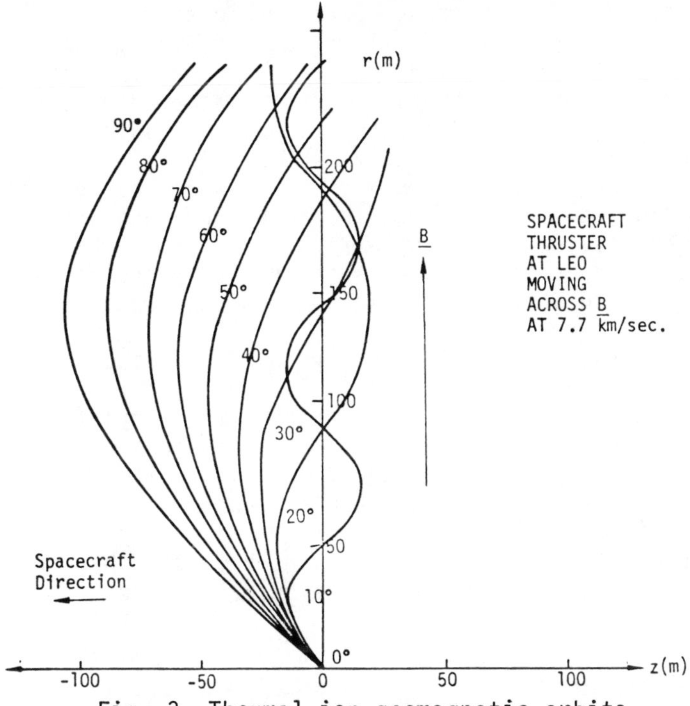

Fig. 3 Thermal ion geomagnetic orbits.

total ion speed, and $\ell_{//}$ is the dimension of the spacecraft parallel to the field. For a typical 100 m extension, less than 0.0005 of the thermal ion charge would be collected directly; the rest enter trapped orbits.

All the other ions travel away along geomagnetic flux tubes until they mirror and return past the plane of the spacecraft. From adiabatic theory for geomagnetically trapped particles,[15,16] the time to return to the spacecraft is on the order of 2 R_e/v, where R_e is the geocentric distance to the spacecraft. For argon ions at 10 eV, this bounce time is $\sim 1.2 \times 10^4$ sec. During this time they drift in longitude at the rate of about 3×10^{-8} rad/sec, which corresponds to an equatorial transit distance of 1500 km, well away from the spacecraft.

IV. Spacecraft Field Effects

Spacecraft that employ ion thrusters for propulsion probably will use solar arrays to generate their electrical power. Since significant propulsion can only be achieved with high power levels, the array will undoubtedly operate at high voltages to reduce conduction losses. Such potentials on exterior spacecraft surfaces produce regions of high electric field which can dramatically alter the thruster plasma. Since the thermal plasma cloud lingers in the vicinity of the spacecraft, the ions and electrons are easily attracted back to the vehicle by these fields. If the surface is insulated, space charge can build up to neutralize the applied fields. If there is access to noninsulated conductors as, for example, exposed solar cell connectors, the return current short-circuits the system, lowering efficiency. Presumably the main thruster beam operates at such an energy and distance from these field sources that it remains unaffected.

A spacecraft driven internally to high voltages behaves like a floating double probe.[17,18] Its potential distribution is biased negative because of the higher mobility of electrons. Experience has shown that a double probe in an ionospheric plasma with an impressed voltage V has its positive end at about 0.1 V above the plasma potential and its negative end around 0.9 V below plasma potential.[19] This large negative field region attracts the thermal ions from the thruster, and their rate of generation limits the return current to the vehicle.

Return Current Theory

The range of spacecraft-generated electric fields determine the volume of plasma that can supply return current. The electric field around a naturally charged spacecraft extends outward only a few Debye lengths because of shielding by the ambient plasma. For high-voltage solar arrays, however, the shielding length is larger, and it is necessary to determine the scaling with array voltage.

When the applied potential energy eV is much greater than the thermal kinetic energy kT, the ions are pulled from the plasma with nearly zero energy (kT), and the electron space charge becomes exponentially small just a few Debye lengths from the plasma edge. Under these conditions the Langmuir-Childs space-charge analysis applies.[20,21] The ion current density flowing from the background plasma to a negative satellite surface is

$$j_{L.C.} = \frac{(2e/M)^{1/2}\, V^{3/2}}{9\pi\beta^2 x^2} \tag{19}$$

where β^2 is a geometry factor and x is the distance between the source plasma and the collector. The ion current j_i collected by the surface is limited to that available from the surrounding environment, including the ambient plasma ions and ram current on forward surfaces (due to spacecraft motion) and ion thruster sources.

The shielding length x_s is determined by limiting $j_{L.C.}$ to the available j_i that can be drawn from the ambient plasma:

$$\beta x_s = \left[\frac{8\pi^{1/2}}{9}\right]^{1/2} \left[\frac{eV}{kT}\right]^{3/4} \left[\frac{kT}{4\pi ne^2}\right]^{1/2} \tag{20}$$

Note that βx_s is independent of particle mass and thus applies to ion- or electron-collecting sheaths. The factor β is unity for plane-parallel geometry (or any curved surfaces with βx much less than a typical dimension), but may vary substantially for other geometries.[21]

For the conditions of geostationary orbit, a plane electrode at 10 kV would have a shielding length of ~7 km. Thermal plasmas produced by ion thrusters on booms even one kilometer away would be within the range of the electric fields from the spacecraft.

Solar Power Satellite Applications

As an illustration of the space charge limit on return current, consider station keeping of the solar power satellite (SPS) at GEO. One version of the SPS system[6] has 25 control thrusters located on a 0.5 km boom extending from each corner of the solar array rectangle. Operating characteristics are those of the 100 cm argon thruster listed in Table 1. The solar panel is designed to operate at 40 kV so that its central power mains are around 4 kV positive and 36 kV negative with respect to the external plasma. At the edge of the array near the boom the panel potential is midway between the mains.

Consider first the case where the coaxial power line to the thrusters is unshielded, with its exterior surface at the negative potential. The thermal ions are attracted to the cable by its external fields. The amount of current that this cable can collect has been estimated using the methods developed by Langmuir[20,21] and treating the cable as a rectangular strip of length L and width πa. The running integral of the charge-limited current, starting from the end of strip away from the source and stopping at the thruster cluster equivalent radius r_0, is given by the expression

$$I(>z) = \frac{4\pi\varepsilon_0 a}{L} \left(\frac{2e}{M}\right)^{1/2} V^{3/2} \left[\frac{\cos\theta}{(1-\cos\theta)} + \ln(1-\cos\theta)\right] \tag{21}$$

where $\theta = \sin^{-1} (z/L) \qquad r_0 \leq z \leq L$ (22)

The plot of I(>z) in Fig. 4 demonstrates the concentration of current density near the thruster. The SPS thruster cluster at the end of the boom produces about 200 A of thermal plasma, and the integral $I(r_0)$ over the 500 m boom reveals that about 300 A could be collected on a 1 m coaxial cable.

This current collection can be eliminated by suitable insulation around the conducting surfaces. But the high-voltage solar panels provide an alternate sink for thermal ions. The space-charge limited current capacity of such surfaces is described by Eq. (20) as well, and the current density has a profile similar to that in Fig. 4. A schematic diagram of this current flow to the SPS is shown in Fig. 5. Most of the charge would be collected in the corner region where the boom attaches to the main structure. The enormous dimensions of the collecting surface imply that all of the thermal ion current would be collected by the exposed solar cell interconnects.

Insulating the entire solar panel surface would eliminate this return current path. However, this insulation membrane must be able to withstand high voltages that are created by charge collection on the outer surface. Laboratory experiments with plastic materials such as kapton have been per-

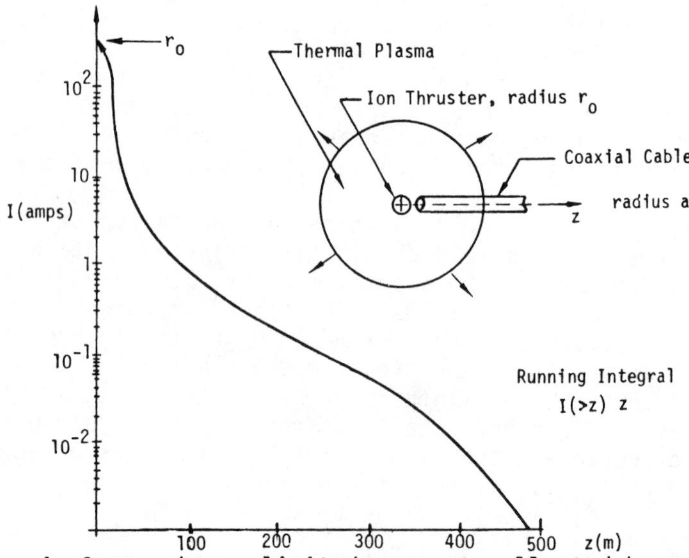

Fig. 4 Space charge limited current collected by exposed boom cable to thruster cluster on SPS.

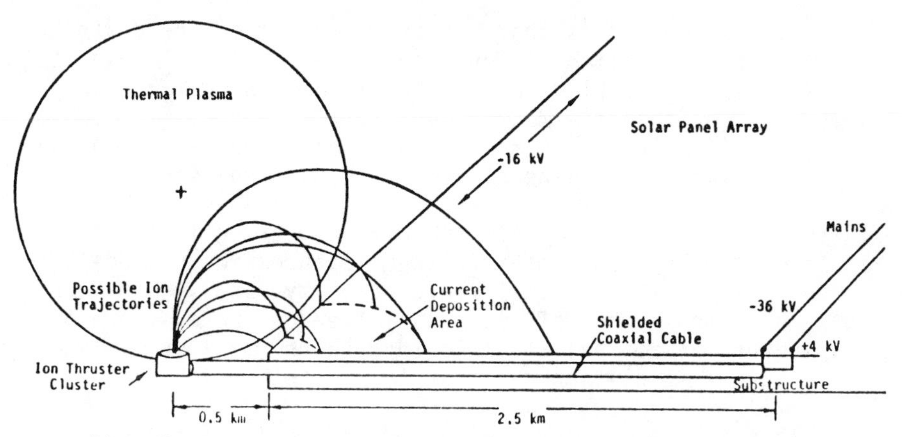

Fig. 5 Thermal ion current collection by unshielded solar cell interconnects on SPS panels.

formed,[22] and the material suffered breakdown and pinhole formation at the edges of conducting elements. Development of new materials or somewhat thicker membranes presumably will be needed to withstand the electric fields. Thus, to avoid solar panel breakdown, collection of thermal return currents on bare coaxial cable appears to be a more desirable procedure.

Appendix 1. Adiabatic Approximation

A description of the expansion process is needed which is tractable and yet systematically related to the exact Klimontovich equation.* The process is well known and will only be summarized here:

1. A continuum kinetic equation (the Vlasov equation) is formed from the Klimontovich equation by smoothing out the discrete particle distribution in phase space. The errors in this process are too small to concern us here.
2. Successive moments of the Vlasov equation are formed. The zeroth moment is the particle conservation equation, the first moment is the momentum conservation equation, and the second moment is the energy conservation equation.
3. The problem at this point is to terminate the series of moment equations. The nth moment of the Vlasov equation always involves the $(n+1)^{th}$ moment of the distribution function, the behavior of which is described by the $(n+1)^{th}$ moment equation. The usual

*The Klimontovich equation is just the ensemble of single particle equations of motion for all particles in the system.

assumption is that the third moment of the distribution function (taken around the mean local velocity) is negligibly small. This is a much weaker assumption than the Maxwellian distribution assumption built into most fluid equations. The resulting equation is the generalized adiabatic equation.

In this description the "temperature" appearing in the fluid equation is the mean kinetic energy measured in the reference frame moving with the local fluid velocity, and $\gamma = (p+2)/p$, where p is the number of degrees of freedom (here, $\gamma = 5/3$). The isothermal approximation would require $\gamma = 1$ or $p \to \infty$, which is not at all accurate for this case.

This is the only systematic procedure for deriving the fluid equations for a collisionless plasma. Sometimes higher moments or the effects of a magnetic field are accounted for in determining the fluid equations; these procedures have the disadvantage that they are not proper approximations--they are not obtained by expansion in a small parameter. However, the general experience in plasma physics is that the fluid equations are quite accurate except for Landau damping and related phenomena which are specific to local features of the distribution function.

Appendix 2. Thermal Ion Generation

Charge-Transfer Plasma

The slow neutrals are assumed to expand spherically as if they originated at a point a distance r_0 behind the accelerating mesh. The fast ions are assumed to flow in a constant radius cylinder over the range where the neutral density is high enough to support charge transfer. Because a significant fraction of the original ion beam undergoes charge transfer, it is necessary to solve the coupled equations for the densities of fast and slow ions.

Momentum transfer during charge transfer collisions can be neglected. However, we also neglect the effects of momentum changing collisions, which is not entirely justified and would have a significant effect on the thermal plasma expansion. In the expansion calculations, we have used spherical expansion rather than cylindrical expansion (as implied by the potential model) as a crude way of accounting for the typical forward motion of the thermal particles acquired in elastic collisions with beam particles.

Slow ion creation and loss:

$$\frac{dn_s^+}{dt} = \nabla \cdot j_s^+ = \frac{\sigma(v_f - v_s)}{v_f\, v_s} (j_f^+ \, j_s - j_s^+ \, j_f) \qquad (A1)$$

where

$$j_s^+ = n_s^+ \, v_s \text{ , etc.,}$$

$$j_f = j_f^+ + j_f^0 \text{ ; } j_s = j_s^+ + j_s^0$$

and σ is the charge transfer cross-section.

Fast ion creation and loss:

$$\frac{dn_f^+}{dt} = -\nabla \cdot j_f^+ = \frac{\sigma(v_f - v_s)}{v_f\, v_s} (j_s^+ \, j_f - j_f^+ \, j_s) \qquad (A2)$$

$$-\frac{dn_f^+}{dt} = \frac{dn_s^+}{dt} \qquad (A3)$$

Taking Z = 0 at the effective center of spherical expansion for the slow particles and letting

$$j_s = I_o/Z^2 \quad , \quad j_s^+ = I_o^+/Z^2 \qquad (A4)$$

we have

$$\nabla \cdot j_s = \frac{1}{Z^2} \frac{\partial}{\partial Z} (Z^2 \, j_s) \text{ when operating} \qquad (A5)$$

on j_s , j_s^+ , j_s^o , and

$$\nabla \cdot j_f = \frac{\partial}{\partial Z} j_f \qquad (A6)$$

Combining Eqs. (A4) and (A5) and defining slow fluences $\overline{v}_s$, $\overline{v}_s^+$ by

$$j_s = \overline{v}_s/Z^2 \text{ , } j_s^+ = \overline{v}_s^+ \, (Z)/Z^2 \qquad (A7)$$

we obtain a differential equation for the slow ion fluence:

$$\frac{\partial^2 j_s^+}{\partial Z^2} + \left(\frac{j_s}{KZ^2} + \frac{j_f}{K}\right) \frac{\partial j_s^+}{\partial Z} = 0 \qquad (A8)$$

where

$$K = \frac{v_f\ v_s}{\sigma(v_f - v_s)} \tag{A9}$$

The solution of Eq. (A8) is

$$\frac{\partial j_s^{+}}{\partial Z} = a\ e^{-p(Z)} \tag{A10}$$

when

$$p(Z) = \frac{j_s}{K}\left(\frac{1}{Z_o} - \frac{1}{Z}\right) + \frac{j_f}{K}(Z - Z_o) \tag{A11}$$

where Z_o is the thruster exit position (i.e., $Z_o \approx r_o$, the exit radius). Since $p(Z_o) = 0$, the constant a is given by the initial value of the charge transfer rate,

$$a = \frac{j_f\ j_s}{K\ Z_o^{2}} \tag{A12}$$

Thus the slow ion source distribution is

$$\frac{\partial j_s^{+}}{\partial Z} = \frac{j_f\ j_s}{K\ Z_o^{2}}\ e^{-p(Z)} \tag{A13}$$

and the net slow ion current is

$$j_s^{+}(Z) = \frac{j_f\ j_s}{K\ Z_o^{2}} \int_{Z_o}^{Z} e^{-p(Z)}\ dZ \tag{A14}$$

Thermal Ion Acceleration

The thruster beam is emitted over an area A_o with average density n_b and speed v_b such that

$$I_b = e\ n_b\ v_b\ A_o \tag{A15}$$

where

$$\frac{1}{2} M v_b^{2} = eV_b - eV_n \tag{A16}$$

and M is the ion mass, V_b is the beam accelerating voltage, and V_n is any ion-retarding potential which builds up between the outer accelerating grid (vehicle skin) and the beam in order to pull in sufficient electrons from the neutralizer to balance the ion space charge.

The region near the vehicle skin where the drop V_n occurs is assumed to be small compared to the charge-transfer mean free path, and we will also assume that relatively few slow-charge transfer ions move backward to the grid. (Both of these assumptions are substantiated by the results.) Then the total positive charge density near the exit grid is e n_b. Overall space charge neutrality requires that

$$n_e = n_b \tag{A17}$$

The electrons are supplied by a plasma source mounted on one side of the thruster. The open area of the plasma source is much smaller than A_o. The outer structure of the neutralizing plasma source is grounded to the outer structure of the thruster. This plasma can be assumed to have an electron temperature of $\approx$1 eV and an ion temperature of perhaps 0.1 eV. Since the net current emitted by the vehicle must be zero, the neutral plasma must be the source of a current I_e.

Drawing this current from the plasma to the beam requires the voltage V_n, i.e.,

$$I_e\ (V_n) = -I_b \tag{A18}$$

The functional form for the neutralization current is the Langmuir-Childs equation.[20]

$$I_e = A\ j_{L.C.} = \frac{(2e/m)^{1/2}\ V_n^{\ 3/2}}{9\pi\beta^2\ell^2}\ A_n \tag{A19}$$

where m is the electron mass, ℓ is the separation between the beam and the thruster plasma, β is the appropriate factor, and A_n is the emitting area of the neutralizing plasma. Equations (A18) and (A19) determine V_n, the difference between the potential at the center of the beam and the vehicle skin potential:

$$V_n = \left\{\frac{9}{\sqrt{\frac{2e}{m}}}\left[\frac{\pi(\beta\ell)^2}{A_n}\right] I_b\right\}^{2/3} \tag{A20}$$

Evaluating the constants,

$$V_n = 5.7 \times 10^3 \left[\frac{I_b(\beta\ell)^2}{A_n}\right]^{2/3} \quad \mathrm{V} \tag{A21}$$

we see that $\beta\ell$ must be very small or A_n must be large in order to explain the observed drops of just a few tens of volts between the beam center and the surrounding space.

It is impossible to obtain a large emitting area with the small emitter presently in use and the low ion temperatures characteristic of these plasma generators.

The gap, however, can be quite small. The charge transfer plasma density is approximately 2.6 x 10^{10} cm^{-3}. The space-charge shielding length for these ions, assuming a temperature of 1000 K, is $\lambda_t \simeq 7(T_t/n_{th})^{1/2}$. The expression for the voltage required to emit I_b across a gap determined by a shielding plasma is

$$V_n = 7.7 \times 10^4 \left[\frac{I_b T_o}{A_n n_{th}}\right]^{1/2} \quad V \qquad (A22)$$

For a current of 80 A and the above plasma characteristics, Eq. (A22) gives

$$V_n = 15.4 \text{ V}$$

Without the charge transfer plasma, V_n would be much larger.

In practice, V_n is just the potential of a typical electron from the neutralizer. The ion beam represents a well of depth V_n in which the electrons are trapped. The ions, on the other hand, are accelerated outward by the electric fields, and consequently, V_n determines the ion expansion speed. This is just another form of ambipolar diffusion, where the diffusion rate is determined by the electron energy and the ion mass.

References

[1] Worlock, R., Trump, G., Sellen, J. M., Jr., and Kemp, R. F., "Measurement of Ion Thruster Exhaust Characteristics and Interaction with Simulated ATS-F Spacecraft," AIAA Paper 73-1101, Nov. 1973.

[2] Bartlett, R. O., DeForest, S. E., and Goldstein, R., "Spacecraft Charging Control Demonstration at Geosynchronous Altitude," AIAA Paper 75-359, March 1975.

[3] Komatsu, G. K., Sellen, J. M., Jr., and Zafran, S., "Ion Beam Plume and E-Flux Measurements of an 8 cm Mercury Ion Thruster," Journal of Spacecraft and Rockets (submitted for publication).

[4] Kaufman, H. R. and Isaacson, G. C., "The Interactions of Solar Arrays With Electric Thrusters," AIAA Paper 76-1051, Nov. 1976.

[5] Austin, R. E., Dod, R. E., and Grim, D., "Solar Electric Propulsion for the Halley's Comet Rendezvous Mission: Foundation for Future Missions," AIAA Paper 78-83, Jan. 1978.

[6] Woodcock, G. R., "Solar Power Satellite System Definition Study, Part III, Preferred Concept System Definition," The Boeing Co., Rept. D180-24071-1, NAS-15196, DRL T-1346, DRD MA-664T, March 1978.

[7] Grim, D., Staff Member, Flight Technology, Boeing Aerospace Company, personal communication, Aug. 1978.

[8] Staggs, J. F., Gula, W. P., and Kerslake, W. R., "Distribution of Neutral Atoms and Charge-Exchange Ions Downstream of an Ion Thruster," Journal of Spacecraft and Rockets, Vol. 5, Feb. 1968, pp. 159-164.

[9] Liemohn, H. B., "Electrical Charging of Shuttle Orbiter," IEEE Transactions, Plasma Science, Vol. PS-4, Dec. 1976, pp. 229-240.

[10] McPherson, D. A., Cauffman, D. P., and Schober, W. R., "Spacecraft Charging at High Altitudes: SCATHA Satellite Program," Journal of Spacecraft and Rockets, Vol. 12, Oct. 1975, pp. 621-626.

[11] Liemohn, H. B., Copeland, R. L., and Leavens, W. M., "Plasma Particle Trajectories Around Spacecraft Propelled by Ion Thrusters," Spacecraft Charging Technology - 1978, edited by R. C. Finke and C. P. Pike, NASA CP-2071/AFGL-TR-79-0082, 1979, pp. 419-436.

[12] Chiu, Y. T., Cornwall, J. R., Luhmann, J. G., and Schultz, M., "Argon-Ion Contamination of the Plasmasphere," Aerospace Corp. Space Sciences Laboratory Rept. No. SSL-79 (7824)-1, June 1979.

[13] Curtis, S. A. and Grebowsky, J. M., "Changes in the Terrestrial Atmosphere-Ionosphere-Magnetosphere System Due to Ion Propulsion for Solar Power Satellite Placement," Space Solar Power Review (submitted for publication).

[14] Longmire, C. L., Elementary Plasma Physics, Interscience Publishers, Wiley, New York, 1963.

[15]Rossi, B. and Olbert, S., Introduction to the Physics of Space, McGraw-Hill, New York, 1970, pp. 157-187.

[16]Liemohn, H. B., "Radiation Belt Particle Orbits," The Boeing Co., Rept. D1-82-0116, June 1961.

[17]Chen, F. F., "Electric Probes," Plasma Diagnostic Techniques, edited by R. H. Huddlestone and S. L. Leonard, Academic Press, New York, 1965, pp. 113-200.

[18]Swift, J. D. and Schwar, M.J.R., Electrical Probes for Plasma Diagnostics, American Elsevier, New York, 1960, pp. 137-155.

[19]Stevens, N. J., "Interactions Between Spacecraft and the Spacecraft Charged-Particle Environment," Spacecraft Charging Technology - 1978, edited by R. C. Finke and C. P. Pike, NASA CP-2071/AFGL-TR-79-0082, 1979, pp. 268-294.

[20]Langmuir, I. and Blodgett, K., "Currents Limited by Space Charge Between Coaxial Cylinders," Physical Review, Vol. 21, Series 2, No. 4, Oct. 1923, pp. 347-356.

[21]Cobine, J. D., Gaseous Conductors: Theory and Applications, Power Publications, New York, 1958, pp. 123-128.

[22]Kennerud, K. L., "High Voltage Solar Array Experiments," NASA/LeRC, CR 12180, March 1974.

Chapter V—Space Environmental Effects on Structures

STRUCTURAL DISTORTIONS OF SPACE SYSTEMS DUE TO ENVIRONMENTAL DISTURBANCES

F. Ayer* and K. Soosaar†
The Charles Stark Draper Laboratory, Cambridge, Mass.

Abstract

It is expected that the size and nature of satellites is likely to change as a result of the greatly enhanced launch capabilities of the space shuttle. Greater payload capability, manned assembly and maintenance will all contribute toward larger space platforms. Such larger platforms are expected to perform to higher levels. As a result, the disturbances, which tend to degrade the performance of these satellites, need to be known to a greater extent.

This paper examines present approaches to environmental disturbance modeling and discusses the methods used to evaluate the resulting structural distortions. For the most critical loadings, such as thermal radiations, solar pressure, gravity gradient, and atmospheric drag, disturbance models and their limitations are described along with suggestions for reducing their impact. The modern discretization of the structures using a finite-element approach is explained as are the solution methods for both static and dynamic responses. Special attention is given to model analysis because of its advantage in interpreting structural behavior and reducing computational difficulties. When the performance goals are not met, the designer is faced with a wide range of options for improvement. Enhancement can be obtained by use of passive or active means of controlling the structural response.

As the performance of new and larger satellites becomes increasingly demanding, it is recognized that both environmental-disturbance and supporting-structure modeling need to be enhanced. New analytical and computational tools must be developed to assess the intricate interactions between spacecraft and environment with increased confidence.

---

*Member of Technical Staff.
†Group Leader.

I. Introduction

The absence of dense atmosphere and strong gravitational attraction makes space an ideal location for conducting many types of highly sensitive experiments and observations which the environment on the Earth's surface does not permit. Because of the delicacy of such experiments, it is important to assess precisely the magnitude of the residual atmosphere, the reduced gravitational forces, and the other unique environmental disturbances encountered in orbit.

A typical satellite consists of one or more sensors or instruments attached to a mechanical/structural platform or spacecraft. In the earlier space missions, these platforms were small, rigid, and relatively insensitive to the environmental disturbances encountered in orbit. The high accelerations of early launch vehicles and their relatively limited payload capabilities rendered them, effectively, point-mass satellites. As the capability of the boosters grew and space missions broadened their goals, larger but more flexible satellites evolved. At the same time, the deformational limits on these satellites became more demanding. Currently, we are on the threshold of the Space Shuttle era which promises greater payloads with softer launches and hints at much larger platforms with yet further enhancements in system performance. To assure these enhanced levels of performance in future missions, it is necessary 1) to refine the models of the environmental sources that tend to degrade that performance, 2) to assess their impact, and then 3) to modify the design of the system to achieve these goals.

The point of view of this paper will be primarily structural since the effects of the space environment will, in many instances, directly or indirectly cause misalignments or vibrations to occur at the payload. It is, of course, recognized that the space environment will also have impacts on other aspects of the satellite, such as electrical components and materials. Some of these are the topics of other papers in this volume.

This paper presents an overview of the major sources of environmental disturbance that effect structural distortions of space systems, attempts to quantify them wherever possible, and indicates how the structural engineer deals with them in assessing, and ultimately assuring, the integrity of the mission.

II. Space System Characteristics

A number of important mission-oriented characteristics must be known before the assessment of the environmental disturbances is begun. For Earth-oriented applications (such as meteorology, Earth resources, and communications), the orbit of the satellite is probably the most fundamental parameter. For astronomical observations in space, the orbit is more a constraint than an objective, since the payload launch limitations and the avoidance of radiation zones will define ranges of orbital operation.

With the orbit defined, the size, shape, and sensitivity of the on-board sensors, antennas, solar panels, etc., can be determined. First-order environmental disturbances can now be assessed since a rough guide exists for the dimensions and configuration of the system. The orbit and attitude maintenance goals can then be compared to the disturbance effects, and development of a preliminary design for the control systems can begin.[1,2,3]

At this point the satellite structure and its performance goals begin to emerge. The satellite structure can be viewed as a mechanical system, usually time-invariant, acted upon by various environmental or on-board disturbances. The performance goals usually can be expressed in terms of static and dynamic displacement perturbations from the nominal design. The structure responds to these disturbances according to its thermoelastic and dynamic properties, and this demonstrates whether the performance goals can be realistically met. In many cases the environmental disturbances directly affect only the rigid-body behavior of the satellite, but the control approaches that are developed to restore the rigid-body behavior have a profound impact on the flexible-body response.

In dealing with the sequential steps of this analysis, great care must be exercised to maintain relative consistency among the various phases of disturbance modeling, structural representation and analysis, and the performance criteria. The level of accuracy must, furthermore, be commensurate with the performance requirements. A solar collector may tolerate large deflections, and thereby, it will demand less accuracy in modeling than a high-performance antenna or astronomical mirror.

As the new classes of large space platforms begin to emerge, it will be found that the great flexibility these systems possess will become the largest source of concern. This will lead to the development of new technologies in structure, dynamics, and control, as well as in new evaluation tools necessary to predict their performance. The importance of accurate knowledge and modeling of the disturbances must be emphasized and will be a major determinant in the success of the large platforms.

III. Disturbance Identification

As stated earlier, the disturbances affecting the structural response of a space system can be broadly divided into external and internal sources. The external sources primarily include solar and terrestrial influences. Solar radiation impinging upon a satellite is absorbed as heat and acts as a source of pressure to alter the orbit of the satellite. Solar wind is occasionally encountered, but its mechanical effects are of a smaller order. Radiation is received by the satellite from the Earth as well. Some of this is a direct reflection of solar radiation, but a small fraction comes from the Earth's thermal emission.

The Earth's gravitational field maintains the orbit of the satellite, but if the satellite configuration departs significantly from a point mass, gravity-gradient effects become important. Solar and lunar gravitational effects can also be of some significance, but these are of small consequence unless the orbits go beyond the geostationary. At lower altitudes, atmospheric drag is an important disturbance, and very accurate models are needed to predict satellite lifetimes and reentry points. In some instances, electrical currents in the satellite will interact with the Earth's magnetic field; this potentially produces torques to misalign the attitude of the satellite.

Very few of these environmental disturbances directly affect the spacecraft structure and its performance. The various attitude, orbit, and thermal control systems installed as a result of the environment will make significant performance contributions, however. The dynamics of the attitude control devices, disturbances from on-board cooling systems, and movement of solar panels and communication antennas will cause significant structural responses from the payload.

The structural analyst prefers to categorize these disturbances primarily by their frequency content. While nothing in orbit is ever truly static, many of the disturbances vary sufficiently slowly so that they do not cause dynamic response to the satellite structure. These quasistatic disturbances tend to be primarily thermal in nature, although it is possible to experience thermal-shock transients during eclipse passage. Attitude control and cooling system effects tend to elicit dynamic responses since they occur quite often within the band of the natural vibration frequencies of the satellite.

In the following sections, a number of environmental disturbances including thermal, radiation pressure, solar wind, gravity-gradient, and atmospheric drag will be examined in more detail, and their effects on structural response will be evaluated.

Radiation Pressure Effects

In addition to causing thermal effects on incident satellite surfaces, the incoming electromagnetic-radiation energy will exert pressure and shear stresses to alter the dynamics of the satellite.[1,2] The primary disturbance source is solar, but the Earth's albedo and emission must also be considered in lower orbital altitudes. Although the combined effects of these pressures do not have a significant direct effect on the elastic distortions of the satellite, they do cause the satellite to wander from its ideal orbit and the pointing of the payload to vary from the desired direction. Attitude and orbit-control devices are therefore needed to restore performance. The effects of such thrusters, and other torquing devices on structural behavior can be significant.

The mechanics of the transmission of force to the satellite depends highly on the properties of the surface intercepted by the radiation. Therefore, the several limiting cases of black, specular, and diffuse surfaces will be examined.

The momentum per unit volume carried by a collimated beam of light is given by H/c^2 where H is the power per unit area and c is the velocity of light. When the light is totally absorbed, the force exerted on the body is the time-rate of momentum deposition. Hence, the pressure exerted in the direction of travel of the light is

$$F = (\text{momentum/volume}) \times (\text{velocity of light}) \quad (1)$$

$$F = H/c^2 \cdot c = H/c \quad (2)$$

For a reflecting surface, the rate of change of momentum of the photons is higher than that of a nonreflective surface because the photons leaving the surface have a component of momentum in a direction opposite to a component of the arriving photons. In computing the radiation pressure on a spacecraft surface, it is particularly important to determine the net resultant pressure when the incident radiation is reflected diffusely. The term "diffuse reflector" implies that the intensity of radiation leaving the surface is independent of

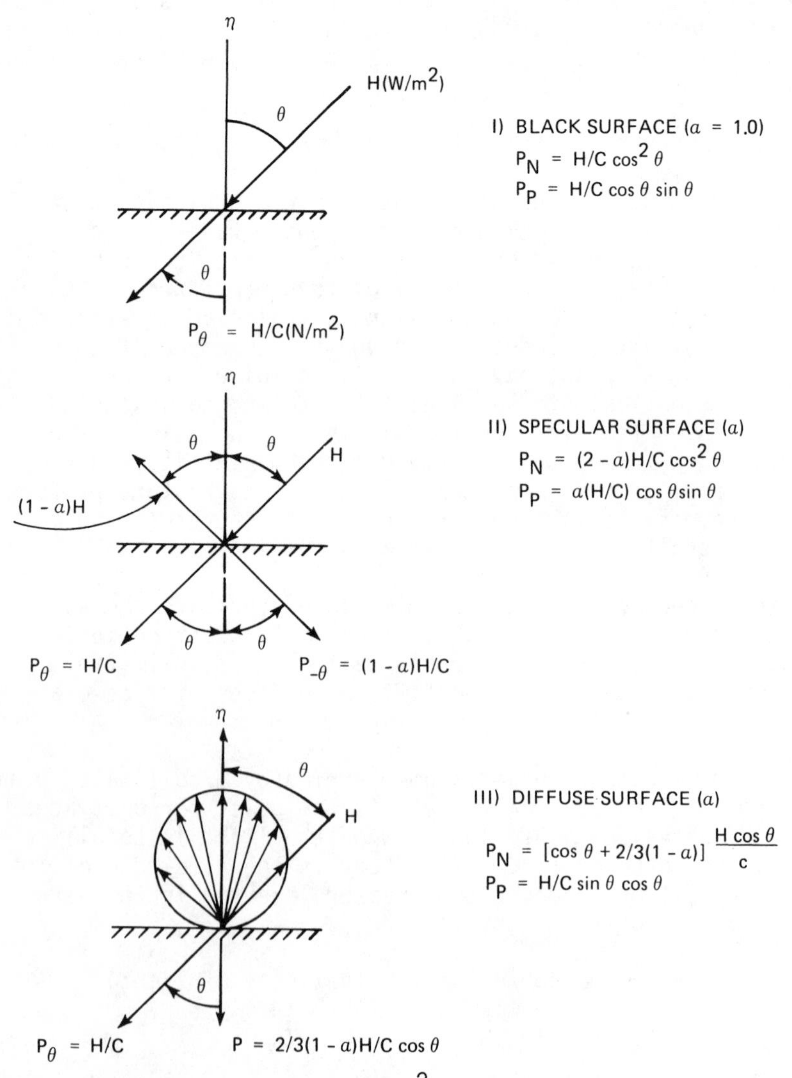

Fig. 1 Radiation pressures (N/m^2) on three different surfaces due to an incident radiation flux intensity $H(W/m^2)$ arriving at an angle to the surface normal.

direction; this is known as Lambert's cosine law. Because of the azimuthal symmetry of diffuse radiation, there exists only a net pressure force in a direction opposite to that of the surface normal (P_N). This component is, for normal incident light

$$P_N = 2/3 \cdot H/c\,(1 - \alpha) \tag{3}$$

where H is the incident radiation flux intensity (W/m^2) and α is the absorptivity of the surface.

Figure 1 examines the three types of pressure forces that can be modeled as part of an orbital heat-flux program. Case I represents a black (totally absorbing) surface and, as such, is just a special case of either Case II (specular surface) or Case III (diffuse surface). Figure 1 provides the resultant force as resolved into components that are perpendicular (P_N) and parallel to the surface (P_p) per unit surface area.

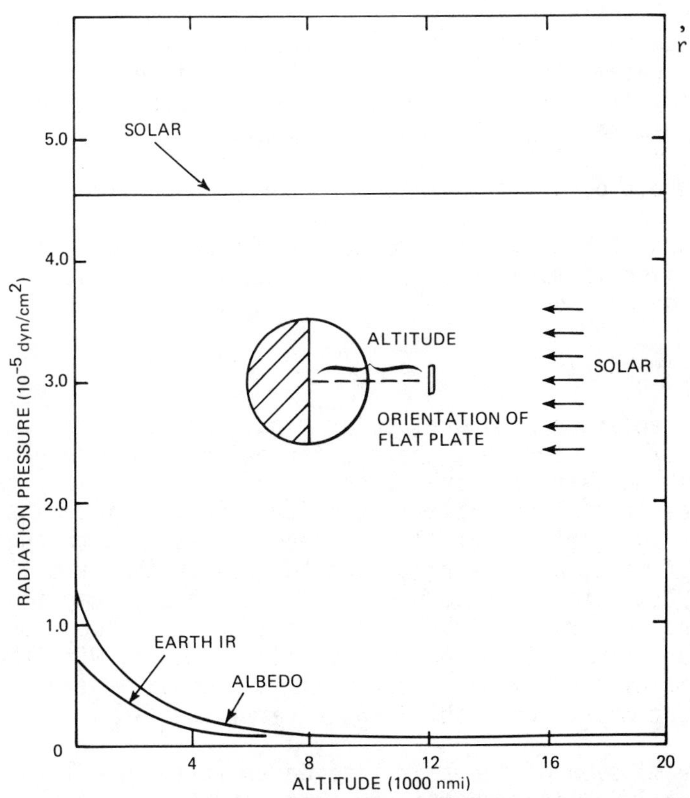

Fig. 2 Solar, maximum albedo, and Earth IR radiation pressures on a flat black plate as a function of altitude.

In Fig. 2, the radiation pressure on a flat black plate viewing the Earth is shown as a function of altitude for three sources of radiation: solar, reflected solar, and Earth infrared (IR). The location of the flat plate over the subsolar point of the Earth was chosen to maximize the pressure force due to the sunlight reflected by the Earth. It is apparent from this figure that the solar pressure forces on a payload can be more than an order-of-magnitude higher than the pressure forces due to Earth IR and reflected sunlight, particularly at higher altitudes.

The actual satellite surfaces will fall somewhere between the absorbing, specular, and diffuse natures. The angles of incidence will depend on the satellite configuration, on its position in orbit, and on functional elements such as Earth-pointing antennas, sun-pointing solar panels, and space-pointing radiators. The procedure for exactly determining the pressure-induced forces and torques will, therefore, be rather complicated, and it will require the use of sophisticated computational software akin to heat-transfer methods. Difficulties arise in accounting for the shadowing effects of surfaces on other satellite components and in determining, as a function of orbital time, the centers of pressure of the partially-shadowed surfaces. With smaller, simpler, "first-generation" satellites, simplifying assumptions generally could be made, and on-orbit behavior could be monitored to provide optimal attitude and orbit-correction strategies. With the larger satellites of the future, a considerable amount of consumables can be saved if more exact computations are performed during the design period.

Gravity Gradient

If radiation pressure, solar wind, and aerodynamic forces are temporarily ignored, the net forces acting on an orbiting body are the Earth's gravitational force and a centripetal force resulting from orbital velocity. For a point-mass or fully axisymmetrical satellite, these forces pass through the center of mass, and the resultant torques on the satellite are zero. Most satellites do not conform to this assumption, however, and the gravitational field will apply torques to the body, tending to align the axis of least inertia with the direction of the field. Often this is used as a simple and convenient means for stabilizing a satellite.[1,2,3]

The mission of the satellite determines, to a large extent, the importance of gravity-gradient effects. If the satellite constantly points at the Earth, as in communications and Earth-resources applications, then the gravity-gradient torques are essentially constant during the orbit. If the goal is astronomy, however, and the spacecraft must remain inertially fixed for long periods of time, then the gravity-gradient torques will have orbital periodicity.

Additional complications will arise if the anomalies due to Earth-mass distribution require recognition. Then a harmonic series representation of the gravitational field must be employed. During the design phase of a satellite, however, these considerations are reduced to first-order simplifications that might account for the Earth's oblateness. The simplest of all, of course, is the performance assessment of a spacecraft in circular orbit about a spherical Earth. The basic relationships for that case follow.

For a body in a circular orbit of radius R_o, about the Earth (see Fig. 3), a reference-coordinate system $(\vec{i}_r, \vec{j}_r, \vec{k}_r)$ is defined such that $\vec{k}_r$ is always directed radially toward the Earth, and its origin coincides with the mass center of the body. The vector radius $\vec{R}$ has components in the reference coordinate frame as follows

$$\begin{aligned} \vec{R} &= R_o \vec{k}_r \\ &= R_x \vec{i}_b + R_y \vec{j}_b + R_z \vec{k}_b \end{aligned} \tag{4}$$

The components of the vector $\vec{R}$ in body axes are obtained from the rotation matrix relating the two systems of axes by use of the Euler angles ψ, θ, ϕ. They are

$$\begin{aligned} R_x &= R_o \sin\theta \\ R_y &= R_o \sin\phi \cos\theta \\ R_z &= R_o \cos\phi \cos\theta \end{aligned} \tag{5}$$

If $\vec{\rho}$ denotes the radius vector from the body's center of mass to a generic mass element dm, it has the form

$$\vec{\rho} = x\vec{i}_b + y\vec{j}_b + z\vec{k}_b \tag{6}$$

and the radius vector from the Earth's center to this mass element becomes

$$\vec{r} = \vec{R} + \vec{\rho} \tag{7}$$

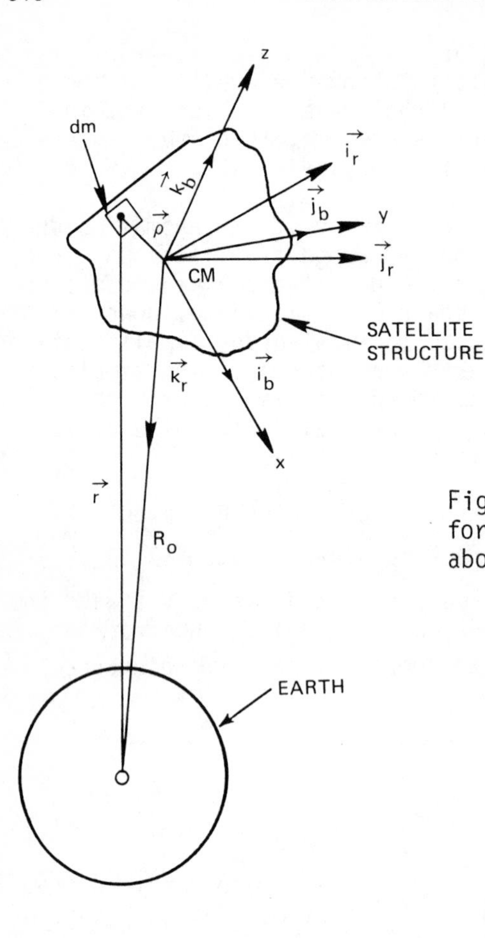

Fig. 3 Reference and body axes for a body in a circular orbit about the Earth.

The force due to gravity on a mass element is

$$d\vec{F} = -\frac{\mu dm}{r^3}\vec{r} \tag{8}$$

where

$$\mu = GM_e$$
$$= g_e R_e^2$$
$$= 398{,}601 \text{ km}^3/\text{s}^2$$

G = universal constant of gravity

M_e = mass of Earth

g_e = gravity acceleration at Earth's surface

R_e = radius of Earth = 6360 km

This causes a torque $\vec{dL}$ about the mass center of the vehicle

$$\vec{dL} = \vec{\rho} \times \vec{dF} = -\frac{\mu dm}{r^3}(\vec{\rho} \times \vec{r}) \tag{9}$$

The equations can be greatly simplified because the origin of the body axes coincides with the center of mass. Further simplifications can be gained if these same axes are the principal axes of inertia. With these assumptions, some algebraic manipulations lead to the final gravitational-torque results in scalar form

$$L_x = \frac{3\mu}{2R_o^3}(I_z - I_y) \sin 2\phi \cos^2 \theta \tag{10}$$

$$L_y = \frac{3\mu}{2R_o^3}(I_z - I_x) \sin 2\theta \cos \phi \tag{11}$$

$$L_z = \frac{3\mu}{2R_o^3}(I_x - I_y) \sin 2\theta \sin \phi \tag{12}$$

Here

$$I_x = \int (y^2 + z^2)\, dm, \text{ etc.}$$

and the product-of-inertia terms

$$I_{xy} = \int xy\, dm$$

are zero, since the principal axes of inertia are chosen to coincide with the body axes. Note that in the expression of the torques, only two Euler angles (θ, ϕ) appear. The third angle ψ, a rotation about the $\vec{k}_r$ vector, has no effect on the gravitational force on a mass element dm.

In an attempt to minimize unwanted gravity-gradient torques, the spacecraft designer must consider the means available to alter the response. The factor $3\mu/2R_o^3$ can only be decreased by moving the satellite to a higher orbit. This possibility is usually undesirable once the mission has been defined. Modifying the principal moments of inertia in order to minimize their differences will, however, affect the torques favorably. This can be achieved by redistributing the existing masses or even by adding a dummy mass at an appropriate location for the sole purpose of shifting axes. At the limit where $I_x = I_y = I_z$ (the case of a homogeneous sphere), for example, there exists no gravity-gradient torque. Finally, the same goal can be obtained if the principal axes of the spacecraft body are made to coincide with those of the reference frame, thus making the Euler angles ϕ, θ, and ψ equal to zero. This requirement is easily fulfilled if there exists mass symmetry with respect to the nadir direction in an Earth-pointing satellite. Although mathematically possible, this approach is never achieved in practice since other disturbances can perturb the spacecraft pointing, thus, nonzero Euler angles arise. The presence, however, of gravity-gradient torques from the Earth gravitational field is unavoidable for an inertially-fixed system in Earth orbit. In this case, the Euler angles ϕ, θ, and ψ (which are constantly changing) lead to disturbed torques that vary with orbital position.

As satellites increase in size and flexibility, the structural and dynamics problems associated with gravity-gradient effects increase in complexity. With highly flexible structures undergoing very large deformations, the inertia properties vary with the dynamic response, and thus, lead to similar high-frequency variations in the gravity-gradient torques. These torques will result in highly distributed force components that need to be resisted by a relatively flexible framework connecting the major mass concentrations. Attitude-control approaches thus need to recognize the high-flexibility and low-resonant frequencies of the satellite. In summary, the design of very large flexible platforms will require a careful assessment of gravity-gradient loads.

Atmospheric Drag

A satellite at an orbit height below 500 km encounters significant aerodynamic forces as it passes through the rarefied upper atmosphere. As a result of the braking effect of the atmosphere, the satellite experiences an orbital decay and usually a deterioration of its pointing accuracy and stability. Although these perturbing forces are sufficiently small that the spacecraft structure is not appreciably deformed in its configuration, the overall effect must be counteracted by an attitude control system. It is, again, the disturbance associated with that control mechanization that must be carefully designed to minimize the dynamic structural response of the delicate spacecraft.[1,2]

This section reviews the important factors entering the calculation of the upper-atmosphere density used in the evaluation of drag forces and torques. The numerous factors affecting the drag coefficient C_D are discussed, and the most appropriate method used to model the entire satellite for drag-force and torque calculations also is presented.

Upper Atmosphere Density Model. The determination of the upper-atmospheric density usually is based on the study of the orbital decay of a large number of Earth satellites.[7-12] The aim of the study is to define a simple analytic relationship for determining the upper-atmospheric density, allowing for its main variations (such as diurnal and semiannual variations) as well as its correlation with the intensity of solar activity at the 10.7-cm wavelength. The density ρ can be computed as a product of several factors, each of them corresponding to one or another form of the known density variations

$$\rho = \rho_H K_1 K_2 K_3 K_4 \qquad (13)$$

where

ρ_H = a nighttime-vertical profile of the atmospheric density

K_1 = a factor describing the density change with the change of solar activity F at 10.7 cm relative to some average level of radiation F_0

K_2 = a coefficient accounting for the diurnal effect of the density distribution

K_3 = a correction for the semiannual effect

K_4 = a factor taking into account the correlation between atmospheric density variations and geomagnetic perturbations.

The coefficients change with a change of the mean level of solar activity F_0 over the 11-year cycle of solar activity. Other density variations are not taken into account since they have a smaller influence. From the observation of the motion of existing Earth satellites, the available experimental material usually is divided in groups corresponding to mean levels of solar activity F of 75, 100, 125, 150, 200, and 250 units of 10^{-22} W/m^2 Hz.

All the K factors are described by involved analytical relationships. With the atmospheric density information and the definition of the orbit parameters, the orbital dynamics and the station-keeping requirements of a satellite can be evaluated with the help of specialized computer programs.

Once the density is estimated, computations to obtain the drag force applied to a satellite can begin. The basic equation for drag is given by the familiar aerodynamic relationship

$$D = 1/2\, \rho V^2 C_D A \tag{14}$$

where D is the drag force opposing the velocity V relative to the ambient gas (whose density is ρ) acting on a body with cross-sectional area A and drag coefficient C_D. In the altitude ranges considered here, the satellite moves in a free molecular-flow regime where the drag coefficient depends on the shape and orientation of the satellite, the angular distribution of reemitted molecules, the "accommodation coefficient", and the molecular speed ratio between free-stream velocity and most probable thermal velocity.

It is important to note that the drag coefficient here is more sensitive to the surface interaction phenomena than to the configuration geometry, making its value very difficult to determine in a laboratory. It is sometimes suggested that the C_D coefficient can be evaluated from the following equation

$$C_D = 2 + 8/9 \sqrt{1 - \alpha} \tag{15}$$

where the coefficient $\alpha = 4\mu/(1 + \mu)^2$ and μ is the ratio of the atomic mass of the gas over the atomic mass of the surface of the satellite.

These difficulties are further increased by the complex geometric shapes of modern satellites. In order to determine their aerodynamic characteristics, it is necessary to take into account the shadowing effects of the components and the multiple collisions of reflected particles. One should also emphasize that the symbol V in the drag equation represents the relative velocity between the satellite and the atmosphere, a value that is highly variable. Finally, if the spacecraft in Earth orbit is inertially fixed (instead of Earth-pointing), the drag force and torque become, in addition, a function of its continuously changing orientation with respect to the nadir.

For the structural designer and analyst to be able to provide an exact evaluation of the drag forces and torques disturbing a large satellite, he should include all the important previously described parameters in his model. A number of simplifying assumptions are imperative to keep the evaluation in line with reasonable computational capabilities. If an upper bound only is desired, the effects of the atmosphere on a satellite can be simply expressed in terms of maximum drag forces and torques. However, when an accurate study of a perturbed orbit is required, the task becomes very involved in order to generate the drag-resultant histories as a function of orbital position and time.

The most appropriate modeling method considers that the spacecraft is made of several major components. The shadowing effects and surface orientation to the incoming stream can be tracked along with other aerodynamic characteristics and orbital data. In a system of axes fixed to the spacecraft center of mass, the drag and torque vectors have the following expressions

$$\text{Drag force vector } \vec{D}(t) = 1/2\ \rho V^2 \sum_i C_{D_i}(t) A_i(t) \vec{N}_i \tag{16}$$

$$\text{Torque vector } \vec{T}(t) = 1/2\ \rho V^2 \sum_i C_{D_i}(t) A_i(t) [\vec{N}_i \times \vec{d}_i] \tag{17}$$

where i extends over all surfaces under consideration. In these expressions, the surface area A_i (the projection of surface i onto a plane perpendicular to velocity vector $\vec{V}$) and the corresponding drag coefficient C_{D_i} might vary with time t. This variation derives mainly from a change in orientation and an evolving shadowing effect. The symbol $\vec{d}_i$ represents the distance between the center of pressure of surface i to the center of mass of the entire satellite, and $\vec{N}_i$ indicates the direction of the unit vector normal to the surface A_i.

For obvious reasons, it is imperative to minimize the drag forces affecting the spacecraft. Once the orbit is defined, the only way to alter these forces is through a change in satellite geometric configurations and, perhaps, a modification of the surface characteristics of its components. At low orbit, especially where the atmospheric drag is more pronounced, it is desirable to consider modifications to the spacecraft geometry to alleviate the effects of drag.

An estimate can be made of the aerodynamic pressures experienced at various altitudes, however, by assuming a spherical satellite in a circular orbit. Then the orbital velocity is

$$V^2 = g/r_o \tag{18}$$

where

V = orbital velocity

r_o = orbital radius

g = gravitational constant (3.9×10^{20} dyn-cm^2/g)

By substituting Eq. (18) into Eq. (14) and dividing the force by the area, the drag pressure p may be computed as follows

$$p = \rho/2(g/r_o)C_D \qquad (19)$$

assuming that $C_D = 2$ and obtaining a suitable model for the variation of p with altitude.

Table 1 shows that when the satellite increases in size, the effects of aerodynamic pressure (especially at lower altitudes) will become more pronounced, and there will be a tendency for the orbits to decay much more rapidly. This fact must be considered for the low Earth-orbit assembly procedures planned for many future large platforms.

Solar Wind

Measurements of the solar-wind phenomenon have found it to be in the range of 10^8 protons/cm^2/s with the energy window ranging from 200 - 640 eV.

Assuming a diffuse collision process, the pressure may be expressed by

$$\text{pressure} = \text{mass flux} \times \text{velocity} \qquad (20)$$

Table 1 Aerodynamic pressures

Altitude (km)	Peak density ρ (gm/cm^3)	Peak aerodynamic pressure p (dyn/cm^2)
150	1.6×10^{-12}	1.2
300	4.3×10^{-15}	3.4×10^{-2}
400	3.6×10^{-16}	7.6×10^{-3}
500	5.0×10^{-17}	2.4×10^{-3}
700	2.2×10^{-18}	3.3×10^{-4}
800	6.8×10^{-19}	1.4×10^{-4}

The proton velocity has been estimated to be

$$V_p^2 = (\text{energy}) \times 1.9 \times 10^{12} \ (\text{cm/s})^2 \tag{21}$$

Using the upper bound of 640 eV for the energy of all particles

$$V_p = 3.5 \times 10^7 \ \text{cm/s} \tag{22}$$

Thus,

$$\text{pressure} = 5.9 \times 10^{-9} \ \text{dyn/cm}^2$$

This is a very small number compared to the solar pressure.

Earth's Magnetic Field

Instances can arise where the electrical currents within a satellite couple with the Earth's magnetic field. This effect is highly dependent on the configuration of the satellite, the generator currents, the circuit currents, and the eddy currents. With proper satellite design and shielding, these phenomena can be considerably reduced. A rough calculation, based on a 200-km orbit with an unshielded satellite containing 10 kg of iron, results in a torque in the range of

$$T = 2.7 \ \text{dyn-cm} \tag{23}$$

In the implementation of solar-power generation megasatellites extending many kilometers, this electromagnetic phenomenon can be expected to increase in importance, and it is necessary to track the phenomenon's growth.

Orbital Thermal Disturbance

An Earth-orbiting satellite is exposed to several external sources of thermal radiation.[4-7] While solar flux is predominant, solar reflection from the Earth, known as albedo, and the IR flux irradiated from a warm Earth may also be significant. Incident flux levels on the satellite vary as a function of orbit geometry, satellite orientation, and time. Such variations give rise to time-varying temperature distributions which, in turn, cause time-varying deformations of the structure. It is the task of the analyst to make suitable assumptions to build a mathematical model capable of adequately predicting the thermal behavior of the spacecraft. As a result of the model analysis, temperature distributions are predicted throughout the structure as a function of both spatial coordinates and time.

The three sources of radiation usually included in the thermal analysis of a space structure have the following characteristics:

1) Direct solar radiation is assumed constant for all practical purposes at 1350 W/m^2, which corresponds to the radiation intensity at the Earth's average distance from the sun.

2) Reflected solar radiation by the Earth (albedo) varies from about 10 - 80% of the direct-sun radiation, depending upon atmospheric conditions as well as the structure of the local Earth surface. A value of 0.30 is usually suggested. Additionally, the albedo depends on the view factor between the satellite and the Earth; thus, it must be multiplied by the factor $1/R^2$ where R is the distance from the center of the Earth to the satellite in Earth radii.

3) The energy-flux magnitude due to direct Earth radiation is a function of Earth's temperature, and it is emitted in the infrared range. An average of 250 W/m^2 is frequently used. Like the albedo, this number is affected by structural and earth view factors and must be multiplied by the same $1/R^2$ factor (R expressed in Earth radii).

The thermal environment must be translated into temperatures experienced by various components on the spacecraft itself. Some of these components, such as detectors, are constrained to operate at specific temperatures while, in other cases, the thermoelastic deflection response needs to be minimized.

In attempting to analyze a complex satellite structure for thermal response, the body under consideration may be idealized as an assemblage of finite elements. These elements (which can be bars, beams, plates, or shells) are so formulated as to characterize their continuum behavior at discrete nodal connections.[13,14]

For steady-state conditions (i.e., the case where time effects on the temperature distribution have not been included) the solution is readily obtainable from the system matrix equation representing the complete structure

$$[M]\vec{\theta} = \vec{Q} \tag{24}$$

where

$[M]$ = conductivity matrix

$\vec{\theta}$ = vector of the nodal point temperatures

$\vec{Q}$ = nodal point heat flow input vector

The vector $\vec{Q}$ may be composed of $\vec{Q}_B$, $\vec{Q}_S$, and $\vec{Q}_C$ where B denotes temperature generated by the body, S by the surface, and C refers to concentrated nodal point heat inflow. The solution vector $\vec{\theta}$ describing the nodal temperatures is obtained by simple inversion of [M], since the system of equations is linear.

With significant heat-flow-input change over time, it is important to include terms that account for the rate at which heat is stored within the material. The rate of heat absorption is proportional to the heat capacity of the material and the rate of temperature change.

The heat flow equilibrium equations corresponding to the <u>transient</u> state are now

$$[C]\dot{\vec{\theta}} + [M]\vec{\theta} = \vec{Q} \tag{25}$$

where [C] is the heat capacity matrix, and $\vec{Q}$ as well as $\vec{\theta}$, is time dependent.

Boundary conditions, such as convection and radiation in which the rate of heat transfer is dependent on the temperature of the surface of the body and its environment, are included through a contribution to both the $\vec{Q}$ vector and the conductivity matrix [M].

The solution to this system of differential equations—one that evaluates the nodal temperatures as a function of time—is most often carried out by iterative techniques, since it is usually nonlinear because of the inclusion of the radiative transfer effects. A number of large engineering-analysis computer programs have been developed for thermal analysis problems.

Current software and hardware capabilities permit the discretization of the continuum in heat-transfer problems to many nodes. (One hundred nodes is typical, but a 1000-node model is not unusual.)[17,20,21] When these techniques are applied to the analysis of a space structure, additional calculations often are needed in the evaluation of the response. The radiation-flux vectors from the sun and the Earth are constantly changing their orientation as the satellite moves along its orbit, and they must be accounted for. In addition, some components of the spacecraft might shield others, and these time-varying effects also are included by generating a table of varying shadows. Other tables also are necessary to express the view factors of various elements radiating to each other. When accounting for thermal conduction and radiation effects between the components of a complex space structure, the thermal model may tend to become very large and possibly unmanageable. The analyst's art, therefore, must enter the assessment of the important factors to be included and in making simplifying assumptions while preserving the required level of accuracy.

As the requirements on dimensional stability become more stringent, it is often necessary to shield the delicate spacecraft from the thermal effects. Several passive means of protecting spacecraft components have been developed (e.g., thermal blankets). By a combination of thin material layers having carefully chosen emissivity and absorptivity coefficients, temperature excursions and gradients can be reduced drastically. These passive methods are preferred to active cooling approaches which often create additional dynamic disturbances on the satellite.

In general, the thermal histories are "slow" compared to the satellite dynamic characteristics (natural periods); thus, they can be treated as quasistatic disturbances. In some cases, especially with large platforms when the structure is extremely flexible with low-heat capacity members, the transition from eclipse to sudden solar exposure could trigger a thermal shock. In this event the vibration modes of the structure could be excited, which could lead to dynamic distortions larger than otherwise expected in the static treatment of that same thermal disturbance.

On-Board Disturbances

The various forces and torques applied to a spacecraft by its environment are resisted by attitude and orbit control

Table 2 On-board disturbances

Source of disturbance	Amplitude range	Frequency range (Hz)
CMG imbalance forces	0.1 - 10 N	10 - 50
Reaction wheel imbalance forces	0.001 - 1 N	1 - 50
Reaction wheel bearing noise	0.0001 - 0.01 N	1 - 100
Solar array drive torques	0.001 - 0.01 N·m	10^{-3} - 10^{-1}

wheels and jets. These devices produce high-frequency mechanical disturbances that can cause a response at natural frequencies of vibration of the satellite. Satellite appendages that vary position with time, such as solar panels and communication antennas, provide disturbances through their drive mechanisms. The magnitudes of these various forces and torques and their spectral content depends, of course, on the specifics of the satellite mission, orbit, and configuration, but some ranges may be estimated as in Table 2.

Thermal effects sometimes are countered by active cooling devices that may provide disturbances through turbulance, mass shifts, or mechanical pump imbalances. Since these applications generally are not common, very little specific information can be provided on them herein.

IV. Structural Distortion Evaluation

The satisfactory performance of a large space system is crucially dependent on an accurate representation of the structure's elastic motion under prescribed forces and temperatures. The study of the expected environmental disturbances, as discussed previously, determines in amplitude and frequency a set of forcing functions that are regarded as input to the structural system. This section discusses problems associated with accurate mathematical modeling of the structure and presents methods of solution.

Thermal Distortions

The results of a thermal study are usually expressed in terms of nodal temperatures from a finite-difference or a finite-element model, which may be either steady-state or transient in nature. The structural distortion calculation requires the development of a further elastic-assembly model, usually also of finite element discretization. Such analysis tools currently are well developed and may be found at almost any computer center. NASTRAN, STARDYNE, ADINA, ANSYS, STRUDL, and SPAR are but a few of the possibilities.[17-21]

Like the thermal finite-element approaches, the structural method discretizes the various struts, plates, shells, and solids into elements over which the internal displacement variation follows an assumed function. The elements are connected to each other at the nodes where equations of continuity assure compatible deformations of adjacent elements. Force equilibrium also is assured at the nodes and the system-level deformations are determined subject to a suitable conservation of energy criterion. In effect, it is highly similar to the solution of a large number of small attached Galerkin or Rayleigh-Ritz problems with mutual boundary conditions.[13-16]

For static mechanical or thermal loads, the system level equations can be expressed as

$$[K]\vec{U} = \vec{R} \tag{26}$$

where

$[K]$ = the stiffness matrix of the structure

$\vec{U}$ = the displacement vector describing the deformed structure

$\vec{R}$ = the vector of applied forces (mechanical or thermally derived)

Under linear behavior conditions or assumptions, this is solved to obtain the set of $\vec{U}$s by inversion of [K] so that

$$\vec{U} = [K]^{-1}\vec{R} \tag{27}$$

Stresses are obtained by further manipulations of the $\vec{U}$ vector at the individual finite element level.

The stiffness matrix [K] is a system-level representation of the materials, dimensions, and mutual interconnections of the structure. The vector $\vec{R}$ includes mechanical forces and torques that are applied to the structure. For thermal loading, the vector $\vec{R}$ is derived from the temperature changes experienced and is proportional to the coefficient of thermal expansion of the structure materials. Having the temperature changes and knowing the thermoelastic properties of the structure, the thermal distortions and stresses can be readily obtained.

Dynamic Distortions

The dynamic analysis takes the same static finite-element approach several steps further in that it develops the entire equations of motion including the inertia forces, damping forces, and time-dependent disturbances. These equations, again in matrix form, can be expressed as follows

$$[M]\ddot{\vec{U}} + [C]\dot{\vec{U}} + [K]\vec{U} = \vec{F}(t) \tag{28}$$

where

$[M]$ = mass matrix of the structure. It includes both the structural masses that resist the deformation as well as the masses derived from nonstructural elements

$\ddot{\vec{U}}$ = acceleration vector at the nodes

$[C]$ = damping matrix. It is generally difficult to determine and computationally cumbersome except in the special cases where it can be diagonalized. Other more convenient approaches use the modal damping method described later

$\dot{\vec{U}}$ = velocity vector at the nodes

$\vec{F}(t)$ = time history of the applied force or torque disturbance on the structure. If this forcing function changes very slowly relative to the natural frequencies of the structure, the acceleration and inertia terms become small in comparison to the elastic term ($[K]\vec{U}$) and reduce the problem to a static one

$[K]$ = stiffness matrix of the structure again

$\vec{U}$ = displacement vector at the nodes

It is possible to develop a very detailed model for static and thermal analysis with many thousands of nodal displacements and rotational degrees of freedom, and have considerable confidence in the validity of the answer. The dynamic problem cannot usually afford this luxury. The main problem with dynamics lies in the proper formulation of an adequate lumped-parameter model which is sufficiently detailed to exhibit all phenomena of interest and yet is not so large that it overwhelms the available computational tools. Problems ordering thousands of degrees of freedom in the static case are computationally equivalent to those with hundreds of degrees of freedom in dynamics. To derive equivalent levels of accuracy in the solution of dynamic problems, additional steps such as component-mode synthesis (substructuring) need to be implemented.

Returning now to the equations of motion, Eq. (28), the first step in a dynamics (vibrational) analysis is the extraction of the natural frequencies and their associated mode shapes. To do this, one obtains the homogeneous solution to the differential equations assuming negligible damping (or proportional for [M] to [K]). Thus

$$[M]\ddot{\vec{U}} + [K]\vec{U} = 0 \tag{29}$$

letting

$$\vec{U} = \vec{\phi}e^{i\omega t}$$

thus

$$\ddot{\vec{U}} = -\omega^2\vec{\phi}e^{i\omega t}$$

Substituting into Eq. (29), one obtains the eigenvalue system

$$([K] - \omega^2[M])\vec{\phi} = 0 \tag{30}$$

This equation is then solved to obtain the N eigenvalues (natural frequencies) ω and N eigenvectors (normal modes) $\vec{\phi}$. The normal modes are, of course, orthogonal (a property responsible for many of the simplification strategies in the dynamics analyses).

The mode shapes are used to transform the physical coordinate vector $\vec{U}$ into the modal coordinate vector $\vec{\xi}$ by a similarity transformation

$$\vec{U}(t) = [\vec{\phi}^{(1)}, \vec{\phi}^{(2)}, \ldots, \vec{\phi}^{(N)}]\vec{\xi}(t) \tag{31}$$

This results in a set of N uncoupled differential equations in modal space

$$M_r\ddot{\xi}_r + \omega_r^2 M_r \xi_r = \Xi_r(t), \quad r = 1,2,\ldots,N \tag{32}$$

where

$M_r = \vec{\phi}^{(r)T}[M]\vec{\phi}^{(r)}$ is the generalized mass of the rth mode

$\Xi_r(t) = \vec{\phi}^{(r)T}\vec{F}(t)$ is the generalized mass force of the rth mode

While adding some complication in computation of the new generalized-force vector, the differential equations are now uncoupled, but wherever necessary, superposition can be used. The advantage here is that now the response calculations can be obtained only for those modes of interest (those likely to be excited) instead of for the entire spectrum. The response is computed in Eq. (32) and transformed back into physical space by Eq. (31). The computational advantages are tremendous. Figure 4 describes the modal analysis process through basic diagrams.

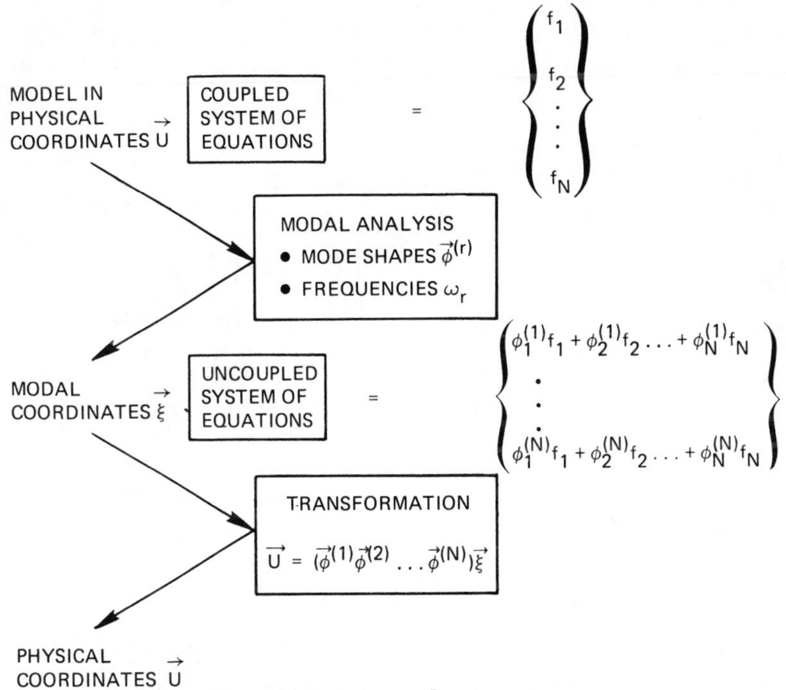

Fig. 4 Modal analysis steps.

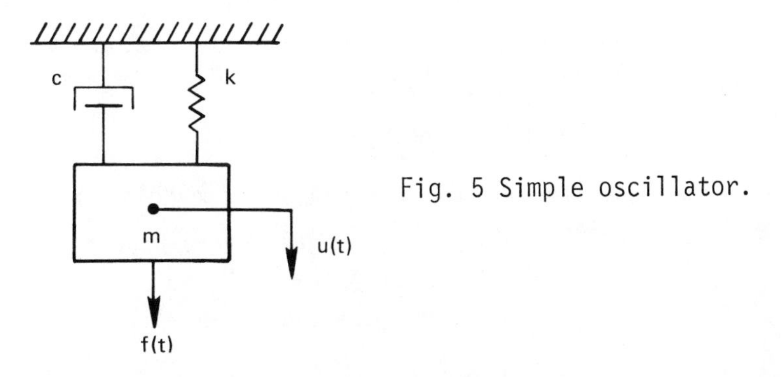

Fig. 5 Simple oscillator.

It should be noted that now each uncoupled modal equation represents a single natural frequency and the mode shape with which it responds. It has become, in effect, a single-degree-of-freedom system. This allows it to be used in highly simplified response calculations. It also permits the introduction of damping through "modal damping" into each equation. Experience with real structures has indicated that this is a more realistic approach than that through the diagonal [C] matrix previously described.

While it is expected that the satellite structure will form a multi-degree-of-freedom system, it is sufficient to examine the response of a number of single-degree-of-freedom systems given the proper modal frequencies.

The disturbances a satellite might experience will vary greatly in form, intensity, and frequency content. Three generic loadings might be sufficient to approximate most of the previously identified disturbances. Therefore, the expressions of dynamic response are examined for 1) impulsive loading (thrusters), 2) harmonic loadings (reaction wheels, control moment gyros), and 3) random input. In the first case the transient response is of greatest interest; whereas with the sinusoidal loading, the steady-state response is needed. This last response is often expressed in terms of resonance peaks. Finally, in random vibration analysis the response of a system to a random input is expressed by a statistical average, most often the mean-square value of displacements. Expressions are given for responses to random forces (or torques) with power spectral densities (PSDs) having the form of pure white noise.

Figure 5 describes a simple oscillator and identifies the parameters entering into the equations of motion as follows.

m = mass of the system

k = spring constant or stiffness

c = viscous damping coefficient

$f(t)$ = applied time-dependent disturbance

$u(t)$ = response of the system

In addition, the following parameters are defined

$\omega = \sqrt{k/m}$ = undamped natural frequency of the system (rad/s)

$C_{cr} = 2m\omega$ = critical damping value

$\zeta = C/C_{cr}$ = damping ratio

$\omega_D = \omega\sqrt{1 - \zeta^2}$ = damped vibration frequency

It should be noted that this could be a single-degree-of-freedom physical system, as described by Eq. (28), or a modal representation, as described by Eq. (32). It should also be noted that the damping ratio ζ must be less than unity (a condition which is very easy for the low-damped ($\zeta = 0.01$) space structures to meet).

Table 3 Simple oscillator responses to three generic loadings (underdamped system $0 < \zeta < 1$)

Impulsive loading	Harmonic loading	Random excitation
Loading:	Loading:	Loading:
$f(t) = \delta(t)$ unit impulse	$f(t) = p_o \sin \Omega t$	$S(\omega)$: PSD; example of white noise: $S(\omega) = S_o$
Response:	Response:	Response:
$h(t) = \frac{1}{m\omega_D} \ell^{-\zeta\omega t} \sin \omega_D t$	$u(t) = \frac{p_o}{m\omega^2}\left[\left(1 - \left(\frac{\Omega}{\omega}\right)^2\right)^2 + \left(2\zeta \frac{\Omega}{\omega}\right)^2\right]^{-1/2} \sin(\Omega t - \theta)$	Mean-square value of response u
where $\omega_D = \omega\sqrt{1 - \zeta^2}$	$\theta = \tan^{-1} \frac{2\zeta\frac{\Omega}{\omega}}{1 - \left(\frac{\Omega}{\omega}\right)^2}$	$s_u = \frac{\pi S_o}{kc}$

Table 3 gives the responses for the single-degree-of-freedom oscillator to the three generic disturbances.

The fact that the large platforms planned for the future have first natural frequencies often starting at 0.01 Hz is of great interest to today's satellite designer. This represents a hundred-fold reduction from today's conventional designs and indicates susceptibility to a far greater range of disturbances.

V. Performance Evaluation and Improvement

The modeling techniques and analysis methods described in the previous sections permit the evaluation of the structural distortions of a complex space structure exposed to the disturbing orbital environment. Proper superposition of the effects of different disturbances, whether quasistatic or dynamic, can provide information on the satellite condition at any desired position or time in orbit. The relative displacements of key components are consequently used in the performance evaluation of the system. For example, the motions of an antenna feed relative to a dish or vibrations of a critical instrument in a payload package may be important measures of performance.

The desired performance goals tend not to be met in all respects at first, since it is a very difficult task to anticipate ultimate structural behavior. The spacecraft designer is faced with a multitude of options and constraints when attempting to improve the performance of this system. The options for system enhancement can be broadly categorized as passive or active means of controlling the structural response.

In the case of passive control, the mathematical expression relating the performance (output) to the disturbances (input) through the structural system illustrates the importance of the interacting parameters. While it is possible to single out those variables that are the predominant contributors to the performance evaluation, one must also recognize that they are subject to constraints of a physical nature or derived from practical requirements. The block diagram in Fig. 6 attempts to group the variables of input and system along with their respective constraints and limits, thereby indicating their relationships. The possibilities of modifying the environmental input disturbance remains quite limited. The geometrical attributes of a satellite could be modified in order to alleviate the effects of such disturbances as atmospheric drag and gravity-gradient. A more effective approach

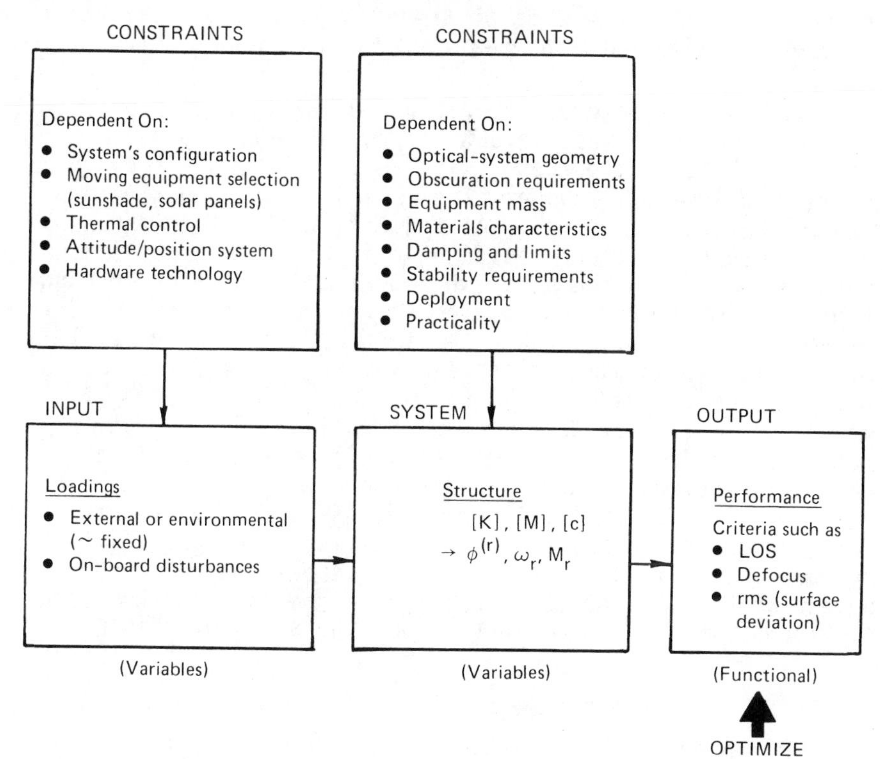

Fig. 6 Interactions of system variables and constraints.

might consist of altering the system by modifying its mechanical, thermal, and dynamic resistance. Improved material stiffness, damping properties, and lower thermal expansion coefficients, when combined with reductions in structural mass, are undeniably most desirable. An unconventional more direct approach may be suggested whereby the ultimate performance is obtained through optimization for highly specific functions. While the complete solution to such optimization formulations is expected to be very complex, it should be explored because of the tremendous efficiencies possible.

When passive means of response minimization remain unsuccessful, active approaches may be investigated. In the early designs of the NASA Large Space Telescope, the thermal environment produced deformations of the mirror surfaces which were unacceptable from a system performance viewpoint. Active optics (consisting of an interferometric error sensor scanning the reflecting surface, displacement actuators behind the mirror bending it into the desired shape, and a controller recognizing the properties of the highly coupled distributed parameter system) showed great promise in solving the problem.

Excessive spacecraft vibration during operation presents an extension of the active-optics problem into the dynamic domain. There, the purpose is to reduce the vibrations to an acceptable level. A number of strategies may be pursued which range from enhanced damping of the entire structure through control means, to the creation of a limited number of control-induced "nodes" in the structure where the vibration is highly attenuated while the rest of the structure is permitted to respond. The first approach tends to be dynamically stable, but the levels of damping achievable are low, and the payload increase due to the controllers may not be acceptable. The second approach provides for very high levels of vibration suppression, but since it applies additional energy into the structure instead of merely absorbing energy, it runs a high risk of destabilization. There are, of course, many intermediate stages that combine the characteristics of both limiting approaches. A full discussion of these new control technologies, which will become increasingly important as the large platforms emerge, is somewhat beyond the scope of this paper.

VI. Conclusions

This paper is a rather brief introduction to the problem of structural distortions in satellites exposed to the orbital environment. Much work remains to define the factors that degrade satellite performance in orbit, and in how these factors are modified as the size of satellites increases from a few meters to many kilometers. New analytical and computational tools must be developed to assess the interaction of the various disturbance sources with the satellite configuration to produce thermal and mechanical loadings. The modeling and computational technologies in the thermal, static, and dynamic areas need enhancement to produce yet finer-grained distortion data over wider bandwidths. Finally, technical means must be developed to enhance the performance of these systems through various control means. The problem, in short, is far from solved.

References

[1]Greensite, A.L., Analysis and Design of Space Vehicles Flight Control System, Spartan Books, New York, 1970.

[2]Jensen, J., Townsend, G., Kork, J., and Kraft, D., Design Guide to Orbital Flight, McGraw-Hill, New York, 1962.

[3] Kaplan, M.H., Modern Spacecraft Dynamics and Control, Wiley, New York, 1976.

[4] Sparrow, E.M. and Cess, R.D., Radiation Heat Transfer, Brooks-Cole Pub. Co., Belmont, Calif., 1966.

[5] Kreith, F., Radiation Heat Transfer, International Textbook Co., Scranton, Penn., 1962.

[6] Hottel, H.C. and Sarofim, A.F., Radiation Heat Transfer, International Textbook Co., Scranton, Penn., 1962.

[7] Slowey, J.W., "Earth Radiation Pressure and the Determination of Density from Atmospheric Drag", Committee on Space Research, Konstanz, 1973.

[8] Jacchia, L.G., "Revised Static Models of the Thermosphere and Exosphere with Empirical Temperature Profiles", Smithsonian Inst., Washington, D.C., Astrophysics SR 332, 1971.

[9] Jacchia, L.G. and Slowey, J.W., "Diurnal and Seasonal Latitudinal Variations in the Upper Atmosphere", Smithsonian Astrophysical Observatory, Washington, D.C., SR 242, 1967.

[10] Jacchia, L.G., "Recent Advances in Upper Atmosphere Structure", Committee on Space Research, Prague, 1969, Space Research X, 1970.

[11] Jacobs, R.L., "Atmospheric Density Derived from the Drag of Eleven Low Altitude Satellites", Journal of Geophysical Research, Vol. 72, June 1967.

[12] Anderle, R.J., "Atmospheric Density Variations at 1000 km Altitude", Naval Surface Weapons Center, Dahlgren, Va., TR 2568, April 1971.

[13] Zienkieweicz, O. and Cheung, U., The Finite Element Method in Structural and Continuum Mechanics, McGraw-Hill, New York, 1967.

[14] Bathe, K.J. and Wilson, E.L., Numerical Methods in Finite Element Analysis, Prentice-Hall, Engelwood Cliffs, N.J., 1976.

[15] Clough, R. and Penzien, J., Dynamics of Structures, McGraw-Hill, New York, 1975.

[16] Gallagher, R., Finite Element Analysis Fundamentals, Prentice-Hall, Englewood Cliffs, N.J., 1975.

[17]MacNeil, R., "The NASTRAN Theoretical Manual (Level 15.0)", NASA SP-221(01), December 1972.

[18]Connor, J., Ferrante, A., and Logeher, R., "ICES STRUBL-II; The Structural Design Language, Engineers Users Manual", Dept. of Civil Engineering, MIT, Cambridge, Mass., 1969.

[19]Bathe, K., "ADINA: A Finite Element Program for Automatic Dynamic Incremental Nonlinear Analysis", Dept. of Mechanical Engineering, MIT, Cambridge, Mass., Rept 82448-5, May 1977.

[20]"ANSYS: Engineering Analysis System Users Manual", Swanson Analysis Systems, Inc., Elizabeth, Penn., E00173-01 (Rev. 1), January 1973.

[21]Bathe, K., "ADINAT: A Finite Element Program for Automatic Dynamic Incremental Nonlinear Analysis of Temperatures", Dept. of Mechanical Engineering, MIT, Cambridge, Mass., Rept. 82498-5, May 1977.

DYNAMICS OF A RIGID BODY IN THE SPACE PLASMA

P. J. L. Wildman*

Air Force Geophysics Laboratory,
Hanscom Air Force Base, Bedford, Mass.

Abstract

The drag and torque forces acting on a large conducting body passing through a partially ionized plasma are calculated over the altitude range 250 km to 36000 km (geosynchronous altitude) for a nonrotating body 2 km long and 10 m wide with mass 2 kg. Drag forces resulting from solar radiation pressure, collisions with neutral particles, collisions and interactions with charged particles, and interactions with the Earth's magnetic field are relatively unimportant. Torques resulting from these same processes are more important. The torque induced by the Earth's gravitational field is the most important of all and dominates all others even at geosynchronous altitudes. The additional forces resulting when the body also has rotational motion are negligible.

I. Introduction

During the decade of the 1980's, many new types of systems will be placed in orbit around the Earth. They will have a number of purposes, the two most probable at the beginning being a) facilities, possibly inhabited, for materials production and processing under 0-g conditions, and b) some form of solar power collector with a means of transmitting that power to Earth. From the point of view of the design and on-orbit operation of these "Space Stations", the most dramatic change from the orbiting objects of the 1970's will be their size. Early examples of these systems will have linear dimensions on the order of hundreds of meters up to several kilometers, with sizes projected to reach 10 km by the mid 1990's (Hagler, et al., 1977). With the exception of thin antennas and booms on scientific satel-

---

*Physicist, Plasmas, Particles and Fields Branch, Space Physics Division; presently Research Physicist, Dynamics Division, British Aerospace, Filton, Bristol, England.

lites, no object so far orbited has exceeded a few tens of meters maximum dimension. The structures will be built initially by deployment of assemblies from the Space Shuttle cargo bay and later by making use of the multiple flight capability of the Space Shuttle to transport construction materials, fabrication equipment, and workers to a low Earth orbit.

In this paper the interaction between large space structures and their environment will be considered insofar as it produces mechanical perturbing forces which might impact orbital lifetime or, possibly more important in practical terms, the attitude stability and control of such a structure. We will not consider possible techniques of attitude control, but simply demonstrate the kinds of forces that will be present. The structure will be assumed to be rigid, nonrotating (see Appendix), and any oscillation or deformation caused by the external forces will not be considered at this time. It should be remembered, however, that the accuracy and reliability of attitude control is crucial in some applications, such as solar power transmission to earth. In those cases the structures will have to exhibit rigidity and/or controllability over a wide frequency range of applied forces, and that requirement may well be the deciding factor in the choice of construction and assembly methods (Oglevie, 1978).

There are many independently acting physical processes that act upon a body, natural or artificial, passing through the space plasma. Frequently, however, many of these processes result in mechanical or electromechanical forces on the body which are many orders of magnitude smaller than one or two dominant mechanisms, and they may, therefore, be neglected. Exactly which mechanism will exert the greatest force on a body in a specific situation depends primarily on the body's location, size, and geometry. Factors such as solar illumination and spacecraft attitude may cause regular or intermittent changes in the balance of forces acting and these changes can have a cumulative effect, due to resonance with the structure of the body or the dynamics of its orbit.

II. Classes of Interaction Between a Moving Body and a Partially Ionized Plasma

In this section, the "body" (spacecraft or structure) is taken to be electrically conducting, nonmagnetic, and to have no rotation about its center of mass (see Appendix). Unless otherwise stated, no specific geometry is implied. Collisions between particles, charged or neutral, are neglected, because of the very low collision frequencies above about 150 km altitude.

A. Charging Effects

Spacecraft Potential (Vehicle Potential or Plasma Potential). Except for transients caused by fluxes of energetic charged particles or by changing solar illumination, there can be no net flux of charge to an isolated body in a partially ionized plasma. Different parts of a spacecraft that are electrically isolated can charge to potentials differing by thousands of volts. This is the case when the spacecraft is in the shadow of the Earth at geosynchronous altitudes, due to the low ambient thermal plasma density. Electrical discharge between surfaces can occur when the breakdown voltage is exceeded. If this happens between thin layers of dielectric coatings separating electrical components, for example, it can damage or destroy critical parts of electrical systems. It is the need to eliminate this damage that has been the driving force behind investigations into spacecraft charging mechanisms over the past decade.

The effect of small potentials of 1 V or less also is of interest, because of the drag and torque forces they cause. These forces are caused by the collision and near collision of the charged spacecraft with the ambient plasma particles, and also by the interaction between the ambient magnetic field and currents flowing to and within the spacecraft.

The equilibrium potential Φ of a body in the space plasma is the potential with respect to the plasma at which there is no net flow of charge to or from the body. At this potential Φ, we have

$$I_- = I_+ + I_S + I_{ph} \tag{1}$$

where I_- = incident electron current
I_+ = incident ion current
I_S = a positive current due to secondary electrons from ion and electron impact and from backscattering of incoming electrons
I_{ph} = photoemission current

In order to determine Φ, expressions must be developed for the individual currents constituting Eq. (1). This requires data on the thermal energy plasma, the energetic particle fluxes, and on the backscattering and secondary emission coefficients (Knott, 1972). In many cases these data are not available, and the usual practice is to use laboratory data or to extrapolate existing space measurements. The full version of Eq. (1) is then solved numerically for Φ. This modeling approach has been used most extensively to establish the equilibrium potential of cylindrical or spherical shaped satellites with dimensions of a few meters (Garrett, this volume). In such models the relatively small disturbing effect of the Earth's magnetic field is often neglected. However, the fact that the electrons generally can travel only parallel to the magnetic field places severe limitations on possible electron fluxes in certain geometries and must be considered in any detailed treatment of the equilibrium potential (Chu and Gross, 1966; Beard and Johnson, 1960). Magnetic field effects will be considered in later sections of this paper.

Under most conditions the polarity and magnitude of Φ depend upon the balance in Eq. (1) between I_{ph}, the total photocurrent, and the portion of I_- that is due to thermal energy electrons reaching the body. There are, however, two important exceptions to this statement: 1) spacecraft in geosynchronous orbit under eclipse conditions (spacecraft in the Earth's shadow), and 2) regions where the thermal plasma density exceeds a value of approximately 5×10^3 cm^{-3}. In the first case, at geosynchronous altitude particles with energies on the order of several tens of keV frequently are present and impinge on the spacecraft. If electrons predominate in this flux, then in shadow, where no photocurrent is generated to balance this negative flux, the spacecraft will rest at a negative potential of many kilovolts in order both to reduce the electron flux and to attract positive ions as a means of restoring zero net charge flow to the spacecraft.

In the second case, at low altitudes in a relatively dense thermal plasma ($N \geq 5 \times 10^3$ cm^{-3}), the isotropic flux of rapidly moving electrons is several orders of magnitude greater than the photocurrent ($I_{ph} \approx 3 \times 10^{-9}$ A cm^{-2}). It is balanced by the flux of relatively slow moving positive ions swept up by the rapid motion of the spacecraft ($v_s \approx 7 \times 10^5$ cm sec^{-1}), and under these conditions the absence of solar illumination to generate a photocurrent makes little or no difference to the value of the equilibrium potential Φ. This is generally the case at altitudes below about 1500 km where Φ has a value of 1- 2 V negative, although plasma densities low enough to allow the photocurrent to predominate in Eq. (1) may be encountered below 1500 km in auroral and polar regions. At plasma densities down to a few cm^{-3} a positive potential of some tens of volts is sufficient to equalize the photocurrent in the absence of high energy particles.

As indicated, extensive studies have been made of the equilibrium conditions of small spherical satellites (Samir and Willmore, 1966; Chang and Smith, 1960; Brundin, 1963). An important factor in those studies is the extent to which the plasma is modified by the electrostatic charge and motion of the spacecraft. This falls under the general category of "sheaths and wakes" phenomena and is of great importance when scientific measurements of the ambient plasma are required; in order to measure charged particles with energies of a few tenths eV the collecting probes must be at the plasma potential (Samir et al., 1973). The electrostatic charge on the spacecraft is screened from the ambient plasma by a sheath region where particles of one sign predominate (positive ions in the case of a negatively charged spacecraft). This sheath is characterized by the Debye length, λ_D which is the distance from the surface at which the potential falls to 1/e:

$$\lambda_D = \left(\frac{kT_e}{4 \pi n_e e^2} \right)^{1/2} \quad \text{cm} \tag{2a}$$

$$\lambda_D = 6.9 \left(\frac{T_e}{n_e} \right)^{1/2} \quad \text{cm} \tag{2b}$$

where T_e = thermal electron temperature, deg K
n_e = electron density, n cm^{-3}
e = electron charge, C
k = Boltzmann constant, 1.38×10^{-16} erg K^{-1}

The degree of electrostatic screening and the effect of ion and electron thermal velocities in "filling in" any wake region left by the rapid passage of the spacecraft through the plasma must be considered in an exact evaluation of equilibrium conditions (Brundin, 1963; Kasha, 1969). However, since they are essentially edge effects, sheaths and wakes have little impact on the mechanical forces acting on an extensive planar structure in the space plasma, except when very high voltages, on the order of thousands of volts, are generated, [e.g., on a solar power station (Parker, 1978; Freeman et al., 1979)].

We note here for completeness the equation for Φ for a conducting sphere, radius R in the 150-1500 km altitude range. At such altitudes Φ is generally negative, and the local Debye length is small compared with the dimensions of the sphere. Neglecting wake effects and the presence of the Earth's magnetic field, Φ is given by (Brundin, 1963)

$$\Phi = \frac{kT_e}{2e}\left[\ln\left(\frac{8kT_e}{\pi m_e v_s^2}\right) - 2\ln\left(1 - \frac{I_{ph}}{v_s \pi R^2 n_e e}\right)\right] \quad (3)$$

where v_s = satellite velocity
m_e = electron mass
I_{ph} = total photoelectron current

Planar Surface Current Flow. When considering large-scale conducting structures, it is necessary to consider the different potentials that are induced at different locations on the structure by its motion through the Earth's magnetic field (Section II E). Currents of different polarities and area densities, therefore, flow through the surface at different locations on the structure. As these currents flow within the structure, they cause drag, and possibly torque, as they interact with the ambient magnetic field. We quote here expressions for the currents flowing from a Maxwellian plasma to a planar surface at known potential for both accelerating and retarding modes, neglecting edge effects (Whipple, 1959). These currents must be evaluated so that the flow pattern within the structure can be used to calculate the resulting magnetically-induced drags and torques.

In an accelerating mode (Φ positive for electrons and negative for ions), the current flowing to area A is given by

$$I_{acc} = \frac{neA}{\sqrt{\pi}\, a}\left\{\frac{\sqrt{\pi}}{2}\gamma\cos\theta\,[1 + \mathrm{erf}(\gamma\cos\theta)] + \frac{1}{2}\exp[-\gamma^2\cos^2\theta]\right\} \quad (4)$$

In a retarding mode (Φ negative for electrons and positive for ions),

$$I_{ret} = \frac{neA}{\sqrt{\pi}\, a}\left\{\frac{\sqrt{\pi}}{2}\gamma\cos\theta\,[1 - \mathrm{erf}(x - \gamma\cos\theta)] + \frac{1}{2}\exp[-(x - \gamma\cos\theta)^2]\right\} \quad (5)$$

$$x = \sqrt{e\Phi/kT} \quad (6)$$

$$a = m/2kT \quad (7)$$

$$\gamma = v_s/\sqrt{a} \quad (8)$$

where e = electron charge
n = density of the particles
m = mass of a particle
T = Maxwellian temperature of the particles
k = Boltzmann constant
Φ = potential difference of surface relative to plasma
A = area of the surface
v_S = velocity of surface relative to plasma
θ = angle between the surface normal and v_S

B. Solar Radiation Presssure

Any object in sunlight will be subjected to a force caused by momentum transfer from solar photons to any surface that they strike. For an Earth satellite in a circular orbit that does not enter the Earth's shadow, the net acceleration due to solar radiation pressure is zero, since the force is accelerating for one-half the orbit and retarding for the other half. For a body in an elliptic orbit that enters the Earth's shadow, there will be a net change in the energy of the body due to solar radiation pressure, unless the accelerating and retarding radiation pressure forces are equal over a complete orbit. This is a geometry that cannot persist for an extended period, because of the shift of the orbit in inertial space; therefore, radiation pressure must be considered in the calculation of most orbital dynamics. For interplanetary missions, the radiation pressure may be in the same sense for extended periods and can be harnessed to accelerate, slow down, or stabilize probes during long, otherwise unpowered, coasting phases (Sohn, 1959). It can even be used as the primary means of propulsion, as in the NASA solar sail proposal, to carry a payload to rendezvous with Halley's comet.

If the area presented to solar radiation is asymmetric about the body's center of mass, a torque will be applied even in orbits where radiation pressure causes no net acceleration of the center of mass. This mechanism can be the principal cause of applied torque, especially in the case of a structure that projects large optically opaque surfaces to solar radiation (as opposed to one with an open structure). It would be possible under some orbital and spacecraft configurations to balance out magnetic and/or aerodynamic torques by careful positioning and trimming of open and closed areas. The feasibility of using solar radiation pressure to trim spacecraft attitude has already been demonstrated on the OTS-2 geosynchronous satellite (Renner, 1979). (A small additional force under conditions of solar illumination arises from the recoil of the solar-induced photoelectrons from the body. This is on the order of 1% of the true radiation pressure and may be neglected.)

For a perfectly absorbing surface, the radiation pressure is equal to the energy density of solar radiation, and the force acting on a body is given by

$$F_R = \frac{L}{4 \pi R_s^2 c} A \cos \alpha \tag{9a}$$

where L = total energy radiated from the sun, 3.86×10^{33} erg cm^{-2} sec^{-1}
R_S = distance from the sun, cm
c = velocity of light, 3×10^{10} cm sec^{-1}
A = area exposed to solar radiation
α = angle between the normal to A and the sun-body line

In the neighborhood of the Earth, R_s = 1 a.u. (1.495×10^{13} cm), and the radiation pressure C_r is given by

$$C_r = 4.6 \times 10^{-5} \cos \alpha \quad \text{dyne cm}^{-2} \tag{9b}$$

C. Neutral Particle Drag

Below about 400 km altitude the largest contribution to spacecraft deceleration comes from collisions with the neutral constituents of the atmosphere which greatly outnumber the charged particles in these regions. Conditions at all practical altitudes for spacecraft, above about 150 km, are those of "free molecular flow", meaning that the mean free path of all particles is much greater than the dimensions of the body with which they interact. Additionally, for regions where neutral particle collisions are significant, their thermal velocities are much lower than the velocity of the spacecraft ($v_s \approx 7 \times 10^5$ cm sec^{-1}). The magnitude of the drag force under these conditions is given by

$$F_{ND} = \frac{1}{2} \rho v_s^2 C_D A \quad \text{dyne} \tag{10}$$

where F_{ND} = drag force
ρ = neutral atmosphere density, g cm^{-3}
v_s = spacecraft velocity, cm sec^{-1}
C_D = drag coefficient
A = cross sectional area projected into the flow, cm^2

The value of C_D depends upon the geometry of the body, the mode of reflection of the incoming particles (specular or diffuse), and the degree of thermal accommodation of the particles to the surface temperature of the body. A value of 2.2 is generally used for C_D for spherical and nearly spherical bodies in the 150-400 km altitude range. Shapes such as flat plates and cylinders have somewhat higher values of C_D, although this is also dependent on the reflection processes (Cook, 1965). At higher altitudes, above about 600 km, the thermal velocity of all particles, neutral and charged, must be considered (Schamberg, 1959); C_D rises to a value of greater than 4.0, and the drag process becomes more efficient (Cook, 1965; Epstein, 1924). Because of the much greater degree of ionization at these higher altitudes, however, the relative contribution of neutral particle collisions to the total drag becomes very small.

D. Charged Particle Drag (Coulomb Drag)

Direct Impact Drag. The potential of the spacecraft modifies the trajectories of charged particles that otherwise would not strike it. Since electrons may be neglected in drag calculations, due to their low mass compared with any ion, the usual situation is that the effective collisional area of the spacecraft is increased by the attraction of positive ions to a negatively-charged body and decreased by a positively-charged body.

In the simplest case, for a spherical body moving at $v_s \gg v_i$ where v_i is the ion thermal velocity and all ions stick to the body after impact and neutralization, the drag is simply

$$F_{ID} = \frac{1}{2} \pi m_i n_i v_s^2 r_e^2 \quad \text{dyne} \tag{11}$$

where m_i = ion mass, g
n_i = ion density, n cm^{-3}
v_s = satellite velocity, cm sec^{-1}
r_e = effective radius for ion collection, cm

From consideration of momentum conservation, r_e is given by:

$$r_e = R\left(1 + \frac{2e\,\Phi}{m_i v_s^2}\right)^{1/2} \quad \text{cm} \tag{12}$$

where R is the radius of the spherical body, and Φ is the spacecraft potential.

The exact determination of the effect of spacecraft charge depends upon the degree to which the charge is screened from the surrounding plasma. This usually is defined in terms of the local Debye length λD, but it must be remembered that this is the distance at which the potential due to spacecraft charge is reduced by 1/e, and it is not the limit of the disturbed region. Where λD is more than a few centimeters, the disturbed region around the spacecraft can be on the order of 10-40% of the total cross section of a spacecraft 1-2 m in diameter (Brundin, 1963).

At impact the ions are neutralized and, if re-emitted, carry momentum that is a further source of drag if the reflection mechanism is diffuse (Schamberg, 1959). This reemission drag can be greater than the initial impact drag of the ions if there is no thermal accommodation of the incident particles to the body surface temperature. This drag has a maximum value F_r, given by (Brundin, 1963)

$$F_R = n_i m_i v_s^2 \pi R^2 \left[1 + \frac{2 e \Phi}{m_i v_s^2}\right] \left\{\frac{4}{9}\left[1 + \left(\frac{2e \Phi}{m_i v_s^2}\right)^{1/2}\right]\right\} \text{dyne} \tag{13}$$

where n_i = ion number density, n cm^{-3}
m_i = ion mass, g
R = spherical body radius, cm
Φ = spacecraft potential, volts
v_s = satellite velocity, cm sec^{-1}

Noncollisional Drag (Maxwell Drag or Dynamic Friction). An additional source of charged particle drag arises from coulomb interaction leading to momentum exchange between a charged body and ions that do not strike it, but whose trajectories are changed. A critical factor in determining this drag is again the extent to which the effect of spacecraft potential penetrates into the ambient plasma. The problem has been approached by a number of investigators (Chang and Smith, 1960; Jastrow and Pearse, 1957; Fournier, 1970), who made varying assumptions about the extent and geometry of the electrostatic screening of a body moving through a plasma. However, in considering large structures we need note only that noncollisional drag is a significant fraction of the total charged particle drag only when the body in question has dimensions no more than an order of magnitude greater than the local Debye length (Knechtel and Pitts, 1964). In regions where this condition is satisfied, for very large structures in geosynchronous orbit outside the

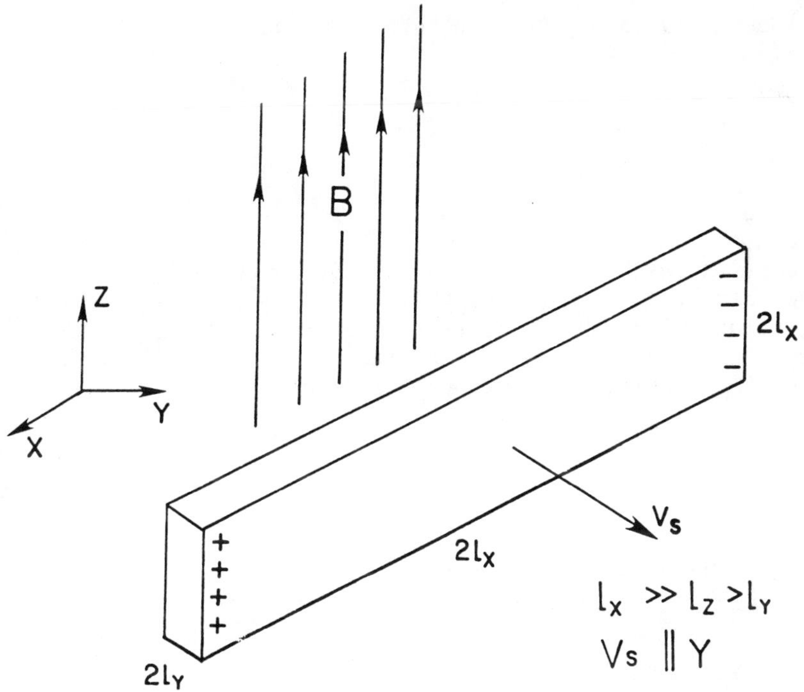

Fig. 1 The basic structure moving perpendicular to the local magnetic field.

plasmapause, for example, ion densities are too low ($\lesssim$ 10 cm^{-3}) to contribute any significant drag.

E. Magnetic Field Drag Effects

The effects of the Earth's magnetic field on the potential distribution and current flows on a conducting body in the space plasma are not significant in practice unless the dimensions of the body exceed approximately 10 m. As linear dimensions increase, the potentials generated on the body by its motion across the magnetic field begin to exceed the energies of the plasma particles with respect to the body. These usually lie within the range ± 25 eV. Ions and electrons flow preferentially to the negative and positive sections of the body; therefore, although there is a zero net current flow to the whole body, positive and negative currents flow to and within specific parts of the body. This results in drag (Lorentz forces) on the body as the various currents interact with the ambient magnetic field.

For a conducting body such as that shown in Fig. 1, with length $2L_X$ perpendicular to v_S and B, a potential E_X

will be set up along the length of the body:

$$E_x = 10^{-8}(v_s \times B) \quad \text{V cm}^{-1} \tag{14}$$

where E_x is in V cm^{-1}
v_s is in cm sec^{-1}
B is in G

Figure 2 shows the distribution of the currents that will flow, partly as a result of this potential and partly due to collisions with ambient ions and electrons, when L_x is on the order of a few meters. The induction drag or Lorentz force (F_L) acting on the body as a result is:

$$F_L = B \int_{-L_x}^{+L_x} I(x) \sin \delta \; dx \quad \text{dyne} \tag{15}$$

where $I(x)$ = net current flowing in an increment dx of the body located at x
δ = angle between the direction of current flow and the magnetic field B

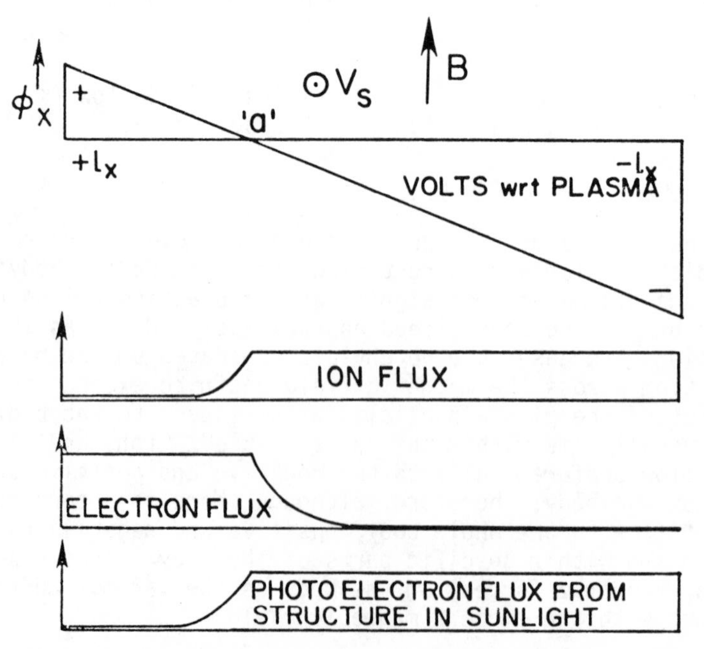

Fig. 2 The current and potential distributions along the x-axis of the structure, when L_x is on the order of a few meters.

For the configuration of Fig. 1, δ = 90 deg, and if the currents are equally distributed about the center of the structure, they flow for an average distance of L_x perpendicular to B, hence

$$F_L = 10^{-1} \text{ I B } L_x \text{ dyne} \tag{16}$$

where I = mean current, A
B is in G
L_x is in cm

The location of a, the point on the structure that rests at plasma potential, is determined by balancing the electron flux to the positive part of the structure with the ion flux to the negative part, and the photocurrent (positive) emitted predominantly from the negative section. Since the electron flow to the structure is parallel to the magnetic field, current densities and distributions depend on the orientation of the structure.

At altitudes where plasma densities are approximately 5×10^3 cm^{-3} or greater, point a for structures no more than a few meters in size is generally located between one-third and one-half of the length of the structure from the positive end. As the linear size is increased along the x-axis, higher potentials are induced across the structure and the positive end is effectively grounded in the local plasma; i.e., $a \geq L_x$ (Chu and Gross, 1966). At higher altitudes, for structures of all sizes, the photocurrent is significantly greater than either the ion or electron flux. The entire structure rests at a positive potential with respect to the plasma and attracts an increased flux of electrons to balance the photocurrent. Electrons still flow preferentially to the less negative end of the structure, and the magnetically induced potential across the structure will still be generated.

F. Torque Processes

All the drag processes outlined so far also will exert a torque on the body if the forces produced have a nonzero total moment about the body's center of mass. Torque also will be generated in the presence of a magnetic field by current loops within the body, but in this case no additional drag results as long as the magnetic field is uniform over the dimensions of the body. This process may have serious consequences when dealing with large structures such as solar power stations, where very large currents are generated. Careful routing of current paths is required to minimize the total magnetic moment of the structure.

Momentum Transfer Torques (Particle Impacts and Radiation Pressure Torques). If a drag process results in a force P per unit area projected in the direction of the body's velocity vector v_s, then the total torque T_p is

$$T_p = P \iint r \quad dA \cos \theta \, dr \qquad \text{dyne cm} \tag{17}$$

where P = drag pressure in dyne cm^{-2}
r = vector distance of dA from center of mass, cm
dA = element of surface area, cm^2
v_s = velocity, cm sec^{-1}
θ = angle between v_s and the normal to dA

Magnetic Torques (Induction Torques). When induction drag results from currents flowing through the body from the space plasma, a torque T_L will result if there is a net nonzero moment of these drag forces about the center of mass. From Eq. (15) we obtain

$$T_L = 10^{-1} \; B \int_{-L_x}^{+L_x} I(x) \, L_x \sin \delta \, dx \qquad \text{dyne cm} \tag{18}$$

where B = magnetic field, G
I(x) = current, A, flowing in an element dx located at L_x cm from the center of mass
δ = angle between the direction of current flow and the magnetic field

In addition, a current flowing around a closed loop in the presence of a uniform magnetic field generates a torque given by

$$T_i = 10^{-1} \; I \; A \; B \sin \alpha \qquad \text{dyne cm,} \tag{19}$$

where I = current around loop, A
A = area enclosed by loop, cm^2
B = magnetic field, G
α = angle between the magnetic field and the normal to the plane of the loop

Gravitational Torque. The gravitational potential gradient of the Earth exerts a torque on a body such that its axis of minimum moment of inertia aligns itself with the radius through the center of the Earth. That is to say, the long axis of a body will line up with the local vertical. Use is made of this torque in the technique of "gravity gradient stabilization" for satellites that are required to

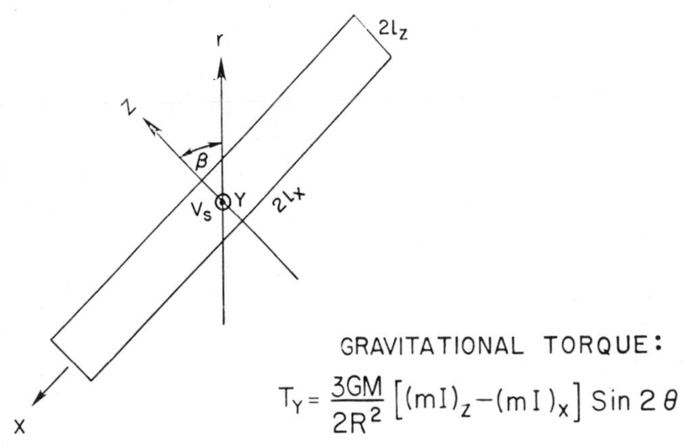

Fig. 3 Geometry to produce a gravitational torque about the y-axis of the structure. The x- and z-axes are coplanar with r, the local vertical.

present the same side toward the Earth at all times. Stabilization is achieved by extending a mass at the end of a long boom erected along the axis required to point downwards. This greatly increases the moments of inertia about the axes perpendicular to the boom, which then aligns itself along the local vertical (Fischell and Mobley, 1964).

Gravitational torques have had little effect on the stability and alignment of many satellites launched to date because of their relatively small overall dimensions and the fact that their moments of inertia about all three axes are frequently of the same order of magnitude. In addition, antennas and solar cell arrays have made large relative contributions to the cross sectional area of the spacecraft, compared with small additions to moments of inertia. Because of this particle impacts, magnetic induction effects, and at higher altitudes solar radiation pressure have been the principal sources of torque. This situation is changed however, when we consider massive, elongated structures with high transparencies for particle collision and solar radiation pressure.

Figure 3 shows the previous sample structure aligned at an angle θ to the local vertical r such that x, z, and r are coplanar. The gravitational torque T_y about the y-axis is given by (Thomson, 1962)

$$T_y = \frac{3GM}{2R^3} \left[(MI)_z - (MI)_x \right] \sin 2\beta \quad \text{dyne cm} \tag{20}$$

Table 1 Environmental Parameters

Altitude h, km	250	500	1000	36000 ($6.6R_e$)
Neutral density ρ, g cm^{-3}	10^{-13}	$3x10^{-15}$	$2x10^{-13}$	...
Neutral density N, cm^{-3}	10^9	10^8	10^5	...
Plasma density n_i, cm^{-3}	$5x10^5$	10^6	$3x10^4$	$5x10^0$ - $5x10^2$
Neutral temperature $T_i \approx T_n$, deg K	$8x10^2$	10^3	$2x10^3$	$5x10^3$ - $2x10^4$
Electron temperature T_e, deg K	10^3	$1.5x10^3$	$2x10^3$	$5x10^3$ - $2x10^4$+ (inside plasmasphere) Several keV (outside plasmasphere)
Mean ion mass $\overline{m}$, amu	24	16	8	1
Debye length λ_D, cm	0.3	1.0	2.0	200+
Geomagnetic field B, G	0.45	0.40	0.33	10^{-3}

where G = universal gravitation constant, 6.67×10^{-8} dyne cm^2 g^{-2}
M = mass of the Earth, 5.973×10^{27} g
R = distance to center of the earth, cm
$(MI)_{x,z}$ = moments of inertia about x- and z-axes, g cm^2
β = the angle between the local vertical r, and the x-axis of the structure

III. Numerical Examples: Drag

In this section we calculate values for the drag forces on a sample structure, which are generated at a number of altitudes by the processes outlined in Section II. Representative values for environmental parameters are given in Table 1 for an exospheric temperature of 1500 deg K. The structure is as shown in Fig. 1 and its dimensions, mass and velocity at the various altitudes are given in Table 2.

A. Solar Radiation Pressure

In the near-Earth region (1 a.u.) the value of solar radiation pressure is effectively constant at all altitudes. For a totally absorbing surface the pressure is

$$C_r = 4.6 \times 10^{-5} \cos \alpha \ \text{dyne cm}^{-2} \tag{9b}$$

Table 2 Structure Details

Dimensions,	Length	$2L_x$, cm	2×10^5
	Height	$2L_z$, cm	10^3
	Thickness	$2L_y$, cm	$< 10^2$
Total projected area $2L_x \times 2L_z$, cm^2			2×10^8
Mass per unit projected area, g cm^{-2}			10^1
Total mass, g			2×10^9
Velocity, cm sec^{-1}	at	250 km	7.80×10^5
		500 km	7.60×10^5
		1,000 km	7.35×10^5
		36,000 km	3.00×10^5

If the structure faces directly toward the sun, $\alpha = 0$ deg and the total force acting on an area of 2×10^8 cm^2 will be

$$C_r = 4.6 \times 10^{-5} \times 2 \times 10^8 = 9.2 \times 10^3 \quad \text{dyne}$$

B. Neutral Particle Drag

The neutral particle drag is a decelerating force acting at all times in the opposite direction to the velocity vector v_s:

$$F_{ND} = \frac{1}{2} \rho v_s^2 C_D A \quad \text{dyne} \tag{10}$$

From Section II C, the drag coefficient $C_D = 2.2$, and the area projected into the velocity vector $A = 2 \times 10^8$ cm^2.

The neutral drag at three altitudes is shown in Table 3. Neutral densities are considered negligible at geosynchronous orbit. Note that at an altitude of 1000 km the neutral component is already less than the solar radiation pressure drag (Section III A) at the time when the structure is facing directly into the sun. However, the net energy transfer due to solar radiation pressure is very close to zero over a complete orbit.

C. Charged Particle Drag

The charged particle drag is a decelerating force acting in the same sense as neutral particle drag:

$$F_{ID} = \frac{1}{2} m_i n_i v_s^2 A, \quad \text{dyne} \tag{21}$$

Table 3 Neutral Drag

h, km	250	500	1000
ρ, g cm^{-3}	10^{-13}	3×10^{-15}	2×10^{-17}
v_s, cm sec^{-1}	7.8×10^5	7.6×10^5	7.35×10^5
F_{ND}, dyne	1.34×10^7	3.81×10^5	2.38×10^3
F_{ND}/A, dyne cm^{-2}	6.69×10^{-2}	1.91×10^{-3}	1.19×10^{-5}

Table 4 Charged Particle Drag

h, km	250	500	1000
m_i, g	4.01×10^{-23}	2.68×10^{-23}	1.34×10^{-23}
n_i, n cm^{-3}	5×10^{5}	10^{6}	3×10^{4}
v_s, cm sec^{-1}	7.8×10^{5}	7.6×10^{5}	7.35×10^{5}
F_{ID}, dyne	1.22×10^{3}	1.54×10^{3}	2.17×10^{1}
F_{ID}/A, dyne cm^{-2}	6.1×10^{-6}	7.73×10^{-6}	1.08×10^{-7}

where m_i = mean mass of ambient ions, g
n_i = number density of ambient ions, n cm^{-3}
v_s = spacecraft velocity, cm sec^{-1}
A = the area projected into v_s, cm^2

The values given in Table 4 are for a structure of cross-sectional area A moving much faster than the thermal velocity of the ions. No allowance is made for increased effective area due to coulomb attraction of nearby ions, since that is most significant when the dimensions of the structure are no more than one order of magnitude greater than the local Debye length. Similarly, no diffuse re-emission of the neutralized ions is considered. Since both of these factors would increase charged particle drag, the figures given in Table 4 for a body withe $A = 2 \times 10^8$ cm^2 are conservative. Since thermal ion velocities at geosynchronous orbit are greater than orbital velocity, Eq. (21) does not apply; however, the ambient density is so low at this altitude that the drag resulting from ion impact is negligible compared with solar radiation pressure forces.

D. Magnetic Drag (Induction Drag)

The three processes dealt with so far all depend on the area that the structure projects into the velocity vector or the sun vector. If an open-girder type of construction is used, the forces resulting from these processes would be reduced in direct proportion to the optical transparency of the area projected. This change in optical transparency must be uniform over the whole structure, however, or a torque will result.

In the case of magnetic field drag (induced drag) it is the linear dimensions and orientation of the structure in

Table 5 Induction Drag

h, km	250	500	1000	36000
I_i, A	$(1.25\text{-}1.31)\times10^{1}$	$(2.43\text{-}2.49)\times10^{1}$	$(7.06\text{-}13.06)\times10^{-1}$	6×10^{-1}
B, G	4.5×10^{-1}	4.0×10^{-1}	3.3×10^{-1}	10^{-3}
F_L max, dyne	5.9×10^{4}	9.96×10^{4}	4.32×10^{3}	6×10^{0}
F_L/A, dyne cm^{-2}	2.94×10^{-4}	4.98×10^{-4}	2.16×10^{-5}	3×10^{-8}

the local magnetic field that are critical. From Eq. (16), the drag for the orientation in Fig. 1 is

$$F_L = 10^{-1} \, I \, B \, L_x \quad \text{dyne} \tag{16}$$

In practice the actual value of I depends upon the local ion density, solar illumination, transparency, velocity of the structure, and orientation in the magnetic field as it affects the potential distribution over the structure. For this example we consider only the photocurrent I_p (due to solar ultraviolet photons) and the ion collection current I_i (swept up by the structure whose velocity is assumed to be large compared with the ion thermal velocities). This assumption is not true at geosynchronous orbit altitudes where the photocurrent dominates. In an equatorial orbit the photocurrent contribution will be zero at 1200 hr local time and maximum at 0600 hr local time and 1800 hr local time if both back and front surfaces are taken to be photoemitters. If almost all the surface of the structure is an ion collector and/or a photoemitter, then the average distance perpendicular to B through which these currents flow will be close to L_x (= 10^5 cm), one half the total length. At each altitude we find I_i using the environmental data in Table 1, and then substitute in Eq. (16) to determine F_L. Photocurrent I_p = 6×10^{-1} A at all altitudes, and its variation around the orbit is reflected in the range of values for I_i given in Table 5.

IV. Numerical Examples: Torque

A. Momentum Transfer Torques

In order to estimate the magnitude of these torques (due to solar radiation pressure and collisions with charged and neutral particles), we modify the structure as shown in Fig. 4. One quarter of the length is covered with conductor as before, but the remainder now has an open construction with an optical transparency of 90%. The total mass and its distribution about the center of gravity remain unchanged. The y-axis of the structure is maintained parallel to v_s at all times.

The applied torque will have a maximum value when the structure is moving directly towards the sun. The solar radiation pressure component varies continuously throughout the orbit and acts to equalize the particle impact torques whenever v_s has a component away from the sun.

If the pressure on the structure at any time is P dyne cm^{-2}, then from Eq. (17), the resulting torque T_p will be

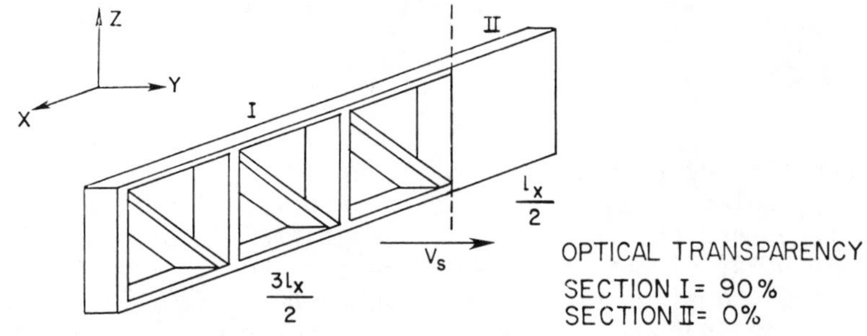

Fig. 4 Modification of the basic structure to demonstrate momentum transfer and solar radiation pressure torques by the imbalance of forces about the center of gravity.

$$T_p = 2\, L_z\, P \int_{-L_x}^{+L_x} L_x\, dL_x \qquad \text{dyne cm} \tag{22a}$$

Due to the basic symmetry about the z-axis, T_p arises from the additional area in the enclosed section of the structure so that:

$$T_p = 0.9 \left[2\, L_z\, P \left(\frac{L_x^2}{2} - \frac{L_x^2}{8} \right) \right]$$

$$= 3.38 \times 10^{12}\ P \qquad \text{dyne cm} \tag{22b}$$

From Section III we obtain maximum values for P given in Table 6 by adding solar radiation pressure and the neutral and charged particle forces.

B. Magnetic Torques (Induction Torques)

As an example of currents flowing in a section of the structure to produce a torque by imbalance about the center of mass, we will use the currents derived in Section III D. We reduce them by a factor of 4 to represent their flowing in the enclosed portion of the structure and ignore currents flowing in the rest of the structure. The mean moment arm of the Lorentz forces produced about the center of mass will be $3L_x/4$ and the currents will flow for a mean distance of $L_x/4$ perpendicular to B. In Table 7 we use these values in Eq. (18) to obtain T_L.

To estimate the torque T_i caused by a flow of current around a closed loop within the structure we postulate a net

Table 6 Pressure Torque

h, km	250	500	1000	36000
P, dyne cm^{-2}	6.69×10^{-2}	1.91×10^{-3}	5.80×10^{-5}	4.6×10^{-5}
T_p, dyne cm	2.26×10^{11}	6.46×10^{9}	1.96×10^{8}	1.56×10^{8}

Table 7 Induction Torque

h, km	250	500	1000	36000
I, A	3.28×10^{0}	6.23×10^{0}	3.27×10^{-1}	1.50×10^{-1}
B, G	4.5×10^{-1}	4.0×10^{-1}	3.3×10^{-1}	10^{-3}
T_L, dyne cm	2.77×10^{8}	4.67×10^{8}	2.03×10^{7}	2.81×10^{4}

Table 8 Current Loop Torque

h, km	250	500	1000	36000
B, G	4.5×10^{-1}	4.0×10^{-1}	3.3×10^{-1}	10^{-3}
T_i, dyne cm	9.0×10^{6}	8.0×10^{6}	6.6×10^{6}	2×10^{4}

current of 1 A flowing around the outermost edges of the structure aligned as in Fig. 1, so that the torque acts about the x-axis of the structure. Note that for this source of torque, internal construction and optical transparency are irrelevent; the torque is determined by the area enclosed by the loop and its alignment in the magnetic field. We use Eq. (19) and Table 1 to obtain the values for T_i given in Table 8.

C. Gravitational Torque (Gravity Gradient)

Starting with the structure in Fig. 4 aligned at $\beta = 45$ deg to the local vertical, we determine the gravitational or gravity gradient torque for the four altitudes.

Taking mass and dimensions from Table 2 we have (since $L_x \gg L_z$ and $L_z > L_y$)

Table 9 Gravitational Torque

h, km	250	500	1000	36000
T_y, dyne cm	1.37×10^{13}	1.23×10^{13}	9.94×10^{12}	5.40×10^{10}

$$(MI)_y \approx (MI)_z = 6.67 \times 10^{18} \text{ g cm}^2$$

$$(MI)_x = 1.67 \times 1014 \text{ g cm2}$$

Substituting in Eq. (20) we obtain the values for T_y given in Table 9.

V. Conclusions

The most striking result from the evaluation of forces acting on a large space structure is the greatly increased importance of gravitational torques. At all altitudes considered this mechanism produces a torque at least two orders of magnitude greater than any other. A gravity gradient stabilization mode is therefore unavoidable unless 1) a continuously operating active attitude control system is used or 2) the structure is designed to have nearly equal moments of inertia about all three axes (Oglevie, 1978).

In the first case, the mass of the structure (10^9-10^{11} g) means that large amounts of power would be required for such a system and that the interaction between the control system and the inevitable flexing of so large a structure would pose totally new problems (Ayer and Soosar, this volume; Ginter and Balas, 1978). The equalization of moments of inertia approach seems more promising if gravitational alignment is undesirable. However, if any other mechanism is employed to orient all or part of a structure (solar radiation pressure using large, light, moveable "sails," for example), the situation is complicated by the need for sensing and moving systems that must operate continuously.

Drag forces are of little significance in practice, except perhaps during the assembly phase in low earth orbit. This is because of the need to ensure an orbital lifetime on the order of tens of years, which means that most structures will be placed in final orbits above 1000 km. Since so many structures, together with an ever increasing number of communications satellites and relay platforms, will be placed in geosynchronous orbit, plans must be made at this early stage concerning the final disposal of the structures at the end of their useful life ("solar sailing" to much higher orbits, for example). Additionally, contingency plans must be made to minimize the possibility of the re-entry of large substructures that experience some malfunction during assembly, since at space shuttle altitudes orbital lifetimes are no more than a few years.

Great care must be exercised in applying the results of Sections III and IV to other structures and orbits. In most examples the orientation of the structure has been chosen to maximize the individual drag and torque forces. It is possible that other combinations of attitude, geometry, and orbit could combine to reduce the very large differences in the magnitude of the various forces.

Appendix: Rotation Effects

Additional mechanisms come into play when a body in the space plasma has rotation about some internal point in addition to translational motion. These mechanisms have been studied principally for spherical bodies (Wood and Hohl, 1965), and are summarized here for completeness.

1) Aerodynamic torque. A spinning body will be decelerated by drag from the neutral particles impinging on its surface (Davis and Harris, 1961). For a spherical body the magnitude of torque resulting is negligible above about 600 km altitude when compared with the other mechanisms listed here.

2) Surface charge torque. The geometry of any bound charge on a conducting body, induced by translational motion across a magnetic field, for example, maintains a fixed orientation in that field. If the body spins, the bound charge flows through it to maintain this fixed orientation. The energy required to overcome the electrical resistance to this current flow is taken from the kinetic energy of the body's rotation, and a deceleration of the spin therefore results. In the case of a sphere, however, this deceleration is many orders of magnitude less than all other torques due to rotation.

3) Reflected ion torque. Ions that strike the body are accelerated as they approach it by the potential of the body with respect to the plasma. If they are neutralized, and undergo diffuse reflection without significant thermal accommodation, they have a net increase in energy. The associated momentum loss to the body applies a force to it at the point of impact. If the ion flow is not symmetric about the center

of mass, because of the potential distribution on the body, for example, then a decelerating torque is applied. For a sphere this torque is some three orders of magnitude smaller than the eddy current and magnetically induced torques.

4) Coulomb torque. Ions that are not reflected but remain on the body after neutralization also will cause a torque if the moments of their impact are not symmetric about the body's center of mass. The magnitude of this torque for a spherical body is some three orders of magnitude less than the eddy current and induced torques which effectively determine the changes to the initial spin rate of an orbiting body.

5) Eddy current torque. The rotation of a body within a magnetic field generates internal eddy currents which interact with the magnetic field via Lorentz forces to give decelerating torques. These are the largest spin-reducing torques above about 600 km altitude. For the Echo II satellite, a 41-meter diameter segmented conducting sphere in a 1400-km circular orbit, the torque had a value on the order of 10^{-11} dyne cm.

6) Magnetically induced torque. As shown in Section II E, the currents flowing between the ambient plasma and a conducting body give rise to Lorentz forces as the body moves across a magnetic field. If the moments of these forces are not equally distributed about the center of mass, an accelerating torque will result, tending to spin up the body. In the case of Echo II this torque was due to nonuniformities in the electrical resistance of the various segments from which the sphere was constructed and was estimated to have a value on the order of 5×10^{-10} dyne cm. Because the induced torque approximately equalized the decelerating eddy current torques, the spin rate of Echo II remained almost constant for many months.

Acknowledgements

I am indebted to Mr. F. Marcos, Dr. M. Smiddy, and Captain H. B. Garrett for several helpful discussions during the preparation of this paper, to Ms. S. Bredesen for help with editing, and to Ms. K. Leccese for her patient typing and retyping.

References

Ayer, F. and Soosar, K., "Structural Distortions of Space Systems due to Environmental Disturbances," published elsewhere in this volume.

Beard, D.B. and Johnson, F.S., "Charge and Magnetic Field Interaction with Satellites," Journal of Geophysical Research, Vol. 65, Jan. 1960, pp. 1-8.

Brundin, C.L., "Effects of Charged Particles on the Motion of an Earth Satellite," AIAA Journal, Vol. 1, Nov. 1963, pp. 2529-2538.

Chang, H.H.C. and Smith, M.C., "On the Drag of a Spherical Satellite Moving in a Partially Ionized Atmosphere," British Interplanetary Society Journal, Vol. 17, Jan.-Feb. 1960, pp. 199-205.

Chu, C.K. and Gross, R.A., "Alfvén Waves and Induction Drag on Long Cylindrical Satellites," AIAA Journal, Vol. 4, Dec. 1966, pp. 2209-2214.

Cook, G.E., "Satellite Drag Coefficients," Planetary and Space Science, Vol. 13, Oct. 1965, pp. 929-946.

Davis, A.H. and Harris, I., "Interaction of a Charged Satellite with the Ionosphere," Rarefied Gas Dynamics, edited by L. Talbot, New York, Academic Press, 1961, pp. 691-699.

Epstein, P.S., "Resistance Experienced by Spheres in their Motion through Gases," Physical Review, Vol. 23, June 1924, pp. 710-733.

Fischell, R.E. and Mobley, F.F., "A System for Passive Gravity-Gradient Stabilization of Earth Satellites," Progress in Astronautics and Aeronautics: Guidance and Control-II, Vol. 13, New York, Academic Press, 1964, pp. 37-71.

Fournier, G., "Electric Drag," Planetary and Space Science, Vol. 18, July 1970, pp. 1035-1041.

Freeman, J.W., Cooke, D., and Reiff, P., "Space Environmental Effects and the Solar Power Satellite," Proceedings of the Spacecraft Charging Technology Conference, USAF Academy, Colorado Springs, Colo., 1978, NASA Conference Publication 2071, 1979, pp. 408-418 (also available as Air Force Geophysics Laboratory report AFGL-TR-79-0082).

Garrett, H.B., "Spacecraft Charging: A Review" published elsewhere in this volume.

Ginter, S. and Balas, M., "Attitude Stabilization of Large Flexible Spacecraft," AIAA Paper 78-1285, Aug. 1978.

Hagler, T., Patterson, H.G., and Nathan, C.A., "Learning to Build Large Structures in Space," Astronautics and Aeronautics, Vol. 15, Dec. 1977, pp. 51-57.

Jastrow, R. and Pearse, C.A., "Atmospheric Drag on the Satellite," Journal of Geophysical Research, Vol. 62, Sept. 1957, pp. 413-423.

Kasha, M.A., The Ionosphere and its Interaction with Satellites, Chapt. 3, New York, Gordon and Breach, 1969.

Knott, K., "The Equilibrium Potential of a Magnetospheric Satellite in an Eclipse Situation," Planetary and Space Science, Vol. 20, Aug. 1972, pp. 1137-1146.

Knechtel, E.D. and Pitts, W.C., "Experimental Investigation of Electric Drag on Satellites," AIAA Journal, Vol. 2, June 1964, pp. 1148-1151.

Oglevie, R.E., "Attitude Control of Large Solar Power Satellites," AIAA Paper 78-1266, Aug. 1978.

Parker, L.W., "Plasma Sheath Effects and Voltage Distributions of Large High-Power Satellite Solar Arrays," Proceedings of the Spacecraft Charging Technology Conference, USAF Academy, Colorado Springs, Colo., 1978, NASA Conference Publication 2071, 1979, pp. 341-357 (also available as Air Force Geophysics Laboratory report, AFGL-TR-79-0082).

Renner, U., "Attitude Control by Solar Sailing - A Promising Experiment with OTS-2," European Space Agency Journal, Vol. 3, No. 1, 1979, pp. 35-40.

Samir, U. and Willmore, A.P., "The Equilibrium Potential of Spacecraft in the Ionosphere," Planetary and Space Science, Vol. 14, Nov. 1966, p. 1131-1137.

Samir, U., Maier, E.J., and Troy, B.E., Jr., "The Angular Distribution of Ion Flux Around an Ionospheric Satellite," Journal of Atmospheric and Terrestrial Physics, Vol. 35, March 1973, pp. 513-519.

Schamberg, R., "Analytic Representation of Surface Interaction for Free Molecule Flow with Application to Drag of Various Bodies," Aerodynamics of the Upper Atmosphere, compiled by D.J. Masson, Rand Corporation Report 339, June 1959, p. 12.

Sohn, M., "Attitude Stabilization by Means of Solar Radiation Pressure," American Rocket Society Journal, Vol. 29, May 1959, pp. 371-373.

Thomson, W.T., "Spin Stabilization of Attitude Against Gravity Torque," Journal of Astronautical Sciences, Vol. 9, Spring 1962, pp. 31-33.

Whipple, E.C., Jr., "The Ion-Trap Results in the 'Exploration of the Upper Atmosphere with the Help of the Third Soviet Sputnik,'" Proceedings of the Institute of Radio Engineers, Vol. 47, Nov. 1959, pp. 2023-2024.

Wood, G.P. and Hohl, F., "Electric Potentials, Forces and Torques on Bodies Moving through Rarefied Plasmas," AIAA Paper 65-628, Aug. 1965.

TORQUING AND ELECTROSTATIC DEFORMATION OF THE SOLAR SAIL

Robert E. LaQuey,* Sherman E. DeForest,† and Marvin Douglas≠

Maya Development Corporation, San Diego, Calif.

Abstract

The impact of natural sources of electrical-mechanical oscillations induced by the environment on the solar sail system is evaluated. The study indicates that, to the level of accuracy (first order) of the analysis, none of the natural sources studied, which range from plasma wave interactions to $\vec{E} \times \vec{B}$ forces, will have a significant impact on the proposed solar sail design. The study is not intended as an exhaustive analysis, and further analysis, particularly in the area of artifically induced oscillations, is needed.

Introduction

The solar sail, an imaginative concept for interplanetary travel, was proposed and seriously studied for a mission to Halley's comet. As part of the study effort, a preliminary analysis of interactions between the solar sail and the natural plasma environment was performed. This study included an examination of torquing and electrostatic deformations of the sail. We thus were led to study the novel problem of electromechanical couplings between the mechanical oscillations of the solar sail and the plasmas both near Earth and in the solar wind.

We have sought to answer questions of the following sort: "Does the interaction between the plasma and the sail cause the sail to develop unstable modes of oscillation?" In sailing parlance, "Does the sail luff?" This question is deceptive in its simplicity, hiding a myriad of possible mechanisms.

---

*President.

†Consultant; also, Physics Dept., University of California, San Diego, Calif.

≠Consultant.

To illustrate, consider the equation of motion for a membrane

$$\rho \frac{\partial^2 y}{\partial t^2} - T\nabla^2 y = f$$

where, for the case at hand,

f =	σE_x	+	P	+	S
=	electrostatic pressure	+	plasma pressure	+	photon pressure

We are interested in perturbations from equilibrium caused by the first two terms. As a worst case, we shall neglect the photon pressure which would appear to be universally stabilizing. Our treatment consists of examining a few cases which are amenable to calculation and then making an informed guess as to the magnitude of the problem.

There is almost no directly relevant literature, and so much of the work must be original. There remain numerous loose ends, and the final answer is far from definitive. Nonetheless, with these substantial qualifications, we find that the answer to the question at hand is "probably not."

Gas Dynamic Coupling

We first illustrate the problem by considering the coupling of oscillations on the surface of the solar sail to sound waves in the surrounding gas— in this instance the solar wind. This case is amenable to calculation and demonstrates one important dimensionless parameter associated with electromechanical oscillations. The result of the calculation is quite clear. This mechanism cannot cause luffing.

The problem is illustrated by analogy. Consider a cylinder, of infinite length, which has a thin membrane (a drum head) stretched across it as shown in Fig. 1. Morse and Ingard[1] have considered the similar problem of a vibrating string coupled to sound waves in a gas. Review of their work indicates the complicated nature of these problems. In this section we demonstrate some of the salient features of such a system. A simple model is used which substantially reduces the algebraic complexity of the problem. The solutions presented here are more easily understood than those in Morse and Ingard and exhibit substantially the same physics.

If the drum head is struck with a hammer, it will oscillate with a characteristic frequency determined by the

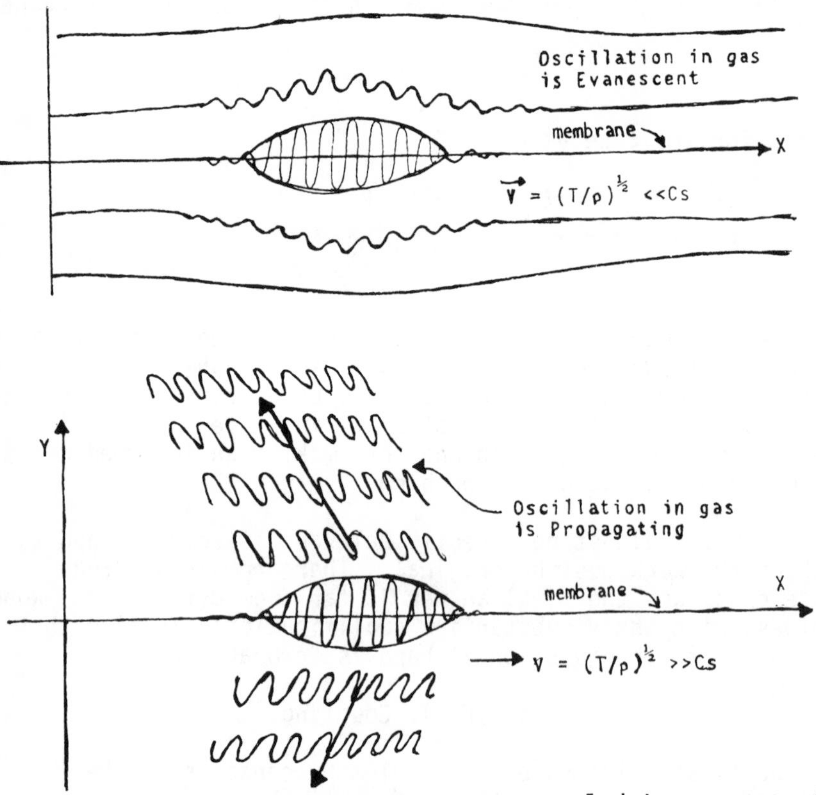

Fig. 1 Waves propagating on a membrane coupled to a surrounding gas. When the wave speed on the membrane $V = (T/\rho)^{\frac{1}{2}} << C_s$, the sound speed in the gas, then the mass of the gas loads the string. When $V >> C_s$, sound waves are radiated from the string in a process analogous to Cherenkov radiation.

parameters of the system. The oscillation of the drumhead will set up sound waves in the cylinder. These sound waves will propagate down the cylinder away from the drumhead. Consider for the moment the modes of oscillation with dimensions that are small compared to that of the cylinder, i.e., wavelengths $\lambda << R$. In this limit one may neglect the details of the boundary conditions and treat the membrane as an infinite sheet embedded in an unbounded medium. What are the modes of oscillation of this system?

Waves propagate on an infinite membrane with the characteristic dispersion relation

$$\omega = (T/\rho)^{1/2} K$$

where

$\omega = 2\pi f$ is the characteristic frequency,

$K = 2\pi/\lambda$ is the wave number for λ the wavelength,

T is the tension in the membrane,

ρ is the density of the membrane.

The characteristic dispersion relation for sound waves in a gas is

$$\omega = C_S K$$

where C_S is the sound speed which depends upon the temperature and molecular composition of the gas.

This is not a complete description of the problem at hand. The presence of the gas around the membrane causes additional forces to act upon the membrane, thus modifying its characteristic dispersion. Similarly the dispersion relationship for waves propagating in the gas is modified by forces exerted by the membrane upon the gas. The oscillations of the membrane and the gas thus are coupled.

Two limits suggest themselves. Whenever the wave speed on the membrane is less than the sound speed in the gas, i.e., $(T/\rho)^{1/2} \ll C_S$, the gas can respond essentially instantaneously to deformations of the membrane. Thus an oscillation propagating slowly across the membrane will simply push the gas back and forth. In the alternative limit $(T/\rho)^{1/2} \gg C_S$, the gas cannot respond instantaneously to the oscillations of the membrane. In this limit an oscillation propagating across the membrane will excite waves in the gas which propagate away from the membrane. The two regimes are illustrated in Fig. 1. These effects will now be illustrated by detailed calculation.

The membrane is described by the linearized wave equation

$$\rho \frac{\partial^2 \eta}{\partial t^2} - T \frac{\partial^2 \eta}{\partial x^2} = p^- - p^+$$

where η is the displacement of the membrane in the y direction, and $p^\pm$ are the pressures exerted by the gas for $y \gtrless 0$. The gas is described by equations for the conservation of number density n and momentum and by an equation of state relating the pressure $p^\pm$ to the number density. These are

$$\frac{\partial n}{\partial t} + n_0 \frac{\partial V_x}{\partial x} + n_0 \frac{\partial V_y}{\partial y} = 0$$

$$n_0 m \frac{\partial V_x}{\partial t} = -\frac{\partial p}{\partial x} \qquad n_0 m \frac{\partial V_y}{\partial t} = -\frac{\partial p}{\partial y}$$

$$p = (\gamma p_0 / n_0) n$$

where m is the mean molecular weight, $\gamma = 1$ for isothermal expansion, and $\gamma = 3$ for adiabatic expansion.

When the gas equations are combined, one finds

$$\frac{\partial^2 p}{\partial t^2} - C_S{}^2 \left(\frac{\partial^2 p}{\partial x^2} + \frac{\partial^2 p}{\partial y^2} \right) = 0$$

which describes the propagation of sound waves in the gas characterized by a sound speed $C_S = (\gamma p_0 / n_0 m)^{\frac{1}{2}} = (\gamma T_0 / m)^{\frac{1}{2}}$, where T is the temperature.

The gas and membrane are further coupled by the requirement that the component of gas velocity in the y direction match the velocity of the y displacement of the membrane. Thus, $V_y = \partial \eta / \partial t$ on the membrane. Using the momentum equation, we relate this to the pressure and find

$$\frac{\partial^2 \eta}{\partial t^2} = - \frac{1}{n_0 m} \frac{\partial p}{\partial y}$$

which, for linear perturbations of the membrane, is to be evaluated on equilibrium surface $y = 0$.

In the limit $(T/\rho)^{\frac{1}{2}} < C_S$, the waves do not propagate into the gas, and the solutions for the pressure may be represented as

$$p = \begin{cases} \hat{p} e^{-\alpha y} & y > 0 \\ \hat{p} e^{\alpha y} & y < 0 \end{cases}$$

where α is a constant to be determined from the gas propagation equation. When these equations are combined, one finds

$$p \pm = \pm \left(\frac{n_0 m}{\alpha} \right) \frac{\partial^2 \eta}{\partial t^2} \Rightarrow p^- - p^+ = - \left(\frac{2 n_0 m}{\alpha} \right) \frac{\partial^2 \eta}{\partial t^2}$$

Thus we find that the equation describing the coupled oscillations is

$$\left[\rho + \frac{2 n_0 m}{\alpha} \right] \frac{\partial^2 \eta}{\partial t^2} - T \frac{\partial^2 \eta}{\partial x^2} = 0$$

This equation can be very simply interpreted. The membrane mass density per unit area ρ is replaced by an effective mass density per unit area $\rho^* = \rho + (2 n_0 m / \alpha)$. Thus we see that the mass of gas contained in the perturbed pressure sheath of thickness $L = 1/\alpha$ simply loads the membrane.

Now let us examine this limit in slightly more detail by considering a specific perturbation of the form

$$\eta = \hat{\eta} \sin (kx - \omega t)$$

$$p = \begin{cases} \hat{p} e^{-\alpha y} \sin (kx - \omega t) & y > 0 \\ \hat{p} e^{\alpha y} \sin (kx - \omega t) & y < 0 \end{cases}$$

The dispersion relation becomes

$$[\rho + (2n_0 m/\alpha)]\omega^2 - TK^2 = 0$$

where

$$\alpha = K\sqrt{1 - (\omega/KC_s)^2}$$

This dispersion relation is cubic in ω^2, thus exhibiting three double modes which correspond to the modified sound waves on either side of the membrane and to the basic membrane mode.

Now, in the limit $(T/\rho)^{\frac{1}{2}} >> C_S$, one takes as the pressure perturbation a representation yielding outgoing waves,

$$p = \begin{cases} \hat{p} e^{i(Ky + Kx - \omega t)} y & y > 0 \\ \hat{p} e^{i(-Ky + Kx - \omega t)} & y < 0 \end{cases}$$

from which we obtain the dispersion relation

$$[\rho + i(2n_0 m/\alpha)]\omega^2 - TK^2 = 0$$

where

$$\alpha = \frac{\omega}{C_s} \sqrt{1 - (KC_s/\omega)^2}$$

Now one finds solutions for the membrane mode of the form

$$\omega = \omega_r + i\gamma$$

where

$$\omega_r = (T/\rho)^{1/2} K$$

$$\gamma = -(n_0 m C_s/\rho) \Rightarrow \text{dampening}$$

Thus the membrane modes are damped by energy radiated into sound waves.

Now let us examine the effect of a gas flowing across the membrane. Although a number of additional complications could be introduced, we shall examine only the effect of Doppler shift which causes the frequency associated with the moving medium to be transformed from ω to $\omega - \underset{\sim}{K}_0 \cdot \underset{\sim}{V}_0$. Thus pressure perturbations associated with a gas moving with velocity $v = v_o e_x$ will obey the equation

$$[(\omega - KV_0)^2 - K^2 C_s^2]p + C_s^2 \frac{\partial^2 \rho}{\partial y^2} = 0$$

which has evanescent solutions

$$p = \begin{cases} \hat{p} e^{-\alpha y} e^{i(Kx - \omega t)} & y > 0 \\ \hat{p} e^{\alpha y} e^{i(Kx - \omega t)} & y < 0 \end{cases}$$

whenever

$$\alpha^2 = K^2 \left[1 - \left(\frac{\omega}{KC_s} - \frac{V_0}{C_s} \right)^2 \right] > 0$$

Now consider the case $\frac{\omega}{K} << V_o \lesssim C_s$. The appropriate solutions are the evanescent ones since $\alpha^2 > 0$. The boundary condition is unmodified, but the momentum equation is Doppler shifted. One obtains

$$\hat{\eta} = \frac{1}{\omega} \left(\frac{1}{\omega - KV_0} \right) \frac{1}{n_0 m} \frac{\partial p}{\partial y}$$

and hence

$$\hat{p}^- - \hat{p}^+ = \omega(\omega - KV_0)(2n_0 m/\alpha)\hat{\eta}$$
$$\simeq -\omega^2 (KV_0/\omega)(2n_0 m/\alpha)\hat{\eta}$$

Thus there results the dispersion relation

$$[p - (KV_0/\omega)(2n_0 m/\alpha)]\omega^2 - TK^2 = 0$$

Note now that the effective mass $\rho^* = \rho - (KV_o/\omega)(2n_o m/\alpha)$ may become negative in which case an instability would develop. The physical mechanism is as follows. The mass displaced by the sail is convected along by the flow, thus shifting by 180 deg the phase with which the mass loading is applied to the membrane. The resulting instability grows by extracting free energy from the flow.

A basic dimensionless parameter $\varepsilon = (n_o mL/\rho)$, where $L = \alpha^{-1}$ is a typical length, characterizes the gas-dynamic coupling. This parameter simply measures the effective mass loading which the gas applies to the membrane.

For the solar wind $n_o m \lesssim 1.6 \times 10^{-16} g/m^3$. For the sail $\rho \simeq 3\ g/m^2$; thus $\varepsilon = 1$ requires $L = (\rho/n_o m) \gtrsim 2 \times 10^{16}$ m. This astronomically large number indicates that gas-dynamic mass coupling is not significant, since L is many orders of magnitude greater than a typical sail dimension. Alternatively, note that in the streaming case $L \simeq (V_o/\omega)$, where $V_o = 800$ km/sec and $\omega \simeq 5 \times 10^{-2}$ rad/sec are typical for the solar wind and the sail, respectively. Thus $L = 1.6 \times 10^7$ m, and hence $\varepsilon \simeq$

10^{-11}, so we are many orders of magnitude away from instability Thus we can say with certainty that the solar wind is of such low mass density that it cannot couple significantly to the sail through these mechanisms.

Resonant Forcing

Obviously we must look elsewhere for potential sources of instability. Treating the solar wind as a simple gas rather than as a plasma neglects the important electrostatic couplings between the solar sail and solar wind. To understand these couplings better we now examine the case of resonant electrostatic forcing of the sail. These are forced oscillations caused by those fluctuations of the solar wind which are of such a frequency as to resonate with the natural frequency of oscillation of the sail.

Consider a membrane subject to external forces. The equation describing linear oscillations of such a membrane is

$$\rho \frac{\partial^2 \eta}{\partial t^2} - T \frac{\partial^2 \eta}{\partial x^2} = \Delta F$$

where ΔF is the perturbing force exerted on the surface of the membrane. In the preceding section we examined perturbing forces caused by a pressure differential across the membrane. Now we examine electrostatic perturbing forces given by

$$\begin{aligned} \Delta F &= \sigma_0 \Delta E_y + \Delta\sigma E_{y0} \\ &= \Delta E_y{}^2 / 4\pi \\ &= \Delta U_\epsilon \end{aligned}$$

where $U_\varepsilon = (E_y{}^2/4\pi)$ is the electrostatic energy density evaluated at the membrane surface.

In equilibrium, a plasma sheath approximately 1 Debye length thick will form about the sail. A typical particle will transit this sheath in times $\tau \sim \omega_\rho{}^{-1}$, which for ions is typically a few microseconds. The natural frequencies of oscillation of the solar sail are many orders of magnitude less than this. Thus the sheath responds quasistatically to fluctuations in the solar wind. Therefore we take

$$\Delta U_\varepsilon = U_\varepsilon(\Delta\alpha/\alpha)$$

where $U_\varepsilon \sim \frac{1}{2} n m_i U_0{}^2 \sim 10^{-7}$ erg/cm^3, which is the wind streaming energy density, and $(\Delta\alpha/\alpha)$ is the relative amplitude of a fluctuation in the solar wind.

Typical unforced oscillations of the membrane are of the form

$$\hat{\eta} = \eta \sin(\pi x/L)\sin\omega_1 t$$

where $\omega_1 = (\pi/L)\ (T/\rho)^{1/2}$. Now if we consider a resonant fluctuation for which

$$\frac{\Delta\alpha}{\alpha}(x,t) = (\Delta\alpha/\alpha)\sin(\pi x/L)\sin\omega_1 t$$

then the sail will respond with a secularly growing oscillation. For an initial displacement of zero, one finds the solution

$$\hat{\eta} = \frac{1}{2\omega_1{}^2}\left(\frac{U_\epsilon}{\rho}\right)\left(\frac{\Delta\alpha}{\alpha}\right)\sin\left(\frac{\pi x}{L}\right)(\sin\omega_1 t - \omega_1 t\cos\omega_1 t)$$

and thus the amplitude of the secular growing term is

$$\eta \sim \frac{1}{2}\left(\frac{U_\epsilon}{\rho}\right)\left(\frac{\Delta\alpha}{\alpha}\right)\left(\frac{t}{\omega_1}\right)$$

Now numbers

$U_\varepsilon = 10^{-7}$ ergs/cm^3 = 10^{-8} J/m^3, $\rho = 3\times10^{-3}$ kg/m^2,
$\omega_1 = 4.9\times10^{-2}$ rad/sec, which result in an amplitude of $\eta \sim 0.06(\Delta\alpha/\alpha)(\omega_1 t)$cm

Thus $(\Delta\alpha/\alpha) \sim 0.5$, indicating a 50% change in amplitude of the solar wind, would require $\omega_1 t \sim 33$, i.e., 33 cycles, in order to produce a wave with amplitude $A \sim 1$ cm on the surface of the sail. Such long trains of large and coherent fluctuations in the solar wind are quite unlikely. Thus we conclude that resonant forcing should not be a serious problem.

To gain insight into this result, the expression for the amplitude is rewritten as

$$\eta \sim \frac{2}{\pi}\left(\frac{U_\epsilon \Sigma}{T}\right)\left(\frac{\Delta\alpha}{\alpha}\right)\left(\frac{\omega_1 t}{2\pi}\right)L$$

Noting that, in equilibrium,

$$-T\partial^2\eta_0/\partial x^2 = U_\nu$$

where U_ν is the photon energy density. Hence, by dimensional analysis, one has the gross estimate

$$T \sim U_\nu L^2/\eta_0$$

and therefore

$$\eta \sim \left(\frac{2}{\pi}\right)\left(\frac{U_\epsilon}{U_\nu}\right)\left(\frac{\Delta\alpha}{\alpha}\right)\left(\frac{\omega_1 t}{2\pi}\right)\eta_0$$

The significant term is $(U_\varepsilon/U_\nu) \lesssim 10^{-3}$, and thus typically many cycles are required before a large oscillation can build up. So once again we see that the sail does not luff. The dimensionless parameter this time involves the energy density ratio, $(U_\varepsilon L^2/T\eta_o) \sim (U_\varepsilon/U_\nu)$, which is quite small.

Debye Sheath Coupling

An alternative approach to the electrostatic perturbing forces is examined next. Again we consider a membrane perturbed by an external electrostatic force and hence governed by the wave equation.

$$\rho \frac{\partial^2 \eta}{\partial t^2} - T \frac{\partial^2 \eta}{\partial x^2} = \Delta F$$

This time we take ΔF to be caused by self-consistent perturbations of the Debye sheath. The rippling of the sail perturbs the sheath, which in turn causes perturbing forces to act back on the sail. Hence this calculation is close in spirit to the earlier calculation of gas dynamic coupling. Now, however, we are examining coupling to the plasma Debye sheath.

We take a simple exponential model for the equilibrium sheath. Forces from either side of the membrane add vectorially. Thus we present the force derivation for a single side and only indicate the result for two sides.

For the front side, y>0, one has an equilibrium sheath

$$\emptyset = \emptyset_o \exp(-K_d Y) \qquad \text{(electrostatic potential)}$$

$$E_{yo} = K_d \emptyset_o \exp(-K_d Y) \qquad \text{(electric field)}$$

$$\sigma_o = (K_d \emptyset_o / 4\pi) \qquad \text{(surface charge density)}$$

where $K_d = (4\pi n_o e^2/T)^{\frac{1}{2}}$.

The perturbing force is

$$\Delta F = \Delta\sigma E_{yo} + \sigma_o \Delta E_y$$

where $E_y = 4\pi\Delta\sigma$ is the perturbed electric field and $\Delta\sigma$ is the perturbed charge density evaluated at the membrane surface. Note that

$$\Delta E_y = \frac{\partial E_{y0}}{\partial y}\delta y + \delta E_y, \qquad \delta y = \eta$$

where the first term represents the contribution to the total perturbed electric field, which results from evaluating the equilibrium electric field on the perturbed boundary, and the second term represents the perturbed fields evaluated at the equilibrium boundary.

To determine the perturbed fields we need a prescription for the plasma. Again, we note that for times t long compared to an ion transit time through the Debye sheath [$\tau_i \sim (\omega_{pi})^{-1} \sim 10^{-6}$sec] that the quasistatic approximation is valid. In this approximation the potential obeys the Debye equation

$$\nabla^2 \delta\phi - K_d^2 \delta\phi = 0$$

which, for perturbations varying as sin kx on the membrane, has the solution

$$\delta\phi = \delta\phi e^{-y\sqrt{K^2+K_d^2}} \sin Kx$$

so that

$$\begin{aligned} \Delta E_y &= 4\pi\Delta\sigma \\ &= -\frac{\partial\delta\phi}{\partial y} + \frac{\partial E_{y0}}{\partial y}\delta y \\ &= \sqrt{K^2+K_d^2}\,\delta\phi - K_d^2\phi_0\delta y \end{aligned}$$

We examine the case of a conducting membrane, which is an appropriate representation of the sail. (For wavelengths >> the sail thickness, the two sides of the sail move together, and no charging of the dielectric takes place in first order.) Thus we demand that the total potential, which again consists of two terms like those in the preceding perturbed electric field equations, be zero on the membrane. Thus one has

$$\frac{\partial\phi_0}{\partial y}\delta y + \delta\phi = 0$$

Hence, using the equilibrium results, one finds that

$$\delta\phi = K_d\phi_0\delta y$$

and so

$$\Delta E_y = \{\sqrt{K^2+K_d^2}\,K_d\phi_0 - K_d^2\phi_0\}\Delta y$$

The force on one side is thus

$$\Delta F = \left(\frac{K_d^2\phi_0^2}{2\pi}\right)[\sqrt{K^2+K_d^2} - K_d]\delta y$$

The total force is obtained by summing the forces acting on the front and back sides. Thus

$$\Delta F = \sum_{f,b} \Delta F, \qquad \text{where } f = \text{front}, \quad b = \text{back}$$

$$= \sum_{f,b} \left(\frac{K^2{}_d \phi_0{}^2}{2\pi} \right) [\sqrt{K^2 + K_d{}^2} - K_d] \delta y$$

and therefore the wave equation may be written as

$$\rho \frac{\partial^2 \delta y}{\partial t^2} = -TK^2 \delta y + \sum_{f,b} \left(\frac{K_d{}^2 \phi_0{}^2}{2\pi} \right) [\sqrt{K^2 + K_d{}^2} - K_d] \delta y$$

in the long wavelength limit $K << K_d$ and this reduces to

$$\rho \frac{\partial^2 \delta y}{\partial t^2} = -TK^2 \left\{ 1 - \sum_{f,b} \left(\frac{K_d{}^2 \phi_0{}^2}{4\pi K_d T} \right) \right\} \delta y$$

A plasma which is the same on the front and back will exhibit unstable modes whenever

$$K_d{}^2 \phi_0{}^2 / 2\pi > K_d T$$

for the solar wind

$$U_\epsilon = (K_d{}^2 \phi_0{}^2 / 4\pi) \sim 10^{-4} \text{ergs/cm}^3$$

whereas

$$U_s = K_d T \sim 10^{-4} \text{ergs/cm}^3$$

Thus no instability will occur. Again we find a dimensionless energy ratio which is quite small. Note again that $T \sim (L^2 U_\nu / \eta_o)$, where η_o is the equilibrium displacement of the sail. Thus the stability parameter can be written as

$$\epsilon = \frac{K_d{}^2 \phi_0{}^2}{4\pi K_d T} = \left(\frac{U_\epsilon}{U_\nu} \right) (K_d L)^{-1} \left(\frac{\eta_0}{L} \right)$$

Each element in this parameter is much less than 1; thus ε is quite small and corresponds to a very stable situation.

Miscellaneous Electromechanical Effects in near Earth Orbit

We also have considered a number of possible electromechanical hazards that could conceivably result from operations near Earth. It appears that none of these poses a threat. Most of these effects are associated with the passage into or out of eclipse. At GSO it takes some 4 s for the heligyro to pass completely into or out of Earth's shadow. We can imagine a pair of opposing blades, one in the sun and one in the shade, for a period of 2 min. We consider the following phenomena that produce mechanical forces that might be detrimental to the spacecraft.

1) Forces caused by different charges on the blades: to bound possible electrostatic forces we imagine one blade in sunlight and one in shade. A voltage of 10 kV on a blade implies a charge on the blade $Q \sim RV \sim (6.25 \times 10^5 \text{cm}) \times (33 \text{ stat V}) = 2 \times 10^7 \text{esu}$. If there are two charges of this magnitude whose separation is on the order of the blade length R, the force between them is only 10^{-2} N. Thus it appears that electrostatic forces on the sail arising from spacecraft charging could not possibly compete with the blade tension 780 N.

2) The current that could flow through the hub section during a charging-eclipse event might be on the order of

$$I = \frac{\Delta Q}{\Delta t} = \frac{2 \times 10^7 esu}{2s} = 3 \times 10^{-3} A$$

This is much smaller than the induction current estimate of the following section, and thus its effects should be negligible.

Two types of nonelectrostatic forces are apparent: 3) The torque τ arising from lack of sunlight pressure on the blade in eclipse and 4) unbalanced momentum arising from photoemission from only the sunlit blades. They are discussed below.

3) To estimate this effect we imagine a system of two opposing blades, one illuminated and one not, and assume that the moment of inertia of the system is given simply by the mass and extent of these two blades. The change in angular momentum is $\Delta L = \tau \cdot (\Delta t \sim 2 \text{ sec})$. The torque is provided by the radiation pressure P_s on one blade

$$\tau = P_s \int_0^R r dr w = w P_s \frac{R^2}{2} = 7.8 \times 10^9 \text{dyne-cm}$$

giving $\Delta L = 1.5 \times 10^{10}$dyne-cm-sec. The moment of inertia of the two blades is

$$I = 2\eta \int_0^R r^2 dr = \frac{2}{3} \eta R^3$$

where η~0.4gm/cm is the linear mass density. We find $I \approx 6.5 \times 10^{16}$gm cm^2, and thus a rotation rate

$$\theta = \Delta L / I = 2.3 \times 10^{-7} \text{rad/sec} = 1.5 \times 10^{-5} \text{deg/sec}$$

It appears that this torque would not be noticed. Another way to examine the effect is to calculate the Δv imparted to a single blade and compare it to the orbital velocity at GSO $v \sim 3 \times 10^5$cm/sec. We find $\Delta v = F\Delta t/m \sim 0.2$ cm/sec.

4) The force resulting from recoil of photo-electrons emitted from a blade at high potential ($\sim$-10 kV) works out to some 1.4×10^{-7}dyne/cm , or less than 1% of the solar pressure, and likewise its contribution to the unbalanced force.

5) Finally we consider the possibility that the electric field associated with some magnetospheric wave disturbance could disrupt the spacecraft. The type of wave in which we are interested will have a period on the order of that of the flapping modes of the heliogyro blade, i.e., ~100 sec. Waves in this range are known as Pc4 waves and are observed in coherent wave trains lasting for some 10-20 cycles. The electric field strength can be as high as $E \sim 0.1$ V/m = 3.33×10^{-6}stat V/cm.

For a simple model we imagine again a single blade charged up to 10 kV, so that the charge estimate of 1) implies a charge density $\sigma \sim 4 \times 10^{-2}$esu/cm^2. This charge density and field imply a force on the blade

$$\sigma E \sim 1.2 \times 10^{-7} \text{dyne/cm}^3 << P_S$$

and thus the whole spacecraft is not moved significantly. We next examine what might happen when this small force is applied in resonance with a blade normal mode. We shall calculate this with a mathematically very simple model. We imagine a blade tied at both ends under tension T=385 N = 3.85×10^7dyne (or about half the blade tension), the average tension along the actual blade. We also neglect damping (which is small: $Q \sim 1000$[6]) and assume E is normal to the blade and in perfect resonance with its lowest mode. The wave equation for the blade displacement is

$$\eta \frac{\partial^2 u}{\partial t^2} - T \frac{\partial^2 u}{\partial x^2} = F_{ext}$$

The lowest mode of the unperturbed blade is

$$u = \sin \pi x / R \sin \omega_l t$$

with

$$\omega_l = \frac{\pi}{R}\sqrt{T/\eta} = c\frac{\pi}{R} = 4.9 \times 10^{-2} \mathrm{rad/sec}$$

(this gives a period of 127 sec). We also assume for mathematical simplicity that the external force has the constant shape sin $\pi x/R$ over the blade. In fact, the E field wavelengths we are considering are much longer than the blade, and the force is effectively constant in space and sinusoidal in time. We are then considering the coupling of this force to the lowest blade mode. The wave equation is thus

$$\frac{\partial^2 u}{\partial t^2} - c^2\frac{\partial^2 u}{\partial x^2} = \beta \sin\frac{\pi x}{R}\sin\omega_l t$$

The constant β is given by the force per unit length divided by the mass per unit length.

$$\beta = F/\eta = \sigma w E/\eta = 2.66 \times 10^{-4} \mathrm{cm/sec^2}$$

If it is assumed that the displacement is zero at time zero, we find

$$u(t) = \frac{\beta}{2\omega_l^{\,2}}\sin\frac{\pi x}{R}(\sin\omega_l t - \omega_l t\cos\omega_l t)$$

and the amplitude grows linearly,

$$A(t) \sim \beta t/2\omega_l$$

after 20 cycles or 2540 sec A~7.0 cm. This appears sufficiently small to be safe. In particular it should be recalled that to compute the force we have used an estimate for the charge density that is rather extreme, and also we neglected damping. In any case, if the heliogyro were developed for use within the magnetosphere, it would be prudent to calculate more realistically the interaction with probable magnetospheric wave modes, using a correct heliogyro blade model.

In summary we have found no special problems of a plasma or electromechanical nature that would prevent the use of the solar sail near Earth.

Induction Current and Induction Force

When a conducting satellite moves through a stationary magnetized plasma, an electric field $\vec{E} = -\vec{V} \times \vec{B}/c$ is set up within it as a result of the polarization of the free charge produced by the motion perpendicular to B(Fig. 2). Likewise, since B field lines can be regarded as frozen into the solar wind, the plasma motion of the solar wind plasma past

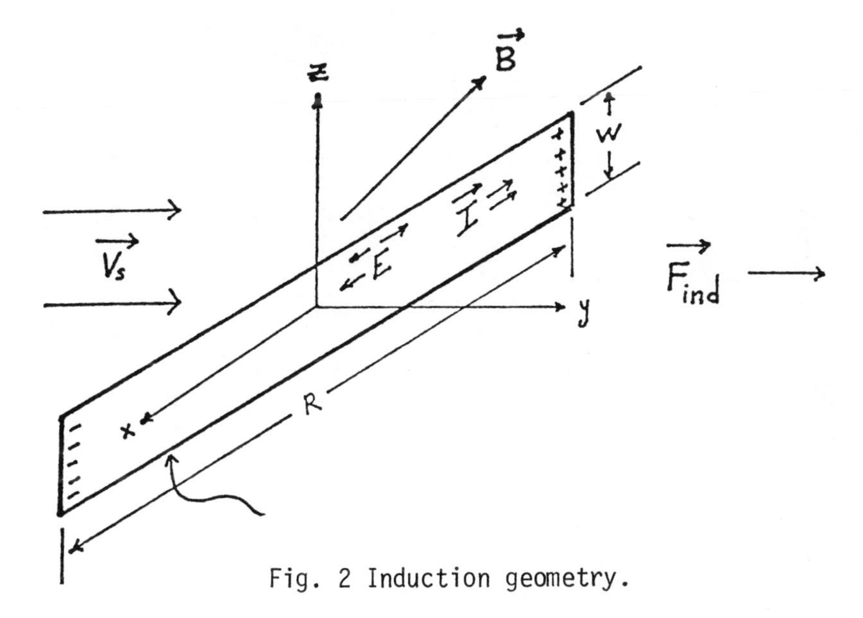

Fig. 2 Induction geometry.

the satellite with velocity V_S produces the field $E = V_S \times B/c$. The potential variation on the satellite surface causes electrons to land preferentially on the more positive portions and protons to land preferentially on the more negative portions. Also, photo-electrons emitted from the positive parts are more likely to be attracted back to the surface.

This situation causes a steady current to flow through the satellite. The current is calculated by using the potential at each point on the surface to calculate the plasma current into the satellite. The potential is a linear function of the dimension x in the $\vec{V}_S \times \vec{B}$ direction. The requirement that the net current into the satellite equal the net current out of the satellite gives the potential. The possible consequences of this current in our view are twofold: 1) the current crossed with the magnetic field produces forces and torques, and 2) the current dissipates energy in the aluminum sail film. We shall argue that the induction forces are far smaller than the radiation pressure and that the energy dissipated by conduction is negligible in comparison to the sunlight energy absorbed by the aluminum.

It is not necessary to perform a detailed calculation to show that the foregoing effects are small. Figure 3 shows the general situation with the currents. For this figure we assumed that the potential changes sign somewhere along the blade. This is not necessarily true. The net current per unit area into or out of the sail at any point x is given by the sum

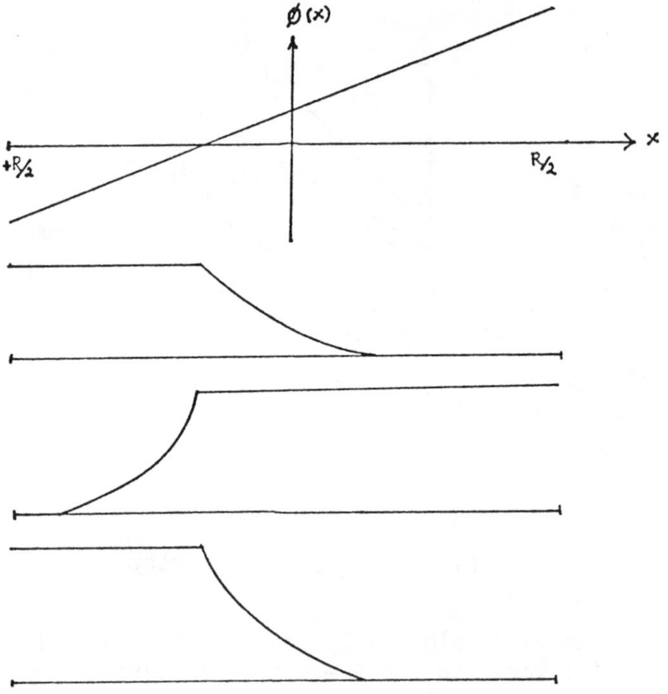

Fig. 3 Sail potential and particle fluxes.

$\Gamma(x)$ of the three fluxes at that point. A crude bound on the maximum current along the sail can be made by taking the largest of the three current densities, $J_{ph}/e = \Gamma_{ph} = 2.6 \times 10^{11}/$ cm^2- ster (at 0.3 a.u.), multiplying by the electronic charge and one-half the area, and assuming this current flows in one-half the blade and out the other half. This current is

$$I_{max} = \tfrac{1}{2} e \Gamma_{ph} \cdot (w = 8 \times 10^2 cm)\ (R = 6.24 \times 10^5 cm) = 3.12 \times 10^{10} esu/sec$$

$$I_{max} = 10.5\ A$$

This total current multiplied by the length and magnetic field and divided by the blade area gives a value for the total induction force per unit area

$$F = \frac{I_{max} \cdot R \cdot (B = 58 \times 10^{-5} G)}{c \cdot R \cdot w}$$

$$F = 7.54 \times 10^{-7} \text{dyne/cm}^2$$

This is about .15% of the solar radiation pressure at 0.3 a.u., and thus is seemingly negligible.

The current density in the VDA implied by I_{max} is

$$J = [I_{max}/w(t = 1000\text{Å}] = 3.9 \times 10^{12}\text{esu/cm}^2\text{-sec}$$

Using the approximate value for the conductivity at 525°K, $\sigma \sim 2.6\times10^{17}\text{sec}^{-1}$, we find the volume energy deposition rate

$$\frac{dw}{dt} = \frac{J^2}{\sigma} = 5.85 \times 10^7 \frac{\text{erg}}{\text{cm}^2 - \text{sec}} = 5.85\text{W/cm}^3$$

This implies a per unit area energy absorption of 5.85 x 10^{-5} W/cm^2. The VDA absorbs about 10% of the incident optical energy or, at 0.3 a.u., about 0.14 W/cm^2. Thus the induction current should not affect the thermal balance of the sail.

Application to Heliogyro

The detailed calculations have used a membrane equation as a description of the basic sail. This is adequate for illustrative purposes and is perhaps a reasonable representation of the square sail. A detailed analysis of the heligyro could be based upon the uncoupled small-motion equations for twist, vertical deflection, and inplane deflection. In view of the strong indication that luffing is not a serious problem, which has been derived using the membrane equation, use of the more elaborate equations seems unjustified. In the event that such analysis were necessary, one would again add the electrostatic and gas dynamic forces to the basic equations.

Concluding Comments

These calculations indicate two dimensionless parameters, 1) $\varepsilon_\rho = (n_\rho m/\rho\alpha)$, the ratio of mass densities; and 2) $\varepsilon_u = (U_\varepsilon/U_\nu)$, the ratio of plasma energy density to photon energy density. Both are much less than unity and thus the gas or plasma forces acting upon the sail are much less than the forces caused by photon pressure. No mechanism has been found which would amplify the impact of the plasma forces significantly. The study is, however, far from exhaustive.

Acknowledgments

This work was supported by NASA Lewis Research Center under Contract NAS3-Z0119. The authors also acknowledge useful discussions with R. Goldstein of the Jet Propulsion Laboratory.

References

[1]Morse, P. M. and Ingard, K. U., Theoretical Acoustics, McGraw Hill, New York, 1968, p. 191-213.

SPACECRAFT CONTAMINATION: A REVIEW

J. M. Jemiola*
Air Force Wright Aeronautical Laboratories/
Wright-Patterson Air Force Base, Ohio

Abstract

Contamination sources include every significant event that occurs in the life of a spacecraft, from manufacturing, through on-orbit operation. To combat these rising contamination concerns, a number of efforts are being directed at controlling the extent and the effects of contamination. Spacecraft materials are selected using screening tests to eliminate high outgassers. Contamination level monitoring procedures and methodology for clean rooms are being developed and the effectiveness of spacecraft protective measures determined. High outgassing components are prebaked and thermal vacuum tested. Contamination models are being developed to predict contaminant levels and effects. Contamination models need to be expanded to include particulate debris. Thruster modeling, though well underway for liquid and electric thrusters, does not yet adequately predict contamination effects. Modeling of solid thrusters is just beginning. Contamination control is currently practiced on a system-by-system basis. Most contamination control technology is developed for a specific system and is not readily transferred to other spacecraft systems.

I. Introduction

The problem of spacecraft contamination has, during the last few years, become very important in space technology. A spacecraft contaminant, for the purpose of this review, will be defined as any foreign material which adversely affects the operation of a spacecraft. The influence of such spacecraft contamination has become increasingly detrimental as spacecraft sensors and other instrumentation have become more complex, the operations of these instruments more critical, and

---

*Satellite Materials Development Engineer.

the operational design lifetimes of spacecraft increasingly long (7-10 years). Available flight evidence of contamination effects includes: a) the measurement of significant mass deposition in the Lincoln Experimental Satellites (LES)-8 and -9 payload area by a quartz crystal microbalance during the firing of the Titan second stage retro rockets (see Fig. 1)[1,2]; b) on-orbit observation of particulates around Skylab (see Fig. 2)[3] and Voyager[4] coupled with evidence of particle interference with star trackers, causing a loss of attitude control; c) sensitivity loss of optical systems probably caused by condensation of outgassing products[5,6] on optical systems; and d) spacecraft heatup due to contamination enhanced degradation of exterior thermal control surfaces (see Fig. 3).[2]

A list of possible contamination sources includes every significant event that occurs in the spacecraft lifetime (see Fig. 4). The prelaunch environment includes all steps in the manufacturing, handling, testing, storage, and transportation of the spacecraft. The complexity of maintaining contamination control during prelaunch is severe even though all steps are within the control of the spacecraft manufacturer and the user. The environment and the organizational control become considerably more complex, however, as the spacecraft inter-

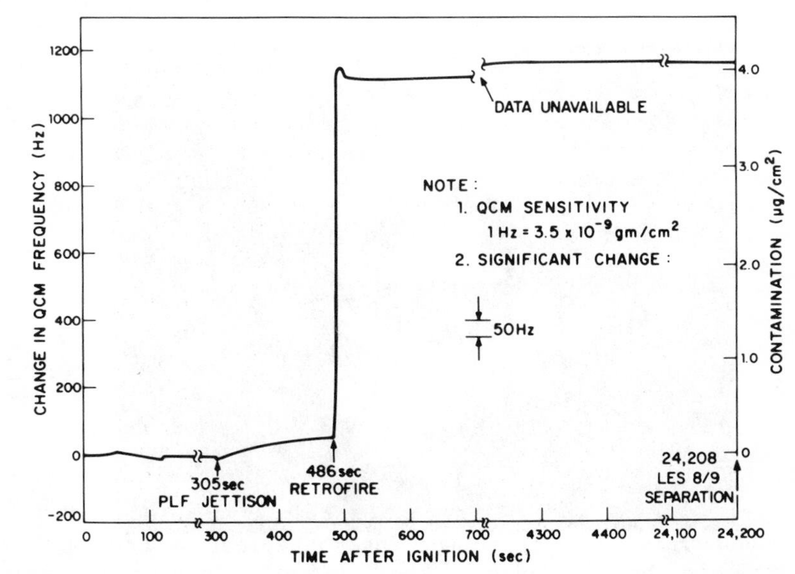

Fig. 1 Quartz crystal microbalance frequency (contaminant deposit thickness) vs time for the entire launch of Lincoln Experimental Satellites -8 and -9 on a Titan III-C rocket; from Ref. 1.

faces with the launch vehicle, and cleanliness concerns meet with the possibly conflicting needs of the launch vehicle and of other spacecraft. During the ascent of present launch vehicles, the spacecraft is subjected to contamination induced by aerodynamic heating of the satellite fairing. Next follows fairing ejection, which allows launch vehicle and satellite outgassing products to be carried away as the launch vehicle speeds through the upper atmosphere.

This scenario is little improved for the Space Transportation System (Space Shuttle). In the case of the Shuttle, both the spacecraft and major components of the launch vehicle vent into the payload bay. The spacecraft will be exposed to contaminants from the payload bay for a period of time (nominally 1 hr.) with the payload bay doors closed, followed by 2 hrs. with the payload bay open.[7] The spacecraft will potentially be exposed to contaminants from rocket plumes during deployment, injection, and on-orbit stabilization. In the case of a Shuttle launch, flash evaporator effluent, outgassing from the crew quarters, and reaction control system thrusters represent additional sources of contamination concern.

Once on-orbit, the spacecraft is exposed to the natural space environment (protons, electrons, ions, ultraviolet

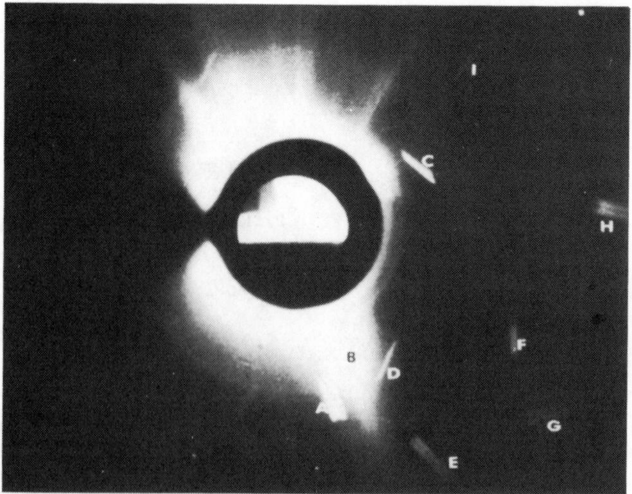

Fig. 2 A typical photograph showing contamination tracks. (A support boom obscures part of the field of view in the middle of the frame taken from the high altitude observatory's white light coronograph [skylab experiment]. The nine contamination tracks analyzed in this photograph are labeled A through I). From Ref. 3.

radiation, and micrometeoroid impact) which can polymerize previously deposited surface contaminant materials or can generate new contaminants, further degrading the operational characteristics of sensors and of thermal control surfaces. Combine these effects with the addition of contaminants from outgassing or particle-generating materials, attitude control thruster products, and evaporated oil from bearing surfaces,[8] and, not surprisingly, on-orbit problems with lifetime and sensor sensitivity become of prime concern.

II. Current Contamination Control Activities and Problem Areas

To combat the rising contamination concerns, a number of efforts are being directed at controlling the extent and effect of contamination. Efforts dealing with material test and selection, contamination monitoring, spacecraft cleaning and testing, contamination modeling, and definition of the space environment are all currently underway. These will be discussed in the following sections.

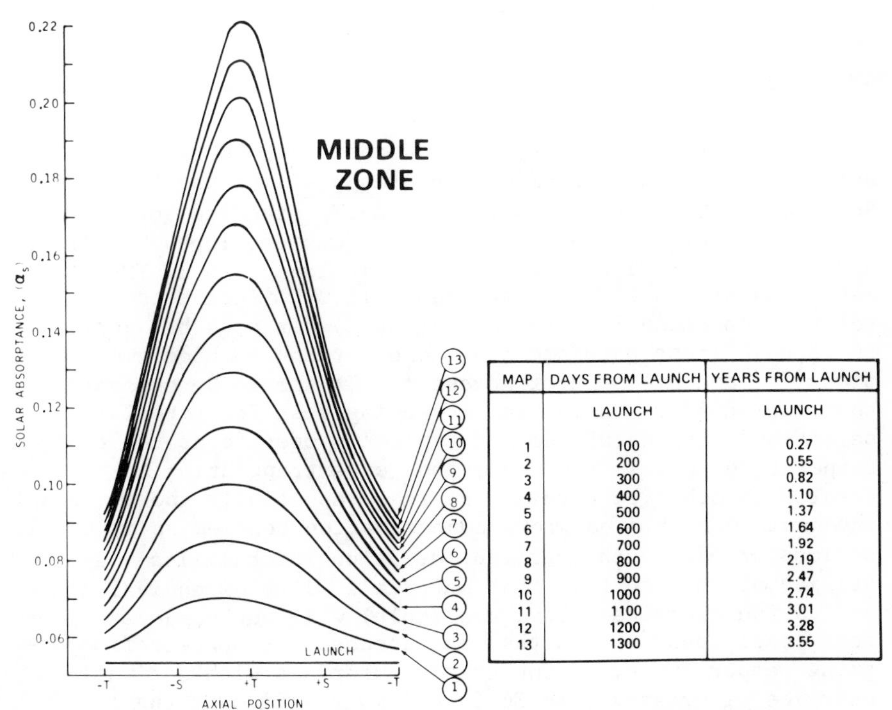

MAP	DAYS FROM LAUNCH	YEARS FROM LAUNCH
	LAUNCH	LAUNCH
1	100	0.27
2	200	0.55
3	300	0.82
4	400	1.10
5	500	1.37
6	600	1.64
7	700	1.92
8	800	2.19
9	900	2.47
10	1000	2.74
11	1100	3.01
12	1200	3.28
13	1300	3.55

Fig. 3 Contamination buildup as evidenced by increased solar absorptance as a function of time on-orbit for the radiator of an operational satellite; from Ref. 37.

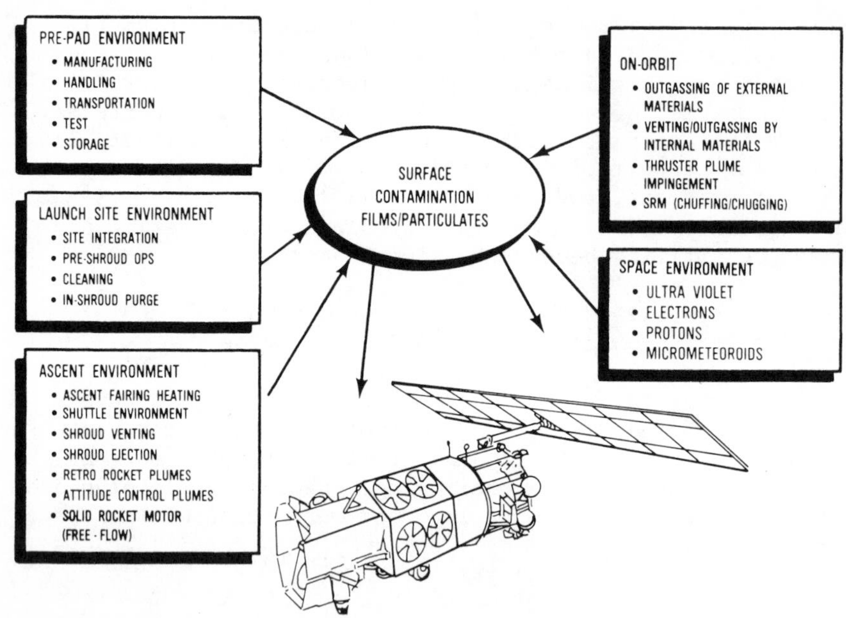

Fig. 4 Sources of spacecraft contamination; from Ref. 82.

A. Materials Testing and Selection

The only generally accepted means of spacecraft materials screening from the contamination standpoint is the American Society for Testing and Materials (ASTM) standard screening test for materials (ASTM E595)[9] which was developed by Miraca and Whittick.[10] The test criterion is less than 1% TML (total mass loss at 125°C) and less than 0.1% CVCM (collected volatile condensable materials at 25°C) for a 24 hr. period in vacuo. These data are available through both documentation[11-13] and by computer access.[14] It should be emphasized that ASTM E595 is an initial screening test for material selection. Spacecraft do not typically operate at the test temperature, nor is there a method of extrapolating test results to other temperature regimes, especially the important cryogenic sensor area where contamination becomes an especially serious problem. Another consideration which limits the utility of this test is that materials with low equivalent activation energies ($\leq$10.1 kcal/mole) will not condense under test conditions, though they may condense at spacecraft operating temperatures.[15] Conversely, materials with activation energies of greater than 36.3 kcal/mole will pass the test, though all of the material may eventually volatilize.[16] The test does not take into consideration space stability of the condensed material[17] nor the effect material-generated CVCM will have upon the operation of a particular spacecraft.[18,19]

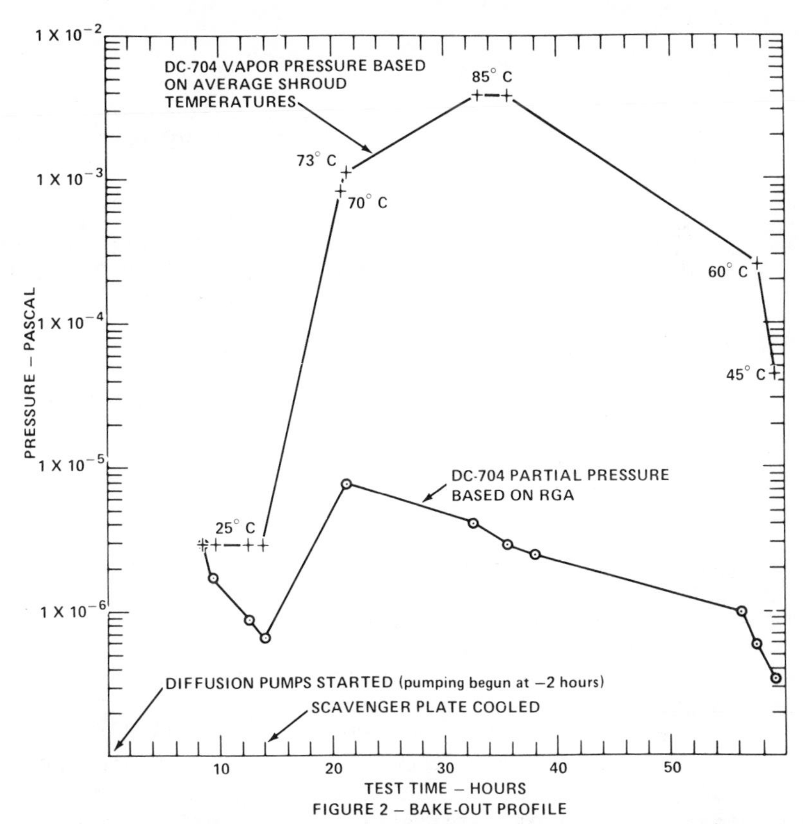

Fig. 5 DC-704 (Dow Corning trade name for tetraphenyltetramethyltrisiloxane) vapor pressure vs thermal vacuum bake-out time of a contamination sensitive spacecraft instrument. (+ is the DC-704 vapor pressure [the pressure at which condensation vapor pressure occurs] based on average instrument shroud temperatures; ʘ is the DC-704 partial pressure as determined by residual gas analysis). From Ref. 32.

The question of the level of actual contamination deposition can be resolved by measuring the light scattered by the collected volatile condensable materials and employing this as an additional acceptance criteria.[20]

B. Contamination Monitoring

To ensure that spacecraft surfaces and materials remain clean, spacecraft are manufactured and stored in clean environments where contaminant levels are monitored. Clean room levels and particulate and nonvolatile residue measurement criteria have been reviewed by Borson[21] whose general conclu-

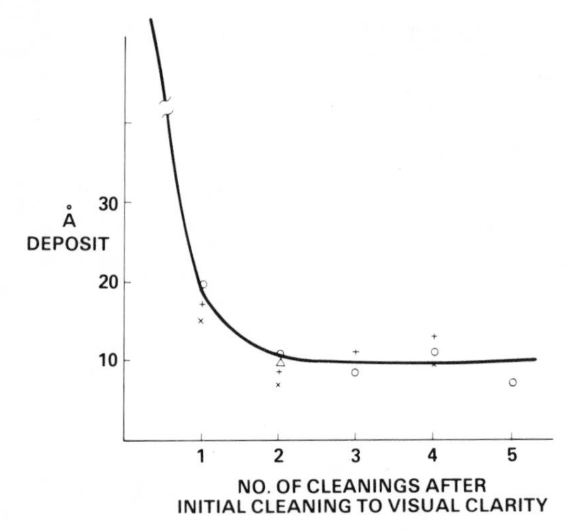

Fig. 6 Effects of repetitive cleanings on a second surface mirror contaminated with RTV-566 (General Electric trade name for an iron oxide filled methylphenylsilicone) adhesive; from Ref. 33.

sions follow. The class of a clean room is based on the design, construction, and operating procedures of that room,[22] whereas actual cleanliness may vary over a wide range depending upon the number of people in the room and the type of operations being performed. If personnel and procedures are rigidly controlled, a clean room can provide a significantly cleaner environment than it was originally designed to provide. Another serious deficiency in clean room standards is that these standards do not consider molecular contaminants that deposit onto exposed surfaces.[22,23] A prime example of this deficiency is the use of dioctylphthalate fogs to test clean rooms and high efficiency particulate air (HEPA) filters[24] for leaks. Dioctylphthalate eventually volatilizes and is carried into the clean room where it can be a serious optical contaminant. Because of the problems with dioctylphthalate, alternate leak check procedures need to be authorized.[21]

Quantitative descriptions of surface cleanliness[25,26] are available to describe particulate and nonvolatile residue (NVR) on surfaces. A serious drawback to implementing these standards is the necessity to define both the number of particles per unit area and the size distribution of these particles.

Normal particle collection procedures for determining particle size and number per unit area are flushing with solvents, photographic techniques,[27] vacuuming, and use of pressure-sensitive tapes. Each of these techniques has sensitivity limitations and particle sizing biases. Another

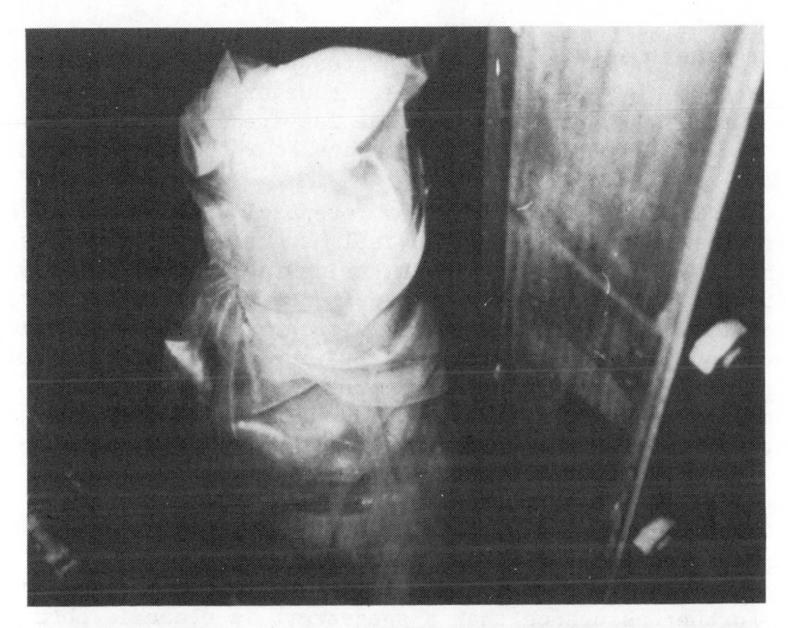

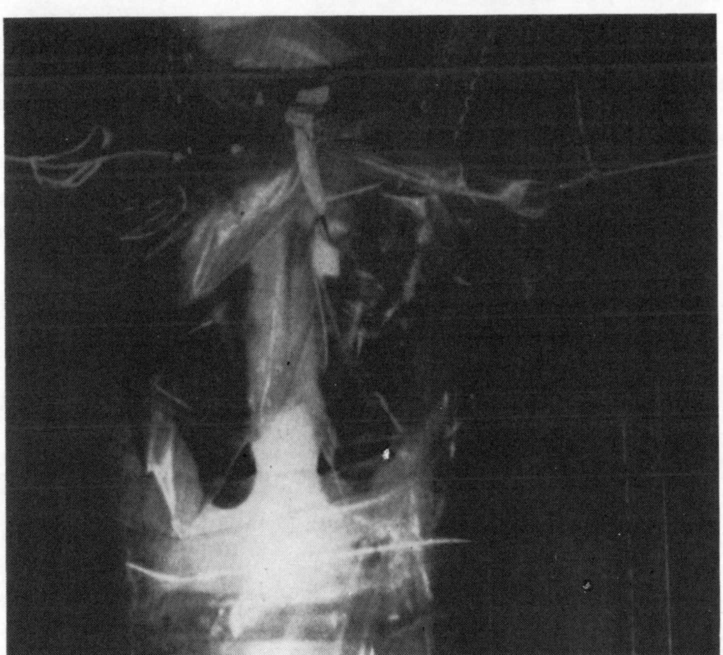

Fig. 7 Infrared payload enshrouded in polyethylene bagging material on the launch tower: top is the payload nosetip, and bottom is the umbilical cable connections to the payload; from Ref. 34.

approach to monitoring clean rooms is to determine particle sizes and numbers from witness plates which are monitored on a regular basis. Questions arise, however, as to whether these witness plates are given special treatment by workers[21] relative to other parts of the spacecraft which are not monitored.

Despite the large number of potential contaminant sources, comprehensive contamination control programs utilizing air cleanliness standards in manufacturing facilities, personnel constraints including garment requirements and limited access to facilities, protective measures such as dust covers, and gaseous purges coupled with monitoring methods for particulate and molecular contaminants have proven successful.[19,28] Assemblage analysis[29] is also being used to identify contaminant sources in clean rooms and a method of relating airborne particle counts to what actually settles out on the spacecraft has been reported.[30]

C. Spacecraft Cleaning and Testing

Further assurance that a spacecraft is clean is obtained by prebaking components to ensure they meet outgassing requirements. This is normally followed by thermal vacuum testing the whole spacecraft while monitoring the chamber for contaminants (see Fig. 5)[31,32] and/or exhaustive cleaning of spacecraft sensitive areas (see Fig. 6)[33,34] prior to bagging (see Fig. 7) or launch. A recommended practice for use of

Table 1 Candidate on-orbit contamination monitors for routine on-orbit use

Contamination Monitor	Operating Principle
Calorimeter	Sample specimen heat up[a]
Photo diode	Light scatter off an optical surface
Quartz crystal microbalance	Oscillator (crystal) frequency change
Solar cells	Decrease in solar cell response[b]

[a]Could be due to contamination buildup or to material degradation (absorptance or emittance change).

[b]Could be due to decrease in light throughput and increased solar cell temperature (both caused by contamination buildup).

residual gas analyzers in thermal vacuum chambers is currently under development.[35]

D. On-orbit Contamination Monitoring

On-orbit contamination levels can be determined only by continued use of on-orbit contamination monitors. Monitoring techniques which are candidates for inclusion in a routine contamination monitoring program are presented in Table 1. In some cases the contamination monitor inadvertently turns out to be the spacecraft's primary sensor[5,6] or a critical thermal control surface.[2] In other cases, the contamination monitors are experiments.[36] Greater use of on-orbit monitors measuring both contamination thickness and contamination effects is needed. Such monitors can provide feedback on the degree of success of a contamination control program and can relate contamination thickness to on-orbit system effects of contamination. Relating contamination levels to effects on-orbit[37] is one of the critical needs of another contamination control area: contamination modeling.

E. Contamination Modeling/Verification

Contamination modeling encompasses a wide area: contamination assessment during ascent and during on-orbit operation (including both material and thruster sources), and verification through flight experiments of such models. Each of these areas is addressed in subsequent sections.

Material Outgassing Modeling/Verification. Contamination modeling of the ascent phase of flight is essentially an extension of vacuum technology. Pressure profiles of a multi-compartment spacecraft can be calculated by a) using standard expressions for viscous, intermediate, and molecular flow regions, b) applying appropriate boundary conditions, such as conductive paths and volumes, and c) solving, via computer, the generated differential equations.[38] Inclusion of external pressure profiles,[39] outgassing characteristics of materials,[40,41] the thermal profile of the spacecraft, and the surface affinity to absorb and desorb gases enables the engineer to predict levels of contaminant deposited during the launch and ascent phases.[42]

It is during the launch and ascent phases of flight that past and present launch vehicle environments diverge significantly from the projected Space Transportation System (Shuttle) environment. On nonreuseable launch vehicles, the payload tends to be isolated from the launch vehicle. As the vehicle accelerates, the fairing temperature rises, causing outgassing contamination and thermal loading of the payload. Shortly after launch, normally 30 sec., the fairing is ejected and the payload rides ahead of the remainder of the launch vehicle. Generated contaminants tend to be swept away by the on-rush of the Earth's upper atmosphere.

For Shuttle-launched payloads, the launch and ascent environment is appreciably different. The Shuttle, because of its size, can accommodate multiple payloads. These payloads are enclosed in the Shuttle bay area with their boosters and kickmotors. The vent paths of the Shuttle's major structures (such as the wings) pass through the same payload area, as do those of a number of Shuttle subsystems such as hydraulic lines which are potential contamination sources. The Shuttle bay is closed during the initial segment of vehicle launch. During the initial venting, outgassing products will be swept away by the venting of the atmosphere enclosed in the Shuttle bay. Soon, however, the mean free path of outgassed molecules approaches the dimensional size of the Shuttle bay. Outgassed products emitted from the payloads, the Shuttle, or less contamination sensitive areas of adjacent payloads, can redistribute themselves in line-of-sight directions from their sources or, by absorption, condensation, and re-evaporation from Shuttle secondary structures, onto more contamination-sensitive portions of other payloads. Because the Shuttle is reuseable, such contamination from dirty payloads, which were launched as many as two or three missions ago, may affect future payloads.

Even after the payload bay doors open, the lower portions of all payloads will be within line-of-sight contamination transport distance of the Shuttle bay. During payload separation from the Shuttle, the payload will be susceptible to the particulate debris cloud associated with manned vehicles and Shuttle-induced contaminant fluxes resulting from the impingement of reaction control system thrusters and the flash evaporator system on the Shuttle's wings. Also during payload separation, the competing interests of the Shuttle to avoid being impinged upon by the spacecraft thrusters (especially solid thrusters, which could erode Shuttle thermal control tiles or cabin windows, having catastrophic effects on the shuttle) and the payload to avoid impingement by Shuttle engines, may be at cross purposes.

The launch and ascent phases merge into the induced orbit environment of material outgassing, meteoroid impact, and thruster impingement products. Sophisticated analytical models and computer programs which can predict the on-orbit environment have been generated.[37,43] These programs take into account the spacecraft geometry and its thermal profile as well as contaminant source, transport, and effects characteristics as a function of time. Numerous experimenters have studied material source transport and effects characteristics resulting in both empirical[16,18,44] and rigorous descriptions[45-47] of the environment. Typically, empirical treatments assume that outgassing can be modeled using the Arrhenius principle, $k=Ae^{-E/RT}$ (where k is the reaction rate of the outgassing process, E is the activation energy, R is the gas constant, and A is a constant) alone or in combination with Langmuir type behavior, $\text{Log } P=A-\frac{B}{T}$ (where P is the vapor pressure of the outgassing material with temperature T, and A and B are constants). Rigorous treatments tend to emphasize diffusion rates, sticking coefficients, evaporation rates, and other parameters that limit the number of materials that can be tested because of time requirements and costs. Less rigorous treatments utilize curve fitting of outgassing rates with empirical equations to generate data at a more reasonable cost. Treatments using line-of-sight deposition,[45] the effects of intermediate surfaces as contaminant transfer agents,[48] the influence of return flux,[49] Monte Carlo simulation of transport characteristics using classical collision dynamics,[50] and the relationship of charged particles and their deposition on spacecraft surfaces by electrostatic attraction,[51] are all reported in the literature. Considerable effort is being devoted to verify these molecular contamination models.

An experiment performed on the Atmospheric Explorer-D satellite, which measured by means of a mass spectrometer the back-scattered flux from neon gas vented in the direction of the satellite velocity vector, compared favorably to calculations based on a single collision theory.[52] The Induced Environmental Contamination Monitor (IECM)[53] will be used to verify the prelaunch, ascent, on-orbit, descent, and post-landing environments of the Shuttle. It will include collection devices, molecular and particulate contamination monitors, and optical systems. One of the experiments on the Spacecraft Charging at High Altitudes (SCATHA) P78-2 satellite (currently on-orbit) will determine whether charging of spacecraft surfaces influences the rate of satellite contamination.[36] Another verification effort has used solar absorptance maps developed from an operational satellite to monitor the progress and to identify the on-orbit sources of contamination (see Fig. 3).[37]

Despite these modeling and verification efforts, there are still areas in molecular contamination modeling that need

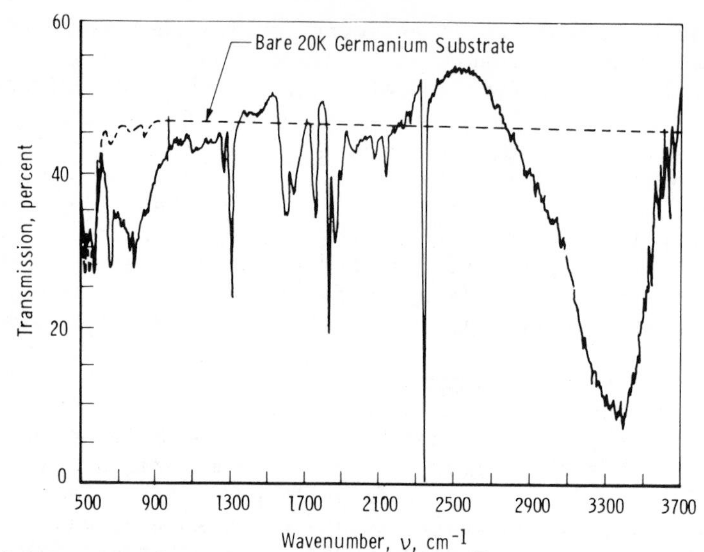

Fig. 8 Transmission spectrum of a 3.62-μm thick deposit of solid bipropellant contamination on a 20K germanium substrate located 90 deg relative to the engine thrust vector. (The thruster was a 5-lb_f nitrogen tetroxide/monomethylhydrazine bipropellant engine.) From Ref. 59.

to be resolved. A way to determine material source and transport characteristics without large expenditures of time and money needs to be found so that large numbers of materials can be evaluated. Another key area is the relationship of contaminant thickness or level to optical and thermal control effects on on-orbit spacecraft systems and subsystems.

Thruster Modeling/Verification. Thruster sources must be included in the on-orbit assessment of spacecraft contamination. Unlike other sources on a spacecraft, thrusters can perform their function only by expelling large quantities of gases and often particulates, either liquid or solid. If even a small portion of this debris finds its way to critical thermal control or optical surfaces of a spacecraft, serious contamination problems could occur.

The CONTAM computer model,[54] developed to model the contamination effects of liquid propellant thrusters, has been verified against chamber experimental data for small monopropellant thrusters.[55-57] Contamination data also are available for bipropellant thrusters (see Fig. 8).[58-60] Mass flux measurements exhibit an exponential falloff from the plume center line (see Fig. 9)[58] rather than the sharp cutoff pre-

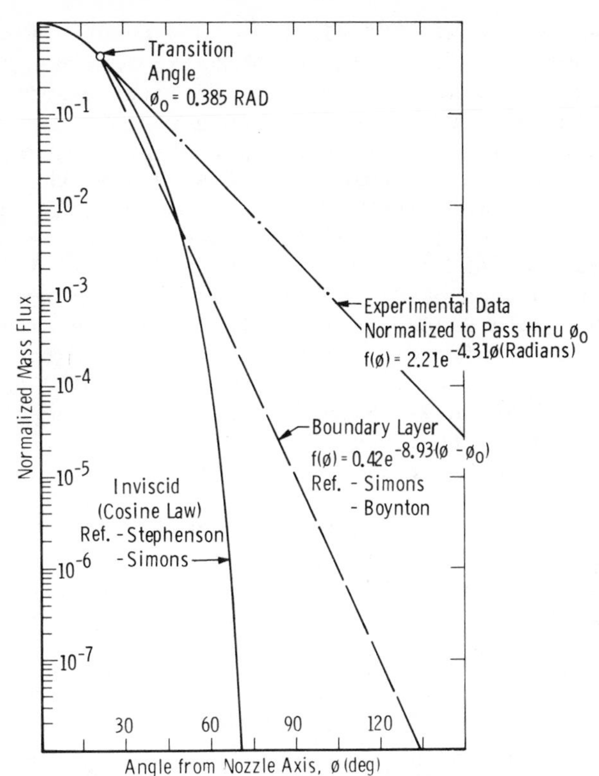

Fig. 9 Comparison of measured and calculated normalized mass flux distributions for a 5-lb_f nitrogen tetroxide/monomethylhydrazine thruster as a function of angle from the thrust vector direction; from Ref. 58.

dicted by present models. Implications are that the cause may be the large radial Mach number variation at the exit of the thruster nozzle which causes the boundary layer near the nozzle wall to expand to large angles with respect to the nozzle center line.[61] Unfortunately, due partially to limitations imposed by the geometry and size of current thruster test chambers, the physics of this boundary layer are not well understood. A number of possible mechanisms which could cause thruster back-flow have been postulated and their magnitude estimated.[62] A more recent update of boundary region theory also is available.[63]

Contamination concerns associated with solid thrusters are similar to those associated with liquids (particulate debris has been observed at 110 deg. and 180 deg. relative to the thrust vector,[64] but the difficulty in modeling and verifying this back-flow is greater. Solid thrusters are loaded with approximately 16% aluminum to prevent catastrophic unstable burning. The main contamination consequence of this aluminum addition is the formation of large quantities of aluminum oxide particles during combustion. Such particles, in combination with the thruster plume, are observed[2] to result in a severe contamination penalty when they contact the spacecraft. Evidence supporting the contention that particle size distribution is more characteristic of the sampling technique used than of any characteristic of the solid thruster also has been presented (see Fig. 10).[65] More important, smaller particle sizes traditionally have been discriminated against because of their limited utility in what here-to-fore has been the particulates prime function: solid thruster plume stabilization.[65] Significantly, a government supported program indicates that when small particles ($\leq 1.0\ \mu m$) are specifically sought, they are found in quantity and may even be the numerically predominant species, although this is not yet certain because of sampling uncertainties. Recent advances in the use of laser fringe anemometry[66-70] give indications that this technique, although size range limited, may be useful as a noninterfering tool for in situ particle size measurements. Finally, another area which deserves increased study is the contamination effect of solid thruster plume debris on optical plates (see Figs. 11 and 12).[64,70,71]

Examples of contamination analyses of ion thrusters on spacecraft and spacecraft components are available in the literature[72-74] as are sputter erosion studies of ion thrusters caused by ion impingement[75] and on-orbit operations of ion thrusters.[76] Operational constraints can be used to prevent direct ion impingement on operational spacecraft surfaces. Contamination problems could arise because of sputtering of material from the ion thruster accelerator grid. Neutralizing the ejected ion beam should reduce the sputtering problem but, in the case of a mercury thruster, amalgamation of spacecraft components could be a problem.[72]

One of the greatest problems in developing thruster models is related to the problem of verification of contamination and plume flow field in the on-orbit environment. Space simulation chambers do not achieve vacuums resembling those which occur in space, yet it is precisely this high vacuum which enables boundary layer expansion and contamination back-flow to occur.[77] Space simulation tests are also limited to tests on mini-thrusters [$\leq$0.9N (0.2 lb_f)]. Whether these test results can be scaled to larger thrust engines of the type used in space is of concern.[78]

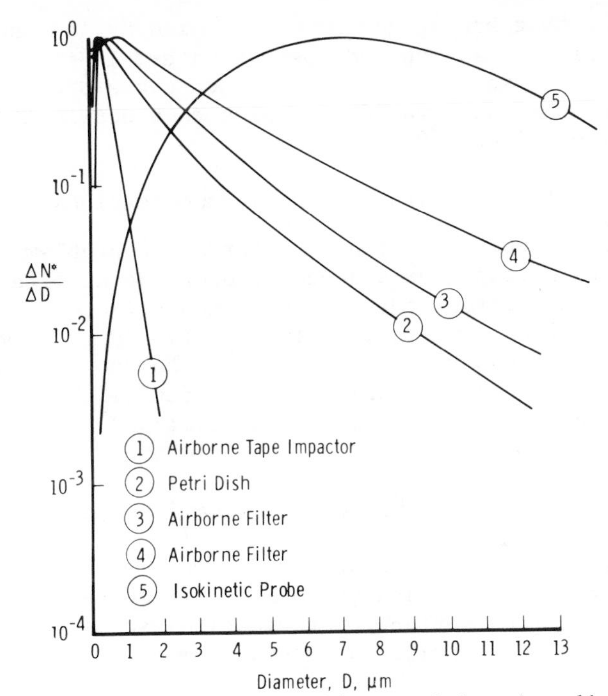

Fig. 10 Normalized aluminum oxide particle size distribution data from various Titan III-C firings using different collection methods; from Ref. 65.

Development and flight of a Propulsion Contamination Effects Module (PCEM)[78] to verify contamination models by flight tests of larger thrust engines has been proposed. Prior to testing of such large thrusters on-orbit, testing and validation of thruster models and development of noninterfering plume diagnostic tools need to be pursued. This is especially true for solid thrusters where diagnostic tools are at a very rudimentary stage of development.

F. Spacecraft Environment Definition

Modeling of the spacecraft charging environment has been reviewed elsewhere[79] and will be given only cursory mention here. These models include modeling of a) ion energies and concentrations in the magnetosphere, b) the influence of the spacecraft plasma sheath on particle trajectories and ions in the vicinity of the spacecraft, and c) discharge or arcing models of the spacecraft exterior surface.[79] Laboratory experiments have shown that arcing is possible under conditions likely to be experienced by satellites in substorm environments,[80] and that arcing can be accompanied by low-energy charged-particle emission and loss of material.[81] The SCATHA satellite, which was launched into synchronous orbit in 1979, will be used for model validation of both electrical[79] and contamination effects.[36]

III. Conclusions and Recommendations

Spacecraft contamination is a pervasive problem which is becoming more critical as spacecraft operational lifetimes and spaceborne instrumentation requirements increase. Contamination comes from a variety of sources (see Fig. 4) which are not within the purview of any one organization. Contamination can be controlled only if a coordinated, unified contamination control program can be developed and implemented by manufacturer, user, and launch vehicle organizations. Such a program would break down into five major areas:[82] 1) development of a generalized spacecraft contamination assessment methodology (including models of material outgassing, thruster effects, and the space environment plus development of cleanliness verification techniques for spacecraft and clean rooms so that surface contamination can be characterized and cleaning technique assessments made, 2) development

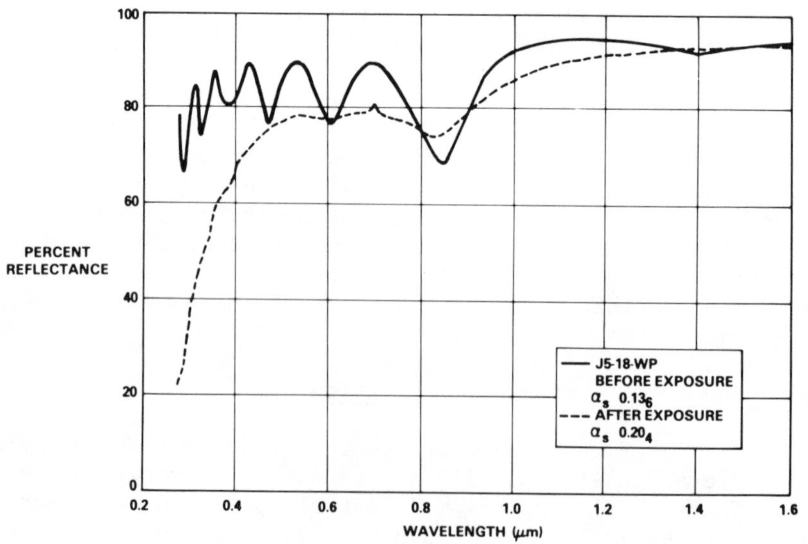

Fig. 11 Spectral reflectance of a quarterwave dielectric coated mirror exposed for 65 sec at 110 deg relative to the thrust vector of a solid thruster (IUS-interim upper stage) in the Arnold Engineering Development Center J-5 test cell; from Ref. 64.

of a data base to support modeling capabilities, 3) spacecraft contamination assessment using the contamination models, 4) development of new materials and material selection procedures to reduce contamination, and 5) flight test validation of contamination control procedures and model predictions by on-orbit contamination monitoring and feedback.

After reviewing the literature regarding spacecraft contamination and spacecraft contamination control, and after postulating the need for a unified spacecraft contamination control program, the following specific conclusions are drawn as to what needs to be added to the current technology to construct such a unified contamination control program:

1) Realistic methods of sampling both the particulate and nonvolatile residue on spacecraft surfaces need to be developed. Such sampling development is critical before cleanliness verification techniques and cleaning methods can be developed or the relationship between the cleanliness level of a clean room, clean room operations, and the debris level on a spacecraft can be determined.

2) Thruster contamination and effects models need to be developed. Probably the most critical area is the development of a contamination model for solid thrusters. Adequate in situ, noninterfering diagnostic tools for measurement of particulates are currently in a very rudimentary stage of development. Such tool development will enable the study of the kinetics of formation of aluminum oxide particulates and the particle size distribution both in the plume and in the back-flow regions. Noninterfering diagnostic tools also need to be developed to measure boundary layer expansion.

3) Current contamination models need to be extended to include the dynamics of particulate formation from nonthruster sources including particle transport through space, the effect of the space environment on particles, and definition of whether particles can stick to and migrate on spacecraft surfaces.

4) Molecular contamination models need basic technique development. Current test techniques for generation of material outgassing data are limited because of the test time required and the cost of data generation. Redesign of current material outgassing models to accept TML/CVCM as currently generated would severely limit the utility of the models and, for this reason, is not a viable alternative. TML/CVCM data contain no time-dependent outgassing information nor any information on the total volatiles capable of being outgassed.

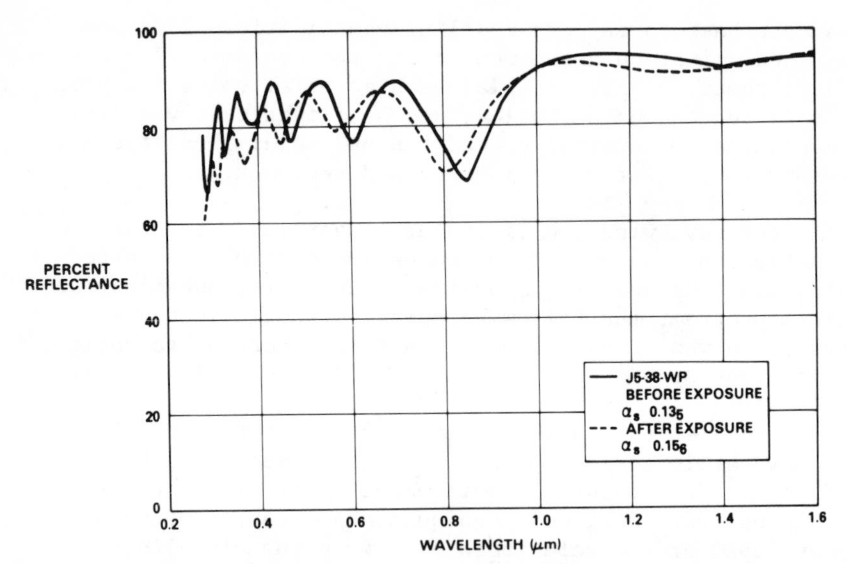

Fig. 12 Spectral reflectance of a quarterwave dielectric coated mirror exposed for 65 sec at 180 deg relative to the thrust vector of a solid thruster (IUS - interim upper stage) in the Arnold Engineering Development Center J-5 test cell; from Ref. 64.

Further limitations are the inability to extend TML/CVCM data to the cryogenic temperature regime where contamination problems become more severe and the lack of data on the effects of space radiation on contaminant films, so that film composition and thickness can be translated into system effects.

5) The existing contamination and effects models need to be verified by on-orbit measurements of contamination and the space environment. Such verified models can then be used to design less contamination-susceptable spacecraft by enabling designers to conduct material and spacecraft geometry trade-offs for sensitive systems, to accurately assess the contamination potential of various launch and deployment profiles, and to design plume shields to prevent back-flow from thrusters.

6) A methodology for acceptance and rejection of materials for spacecraft use needs to be developed. The current TML/CVCM test is useful only as a screening test. A unified materials test methodology needs to be devised enabling the assessment of low-temperature effects on outgassing, the assessment of the synergistic effect of ultraviolet and ions on material stability and degradation, and the resolution of

which materials may outgas but do not degrade spacecraft performance.

7) The overall conclusion of this review is that contamination control is currently practiced on a system-by-system basis. Those problems determined to be critical to the performance of that system are solved on an empirical basis. Little basic technology in contamination control is being developed, and system specific technology, which is being developed, is not being transfered to other systems.

References

[1]Lynch, J. T., "QCM Monitor of Contamination for LES 8/9," 9th Space Simulation Conference, Los Angeles, Calif, April 1977, NASA CP-2007, pp. 211-219.

[2]Maag, C. R., "Backflow Contamination from Solid Rocket Motors," Proceedings of the USAF/NASA International Spacecraft Contamination Conference, Colorado Springs, Colo., March 1978, NASA CP-2039/AFML-TR-78-190, 1978, pp. 332-346.

[3]Schuermen, D. W., Beeson, D. E., and Giovane, F. G., "Coronagraphic Technique to Infer the Nature of the Skylab Particulate Environment," Applied Optics, Vol. 16, June 1977, pp. 1591-1597.

[4]Lenorovitz, J. M., "Voyager 2 Boom Appears Operational," Aviation Week and Space Technology, Sept. 5, 1977, pp. 19-21.

[5]"Malfunction of Infrared System Reduces Landsat's Data Return," Aviation Week and Space Technology, Aug. 28, 1978, pp. 49-50.

[6]"Report of the Findings of the Radiation Cooler Task Group," NASA DIRS 02273-I-2-TR-239-032-215, Sept. 1970.

[7]"Shuttle Orbiter - IUS - DSP Satellite Interface Contamination Study, Final Report," NASA Contract NAS9-15335, March 1978.

[8]Dormont, L.M. and Feuerstein, S. F., "Nylon Pore System," Journal of Spacecraft and Rockets, Vol. 13, June 1976, pp. 306-309.

[9]"Standard Test Method for Total Mass Loss and Collected Volatile Condensable Materials from Outgassing in a Vacuum Environment," American Society for Testing and Materials, Annual Report, ASTM E595-77, Part 41, 1977, pp. 761-769.

[10]Miraca, R. F. and Whittick, J. S., "Polymers for Spacecraft Applications," Stanford Research Institute, Stanford, Calif., SRI Project ASD-5046, NASA CR-89557, Sept. 1967.

[11]"Compilation of CVCM Data of Nonmetallic Materials," NASA Johnson Space Center, JSC 08962, Revision U, May 1979, plus addendum.

[12]"Listing of Materials Meeting JSC Vacuum Stability Requirements," NASA Johnson Space Center, JSC 07572, April 1978.

[13]Campbell, W. A., Jr., Marriott, R. S., and Park, J. J., "An Outgassing Data Compilation of Spacecraft Materials," NASA Ref. Pub. 1014, Jan. 1978.

[14]Product Assurance Division, European Space Research and Technology Centre, European Space Agency, Noordwijk, Netherlands.

[15]Scialdone, J. J., "An Equivalent Energy for the Outgassing of Space Materials," NASA TN D-8294, Aug. 1976.

[16]Hughes, T. A., "Outgassing of Materials in the Space Environment," Proceedings of the USAF/NASA International Spacecraft Contamination Conference, Colorado Springs, Colo., March 1978, NASA CP-2039/AFML-TR-78-190, 1978, pp. 14-33.

[17]Phillips, R. W., Tolentino, L. U., and Feuerstein, S., "Spacecraft Contamination Under Simulated Orbital Environment," Journal of Spacecraft and Rockets, Vol. 14, Aug. 1977, pp. 501-508.

[18]Dauphin, J. and Zwaal, A., "The Outgassing of Space Materials and Its Measurement," European Space Agency, ESRO IN-124, Feb. 1975.

[19]Hunter, W. R., "Optical Contamination: Its Prevention in the XUV Spectrographs Flew by the U. S. Naval Research Laboratory in the Apollo Telescope Mount," Applied Optics, Vol. 16, April 1977, pp. 909-916.

[20]Poehlmann, H. C., International Seminar on Simulation in Space, Toulouse, France, Sept. 1973.

[21]Borson, E. N., "The Control of Spacecraft Contamination -- Where Are We Going?" Proceedings of the USAF/NASA International Spacecraft Contamination Conference, Colorado Springs, Colo., March 1978, NASA CP-2039/AFML-TR-78-190, 1978, pp. 1129-1135.

[22]"Clean Room and Work Station Requirements, Controlled Environments," General Services Administration, FED STD No. 209B, April 24, 1973.

[23]"Criteria for Air Force Clean Room Facility Design and Construction," U.S. Air Force, Air Force Manual 88-4, Nov. 13, 1964, Ch. 5.

[24]"Military Specification, Filter, Particulate, High Efficiency, Fire Resistant," Department of Defense, MIL-F-51068C, June 8, 1970.

[25]"Product Cleanliness Levels and Contamination Control Program," Department of Defense, MIL-STD-1246A, Aug. 18, 1967.

[26]"Specification, Contamination Control Requirements for the Space Shuttle Program," NASA Johnson Space Center, Texas, SN-C-0005, March 1974.

[27]Hamberg, O., "Photographic Measurement of Particulate Surface Contamination," Proceedings of the 7th Space Simulation Conference, NASA SP-336, Nov. 1973, pp. 379-400.

[28]Hoffman, A. R. and Koukol, R. C., "Particulate Contamination Control for Viking and Voyager Unmanned Planetary Spacecraft," Proceedings of the USAF/NASA International Spacecraft Contamination Conference, Colorado Springs, Colo., March 1978, NASA CP-2039/AFML-TR-78-190, 1978, pp. 899-925.

[29]Crutcher, E. R., "Assemblage Analysis-Identification of Contaminant Sources," Proceedings of the USAF/NASA International Spacecraft Contamination Conference, Colorado Springs, Colo., March 1978, NASA CP-2039/AFML-TR-78-190, 1978, pp. 763-778.

[30]Davis, R. W., "Evaluation and Comparison of Airborne and Settled Particulate for Controlled Equipment Work Stations," 24th National SAMPE Meeting, May 1979, Society for the Advancement of Materials and Process Engineering, Vol. 24, pp. 707-716.

[31]Horr, K. S., "Contamination Monitoring for Aerospace Instruments During Testing," Symposium for Optical Contamination in Space, Aspen, Colo., Aug. 1969.

[32]Kruger, R., "Evaluating a Contamination Hazard with a Residual Gas Analyzer," Proceedings of the USAF/NASA International Spacecraft Contamination Conference, Colorado Springs,

Colo., March 1978, NASA CP-2039/AFML-TR-78-190, 1978, pp. 644-656.

[33]Barsh, M. K., "2nd Surface Mirror Cleaning and Verification," Proceedings of the USAF/NASA International Spacecraft Contamination Conference, March 1978, NASA CP-2039/AFML-TR-78-190, 1978, pp. 927-944.

[34]Cunniff, C. V., "AFGL Infrared Survey Experiments Cleaning Procedure," Proceedings of the USAF/NASA International Spacecraft Contamination Conference, Colorado Springs, Colo., March 1978, NASA CP-2039/AFML-TR-78-190, 1978, pp. 880-888.

[35]American Society for Testing and Materials, ASTM E21.05 subcommittee on spacecraft contamination, personal communication.

[36]Hall, D. F., Borson, E. N., Winn, R. A., and Lehn, W. L., "Experiment to Measure Enhancement of Spacecraft Contamination by Spacecraft Charging," 8th Conference on Space Simulation, Silver Springs, Md., Nov. 1975, NASA SP-379, 1975, pp. 85-107.

[37]Millard, J. M. and Maag, C. R., "Spacecraft Contamination Model Development," Proceedings of the USAF/NASA International Spacecraft Contamination Conference, Colorado Springs, Colo., March 1978, NASA CP-2039/AFML-TR-78-190, 1978, pp. 208-229.

[38]Scialdone, J. J., "Internal Pressures of a Spacecraft or Other System of Compartments, Connected in Various Ways and Including Outgassing Materials, in a Time-Varying Pressure Environment," NASA X-327-69-524, Aug. 1969.

[39]Neff, W. J. and Montes de Oca, R. A., "Launch Environment Profiles for Sounding Rockets and Spacecraft," NASA-TN-D-1916, Jan. 1964.

[40]Dayton, B. B., "Relations Between Size of Vacuum Chamber, Outgassing Rates and Required Pumping Speed," 6th National Symposium on Vacuum Technology Transfer, Pergamon Press, New York, 1959.

[41]Dayton, B. B., "Outgassing Rate of Contaminated Metal Surfaces," 8th National Vacuum Symposium and 2nd International Congress on Vacuum Science, Pergamon Press, New York, 1961.

[42]Hale, R. R., "An Improved Analytic Technique to Predict Space System Contamination," Proceedings of the USAF/NASA International Spacecraft Contamination Conference, Colorado Springs, Colo., March 1978, NASA CP-2039/AFML-TR-78-190, 1978, pp. 230-249.

[43]Bareiss, L. E., "Spacelab Induced Environment Technical Overview," Proceedings of the USAF/NASA International Spacecraft Contamination Conference, Colorado Springs, Colo., March 1978, NASA CP-2039/AFML-TR-78-190, 1978, pp. 152-175.

[44]Heslin, T. M., "An Equation that Describes Material Outgassing for Contamination Modeling," NASA TN D-8471, May 1977.

[45]Harvey, R. L., "Spacecraft Neutral Self Contamination by Molecular Outgassing," Journal of Spacecraft and Rockets, Vol. 13, May 1976, pp. 301-305.

[46]Zeiner, E. A., "A Multinodal Model for Surface Contamination Based Upon the Boltzman Equation of Transport," Proceedings of the USAF/NASA International Spacecraft Contamination Conference, Colorado Springs, Colo., March 1978, NASA CP-2039/AFML-TR-78-190, 1978, pp. 34-82.

[47]Zeiner, E. A., "AESC Multinodal Free Molecular Contamination Transport Model," 8th Conference on Space Simulation, Silver Springs, Md., Nov. 1975, NASA SP-379, 1978, pp. 1-32.

[48]Kan, H. K. A., "Desorptive Transfer: A Mechanism of Contaminant Transfer in Spacecraft," Journal of Spacecraft and Rockets, Vol. 12, Jan. 1975, pp. 62-64.

[49]Scialdone, J. J., "Self Contamination and Environment of an Orbiting Satellite," Journal of Vacuum Science and Technology, Vol. 9, March-April 1972, pp. 1007-1015.

[50]Heuser, J. E. and Brock, F. J., "Shuttle Flow Field Analysis Using the Direct Simulation Monte Carlo Technique," Proceedings of the USAF/NASA International Spacecraft Contamination Conference, Colorado Springs, Colo., March 1978, NASA CP-2039/AFML-TR-78-190, 1978, pp. 250-273.

[51]Stevens, N. J., Roche, J. C., and Mandell, M. J., "NASA Charging Analyzer Program - A Computer Tool that can Evaluate Electrostatic Contamination," Proceedings of the USAF/NASA International Spacecraft Contamination Conference, Colorado Springs, Colo., March 1978, NASA CP-2039/AFML-TR-78-190, 1978, pp. 274-289.

[52]Scialdone, J. J., "Correlation of Self-Contamination Experiments in Orbit and Scattering Return Flux Calculations," NASA TN D-8438, March 1977.

[53]"An Induced Environment Contamination Monitor for the Space Shuttle," NASA TM-78193, Aug. 1978.

[54]Hoffman, R. J., Webber, W. T., Oeding, R. G., and Nunn, J. R. "An Analytical Model for the Prediction of Liquid Rocket Plume Contamination Effects on Sensitive Surfaces," AIAA Paper No. 72-1172, 8th Joint Propulsion Specialist Conf., New Orleans, La., Nov. 1972.

[55]Davis, L. P. and Wax, S. G., "Verification of Contamination Predictions for Monopropellant Thrusters," Air Force Rocket Propulsion Laboratory, Edwards Air Force Base, Calif., AFRPL-TR-77-56, Oct. 1977.

[56]Passamaneck, R. S. and Chiravella, J. E., "Small Monopropellant Thruster Contamination Measurement in a High-Vacuum, Low-Temperature Facility," Journal of Spacecraft and Rockets, Vol. 14, July 1977, pp. 419-426.

[57]McKay, T. D., Weaver, D. P., Williams, W. D., Powell, H. M., and Lewis, J. W. L., "Exhaust Plume Contaminants from an Aged Hydrazine Monopropellant Thruster," Proceedings of the USAF/NASA International Spacecraft Contamination Conference, Colorado Springs, Colo., March 1978, NASA CP-2039/AFML-TR-78-190, 1978, pp. 456-517.

[58]Scott, H. E., Frazine, D. F., and Lund, E. G., "Bipropellant Engine Plume Study," Proceedings of the USAF/NASA International Spacecraft Contamination Conference, Colorado Springs, Colo., March 1978, NASA CP-2039/AFML-TR-78-190, 1978, pp. 682-740.

[59]Roux, J. A., Wood, B. E., Smith, A. M., and Pipes, J. G., "IR Properties of Bipropellant Cryo-contaminants," Proceedings of the USAF/NASA International Spacecraft Contamination Conference, Colorado Springs, Colo., March 1978, NASA CP-2039/AFML-TR-78-190, 1978, pp. 412-455.

[60]Etheridge, F. G. and Boudreaux, R. A., "Attitude Control Rocket Exhaust Plume Effects on Spacecraft Functional Surfaces," Journal of Spacecraft and Rockets, Vol. 7, Jan. 1970, pp. 44-48.

[61]"Executive Summary - Session III. Rocket Exhaust Plumes," Proceedings of the USAF/NASA International Spacecraft Contamination Conference, Colorado Springs, Colo., March 1978, NASA CP-2039/AFML-TR-78-190, 1978, pp. 5-7.

[62]Barsh, M. K., Jeffrey, J. A., Brestyanszky, P., and Adams, K. J., "Contamination Mechanisms of Solid Rocket Motor Plumes," Proceedings of the USAF/NASA International Contamination Conference, Colorado Springs, Colo., March 1978, NASA CP-2039/AFML-TR-78-190, 1978, pp. 347-384.

[63]Cooper, B. P., Jr., "A Computational Scheme Useable for Calculating the Plume Back Flow Region," AIAA Paper No. 78-1631, 10th Space Simulation Conference, Bethesda, Md., Oct. 1978.

[64]Maag, C. R. and Scott, R. R., "Ground Contamination Monitoring Methods," Proceedings of the USAF/NASA International Spacecraft Contamination Conference, Colorado Springs, Colo., March 1978, NASA CP-2039/AFML-TR-78-190, 1978, pp. 741-762.

[65]Dawbarn, R., "Aluminum Oxide Particles Produced by Solid Rocket Motors," Proceedings of the USAF/NASA International Spacecraft Contamination Conference, Colorado Springs, Colo., March 1978, NASA CP-2039/AFML-TR-78-190, 1978, pp. 809-845.

[66]Hong, N. S. and Jones, A. R., "A Light Scattering Technique for Particle Sizing Based on Laser Fringe Anemometry," Journal of Physics, Vol. D9, Sept. 1976, pp. 1839-1848.

[67]Roberds, D. W., "Particle Sizing Using Laser Interferometry," Applied Optics, Vol. 16, Aug. 1977, pp. 1861-1868.

[68]Adrian, R.J. and Orloff, K. L., "Laser Anemometer Signals: Visibility Characteristics and Application to Particle Sizing," Applied Optics, Vol. 16, March 1977, pp. 677-684.

[69]Chu, W. P. and Robinson, D. M., "Scattering from a Moving Particle by Two Crossed Coherent Plane Waves," Applied Optics, Vol. 16, March 1977, pp. 619-626.

[70]Young, R. P., "Degradation of Low Scatter Mirrors by Particle Contamination," AIAA Paper No. 75-667, 10th Thermophysics Conference, Denver, Colo., May 1975.

[71]Dowling, J. M. and Randall, C. M., "Infrared Emissivities of Micron-Sized Particles of C, MgO, Al_2O_3 and ZnO_2," Air Force Rocket Propulsion Lab., AFRPL-TR-77-14, April 1977.

[72]Rees, T. and Fern, D. G., "N-S Station Keeping by 10-cm Ion Thruster," Journal of Spacecraft and Rockets, Vol. 15, May 1978, pp. 147-153.

[73]Parks, D. E. and Katz, I., "Solar Electric Propulsion Interactions with Solar Arrays," NASA CR-135257, Aug. 1977.

[74]Liemohn, H. B., Copland, R. L., and Leavens, W. M., "Plasma Particles Trajectories around Spacecraft Propelled by Ion Thrusters," Spacecraft Charging Technology Conference - 1978, Colorado Springs, Colo., Oct. 1978, NASA CP-2071/AFGL-TR-79-0082, 1979, pp. 419-436.

[75]Williamson, W. S. and Hyman, J., Jr., "Discharge Chamber in Mercury Ion Thrusters," Journal of Spacecraft and Rockets, Vol. 15, Nov. 1978, pp. 375-380.

[76]Olson, R. C. and Whipple, E. C., "Operations of the ATS-6 Ion Engine," Spacecraft Charging Technology Conference - 1978, Colorado Springs, Colo., Oct. 1978, NASA CP-2071/AFGL-TR-79-0082, 1979, pp. 59-68.

[77]Lyon, W. C., "Thruster Exhaust Effects upon Spacecraft," NASA TM X-65427, Oct. 1970.

[78]Molinari, L. F., "Design of a Propulsion System Facility to Study Rocket Plumes in the Space Environment," Proceedings of the USAF/NASA International Spacecraft Contamination Conference, Colorado Springs, Colo., March 1978, NASA CP-2039/AFML-TR-78-190, 1978, pp. 518-532.

[79]O'Donnell, E. E., "Spacecraft Charging Modeling Development and Validation Study," Spacecraft Charging Technology Conference - 1978, Colorado Springs, Colo., Oct. 1978, NASA CP-2071/AFGL-TR-79-0082, 1979, pp. 797-816.

[80]Stevens, N. J., Berkopec, F. D., Staskus, J. V., Blech, R. A., and Narcisco, S. J., "Testing of Typical Spacecraft Materials in a Simulated Substorm Environment," Proceedings of Spacecraft Charging Technology Conference, Colorado Springs, Colo., Oct. 1976, NASA TMX-73537/AFGL-TR-77-0051, 1977, pp. 431-458.

[81]Yadlowsky, E. J., Hazelton, R. C., and Churchill, R. J., "Puncture Discharges in Surface Dielectrics as Contaminant Sources in the Spacecraft Environment," Proceedings of the USAF/NASA International Spacecraft Contamination Conference, Colorado Springs, Colo., March 1978, NASA CP-2039/AFML-TR-78-190, 1978, pp. 945-969.

[82]Jemiola, J. M., "Contamination Control Program for Spacecraft," 24th National SAMPE Symposium, San Francisco, Calif., May 1979, Society for the Advancement of Material and Process Engineering, Vol. 24, pp. 699-706.

COLLISION FREQUENCY OF ARTIFICIAL SATELLITES: CREATION OF A DEBRIS BELT

Donald J. Kessler* and Burton G. Cour-Palais+
NASA Johnson Space Center, Houston, Texas

Abstract

As the number of artificial satellites in Earth orbit increases, the probability of collisions between satellites also increases. Satellite collisions would produce orbiting fragments, each of which would increase the probability of further collisions, leading to the growth of a belt of debris around the Earth. This process parallels certain theories concerning the growth of the asteroid belt. The debris flux in such an Earth-orbiting belt could exceed the natural meteoroid flux, affecting future spacecraft designs. A mathematical model was used to predict the rate at which such a belt might form. Under certain conditions the belt could begin to form within this century and could be a significant problem during the next century. The possibility that numerous unobserved fragments already exist from spacecraft explosions would decrease this time interval. However, early implementation of specialized launch constraints and operational procedures could significantly delay the formation of the belt.

Introduction

Since the beginning of the space age, thousands of satellites have been placed in Earth orbit by various nations. These satellites may be grouped into three categories: payloads, rocket motors, and debris associated with the launch or breakup of a particular payload or rocket; most satellites fall into the last category. Because many of these satellites are in orbits which cross one another, there is a finite probability of collisions between them. Satellite

---

 Published in 1978 by the American Geophysical Union.

*Astrophysicist, NASA Johnson Space Center.
+Aerospace Engineer, NASA Johnson Space Center.

collisions will produce a number of fragments, some of which may be capable of fragmenting another satellite upon collision, creating even more fragments. The result would be an exponential increase in the number of objects with time, creating a belt of debris around the Earth.

This process of mutual collisions is thought to have been responsible for creating most of the asteroids from larger planetlike bodies. The time scale in which this process is taking place in the asteroid belt is of the order of billions of years. A much shorter time scale in Earth orbit is suggested by the much smaller volume of space occupied by Earth-orbiting satellites compared to the volume of space occupied by the asteroids.

Conceivably, a significant number of small satellite fragments already exist in Earth orbit. Fragments which are undetected by radar are likely to have been produced from "killer satellite" tests and the accidental explosions of rocket motors. Although some work already has been completed to estimate the number of these fragments, further investigations in this area are still required.

This paper will determine possible time scales for the growth of a "debris belt" from collision fragments and will predict some of the possible consequences of continued unrestrained launch activities. This will be accomplished by applying techniques formerly developed for studying the evolution of the asteroid belt. A model describing the flux from the known Earth-orbiting satellites will first be developed. The results from this model will then be extrapolated in time to predict the collision frequency between satellites. The hypervelocity impact phenomena will then be examined to predict the debris flux resulting from collisions. Other sources and sinks for debris will be discussed, and the effects of atmospheric drag will be predicted. These results will be applied to design requirements for three types of space missions in the future. The potential, or upper limit, debris flux will then be discussed. Although further studies are recommended, the conclusion is reached by this study that over the next few decades a significant amount of debris could be generated by collisions, affecting future spacecraft designs.

Satellite Environment Model

A model describing the environment resulting from orbiting satellites was constructed by first calculating the spatial density (average number of satellites per unit volume) as a function of distance from the Earth and geocentric latitude. Flux (number of impacts per unit area per unit time) was then related to spatial density through the relative impact velocities. This technique was also used to model the collision frequency in the asteroid belt (Kessler, 1971).

Orbital perturbations can be expected to cause the orbital argument of perihelion and right ascension of ascending node to change fairly rapidly, causing these two distributions to be nearly random. This randomness was observed (Brooks et al., 1975) and led to a uniform distribution in the spatial density as a function of geocentric longitude. The model was thus reduced to determining the spatial density S as a function of distance from the surface of the Earth R and geocentric latitude β. To construct the model, volume elements were defined as $\Delta R = 50$ km and $\Delta\beta = 3$ deg. The spatial density in each of these volume elements was found by calculating the probability of finding each satellite in a particular volume element and then summing these probabilities. Spatial density is then this sum divided by the volume of the volume element.

The April 30, 1976, Satellite Situation Report (NASA, 1976) contains a total of 3866 satellites, indicating that as of that date a total of 3866 were being tracked, most by radar in low Earth orbit. The Satellite Situation Report is compiled from data provided by the Space Defense Center (SDC) and represents the most complete data available. Even so, it was found to be significantly incomplete by a test performed in 1976 (Hendren and Anderson, 1976), especially at altitudes below 500 km. In addition, the drop in radar sensitivity for objects smaller than 10 cm (Brooks et al, 1975) and for objects at higher altitudes produces another bias in these data. Such deficiencies caused the calculated spatial densities to be too low; the implications of this result will be discussed in later sections. Since only a statistical solution was required, sufficient accuracy was maintained by performing all calculations with a random sample consisting of 125 of the 3866 satellites.

The resulting spatial densities are given in Figs. 1 and 2. Figure 1 shows the spatial density as a function of distance from the Earth (averaged over latitude), while Fig. 2 shows the spatial density as a function of latitude for a few selected distances. Of particular note is that most of the satellites were found within about 2000 km of the Earth, the peak density being found at about 900 km. Significant peaks were found at 1500 and 3700 km. In latitude, significant peaks were found between 30 deg. and 35 deg. and at >60 deg. For most latitudes the spatial density was found to be well within a factor of 2 of the average for that distance, although specific peaks were a factor of 3 or 4 from the average.

As a note of interest at this point, the impact rate on a particular spacecraft, dI/dt, can be approximated quickly from Fig. 1 by using the equation

$$dI/dt = S\bar{V}_s A_c \tag{1}$$

where $\bar{V}_s$ is an average relative velocity, A_c is the cross-sectioned area of the spacecraft, and I is the total number of impacts with the spacecraft at time t. As will be shown later, $\bar{V}_s \approx 7$ km/sec. Therefore for a space station of 50-m radius, at 500-km alt ($S \approx 2.8\text{x}10^{-9}/\text{km}^3$ and $A_c = 7.9\text{x}10^{-3}\text{km}^2$) the impact rate is $1.5\text{x}10^{-10}$/sec, or $4.9\text{x}10^{-3}$/yr. This compares to about $3\text{x}10^{-3}$/yr found by McCarter (1972) using a 1971 Satellite Situation Report containing 1805 objects. This impact rate is only slightly dependent on the space station orbital inclination: for inclinations less than 90 deg the rate varies by less than a factor of 2.

The collision rate between all satellites, dC/dt, is given by

$$\frac{dC}{dt} = \frac{1}{2}\int S^2 \bar{V}_s \bar{A}_{cc} dU \tag{2}$$

where C is the number of collisions between satellites, $\bar{A}_{cc}$ is an average collision cross-sectional area of the satellites, and dU is an element of volume. Thus both an average velocity and a collision cross-sectional area are required. These distributions and their resulting averages will be discussed now.

Velocity Distribution

The velocity distribution in each volume element was calculated by computing the relative velocity between each of the satellites in the random sample and then weighting the number having this velocity by the probability of finding the two satellites in the volume element. Two average velocities, the average relative velocity $\bar{V}_s$ and the average collision velocity $\bar{V}_c$, were found from these distributions. To illustrate the difference between these averages, assume that there are three objects in a unit volume having velocities of 1, 2 and 3 km/sec relative to one another. The average relative velocity would be 2 km/sec. However,

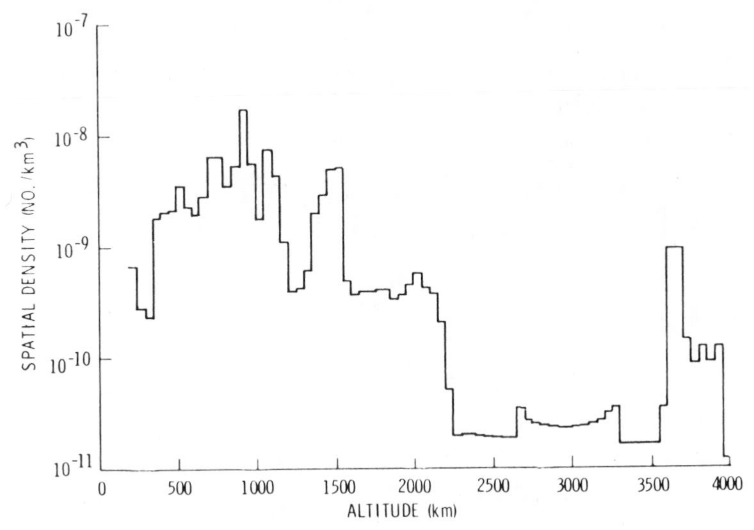

Fig. 1 Current distribution of satellites in Earth orbit as observed by radar. A total of 3866 satellites are in the April 1976 catalog and are represented here.

since the probability of collision is proportional to the relative velocity, for every collision at 1 km/sec there will be two collisions at 2 km/sec and three collisions at 3 km/sec. The average collision velocity is thus 2.33 km/sec. The average relative velocity can be shown to be the proper average to transform spatial density to flux and collision frequency, while the average collision velocity describes the average velocity at which objects would be observed to collide.

The results of these calculations for altitudes less than 2000 km were that $\bar{V}_S \simeq 7$ km/sec and $\bar{V}_C \simeq 10$ km/sec; these averages were found to be nearly independent of the volume element. At latitudes where the spatial density was large, these average velocities sometimes were slightly smaller (by 1 or 2 km/sec); however, this effect was small, so that $\bar{V}_S$ and $\bar{V}_C$ could be considered constant over all space below 2000 km.

Size Distribution

The size distribution of satellites was obtained from the radar cross-section measurements. The data in Fig. 3 were obtained during a 12-hr test of the perimeter acquisition radar (PAR) on July 31, 1976 (Hendren and Anderson, 1976). The purpose of this test was to determine PAR's

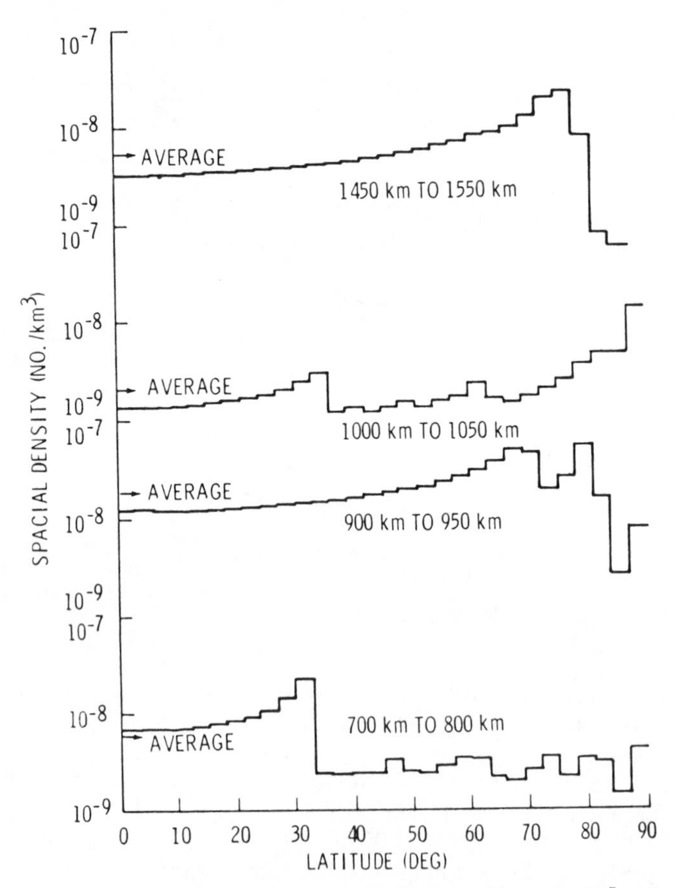

Fig. 2 Latitude variation in spatial density at selected altitudes. At most latitudes the spatial density is within a factor of 2 of the average at that distance.

capability to detect and track objects that were not in the Space Defense Center catalog. The results of the test concluded that the SDC catalog was 18% deficient, and the implications of these results will be discussed later. However, the PAR data have other applications, since they represent a "point-in-time" sampling of the satellite data and were gathered by a single instrument. Thus these data were used to analyze the satellite size distribution.

In an individual case, the physical cross section A_c of an object may be orders of magnitude different from its radar cross section σ (Ruck et al., 1970). However, the difference may be small (less than a factor of 2) when an average physical cross section is compared to an average

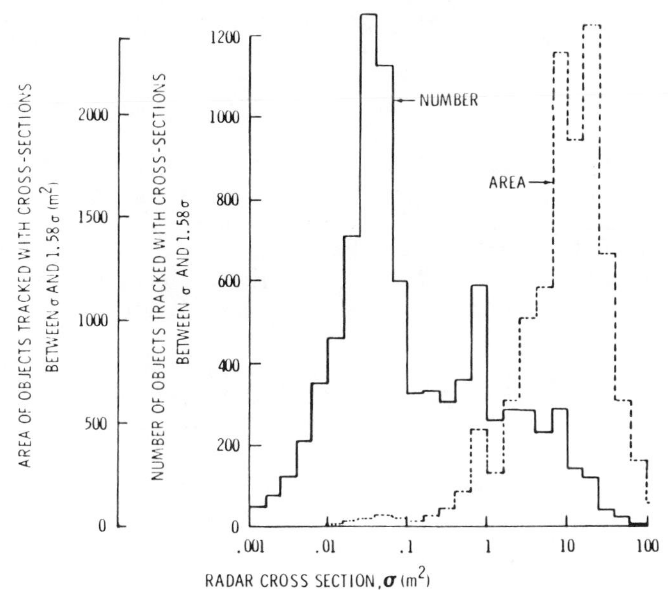

Fig. 3 Size distribution of Earth-orbiting satellites observed by radar. The largest number of satellites have a radar cross section of about 0.04 m^2, while the largest area contribution is around a radar cross section of 10 m^2.

radar cross section resulting from the observation of many objects. But when the linear dimensions of an object become small compared to the wavelength, Rayleigh scattering causes the radar cross section to be much smaller than the physical cross section. This drop in radar sensitivity was at least partially responsible for the drop in number for sizes smaller than 0.03 m^2 shown in Fig. 3. Hence the assumption was made that for $\sigma \geq 0.02\ m^2$, $A_c = \sigma$. For $\sigma < 0.02\ m^2$, $A_c \approx k\sigma$, where k approaches 6 at the smaller values of σ in Fig. 3. This correction for Rayleigh scattering produced only minor changes in the overall collision process.

Also plotted in Fig. 3 is the distribution of total area of satellites. This distribution is of interest, since collision probabilities are related to area. The distribution suggests that most collisions will involve objects between the sizes of 1- and 100-m^2 cross section. The average cross section $\bar{A}_c = 1.5\ m^2$ was found by dividing the integral of the area curve by the integral of the number curve.

However, the average collision cross section A_{cc} must include the finite size of both objects. Collision cross section is related to the physical cross section of two objects by

$$A_{cc} = (A_{ci}^{1/2} + A_{cj}^{1/2})^2 \tag{3}$$

where A_{ci} and A_{cj} are the physical cross sections of objects i and j, respectively. The collision cross section was calculated between each of the sizes in Fig. 3, weighted according to the probability of those sizes colliding, and then averaged. The results were that $\bar{A}_{cc} \simeq 4\ m^2$ and will be assumed to be independent of the volume element.

When the data in Fig. 3 were compared with the SDC catalog data, the PAR radar cross sections were found to be about 50% less, on the average, than those of the SDC. This indicates an additional calibration uncertainty between the two radar systems. When combined with the uncertainty between radar and physical cross sections, the value of $\bar{A}_{cc}$ may range from about 2 m^2 to 12 m^2.

Thus, Eq. (2) was integrated over the space between 150- and 4000-km alt by using $\bar{V}_s$ = 7 km/sec and $\bar{A}_{cc}$ = 4 m^2 to give a collision rate of 0.013 collisions/yr, but this rate could range from 0.007 to 0.039 collisions/yr, because of the uncertainty in the relationship between the real and radar cross sections. These collision rates could be significantly higher if a large number of unobserved satellites exist.

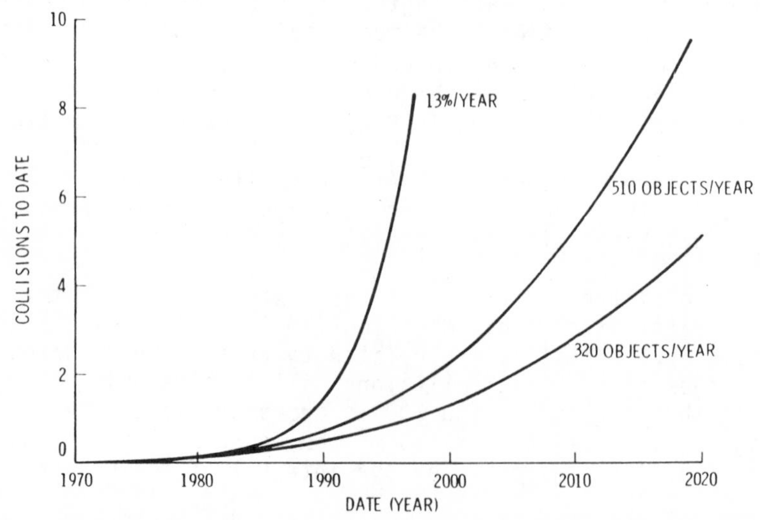

Fig. 4 Total collisions by the given date under various growth assumptions. The first collision is expected between 1989 and 1997.

Extrapolation into the Future

Between 1968 and 1974 the net number of trackable objects in space increased at the rate of about 320 objects/yr (Brooks et al., 1975). From 1975 to the present that rate increased to 510/yr (NASA, 1974, 1975, 1976). The period from 1966 to the present could be summarized by an increase of 13%/yr.

If it is assumed that the same pattern of debris buildup will continue, then the number of collisions C by time t is found by integrating Eq. (2) over time, where the spatial density is then a function of time and proportional to the number of objects in space.

Figure 4 shows the results of such an integration, using the three different buildup rates. Note that under the more conservative assumption (320 objects/yr) the first collision would be expected around 1997. However, at a growth rate of 13%/yr this collision would occur around 1989. If the average collision cross section is overestimated by a factor of 2, the first collision could be as late as the year 2005, while an underestimation by a factor of 3 results in a first collision between 1985 and 1990 under any of the growth assumptions. The presence of unobserved satellites would move these dates even closer to the present. Thus, unless significant changes are made in the method of placing objects into space, fragments from inter-collisions probably will become a source of additional space debris by the year 2000 and perhaps much earlier. The significance of this new source is seen by taking a closer look at the hypervelocity impact phenomenon.

Hypervelocity Impacts

The average impact velocity of 10 km/sec insures that almost all of the Earth-orbiting objects will exhibit hypervelocity impact characteristics when they collide. Both objects will be subjected to very high instantaneous pressures ($\geq 10^{12}$ dyne/cm^2), the strong shock waves causing melting and possible vaporization in the immediate region of the impact. A crater, or hole, will be formed, the molten ejected mass coalescing into more or less spherical particles. In addition, the shock waves, particle fragments hitting other surfaces, and vapor pressure may cause fragmentation outside the cratered region, possibly resulting in the catastrophic disruption of both objects. This process has been studied for some time, mostly from the standpoint of protection of spacecraft from meteoroids, crater formation on the moon, and fragmentation of rocks on the lunar surface or in the asteroid belt. Because of the parallel between the potential formation of an Earth-orbiting debris belt and the hypothetical formation of the asteroid belt, and because of the availability of data concerning impacts into solid, homogeneous objects, these data will be discussed first.

Impacts into Solid Structures and Basalt

Hypervelocity impacts into solid structures can be divided into two groups: catastrophic and noncatastrophic. A noncatastrophic collision results from the collision of two masses M_1 and M_2, where M_1 is much smaller than M_2 by an amount

$$M_2 > \Gamma' M_1 \tag{4}$$

where Γ' is a function of the impact velocity and the structure and materials of M_1 and M_2. In noncatastrophic collisions, only M_1 is destroyed, and a crater is produced in M_2, ejecting a total mass of M_e, which may be expressed as

$$M_e = \Gamma M_1 \tag{5}$$

where Γ is also a function of the impact velocity and the structure and materials of M_1 and M_2.

If M_1 is larger than the amount given in Eq. (4), then not only is a crater produced in M_2, but the entire structure of M_2 begins to fragment. This process is referred to as a catastrophic collision. These additional fragments usually are larger than the fragments from the crater and are ejected at a much slower velocity. The mass ejected from a catastrophic collision is

$$M_e = M_1 + M_2 \tag{6}$$

The ejected mass also has been shown to be proportional to the impact kinetic energy (Moore et al., 1965; Dohnanyi, 1971). Thus the values for Γ and Γ' will vary as V^2. At 10 km/sec, the values of Γ and Γ' for basalt are 500 and 25,000, respectively (Dohnanyi, 1971). That is, for basalt, if M_2 is greater than 25,000 times M_1, then the ejected mass resulting from a collision between M_1 and M_2 at 10 km/sec is 500 times M_1. If M_2 is less than 25,000 times M_1, then the collision is catastrophic. These results are summarized in Table 1, along with the results for glass and 1100-0 aluminum (low strength, high ductility). The glass and aluminum tests were performed by the Ames Research Center and the General Motors Defense Research Laboratories for the Johnson Space Center.

The number of small fragments of mass M and larger ejected from a noncatastrophic collision can be expressed as

$$N = K(M/M_e)^{\eta} \qquad (7)$$

where K and η are constants. From tests performed on basalt, $K = 0.4$ and $\eta = 0.8$ (Gault et al., 1963; Dohnanyi, 1971). From these results and associated modeling (Kessler, 1971) it was concluded that the asteroid belt must include particles as small as dust grains. Of course, the objects in the asteroid belt are solid chunks of material, unlike the anthropogenic satellites in Earth orbit.

Impacts into Spacecraft Structures

The objects expected to collide in Earth orbit consist of predominantly scientific and military satellites, rocket motors, and fragments of the same caused by malfunctions or deliberate destruction. The proportion of solid chunks of material will be very small, and the majority of collisions will be between open and closed structures filled with equipment. Thus the typical object will be a nonhomogeneous mass with discontinuities and many voids. The only reported hypervelocity tests where fragment distributions were

Table 1 Hypervelocity impact parameters

Material	Γ' [a]	Γ [b]
Basalt	25,000	500
Glass	120,000	2,000
1100-0 aluminum	2,600	130
Spacecraft structure	>115[c]	115

[a] Γ' is the minimum ratio of target mass to projectile mass causing catastrophic disruption at 10 km/sec.

[b] Γ is the ratio of ejected mass to projectile mass in a noncatastrophic collision at 10 km/sec.

[c] No tests have been performed to obtain this value. This lower limit follows from the definitions of Γ' and Γ.

obtained for "typical" space structures, with internal components, were performed by Langley Research Center (Bess, 1975). The tests showed that for these particular configurations the same general ejected mass and fragment size distribution laws established for the solid objects also apply to spacecraft structures.

When these results were scaled to 10 km/sec, a value of $\Gamma = 115$ was obtained for spacecraft structures. That is, in a noncatastrophic collision between a spacecraft structure and smaller object at 10 km/sec, the ejected mass would be 115 times the mass of the smaller object. Note from Table 1 that this value is not too different from that for solid 1100-0 aluminum. No tests have been performed to try to duplicate a catastrophic collision involving a spacecraft structure. Obviously, if the crater produced in a "noncatastrophic" collision is larger than the satellite, then the collision is catastrophic; thus Γ' must be greater than 115 and, by analogy to solid objects, is likely to be much larger than 115.

When normalized by the total ejected mass, as in Eq. (7), the distribution of fragments from spacecraft structures looks very similar to that of basalt. Values of $K = 0.89$ and $\eta = 0.8$ were obtained from one test into the spacecraft structure, and $K = 0.69$ and $\eta = 0.84$ were obtained from the other test (Bess, 1975). The velocities of fragments, measured from a 400-frame/sec film, were found to be very slow, about 10-30 m/sec. Most of the fragment mass from basalt targets is slower than 100 m/sec (Zook, 1967).

Thus, for collisions between Earth-orbiting objects the following relationship was adopted:

$$M_e = 115M_1 \tag{8}$$

when $M_2 > 115M_1$. If $M_2 < 115M_1$, then $M_e = M_2$. The mass of M_1 was assumed to be small or lost. The number of fragments of mass M and larger resulting from the collision is given by

$$N = 0.8(M/M_e)^{-0.8} \tag{9}$$

Satellite Mass

Before these impact equations can be used, the individual mass of each satellite must be known. Some of these masses are available, but not in one source. Mass and areas for a few payloads and rocket motors were found in several

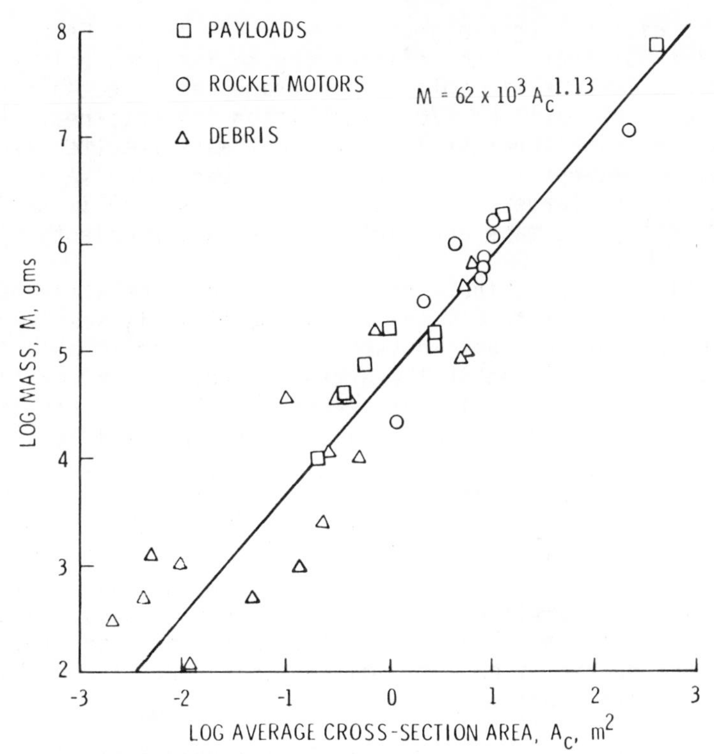

Fig. 5 Satellite mass vs average cross section.

publications (Bowman, 1963; Corliss, 1967; Martin, 1967; Von Braun and Ordway, 1975; Aerospace Defense Command, 1977), and these data are plotted in Fig. 5. Also plotted is the debris resulting from the breakup of a Centaur D-IT rocket (Drago and Edgecombe, 1974). The data were found to fit the relationship

$$M = 62 \times 10^3 A_c^{1.13} \tag{10}$$

where M is the mass of the satellite in grams and A_c is its average cross section in square meters. The slope of this fit was expected to be between 1.0 (which would be true of hollow structures with constant thickness walls) and 1.5 (which would be true of structures having a constant mass density). The value of 1.13 fell within this range.

Therefore, masses were assigned to each of the sizes shown in Fig. 3, according to Eq. (10). The amount of mass ejected in collisions between size was calculated by using Eq. (8); however, the ejected mass was not allowed to exceed the mass of the larger object (the target mass). Each collision was then weighted according to its probability of occurrence in order to obtain an average mass ejected per collision. A result of 870 kg/collision was obtained. A detailed examination revealed that most mass resulted from satellites between 16 and 40 m^2 being impacted by satellites of 0.25 m^2 and larger; thus the ejected mass was limited in many cases by the mass available in the larger object.

This suggested that the average ejected mass may be fairly insensitive to Eq. (8). This sensitivity was tested by allowing the constant to vary by a factor of 10 on either side of 115, resulting in the average ejected mass varying from 440 to 1190 kg/collision. A more realistic lower limit for the constant in Eq. (8) is about a factor of 4 less (resulting in an average of 600 kg/collision), while the upper limit may be a factor of 10 or more larger, owing to the concept of catastrophic collisions (i.e., the value of Γ' in Eq. (4) could be 1150 or larger). Thus the measured size distribution of debris causes the average ejected mass to be fairly insensitive to the uncertainty in mass ejected during a collision. The sensitivity would increase if a sufficient number of objects smaller than 0.25 m^2 were known to be in Earth orbit.

Thus on the basis of currently observed distribution of satellites, an average of $M_e = 8.7 \times 10^5$ g is ejected in each collision shown in Fig. 4.

Average Debris Flux Between 700 and 1200 km

By integrating Eq. (2) over the region of space between 700 and 1200 km, a current collision rate of 0.01/yr was obtained, compared to 0.013/yr for all of space. Thus, about 77% of the collisions were found to occur within this volume. The near-circular orbit of most objects found within this volume combined with the low ejection velocity of most fragments was justification for assuming that 77% of the collisional fragments also would be found within this volume.

Thus, an average debris flux for this volume of space was found by

$$F = (N/U)\bar{V}_s \qquad (11)$$

where N is the average number of objects of mass M and larger found within volume U, and F is the flux of debris of mass M and larger. This flux was computed as a function of time by increasing the 3866 satellites (1976 number) at the "nominal" rate of 510/yr. It was calculated that 51% of these satellites are normally found within the volume of space between 700 and 1200 km. The collision rate within this volume was assumed to be 77% of that shown in Fig. 4. The number of fragments generated by each collision is given by Eq. (9), where $M_e = 8.70 \times 10^5$.

The projected debris flux is shown in Fig. 6 for the years 1990, 2020, and 2100, compared with the natural meteoroid flux (Cour-Palais, 1969) and the current (1976) debris flux. The curve for the current debris flux is flat for debris masses less than 200 g only because these sizes cannot currently be observed. Even so, for meteoroid mass greater than 0.3 g the current debris flux between 700 and 1200 km already exceeds the natural meteoroid flux. As is illustrated in Fig. 6, a significant number of debris fragments smaller than 200 g will be generated in the future, further exceeding the meteoroid flux.

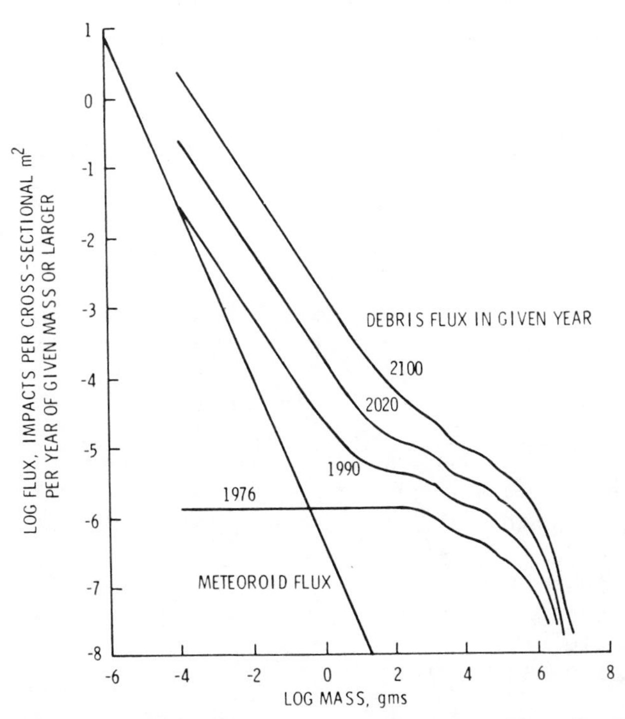

Fig. 6 Average debris flux between 700 and 1200 km; assumes that the net satellite input rate is always 510/yr, and there is no atmospheric drag.

To illustrate the effect on future space missions, consider three types of missions. The first is an unmanned satellite having an average cross section of 10 m^2 (i.e., about 3.6 m diam) and a desired average lifetime of 10 yrs. The area-time product of such a satellite would be $10^2 m^2$-yr, so that the design flux, to ensure an average 10-yr lifetime, would be $10^{-2}/m^2$-yr. If meteoroids are the only hazard, then Fig. 6 predicts that the satellite should be designed to survive a 2×10^{-4} g meteoroid impact. The second type is a manned spacecraft having an average cross section of 100 m^2, a mission duration of 1 yr, and a desired probability of impact damage of less than 0.01. In this case the design flux would be $10^{-4}/m^2$-yr, or about the same as the 1973-1974 Skylab mission. A meteoroid shield, weighing over 300 kg, was added to the Skylab structure in order to protect it against 10^{-2} g meteoroid impacts. The third type of mission is a large space station having an average cross section of 10,000 m^2 (i.e., a little over 100 m diam), a mission duration of 10 yr, and a desired probability of impact damage of less than 0.1. The design flux would be $10^{-6}/m^2$-yr, requiring protection against a 0.4-g meteoroid. These conditions are summarized in Table 2.

Table 2 Design requirements for three hypothetical missions

Mission type	Unmanned	Skylab	Space Station
Design flux, impacts/ m^2-yr	10^{-2}	10^{-4}	10^{-6}
Design meteoroid impact mass, g	2×10^{-4}	1×10^{-2}	4×10^{-1}
Year 2020, 1200-km altitude, debris design impact mass, g	1×10^{-2}	2.0	5×10^{5}
Altitude band where equilibrium debrus flux exceeds meteoroid flux (assuming change to zero net satellite input after given year), km			
1980	None	750-1200	500-1200
2020	800-1200	550-1200	400-1500

However, under certain conditions, as illustrated in Fig. 6, the debris flux for these missions may exceed the meteoroid flux. By the year 2020 the unmanned Skylab, and space-station-type missions may require protection against a 1×10^{-2} g, 2 g, and 5×10^5 g debris particle, respectively. These missions would require more weight for impact protection. However, protection requirements against even a 100-g impact are so severe that a space station may have to either accept a much higher probability of impact damage or be restricted to altitudes where the debris flux is lower.

The increased risk of impact damage may lead to certain constraints being placed on launched satellites in order to reduce the projected debris hazard. For purposes of illustration it was assumed that beginning in 2020, the net satellite input rate of 510/yr is changed to zero. A zero rate can be maintained by ceasing all launch activity, returning a similar object for every object placed into orbit, or causing the reentry of unused objects. The results of

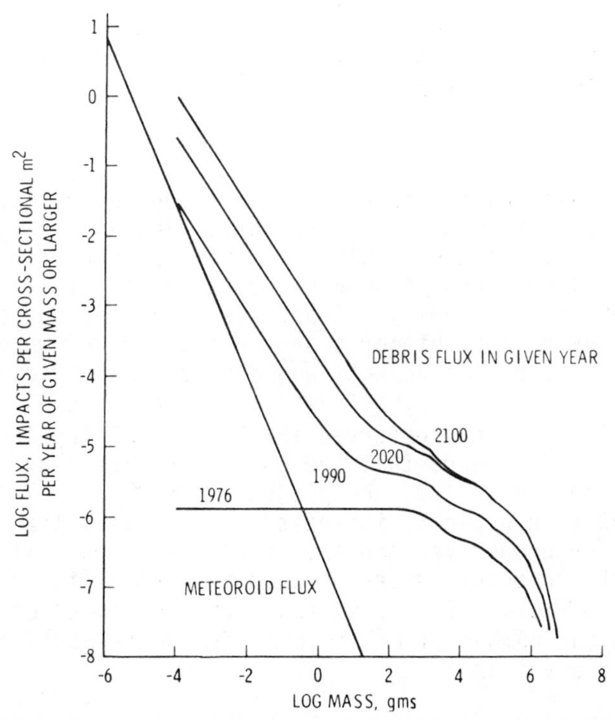

Fig. 7 Average debris flux between 700 and 1200 km; assumes that the net satellite input rate changes from 510/yr to zero in the year 2020, and there is no atmospheric drag.

this assumption are shown in Fig. 7. Notice that the flux of fragments continues to increase after the year 2020.

Other Sources and Sinks for Debris

The effects of catastrophic collisions, collisions involving fragments from previous collisions, and the current number of unobserved small fragments all represent other sources of debris, increasing the results presented thus far. Each source has been looked at in some detail, and each is found to have a relatively small effect:

1)As was previously stated, the amount of mass produced per collision is mostly limited by the amount of mass available; thus the concept of catastrophic collisions could only about double the amount of fragment mass per collision.

2)With time, enough collisional fragments could be produced to become important in producing new collisional fragments. When these conditions apply, the number of objects will increase exponentially with time, even though no new objects may be placed into orbit by man. Some preliminary analysis indicated that the uncertainty in timing of this phenomenon is large but probably of the order of several hundred years.

3)The presence of objects that are too small to be detected by ground radar would imply that the current debris flux should be increased correspondingly. These objects must already exist as the results of numerous satellite explosions and other types of debris (Neste et al., 1976). However, a preliminary analysis concluded that the current number would have to be higher than the 1990 or 2020 projected number before the objects would become significant contributers to the collision-fragmentation process. Thus while the 18% deficiency observed in the SDC catalog (Hendren and Anderson, 1976) means that the current debris flux should be increased by 18%, the projected debris resulting from fragmentation would be increased by much less than 18%. In fact, if the unobserved population of small fragments requires that the total number of observed satellites be increased by a factor of 2.5, as suggested by Brooks et al., (1975), then the number of collisional fragments would be increased by less than 3%.

While these three additional sources may represent a small effect when they are taken individually, they may combine to produce a significant effect. For example, a 2.5

factor increase in the observed population, combined with a more realistic value of catastrophic disruption mass ratio Γ', could double the number of collision fragments. Also, the exponential increase in fragments with time (source 2) could be observed much earlier than several hundred years, depending on the nature of catastrophic collisions, the projected number and size of future satellites, and the current number of unobserved small fragments. Meaningful analysis of this type must await more data.

Only a few "sinks", or removal mechanisms, exist for Earth-orbiting satellites. They are basically only retrieval and atmospheric re-entry. One could argue that as a result of catastrophic collisions, large objects disappear, and thus collision is a sink for these objects. However, hundreds of years are required before this becomes a significant sink, whereas collision is a much more important source of small fragments at a much earlier date. Thus, for the near future it is accurate to think of collisions as only a source.

If retrieval is implemented, it could significantly alter the conclusions reached thus far. Collision rates are proportional to the collision cross section of satellites. Figure 3 reveals that 90% of the satellite area is contained in 20% of the satellites. Thus, removal of large satellites could effectively slow down the collision rate. However, as the number of collision fragments increases, the concentration of area will move toward the more numerous, smaller objects, making it more difficult to slow down the fragmentation process by retrieval.

Atmospheric drag eventually will cause the re-entry of many objects in orbit. Drag acts most quickly on small, low-altitude objects and is a significant factor in reducing the number of satellites below 400 km, as is shown in Fig. 1. The projected number of small fragments between 700 and 1200 km will be reduced from that shown in Figs. 6 and 7 by the effects of atmospheric drag. However, since the collision frequency is low at altitudes less than 700 km, atmospheric drag will act as the primary source of fragments at these lower altitudes. That is, drag will act to remove collision fragments from the 700- to 1200-km region and drag them through these lower altitudes. Thus since atmospheric drag is the only natural sink for debris and since it may be an important source of debris at lower altitudes, it deserves a more detailed analysis.

Atmospheric Drag

Atmospheric drag will act to first circularize an orbit, then cause the object to spiral into the atmosphere. The speed at which this process works is proportional to the area-to-mass ratio of the object, as well as the atmospheric density at a given altitude (Martin, 1967). With the use of the atmospheric model of Lou (1973), energy loss rates were calculated to obtain orbital decay times. The time for a 1-cm-radius sphere of mass density 2 g/cm^3 (mass of 8.4 g) to change its altitude by 100 km for circular orbits of 800 and 1200 km was calculated to be 32 and 455 yr, respectively. This compares to 110 and 2000 yr, respectively, calculated by Martin (1967) and 20 and 100 yr, respectively, calculated by Brooks et al. (1975). The large range in values results from uncertainties in the atmospheric model and drag coefficients.

Lifetimes at a particular altitude were assumed to be the same as the calculated values for a circular orbit to

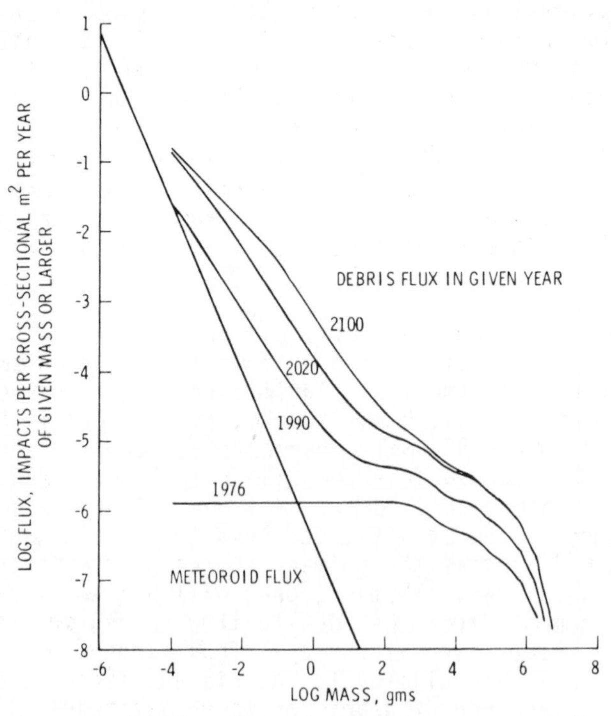

Fig. 8 Average debris flux at 1200 km; assumes that the net satellite input rate changes from 510/yr to zero in the year 2020.

decrease by 100 km. Of course, actual lifetimes could be longer or shorter, depending on orbital perigee or apogee. However, since most of the satellites within the 700- to 800-km band are in nearly circular orbits, this approximation should be fairly accurate. Lifetime as a function of particle size varies as particle radius, assuming a constant mass density.

Thus a model was developed which calculated the number of collisional fragments of a particular size which are produced during the lifetime of that size. Figure 8 gives the results of that model at 1200-km altitude. By comparison to Fig. 7, drag had no effect on the debris design impact masses given in Table 1 for the year 2020. The change to a zero net satellite input rate in 2020 still led to an increased debris flux in 2100, although the flux of debris particles of less than 0.1 g was reduced by atmospheric drag.

Figure 9 gives the results of atmospheric drag at 800 km. In comparison with Fig. 7 the debris design impact

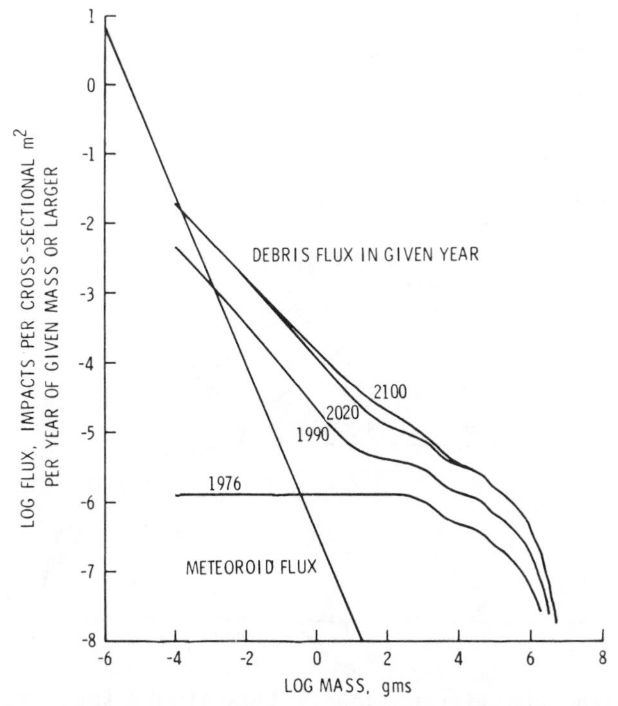

Fig. 9 Average debris flux at 800 km; assumes that the net satellite input rate changes from 510/yr to zero in the year 2020.

mass for an unmanned mission in the year 2020 was reduced, although it was still higher than the meteoroid design impact mass. For Skylab and space-station-type missions the debris flux was essentially unchanged by atmospheric drag. Of particular interest in Fig. 9 is that a change in the net satellite input rate to zero in the year 2020 resulted in a near-equilibrium being established for times after that date. That is, collisional fragments are being generated at the same rate as they are being removed by atmospheric drag. This equilibrium was reached almost immediately after the year 2020 for sizes smaller than about 0.1 g, while the sizes between 0.1 and 10^4 g were near equilibrium in 2020, reaching it by the year 2100. By comparing Figs. 8 and 9, an equilibrium at 1200 km was reached for sizes smaller than 0.1 g by the year 2100 but at a flux level about a factor of 10 larger than that at 800 km.

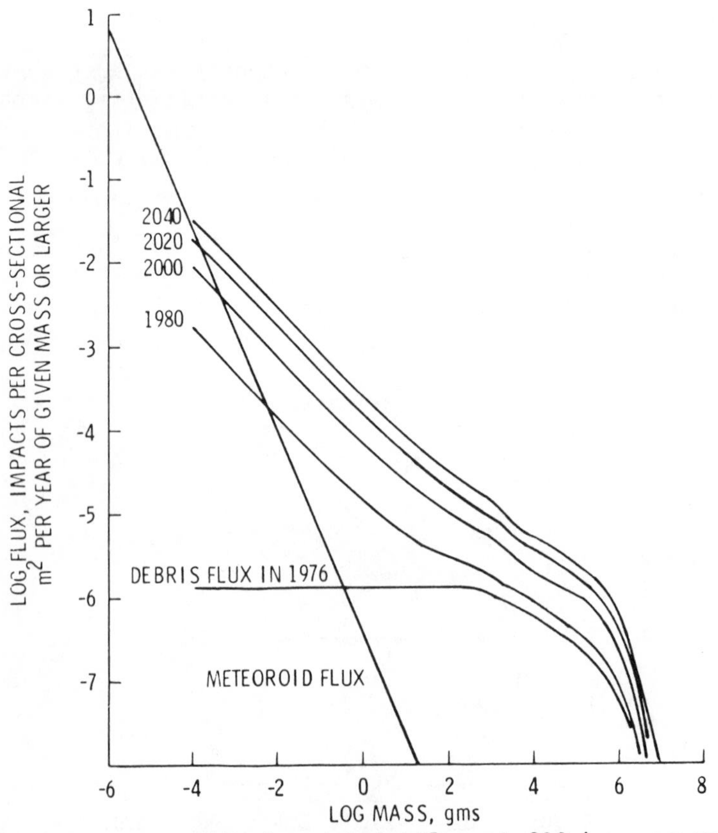

Fig. 10 Average equilibrium debris flux at 800 km; assumes that the net satellite input rate changes from 510/yr to zero in given year.

These equilibrium debris levels suggest another way of looking at the future debris fluxes: that is, given a date to change from a net satellite input rate of 510/yr to zero, what will the equilibrium debris flux eventually become? For altitudes below 800 km the equilibrium will be reached almost immediately after the change to zero input date, whereas for altitudes up to 1200 km, several hundred years may be required, especially for the larger debris sizes. Figure 10 shows the average equilibrium debris flux at 800 km for various change input rate dates. Notice that even an early (1980) change to zero net input rate would eventually affect Skylab-type missions at 800 km, whereas unmanned missions would not be affected at this altitude until a change after 2020. A similar curve for 1200-km alt would be about a factor of 10 higher. The time required for atmospheric drag to drag a fragment through a particular altitude is inversely proportional to the atmospheric density at that altitude. Thus the average equilibrium debris flux at altitudes below 800 km (and actually above 800 km, up to 1200 km), was found by taking the ratio of the atmospheric density at that altitude to the atmospheric density at 800 km. Figure 11 gives that ratio of equilibrium fluxes. Thus for example, the equilibrium flux at 500 km was found by reducing the debris fluxes in Fig. 10 by a factor of 30.

From this type of analysis the region of space where the debris flux could exceed the natural meteoroid flux was determined for the three types of missions. Table 2 sum-

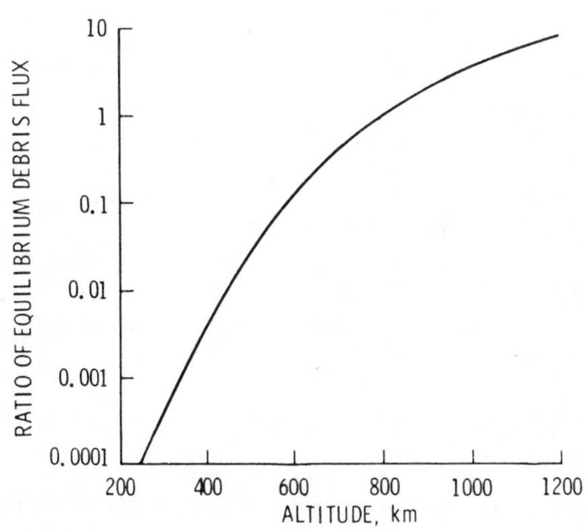

Fig. 11 Ratio of equilibrium debris flux at given altitude to flux at 800 km.

marizes these altitude bands, assuming a change to the zero net input rate in 1980 or 2020. An early (1980) change would prevent unmanned satellites from being affected by the debris flux anywhere in space. Skylab and space-station-type missions would be affected between 750 and 1200 km and 500 and 1200 km, respectively. However, if the change does not occur until the year 2020, the region of space where unmanned missions would be affected becomes 800-1200 km, and the Skylab and space stations regions affected expand to 550-1200 and 400-1500 km, respectively. Thus an increase in the change date leads to larger regions of space where the debris flux could eventually exceed the meteoroid flux. The only obvious method of lowering this eventual equilibrium flux is to minimize the number of large satellites, either by a change in launch practices or by retrieval.

The "Potential" Debris Environment

Thus far, the physical processes of collision and atmospheric drag have been applied to the observed distribution of satellites in Earth orbit in an attempt to predict the time history of debris in Earth orbit. However, it is known that satellites can fragment from other sources and that attempted modeling of incomplete data may predict the future debris flux inadequately.

Another approach might be to define the "potential" debris flux and then develop arguments which would prevent this potential from being realized. This type of analysis would produce flux curves much higher than the previous curves, but the analysis would predict accurately what is possible, although not necessarily what is probable. A beginning point would be to assume that all satellites, through some unknown mechanism, become fragmented into some preferred but yet to be determined fragment size. The potential flux as a function of mass is then found by assuming that all of these fragments are of the preferred size. This potential flux is actually an envelope of single-size fluxes. As data become available that indicate that only some fraction of the total mass goes into a particular size interval, then the potential flux could be lowered appropriately in that size interval.

The average satellite mass was found [by using Fig. 3 and Eq. (10)] to be 1.3×10^5 g, or a total 1976 satellite mass of 5×10^8 g. If this total mass were to fragment into debris of mass M, then the number of fragments would be

$$N = 5 \times 10^8 M^{-1} \quad (12)$$

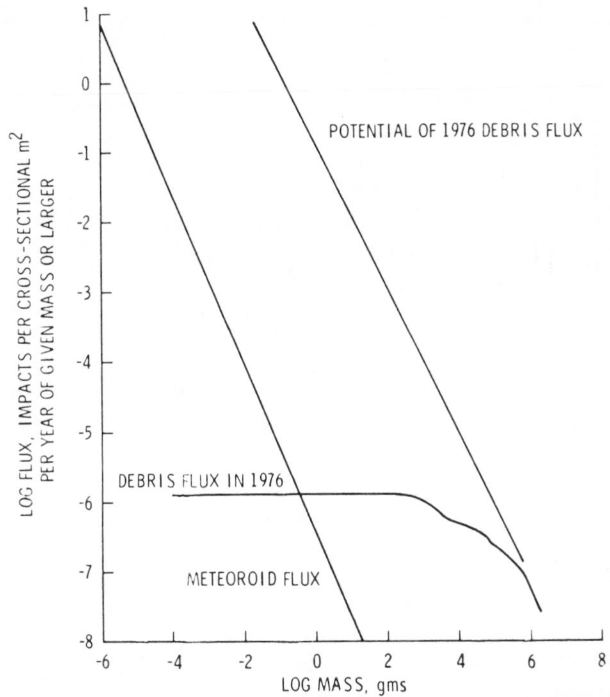

Fig. 12 Potential debris flux between 700 and 1200 km. Potential flux assumes that all satellites are fragmented into given mass.

If these fragments maintained the same orbits as the original 3866 satellites, then by using the same techniques as the original model, the "average potential flux" between 700 and 1200 km was found and is shown in Fig. 12. Note that if this potential were ever realized, all types of missions into this region would experience a flux level far in excess of the natural meteoroid flux. In fact, since it becomes impractical to protect against impacts larger than about 100 g, all missions would have to expect damage in certain regions of space.

The potential flux for other distances may be scaled from Fig. 1. For example, the spatial density at 300 km was found from Fig. 1 to be about a factor of 30 lower than the average between 700 and 1200 km; thus the debris curves in Fig. 12 could be lower by this amount to obtain the 300-km fluxes.

As more mass is added to the system with time, the potential flux will increase correspondingly. The potential

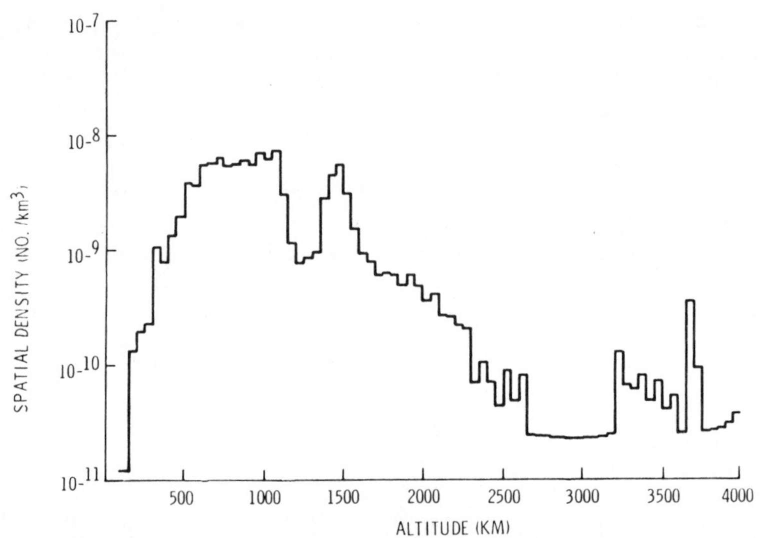

Fig. 13 Distribution of the 4762 satellites in Earth orbit which were maintained by NORAD on Oct. 1, 1978.

flux can be decreased as data become available. For example, at least 31 larger satellites, or about 1% of the total number of satellites, already have exploded in orbit (Neste et al., 1976). If fragmentation could be limited to these 31 explosions, then the potential flux could be reduced by a factor of 100. In addition, if it were found that only a small fraction of the fragmented mass were of a particular size or larger, then the potential flux for that size could be reduced further.

However, if fragmentation is maintained at 1%, or about 5 explosions per year, the potential flux would continue to increase until an equilibrium is reached with atmospheric drag. At 800 km this equilibrium would be about 4% of the potential flux shown in Fig. 12 for fragments of 8.4 g and 0.4% for fragments of 8.4×10^{-3} g. Thus without meaningful data on the actual size distribution of fragments, the expected fragmentation rate, and the lifetime of these fragments, it is difficult to lower the potential debris flux significantly.

Future Analysis Required

In order to reduce the uncertainty in the projected debris environment, additional data will be required on the effects of hypervelocity collisions between spacecraft, as well as the effects of other spacecraft fragmentation processes. Additional data on the number of small objects in space can be obtained by a detailed examination of individual launch and orbit injection procedures. The results of these studies should be tested by experiments designed to detect objects in orbit smaller than 10 cm. An optical experiment is described by Neste et al. (1976) which would detect debris in the 1-mm to 10-cm range. With the availability of improved input data, a model could be developed to include the results from explosions, the effects of collisions resulting from collision fragments, the effects of catastrophic collisions, and orbital changes resulting from collisions.

Various methods to stop or slow the formation of a debris belt should be studied. The model suggests that the most effective way would be to keep the number of large objects as small as practical. This could be accomplished by planning launches so that large objects can be caused to reenter when their usefulness is complete or by using the space shuttle concept to retrieve objects in orbit which no longer serve a useful function. Since it is impractical to retrieve the much larger number of large and small fragments, every effort should be made to prevent their production in space, either by explosion or by collision.

The evolution of the debris belt should be followed to its conclusion. As was pointed out by Alfven and Arrhenius (1976), the consequence of many collisions is to change the orbits of objects to be more alike. This process may be responsible for creating "jet streams" in the asteroid belt. With time, according to Alfven, these jet stream orbits will become identical, and a single planet will accrete (billions of years in the future) at 2.8 AU. However, in the case of the Earth, the debris belt will be within the Roche's limit (less than 9000-km alt), preventing accretion into a single object. Thus the end result, assuming that drag does not act fast enough, could be a ring system, similar to that around Saturn and Uranus.

Conclusions

A model has been developed which considers the major source and sink terms for the growth of the satellite population in Earth orbit. While significant uncertainties exist, the following conclusions, if current trends continue, seem unavoidable:

1) Collisional breakup of satellites will become a new source for additional satellite debris in the near future, possibly well before the year 2000.

2) Once collisional breakup begins, the debris flux in certain regions near Earth may quickly exceed the natural meteoroid flux.

3) Over a longer time period the debris flux will increase exponentially with time, even though a zero net input rate may be maintained.

4) The processes which will produce these fragments are totally analogous to the processes that probably occurred in the formation of the asteroid belt but require a much shorter time.

Effective methods exist to alter the current trend without significantly altering the number of operational satellites in orbit. These methods include reducing the projected number of large, nonoperational satellites and improved engineering designs which reduce the frequency of satellite breakups from structural failure and explosions in space. Delay in implementation of these methods reduces their effectiveness.

Appendix: Update

Since the original publication of the previous paper, the following additional information has been obtained:

1) Figure 13 gives the October 1978 distribution of tracked satellites, based on an Aerospace Defense Command paper (1978). This figure would update Fig. 1 in the paper. Figure 13 is smoother than Fig. 1 because of additional orbital information available in the reference, and also because a larger random sample was used to do the analysis.

2) An experiment was performed by the Aerospace Defense Command for NASA in August 1978 to determine the "completeness" of the official catalogue of space objects. Although analysis of this experiment are not yet complete, the experiment does suggest a large population of objects larger than 1 cm diam, and not represented in Fig. 13. The obvious source of these objects is the more than 50 satellite explosions or break-ups. Approximately 60% of the objects represented in Fig. 13 are explosion fragments.

Acknowledgments

We thank Joe M. Alvarez (NASA) for helpful discussions of past work in these areas, leading to further helpful discussion with Preston M. Landry (NORAD) and Larry Rice (SAI). We acknowledge suggestions by Herbert A. Zook and Andrew E. Potter (NASA) which led to changes in the content of this paper.

References

Aerospace Defense Command, "Average Radar Cross Section (sq. m.) as of 1 July 1977," Peterson Air Force Base, Colo., 1977.

Aerospace Defense Command, "Average Radar Cross Section (sq. m.) as of 1 Oct 1978," Peterson Air Force Base, Colo., 1978.

Alfven, H. and Arrhenius, G., "Evolution of the Solar System," NASA SP-345, 1976.

Bess, T.D., "Mass Distribution of Orbiting Man-Made Space Debris," NASA TND-8108, 1975.

Bowman, N.J., The Handbook of Rockets and Guided Missiles, Perastadion Press, Newtown Square, Pa., 1963.

Brooks, D.R., Gibson, G.G., and Bess, T.D., "Predicting the Probability that Earth-Orbiting Spacecraft will Collide with Man-Made Objects in Space," Space Rescue and Safety, American Astronautical Society Publications Office, Tarzana, Calif., pp. 79-139, 1975.

Corliss, W.R., "Scientific Satellites," NASA SP-133, 1967.

Cour-Palais, B.G., "Meteoroid Environment Model - 1969 (near earth to lunar surface)," NASA SP-8013, 1969.

Dohnanyi, J.S., "Fragmentation and Distribution of Asteroids, Physical Studies of Minor Planets," NASA SP-267, 263-295, 1971.

Drago, V.J. and Edgecombe, D.S., "A Review of NASA Orbital Decay Reentry Debris Hazard," Battelle Memorial Inst., Columbus, Ohio, Rept. BMI-NVLP-TM-74-1, 1974.

Gault, D.E., Shoemaker, E.M., and Moore, H.J., "Spray Ejected from the Lunar Surface by Meteoroid Impact," NASA TND-1767, 1963.

Hendren, J.K. and Anderson, A., "Comparison of the Perimeter Acquisition Radar (PAR) Satellite Track Capability to the Space Defense Center (SDC) Satellite Catalogue - Unknown satellite track experiment," Sci. Appl., Inc., Huntsville, Ala., Rept. SAI-77-701-HU, 1976.

Kessler, D.J., "Estimate of Particle Densities and Collision Danger for Spacecraft Moving through the Asteroid Belt," Physical Studies of Minor Planets, NASA SP-267, 1971, pp. 595-605.

Lou, G.Y., "Models of Earth's Atmosphere (90 to 2500 km), NASA SP-8021, 1973.

Martin, C.N., Satellites Into Orbit, trans. from French by T. Schoeters, George G. Harrap Co. Ltd., London, 1967.

McCarter, J.W., "Probability of Satellite Collision," NASA TMX-64671, 1972.

Moore, H.J., Gault, D.E., and Heitowit, E.D., "Change of Effective Target Strength with Increasing Size of Hypervelocity Impact Craters," Proceedings of 7th Hypervelocity Impact Symposium, Vol. IV: Theory, p. 35, U.S. Army, Navy, Air Force, 1965. (Available from National Technical Information Service, Springfield, Va.)

NASA, Satellite Situation Report, Vol. 14, NASA Goddard Space Flight Center, Greenbelt, Md., 1974.

NASA, Satellite Situation Report, Vol. 15, NASA Goddard Space Flight Center, Greenbelt, Md., 1975.

NASA, Satellite Situation Report, Vol. 16, NASA Goddard Space Flight Center, Greenbelt, Md., 1976.

Neste, S.L., Soberman, R.K., Lichtenfield, K., and Eaton, L.R., "The Sisyphus System - Evaluation, Suggested Improvements, and Application to Measurements of Space Debris, Final Report," Contract NAS1-13407, General Electric Co., Philadelphia, Pa., 1976.

Ruck, G.T., Barrick, D.E., Stuart, W.D., and Krichbaum, C.K., Radar Cross Section Handbook, Plenum, New York, 1970.

Von Braun, W., and Ordway, F.I., III, History of Rockets and Space Travel, Thomas Y. Crowell Co., New York, 1975.

Zook, H.A., "The Problem of Secondary Ejecta Near the Lunar Surface," Saturn V/Apollo and Beyond, American Astronautical Society Publications Office, Tarzana, Calif., 1967, pp. EN-8 to 1-24.

Index to Contributors to Volume 71